ANNALS OF THE NEW YORK ACADEMY OF SCIENCES

Volume 305

SEROTONIN NEUROTOXINS

Edited by J. H. Jacoby and L. D. Lytle

The New York Academy of Sciences
New York, New York
1978

Library of Congress Cataloging in Publication Data

Main entry under title:

Serotonin neurotoxins.

(Annals of the New York Academy of Sciences; v. 305)
"This series of papers is the result of a conference . . . held by the New York Academy of Sciences on March 30 and 31 and April 1, 1977."
Includes index.
1. Serotonin metabolism—Congresses. 2. Serotonin antagonists—Physiological effect—Congresses. I. Jacoby, Jacob H. II. Lytle, Loy D. III. New York Academy of Sciences. IV. Title. V. Series: New York Academy of Sciences. Annals; v. 305. [DNLM: 1. Serotonin—Toxicity—Congresses. 2. Nervous system—Drug effects—Congresses. W1 AN626YL v. 305 / QV126 S486]
Q11.N5 vol. 305 [QP801.S4] 508'.1s [599'.02'188]
ISBN 0–89072–061–4 78–9143

PCP
Printed in the United States of America
ISBN 0-89072-061-4

ANNALS OF THE NEW YORK ACADEMY OF SCIENCES

VOLUME 305

June 12, 1978

SEROTONIN NEUROTOXINS *

Editors and Conference Chairmen
J. H. JACOBY AND L. D. LYTLE

CONTENTS

Introductory Remarks. *By* JACOB H. JACOBY AND LOY D. LYTLE 1

Part I. Mechanisms of Action

Mode and Mechanism of Action of Neurotoxic Indoleamines: A Review and a Progress Report. *By* H. G. BAUMGARTEN, H. P. KLEMM, L. LACHENMAYER, A. BJÖRKLUND, W. LOVENBERG, AND H. G. SCHLOSSBERGER 3

Synthesis and Chemical Properties of Some Indole Neurotoxins. *By* H. G. SCHLOSSBERGER 25

Interaction of 5,6- and 5,7-Dihydroxytryptamine with Tissue Monoamine Oxidase. *By* H. P. KLEMM AND H. G. BAUMGARTEN 36

Mechanism of Action of Dihydroxytryptamines. *By* C. R. CREVELING AND A. ROTMAN 57

Mechanisms of Action of Hydroxylated Phenylethylamine and Indoleamine Neurotoxins. *By* GERALD COHEN AND RICHARD E. HEIKKILA 74

Effects of Neurotoxins *in Vitro* on the Binding of Serotonin to Serotonin-Binding Protein. *By* HADASSAH TAMIR AND MAURICE M. RAPPORT 85

Site of Action of Neurotoxins on Protein Synthesis. *By* MICHAEL A. MOSKOWITZ, DEBORAH RUBIN, THADDEUS S. NOWAK, JR., B. SUREN BALIGA, AND HAMISH N. MUNRO 96

Part II. Pharmacodynamics

Effect of Drugs on the Distribution of [^{14}C]-5,6-Dihydroxytryptamine and [^{14}C]-5,7-Dihydroxytryptamine in Rat Brain. *By* H. G. BAUMGARTEN, H. P. KLEMM, L. LACHENMAYER, AND H. G. SCHLOSSBERGER 107

Cerebral Metabolism of Some Serotonin Depletors. *By* E. M. GÁL AND A. D. SHERMAN 119

Structure-Activity Relationships of Serotonin Neurotoxins: Effects on Serotonin Uptake. *By* A. S. HORN 128

Structure-Activity Relationships in Compounds Analogous to 5,6-Dihydroxytryptamine. *By* ROGER P. MAICKEL, TALMAGE R. BOSIN, ALAN C. DONELSON, ERNEST CAMPAIGNE, AND RICHARD B. ROGERS 134

Structure-Activity Relationships among the Halogenated Amphetamines. *By* RAY W. FULLER 147

* This series of papers is the result of a conference entitled Serotonin Neurotoxins, held by The New York Academy of Sciences on March 30 and 31 and April 1, 1977.

Alterations in the Neurocytotoxicity of 5,7-Dihydroxytryptamine by Pharmacologic Agents in Adult and Developing Rats. *By* G. R. BREESE AND R. A. MUELLER 160
General Discussion 175

Part III. Neurochemical Effects

Neurochemical Effects of Serotonin Neurotoxins: An Introduction. *By* RAY W. FULLER 178
Enzyme Changes as an Index of Neurotoxin Specificity. *By* WALTER LOVENBERG AND ELEANOR BRUCKWICK 182
Studies of Serotonin Neurotoxins in Discrete Brain Nuclei. *By* JAMES L. MEEK 190
Biochemical, Behavioral, and Pharmacologic Studies of the Effects of Dihydroxytryptamines in the Rodent Brain. *By* R. MALCOLM STEWART, SYLVIA C. GERSON, GÜNTHER SPERK, ALEXANDER CAMPBELL, AND ROSS J. BALDESSARINI 198
Immediate and Long-Term Effects of *p*-Chloroamphetamine on Brain Amines. *By* ELAINE SANDERS-BUSH AND LARRY R. STERANKA 208
Fenfluramine and Brain Serotonin. *By* BRADLEY V. CLINESCHMIDT, ANTHONY G. ZACCHEI, JAMES A. TOTARO, A. BARBARA PFLUEGER, JODIE C. MCGUFFIN, AND THEODORE I. WISHOUSKY 222
Effect of Central Nervous System-Acting Drugs After Selective Destruction by Neurotoxins of 5-Hydroxytryptamine Fibers in the Brain. *By* MICHELE O. CARRUBA, GIUSEPPE NISTICÒ, AND GIULIANA GARGIULO 242
General Discussion 259

Part IV. Neuroanatomic Correlates

Morphologic Consequences of Serotonin Neurotoxin Administration: Neuron-Target Cell Interaction in the Rat Subcommissural Organ. *By* K. MØLLGÅRD, J. J. LUNDBERG, L. WIKLUND, L. LACHENMAYER, AND H. G. BAUMGARTEN 262
Neurotoxic Action of Halogenated Amphetamines. *By* JOHN A. HARVEY 289
Degenerative Effects of Various Neurotoxic Indoleamines on Central Monoamine Neurons. *By* ANDERS NOBIN AND ANDERS BJÖRKLUND 305
Developmental Plasticity of Central Serotonin Neurons After 5,7-Dihydroxytryptamine Treatment. *By* G. JONSSON, T. POLLARE, H. HALLMAN, AND CH. SACHS 328
5,7-Dihydroxytryptamine as a Tool to Study the Functional Role of Central 5-Hydroxytryptamine Neurons. *By* KJELL FUXE, SVEN-OVE ÖGREN, LUIGI F. AGNATI, GÖSTA JONSSON, AND JAN-ÅKE GUSTAFSSON 346
Regeneration of Serotonin Neurons in the Rat Brain After 5,6-Dihydroxytryptamine-Induced Axotomy. *By* LEIF WIKLUND, ANDERS BJÖRKLUND, AND ANDERS NOBIN 370
General Discussion 385

Part V. Physiologic Effects

Neuroendocrine Consequences of Chemical Indolectomy. *By* JACOB H. JACOBY AND RONALD F. THOMAS 387
Effects of Serotonin Neurotoxins on Pituitary Hormone Release. *By* JAMES A. CLEMENS 399
Endocrine Effects of Brain Serotonin Depletion by 5,6-Dihydroxytryptamine in Prepubertal Male Rats. *By* ROBERT COLLU 411
Effect of Intraventricular Injection of 5,7-Dihydroxytryptamine on Serum Gonadotropins and Prolactin. *By* W. WUTTKE, J. L. HANCKE, K. G. HÖHN, AND H. G. BAUMGARTEN 423

Changes in Seizure Susceptibility After Intracerebral Treatment with 5,7-Dihydroxytryptamine: Role of Serotonergic Neurons. *By* RONALD A. BROWNING, WILLIAM E. HOFFMANN, AND RONALD L. SIMONTON 437
Cardiovascular Effects of Neurotoxic Indolethylamines. *By* MANFRED GÖTHERT AND NIELS KLUPP 457
General Discussion 478

Part VI. Behavioral Effects

Behavioral Effects of Serotonin Neurotoxins: An Overview. *By* RITA B. MESSING, DOUGLAS J. PETTIBONE, NATHAN KAUFMAN, AND LOY D. LYTLE .. 480
Behavioral Evidence for Denervation Supersensitivity After Destruction of Central Serotonergic Nerve Terminals. *By* MICHAEL E. TRULSON AND BARRY L. JACOBS 497
Changes in Motor Behavior Following the Administration of Serotonin Neurotoxins. *By* JOHN H. GROWDON 510
Effects of Serotonin Neurotoxins on Rotational Behavior in the Rat. *By* C. T. GIAMBALVO, S. R. SNODGRASS, AND N. J. URETSKY 524
Some Behavioral Effects of Serotonin Depletion Depend on Method: A Comparison of 5,7-Dihydroxytryptamine, *p*-Chlorophenylalanine, *p*-Chloroamphetamine, and Electrolytic Raphe Lesions. *By* STANLEY A. LORENS 532
Hypothalamic Actions of 5-Hydroxytryptamine Neurotoxins: Feeding, Drinking, and Body Temperature. *By* R. D. MYERS 556
5,6-Dihydroxytryptamine as a Tool for Studying Sleep Mechanisms and Interactions between Monoaminergic Systems. *By* J. F. PUJOL, P. KEANE, P. BOBILLIER, B. RENAUD, AND M. JOUVET 576
Differential Effects of *p*-Chlorophenylalanine and 5,7-Dihydroxytryptamine on Feeding in Rats. *By* B. G. HOEBEL, F. P. ZEMLAN, M. E. TRULSON, R. G. MACKENZIE, R. P. DUCRET, AND C. NORELLI 590
Effects of 5,6-Dihydroxytryptamine and 5,7-Dihydroxytryptamine on the Rat Central Nervous System After Intraventricular or Intracerebral Application and on Blood Platelets *in Vitro*. *By* M. DA PRADA, L. PIERI, H. H. KELLER, M. PIERI, AND E. P. BONETTI 595
Influence of *p*-Chloroamphetamine and *p*-Chlorophenylalanine on Female Mating Behavior. *By* FRANK P. ZEMLAN 621
Effects of Central 5,7-Dihydroxytryptamine on the Medial Hypothalamic Syndrome in Rats. *By* DONALD V. COSCINA 627
Long-Term Biochemical and Behavioral Effects of *p*-Chloroamphetamine in the Rat. *By* CHRISTER KÖHLER, SVANTE B. ROSS, BOLEK SREBRO, AND SVEN-OVE ÖGREN 645
General Discussion 664

Author Index 667
Subject Index 669

Financial assistance was received from:

- CIBA-GEIGY CORPORATION
- ELI LILLY AND COMPANY
- FARMITALIA, MONTEDISON GROUP
- HOECHST-ROUSSEL PHARMACEUTICALS INC.
- HOFFMANN-LA ROCHE INC.
- MCNEIL LABORATORIES, INC.
- MERCK SHARP & DOHME
- NEW ENGLAND NUCLEAR CORPORATION AND SUBSIDIARIES
- NORWICH PHARMACAL COMPANY
- REGIS CHEMICAL
- SCHERING-PLOUGH CORPORATION
- SMITH KLINE & FRENCH LABORATORIES
- THE UPJOHN COMPANY
- WARNER-LAMBERT RESEARCH INSTITUTE

INTRODUCTORY REMARKS

Jacob H. Jacoby
Department of Pharmacology
College of Medicine and Dentistry of New Jersey
New Jersey Medical School
Newark, New Jersey 07103

Loy D. Lytle
Department of Nutrition and Food Science
Massachusetts Institute of Technology
Cambridge, Massachusetts 02139

It has been almost 30 years since Maurice Rapport, Arda Green, and Irvine Page, working with serum at the Cleveland Clinic, isolated and characterized serotonin, a factor that was found to be identical to the potent uterine-contracting substance extracted from the gastrointestinal tract by Erspamer 12 years earlier.

Because of the marked cardiovascular and smooth muscle effects of enteramine, or serotonin, or 5-hydroxytryptamine, or 5-HT (a term first introduced by John Gaddum), earlier research efforts were directed solely at the peripheral effects of this amine.

Not until its demonstration in brain tissue, again in Page's laboratory, the further observation by British investigators that this amine appeared to be discretely localized in different areas of the central nervous system, and the actual visualization of the distribution of serotonin-containing neurons, thanks to the development and practical utilization of selective histofluorescent staining by Swedish investigators, did the neurobiology of serotonin begin in earnest.

The relatively recent discoveries of several pharmacologic agents that appear to have specific neurotoxic effects on various brain and peripheral neurotransmitter systems have expanded greatly our understanding of the biochemical, physiologic, and behavioral importance of these transmitter systems.

With the introduction of 6-hydroxydopamine, a drug that selectively destroys brain or peripheral catecholamine-containing neurons, one could have predicted that it would only be a matter of time, and a short time at that, until drugs with a similar effect on serotonin-containing neurons would be devised. Well, thanks to the synthesis of 5,6-dihydroxytryptamine, the first such serotonin neurotoxic agent, by Dr. Hans Schlossberger, and the prolific efforts of Dr. Hans Baumgarten and his coworkers, we now have the means of introducing a chemical indolectomy and a potent new tool for studying and elucidating the role of serotonin in neuronal and, possibly, metabolic processes. We hope that this conference will be of some help in consolidating and more clearly defining our understanding of what these roles may be. The attainment of such a desirable state of affairs can only help future investigations in this multifaceted area. But who would have believed that so quickly after their introduction as experimental tools, we would already have, as recent findings from John Harvey's laboratory indicate, a captive population of human subjects ripe for studying the clinical effects of serotonin deficiency, since these individuals have possibly had more than their appetites reduced with the use of fenfluramine.

We would like to express our appreciation to Drs. Srinivasan and Banay-Schwartz for their help in the organization of this conference, to The New York Academy of Sciences Conference Director Ellen Marks, and to Reneé Wilkerson, Ann Collins, and Heidi Abrams. A more efficient and helpful group is hard to find.

We also gratefully acknowledge our supporters from the pharmaceutical industry and from the National Science Foundation, without which this gathering of such a distinguished group of investigators would not have been possible.

The following papers describe in great detail how men and women of good will have expended their efforts in attempting to more effectively destroy, and to study the effects of such destruction, of neurons whose sole candidacy for such a dismal fate is due to the fact that they contain the indoleamine first isolated by Dr. Rapport.

Part I. Mechanisms of Action

MODE AND MECHANISM OF ACTION OF NEUROTOXIC INDOLEAMINES: A REVIEW AND A PROGRESS REPORT

H. G. Baumgarten, H. P. Klemm, and L. Lachenmayer

Department of Functional Anatomy
University of Hamburg
2000 Hamburg, Federal Republic of Germany

A. Björklund

Department of Histology
University of Lund
S-223 62 Lund, Sweden

W. Lovenberg

Section on Biochemical Pharmacology
National Heart, Lung, and Blood Institute
Bethesda, Maryland 20014

H. G. Schlossberger

Department of Organic Chemistry and Spectroscopy
Max Planck Institute for Biochemistry
8033 Martinsried, Munich, Federal Republic of Germany

The first description of the rather selective reduction in regional brain 5-hydroxytryptamine (5-HT) content after injection of 25–75 μg of 5,6-dihydroxytryptamine (5,6-DHT) into the lateral ventricle of the adult rat was provided by Baumgarten *et al.*[1] in 1971. These results were soon confirmed by two studies,[2, 3] one of which pointed to serious limitations in the applicability of 5,6-DHT as a tool for selective degeneration of serotonin (5-HT) neurons by demonstrating that there was a partial atrophy of the septum and caudate on the injection side.[2] Electron microscopic findings of Baumgarten *et al.*[4] revealed that, in addition to injuring nonmyelinated presumptive monoaminergic axons and terminals, 5,6-DHT also lesioned glial cells and myelinated fibers in the head of the caudate and the septum on the injection side. The potential of 5,6-DHT to nonselectively damage both catecholamine-containing and serotonin-containing neurons was substantiated by four findings: the reduction of striatal dopamine (DA) and forebrain noradrenaline (NA) content after a high intraventricular dose of 5,6-DHT,[2] the destruction of peripheral sympathetic noradrenergic fibers by 5,6-DHT, particularly after prior depletion of NA by reserpine and subsequent inhibition of monoamine oxidase (MAO),[5] the destruction of subependymally located DA terminals in the head of the left caudate nucleus after left lateral ventricular injections of low to high doses of 5,6-DHT (25–75 μg),[6] and the destruction of DA-containing axons by small intracerebral injections of 5,6-DHT.[7, 8] The need for a less generally cytotoxic serotonin neurotoxin was thus clearly evident, and in 1972 Baumgarten and Lachenmayer[9] described the effects of intraventricular administration of 5,7-dihydroxytryptamine (5,7-DHT) on central 5-HT neurons by use of the conventional Falck-Hillarp method. This dihydroxyindoleamine appears to have

0077–8923/78/0305–0003 $01.75/0

more pronounced lesioning actions on 5-HT axons than does 5,6-DHT. This drug was tolerated in much higher doses than was 5,6-DHT but damaged NA fibers, in addition to serotonergic fibers. This effect was particularly evident at higher dose levels (more than 50 μg of free base, intraventricularly).[9] The neurochemical and morphologic consequences of different intraventricular doses of 5,7-DHT on central nervous system levels of 5-HT, NA, and DA were studied in 1973 by Baumgarten *et al.*[10] 5,7-Dihydroxytryptamine was reported to cause dose-related decreases in regional 5-HT levels but nondose-related changes in whole brain and spinal NA content. This study received confirmation through the careful evaluation of changes in central nervous system monoamine, 5-hydroxyindoleacetic acid (5-HIAA), and tryptophan levels.[11, 12] Serotonin synthesis was also decreased in rat brain after intraventricular or intracisternal administration of 5,7-DHT.[11, 12]

In developing rats, Lytle *et al.*[13] obtained dose-related serotonin depletions in the brain after intracisternal injections of varying doses (12.5–50 μg) of 5,7-DHT, given at birth. Similar effects were observed by Breese and Cooper,[14] who used a high intracisternal dose of 5,7-DHT on Day 7 or 14 after birth. Three independent studies[14-16] established that the nondose-related decreases of cerebral NA that impair the specificity of action of 5,7-DHT could be antagonized by pretreatment of the animals with desmethylimipramine (DMI). Breese and Cooper[14] noted that the central NA fibers could also be protected from 5,7-DHT-induced toxicity by pretreatment with various MAO inhibitors. They also found that the anorectic side effects in newborn rats after DMI-5,7-DHT combinatioin treatment were not observed in the pargyline-5,6-DHT-treated animals. This observation can be interpreted as indicating that the latter treatment can be more easily tolerated by the developing animals. However, it remains to be shown conclusively that MAO inhibition improves the selectivity of 5,7-DHT. Sachs and Jonsson[17] showed that regional differences in the degree of 5-HT depletion in the central nervous system of the developing rat occur after systemic injection of 5,7-DHT during the first postnatal week. The depletion of 5-HT that occurs in individual brain regions after systemic 5,7-DHT treatment of neonatal rats differs significantly from that obtained after intraventricular or intracisternal injection of 5,7-DHT, in that systemic injection leads to a decrease of 5-HT in the forebrain and an increase of 5-HT in the brain stem. It is to be expected that the pattern and time course of recovery of 5-HT that will occur after intracisternal, intraventricular, or systemic 5,7-DHT treatment will also be very different.

More recent developments concern the search for more selective serotonin neurotoxins. Of all of the derivatives tested, only two compounds proved more selective than 5,6-DHT or 5,7-DHT when injected at a dose of 50 μg, namely, 5,6-diacetoxytryptamine and α-methyl-5,7-DHT.[18, 19] However, both analogs have the disadvantage that they are less potent than is 5,6-DHT or 5,7-DHT in destroying 5-HT axons. They also tended to be less selective and to have toxic effects on central catecholamine fibers when the intraventricular dose was increased above 50 μg.[18] The search for a more selective serotonin neurotoxin will therefore continue despite the fact that substantial and remarkably selective 5-HT depletions can be accomplished in the neonatal and adult rat brain with combined 5,7-DHT-DMI or 5,7-DHT-pargyline treatment.

In this paper, we will present new data on the potency and selectivity of action of two serotonin neurotoxins, 5,7-DHT and α-methyl-5,7-DHT, and will compare the intracisternal and intraventricular routes of administration. This review also summarizes some more recent findings relevant to the question of the conditions that determine the mode and mechanism of action of 5,6-DHT and 5,7-DHT on central monoaminergic and nonmonoaminergic structures.

In Vivo Binding of Radioactivity to Brain Proteins After Intraventricular Injection of [^{14}C]-5,6-DHT or [^{14}C]-5,7-DHT and Excretion of Radioactivity into the Urine

Accumulation and disappearance of radioactivity were analyzed in eight separate regions of the rat central nervous system at various times after injection of [1-^{14}C]-5,6-DHT [37] and in 13 central nervous system regions at various intervals after injection of [1-^{14}C]-5,7-DHT into the left lateral forebrain ventricle (Figures 1–6).

The radioactivity retained in the central nervous system was measured as free [perchloric acid (PCA) supernatant fraction] or bound (PCA-insoluble) material. As seen in Figures 1–6, significant amounts of [^{14}C]-5,6-DHT or [^{14}C]-5,7-DHT-derived radioactivity were bound to the protein-containing pellet in all regions analyzed and at all time points assayed. In the [^{14}C]-5,6-DHT-injected specimens, radioactivity increased during the first hour in all central nervous system regions that border on the lateral and third ventricles and on the aqueduct (Figure 1), whereas the [^{14}C]-5,7-DHT-derived radioactivity was at peak levels 30 min after injection in all regions (Figures 4 & 5), except the spinal cord (cf. Figure 6). This finding suggests that penetration of [^{14}C]-5,6-DHT into the brain parenchyma is retarded and may in part be dependent on transport through ependymal cells and that 5,7-DHT equilibrates more rapidly with brain tissue, probably because of its higher lipid solubility and lesser tendency to undergo nonspecific protein binding. Regions located at increasing distances from the injection site accumulated smaller amounts of radioactivity derived from [^{14}C]-5,6-DHT. This craniocaudal concentration gradient was less apparent in the [^{14}C]-5,7-DHT-treated brain regions and was clearly observed only in the three spinal cord portions.

Much smaller amounts of radioactivity were recovered from the PCA supernatant fraction than from the pellet in all of the periventricular forebrain regions and midbrain parts after injection of [^{14}C]-5,6-DHT. This soluble radioactivity had largely disappeared by 24 hr after injection (Figures 1–3). This observation contrasts with the finding of higher levels of radioactivity in the PCA-soluble supernatant fractions of all central nervous system regions in the [^{14}C]-5,7-DHT-treated animals (Figures 4–6) and also contrasts with respect to the time course of disappearance of radioactivity from the PCA-soluble fraction. The levels of [^{14}C]-5,7-DHT-derived radioactivity declined slowly and reached values close to background only by 4–8 days (cf. Figures 4–6). Approximately 50% of the administered radioactivity (2×10^6 dpm) from [1-^{14}C]-5,6-DHT was excreted in the urine during the first week (Figure 7). The majority of this radioactivity appeared during the first 2 days. After intraventricular injection of [^{14}C]-5,7-DHT, an equivalent amount of radioactivity appeared to have been excreted into the urine during the first few hours after drug administration (in about 3–6 hr). Logarithmic plots of the time-dependent accumulation of radioactivity in the urine suggests that [^{14}C]-5,7-DHT is excreted from different brain compartments at considerably different velocities (cf. Figure 8).

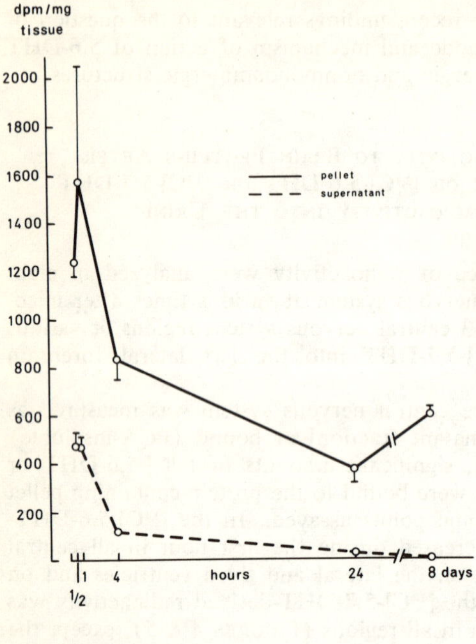

FIGURE 1. Hypothalamic acid-soluble (supernatant) and acid-insoluble (pellet) radioactivity at various times after intraventricular injection of [1-^{14}C]-5,6-DHT (75 μg of free base; 2×10^6 dpm); $n=3$. Regions from two animals were pooled.

FIGURE 2. Pons-medulla oblongata radioactivity. See the legend to FIGURE 1 for details.

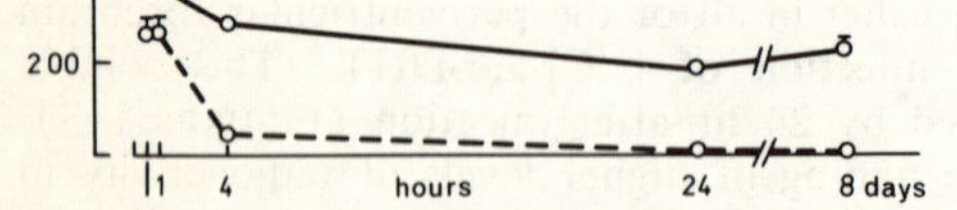

FIGURE 3. Spinal cord radioactivity. See the legend to FIGURE 1 for details.

For comparison with the above data, the accumulation and disappearance of brain radioactivity were analyzed 1 and 24 hr after intraventricular administration of [^{14}C]-5-HT. The activity in the supernatant fraction was about twice that recovered from the pellet by 1 hr in all central nervous system regions analyzed, and the radioactivity largely disappeared from both the pellet and the supernatant fractions within 1 day. Accordingly, by 24 hr, about 65% of the injected radioactivity was found in the urine. This amount is twofold higher than that seen after [^{14}C]-5,6-DHT injection but is close to that recovered from the 24-hr urine sample after administration of [^{14}C]-5,7-DHT. These *in vivo*

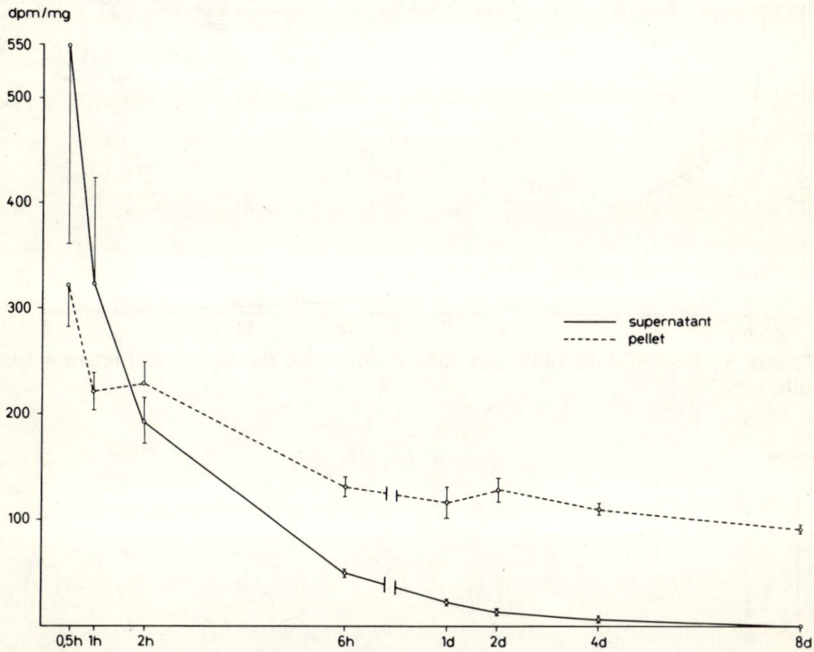

FIGURE 4. Hypothalamic acid-soluble (supernatant) and acid-insoluble (pellet) radioactivity at various times after intraventricular injection of [1-^{14}C]-5,7-DHT (94 μg of free base; 1.3×10^6 dpm); $n=3$. Regions from two animals were pooled.

differences in the binding of 5-HT, 5,6-DHT, and 5,7-DHT agree well with the relative binding of these compounds observed after a 1-hr incubation of brain slices *in vitro*.[19]

To ascertain whether the precipitation of radioactivity with the insoluble pellet fraction of PCA-homogenized brain tissue reflected irreversible protein binding, a differential extraction of brain tissue with polar and nonpolar solvents was performed. Control samples were obtained by homogenization of untreated brain regions in PCA that contained the same amount of [^{14}C]-5,6-DHT or [^{14}C]-5,7-DHT as that used for the *in vivo* injections. The supernatants of such control brain samples (washed five times in cold PCA) contained 94.5 and

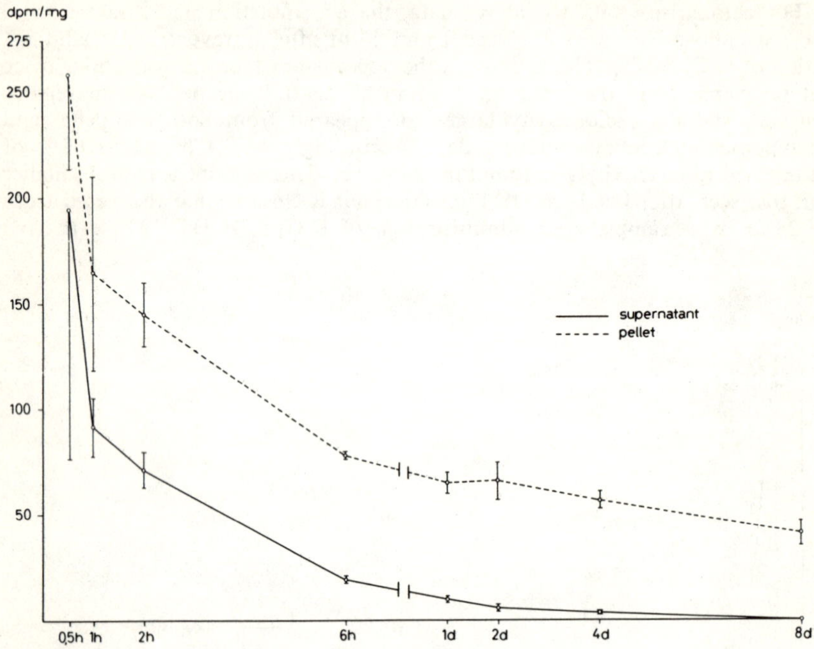

FIGURE 5. Pons-medulla oblongata radioactivity. See the legend to FIGURE 4 for details.

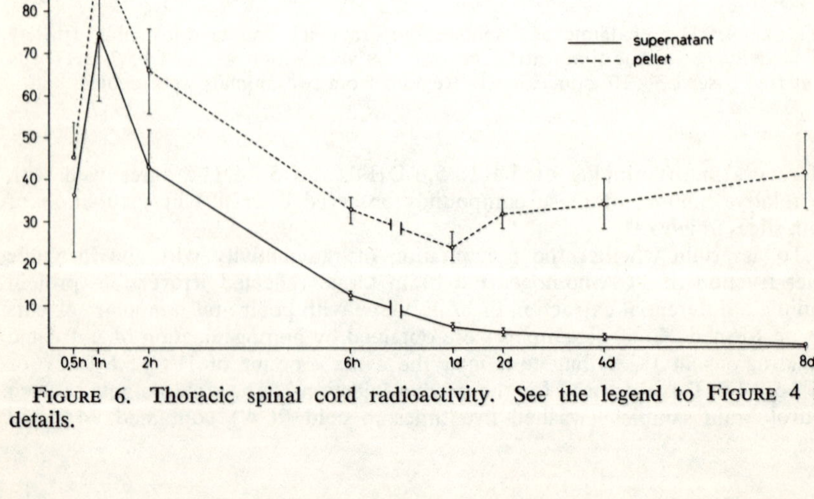

FIGURE 6. Thoracic spinal cord radioactivity. See the legend to FIGURE 4 for details.

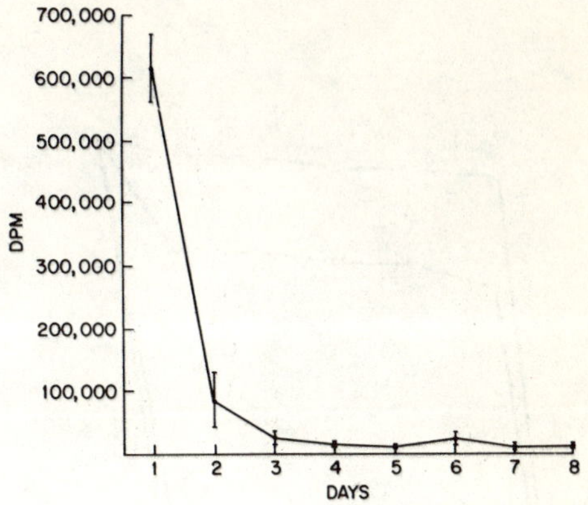

FIGURE 7. Urinary excretion of radioactivity from rats after intraventricular injection of 2×10^6 dpm of [1-^{14}C]-5,6-DHT (75 μg of free base); $n=6$.

78.5%, respectively, of the amount of [^{14}C]-5,6-DHT and [^{14}C]-5,7-DHT added to the homogenization medium (0-hr control in FIGURES 9 & 10). Further extraction of the PCA pellet with buffer, water, or Folch extraction medium (chloroform-methanol) failed to remove significant amounts of radioactivity from the [^{14}C]-5,6-DHT-treated control homogenate. The final pellet thus contained about 2.5% of the total radioactivity. Extraction of the PCA-insoluble pellet from the [^{14}C]-5,7-DHT-treated control homogenates with Folch medium resulted in a loss of 11.3% of the pellet-bound activity. The lipid-free pellet contained 5.7% of the total radioactivity added to the homogenization medium. This value is twice as much as that obtained after extraction of the [^{14}C]-5,6-DHT-treated control sample.

FIGURES 9 and 10 demonstrate that after intraventricular injection, the

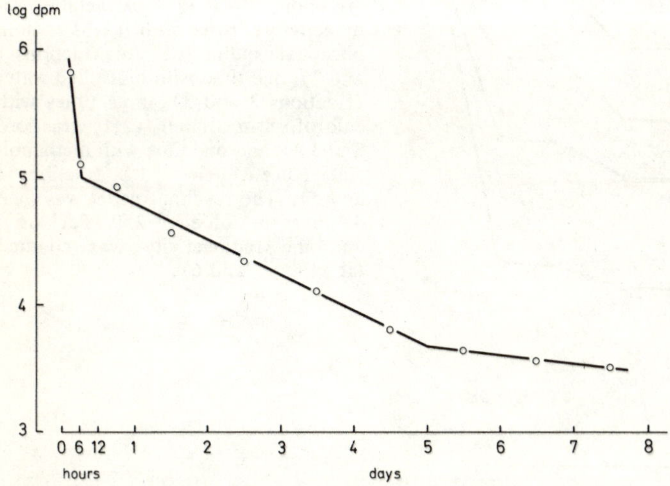

FIGURE 8. Urinary excretion of radioactivity (log dpm) from rats after intraventricular injection of 1.3×10^6 dpm of [1-^{14}C]-5,7-DHT (94 μg of free base); $n=6$.

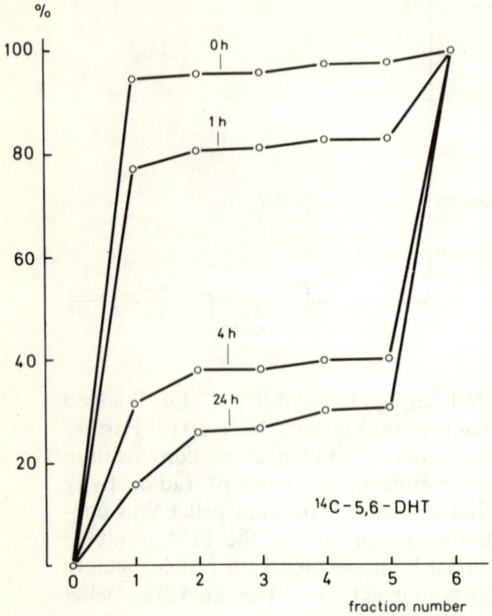

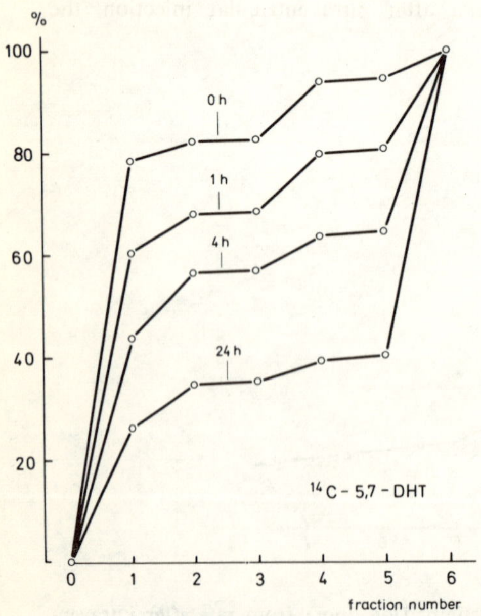

FIGURES 9 & 10. Differential, exhaustive extraction of PCA-homogenized combined central nervous system regions of the rat (hypothalamus, tectum, pons-medulla oblongata; pooled from two animals) injected with 3.6×10^6 dpm of [1-^{14}C]-5,6-DHT (75 μg of free base) (FIGURE 9, *top*) or 4×10^6 dpm of [1-^{14}C]-5,7-DHT (FIGURE 10, *bottom*) (1, 4, and 24 hr before homogenization in cold PCA). Blank values (0-hr control in FIGURES 9 and 10) were obtained by the addition to the first PCA homogenate of pooled control central nervous system regions of 3.6×10^6 dpm of [^{14}C]-5,6-DHT or 4×10^6 dpm of [^{14}C]-5,7-DHT. All data are expressed as the percentage of radioactivity contained in the first PCA homogenate (100%). The tissues were homogenized in PCA (0.4 N; 1:10, w/v), and the pellet was washed five times with PCA (fractions 0–1, PCA-extractable radioactivity), twice with 0.5 M sodium phosphate buffer (pH 7.4; fractions 1 and 2), one time with bidistilled water (fractions 2 and 3), three times with chloroform-methanol (2:1; fractions 3 and 4), and one time with methanol-glacial acetic acid (10:1; fractions 4 and 5). The resultant pellet was then dissolved in Soluene® 250 (Packard), and the radioactivity was counted (fractions 5 and 6).

PCA-extractable radioactivity (expressed as a percentage of the total radioactivity in the first PCA homogenate) decreased with the time between intraventricular injection of [^{14}C]-5,6-DHT or [^{14}C]-5,7-DHT and sacrifice. As indicated in FIGURES 9 and 10, there was a time-dependent relative increase of bound radioactivity in the final pellet. The pellet-bound radioactivity appeared more quickly in tissue from [^{14}C]-5,6-DHT-injected brains than from [^{14}C]-5,7-DHT treated ones. The Folch medium-extractable fraction of radioactivity from the pellet of [^{14}C]-5,7-DHT-injected brains (i.e., the difference between fractions 3 and 4 in FIGURES 9 and 10) showed relative diminution with time. Conversely, there was a time-dependent increase in the relative percentage of phosphate buffer-extractable radioactivity in both the [^{14}C]-5,6-DHT and the [^{14}C]-5,7-DHT-injected specimens. This increase was somewhat larger in the [^{14}C]-5,6-DHT-treated brain tissue.

The results described above indicate that, at least in the early hours after intraventricular injection of [^{14}C]-5,7-DHT, part of the PCA-precipitable radioactivity is not covalently bound to protein and that protein binding of radioactivity occurs somewhat faster and also more extensively after treatment with 5,6-DHT. The amount of PCA-insoluble radioactivity recovered from the [^{14}C]-5,7-DHT-treated pellet fractions 24 hr or more after injection may thus be roughly equal to the amount of protein-bound activity.

BINDING OF RADIOACTIVITY TO MITOCHONDRIAL FRACTIONS OF BRAIN TISSUE AFTER INTRAVENTRICULAR INJECTION OF [^{14}C]-5,6-DHT OR [^{14}C]-5,7-DHT

Three lines of evidence have suggested that mitochondria may be a site of accumulation of 5,6-DHT and 5,7-DHT and, therefore, somehow involved in mediating the neurotoxicity of these drugs: the appearance of electron-dense material in the intermembraneous compartment of mitochondria soon after intraventricular injection of 5,6-DHT;[4] the development of ultrastructural signs of mitochondrial damage after intraventricular or systemic 5,6-DHT or 5,7-DHT injections;[4,5,20,21] and the localization of much of the intraventricularly applied [^{14}C]-5,6-DHT- and [^{14}C]-5,7-DHT-derived radioactivity in the mitochondria-rich fractions of brain homogenates separated on a discontinuous density gradient.[22] From the observations that MAO inhibition prevents the degenerative actions of 5,7-DHT on cardiac sympathetic nerves and that the MAO-resistant analog α-methyl-5,7-DHT lacks long-term depletory effects on these nerves,[23,24] Creveling *et al.*[23] and Rotman *et al.*[24] have concluded that metabolism of 5,7-DHT by MAO is required for the expression of its toxicity in this system *in vivo.* The data reported here and those of Baumgarten *et al.*[18] and Björklund *et al.*,[19] however, suggest that this is not the case in the brain, because α-methyl-5,7-DHT exerts toxic effects on both NA- and 5-HT-containing axons in the brain. It was, therefore, of particular interest to determine how much of the [^{14}C]-5,6-DHT- and [^{14}C]-5,7-DHT-derived radioactivity in the brain is associated with MAO prepared from mitochondrial fractions of brain tissue.

For isolation of MAO, we employed the procedure of Shih and Eiduson.[25] Animals were processed 4 hr after intraventricular injection of either drug. The initial sucrose homogenate of periventricular brain regions contained 5.56×10^6 dpm of radioactivity (in the [^{14}C]-5,6-DHT-injected animals) and 3.69×10^6

dpm (in the [^{14}C]-5,7-DHT-injected animals i.e., 100%). The debris-free homogenate retained 40 and 48%, respectively, of the activity found in the sucrose homogenates. The crude mitochondrial pellet obtained by centrifugation of the debris-free homogenate was found to store about 21% of the initial activity in the sucrose homogenate. The pellet was dissolved in Triton® X-100 that contained phosphate buffer, and a subsequent 20% ammonium sulfate precipitation gave a supernatant from which MAO was precipitated by 40% saturation with ammonium sulfate (which contained only 0.3% of the initial radioactivity in the [^{14}C]-5,6-DHT-treated sample and 0.47% in the [^{14}C]-5,7-DHT-treated sample). The MAO fraction was dialyzed overnight. After centrifugation, the supernatant fraction (which contained the MAO) was found to contain 0.17 and 0.32%, respectively, of the initial radioactivity of the [^{14}C]-5,6-DHT- and [^{14}C]-5,7-DHT-treated sucrose homogenates.

Thus, this crude MAO bound twice as much [^{14}C]-5,7-DHT-derived radioactivity. This finding agrees with the observation that 5,7-DHT has a higher affinity than does 5,6-DHT for this enzyme.[38] The results demonstrate that only a very small part of the mitochondria-confined radioactivity is bound to the MAO-enriched fraction and that most of the radioactivity retained by mitochondria must be associated with mitochondrial constituents other than MAO.

Comparison of Intraventricular and Intracisternal Routes of Administration

As suggested by previous work, the extent, regional pattern, time course, and, possibly, also the selectivity of action of 5,6-DHT or 5,7-DHT on cerebral monoamines may vary with the site of administration of the neurotoxins into the cerebrospinal fluid. The injection site may also affect the subsequent regional and time course patterns of recovery of 5-HT and NA in the brain.[26]

In the present experiments, the intraventricular and intracisternal routes of administration of 5,7-DHT and α-methyl-5,7-DHT have been compared with respect to the pattern of depletion and recovery of monoamines in different regions of the central nervous system. In addition, the distribution of radioactivity in the pellet and supernatant fractions of different brain regions was investigated after intraventricular and intracisternal injections of [^{14}C]-5,7-DHT.

Distribution of Radioactivity

Twenty-four hours after a single intraventricular or intracisternal injection of 94 μg of [^{14}C]-5,7-DHT (1.3×10^6 dpm), retention of radioactivity occurs mainly in the pellet fraction of the PCA-homogenized brain regions (cf. Table 1). In intraventricularly injected brains, the highest amounts of radioactivity are contained in the PCA-insoluble pellets of the septum, left striatum, and hippocampus, which border on the injected left lateral ventricle. The unusually high amounts of radioactivity in the hypothalamus are undoubtedly due to the fact that this region has access to the third ventricle and protrudes into the cisternae intrapeduncularis, basalis, and chiasmatis. A craniocaudal gradient in the amounts of radioactivity was seen in the regions that surround the fourth ventricle and the central canal. Significant activity was also detected in the spinal cord, which is most distant from the injection site. As expected, the amounts of radioactivity accumulated by lower brain stem structures and by the spinal cord are higher in intracisternally injected animals, and the decrease in activity found with increasing distance from the injected fourth ventricle is striking in the rostral brain regions; the septum and caudate show only traces of label incorporation. From these findings, it must be concluded that 5,7-DHT injected intracisternally (provided that there is a relationship between the extent of protein binding and neurotoxic potency) will indirectly affect the forebrain serotonin terminals almost exclusively by an action on the most proximal parts of the nonterminal axons in the upper mesencephalon and/or by a direct action on the surface near 5-HT terminals in the basal forebrain regions.

TABLE 1

EFFECT OF INJECTION ROUTE ON REGIONAL RETENTION OF RADIOACTIVITY DERIVED FROM 5,7-DHT

	Radioactivity in Acid-Insoluble Fraction (dpm/mg tissue)	
	Intraventricular	Intracisternal
Left striatum	42.1±1.2	1.9±0.4
Right striatum	19.8±5.5	1.4±0.2
Hypothalamus	117.9±13.6	93.3±16.7
Pons-medulla oblongata	65.2±5.4	58.5±7.2
Cervical spinal cord	41.5±3.2	54.0±6.1
Thoracic spinal cord	24.3±3.7	38.0±7.8
Lumbar spinal cord	8.4±1.3	15.7±5.8

Brain Monoamine Content

The effect of 100 or 150 μg (free base) of 5,7-DHT, given either intracisternally or intraventricularly, with or without DMI pretreatment, on the monoamine content in the forebrain, spinal cord (rich in 5-HT terminals), and the brain stem (which contains all of the 5-HT cell bodies) of the adult rat was compared to monoamine levels in age-matched, vehicle-injected controls. Intraventricular injection caused larger initial degrees of 5-HT depletion in the forebrain, with almost no time-dependent recovery. By contrast, intracisternal injection resulted in a slower decline of forebrain 5-HT to minimum levels at 12 days and a clear-cut, progressive recovery in the 5-HT thereafter. Depletion in the spinal cord is similar with either route of administration. The percentage decrease in brain stem 5-HT was lower in intracisternally than in intraventricularly injected animals. The levels were transiently reduced to about 50% at Day 12 but recovered to values slightly above control by Day 42 (TABLE 2). Intraventricularly injected specimens, on the other hand, showed brain stem 5-HT depletion for up to 55 days (about 52–54% of control values). These important differences between the recovery patterns of 5-HT levels after both

TABLE 2
INFLUENCE OF 5,7-DHT AND α-METHYL-5,7-DHT ON REGIONAL 5-HT CONTENT AFTER VARIOUS TREATMENT REGIMENS

Treatment	Regional 5-HT Content (%) Forebrain	Brain Stem	Spinal Cord
Control *	100.0	100.0	100.0
5,7-DHT, 150 μg, intraventricular, 10 days	11.2	24.9	9.2
5,7-DHT, 150 μg, intracisternal, 10 days	40.2	40.2	7.5
α-Methyl-5,7-DHT, 150 μg, intraventricular, 10 days	38.9	41.6	15.3
α-Methyl-5,7-DHT, 150 μg, intracisternal, 10 days	52.5	62.6	17.2
DMI-5,7-DHT, 150 μg, intraventricular 10 days	15.8	33.0	9.1
DMI-5,7-DHT, 150 μg, intracisternal, 10 days	41.3	50.2	5.6

* The control content of 5-HT was as follows: forebrain, 0.267±0.028; brain stem, 0.402±0.027; spinal cord, 0.31±0.01 μg/g. The control group consisted of 36 samples, whereas each experimental result is the mean of six values.

routes of administration of 5,7-DHT should be taken into consideration when changes in serotonin-modulated functions are being investigated. Wuttke *et al.*[27] have recently given an example of such correlations and have shown that recovery of function (in this case, luteinizing hormone secretion) follows the time course of recovery of hypothalamic [^{3}H]-5-HT uptake and synthesis.

After intraventricular injection of 5,7-DHT in DMI-pretreated rats, brainstem NA and forebrain NA and DA content did not differ significantly from control values at any time point assayed, thus substantiating earlier findings that DMI protects NA neurons against the action of 5,7-DHT[15] (TABLE 3).

A more detailed evaluation of the effects of different intracisternal and intraventricular doses of α-methyl-5,7-DHT on forebrain, brain stem, and spinal cord levels of serotonin and catecholamines was conducted on animals assayed 10 days after drug administration (see TABLE 2). Intraventricular injection of 75 μg of α-methyl-5,7-DHT reduced brain stem and forebrain 5-HT by about 35% but decreased spinal cord 5-HT content to about 20% of control levels. Doubling the dose produced a further reduction of brain stem and forebrain 5-HT to about 40% of control values, but this increase in dose did not potentiate the effect on spinal cord 5-HT. The extent of the decrease in forebrain and brain stem 5-HT produced by 150 μg of α-methyl-5,7-DHT given intracisternally did not exceed that produced by half of that amount given intraventricularly (see above). Increasing the dose of α-methyl-5,7-DHT to 300 μg caused a further reduction of forebrain 5-HT (to about 30% of control) but did not enhance the depletion in brain stem and spinal cord 5-HT content. Therefore, with both routes of administration, 5,7-DHT-DMI is more powerful than is α-methyl-5,7-DHT with regard to long-term 5-HT reduction. All doses of α-methyl-5,7-DHT tested (both routes) caused decreases in forebrain, brain stem, and spinal cord NA, but significant depletion of forebrain DA content was evident only with intracisternal injection of 300 μg of α-methyl-5,7-DHT, clearly indicating that there is no gain in specificity of α-methyl-5,7-DHT over

5,7-DHT. This is particularly true if one takes into account the observation that the potency of α-methyl-5,7-DHT in depleting cerebral 5-HT is clearly lower than that of 5,7-DHT (TABLE 3).

Fluorescence Histochemistry

Coronal sections of rat brain perfused with formaldehyde and glyoxylic acid, according to the method of Loren *et al.*,[35] on Day 3 after administration of 5,7-DHT or α-methyl-5,7-DHT were analyzed for signs of damage to catecholamine fibers, particularly the NA and DA systems in the telencephalon and diencephalon. As reported previously,[27] 5,7-DHT injected intraventricularly at a dose of 100 μg into animals treated with DMI (intraperitoneally) caused some scattered lesions of DA fibers in the right and left caudates, in the supracommissural part of the bed nucleus of the stria terminalis, and in the medial parts of the olfactory tubercle. Administration of 150 μg of 5,7-DHT, with or without DMI pretreatment, produced a slightly higher frequency of lesions in DA fibers in the regions mentioned. Desmethylimipramine fully antagonized the damage exerted by 100 μg of 5,7-DHT, injected intraventricularly, on the NA fiber systems and almost completely prevented the NA damage caused by 150 μg of 5,7-DHT.[15] Intraventricular injection of α-methyl-5,7-DHT (150 μg) caused even greater damage to periventricular and basal forebrain DA fibers.

TABLE 3

INFLUENCE OF 5,7-DHT AND α-METHYL-5,7-DHT ON REGIONAL CATECHOLAMINE CONTENT AFTER VARIOUS TREATMENT REGIMENS

	Regional Nonadrenaline Content (%)			Forebrain Dopamine Content (%)
	Forebrain	Brain Stem	Spinal Cord	
Control *	100.0	100.0	100.0	100.0
α-Methyl-5,7-DHT, 150 μg, intraventricular, 10 days	54.2	64.2	36.9	87.2
α-Methyl-5,7-DHT, 150 μg, intracisternal, 10 days	77.6	80.7	26.6	104.3
α-Methyl-5,7-DHT, 300 μg, intracisternal, 10 days	53.1	54.3	10.7	70.2
DMI-5,7-DHT, 100 μg, intraventricular, 12 days	100.0	106.5	—	124.6
DMI-5,7-DHT, 100 μg, intraventricular, 55 days	96.0	110.5	—	101.1
DMI-5,7-DHT, 100 μg, intracisternal, 12 days	91.0	100.3	—	90.2
DMI-5,7-DHT, 100 μg, intracisternal, 42 days	99.7	107.3	—	108.8

* The control content of NA was as follows: forebrain, 0.277±0.012; brain stem, 0.405±0.034; spinal cord, 0.233±0.042. The DA content for the forebrain was 0.66±0.13 μg/g. The control group consisted of 36 samples, whereas each experimental result is the mean of six values.

The depletion of DA from the periventricular caudate was thus similar to that observed after intraventricular treatment with 5,6-DHT (75 μg).[6] α-Methyl-5,7-DHT also caused substantial damage to the NA fiber systems, which was partially antagonized by DMI pretreatment.

After administration of 5,7-DHT (150 μg) or α-methyl-5,7-DHT (150 or 300 μg) via the intracisternal route, there was little, if any, injury to the forebrain DA systems. Some scattered damaged fibers were observed in the medial medial part of the olfactory tubercle, but none were found in the caudate or septum. The damage inflicted on the NA fiber systems in the lateral hypothalamus was markedly counteracted by DMI pretreatment. These observations demonstrate that intracisternal injections of 5,7-DHT or α-methyl-5,7-DHT exert less toxic side effects on the forebrain DA innervations than do intraventricular injections.

AFFINITY OF [^{14}C]-5,7-DHT FOR THE 5-HT AND NA UPTAKE SITES IN BRAIN

Previous work[19, 28, 29] has shown that uptake of [^{3}H]-5-HT and [^{3}H]-NA into synaptosome-rich homogenates of rat hypothalamus is inhibited by 5,7-DHT to almost the same extent, and the tentative conclusion has been made that the affinity of 5,7-DHT for the NA and 5-HT uptake sites might be similar. If so, it remains unclear why 5,7-DHT should have a divergent toxic potency on noradrenergic and serotonergic fibers in the brain, particularly in doses higher than 50 μg (free base).[10, 18] Björklund *et al.*[19] attempted to resolve this obvious discrepancy by proposing that there might be differences in the rate of uptake of 5,7-DHT into the two types of monoaminergic neurons. We have now followed this line of reasoning and determined the actual affinity of [^{14}C]-5,7-DHT to the NA transport sites *in vitro* in serotonin-denervated homogenates of pooled hypothalamus, hippocampus, and tectum, to the 5-HT transport sites in NA-denervated brain regions, and to the NA and 5-HT transport sites in the untreated brain *in vitro*. The results are different from those of previous uptake inhibition experiments, in that the K_m of 5,7-DHT for 5-HT uptake (6.0×10^{-7} M) is about 1/16 of the K_m of 5,7-DHT for NA uptake (10.0×10^{-6} M). Therefore, 5,7-DHT has a 16-fold higher affinity for the 5-HT transport mechanism. The V_{max} of the two uptake systems is also different ($\sim 0.08 \times 10^{-12}$ mol of [^{14}C]-5,7-DHT/min for 5-HT uptake and 0.6×10^{-12} mol/min for NA uptake *). If these K_m values can be confirmed, the difference in the affinity of 5,7-DHT for the two transport systems is sufficiently large to explain the differences in the neurotoxic potency observed *in vivo,* and this difference also helps to understand why highly potent blockers of NA uptake, such as DMI, readily antagonize the accumulation of 5,7-DHT into NA neurons to concentrations sufficient to induce neuronal damage. These findings may also

* The K_m values given here must be considered preliminary, because the specific activity of the [^{14}C]-5,7-DHT used (25 mCi/mmol) is rather low.

have some bearing on the role of MAO inhibition in the counteraction of the toxicity of 5,7-DHT in central and peripheral NA fibers (see DISCUSSION).

METABOLISM OF [^{14}C]-5,6-DHT IN RAT BRAIN *in Vivo*

To understand the mechanism of action of 5,6-DHT on central monoaminergic neurons, the distribution of radioactivity in the brain after intraventricular injection of [^{14}C]-5,6-DHT, and the effects of drugs on the toxicity and distribution of 5,6-DHT in the brain, information about the fate of intraventricularly injected [^{14}C]-5,6-DHT is of great value. We therefore attempted to identify some metabolites of [^{14}C]-5,6-DHT 30 and 60 min after intraventricular injection of the labeled drug. Thin-layer chromatograms of PCA-soluble fractions from untreated and probenecid-treated whole rat brain were prepared. Radioactivity peaks were detected in spots that cochromatographed with authentic 5,6-DHT, 5-methoxy-6-hydroxytryptamine, and/or 6-methoxy-5-hydroxytryptamine and 5,6-dihydroxyindoleacetic acid (5,6-DHIAA) (in one case). Two of the radioactive spots were not isographic with, but localized just in front of, the following reference substances: 5-methoxy-6-hydroxyindoleacetic acid and/or 5-hydroxy-6-methoxyindoleacetic acid, and 5,6-DHIAA (in five cases). After pargyline pretreatment, the amount of radioactivity isographic with the methoxyhydroxytryptamines increased, whereas no radioactivity was detected at the reference spots of the acid metabolites, suggesting that the latter peaks, which were not strictly isographic with the methoxyhydroxyindoleacetic acids and 5,6-DHIAA, are formed in the brain by interaction with MAO. The results obtained by Klemm and Baumgarten[38] support this notion, because only a fraction of the radioactivity extractable from homogenates incubated with [^{14}C]-5,6-DHT was recovered as 5,6-DHIAA, and part of it represented nonidentifiable activity running close to the authentic 5,6-DHIAA on the chromatograms. Inhibition of MAO, however, prevented the formation of this radioactive spot. It is clear from these findings that both MAO and COMT (catechol-*O*-methyltransferase) are involved in the acute metabolism of intraventricularly injected 5,6-DHT.

DISCUSSION

Despite intense research conducted during the past few years, there remain unexplained several questions concerning the selectivity of action of 5,6-DHT and 5,7-DHT on central monoaminergic neurons, the discrepancies between the behavior of some indole neurotoxins *in vitro* and *in vivo,* and the differences in the suspected mechanism of action of 5,6-DHT and 5,7-DHT (the two most frequently used serotonin neurotoxins; for details, see Reference 19). Following several lines of research, we have attempted to provide an answer to these questions and to propose a unifying concept as to the molecular mechanism of toxicity of this class of compounds.

The dose of 5,6-DHT tolerated by adult rats without signs of unspecific cytotoxicity and general brain damage was much smaller than that of 5,7-DHT.

These differences in the tolerated dose range appear to depend on differences in molecular structure and reactivity of the drugs and on their metabolism. After administration of the labeled compounds into the cerebrospinal fluid, a considerably higher proportion of radioactivity from 5,6-DHT than from 5,7-DHT is bound to the acid-insoluble pellet. This radioactivity is not extractable in water, buffer, or lipid extraction medium, indicating that most, if not all, of the radioactivity is irreversibly bound to protein. Model experiments with homogenates of brain tissue incubated *in vitro* [38] have shown that both [^{14}C]-5,6-DHT and [^{14}C]-5,6-DHIAA undergo a rapid, irreversible binding to tissue proteins and also precipitate from the incubation medium as insoluble, colored polymers. Conversely, [^{14}C]-5,7-DHT is bound more slowly and to a lesser extent. The binding of [^{14}C]-5,7-DHT *in vitro* does not appear to be concentration dependent, and only a fraction of the radioactivity is irreversibly bound to protein. This phenomenon is also evident *in vivo,* because 10% or more of the acid-precipitable radioactivity can be extracted from the pellet by methanol-chloroform. The rapid and irreversible protein binding of [^{14}C]-5,6-DHT and of its acid metabolite [^{14}C]-5,6-DHIAA does not occur in the absence of oxygen. The *o*-dihydroxy configuration of 5,6-DHT and of some of its metabolic products indicates that they probably interact with molecular oxygen to form highly reactive *o*-quinones that have the capability of either undergoing polymerization (to establish melanin-like structures) or of reacting with sulfhydryl groups in peptides and proteins (see the recent paper by Rotman *et al.*[24]). This hypothesis implies that much of the intracerebrally injected 5,6-DHT binds nonselectively to susceptible proteins (those that have free sulfhydryl groups or accessible disulfide bonds) or precipitates as a polymer (indicated by the macroscopically visible discoloration of ventricle bordering brain structures; cf. Reference 2). This readily bound fraction of 5,6-DHT may not gain access to the 5-HT uptake sites in the brain and will probably contribute to its nonspecific cytotoxicity. Furthermore, the metabolism of 5,6-DHT by MAO gives rise to another fraction of intermediates that share with 5,6-DHT the property of reacting with molecular oxygen to form quinoid structures that also attack nucleophiles in peptides and proteins and thus participate in the deleterious actions of this compound.

As clearly demonstrated by the studies described by Klemm and Baumgarten,[38] the site of covalent attack may be the enzymes that metabolize 5,6-DHT, such as MAO, and possibly also aldehyde dehydrogenase. The formation of an aldehyde intermediate of 5,6-DHT provides a structure with an additional reactive site that might be involved in the inactivation of MAO *in vitro* and *in vivo.*

Metabolism of an *o*-methylated derivative of 5,6-DHT, for example, 5-hydroxy-6-methoxytryptamine, by MAO would result in the formation of another reactive aldehyde capable of interacting with molecular oxygen to form a quinoid system that may attack nucleophiles. It is at least conceivable that MAO itself participates in the detoxification of 5,6-DHT by transforming it into a more reactive species that becomes in part irreversibly bound to the enzyme itself. This binding of the 5,6-hydroxyindoleacetaldehyde or its autoxidation products to MAO would prevent the damage produced by this metabolite of 5,6-DHT on other susceptible structures of brain cells. The dramatic increase of nonspecific toxicity of 5,6-DHT in MAO-inhibited animals may be considered evidence for this paradoxic role of MAO.

Another interesting aspect of the nonselective cytotoxicity of 5,6-DHT relates to the role of the monoaminergic neurons as a sink for this drug. In autoradiograms of rat brain processed 1 hr after intraventricular injection of labeled 5,6-DHT,[39] a considerable fraction of label is observed confined to dopaminergic fibers and cell bodies. After a chemically induced degeneration of the central catecholaminergic fibers by intraventricular and intracisternal injection of 6-OHDA (combined with a MAO inhibitor), doses of 5,6-DHT become significantly more toxic than they are in nontreated animals.[39] This finding suggests that a significant fraction of unchanged 5,6-DHT is removed from the extracellular space by active uptake into catecholaminergic fibers, an action that thereby protects other structures from damage by the autoxidation products of 5,6-DHT.[36]

From test tube experiments, it is evident that both 5,6-DHT and 5,7-DHT undergo time-dependent reactions with molecular oxygen at biologic pH at 37° C (ultraviolet spectroscopic measurements). There are two important differences in the aerobic oxidation of 5,6-DHT and 5,7-DHT. For 5,6-DHT, the spectral changes indicate that the amount of unchanged 5,6-DHT decreases with time until finally 5,6-DHT is removed from the buffer by precipitation of a colored polymer, most probably formed via *o*-indolequinone intermediates. This aerobic oxidation of 5,6-DHT occurs rapidiy at 37° C. By contrast, there is a much slower, time-dependent interaction of 5,7-DHT with molecular oxygen in solution that results in the appearance of a new absorption maximum that may well be due to the formation of the *p*-quinoneimine of 5,7-DHT. There is no evidence for any polymerization of this stable autoxidation product of 5,7-DHT. The postulated *p*-quinoneimine of 5,7-DHT remains to be identified. If the *p*-quinoneimine is the species that is formed by interaction of 5,7-DHT with molecular oxygen *in vivo,* it would undergo covalent binding at a much reduced rate because of the negative charge on the oxygen in position 7 of the indole nucleus (as discussed by Rotman *et al.*[24]) and would be a less effective cross-linking agent.

Both 5,6-DHT and 5,7-DHT are substrates for MAO *in vitro* and *in vivo,*[38] but despite a higher affinity for this enzyme, 5,7-DHT is metabolized by MAO at a lower rate. However, we can postulate that much greater amounts of 5,7-DHT will finally be converted to the corresponding aldehyde and acid metabolites *in vivo* because of a prolonged availability of 5,7-DHT to this enzyme due to the slow formation of the less reactive *p*-quinoneimine, because of the absence of metabolism of 5,7-DHT by enzymes that compete with MAO for 5,7-DHT (e.g., COMT), and because MAO is less likely to be inactivated by 5,7-dihydroxyindoleacetaldehyde.[38] Thus, a much higher percentage of [^{14}C]-5,7-DHT-derived radioactivity is excreted from the brain and recovered in the urine during the first 24 hr than occurs with [^{14}C]-5,6-DHT.

The properties of 5,7-DHT described and discussed above greatly favor the accumulation of this form of DHT in monoaminergic neurons despite a lower affinity for the membrane-bound amine uptake mechanism when compared to 5,6-DHT (cf. References 19 and 28).

As outlined above for the monomethoxylated tryptamines, the interaction with MAO provides these relatively inert structures with a reactive site. The same would be true for 5,7-DHT. MAO catabolism appears to be the major mechanism of enzymatic metabolism of 5,7-DHT. If this assumption can be confirmed in ongoing investigations on the metabolism of [^{14}C]-5,7-DHT *in vivo,* one must conclude that the metabolism of 5,7-DHT increases the neurotoxic potential by forming a reactive aldehyde. Conversely, one would have to postulate that inhibition of MAO should diminish the molar neurotoxicity of 5,7-DHT and the extent of its irreversible protein binding. A similar effect should result from α-methylation of 5,7-DHT, whereby the drug would become resistant to metabolism by MAO. In fact, our comparative studies on long-term depletion of brain 5-HT, NA, and DA reveal that the molar toxicity of 5,7-DHT is higher than that with α-methyl-5,7-DHT (by 30–40%). These studies on NA and 5-HT neurons also suggest that the injury exerted on the central DA fibers is greater with the α-methyl analogs. The same tendency is observed *in vitro.*[19] It is also relevant to point out that a dose as small as 75 μg of α-methyl-5,7-DHT given intraventricularly causes significant depletion of NA in the forebrain (by 25%) and that 150 μg of α-methyl-5,7-DHT reduces NA even further (to about 55% of control). Thus, contrary to the findings of Daly *et al.*[30] and Creveling *et al.*,[23] who did not obtain long-lasting depletion of cardiac NA in mice treated with α-methyl-5,7-DHT, the MAO-resistant 5,7-DHT analog proved toxic in central 5-HT and NA neurons, though less potent than 5,7-DHT. This finding is remarkable, because inhibition of MAO prior to intracisternal administration of 5,7-DHT has been found to antagonize the long-term decrease of brain NA in adult and newborn rats.[14] However, it did not influence the dose-dependent long-lasting reduction of brain 5-HT. Pargyline increased, rather than decreased, the amount of protein-bound radioactivity in pellet fractions of PCA-homogenized brain regions by 24 hr and to even a greater extent by 48 hr after intraventricular administration of [^{14}C]-5,7-DHT.[39] This finding suggests that MAO is probably not significantly involved in the irreversible binding of 5,7-DHT to brain proteins, nor is it required for 5,7-DHT to injure central 5-HT fibers. Thus, it appears doubtful if cross-linking of proteins by the 5,7-dihydroxyindoleacetaldehyde or its *p*-quinoneimine is required for the neurotoxicity of 5,7-DHT in the central nervous system, as proposed by Rotman *et al.*,[24] for peripheral noradrenergic axons. As an alternative, it is conceivable that nucleophilic 4-addition of the *p*-quinoneimine of 5,7-DHT to functionally indispensable groups in proteins without substantial cross-linking may be sufficient for this drug to become neurotoxic.

Our recent data, obtained by measuring the incorporation of [^{14}C]-5,7-DHT into rat brain homogenates of animals subjected to selective degeneration of their central 5-HT or NA system (see above), indicate that the affinity of 5,7-DHT for 5-HT and NA uptake differs by a factor of about 16. This observation is in contrast to what is suggested by the IC_{50} values of 5,7-DHT as inhibitors of 5-HT and NA uptake in hypothalamic homogenates,[19, 28] namely, that 5,7-DHT should have equal affinity for the 5-HT and NA transport sites.

The difference in the affinity of the two transport systems is sufficiently large to explain why central 5-HT neurons are always more profoundly affected by 5,7-DHT than are NA neurons. The results of our autoradiographic investigations[39] give direct support for this assumption. Chloroimipramine is somewhat more potent in inhibiting 5-HT transport into 5-HT neurons that NA transport into NA neurons (cf. References 31 and 32). This drug has, in fact, been found to attenuate the damaging effect of 100 μg of 5,7-DHT (given intraventricularly) on NA, but not on 5-HT, neurons.[40]

If we now consider the influence of MAO inhibition on the uptake and accumulation of 5,7-DHT in noradrenergic neurons, the following picture emerges: By inhibition of the catabolism of NA, MAO inhibition can be assumed to result in an accumulation of NA at the synapse, for example, by increasing cytoplasmic NA and increasing efflux of NA from the neuron but decreasing the net uptake of NA into the neuron. The fraction of NA that effluxes from the neuron will compete with 5,7-DHT for uptake into the NA nerve terminals. The difference in the affinity of NA for NA uptake ($K_m = 4 \times 10^{-7}$ M[34]) and of 5,7-DHT for NA uptake (10×10^{-6} M) is, considering the decreased rate of uptake of amines into NA neurons under the influence of MAO inhibition, large enough to prevent accumulation of concentrations of 5,7-DHT that exert toxic effects on NA axons. Therefore, the net uptake of 5,7-DHT is decreased in MAO-treated animals. In addition, formation of the more reactive aldehyde of 5,7-DHT is prevented, and the molar toxicity of 5,7-DHT is reduced by about 30–40% (see comparison of the neurotoxic potency of 5,7-DHT and α-methyl-5,7-DHT). In concert, these changes in the distribution of endogenous NA and their consequences for 5,7-DHT uptake and the lack of toxicity reinforcement by MAO catabolism explain the absence of toxicity of intraventricular 5,7-DHT in MAO-inhibited animals. Due to the higher affinity of 5,7-DHT for the 5-HT uptake sites, the consequences of MAO inhibition for the net transport of 5,7-DHT into the serotonin axons may not be as deleterious as for 5,7-DHT transport into NA neurons.

In conclusion, it appears that 5,6-DHT and 5,7-DHT become neurotoxic *in vivo* by different mechanisms, mainly because of differences in the nature and reactivity of the products formed by interaction with molecular oxygen (*o*-quinone and *p*-quinoneimine, respectively). The concept that 5,7-DHT attains neurotoxic potency only if metabolized by MAO *in vivo*[24] is difficult to reconcile with the observation that α-methyl-5,7-DHT is toxic to central 5-HT, NA, and DA axons. If α-methyl-5,7-DHT accumulates in central 5-HT and NA axons at the same rate as does 5,7-DHT, the decrease in neurotoxic potency that follows from α-methylation is evidence for an enhancement of the neurotoxic potency of 5,7-DHT through metabolism by MAO. The differences in the potency of 5,7-DHT as a neurotoxin in central NA and 5-HT fibers may be explained by its higher affinity for the 5-HT transport sites in the rat central nervous system.

References

1. Baumgarten, H. G., A. Björklund, L. Lachenmayer, A. Nobin & U. Stenevi. 1971. Long-lasting selective depletion of brain serotonin by 5,6-dihydroxytryptamine. Acta Physiol. Scand. Suppl. **373:** 1–15.
2. Baumgarten, H. G., K. D. Evetts, R. B. Holman, L. L. Iversen, M. Vogt & G. Wilson. 1972. Effects of 5,6-dihydroxytryptamine on monoaminergic neurones in the central nervous system of the rat. J. Neurochem. **19:** 1587–1597.
3. Costa, E., H. Lefevre, J. Meek, A. Revuelta, F. Spano, S. Strada & J. Daly. 1972. Serotonin and catecholamine concentrations in brain of rats injected intracerebrally with 5,6-dihydroxytryptamine. Brain Res. **44:** 304–308.

4. Baumgarten, H. G., A. Björklund, A. F. Holstein & A. Nobin. 1972. Chemical degeneration of indoleamine axons in rat brain by 5,6-dihydroxytryptamine, an ultrastructural study. Z. Zellforsch. **129:** 256–271.

5. Baumgarten, H. G., M. Goethert, A. F. Holstein & H. G. Schlossberger. 1972. Chemical sympathectomy induced by 5,6-dihydroxytryptamine. Z. Zellforsch. **128:** 115–134.

6. Björklund, A., A. Nobin & U. Stenevi. 1973. Effects of 5,6-dihydroxytryptamine on nerve terminal serotonin and serotonin uptake in the rat brain. Brain Res. **53:** 117–127.

7. Björklund, A., A. Nobin & U. Stenevi. 1973. The use of neurotoxic dihydroxytryptamines as tools for morphological studies and localized lesioning of central indoleamine neurons. Z. Zellforsch. **145:** 479–501.

8. Saner, A., L. Pieri, J. Moran, M. Da Prada & A. Pletscher. 1975. Decrease of dopamine and 5-hydroxytryptamine after intracerebral application of 5,6-dihydroxytryptamine. Brain Res. **76:** 109–117.

9. Baumgarten, H. G. & L. Lachenmayer. 1972. 5,7-Dihydroxytryptamine: improvement in chemical lesioning of indoleamine neurons in the mammalian brain. Z. Zellforsch. **135:** 399–414.

10. Baumgarten, H. G., A. Björklund, L. Lachenmayer & A. Nobin. 1973. Evaluation of the effects of 5,7-dihydroxytryptamine on serotonin and catecholamine neurons in the rat CNS. Acta Physiol. Scand. Suppl. **391:** 1–19.

11. Gerson, S., R. J. Baldessarini & S. C. Wheeler. 1974. Biochemical effects of dihydroxylated tryptamines on central indoleamine neurons. Neuropharmacology **13:** 987–1004.

12. Jacoby, J. H., L. D. Lytle & M. F. Nelson. 1974. Long-term effects of 5,7-dihydroxytryptamines on brain monoamines. Life Sci. **14:** 909–919.

13. Lytle, L. D., J. H. Jacoby, M. F. Nelson & H. G. Baumgarten. 1975. Long-term effects of 5,7-dihydroxytryptamine administered at birth on the development of brain monoamines. Life Sci. **15:** 1203–1217.

14. Breese, G. R. & B. R. Cooper. 1975. Behavioral and biochemical interactions of 5,7-dihydroxytryptamine with various drugs when administered intracisternally to adult and developing rats. Brain Res. **98:** 517-528.

15. Björklund, A., H. G. Baumgarten & A. Rensch. 1975. 5,7-Dihydroxytryptamine: improvement of its selectivity for serotonin neurons in the CNS by pretreatment with desipramine. J. Neurochem. **24:** 833–835.

16. Gerson, S. & R. J. Baldessarini. 1975. Selective destruction of serotonin terminals in rat forebrain by high doses of 5,7-dihydroxytryptamine. Brain Res. **85:** 140–145.

17. Sachs, Ch. & G. Jonsson. 1975. 5,7-Dihydroxytryptamine induced changes in the postnatal development of central 5-hydroxytryptamine neurons. Med. Biol. **53**: 156–164.

18. Baumgarten, H. G., A. Björklund, A. Nobin, E. Rosengren & H. G. Schlossberger. 1975. Neurotoxicity of hydroxylated tryptamines: structure-activity relationships. 1. Long-term effects on monoamine content and fluorescence morphology of central monoamine neurons. Acta Physiol. Scand. Suppl. **429:** 1–27.

19. Björklund, A., A. S. Horn, H. G. Baumgarten, A. Nobin & H. G. Schlossberger. 1975. Neurotoxicity of hydroxylated tryptamines: structure-activity relationships. 2. In vitro studies on monoamine uptake inhibition and uptake impairment. Acta Physiol. Scand. Suppl. **429:** 29–60.

20. Baumgarten, H. G., H. P. Groth, M. Goethert & A. A. Manian. 1974. The effect of 5,7-dihydroxytryptamine on peripheral adrenergic nerves in the mouse. Arch. Exp. Pathol. Pharmakol. **282:** 245–254.

21. MØLLGARD, K., J. J. LUNDBERG, L. LACHENMAYER, L. WIKLUND & H. G. BAUMGARTEN. This monograph.

22. BAUMGARTEN, H. G., A. BJÖRKLUND & D. F. BOGNASKI. 1975. Similarities and differences in the mode of action of 6-hydroxydopamine and neurotoxic indoleamines. *In* Chemical Tools in Catecholamine Research. G. Jonsson, T. Malmfors & Ch. Sachs, Eds. : 59–66. North Holland Publishing Company. Amsterdam, The Netherlands.

23. CREVELING, C. R., J. LUNDSTRÖM, E. T. MCNEAL, L. TICE & J. W. DALY. 1975. Dihydroxytryptamines: effects on noradrenergic function in mouse heart in vivo. Mol. Pharmacol. **11:** 211–222.

24. ROTMAN, A., J. W. DALY & C. R. CREVELING. 1976. Oxygen-dependent reaction of 6-hydroxydopamine, 5,6-dihydroxytryptamine, and related compounds with proteins in vitro: a model for cytotoxicity. Mol. Pharmacol. **12:** 887–899.

25. SHIH, J. C. & S. EIDUSON. 1973. Monoamine oxidase (EC 1.4.3.4.): isolation and characterization of multiple forms of the brain enzyme. J. Neurochem. **21:** 41–49.

26. BAUMGARTEN, H. G., L. LACHENMAYER & A. BJÖRKLUND. 1977. Chemical lesioning of indoleamine pathways. *In* Methods in Psychobiology. R. D. Myers, Ed. Vol. **III:** 47–98. Academic Press. London and New York.

27. WUTTKE, W., A. BJÖRKLUND, H. G. BAUMGARTEN, L. LACHENMAYER, M. FENSKE & H. P. KLEMM. 1977. De- and regeneration of brain serotonin neurons following 5,7-dihydroxytryptamine treatment: effects on serum LH, FSH and prolactin levels in male rats. Brain Res. **134:** 317–331.

28. HORN, A. S., H. G. BAUMGARTEN & H. G. SCHLOSSBERGER. 1973. Inhibition of the uptake of 5-hydroxytryptamine, noradrenaline and dopamine into rat brain homogenates by various hydroxylated tryptamines. J. Neurochem. **21:** 233–236.

29. BAUMGARTEN, H. G., A. BJÖRKLUND, A. S. HORN & H. G. SCHLOSSBERGER. 1974. Studies on the neurotoxic properties of hydroxylated tryptamines. *In* Dynamics of Degeneration and Growth in Neurons. K. Fuxe, L. Olson & Y. Zotterman, Eds. : 153–167. Pergamon Press. Oxford, England.

30. DALY, J., J. LUNDSTRÖM & C. R. CREVELING. 1974. Structure-activity correlations in trihydroxyphenylethylamines and dihydroxytryptamines, relationship to cytotoxicity in adrenergic and serotonergic neurons. *In* Dynamics of Degeneration and Growth in Neurons. K. Fuxe, L. Olson & Y. Zotterman, Eds. : 29–42. Pergamon Press. Oxford, England.

31. IVERSEN, L. L. 1975. Uptake processes for biogenic amines. *In* Handbook of Psychopharmacology. L. L. Iversen, S. D. Iversen & S. H. Snyder, Eds. Vol. 3: 381–442. Plenum Press. New York and London.

32 HORN, A. S. 1976. Characteristics of transport in dopaminergic neurons. *In* The Mechanism of Neuronal and Extraneuronal Transport of Catecholamines. D. M. Paton, Ed. : 195–214. The Raven Press. New York, N.Y.

33. GRAEFE, K. H. 1976. Methodology of catecholamine transport studies: definition of terms. *In* The Mechanism of Neuronal and Extraneuronal Transport of Catecholamines. D. M. Paton, Ed. : 7–35. The Raven Press. New York, N.Y.

34. SNYDER, S. H. & J. T. COYLE. 1969. Regional differences in ^{3}H-norepinephrine and ^{3}H-dopamine uptake into rat brain homogenates. J. Pharmacol. Exp. Ther. **165:** 78–86.

35. LOREN, J., A. BJÖRKLUND, B. FALCK & O. LINDVALL. 1976. An improved histofluorescence procedure for freeze-dried paraffin-embedded tissue based on combined formaldehyde-glyoxylic acid perfusion with high magnesium content and acid pH. Histochemistry **49:** 177–192.

36. BJÖRKLUND, A., H. G. BAUMGARTEN & A. NOBIN. 1974. Chemical lesioning of central monoamine axons by means of 5,6-dihydroxytryptamine and 5,7-dihydroxytryptamine. Advan. Biochem. Psychopharmacol. **10:** 13–33.

37. Baumgarten, H. G., S. J. Victor, W. Lovenberg & H. G. Schlossberger. 1977. To be published.
38. Klemm, H. P. & H. G. Baumgarten. This monograph.
39. Baumgarten, H. G., H. P. Klemm, L. Lachenmayer & H. G. Schlossberger. This monograph.
40. Baumgarten, H. G. 1977. Unpublished results.

SYNTHESIS AND CHEMICAL PROPERTIES OF SOME INDOLE DERIVATIVES

H. G. Schlossberger

Department of Organic Chemistry and Spectroscopy
Max Planck Institute for Biochemistry
8033 Martinsried, Munich, Federal Republic of Germany

Ortho-dihydroxytryptamines (*o*-DHT) have been claimed to occur in biologic tissues (e.g., in the crustacean heart[1]) and hypothesized to represent "active" metabolites of serotonin.[2] These assumptions prompted us to synthesize 5,6-DHT[3,4] and 4.5-DHT.[5] More recent work by Baumgarten *et al.*[6,7] renders these earlier assumptions unlikely, since both drugs damage central serotonin (5-HT) neurons when introduced into the cerebrospinal fluid. Both 5,6-DHT and 4,5-DHT have limited applicability as tools for induction of serotonin fiber degeneration because of unspecific cytotoxicity. These findings stimulated a search for neurotoxic indole derivatives with lesser side effects. Isomeric dihydroxytryptamines, in addition to N^{ω}-monomethylated, N^{ω}-dimethylated, α-*C*-alkylated, and *O*-methylated derivatives, and otherwise substituted dihydroxytryptamines appeared to be of special interest in this context.[8–11]

SYNTHESES

The synthesis of tryptamines modified with two substituents on the benzene nucleus of the indole ring is generally performed according to methods for synthesis of simple or nonsubstituted tryptamines; that is, in general, the side chain is attached to an indole derivative.[12] For the synthesis of dihydroxy- and hydroxymethoxytryptamines, the indoles required were the corresponding dibenzyloxy- resp. benzyloxymethoxyindoles. The free hydroxyl groups in the tryptamines to be synthesized must be protected in this stage of synthesis. We obtained these indoles by two well-established methods: reductive cyclization of *o*,ω-dinitrostyrenes[13] or according to the method of Reissert,[14] namely, condensation of an *o*-nitrotoluene with diethyl oxalate to the corresponding ethyl *o*-nitrophenylpyruvate and, after cyclization and saponification, decarboxylation of the indole-2-carboxylic acid.

The accessibility of the starting material, either *o*,ω-dinitrostyrene or *o*-nitrotoluene, is of importance for the choice of the method of synthesis of the indole derivative. The last step in the synthesis, according to Reissert, decarboxylation, is the critical step; depending on the substitution, the yield of indole varies greatly.[15–17]

FIGURE 1 shows the different methods we used to synthesize 4,5-dibenzyloxyindole, the starting indole for the preparation of 4,5-DHT. This synthesis has not been published before but will be outlined below.

6-Nitrovanillin (**II**)[18] obtained from o-vanillin (**I**) could not be demethylated to 6-nitro-2,3-dihydroxybenzaldehyde (**III**), so that 6-nitro-2,3-dibenzyloxybenzaldehyde (**XV**) was not accessible by this method. Reduction of **I** yields 2-hydroxy-3-methoxytoluene (**VII**), the 6-nitro derivative of which, **VIII**,

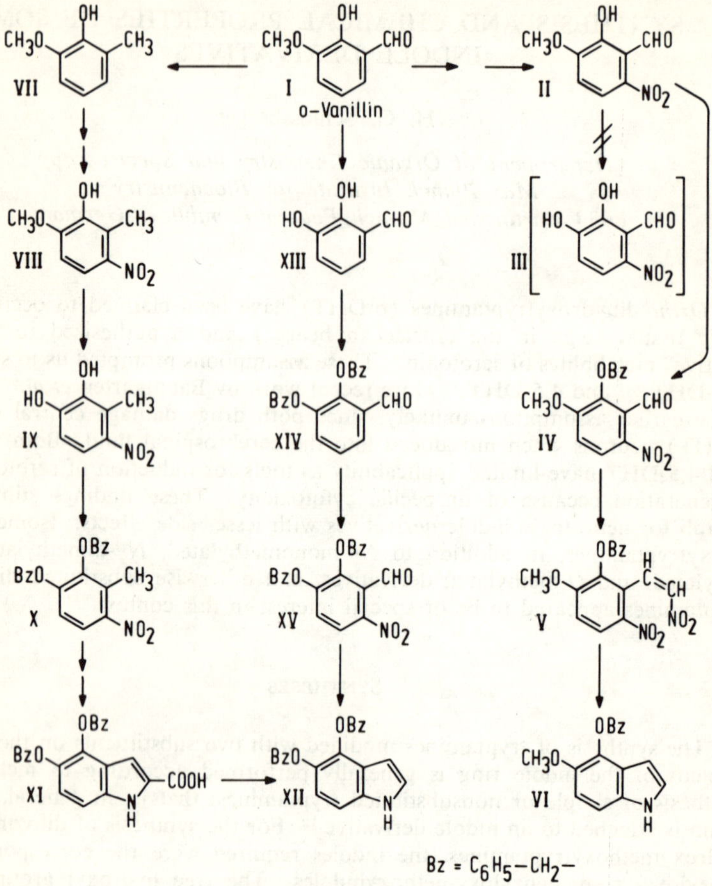

FIGURE 1. Synthesis of 4,5-dibenzyloxyindole (**XII**), and 4-benzyloxy-5-methoxyindole (**VI**).

gives 6-nitro-2,3-dihydroxytoluene (**IX**) during demethylation. From 6-nitro-2,3-dibenzyloxytoluene (**X**), ethyl 6-nitro-2,3-dibenzyloxyphenylpyruvate is obtained, and by its cyclization, ethyl 4,5-dibenzyloxyindole-2-carboxylate is produced. The yield of 4,5-dibenzyloxyindole (**XII**) by decarboxylation of 4,5-dibenzyloxyindole-2-carboxylic acid (**XI**) [19] is so poor that for the preparation of **XII** in larger quantities, another method had to be chosen. Compound **I** was demethylated to 2,3-dihydroxybenzaldehyde (**XIII**),[20] the dibenzyl derivative (**XIV**) of which could be nitrated to 6-nitro-2,3-dibenzyloxybenzaldehyde (**XV**) in a yield of 40%. Condensation with nitromethane and subsequent reductive cyclization give **XII** in an overall yield of 10%. Compound **II** reacts with benzylchloride to 6-nitro-2-benzyloxy-3-methoxybenzaldehyde (**IV**); via the stage of the *o,ω*-dinitrostyrene **V**, 4-benzyloxy-5-methoxyindole (**VI**) [19] is obtained, the indole derivative for the synthesis of 4-hydroxy-5-methoxytryptamine.

Among the different methods that exist for attaching side chains to an indole, we chose the oxalyl chloride method of Speeter and Anthony[21] for the preparation of N^{ω}-monomethylated and N^{ω}-dimethylated tryptamine derivatives and also for tryptamines with a free amino group. For α-alkylated derivatives, the synthesis must proceed via the indole-3-carboxaldehyde and the corresponding nitrovinylindole,[22] but α-unsubstituted tryptamines are also available by this method.

FIGURE 2 shows all of the reaction steps that proceed via the gramine (**XVI**)[23] and that we used successfully for preparation of numerous dihydroxy- and hydroxymethoxyindole derivatives. The corresponding tryptophan (**XIX**) is obtained according to Snyder and Smith.[24] Via the stages of the nitrile (**XVII**), the tryptamines[25, 26] (**XVIII**), indole acetic acids[27] (**XX**), and tryptophols[27] (**XXI**) are accessible. By means of catalytic debenzylation, the end products are obtained. The stage of indoleacetaldehyde (**XXII**) would be the intermediate for dihydroxy- and hydroxymethoxyindoleacetaldehydes.

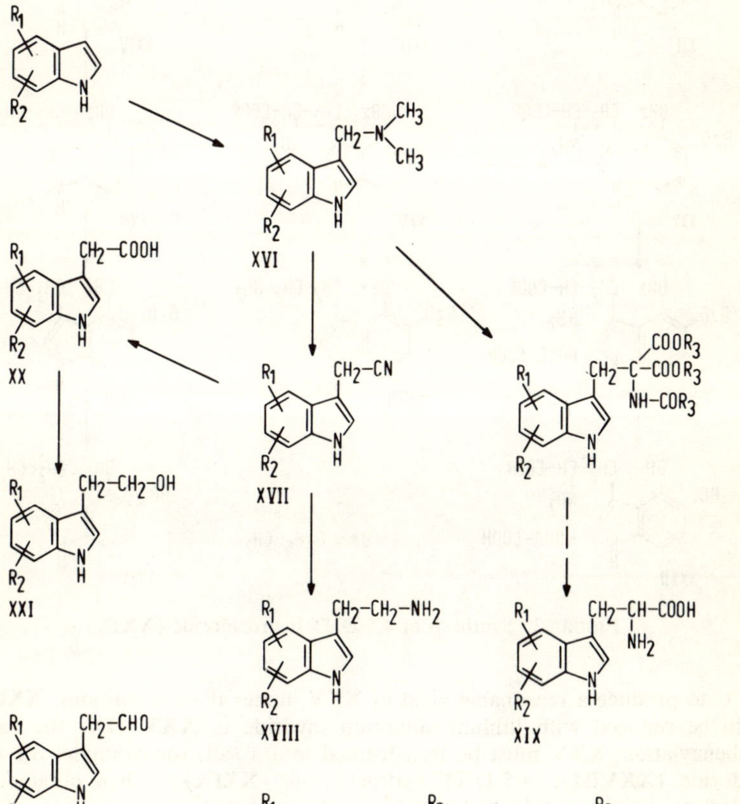

R_1	R_2	R_3
$C_6H_5-CH_2O-$	$C_6H_5-CH_2O$	i.e. CH_3
$C_6H_5-CH_2O-$	CH_3O-	

FIGURE 2. Syntheses of indole derivatives via the stage of gramine.

In FIGURE 3, the synthesis of 4,5-DHT is shown. From 4,5-dibenzyloxygramine (**XXIII**), 4,5-dibenzyloxyindoleacetonitrile (**XXIV**) is obtained. However, attempts to reduce this compound to 4,5-dibenzyloxytryptamine (**XXV**) with lithium aluminum hydride were unsuccessful at first; because large quantities of dark-colored products were always obtained; it was not possible to isolate **XXV** from the reaction mixture. Therefore, we condensed **XXIII** with ethyl nitroacetate to ethyl α-nitro-β-4,5-dibenzyloxyindole-3-propionate (**XXVI**) according to Lyttle and Weisblat;[28] after saponification, the nitropropionic acid derivative is decarboxylated to give 1-nitro-2-(4,5-dibenzyloxyindole-3)-ethane (**XXVII**). After many attempts to reduce **XXVII** to **XXV** with lithium aluminum hydride, we found it necessary to keep the reaction temperature at

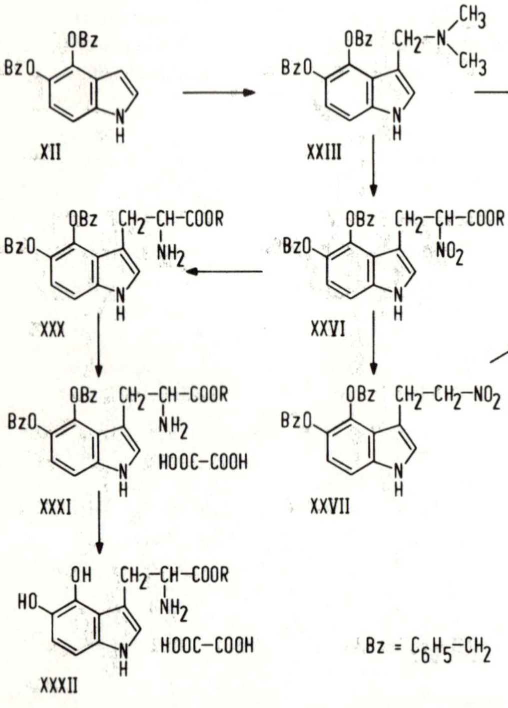

FIGURE 3. Synthesis of 4,5-DHT hydrochloride (**XXIX**).

0° C to produce a reasonable yield of **XXV**; under these conditions, **XXIV** can also be reduced with lithium aluminum hydride to **XXV**. For the catalytic debenzylation, **XXV** must be transformed into a salt, for example, the hydrochloride (**XXVIII**). 4,5-DHT hydrochloride (**XXIX**) is then obtained as a colorless substance, which, however, in the presence of oxygen becomes blue within a few hours. Analogously, 4-hydroxy-5-methoxytryptamine[5, 29] hydrochloride is obtained from 4-benzyloxy-5-methoxyindole (**VI**) via the stages of the gramine, nitrile, and the 4-benzyloxy-5-methoxytryptamine. This compound is somewhat more stable than **XXIX**; however, with time, it also becomes blue when exposed to air. The ultraviolet absorption spectrum of 4-hydroxy-5-

methoxytryptamine is seen in FIGURE 4; the spectra of **XXV** and 4-benzyloxy-5-methoxytryptamine are shown in FIGURE 5. Compound **XXVI** is reduced, according to Lyttle and Weisblat,[28] to ethyl-D,L-4,5-dibenzyloxytryptophan, the hydrogen oxalate, of which **XXXI** is debenzylated to ethyl-D,L-4,5-dihydroxytryptophan hydrogen oxalate (**XXXII**).

The rapid autoxidation of both 4,5-DHT and 4-hydroxy-5-methoxytryptamine in a buffer at physiologic pH prevents the application of these two tryptamines as selective neurotoxic pharmacologic agents.

The most potent and useful indole neurotoxins are 5,6-DHT and 5,7-DHT.[30, 31] 5,7-DHT, first synthesized by Lee *et al.*,[10] exhibits effects on serotonergic neurons that are qualitatively similar to those produced by 5,6-DHT. For autoradiographic studies and for *in vivo* and *in vitro* metabolism, both 5,6-DHT and 5,7-DHT were synthesized as ^{14}C-labeled compounds. The reaction of 5,6- and 5,7-dibenzyloxygramine methiodide with [^{14}C]NaCN gives the two ^{14}C-labeled dibenzyloxyindoleacetonitriles. This reaction, which is normally performed with a great excess of NaCN,[25] is achieved with stoichiometric cyanide quantities in a yield of approximately 60%. [ω-^{14}C]-5,6-DHT and [ω-^{14}C]-5,7-DHT and [*carboxy*-^{14}C]-5,6- and [*carboxy*-^{14}C]-5,7-dihydroxyindoleacetic acid are obtained from the nitriles. The specific activities of the ^{14}C-labeled DHTs amounted to up to 25 mCi/mmol.

Chemical Properties of 5,6- and 5,7-Dihydroxytryptamine

Both 5,6-DHT and 5,7-DHT are capable of damaging serotonergic neurons, but the chemical properties of the two compounds differ markedly in some respects.

One hour after intraventricular administration of [^{14}C]-5,6-DHT, twice as much radioactivity is found in the insoluble residue of perchloric acid-homogenized brain tissue than in the case of animals treated with [^{14}C]-5,7-DHT.[32] In *in vitro* studies, Rotman *et al.*[33] found even a higher proportion of 5,6-DHT-derived radioactivity irreversibly bound to bovine serum albumin.

Only after intraventricular injection of 5,6-DHT are melanin-like pigments deposited in the brain;[34] this finding can be interpreted as an indication for the *in vivo* formation of quinoid intermediates in the case of 5,6-DHT metabolism. However, the possibility also exists that 5,7-DHT is oxidized to quinoid products.[31, 33]

In addition, spectroscopic investigations point to another form of reactivity in the case of 5,7-DHT. Unlike 5,6-DHT, 5,7-DHT shows in buffer at pH 7.0 a relatively long-wave ultraviolet absorption maximum at 350 nm (FIGURE 4); this absorption does not occur at low pH and if the two hydroxyl groups are etherified (FIGURE 5). The intensity of the maximum at 350 nm increases with pH (FIGURE 6). The existence of isosbestic points can be explained by a pH-dependent equilibrium between different keto-enol tautomeric forms of 5,7-DHT (FIGURE 7). The mesomeric β-diketo anion, which is formed from the tautomeric species b and d through proton removal, can be expected to undergo nucleophilic reactions. 5,6-DHT does not exhibit such spectroscopic behavior, because it has no resorcinol-like structure. The concept of keto-enol tautomerism in 5,7-DHT is supported by nuclear magnetic resonance spectroscopic data. In deuterium oxide, the two aromatic protons in positions 4 and 6 of 5,7-DHT creatinine sulfate are immediately replaced by deuterium; only

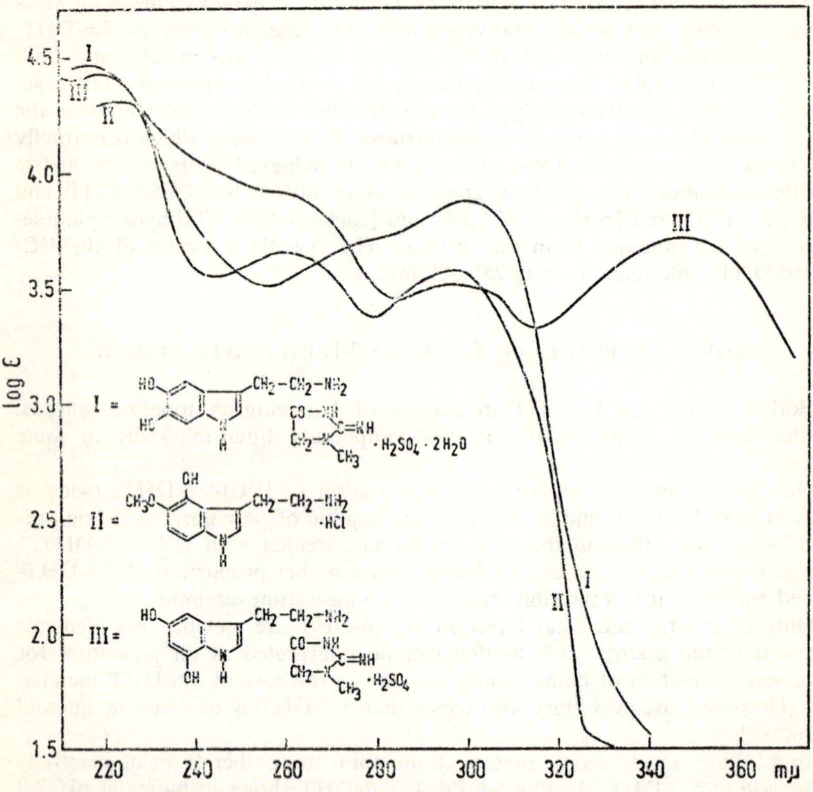

FIGURE 4. Ultraviolet spectra of 5,6-DHT creatinine sulfate·$2H_2O$ (**I**), 4-hydroxy-5-methoxytryptamine hydrochloride (**II**), and 5,7-DHT creatinine sulfate (**III**), in buffer at pH 7.0.

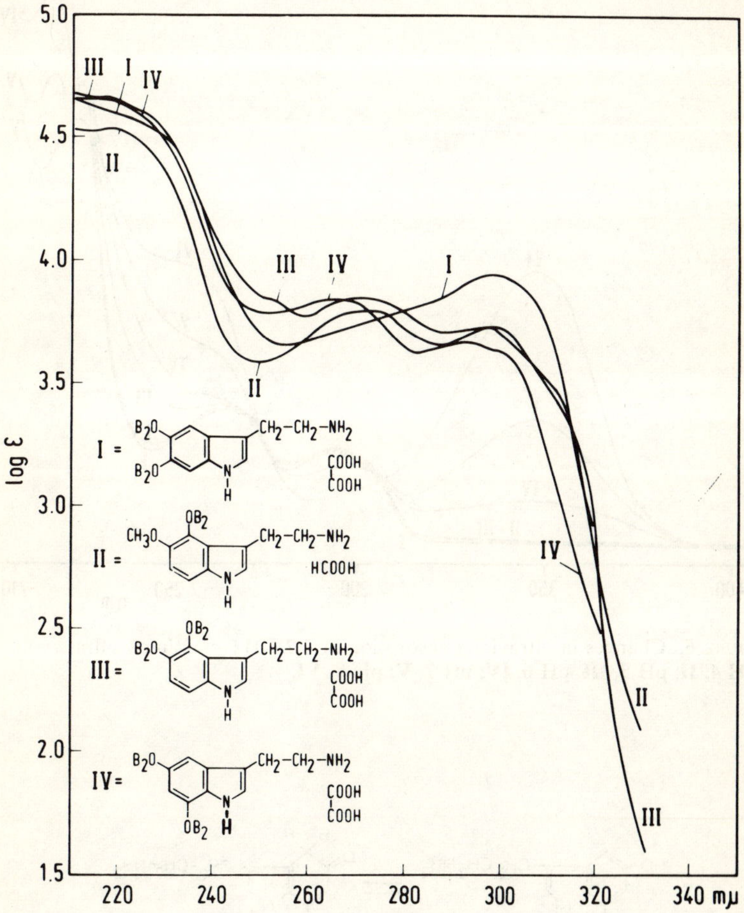

FIGURE 5. Ultraviolet spectra of 5,6-dibenzyloxytryptamine hydrogen oxalate (**I**), 4-benzyloxy-5-methoxytryptamine formiate (**II**), 4,5-dibenzyloxytryptamine hydrogen oxalate (**III**), and 5,7-dibenzyloxytryptamine hydrogen oxalate (**IV**) in ethanol.

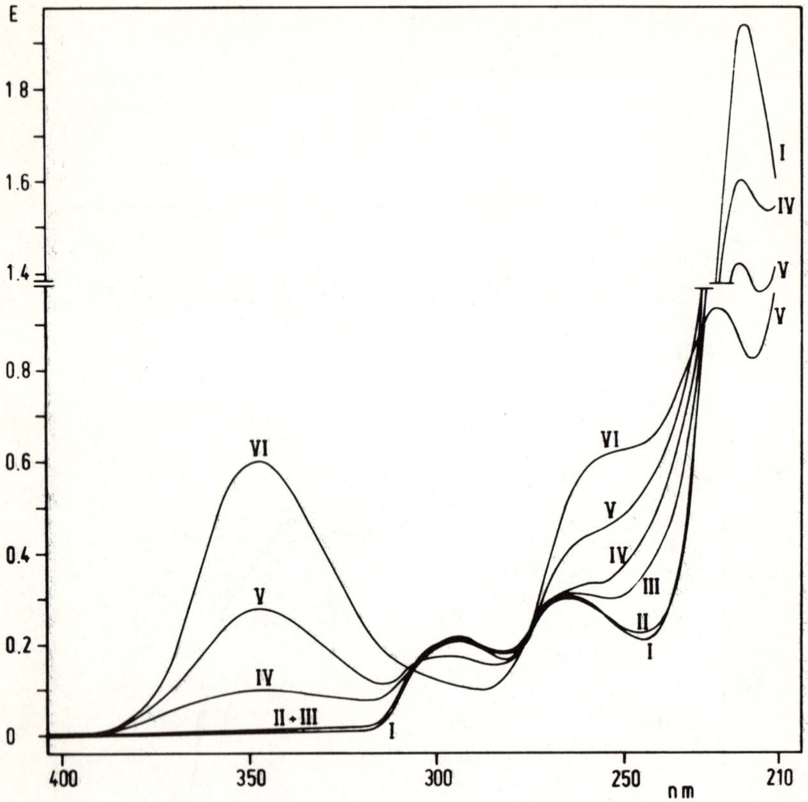

FIGURE 6. Changes in ultraviolet absorption of 5,7-DHT creatinine sulfate with pH: 2,**I**; pH 4, **II**; pH 5, **III**; pH 6, **IV**; pH 7, **V**; pH 8, **VI**.

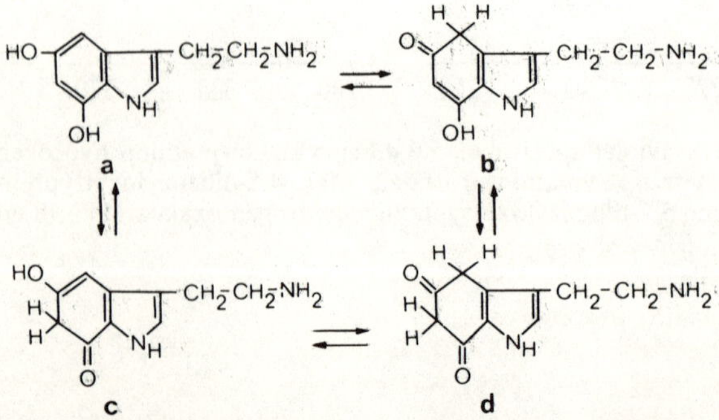

FIGURE 7. Keto-enol tautomeric forms of 5,7-DHT.

the proton in position 2 is detectable. Neither with 5,6-DHT nor with 5-HT creatinine sulfate was any exchange of aromatic protons detected in deuterium oxide, even after 24 hr.

Another difference in the behavior of the two dihydroxytryptamines is observed when the two compounds are dissolved in buffer solutions at physiologic pH at 37° C. While a dark precipitate becomes visible in the 5,6-DHT-containing solution, the 5,7-DHT-containing solution slowly attains a blue-violet color. The same result is observed when oxygen is passed through such solutions in buffer at pH 7.4. When the ultraviolet absorption of these solutions

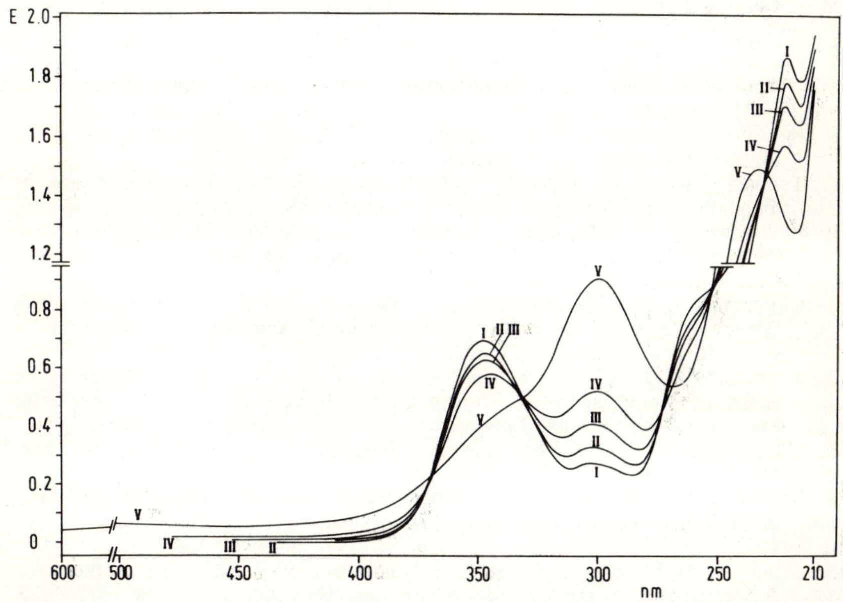

FIGURE 8. Reaction of 5,7-DHT creatinine sulfate in buffer at pH 7.4 with oxygen: ultraviolet absorption spectra after 0 min (**I**), 40 min (**II**), 80 min (**III**), 160 min (**IV**), and 10 hr (**V**) reaction times.

is studied over a longer period of time, initially a decrease of the extinction of the absorption maximum at 299 nm is seen in case of 5,6-DHT, followed by a broadening and leveling off of the absorption curve. After a few hours, only the spectrum of creatinine is visible. In the case of 5,7-DHT, the absorption maximum at 350 nm disappears with time, whereas the intensity of the maximum at 300 nm increases (FIGURE 8), and a low and flat maximum is formed in the visible range of the spectrum at 520 nm. No precipitate can be seen in such solutions. The ultraviolet spectra of 5,7-DHT measured at different times show isosbestic points. We are now testing to determine whether this reaction leads to defined products.

References

1. Carlisle, D. B. 1956. An indole-alkylamine regulating heartbeat in Crustacea. Biochem. J. **63:** 32P–33P.
2. Dalgliesh, C. E. 1958. Comments on the biochemistry of 5-hydroxtryptamine. *In* 5-Hydroxtryptamine. G. P. Lewis, Ed. : 58–61. Pergamon Press. London, New York, Paris, Los Angeles.
3. Schlossberger, H. G. & H. Kuch 1960. Synthese des 5.6-Dihydroxy-tryptamins. Chem. Ber. **93:** 1318–1323.
4. Schlossberger, H. G. & H. Kuch. 1963. Synthese von 5.6-Dihydroxytryptophan und verwandten Verbindungen. Liebigs Ann. Chem. **662:** 132–138.
5. Schlossberger, H. G. & H. Kuch. 1977. In preparation.
6. Baumgarten, H. G., A. Björklund, L. Lachenmayer, A. Nobin & U. Steveni. 1971. Long-lasting selective depletion of brain serotonin by 5.6-dihydroxytryptamine. Acta Physiol. Scand. Suppl. **373:** 1–15.
7. Baumgarten, H. G. & H. G. Schlossberger. 1973. Effects of 5.6-dihydroxytryptamine on brain monoamine neurons in the rat. *In* Serotonin and Behavior. J. Barchas & E. Usdin, Eds. : 209–224. Academic Press, Inc. New York, N.Y.
8. Benigny, J. D. & R. L. Minnis. 1965. The synthesis of 5.6-dihydroxyindole and some of its derivatives (1). J. Heterocyclic Chem. **2:** 387–392.
9. Taborsky, R. G., P. Delvigs & I. H. Page. 1965. 6-Hydroxy-indoles and the metabolism of melatonin. J. Med. Chem. **8:** 855–858.
10. Lee, F. G. H., D. E. Dickson, J. Suzuki, A. Zirnis & A. A. Manian. 1973. Synthesis of 5.7- and 6.7-disubstituted tryptamines and analogs. J. Heterocyclic Chem. **10:** 649–654.
11. Creveling, C. R., J. Lundstrom, E. T. McNeal, L. Tice & J. W. Daly. 1975. Dihydroxytryptamines: effects on noradrenergic function in mouse heart in vivo. Mol. Pharmacol. **11:** 211–222.
12. Sundberg, R. J. 1970. The chemistry of indoles. *In* Organic Chemistry, A Series of Monographs. A. T. Blomquist, Ed. Vol. 18: 218–230. Academic Press, Inc. New York and London.
13. Nenitzescu, C. 1925. Über eine neue Indol-Synthese. Chem. Ber. **58:** 1063–1064.
14. Reissert, A. 1897. Einwirkung von Oxalester und Natriumäthylat auf Nitrotoluole. Synthese nitrierter Phenylbrenztraubensäuren. Chem. Ber. **30:** 1030–1053.
15. Stoll, A., F. Troxler, J. Peyer & A. Hofmann. 1955. Eine neue Synthese von Bufotenin und verwandten Oxy-tryptaminen. Helv. Chim. Acta **38:** 1452–1472.
16. Mohr, G. 1952. Versuche zur Synthese des 7-Oxytryptophans. Ph.D. Thesis. Tübingen, Federal Republic of Germany.
17. Piers, E. & R. K. Brown. 1962. The decarboxylation of ring-substituted indole-2 (and 3) -carboxylic acids. Can. J. Chem. **40:** 559–561.
18. Ried, W. & H. Schiller. 1952. Über das 5.6-Dioxychinolin und einige seiner Abkömmlinge. Chem. Ber. **85:** 216–224.
19. Kuch, H. 1962. Synthese des 5.6-Dihydroxy-tryptamins und verwandter Verbindungen. Ph.D. Thesis. Munich, Federal Republic of Germany.
20. Pauly, H., K. Schübel & K. Lockemann. 1911. Über die Reaktionsfähigkeit der Phenolgruppen in den Phenolaldehyden. Liebigs Ann. Chem. **383:** 288–337; Gerngross, O. 1920. Die Fähigkeit der tierischen Haut zur Reaktion mit Phenolaldehyden. Biochem. Z. **108:** 82–97.
21. Speeter, M. E. & W. C. Anthony. 1954. The action of oxalyl chloride on indoles: a new approach to tryptamines. J. Amer. Chem. Soc. **76:** 6208–6210.
22. Young, E. H. P. 1958. The synthesis of 5-hydroxytryptamine (serotonin) and related tryptamines. J. Chem. Soc. : 3493–3496.
23. Kühn, H. & O. Stein. 1937. Über Kondensationen mit Aldehyden und sekundären Aminen. I. Mitteil: Eine neue Gramin-Synthese. Chem. Ber. **70:** 567–569.

24. Snyder, H. R. & C. W. Smith. 1944. A convenient synthesis of d,l-tryptophan. J. Amer. Chem. Soc. **66:** 350–351.
25. Thesing, J. & F. Schülde. 1952. Notiz über die präparative Darstellung von Heteroauxin und Tryptamin. Chem. Ber. **85:** 324–327.
26. Speeter, M. E., R. V. Heinzelmann & D. I. Weisblat. 1951. The synthesis of the blood stream vasoconstrictor principle serontonin creatinine sulfate. J. Amer. Chem. Soc. **73:** 5514–5515.
27. Snyder, H. R. & F. J. Pilgrim. 1948. The preparation of 3-indoleacetic acid; a new synthesis of tryptophol. J. Amer. Chem. Soc. **70:** 3770–3771.
28. Lyttle, D. A. & D. I. Weisblat. 1947. The chemistry of nitroacetic acid and its esters. I. The alkylation of alkylnitroacetates with gramine. J. Amer. Chem. Soc. **69:** 2118–2119.
29. Julia, M., P. Manoury & Ch. Voillaume. 1965. Recherches en série indolique. XIV. Sur des méthoxy-5 hydroxy-4, méthoxy-5 hydroxy-6 et méthoxy-7 hydroxy-6 tryptamines. Bull. Soc. Chim. France: 1417–1423.
30. Baumgarten, H. G., A. Björklund, A. Nobin, E. Rosengren & H. G. Schlossberger. 1975. Neurotoxicity of hydrolated tryptamines: structure-activity relationships. 1. Long-term effects on monoamine content and fluorescence morphology of central monoamine neurons. Acta Physiol. Scand. Suppl. **429:** 1–27.
31. Björklund, A., A. S. Horn, H. G. Baumgarten, A. Nobin & H. G. Schlossberger. 1975. Neurotoxicity of hydroxylated tryptamines: structure-activity relationships. 2. In vitro studies on monoamine uptake inhibition and uptake impairment. Acta Physiol. Scand. Suppl. **429:** 29–60.
32. Baumgarten, H. G., H. P. Klemm, L. Lachenmayer & H. G. Schlossberger. This monograph.
33. Rotman, A., J. W. Daly & C. R. Creveling. 1976. Oxygen-dependent reaction of 6-hydroxydopamine, 5.6-dihydroxytryptamine, and related compounds with proteins in vitro: a model for cytotoxicity. Mol. Pharmacol. **12:** 887–899.
34. Baumgarten, H. G., K. D. Ivett, L. L. Iversen, M. Vogt & G. Wilson. 1972. Effects of 5.6-dihydroxytryptamine on monoaminergic neurons in the central nervous system of the rat. J. Neurochem. **19:** 1587–1597.

INTERACTION OF 5,6- AND 5,7-DIHYDROXYTRYPTAMINE WITH TISSUE MONOAMINE OXIDASE *

H. P. Klemm and H. G. Baumgarten

Department of Functional Anatomy
University of Hamburg
2000 Hamburg, Federal Republic of Germany

Since the discovery by Thoenen and Tranzer [1, 2] that 2,4,5-trihydroxyphenylethylamine ("6-hydroxydopamine;" 6-OH-DA) caused a long-lasting depletion of noradrenaline (NA) in peripheral organs by destruction ("chemical sympathectomy") of noradrenergic axons, there has been continuing interest in the mechanism that is responsible for its unique selective neurotoxicity. Based on experimental evidence, two main theories for the molecular mechanism of 6-OH-DA toxicity have been proposed: the theory that oxidation products of 6-OH-DA undergo irreversible, covalent binding to functionally indispensable nucleophiles in proteins [3] and the postulate that the autoxidation of 6-OH-DA gives rise to oxygen and hydroxyl radicals and to hydrogen peroxide, which are the actual species responsible for neuronal damage.[4] The former theory has received considerable support by three recent studies,[5-7] one of which has impressively demonstrated that covalent attack of quinoid 6-OH-DA intermediates to the active site of a defined enzyme in catecholamine metabolism, catechol-*O*-methyltransferase, is correlated with and probably responsible for its capability to irreversibly inactivate the enzyme.[8] The suggestion by Borchardt [6] that an aminochrome-like intermediate of 6-OH-DA (5,6-*o*-indolquinone) was one of the more important candidates for such a role in inactivating functional groups in this enzyme focused our attention on the possibility that the neurotoxic potential of 5,6-dihydroxytryptamine (5,6-DHT), a drug with known preferential action on central 5-hydroxytryptamine (5-HT) neurons,[9, 10] might be due to an analogous mechanism. Therefore, its interaction with the most important enzyme in serotonin catabolism, monoamine oxidase (MAO), was studied *in vitro* and *in vivo* and correlated with its capability to undergo protein binding.

From the beginning of these studies, it was anticipated that 5,7-dihydroxytryptamine (5,7-DHT), which has a selective damaging action on central 5-HT fibers similar to that of 5,6-DHT but less unselective cytotoxic side effects,[11] should have qualitatively different effects on this enzyme due to its differing physicochemical properties, despite its similarity in causing neuronal damage. The results of the present study substantiate these suspected differences in the behavior of 5,6-DHT and 5,7-DHT but suggest at the same time that the final process that results in neurotoxicity is similar for both drugs, namely, covalent binding of reactive intermediates to nucleophiles in proteins, disregarded important differences in the initial oxidation and metabolism of 5,6-DHT and 5,7-DHT *in vivo*.

* Supported by a grant from the Deutsche Forschungsgemeinschaft (H.G.B.).

0077–8923/78/0305–0036 $01.75/0

Materials and Methods

Female albino rats, Sprague-Dawley strain (Wiga, Sulzfeld, Federal Republic of Germany), 180–220 g body weight, were used throughout the present study. Animals were killed by decapitation and quickly dissected. The dissected brains or brain regions were stored in liquid nitrogen prior to use. Intraventricular injections were performed in Brevimytal® (Eli Lilly, Giessen, Federal Republic of Germany) anesthetized rats by the technique devised by Noble *et al.*[12] Injections were made with a Hamilton microliter syringe, generally 20 μl per animal. 5,6-Dihydroxytryptamine creatinine sulfate·$2H_2O$ (5,6-DHT) or 5,7-dihydroxytryptamine creatinine sulfate·½H_2O (5,7-DHT) was obtained from Regis Chemical Company (Morton Grove, Illinois) and dissolved in 20 μl of sterile physiologic saline, which contained 1 mg/ml of ascorbic acid. [1-^{14}C]-5,6-DHT and [1-^{14}C]-5,7-DHT creatinine sulfate (5.5, 10, or 25 mCi/mmol) were synthesized by Dr. H. G. Schlossberger (MPI-Biochemie, Martinsried, Federal Republic of Germany) and dissolved as described above. *In vivo* inhibition of monoamine oxidase (MAO) was achieved by intraperitoneal treatment of rats with 200 mg/kg of nialamide (Niamid®, Pfizer, Karlsruhe, Federal Republic of Germany); the animals were killed 6 hr after injection of the drug.

MAO Determinations (using [^{14}C]-5,6-DHT and [^{14}C]-5,7-DHT as substrates)

A modification of the method of Wurtman and Axelrod[13] was used. Hypothalami were homogenized in cold physiologic potassium chloride (1:50, w/v); an aliquot of this homogenate (50 μl = 1 mg tissue) was incubated with labeled 5,6-DHT or 5,7-DHT or with 5-[1,2-^{3}H(*N*)]hydroxytryptamine binoxalate (21.4 Ci/mmol; NEN, Dreieichenhain, Federal Republic of Germany). The incubation time was 20 min for [^{14}C]-5,6-DHT and [^{3}H]-5-HT but 60 min for [^{14}C]-5,7-DHT. The deaminated radioactive material was extracted with 6 ml of toluene-isoamylalcohol (5:1, v/v) by vigorous shaking for 10 min. For [^{3}H]-5-HT, 6 ml of toluene-isoamylalcohol (10:1; v/v) were used for extraction. The recovery of the extraction step was evaluated by the internal standards [1-^{14}C]-5,6-dihydroxyindoleacetic acid ([^{14}C]-5,6-DHIAA) (0.55 mCi/mmol) and [1-^{14}C]-5,7-dihydroxyindoleacetic acid ([^{14}C]-5,7-DHIAA) (1.3 mCi/mmol; kindly donated by Dr. H. G. Schlossberger). Four milliliters of the organic layer were counted in 10 ml of Triton X-100 cocktail in a Packard Tri-Carb liquid scintillation spectrometer (model C 2405) (LSC).

Five different substrate concentrations (which ranged from 10^{-6} to 10^{-4} M) were employed, and the linearity of the reaction was tested in time course experiments.

Inhibition of MAO by 5,6-DHT, 5,7-DHT, and 5-HT

[2-^{14}C]Tryptamine bisuccinate (48.5 mCi/mmol) was used as the substrate (at concentrations of 10^{-7}–10^{-4} M; according to the technique of Wurtman and Axelrod[13]). Test tubes were preincubated with 5,6-DHT, 5,7-DHT, or 5-HT dissolved in bidistilled water (inhibitory concentrations in the final reaction mixture: 10^{-7}–10^{-4} M). In some incubation experiments, catalase (Boehringer, Mannheim, Federal Republic of Germany) and superoxide dismutase (Sigma, Munich, Federal Republic of Germany) dissolved in 0.5 M phosphate buffer (pH 7.4) were added to the incubation mixture. MAO activity was assayed in potassium chloride homogenates of rat pons-medulla oblongata (1:100, w/v).

Inactivation of MAO in Vitro

Homogenates of rat pons-medulla oblongata (1:100, w/v) were incubated with 2.5×10^{-4} M 5,6-DHT or 5,7-DHT for 15, 30, 45, or 60 min at 37° C; at the end of each incubation experiment, MAO activity was assayed with 3.4×10^{-5} M [^{14}C]tryptamine by use of 50 μl of the preincubated homogenate. In two series of experiments, all solutions used were gassed with nitrogen, and the preincubation was performed with air sealed off as completely as possible. Further series of incubations were run in the presence of 4 mM dithiothreitol (Sigma, Munich), glutathione (Boehringer, Mannheim), and ascorbate (Merck, Darmstadt, Federal Republic of Germany), 0.125 mg of superoxide dismutase, or 0.125 mg of catalase.

Inactivation of MAO in Vivo

The basal hypothalamus of rats injected with 150 μg of 5,6-DHT or 150 μg of 5,7-DHT (1.5, 3, 6, or 24 hr before killing) or with vehicle was homogenized in potassium chloride, and MAO activity was assayed as described above (by use of 3.6×10^{-6} M [^{14}C]tryptamine).

In Vitro *Binding of [^{14}C]-5,6-DHT, [^{14}C]-5,7-DHT, [^{14}C]-5,6-DHIAA, or [^{14}C]-5,7-DHIAA to Homogenates of Rat Hypothalami*

Concentrations of 10^{-6}, 10^{-5}, or 10^{-4} M [^{14}C]-5,6-DHT or [^{14}C]-5,7-DHT and 10^{-5} or 10^{-4} M [^{14}C]-5,6-DHIAA or [^{14}C]-5,7-DHIAA were dissolved in 0.25 ml of 0.5 M phosphate buffer (pH 7.4) and incubated with 50 μl of homogenate from rat hypothalami (1:50, w/v). The incubation was terminated by the addition of 2 ml of 0.4 N perchloric acid (PCA) and by cooling down to 0° C. Subsequently, the samples were centrifuged at 10.000 rpm for 10 min. One milliliter of the PCA supernatant was diluted in 10 ml of Triton X-100 cocktail and counted by liquid scintillation counting (LSC). Binding of radioactivity to the acid-insoluble pellet was calculated by measuring the decrease of radioactivity in the supernatant fraction.

In Vivo *Binding of [^{14}C]-5,6-DHT or [^{14}C]-5,7-DHT to Acid-Insoluble Homogenate Pellets*

Hypothalamus, tectum, and pons-medulla oblongata of two rats injected with 4×10^{6} dpm of [^{14}C]-5,6-DHT (75 μg of free base) or 3.6×10^{6} dpm of [^{14}C]-5,7-DHT (94 μg of free base) 1, 4, or 24 hr before processing were homogenized in ice-cold 0.4 N PCA (1:10, w/v) and centrifuged. The acid-insoluble pellet was washed five times with 5 ml of ice-cold PCA, two times with 5 ml of cold 0.5 M phosphate buffer (sodium, pH 7.5), three times with chloroform-methanol (2:1, v/v), and two times with 5 ml of methanol-glacial acetic acid (10:1, v/v). The resultant pellet was suspended in 5 ml of 0.4 N PCA, centrifuged, and dissolved in Protosol® (NEN, Dreieichenhain; 1 g/10 ml). The radioactivity in the supernatant fractions was measured in Triton cocktail; radioactivity in the dissolved pellet was determined by counting in toluene-bis-MSB (Packard) cocktail.

Thin-Layer Chromatography of MAO-Catalyzed Metabolites from [^{14}C]-5,6-DHT and [^{14}C]-5,7-DHT

The following reference substances were available: 5,6-DHT, 5,7-DHT, 5,6-DHIAA, and 5,7-DHIAA. About 3×10^6 dpm of 5,6-DHT or 3×10^6 dpm of 5,7-DHT (10 mCi/mmol) were incubated in homogenates of rat hypothalamus (1:10, w/v) for 10 min (5,6-DHT) or 30 min (5,7-DHT) and extracted as described above. Another series of incubations was run in the presence of 2 U of aldehyde dehydrogenase (Boehringer, Mannheim) and 10^{-6} M NAD (Boehringer, Mannheim). To the extracts, pooled from 10 tubes, 0.1 mg of 5,6-DHT and 0.1 mg of 5,6-DHIAA or 0.1 mg of 5,7-DHT and 0.1 mg of 5,7-DHIAA were added, and the extract was evaporated to dryness in a nitrogen atmosphere. The residue was taken up in butanol-glacial acetic acid-water (12:3:5, v/v) and chromatographed on Cellulose-Fertigplatten (Merck, Darmstadt). The reference compounds were identified by spraying *p*-dimethylaminocinnamaldehyde onto the chromatograms. One-centimeter strips were scraped off the plates and suspended in 3 ml of H_2O and 10 ml of Instagel® (Packard, Frankfurt, Federal Republic of Germany) and counted by LSC. Counting efficiency was monitored by internal standards that contained [^{14}C]-toluene (Packard, Frankfurt).

Results

Time Course of Conversion of [^{14}C]-5,6-DHT and [^{14}C]-5,7-DHT by MAO

Figure 1 illustrates the time course characteristics of the MAO-catalyzed conversion of [^{14}C]-5,6-DHT and [^{14}C]-5,7-DHT (*in vitro* incubation of the labeled amines in homogenates of rat hypothalamus). The toluene-isoamylalcohol extractable fraction of the reaction products shows linearity over time with 5,7-DHT; with 5,6-DHT as the substrate, linearity is evident only during the first 10 min of the incubation

In homogenates of hypothalamus from rats treated intraperitoneally with 200 mg/kg of nialamide (1 hr prior to sacrifice) and incubated for 10 min in the presence of 0.6×10^{-3} M nialamide, there was no apparent conversion of [^{14}C]-5,6-DHT to solvent-extractable products. With [^{14}C]-5,7-DHT as the substrate, processed under identical conditions, the MAO-catalyzed conversion was reduced to 40% of control values in a 60-min incubation.

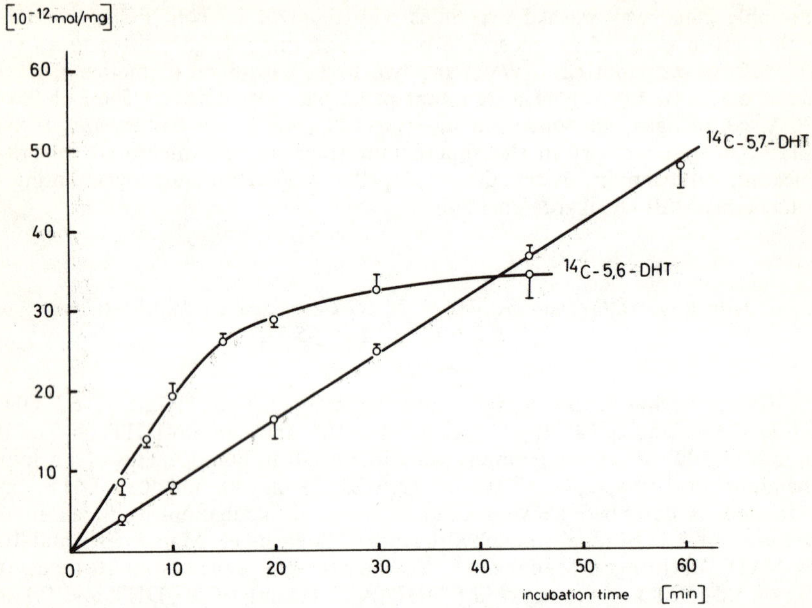

FIGURE 1. Time course characteristics of the MAO-catalyzed conversion of [14C]-5,6-DHT (2.0×10^{-5} M; 5.5 mCi/mmol) and [14C]-5,7-DHT (1.9×10^{-5} M; 10.8 mCi/mmol). Homogenate aliquots of rat hypothalamus (50 μl=1 mg) were incubated with 0.25 ml of 0.5 M phosphate buffer (pH 7.4) at 37°C. $n=3$; mean±SD.

Kinetics of Conversion of [14C]-5,6-DHT and [14C]-5,7-DHT by MAO

FIGURES 2–4 depict double reciprocal plots of the initial linear reaction velocities of MAO versus different substrate concentrations of [14C]-5,6-DHT, [14C]-5,7-DHT, and [3H]-5-HT, respectively. The approximate values for the Michaelis constants ($K_{m\ app}$) and for the maximal velocities (V_{max}) (see TABLE 1) were determined graphically from the Lineweaver-Burk diagram.

TABLE 1

MICHAELIS CONSTANTS ($K_{m\ app}$) MAXIMAL VELOCITIES (V_{max}) FOR THE CONVERSION OF [14C]-5,6-DHT, [14C]-5,7-DHT, AND [3H]-5-HT BY MAO IN RAT BRAIN HYPOTHALAMUS HOMOGENATES

	[14C]-5,6-DHT	[14C]-5,7-DHT	[3H]-5-HT
$K_{m\ app}$ (10^{-3} M)	1.25	0.50	0.20
V_{max} (10^{-9} mol/min/mg^{-1})	0.30	0.07	0.25

Kinetics of Inhibition of MAO by 5,6-DHT and 5,7-DHT

The conversion of [^{14}C]tryptamine in homogenates of rat pons-medulla oblongata was found to be inhibited when 5,6-DHT or 5,7-DHT was present in the incubation mixture at a concentration of 10^{-5} M or more. FIGURE 5 illustrates the Lineweaver-Burk representation of [^{14}C]tryptamine conversion under control conditions, in the presence of different concentrations of 5,6-DHT, and with catalase added to the reaction mixture that contained 5,6-DHT. The straight lines intersect at a point to the left of the origin but above the $1/s$ axis. Inhibition of [^{14}C]tryptamine conversion was not antagonized by either catalase or superoxide dismutase. In FIGURE 6, the $1/v$ versus i plot, directly obtained from the kinetic characteristics in the Lineweaver-Burk diagram of

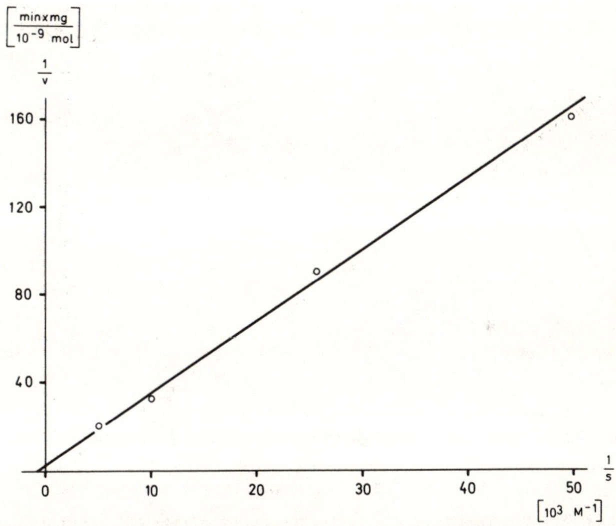

FIGURE 2. Lineweaver-Burk representation of [^{14}C]-5,6-DHT conversion by MAO *in vitro* (cf. legend to FIGURE 1 for details).

FIGURE 5, is presented; this plot permits easy recognition of the type of MAO inhibition by 5,6-DHT (noncompetitive type of inhibition). In this form of inhibition, the inhibitor not only interacts with a second binding site of the enzyme but also interferes with the affinity of the substrate to its binding site.

The kinetics of inhibition of MAO by 5,7-DHT (FIGURES 7 & 8) resemble those of 5,6-DHT (noncompetitive type of inhibition). By contrast, the inhibition of conversion of [^{14}C]tryptamine by 5-HT conforms to a purely competitive type of inhibition, in which all straight lines intersect on the $1/v$ axis (FIGURE 9) at $1/V_{max}$. When $1/v$ is plotted against i, the straight lines intersect at a point above the x axis (FIGURE 10).

Graphic reconstruction of $K_{m\ app}$ for [^{14}C]tryptamine in homogenates of rat pons-medulla oblongata yielded a value of $6.9 \pm 1.2 \times 10^{-6}$ M and a V_{max} of $79.3 \pm 5.0 \times 10^{-12}$ mol/min/mg^{-1}. The inhibition constants (K_i) were

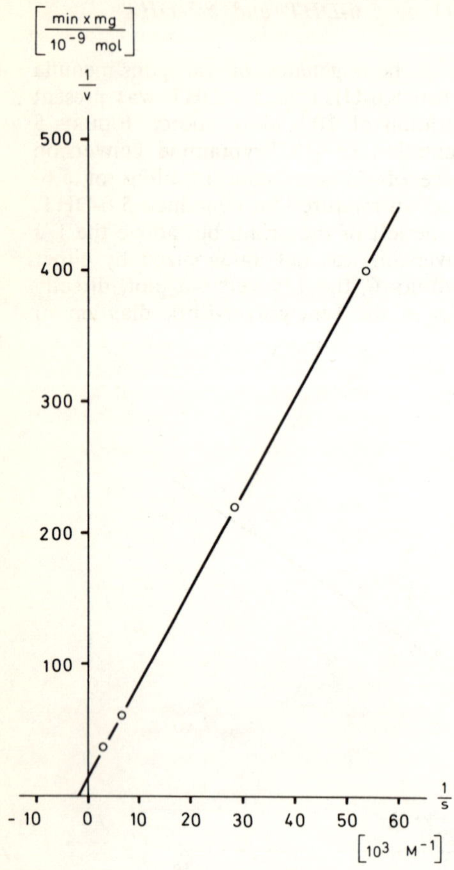

FIGURE 3. Lineweaver-Burk representation of [^{14}C]-5,7-DHT conversion by MAO *in vitro* (cf. legend to FIGURE 1 for details).

0.9×10^{-4} M for 5,6-DHT, 5.6×10^{-4} M for 5,7-DHT, and 2.4×10^{-4} M for 5-HT.

In Vitro *Inactivation of MAO by 5,6-DHT and 5,7-DHT*

Preincubation of pons-medulla oblongata homogenates with 5,6-DHT or 5,7-DHT and subsequent coincubation (generally for 6 min) of an aliquot of the homogenate with [^{14}C]tryptamine allowed a distinction between reversible enzyme inhibition and time-dependent irreversible MAO inactivation. FIGURE 11 shows the time-dependent loss of enzymatic activity due to 5,6-DHT. After a 30-min preincubation, only 40% of the initial activity is detected in the homogenates.

In prolonged preincubations, there is only a slight increase in the time-dependent enzyme inactivation. The extent of this late MAO inactivation by 5,6-DHT roughly corresponds to the time-dependent loss of MAO activity caused by the incubation procedure itself.

The inactivation of MAO by 5,6-DHT can be partially prevented by the addition of dithiothreitol, glutathione, or reduced oxygen to the preincubation medium. There was no effect on MAO activity by catalase or superoxide dismutase.

By contrast, there is no clear-cut effect of equimolar amounts of 5,7-DHT on MAO activity. The loss of activity by the 60-min preincubation with 5,7-DHT exceeded that in control incubations by only 10%. The initial fall of MAO activity by 5,7-DHT is readily explained by the transfer from the preincubate of small amounts of 5,7-DHT into the reaction mixture. The addition of ascorbate, glutathione, or dithiothreitol had no effect on enzymatic activity (FIGURE 12).

In Vivo *Inhibition of MAO by 5,6-DHT or 5,7-DHT*

The effects of intraventricular injection of 5,6-DHT or 5,7-DHT on MAO activity of the basal hypothalamus are illustrated in FIGURE 13. Following a transient mild inhibition of MAO activity by 5,7-DHT (15% reduction at 1.5 hr) or by 5,6-DHT (20% reduction at 3 hr), the activity turned to control levels by 6 hr.

In Vitro *Binding of [^{14}C]-5,6-DHT, [^{14}C]-5,7-DHT, [^{14}C]-5,6-DHIAA, and [^{14}C]-5,7-DHIAA to Homogenates of Rat Hypothalamus*

Time-dependent binding of all four dihydroxyindoles to components of hypothalamic homogenates was revealed by *in vitro* studies (FIGURES 14 & 15). Homogenates of rat hypothalamus were incubated in 0.5 M phosphate buffer

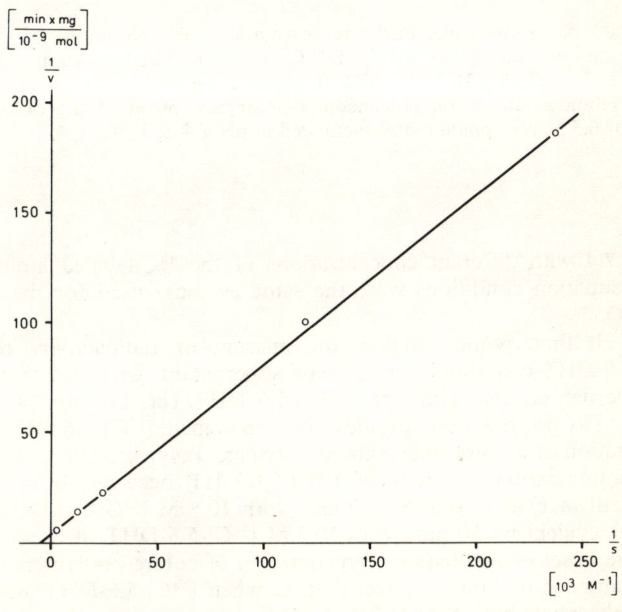

FIGURE 4. Lineweaver-Burk representation of [^3H]-5-HT conversion by MAO *in vitro* (cf. legend to FIGURE 1 for details).

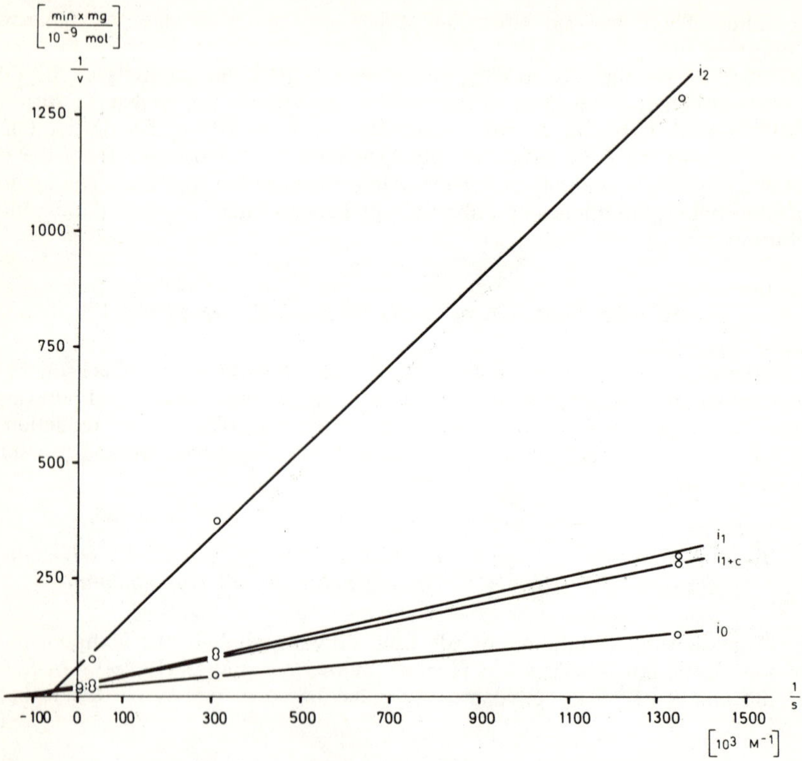

FIGURE 5. Lineweaver-Burk representation of inhibition of MAO-catalyzed [^{14}C]tryptamine conversion by 5,6-DHT. i_0=uninhibited reaction; $i_1 = 2.78 \times 10^{-4}$ M 5,6-DHT; $i_{1+c} = 2.78 \times 10^{-4}$ M 5,6-DHT plus 0.1 mg of catalase; $i_2 = 1.39 \times 10^{-3}$ M 5,6-DHT. Homogenate of rat ponsmedulla oblongata, 50 μl (1 mg of tissue) added to 250 μl of 0.5 M phosphate buffer incubated at pH 7.4 at 37°C.

at pH 7.4 with different concentrations of the ^{14}C-labeled amines and acids. The incubation conditions were the same as those used for the *in vitro* assay of MAO.

At all time points studied, the amount of radioactivity removed from [^{14}C]-5,6-DHT-containing homogenate supernatants exceeded that cleared from the supernatants that contained [^{14}C]-5,7-DHT (cf. FIGURE 14). Part of the radioactivity in the homogenates that contained [^{14}C]-5,6-DHT was lost by precipitation of an insoluble colored polymer. Polymerization of what appeared to be autoxidation products of [^{14}C]-5,6-DHT occurred in a concentration-dependent manner at different times: with 10^{-4} M [^{14}C]-5,6-DHT, polymerization was evident by 30 min; with 10^{-5} M [^{14}C]-5,6-DHT, it occurred by 60 min after the onset of incubation. Precipitation of colored polymers occurred even in the absence of homogenates, that is, when [^{14}C]-5,6-DHT was dissolved in 0.5 M phosphate buffer at pH 7.4 at 37° C.

The removal of radioactivity from the [^{14}C]-5,7-DHT-containing homogenate supernatants progressed slowly with time and did not exceed 10% of the amount of radioactivity added to the homogenates by 90 min after the onset of incubation. There was no indication of formation of an insoluble colored polymer by 5,7-DHT.

It is important to note that there was an initial lag phase of about 10 min after the onset of incubation, during which time little, if any, radioactivity was found in the acid-insoluble homogenate-pellet fractions.

The ^{14}C-labeled dihydroxyindoleacetic acids reacted similarly to the ^{14}C-labeled amines, and the difference in percentage removal from the homogenate supernatants of [^{14}C]-5,6-DHIAA versus [^{14}C]-5,7-DHIAA was also similar to the difference noted between the amines (see above), and part of the [^{14}C]-5,6-DHIAA precipitated from the solutions in the form of colored insoluble polymers.

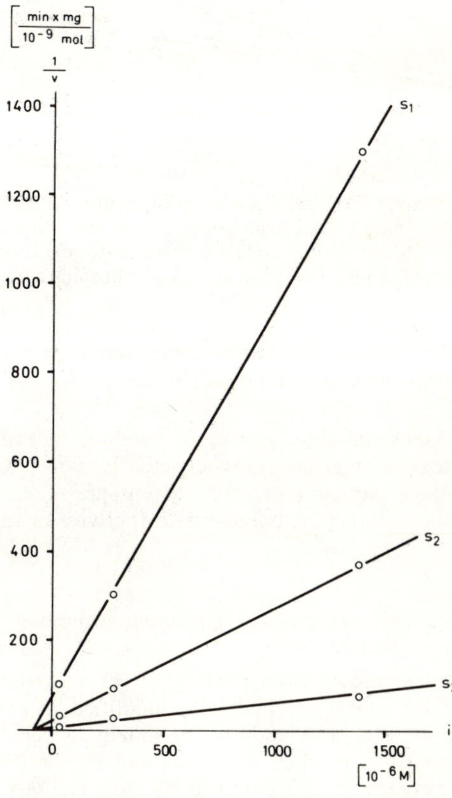

FIGURE 6. $1/v$ versus i plot of inhibition of MAO-catalyzed [^{14}C]tryptamine conversion by 5,6-DHT. $s_1 = 0.74 \times 10^{-6}$ M [^{14}C]tryptamine; $s_2 = 3.24 \times 10^{-6}$ M [^{14}C]tryptamine; $s_3 = 32.4 \times 10^{-6}$ M [^{14}C]tryptamine. Homogenate of rat pons-medulla oblongata, 50 μl (1 mg of tissue), added to 250 μl of 0.5 M phosphate buffer at pH 7.4 at 37° C.

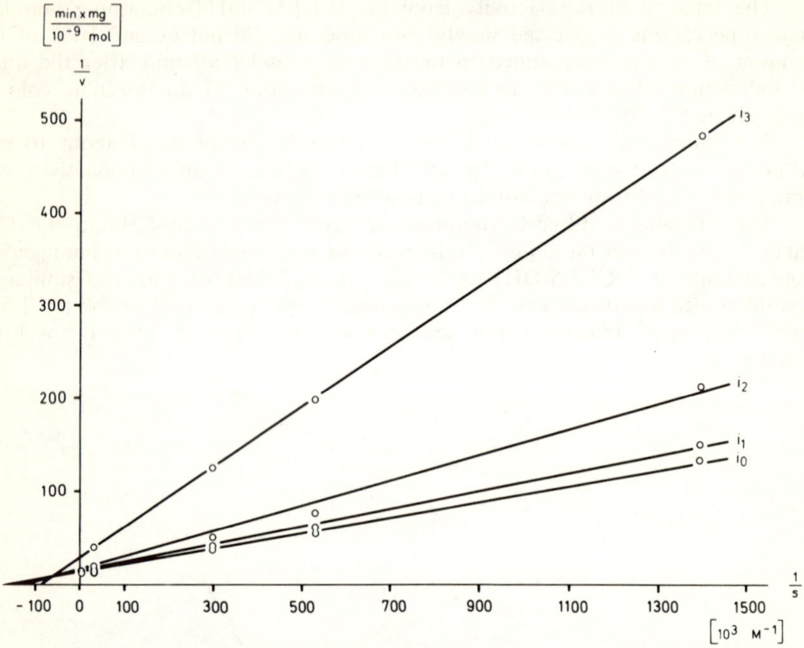

FIGURE 7. Lineweaver-Burk representation of inhibition of [^{14}C]tryptamine conversion by 5,7-DHT. i_0=uninhibited reaction; $i_1=2.94\times10^{-5}$ M 5,7-DHT; $i_2=2.94\times10^{-4}$ M 5,7-DHT; $i_3=1.47\times10^{-3}$ M 5,7-DHT. Homogenate of rat pons-medulla oblongata, 50 μl (1 mg of tissue), added to 250 μl of 0.5 M phosphate buffer at pH 7.4 at 37°C.

Protein Binding of Radioactivity After Intraventricular Injection of [^{14}C]-5,6-DHT or [^{14}C]-5,7-DHT

FIGURE 16 shows the time course of binding of radioactivity to protein pellets of brain regions injected intraventricularly with [^{14}C]-5,6-DHT or [^{14}C]-5,7-DHT. The data plotted represent percentage values of the radioactivity recovered from the initial PCA homogenate (activity in supernatant and pellet; cf. FIGURE 16).

Thin-Layer Chromatography

In addition to nonmetabolized [^{14}C]-5,6-DHT or [^{14}C]-5,7-DHT, [^{14}C]-5,6-DHIAA and [^{14}C]-5,7-DHIAA were also identified in organic extracts of the MAO reaction mixtures. The addition of aldehyde dehydrogenase and NAD to the reaction mixtures resulted in triplication of the [^{14}C]-5,6-DHIAA spot, but little or no increase was observed in the radioactivity that accumulated at the R_F of 5,7-DHIAA. Though the background activity on the chromatograms was considerable, the amines and acids appeared as distinct radioactivity peaks. A considerable fraction of the radioactivity moved with the solvent front and was not identified.

Discussion

The results of the present radiometric study show that both 5,6-DHT and 5,7-DHT are metabolized by MAO in homogenates of rat hypothalamus. The conversion of both drugs by MAO was inhibited by nialamide and obeyed, at least initially, Michaelis-Menten kinetics. 5,6-DHIAA and 5,7-DHIAA were identified as part of the deamination products by thin-layer chromatography.

A comparison of the kinetic data of 5,6-DHT and 5,7-DHT with those of 5-HT obtained from Lineweaver-Burk diagrams permits the following conclusions regarding enzyme-substrate interaction: 1. The K_m of 5,6-DHT and 5,7-DHT is on the order of 10^{-4} M and thus in a concentration range that corresponds to the postulated concentration of extravesicular, cytoplasmic monoamines.[14] The affinity of MAO for 5,6-DHT and 5,7-DHT is thus high enough to ensure metabolism of these false neurochemicals following accumulation in the neuron mediated by the membrane-bound amine uptake mechanism.

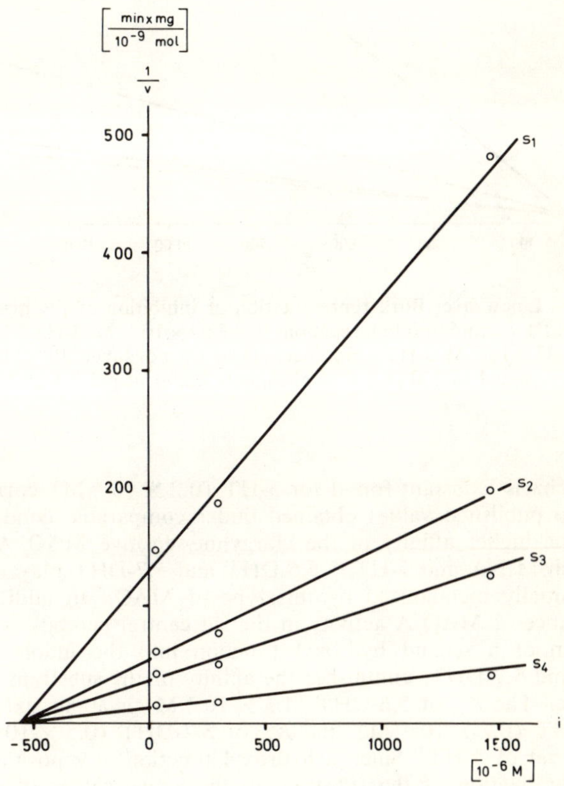

FIGURE 8. $1/v$ versus i plot of inhibition of MAO-catalyzed [^{14}C]tryptamine conversion by 5,7-DHT. $s_1 = 0.72 \times 10^{-6}$ M[^{14}C]tryptamine; $s_2 = 1.9 \times 10^{-6}$ M [^{14}C] tryptamine; $s_3 = 3.4 \times 10^{-6}$ M [^{14}C]tryptamine; $s_4 = 33.6 \times 10^{-6}$ M [^{14}C]tryptamine. Homogenate of rat pons-medulla oblongata, 50 μl (1 mg of tissue), 250 μl of 0.5 M phosphate buffer at pH 7.4 at 37°C.

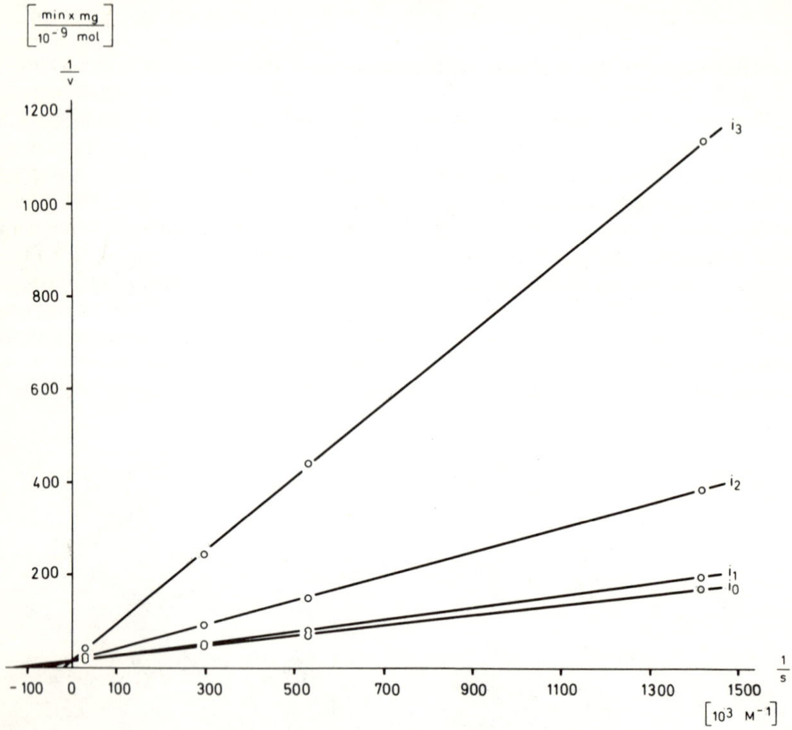

FIGURE 9. Lineweaver-Burk representation of inhibition of [^{14}C]tryptamine conversion by 5-HT. i_0=uninhibited reaction; $i_1=2.94\times10^{-5}$ M 5-HT; $i_2=2.94\times10^{-4}$ M 5-HT; $i_3=1.47\times10^{-3}$ M 5-HT. Homogenate of rat pons-medulla oblongata, 50 μl (1 mg of tissue), added to 250 μl 0.5 M phosphate buffer at pH 7.4 at 37° C.

2. The Michaelis constant found for 5-HT (0.2×10^{-3} M) corresponds rather precisely to published values obtained under comparable conditions.[15, 16] Because of the higher affinity of the clorgyline-sensitive MAO A to polar substrates, such as NA and 5-HT,[17] 5,6-DHT and 5,7-DHT may be presumed to be preferentially metabolized by this type of MAO. In addition, there is a preponderance of MAO A activity in the rat central nervous system.[18] 3. The introduction of a second hydroxyl function into the indole nucleus, as in 5,6-DHT and 5,7-DHT, diminishes the affinity of the substrate to the catalytic binding site. The K_m of 5,6-DHT (1.3×10^{-3} M) is about sixfold higher than that of 5-HT (0.2×10^{-3} M); the K_m of 5,7-DHT (0.5×10^{-3} M) is about 2.5 times that of 5-HT. Since a hydroxyl function in *p*-position seems to be required for fixation of the substrate in the active center of MAO A,[19] the additional hydroxyl group in the indole nucleus could interfere with the binding by steric hindrance or by electrostatic repulsion. The latter mechanism appears to be more important for *o*-dihydroxylated derivatives, such as 5,6-DHT. 4. The V_{max} of 5,6-DHT (0.3×10^{-9} mol/min/mg^{-1}) and of 5-HT (0.25×10^{-9} mol/min/mg^{-1}) are rather similar, whereas that of 5,7-DHT is about four times less

(0.07×10^{-9} mol/min/mg^{-1}); the thermodynamic reason for this difference is unknown at present. When homogenates of rat pons-medulla oblongata were incubated with [^{14}C]tryptamine, a substrate for both MAO A and B,[17] at a concentration of 10^{-7} M, a 100-fold excess of 5,6-DHT, 5,7-DHT, or 5-HT was required to produce inhibition of tryptamine conversion. This finding is in accord with those of Ekstedt,[20] indicating that substrates preferentially metabolized by MAO A must be present in rather substantial excess to antagonize the metabolism of nonpolar amines that are preferential substrates of MAO B. The type of inhibition that results from such interaction (e.g., inhibition of tryptamine oxidation by 5-HT, as studied by us) follows competitive inhibition patterns. The observed noncompetitive inhibition of MAO by 5,6-DHT and 5,7-DHT appears to be in contradiction to these observations but seems to correlate with the findings of Meyerson *et al.*,[15] which show that certain neuroamine-derived alkaloids inhibited the metabolism of MAO A-specific substrates in a competitive manner but inhibited MAO B-specific substrates in a non-

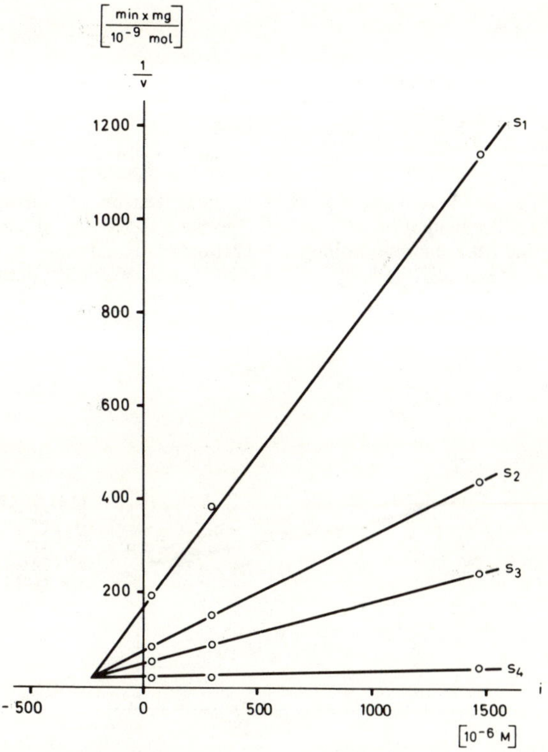

FIGURE 10. $1/v$ versus i plot of inhibition of MAO-catalyzed [^{14}C]tryptamine conversion by 5-HT. $s_1 = 0.72 \times 10^{-6}$ M [^{14}C]tryptamine; $s_2 = 1.88 \times 10^{-6}$ M [^{14}C] tryptamine; $s_3 = 3.35 \times 10^{-6}$ M [^{14}C]tryptamine; $s_4 = 3.36 \times 10^{-5}$ M [^{14}C]tryptamine. Homogenate of rat pons-medulla oblongata, 50 μl (1 mg of tissue), added to 250 μl 0.5 M phosphate buffer at pH 7.4 at 37°C.

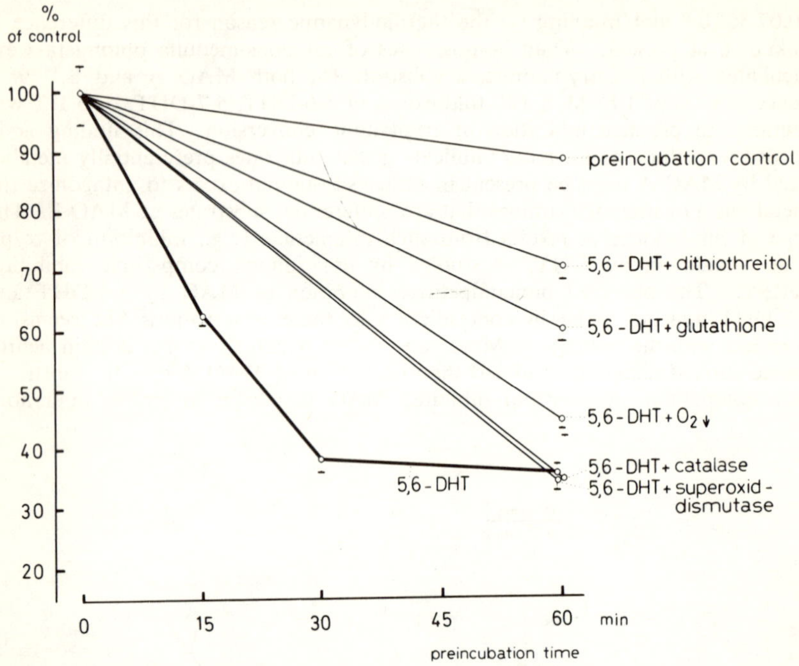

FIGURE 11. *In vitro* inactivation of MAO in homogenates of pons-medulla oblongata by 5,6-DHT. Preincubation of homogenates with 2.5×10^{-4} M 5,6-DHT (in part with 4 mM dithiothreitol or glutathione, 0.12 mg/ml of catalase or superoxide dismutase). $n=4$; mean±SD; 100% = 53.6×10^{-12} mol of [^{14}C]tryptamine/min/mg^{-1}.

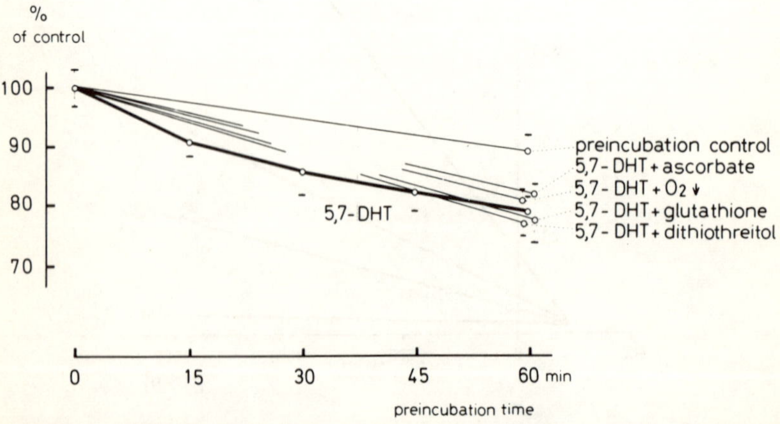

FIGURE 12. *In vitro* inactivation of MAO in homogenates of pons-medulla oblongata by 5,7-DHT. Preincubation of homogenates with 2.5×10^{-4} M 5,7-DHT (in part with 4 mM ascorbate, glutathione, or dithiothreitol). $n=4$; mean±SD; 100% = 60.0± 1.7×10^{-12} mol of [^{14}C]tryptamine/min/mg^{-1}.

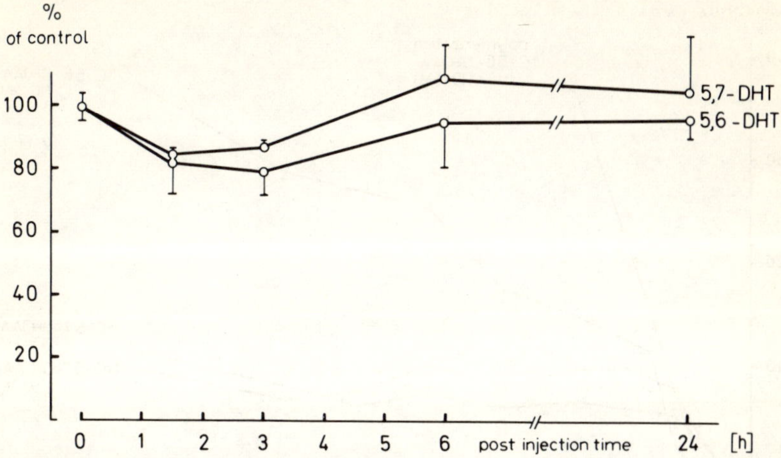

FIGURE 13. *In vivo* inhibition of MAO by 5,6-DHT or 5,7-DHT. Homogenates of rat basal hypothalamus (50μl=0.5 mg of tissue) incubated with 36.2×10^{-6} M [^{14}C]-tryptamine plus 250 μl of 0.5 M phosphate buffer at pH 7.4 at 37°C. Rats were pretreated with 150 μg (free base) of 5,6-DHT or 5,7-DHT and killed at 1.5, 3, 6, or 24 hr after intraventricular administration of the drugs. 100% $=153\pm6.6\times10^{-12}$ mol of [^{14}C]tryptamine. $n=4$; mean±SD.

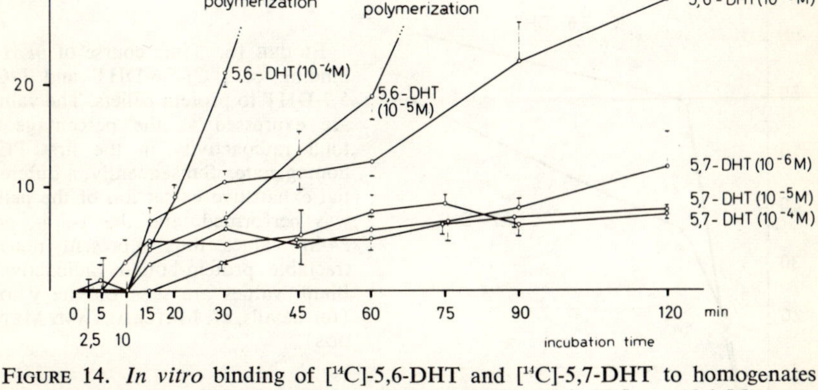

FIGURE 14. *In vitro* binding of [^{14}C]-5,6-DHT and [^{14}C]-5,7-DHT to homogenates of rat hypothalamus. [^{14}C]-5,6-DHT or [^{14}C]-5,7-DHT (10^{-4}, 10^{-5}, or 10^{-6} M) were dissolved in 250 μl of 0.5 M phosphate buffer at pH 7.4 at 37°C; 50 μl (1 mg of tissue) of homogenate of rat hypothalamus were added to the amine-containing solutions. $n=4$; mean±SD.

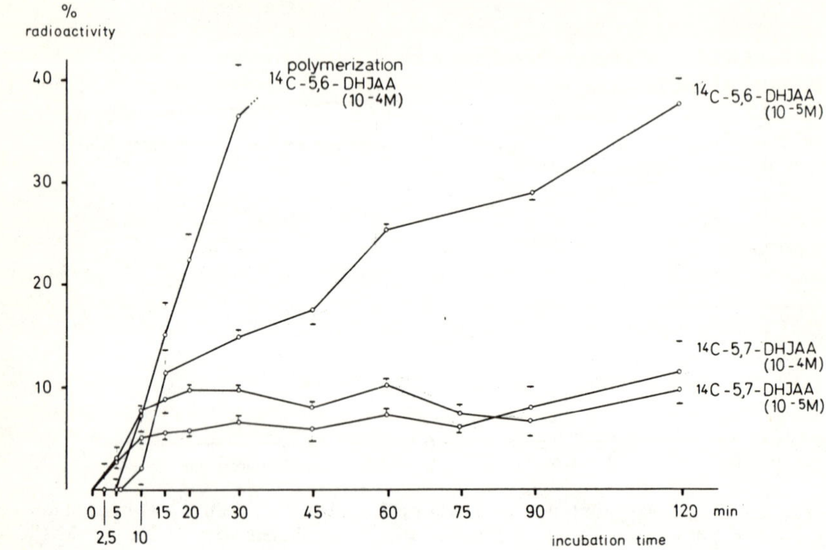

FIGURE 15. *In vitro* binding of [^{14}C]-5,6-DHIAA and [^{14}C]-5,7-DHIAA to homogenates of rat hypothalamus. [^{14}C]-5,6-DHIAA or [^{14}C]-5,7-DHIAA (10^{-4} or 10^{-5} M) was dissolved in 250 μl of 0.5 M phosphate buffer at pH 7.4 at 37°C; 50 μl (1 mg of tissue) of homogenate of rat hypothalamus were added to the dihydroxyindoleacetic acid-containing solutions. $n=4$; mean$\pm$SD.

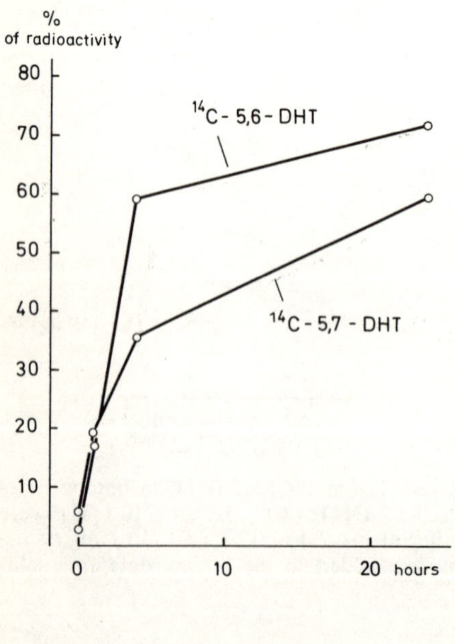

FIGURE 16. Time course of *in vivo* binding of [^{14}C]-5,6-DHT and [^{14}C]-5,7-DHT to protein pellets. The values are expressed as the percentage of total radioactivity in the first PCA homogenate. Subsequently, a differential exhaustive extraction of the pellet was performed, and the 1-, 4-, and 24-hr values thus represent nonextractable protein-bound radioactivity. Blank values are seen on the y axis (for details, cf. MATERIALS AND METHODS).

competitive manner. The early (irreversible?) inactivation of MAO by 5,6-DHT or 5,7-DHT is difficult to distinguish from a noncompetitive type of inhibition because of the lowering of V_{max}. This uncertainty applies at least for 5,7-DHT, because there was linearity in the conversion of this substrate for as long as 60 min.

The noncompetitive inhibition of [^{14}C]-tryptamine by 5,6-DHT or 5,7-DHT indicates that in addition to the enzyme-substrate complex there exists an inhibitor-enzyme-substrate complex, which results in formation of product at a reduced rate. This complex requires binding of 5,6-DHT or 5,7-DHT to a second binding site of the enzyme molecule. Small amounts of quinones or quinoneimines formed from 5,6-DHT or 5,7-DHT by interaction with oxygen that preferentially react with nucleophiles in proteins [7] might well be initially fixed at a second binding site of the enzyme molecule without denaturation (e.g., by cross-linking). This ambivalent nature of 5,6-DHT and 5,7-DHT (substrate/inhibitor of MAO) leads to the possibility of an affinity labeling of MAO, similar to the inactivation of COMT by autoxidation products of 6-OH-DA.[6, 8]

Ultraviolet absorption spectrographic analysis revealed a rapid, time-dependent breakdown of 5,6-DHT at biologic pH and 37° C. By contrast, 5,7-DHT appeared to be more stable and slowly formed an apparently stable oxidation product. In correlation to the time course of oxidation in tissue-free medium, an initially slowly and then rapidly developing binding of [^{14}C]-5,6-DHT-derived radioactivity to PCA-precipitable tissue components was observed in homogenates of brain tissue. This rapidly occurring binding process might well have been autocatalytically promoted by hydrogen peroxide formed during the oxidation of 5,6-DHT, and this binding process stopped only after precipitation of an insoluble colored polymer. Heikkila and Cohen [4] presented evidence for a similar autocatalytic effect of hydrogen peroxide in the air oxidation of 6-OH-DA.

The binding of [^{14}C]-5,7-DHT-derived radioactivity to PCA-insoluble homogenate pellets occurred slowly and to a much lesser extent. An analogous pellet binding was monitored after incubation of homogenates with [^{14}C]-5,6-DHIAA and [^{14}C]-5,7-DHIAA, the potential MAO metabolites of 5,6-DHT and 5,7-DHT, respectively.

These differences in the binding of labeled 5,6-DHT and 5,7-DHT may correspond to the time course of inactivation of MAO by these substrates. After an interval of about 10 min, 5,6-DHT drastically reduced MAO activity, but after 30 min, when formation of the colored polymer had become observable, the inactivation of MAO terminated. There was, by contrast, little, if any, time-dependent inactivation of MAO by 5,7-DHT. There was no effect of catalase or superoxide dismutase on the inactivation of MAO by 5,6-DHT, which parallels findings by Borchardt,[6, 8] who used 6-OH-DA and 5,6-dihydroxyindole as substrates for COMT. This finding casts doubt on the postulated role of hydrogen peroxide, oxygen, or hydroxyl radical in the neurotoxicity of 5,6-DHT and 5,7-DHT. Dithiothreitol and glutathione, compounds that have a high reduction potential and reactive, easily accessible thiol groups, as well as reduced oxygen in the incubation medium, antagonized the toxic action of 5,6-DHT on MAO.

The prevention of the inactivation of MAO by glutathione and dithiothreitol could be due to formation of conjugates between 5,6-DHT and these agents. Analogous mechanisms of protection have been described for 6-OH-DA by

Liang *et al.*[21] The protection of the enzyme by maintenance of the "reductive milieu" for oxidation-prone sulfhydryl groups appears less likely because sulfhydryl groups are the preferred partners for nucleophilic addition of oxidation products of 5,6-DHT.[7] On the other hand, both glutathione and dithiothreitol could prevent the oxidation of 5,6-DHT and thereby the formation of reactive quinoid structures. This prevention is possible only if the redox potential of both thiols is more negative than that of the quinone system of 5,6-DHT.

In contrast to 5,6-DHT, there was no clear protective effect of glutathione and dithiothreitol on MAO inactivation by 5,7-DHT, nor was there any influence of ascorbate or reduced oxygen. This finding suggests that the time-dependent inactivation of MAO by 5,7-DHT was minimal during the time period studied or that the loss of activity observed was solely due to the incubation inactivation.

Exhaustive extraction of PCA-insoluble pellet fractions of rat brain tissue injected *in vivo* with [^{14}C]-5,6-DHT or [^{14}C]-5,7-DHT indicated that as early as 1 hr, a considerable proportion of radioactivity from either compound was tightly bound to proteins. A comparison of the time courses of the *in vivo* with *in vitro* binding of [^{14}C]-5,7-DHT-derived radioactivity reveals a manyfold higher binding *in vivo*. This finding points to the possibility of enhancement of the toxicity of 5,7-DHT by metabolic processes.[7] The potential role of the 5,7-dihydroxyindoleacetaldehyde in this augmentation of toxicity needs to be evaluated in the future.

Conclusions

When [^{14}C]-5,6-DHT or [^{14}C]-5,7-DHT was incubated with homogenates of rat brain, both drugs were metabolized by MAO, as evidenced by radiometric measurements. The deamination proceeded via Michaelis-Menten kinetics and was inhibited by nialamide. 5,6-DHIAA and 5,7-DHIAA were identified as metabolites by thin-layer chromatography; unidentified products were also found. The Michaelis constants were found to be in the range of 10^{-4} M, suggesting effective metabolism also *in vivo*. The higher polarity of 5,6-DHT and 5,7-DHT, which is due to the additional hydroxyl function in the indole molecule (when compared with 5-HT), may be responsible for the increased K_m values of the dihydroxy analogs. The following circumstances argue in favor of a preferential metabolism of 5,6-DHT and 5,7-DHT by MAO A: the high concentration of 5,6-DHT and 5,7-DHT required for inhibition of [^{14}C]-tryptamine oxidation by MAO, the serotonin-like properties of 5,6-DHT and 5,7-DHT, and the preponderance of MAO A in the rat central nervous system. The correlation between autoxidation of 5,6-DHT and 5,7-DHT, the irreversible binding of radioactivity from both drugs to tissue components, and the time-dependent inactivation of MAO were investigated *in vitro*. In contrast to 5,7-DHT, 5,6-DHT was rapidly and extensively irreversibly bound to tissue components and inactivated MAO within 30 min after the onset of incubation at 37° C. It is possible that the inactivation of MAO by 5,6-DHT is accomplished through "affinity labeling" of the enzyme by aminochrome-like autoxidation products of 5,6-DHT. A similar time course of protein binding in homogenates was demonstrated with 5,6-DHIAA and 5,7-DHIAA. The inactivation of MAO by 5,6-DHT was antagonized by reduced glutathione, dithiothreitol, and reduced oxygen. The inhibition of MAO in homogenates in rat brain treated *in vivo* with 5,6-DHT or 5,7-DHT was transient, despite the fact that radioactivity derived from either amine was bound to protein. The amount of [^{14}C]-5,7-DHT-derived radioactivity bound to protein *in vivo* was larger than that bound *in vitro,* which may be interpreted as signifying a metabolism-dependent enhancement of the toxicity of 5,7-DHT. The postulate is raised that the MAO-catalyzed formation of the 5,7-dihydroxyindoleacetaldehyde plays a role in this toxicity reinforcement, in addition to the dominant role of the major autoxidation product of 5,7-DHT, the *p*-quinoneimine of 5,7-DHT, in the neurotoxicity of this *m*-substituted DHT.

Acknowledgments

The authors express their appreciation to Mrs. E. Lachenmayer, Mrs. U. Jakubowsky, and Mr. G. Nohl for technical assistance.

References

1. Tranzer, J. P. & H. Thoenen. 1967. Ultramorphologische Veränderung der sympathischen Nervenendigungen der Katze nach Vorbehandlung mit 5- und 6-Hydroxy-Dopamin. Arch. Exp. Pathol. Pharmakol. **257:** 343–344.
2. Thoenen, H. & J. P. Tranzer. 1968. Chemical sympathectomy by selective destruction of adrenergic nerve endings with 6-hydroxydopamine. Arch. Exp. Pathol. Pharmakol. **261:** 271–288.
3. Saner, A. & H. Thoenen. 1971. Model experiments on the molecular mechanism of action of 6-hydroxydopamine. Mol. Pharmacol. **7:** 147–154.
4. Heikkila, R. & G. Cohen. 1971. Inhibition of biogenic amine uptake by hydrogen peroxide: a mechanism for toxic effects of 6-hydroxydopamine. Science **172:** 1257–1258.
5. Jonsson, G. & Ch. Sachs. 1975. On the mode of action of 6-hydroxydopamine. *In* Chemical Tools in Catecholamine Research. G. Jonsson, T. Malmfors & Ch. Sachs, Eds. Vol. I: 41–50. North Holland Publishing Company. Amsterdam and Oxford.
6. Borchardt, R. T. 1975. Catechol O-methyltransferase: a model to study the mechanism of 6-hydroxydopamine interaction with proteins. *In* Chemical Tools in Catecholamine Research. G. Jonsson, R. Malmfors & Ch. Sachs, Eds. Vol. I: 33–40. North Holland Publishing Company. Amsterdam and Oxford.
7. Rotman, A., J. W. Daly & C. R. Creveling. 1976. Oxygen-dependent reaction of 6-hydroxydopamine, 5,6-dihydroxytryptamine, and related compounds with proteins in vitro: a model for cytotoxicity. Mol. Pharmacol. **12:** 887–899.
8. Borchardt, R. T. 1975. Affinity labeling of catechol O-methyltransferase by the oxidation products of 6-hydroxydopamine. Mol. Pharmacol. **11:** 436–449.
9. Baumgarten, H. G., A. Björklund, L. Lachenmayer, A. Nobin & U. Stenevi. 1971. Long-lasting selective depletion of brain serotonin by 5,6-dihydroxytryptamine. Acta Physiol. Scand. Suppl. **373:** 1–16.
10. Baumgarten, H. G., K. D. Evetts, R. B. Holman, L. L. Iversen, M. Vogt & G. Wilson. 1972. Effects of 5,6-dihydroxytryptamine on monoaminergic neurons in the central nervous system of the rat. J. Neurochem. **19:** 1587–1597.
11. Baumgarten, H. G., A. Björklund, L. Lachenmayer & A. Nobin. 1973. Evaluation of the effects of 5,7-dihydroxytryptamine on serotonin and catecholamine neurons in the rat CNS. Acta Physiol. Scand. Suppl. **391:** 1–19.
12. Noble, E. P., R. J. Wurtman & J. Axelrod. 1967. A simple and rapid method for injecting ^{3}H-norepinephrine into the lateral ventricle of rat brain. Life Sci. **6:** 281–291.

13. WURTMAN, R. J. & J. AXELROD. 1963. A sensitive and specific assay for the estimation of monoamine oxidase. Biochem. Pharmacol. **12:** 1439–1440.
14. VON EULER, U. S. 1967. Cited in: TIPTON, K. F. 1973. Biochemical aspects of monoamine oxidase. Brit. Med. Bull. **29:** 116–119.
15. Meyerson, L. R., K. D. MCMURTREY & V. A. DAVIS. 1976. Neuroamine-derived alkaloids: substrate-preferred inhibitors of rat brain monoamine oxidase in vitro. Biochem. Pharmacol. **25:** 1013–1020.
16. BOURGOIN, S., F. ARTAUD, J. ADRIEN, F. HERY, J. GLOWINSKI & M. HAMON. 1977. 5-Hydroxytryptamine catabolism in the rat brain during ontogenesis. J. Neurochem. **28:** 415–422.
17. NEFF, N. H. & H. Y. T. YANG. 1974. Monoamine oxidases and monoamine oxidase inhibitor drugs. Life Sci. **14:** 2061–2074.
18. SUZUKI, O. & K. YAGI. 1976. Distribution of type A and B monoamine oxidase activities in the central nervous system of rat and chick. Separatum Experientia **32:** 13–14.
19. SEVERINA, I. S. 1973. On the substrate-binding sites of the active centre of mitochondrial monoamine oxidase. Eur. J. Biochem. **38:** 239–246.
20. EKSTEDT, B. 1976. Substrate specificity of the different forms of monoamine oxidase in rat liver mitochondria. Biochem. Pharmacol. **25:** 1133–1138.
21. LIANG, Y.-O., R. M. WIGHTMAN, P. PLOTSKY & R. N. ADAMS. 1975. Oxidative interactions of 6-hydroxydopamine with CNS constituents. *In* Chemical Tools in Catecholamine Research. G. Jonsson, T. Malmfors & Ch. Sachs, Eds. Vol. I: 15–22. North Holland Publishing Company. Amsterdam and Oxford.

DISCUSSION

DR. H. TAMIR: Why do MAO inhibitors affect the cytotoxicity of 5,7-DHT and 5,6-DHT differently?

DR. BAUMGARTEN: The subject is complex. 5,6-Dihydroxytryptamine and its end products produced by MAO action form compounds that rapidly react with sulfhydryl groups and cause inactivation of MAO. 5,7-Dihydroxytryptamine, in contrast, is a good substrate for MAO, and its products do not inhibit MAO or attack sulfhydryl groups. The complete explanation requires a consideration of the dynamics of neurotransmitter storage and release.

There is possibly a high-affinity uptake of 5,7-DHT into certain adrenergic sites that can be blocked with low concentrations of DMI, whereas the uptake of 5,7-DHT into serotonergic sites can only poorly be blocked with chlorimipramine.

DR. M. DA PRADA: Regarding Dr. Baumgarten's hypothesis that chlorimipramine blocks the uptake of 5,7-DHT, we found that LU-10,171, a new compound that is quite selective and strong in blocking serotonin uptake, can prevent the destruction of the serotonergic system when injected intraperitoneally at 50 mg/kg.

DR. C. R. CREVELING: The C cell of the thyroid, which takes up norepinephrine and dopamine, is interesting in regard to neurotoxins. It takes up 6-hydroxydopamine, but it doesn't undergo any visible degeneration. The only changes observed were a marked increase in the rate of protein synthesis and an increased number of ribosomes.

MECHANISM OF ACTION OF DIHYDROXYTRYPTAMINES

C. R. Creveling and A. Rotman *

Laboratory of Chemistry
National Institute of Arthritis, Metabolism, and Digestive Diseases
Bethesda, Maryland 20014

Two of the autoxidizable dihydroxytryptamines, 5,6-dihydroxytryptamine (5,6-DHT) and 5,7-dihydroxytryptamine (5,7-DHT), have been shown to have a relatively selective cytotoxic action on central serotoninergic neurons.[1-3] In addition, 5,7-DHT has been shown to have a cytotoxic action on peripheral noradrenergic neurons.[4,5] The cytotoxic actions of these tryptamine derivatives are similar in many respects to the actions of autoxidizable phenethylamine derivatives, of which 6-hydroxydopamine (6-HD) is the best known example.[5,6] It is now clear that the selectivity of the cytotoxic effect of 6-HD is derived from the specificity of the active uptake mechanism of 6-HD at the plasma membrane of the nerve terminal. The cytotoxic action occurs when a critical intraneuronal concentration of the amine is achieved through the action of the transport mechanism located in the plasma membrane. The subsequent events that result in the cytotoxic effect are as yet unclear. Two mechanisms have been proposed, both of which invoke the rapid autoxidation of 6-HD to form an initial oxidation product, the highly reactive paraquinone species. One reaction mechanism involves the rapid generation of toxic agents, such as hydrogen peroxide, superoxide ion, and hydroxyl radical, during the autoxidation of 6-HD;[7,8] the other mechanism implicates the covalent interaction of one or more of the oxidation products of the amine with intraneuronal proteins, leading to an extensive cross-linking of the proteins responsible for the cytotoxicity.[9,10]

However, the actions of both 5,6-DHT and 5,7-DHT differ in several respects from that of 6-HD. While the selectivity of the cytotoxic action of 5,7-DHT and 5,6-DHT appears to depend on an affinity for serotonin-containing neurons,[3,11] such effects are not limited to serotonin-containing structures [5] but also occur in both central and peripheral catecholamine-containing neurons. The relative selectivity results from a higher affinity of the uptake processes at serotonin-containing terminals than at catecholamine-containing terminals.[12] The specific neurotoxic action of 5,6-DHT appears to be obscured by a marked nonspecific toxicity both centrally and peripherally. 5,6-DHT induces the formation of invasive lesions of the epimyocardium without showing any effect on the noradrenergic nerve terminals of the heart.[13] This finding is surprising since a variety of evidence indicates that 5,6-DHT is accumulated within noradrenergic neurons in the heart. Both 5,7-DHT and 6-HD produce neurodegenerative toxicity in noradrenergic terminals in cardiac tissue but show no tendency to cause epimyocardial lesions.[13] A major difference between the properties of

* Present address: Department of Biochemistry, Hebrew University-Hadassah Medical School, Jerusalem, Israel.

0077–8923/78/0305–0057 $01.75/0 © 1978, The New York Academy of Sciences

6-HD and those of 5,7-DHT is the apparent role of monoamine oxidase in their cytotoxic action. Prior inhibition of monoamine oxidase, *in vivo,* potentiates the cytotoxic action of 6-HD but prevents that of 5,7-DHT.[5, 14] Furthermore, although the α-methyl derivative of 5,7-DHT accumulates in noradrenergic neurons, it is devoid of cytotoxic action.[14] The studies presented in this report utilized isotopically labeled 5,6-DHT and 5,7-DHT in an attempt to provide more direct evidence concerning the mechanism of action of the cytotoxicity of these compounds. The binding and ability to induce cross-linking of model proteins *in vitro* and the effects on the uptake mechanism in isolated atria by 5,6-DHT and 5,7-DHT are presented.

Materials and Methods

[^{3}H]-5,6-Dihydroxytryptamine (6.6 mCi/mmol) and [^{3}H]-5,7-dihydroxytryptamine (45 mCi/mmol) were prepared according to the procedure of Lee *et al.*,[15] as modified by Rotman *et al.*[9] 5,6-DHT creatinine sulfate and 5,7-DHT creatinine sulfate were obtained from Regis Chemical Company and were provided by Dr. A. A. Manian, Psychopharmacology Service Center, National Institute of Mental Health. Bovine serum albumin (recrystallized six times) was obtained from Armour Pharmaceutical Company. Horse and human hemoglobin (recrystallized two times) were obtained from Dr. Harry A. Saroff, Laboratory of Biophysical Chemistry, National Institute of Arthritis, Metabolism, and Digestive Diseases. Acetylated bovine serum albumin was prepared as described by Sander and Thoenen,[16] as modified by Rotman *et al.*[9] [^{3}H]-DL-Norepinephrine (10 Ci/mmol) was purchased from New England Nuclear Corporation. The *N*-ethylmaleimide derivative of horse hemoglobin was prepared as previously described.[9] Animals used were National Institutes of Health general-purpose, male mice weighing 18–24 g. All other materials was obtained from standard commercial sources.

Determination of the extent of covalent binding of [^{3}H]amines to proteins *in vitro* was performed as previously described.[9] Polyacrylamide gel electrophoresis of amine-protein complexes was performed as previously described,[9] according to the chromatography systems of Neville.[17] Protein was determined by the procedure of Lowry *et al.*[18] In all cases, radioactivity was measured by scintillation spectrophotometry with a xylene-based scintillation fluid (Aquasol®). Values were corrected for quenching.

Uptake of amines by mouse atria was determined by the following technique. Mice were sacrificed by cervical dislocation, and both atria were removed, cut in half, and washed in several changes of warm (37° C), oxygenated (95% O_2, 5% CO_2) Krebs-Ringer bicarbonate buffer. Atria maintained their ability to take up amines for several hours under these conditions. Media were decanted, and approximately equal portions of atrial fragments, which represented four to eight atria, were transferred to small test tubes that contained Krebs-Ringer solution (0.3 ml) and amine. Tubes were prepared by evaporating, under nitrogen, the appropriate quantity of an ethanol solution of the dihydroxytryptamines prior to the addition of Krebs-Ringer bicarbonate buffer. Incubation was continued for 1–120 min and terminated by decanting the mixture onto a Millipore® filter. The atrial fragments were quickly washed with cold buffer (5 ml) and transferred to a conical glass homogenizing tube (3 ml) that contained 0.4 N perchloric acid. After homogenization, the mixture was centrifuged (1000*g*) for 10 min, and the radioactivity in the soluble supernatant fraction was measured. The sediment was rehomogenized in perchloric acid, and the procedure was then repeated until no additional radioactivity was present in the acid-soluble fraction. The sediment was suspended in perchloric acid, and the protein-bound radioactivity and the total protein of the sample were determined. The mean protein per atrium was 0.46 mg, and the total sample ranged from 1.8 to 3.7 mg of protein. Control, or "zero degree," incubations were performed by adding atrial preparations to tubes precooled and maintained in an ice-water bath. In practice, uptake was measured for 10 min. Results were expressed as picomoles of amine taken up or bound per milligram of protein, corrected for the uptake at "zero degrees."

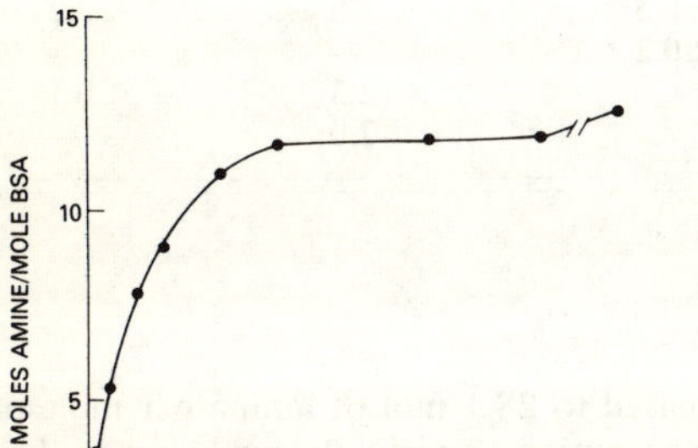

FIGURE 1. Time course of the binding of [^{3}H]-5,6-DHT (●) and [^{3}H]-5,7-DHT (○) to native bovine serum albumin.

RESULTS

Binding of 5,6-DHT and 5,7-DHT to Proteins in Vitro

The maximum rate of irreversible binding of 5,6-DHT to native bovine serum albumin at pH 7.2 was achieved with a 50-fold excess of amine (8.0 mM) with respect to the concentration of protein (0.16 mM). The reaction was a relatively slow process, exhibiting an initial rate of 5.0 mol of amine bound per mole of protein. Half-maximal binding was attained in 1.5 hr and apparent saturation at 18 hr, with 12.6 mol of amine bound per mole of protein (FIGURE 1). In marked contrast, the irreversible binding of 5,7-DHT was much slower and failed to reach an apparent saturation (FIGURE 1). The maximum binding achieved with 5,7-DHT in 24 hr was 1.2 mol of amine per mole of protein. Amine concentrations greater than 8 mM or reaction times as long as 7 days failed to significantly increase the total binding. The time course of irreversible binding of 5,6-DHT to horse or human hemoglobin or to acetylated bovine serum albumin was similar to that of native bovine serum albumin, except that the maximum binding results achieved were 3.6, 5.0, and 26.2 mol of amine per mole of protein, respectively (TABLE 1). However, when the reaction was performed in the presence of 6 M urea (TABLE 1), the total bind-

TABLE 1

IRREVERSIBLE BINDING OF [³H]-5,6-DHT AND [³H]-5,7-DHT TO PROTEINS *in Vitro*

Protein	Extent of Binding * of [³H]-5,6-DHT (moles of amine/moles of protein)		
	Native	6 M Urea	Anaerobic
Serum albumin (bovine)	11.9±2.1 [b]	28.1	<0.1
Acetyl albumin (bovine)	26.2	29.0	<0.1
Hemoglobin (horse)	3.6	4.1	<0.1
NEM hemoglobin (horse)	0.3	—	—
Hemoglobin (human)	5.0	6.3	<0.1
	[³H]-5,7-DHT		
Serum albumin (bovine)	1.9±0.5 †	2.0	<0.2
Acetyl albumin (bovine)	1.8±0.2 †	—	—
Hemoglobin (horse)	2.0	2.3	<0.1
Hemoglobin (human)	1.7	2.0	—

* Reaction time, 24 hr.

† Mean ± SD for four samples.

ing to native bovine serum albumin increased to 28.1 mol of amine per mole of protein, whereas binding to other proteins remained virtually unchanged. The binding of 5,7-DHT was not significantly different with the other proteins tested from that observed with native bovine serum albumin nor was any significant increase in binding observed in the presence of 6 M urea. Virtually no binding of 5,6-DHT or of 5,7-DHT occurred in the absence of oxygen or with the *N*-ethylmaleimide derivative of horse hemoglobin. The binding of 5,6-DHT to native bovine serum albumin in the presence of a twofold excess of mercaptoethanol, dithiothreitol, or ascorbate with respect to the amine was partially blocked, reducing the binding to 2–6 mol of amine per mole of protein (data not shown).

Cross-linking of Proteins by Amines

Examination of the reaction products of 5,6-DHT with bovine serum albumin by chromatography on polyacrylamide gels suggested that in addition to an irreversible binding of the amine to protein, a cross-linking of protein was occurring, yielding polymers of progressively increasing size. As indicated in FIGURE 2, chromatography of the reaction mixture of 5,6-DHT and bovine serum albumin, after allowing the reaction to proceed for 60 min, revealed the presence of protein species that remained at the origins of the 3% stacking and 5% separating gels. Intermediate-sized polymers of the monomeric form of bovine serum albumin were present in smaller quantities. The protein species that remained at the gel origin retained 72% of the radioactivity derived from 5,6-DHT, whereas 12% was bound to the monomeric bovine serum albumin. The formation of these higher-molecular-weight species was not observed when the reaction was performed anaerobically or with 5,7-DHT (FIGURE 2). When the reaction between 5,6-DHT and bovine serum albumin was allowed to proceed to completion, a dark brown precipitate gradually appeared. The material was cross-linked protein and could not be dissociated by dialysis against buffer or by treatment with mercaptoethanol, sodium bisulfite urea, or detergents. The 5,6-DHT bound to the cross-linked protein could not be removed by prolonged washing with perchloric acid (0.4 N) or trichloroacetic acid (6%).

Uptake of 5,7-DHT and 5,6-DHT by Mouse Atria in Vitro

Mouse atrial preparations displayed an active uptake process for both 5,6-DHT (FIGURE 3) and 5,7-DHT (FIGURE 4). At amine concentrations of 30 μM, the initial uptakes, corrected for the accumulation at 0° C, for 5,7-DHT and 5,6-DHT were 1.3 and 7.8 pmol/mg/min, respectively. The uptake of 5,7-DHT was linear with time for 20–30 min and leveled off after 60 min. The uptake of 5,6-DHT was linear for only 5–10 min; however, the accumulation of the amine continued at a slower rate for at least 120 min. The uptake of 5,7-DHT by atria was a saturable process, with an apparent K_m of 39.5 μM and V_{max} of 9.8 pmol/mg/min (FIGURE 5). No evidence for a high-affinity component for the uptake of 5,7-DHT was observed. Attempts to determine the K_m for the uptake of 5,6-DHT were not successful because of the low specific

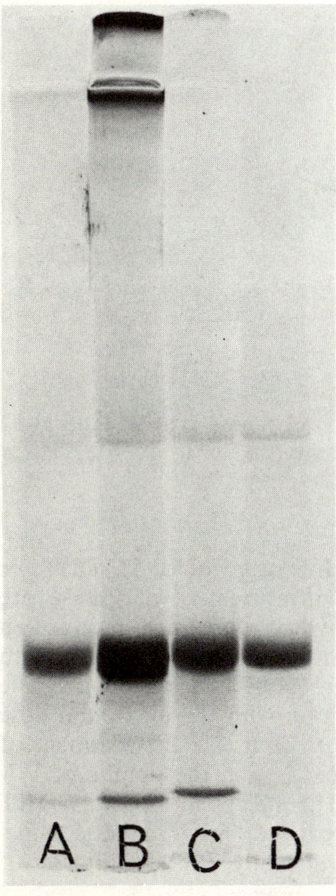

FIGURE 2. Chromatography of the soluble reaction products of bovine serum albumin with [^{3}H]-5,6-DHT a sodium dodecyl sulfate-polyacrylamine gel: stacking gel, 3%; separating gel, 5%. A, 5,7-DHT; B, 5,6-DHT; C, 5,6-DHT, anaerobic; D, bovine serum albumin control.

activity of the amine preparation; however, the uptake curve at 30 μM amine suggests the presence of a two-component system. The initial uptake of 5,6-DHT is at least fourfold to 20-fold greater than that of 5,7-DHT.

The initial uptake of 5,7-DHT was sensitive to inhibition by cocaine and desmethylimipramine (DMI), each exhibiting an IC_{50} in the range of 0.5 mM (TABLE 2). Only marginal reductions of the uptake of 5,6-DHT were observed in the presence of 0.5 mM cocaine or DMI. The uptake of 5,7-DHT was also inhibited by the presence of α-methyltyramine and serotonin (TABLE 2). The presence of these amines, even at a concentration of 1.0 mM, had no significant effect on the uptake of 5,6-DHT.

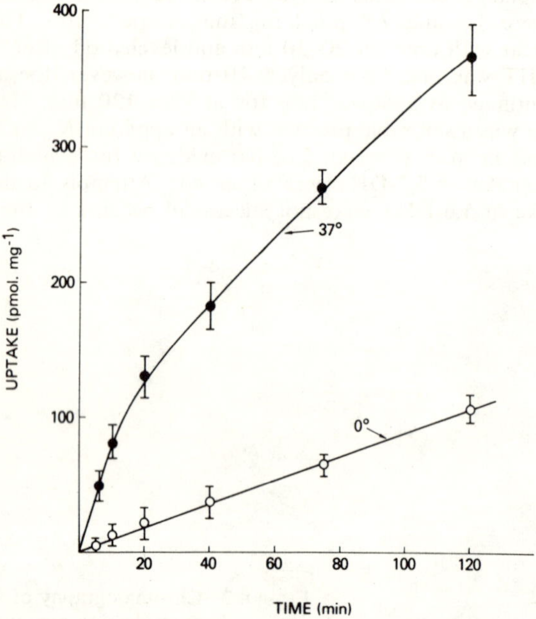

FIGURE 3. Uptake of [^{3}H]-5,6-DHT by a preparation of mouse atria *in vitro* at 37 and 0° C. Each point (pmol/mg of protein/ represents the mean ± SD for the equivalent of four to eight atria in FIGURES 1 and 7.

The effect of pretreatment of mice with 6-hydroxydopamine, 5,7-DHT, or 5,6-DHT 18 hr before removal of atria and the measurement of uptake are shown in TABLE 3. Treatment of mice with 6-hydroxydopamine produces a dose-dependent loss in the uptake of 5,7-DHT and norepinephrine (NE), with a dose of 90 mg/kg reducing the uptake to 12–15% of control values. The effect on the uptake of 5,6-DHT was less pronounced, with the maximum dose producing a reduction in uptake of a little more than 50%. Pretreatment of mice with 5,7-DHT also produced a dose-dependent loss in the uptake of 5,7-DHT, 5,6-DHT, and NE, as measured *in vitro*. The effectiveness of 5,7-DHT in this regard was less than half of that observed with 6-hydroxydopamine. Pretreatment of mice with 5,6-DHT failed to significantly alter the uptake of

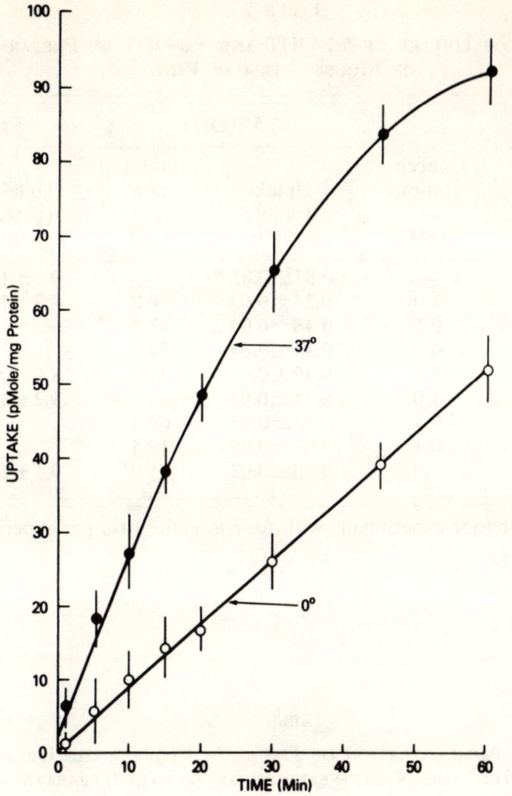

FIGURE 4. Uptake of [³H]-5,7-DHT by mouse atria at 37 and 0° C *in vitro*.

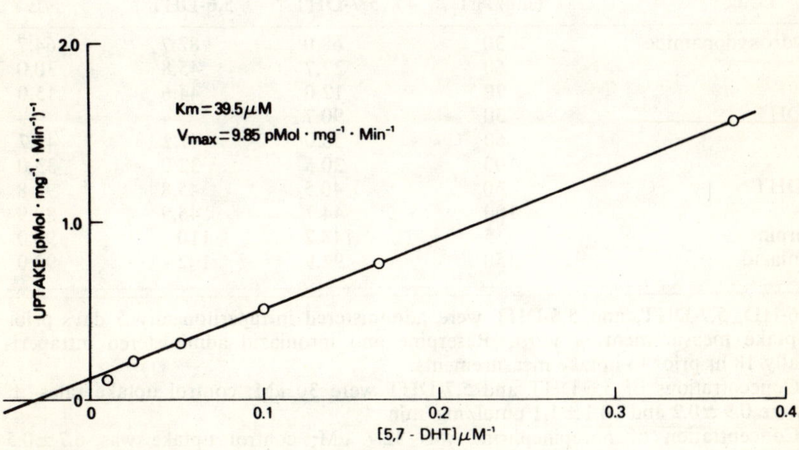

FIGURE 5. Double reciprocal plot of the temperature-dependent uptake of [³H]-5,7-DHT by mouse atria *in vitro*. Incubation time, 10 min.

TABLE 2

INHIBITION OF UPTAKE OF 5,7-DHT AND 5,6-DHT BY PREPARATIONS OF MOUSE ATRIA *in Vitro*

Agent	Concentration (mM)	5,7-DHT Uptake (pmol/mg/min)	5,7-DHT Inhibition (%)	5,6-DHT Uptake (pmol/mg/min)	5,6-DHT Inhibition (%)
None	—	0.83±0.05 *	—	9.0±1.4	—
Cocaine	1.0	0.24±0.02	70.2	8.7±1.2	3
	0.5	0.46±0.03	43.6		
	0.1	0.54±0.04	34.3		
DMI	0.5	0.49±0.03	41.0	8.6±0.7	4
α-Methyltyramine	1.0	0.14±0.01	82.7	9.2±1.0	0
	0.5	0.25±0.03	69.8		
	0.1	0.67±0.08	19.4		
Serotonin	1.0	0.20±0.02	75.0	9.1±1.1	0

* Mean ± SD for four experiments with four to eight atria per experiment.

TABLE 3

EFFECT OF ADMINISTRATION OF DRUGS *in Vivo* ON THE UPTAKE OF 5,7-DHT, 5,6-DHT, AND NOREPINEPHRINE BY ATRIAL PREPARATIONS *in Vitro*

Drug *	Dose (mg/kg)	Uptake (percent of control) 5,7-DHT †	5,6-DHT †	NE ‡
6-Hydroxydopamine	30	68.0	82.7	64.7
	60	22.7	45.8	30.0
	90	12.0	44.6	15.0
5,7-DHT	30	90.7	—	—
	60	50.0	47.2	45.7
	100	30.5	32.0	33.0
5,6-DHT	60	40.5	45.8	92.8
	100	44.7	46.9	89.9
Reserpine	5	148.2	110	96.0
Iproniazid	150	92.9	102	98.0

* 6-HD, 5,7-DHT, and 5,6-DHT were administered intraperitoneally 5 days prior to uptake measurement *in vitro*. Reserpine and iproniazid administered intraperitoneally 18 hr prior to uptake measurements.

† Concentrations of 5,6-DHT and 5,7-DHT were 30 μM; control uptake rates ± SD were 0.9±0.2 and 10.1±1.1 pmol/mg/min.

‡ Concentration of norepinephrine was 32.7 μM; control uptake was 6.7±0.5 pmol/mg/min. Total incubation time 10 min. All uptake values were corrected for uptake at "zero degrees."

NE. The uptake of both 5,7-DHT and 5,6-DHT was reduced nearly 50%; however, the reduction did not appear to be dose dependent. Reserpine pretreatment had no effect on the uptake of NE or 5,6-DHT and appeared to result in a modest increase in the uptake of 5,7-DHT. Inhibition of monoamine oxidase activity by pretreatment with iproniazid did not alter the uptake.

Covalent Binding of 5,7-DHT and 5,6-DHT to Atrial Protein in Vitro

The binding of both 5,7-DHT and 5,6-DHT and/or their oxidation products was measured at various times during the uptake of these amines. As seen in FIGURE 6, the amount of 5,7-DHT or its oxidation products that is tightly bound to atrial constituents is small and represents only 7–9% of the net uptake of 5,7-DHT. The rate at which this binding occurred was estimated as 0.1 pmol/

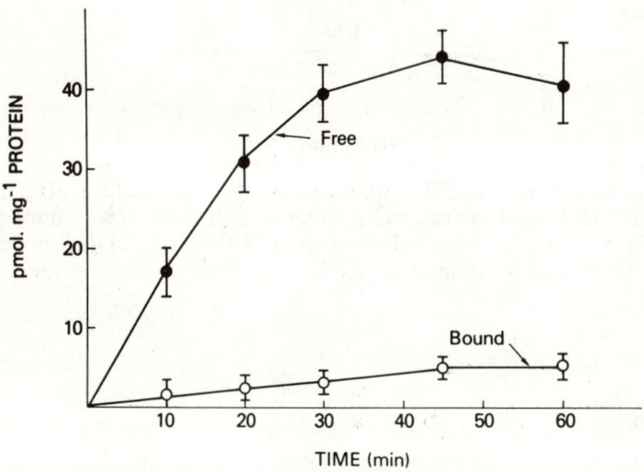

FIGURE 6. Time course of the temperature-dependent uptake (●) and irreversible binding (○) of [^{3}H]–5,7-DHT to mouse atria *in vitro*.

mg/min. The efflux of 5,7-DHT after preloading atria with 5,7-DHT for 30 min is shown in FIGURE 7. After an initial outflow of approximately 58% of the 5,7-DHT in the first 20 min of the postincubation period, the outflow stopped, with 42% of the radioactivity derived from 5,7-DHT remaining in the atria. Further postincubation in amine-free media failed to release this fraction of 5,7-DHT. The amount, approximately 8% of the total uptake, that was covalently bound remained unchanged throughout the postincubation period. Thus, it appeared that approximately 30% of the 5,7-DHT remained sequestered in the atria in some form not freely diffusable with the media.

In contrast to the limited irreversible binding of 5,7-DHT, the uptake of 5,6-DHT by atria was accompanied by extensive, irreversible binding of 5,6-DHT and/or its oxidation products to atrial protein. As shown in FIGURE 8, the binding that occurred during the initial 10 min of uptake was approximately 13% of the net uptake and progressively increased to greater than 60% after 2 hr. Postincubation of atria, prelabeled with 5,6-DHT for 60 min, in amine-

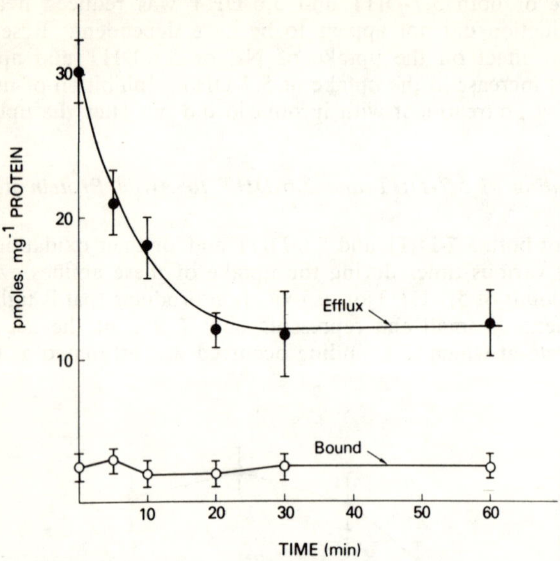

FIGURE 7. Time course of the efflux of radioactivity from mouse atria prelabeled with [³H]-5,7-DHT. Unbound radioactivity remaining in atria (●); bound radioactivity remaining in atria (○). Preincubation time, 30 min; 5,7-DHT concentration, 30 μM. Postincubation media, amine-free Krebs-Ringer bicarbonate buffer.

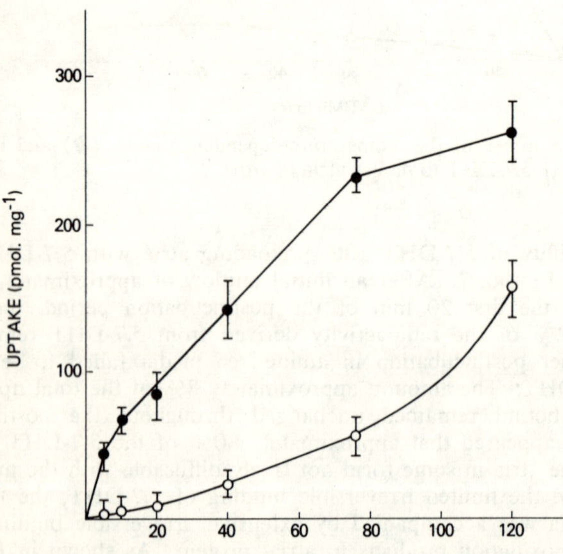

FIGURE 8. Time course of the temperature-dependent uptake (●) and the irreversible binding (○) of [³H]-5,6-DHT by mouse atria *in vitro*.

free media resulted in the rapid efflux of nearly all of the unbound radioactivity (78%) within the first 20 min. The remaining 22% was irreversibly bound. No evidence for retention of 5,6-DHT other than the irreversibly bound fraction could be demonstrated.

Subcellular Distribution of 5,7-DHT in Mouse Atria

A study of the subcellular distribution of 5,7-DHT, shown in TABLE 4, confirmed the observation shown in FIGURES 5 and 6 that only a small fraction (3.8%) of the total 5,7-DHT accumulated by atria was present in an irreversibly bound form. It is of interest that virtually all of the radioactivity in the

TABLE 4

SUBCELLULAR DISTRIBUTION OF 5,7-DHT IN MOUSE ATRIA *

Fraction	Total Content (pmol)	Percent	
		Bound	Free
Total uptake	247	3.8	96.2
Soluble	179	0	100
Particulate	68	13.8	86.2
Particulate			
Cell debri	32.7	>0.1	98
Mitrochondrial	9.2	99	>0.1
Microsomal	26.0	>0.1	96

* Uptake was allowed to proceed for 30 min with an atrial preparation from 18 atria in [^{3}H]-5,7-DHT (30 μM). The washed fragments were homogenized in phosphate-buffered saline (PBS), in a conical glass-glass homogenizer, and a portion centrifuged for 60 min at 105,000*g* to give the total soluble and particulate fractions; a portion was centrifuged successively at 600, 11,000, and 105,000*g* to give the particulate fractions "cell debri, mitochondria, and microsomes," respectively. Each sediment was washed in PBS (2 ml) and resedimented. "Bound" radioactivity was determined after the addition of bovine serum albumin (5.0 mg) and 0.4 N perchloric acid and homogenization and washing with 0.4 N perchloric acid, as described in MATERIALS AND METHODS, and expressed in percent for each fraction.

soluble fraction was present in an unbound form. Approximately 38% of the radioactivity derived from 5,7-DHT was associated with the particulate fraction and was divided almost equally between the cell debri and the microsomal fractions, with a lesser amount in mitochondria. The portion bound to the cell debri and microsomes was present as an acid-soluble form or "free," whereas the fraction bound to the mitochondrial sediment was present in an acid-insoluble or irreversibly bound form.

DISCUSSION

The properties of 5,7-DHT and 5,6-DHT and those of 6-HD are dissimilar in many respects, so that a unifying mechanism of action that would account for the cytotoxic properties *in vivo* must be quite speculative at present. However, the oxygen-dependent binding of 5,6-DHT and 6-HD to protein and the subsequent cross-linking of protein are clearly relevant to considerations of the mechanism of cytotoxicity *in vivo*. Studies of the interaction of 6-HD [9, 10] with proteins *in vitro* have shown that this reaction has an absolute requirement for oxygen and that it is blocked in the presence of reducing agents. The initial event in the binding of 6-HD is the oxidative formation of the electron-deficient [9, 19] 1,4-*p*-quinone derivative. This quinone is highly reactive with respect to nucleophilic groups and reacts readily with free sulfhydryl groups on proteins. Subsequently, a second oxygen-dependent reaction occurs, which results in the cross-linking of proteins to form an insoluble complex protein matrix.[9] Amino acid analysis of this protein matrix revealed the presence of a stable, ninhydrin-positive substance derived from [^{3}H]-6-HD that was tentatively identified as 2,5-dicysteinyldopamine.[9] In addition, it was recently reported by Rotman *et al.*[20] that 6-HD not only bound covalently to the intracellular proteins of a clonal line of cultured neuroblastoma cells but that it also resulted in the formation of high-molecular-weight proteins that contained irreversibly bound isotope derived from radioactive 6-HD.[20, 21] This observation provided direct evidence that 6-HD can induce cross-linking of proteins in cultured cells.

The interaction of 5,6-DHT with proteins shows an absolute dependence on oxygen (TABLE 1) and diminished binding in the presence of reducing agents. The approximate equivalency between 5,6-DHT bound and the total number of potential sulfhydryl groups present, both free and in disulfide bonds in acetylated bovine serum albumin and in native bovine serum albumin (34 potential sulfhydryl residues), suggest that the reactive nucleophiles are sulfhydryl groups. Further evidence is the equivalency between the binding of 5,6-DHT and the number of free sulfhydryl groups in horse (four sulfhydryl residues) and human (six sulfhydryl residues) hemoglobin and the reduced binding that occurs when the sulfhydryl groups of horse hemoglobin are blocked with *N*-ethylmaleimide. Thus, it appears that the initial event in the covalent binding of 5,6-DHT is the oxidative formation of an aminochrome derivative(s), followed by a nucleophilic attack by sulfhydryl residues. 5,6-DHT has three resonance forms,[9] which are activated toward nucleophilic attack in positions 4 and 7, as outlined in FIGURE 9. It is suggested that this sequence results in the observed binding and cross-linking of proteins (FIGURE 2). Efforts are in progress to define the structure of the product(s) obtained through the interaction of 5,6-DHT with protein.

To define this reaction mechanism in a system that more closely approximates the conditions seen *in vivo*, the uptake and binding of 5,6-DHT were studied in preparations of mouse atria. As indicated in FIGURES 3 and 8, 5,6-DHT is accumulated by a temperature-dependent mechanism in atrial preparations *in vitro*. This uptake is accompanied by the covalent binding of 5,6-DHT or its oxidation products to atrial protein. The uptake of 5,6-DHT is not sensitive to cocaine or DMI nor is it inhibited by α-methyltyramine, which suggests that the primary uptake site is not noradrenergic neurons. However, prior treatment of animals with 6-HD does reduce the uptake of 5,6-DHT observed *in vitro*, as does pretreatment with 5,7-DHT or 5,6-DHT itself (TABLE 3). It is noteworthy, however, that pretreatment of animals with 5,6-DHT fails to block the subsequent uptake of NE (TABLE 3). This observation agrees with an earlier study performed *in vivo* [14] that demonstrated that no alteration in NE uptake followed the administration of 5,6-DHT or α-methyl-5,6-DHT. A curious discrepancy exists, in that 5,6-DHT was reported to displace NE from mouse heart. This displacement was enhanced following inhibition of monoamine oxidase.[17] However, α-methyl-5,6-DHT, unlike other α-methyltryptamines, which, in general, are more efficacious in the displacement of NE than are the parent tryptamines, was a remarkable exception, producing little, if any, displacement of NE from mouse heart. Thus, there is some question regarding the localization of the uptake of 5,6-DHT (FIGURE 3) and the site of the extensive irreversible binding of 5,6-DHT (FIGURE 8). Recently, we reported[13] a time- and dose-dependent induction of subepicardial lesions in the mouse after the administration of 5,6-DHT. The subepicardial localization of these lesions indicates that the effect on the myocardium is due to a direct action of 5,6-DHT or its oxidation products or metabolites rather than a consequence of ischemia. The uptake of NE by mouse heart *in vivo* is completely unimpaired in animals with massive subepicardial lesions induced by 5,6-DHT.[13, 14] It is suggested,

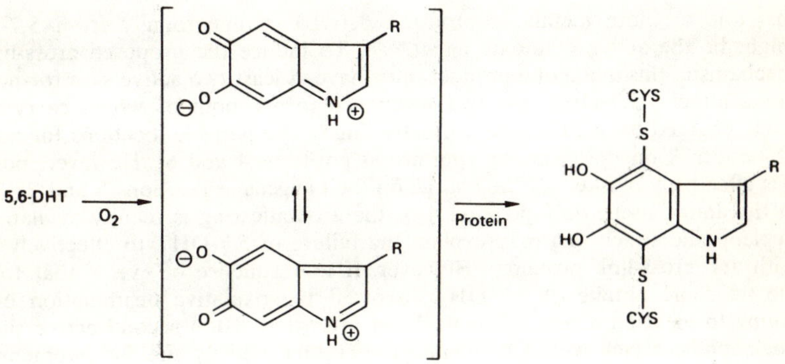

FIGURE 9. Proposed mechanism for the reaction and cross-linking of protein by 5,6-DHT.

therefore, that the irreversible binding of 5,6-DHT is related to some element in the myocardium rather than noradrenergic neurons.

Unlike 5,6-DHT or 6-HD, 5,7-DHT reacts with proteins *in vitro* to a very limited extent (FIGURE 1) and fails to induce the cross-linking process (FIGURE 2).[9] Thus, if the cross-linking of proteins is operant in the cytotoxic action of 5,7-DHT, some further explanation is required. It is unlikely that the failure of 5,7-DHT to effectively bind or cross-link proteins can be attributed to a reduced rate of autoxidation, because Rotman *et al.*[9] have shown that 5,7-DHT undergoes autoxidation at nearly the same rate as does 5,6-DHT. 5,7-DHT differs from 6-HD in that the neurodegenerative effects of 5,7-DHT are blocked by prior inhibition of monoamine oxidase,[5, 14] whereas the effects of 6-HD are potentiated by inhibition of monoamine oxidase.[22] Furthermore, the α-methyl derivative of 5,7-DHT is devoid of neurotoxicity, even though it is taken up and accumulated in noradrenergic neurons,[14] while the neurotoxic action of α-methyl-6-HD is equal to or greater than that of 6-HD.[14, 23] The apparent role of monoamine oxidase in the cytotoxic action of 5,7-DHT led to the suggestion

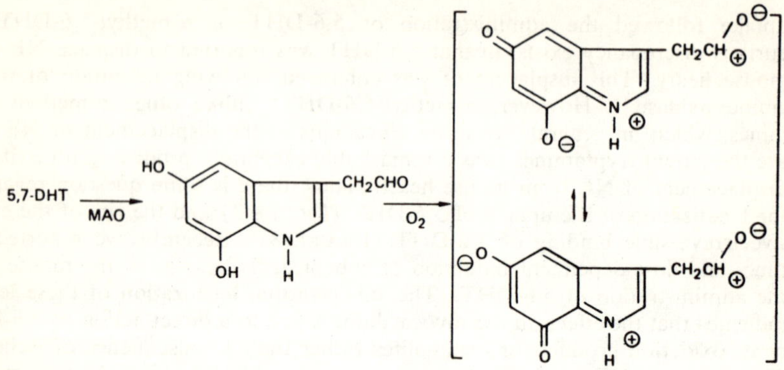

FIGURE 10. Proposed intermediates for the reaction of 5,7-DHT with protein *in vivo*.

that one or more metabolic products derived intraneuronally from 5,7-DHT might be the active cytotoxic agent.[9, 14] To induce the proposed cross-linking mechanism, this metabolic product must have at least two active sites for nucleophilic attack. 5,7-DHT has two resonance forms, both of which carry negatively charged oxygens on the aromatic ring.[9] The possible locations for nucleophilic attack on the aromatic ring are at positions 4 and 6. However, position 6 is effectively masked by the charge on the oxygens at positions 5 and 7. Thus, in the amine itself, only position 4 of the aromatic ring is readily available for nucleophilic attack and may explain the failure of 5,7-DHT to effectively react with and cross-link proteins. However, if the sequence of events that follows the neuronal uptake of 5,7-DHT involved the oxidative deamination of the amine to yield an aldehyde, as depicted in FIGURE 10, a second active site for nucleophilic attack would be available. The most likely site for interaction of this aldehyde with protein is at a tyrosyl residue, giving rise to the proposed structure illustrated in FIGURE 11. Thus, 5,7-DHT may, following the intermediate formation of an aldehyde, cross-link proteins by forming a bridge between cysteinyl and tyrosyl residues. However, no direct evidence for this proposal can be cited at present.

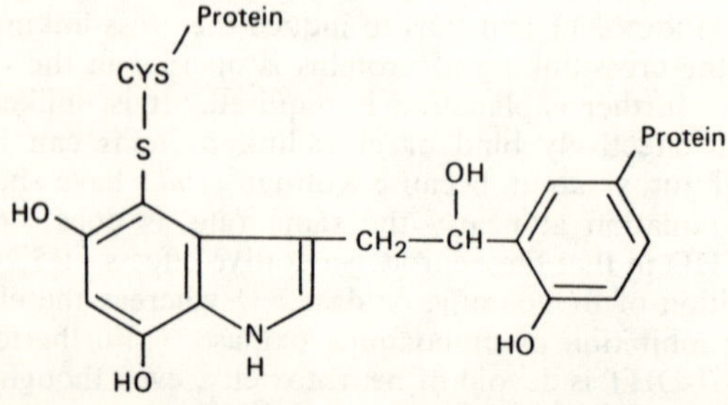

FIGURE 11. Proposed reaction product of 5,7-dihydroxyindolaldehyde with cysteine and tyrosine residues in protein.

The uptake of 5,7-DHT by atrial preparations is, however, accompanied by the irreversible binding of radioactivity derived from [^{3}H]-5,7-DHT (FIGURES 6 & 7). The binding parallels the active uptake and does not show the marked increase in binding on continued incubation of the atrial preparation seen with 5,6-DHT (FIGURE 8). The extent of binding of 5,7-DHT is relatively low (4% of the total uptake in 30 min). The present study indicates that the uptake of 5,7-DHT by mouse atria, unlike that of 5,6-DHT, is mediated primarily by noradrenergic terminals in atria. The uptake is temperature dependent (FIGURE 4), sensitive to cocaine and DMI, is inhibited by competitive substrates like α-methyltryamine and serotonin (TABLE 2) and is a saturable system (FIGURE 5). Furthermore, pretreatment of mice with 6-HD or 5,7-DHT results in a dose-dependent loss of the uptake of 5,7-DHT measured *in vitro* (TABLE 3). The effects on the uptake of 5,7-DHT are paralleled by similar effects on the uptake of NE, except as shown in TABLE 3, where pretreatment with reserpine results in an apparent increase in the uptake of 5,7-DHT. Previous studies have shown that uptake of 5,7-DHT by atria *in vivo* produces degenerative changes in the ultrastructure of neuronal terminals and axons in atria that are similar to those seen after administration of 6-HD.[4, 9, 24] It is conceivable that the limited binding of 5,7-DHT is due to both the low affinity of 5,7-DHT for the uptake system ($K_m = 40\ \mu M$) and to the *in vitro* nature of the atrial preparation. Similar studies with [^{3}H]-6-HD *in vitro* result in the irreversible binding of 15–20% of the total uptake in 30 min,[25, 26] whereas the binding observed *in vivo* by Jonsson and Sachs [^{3}H]-6-HD[27, 28] was approximately 40% after 60 min. The affinity of 6-HD for the uptake system, *in vitro* ($K_m = 4\ \mu M$),[25] like other phenethylamine substrates is much greater than that observed for indolamines. Thus, the binding observed in atria *in vitro* may represent a minimal value. It may also be of importance that the primary site of irreversible binding of 5,7-DHT is associated with the mitochondrial fraction (TABLE 4). Certainly, formation of the proposed aldehyde (FIGURE 10) could lead to either inactivation of monoamine oxidase at the active center of the enzyme or local damage within the mitochondrion. As shown in FIGURE 7, approximately 30% of the radioactivity taken up by atria is sequestered in some *non*covalently bound form. This fraction may represent the intraneuronal binding of 5,7-DHT at some site protected from mitochondrial activity. It would be of interest to determine the effect of 5,7-DHT on oxidative phosphorylation or other mitochondrial functions. Preliminary studies have indicated that 5,7-DHT is bound irreversibly to mitochondrial preparations. However, the presence of a monoamine oxidase inhibitor (harmaline) failed to reduce this binding.[25] In this regard, it should be noted that Wagner and Trendelenburg[29] suggested that the cytotoxic actions of 6-HD may be related to either the potent ability of 6-HD to uncouple oxidative phosphorylation or to its effects on monoamine oxidase. The possibility suggested earlier by Tranzer[30] that mitochondria within noradrenergic terminals could be selective target organelles for 6-HD may well apply to the mode of action of 5,7-DHT and 5,6-DHT.

ACKNOWLEDGMENTS

The authors express their appreciation to Ms. E. T. McNeal and Mr. W. Padgett for their expert technical assistance in this study.

REFERENCES

1. BAUMGARTEN, H. G., A. BJÖRKLUND, L. LACHENMAYER & A. NOBIN. 1973. Evaluation of the effects of 5,7-dihydroxytryptamine on serotonin and catecholamine neurons in the rat CNS. Acta Physiol. Scand. Suppl. **391:** 1–19.
2. BAUMGARTEN, H. G., A. BJÖRKLUND, A. NOBIN, E. ROSENGREN & H. G. SCHLOSSBERGER. 1975. Neurotoxicity of hydroxylated tryptamines: structure activity relationships, 1. Long-term effects on monoamine content and fluorescence morphology of central monoamine neurons. Acta Physiol. Scand. Suppl. **429:** 1–27.
3. BAUMGARTEN, H. G., K. D. EVETTS, R. B. HOLMAN, L. L. IBERSEN, M. VOGT & G. WILSON. 1972. Effects of 5,6-dihydroxytryptamine on monoaminergic neurones in the central nervous system of the rat. J. Neurochem. **19:** 1587–1597.
4. BAUMGARTEN, H. G., H. P. GROTH, M. GOTHERT & A. A. MANIAN. 1974. The effect of 5,7-dihydroxytryptamine on peripheral adrenergic nerves in the mouse. Arch. Exp. Pathol. Pharmakol. **282:** 245–254.
5. DALY, J. W., J. LUNDSTROM & C. R. CREVELING. 1974. Structure-activity correlations in trihydorxyphenethylamines and dihydroxytryptamines. *In* Dynamics of Degeneration and Growth in Neurons. K. Fuxe, L. Olson & Y. Zotterman, Eds. : 29–41. Pergamon Press. Oxford and New York.
6. SACHS, C. & G. JONSSON. 1975. Mechanisms of action of 6-hydroxydopamine. Biochem. Pharmacol. **24:** 1–8.
7. COHEN, G. & R. E. HEIKKILA. 1974. The generation of hydrogen peroxide, superoxide radical, and hydroxylradical by 6-hydroxydopamnie, dialuric acid, and related cytotoxic agents. J. Biol. Chem. **249:** 2447–2452.
8. COHEN, G., B. ALLIS, W. BRADLY, C. MYTILINEOU & R. HEIKKILA. 1975. Prevention of 6-hydropamine toxicity. Eur. J. Pharmacol. **33:** 217–221.
9. ROTMAN, A., J. W. DALY & C. R. CREVELING. 1976. Oxygen-dependent reaction of 6-hydroxydopamine, 5,6-dihydroxytryptamine, and related compounds with proteins *in vitro:* a model for cytotoxicity. Mol. Pharmacol. **12:** 887–899.
10. ROTMAN, A. & C. R. CREVELING. 1977. A rationale for the design of cell-specific toxic agents: the mechanism of action of 6-hydroxydopamine. FEBS Lett. **72:** 227–230.
11. DALY, J., K. FUXE & G. G. JONSSON. 1973. Effects of intracerebral injections of 5,6-dihydroxytryptamine on central monoamine neurons: evidence for selective degeneration of central 5-hydroxytryptamine neurons. Brain Res. **49:** 476–482.
12. BJÖRKLUND, A., A. S. HORN, H. G. BAUMGARTEN, A. NOBIN & H. G. SCHLOSSBERGER. 1975. Neurotoxicity of hydroxylated tryptamines: structure-activity relationships, 2. *In vitro* studies on monoamine uptake and uptake impairment. Acta Physiol. Scand. Suppl. **429:** 30–60.
13. CREVELING, C. R., J. LUNDSTROM & B. HIGHMAN. 1976. Cardiovascular toxicity of dihydroxytryptamines. Res. Commun. Chem. Pathol. Pharmacol. **14:** 1–11.
14. CREVELING, C. R., J. LUNDSTROM, E. T. MCNEAL, L. TICE & J. W. DALY. 1975. Dihydroxytryptamines: effects on noradrenergic function in mouse heart *in vivo.* Mol. Pharmacol. **11:** 211–222.
15. LEE, F. G. N., D. E. DICKSON, J. SUZUKI, A. ZIRNIS & A. A. MANIAN. 1973. Synthesis of 5,7- and 6,7-disubstituted tryptamines and analogs. J. Heterocyclic Chem. **10:** 649–654.
16. SANDER, A. & H. THOENEN. 1971. Model experiments on the molecular mechanism of action of 6-hydroxydopamine. Mol. Pharmacol. **7:** 147–154.
17. NEVILLE, D. M. 1971. Molecular weight determination of protein-dodecylsulfate complexes by gel electrophoresis in a discontinuous buffer system. J. Biol. Chem. **246:** 6328–6334.
18. LOWRY, O. H., N. J. ROSEBROUGH, A. L. FARR & R. J. RANDALL. 1951. Protein measurement with the Folin phenol reagent. J. Biol. Chem. **193:** 265–275.
19. BLANK, C. L., R. L. MCCREERY, R. M. WRIGHTMAN, W. CHEY, R. N. ADAMS,

J. R. Ried & E. E. Smissman. 1976. Intracyclization rates of 6-hydroxydopamine and 6-aminodopamine analogs under physiological conditions. J. Med. Chem. **19:** 178–180.
20. Rotman, A., J. W. Daly, C. R. Creveling & X. O. Breakfield. 1976. Uptake and binding of dopamine and 6-hydroxydopamine in murine neuroblastoma and fibroblast cells. Biochem. Pharmacol. **25:** 383–388.
21. Creveling, C. R., A. Rotman & J. W. Daly. 1975. Interactions of 6-hydroxydopamine and related compounds with proteins: a model for the mechanism of cytotoxicity. *In* Chemical Tools in Catecholamine Research. G. Jonsson, T. Malmfors & C. Sachs, Eds. Vol. **1:** 23–32. North Holland Publishing Company. Amsterdam, The Netherlands.
22. Jonsson, G., T. Malmfors & C. Sachs. 1972. The effects of drugs on the 6-hydroxydopamine induced degeneration of adrenergic nerves. Res. Commun. Chem. Pathol. Pharmacol. **3:** 543–555.
23. Lundstrom, J., H. Ong, J. Daly & C. R. Creveling. 1973. Isomers of 2,4,5-trihydroxyphenethylamine (6-hydroxydopamine): long-term effects on the accumulation of [^{3}H]-norepinephrine in mouse heart *in vivo*. Mol. Pharmacol. **9:** 505–513.
24. Tranzer, J. P. & J. G. Richard. 1971. Fine structural effects of 6-OH-DA in the periphery. *In* 6-Hydroxydopamine and Catecholamine Neurons. T. Malmfors & H. Thoenen, Eds. : 15–22. American Elsevier. New York, N.Y.
25. Rotman, A. & C. R. Creveling. 1976. Unpublished results.
26. Rotman, A. 1977. The mechanisms of action of neurotoxic compounds. Life Sci. In press.
27. Jonsson, G. & C. Sachs. 1975. On the mode of action of 6-hydroxydopamine. *In* Chemical Tools in Catecholamine Research. G. Jonsson, T. Malmfors & C. Sachs, Eds. Vol. **1:** 41–50. North Holland Publishing Company. Amsterdam, The Netherlands.
28. Jonsson, G. 1976. Studies on the mechanism of action of 6-hydroxydopamine. Med. Biol. **54:** 406–420.
29. Wagner, K. & U. Trendelenburg. 1971. Effect of 6-hydroxydopamine on oxidative phosphorylation and on monamine oxidase activity. Arch. Exp. Pathol. Pharmakol. **269:** 112–116.
30. Tranzer, J. P. 1971. Fine structural mechanism of action of 6-OH-DA. *In* 6-Hydroxydopamine and Catecholamine Neurons. T. Malmfors & H. Thoenen, Eds. : 257–264. North Holland Publishing Company. Amsterdam, The Netherlands.

Discussion

Dr. H. Tamir: Were you able to enhance the release of 5,7-DHT with serotonin?

Dr. Creveling: No.

MECHANISMS OF ACTION OF HYDROXYLATED PHENYLETHYLAMINE AND INDOLEAMINE NEUROTOXINS *

Gerald Cohen and Richard E. Heikkila

Department of Neurology
Mount Sinai School of Medicine
New York, New York 10029

Among the neurotoxic dihydroxytryptamines,[1,2] 5,7-dihydroxytryptamine (5,7-DHT) is unusual in that its destructive actions *in vivo* on both central and peripheral noradrenergic neurons can be blocked by inhibitors of monoamine oxidase (MAO).[3,4] The destruction of central serotonergic neurons *in vivo* is not similarly blocked.[3,5] In contrast, the toxicity of 6-hydroxydopamine (6-OHDA) on central or peripheral noradrenergic neurons is enhanced by MAO inhibitors.[6,7]

In considering the status of MAO in the neurotoxic action of 5,7-DHT, some investigators have expressed a belief that the aldehyde product derived from 5,7-DHT (FIGURE 1) may play a critical role.[4] Alternatively, another anticipated product is hydrogen peroxide,[8] an agent that has been linked to the neurotoxic action of 6-OHDA.[7,9,10] The relatively low levels of catalase and glutathione peroxidase in neural tissue [11–13] provide a mechanism for the toxicity of hydrogen peroxide to be readily expressed. Hydrogen peroxide may act either directly or indirectly via the generation of hydroxyl radicals.[10]

This report deals with two separate kinds of experimental observations. First, we review recent data that indicate a critical role for hydroxyl radicals in the neurotoxicity of 6-OHDA. The prevention of the neurotoxicity of both 6-OHDA and 5,7-DHT on mouse heart by scavengers of hydroxyl radicals tends to point to some similarity in mechanism for these two neurotoxins on noradrenergic nerves. Second, we describe preliminary studies in which rat brain mitochondria catalyzed the oxidation of 5,6-DHT and 5,7-DHT. This phenomenon was not blocked by MAO inhibitors.

MATERIALS AND METHODS

Detailed descriptions of the *in vivo* studies with 6-OHDA and 5,7-DHT are provided elsewhere.[10,14] The integrity of the sympathetic (noradrenergic) nerve plexus of the left atria of control and drug-injected Swiss-Webster mice was evaluated by measuring the uptake and accumulation of [^{3}H]norepinephrine ([^{3}H]-NE) *in vitro*. In brief, the procedure consisted of incubating groups of atria at 37° C in a Krebs-Ringer phosphate buffer that contained glucose, ascorbate, and EDTA. [^{3}H]Norepinephrine was added to 2.5 μCi/10 ml (sp act, 5.9–9.8 Ci/mmol), and samples were incubated for 10 min. The atria were then isolated, rinsed, and reincubated with fresh buffer to remove loosely bound

* Supported by United States Public Health Service Grant NS-11631 (Clinical Research Center for the study of Parkinson's and Allied Diseases).

$$\text{6-OHDA} + O_2 \xrightarrow{\text{spontaneous}} H_2O_2 + \text{6-OHDA-quinone}$$

$$\text{5,7-DHT} + O_2 \xrightarrow{\text{MAO}} H_2O_2 + \text{5,7-Dihydroxyindoleacetaldehyde}$$

FIGURE 1. Formation of hydrogen peroxide from the spontaneous reaction of 6-hydroxydopamine with molecular oxygen (*top*) and from the oxidative deamination of 5,7-dihydroxytryptamine, catalyzed by monoamine oxidase (*bottom*).

or extraneuronal tritium. Finally, the remaining tritium, of which 90% or more represented unmetabolized [^{3}H]NE, was extracted from individual atria with ethanol and measured by liquid scintillation spectrometry. For convenience, experimental results for individual atria were expressed as percentages of the mean value seen in control atria for each experiment.

The ethylene-generating system for measuring a flux of hydroxyl radicals was based on the prior observations of Beauchamp and Fridovich[15] and has been described in detail.[10] The system consisted of autoxidizing 6-aminodopamine (6-ADA) in a 0.2 M phosphate buffer (final pH, 8.4) that contained H_2O_2 (1 mM), EDTA (0.1 mM), and 2-keto-4-thiomethylbutyric acid (1 mM). The generation of hydroxyl radicals was initiated by the addition of 6-ADA (0.1 mM). Ethylene production was measured after 6 min by gas chromatography.

Oxygen consumption by rat brain mitochondria in the presence of 5,6-DHT and 5,7-DHT (0.5 mM) was measured on a Biological Oxygen Monitor at 35° C in a closed system. The methodology was essentially the same as that described for the autoxidation reactions of 6-OHDA and 6,7-DHT,[16] except for the use of another buffer and the addition of rat brain mitochondria. The rat brain mitochondria were prepared by the method of Clark and Nicklas,[17] and we used their buffer that contained 5 mM potassium.

FREE RADICALS DERIVED FROM OXYGEN DURING OXIDATIVE REACTIONS

During spontaneous or enzymatically catalyzed oxidation reactions, oxygen can accept either one or two electrons at a time. The one-electron transfer product is the superoxide radical, while the two-electron product is an ionized form of hydrogen peroxide (FIGURE 2, *top*). Water results from the transfer of four electrons.

In biologic systems, superoxide radicals are removed by a broadly distributed enzyme, superoxide dismutase. This enzyme catalyzes the dismutation of two

$$O_2 \xrightarrow{+1e^-} O_2^{\overline{\cdot}} \xrightarrow{+1e^-} O_2^{=}$$

$$O_2^{\overline{\cdot}} + H_2O_2 \longrightarrow \cdot OH + O_2 + OH^-$$

FIGURE 2. Reduction of oxygen in one-electron steps to form the superoxide radical and ionized peroxide (*top*). The Haber-Weiss reaction[21] for production of hydroxyl radicals (·OH) from superoxide and hydrogen peroxide (*bottom*).

superoxide radicals to yield oxygen and hydrogen peroxide.[18] In addition, catecholamines are good scavengers of superoxide radicals.[19, 20]

When superoxide radicals are generated in the presence of hydrogen peroxide, an interaction between the two (FIGURE 2, *bottom*) can lead to formation of the hydroxyl radical.[15, 21, 22] Hydroxyl radicals are extremely reactive, and they have an acknowledged cytotoxic role in biologic systems, such as during the exposure of tissues to high-energy radiation.[23]

STUDIES WITH 6-HYDROXYDOPAMINE AND 6-AMINODOPAMINE

The rapid, spontaneous autoxidation of 6-OHDA, 6-ADA, or related hydroxylated phenylethylamine neurotoxins generates both quinoidal products and hydrogen peroxide.[9, 24, 25] A precursor of the latter is the superoxide radical,[26] and a by-product of the reaction is the hydroxyl radical.[16] Superoxide radicals have been implicated in the neurotoxicity of 6-OHDA.[19]

The generation of hydroxyl radicals *in vitro* can be conveniently studied in the system described by Beauchamp and Fridovich.[15] In this system, hydroxyl radicals react with (are scavenged by) either methional [15] or 2-keto-4-thiomethyl butyric acid [10] to yield ethylene gas. The competitive removal of hydroxyl radicals by scavengers suppresses ethylene production and provides a means for evaluating the relative scavenging action of added substances.

The system is illustrated in TABLE 1. In these experiments, H_2O_2 (1 mM) was added prior to 6-ADA. Although 6-ADA itself generates hydrogen peroxide, the production of hydroxyl radicals (production of ethylene) was aug-

TABLE 1

SUPPRESSION OF ETHYLENE PRODUCTION BY CATALASE, SUPEROXIDE DISMUTASE, AND HYDROXYL RADICAL SCAVENGERS *

Sample	Ethylene † (% control ± SD)	(n)
No 6-ADA (H_2O_2 alone)	4.4±0.6	(9)
Superoxide dismutase (50 μg/ml)	51.3±7.8	(10)
Catalase (33 μg/ml)	3.9±0.1	(7)
Ethanol (0.5 mM)	90.1±7.2	(10)
Ethanol (5.0 mM)	52.2±4.8	(10)
Thiourea (0.5 mM)	66.8±5.6	(10)
Thiourea (5.0 mM)	15.5±1.0	(4)
Methimazole (0.5 mM)	58.5±6.3	(10)
PTTU (0.5 mM)	57.9±9.4	(10)
Urea (0.5 mM)	97.6±7.7	(10)

* The system consisted of 0.2 M phosphate buffer (final pH, 8.4) that contained 0.1 mM EDTA, 1 mM H_2O_2, and 1 mM 2-keto-4-thiomethylbutyric acid at 37° C. The reaction was initiated by the addition of 6-ADA (0.1 mM). The total volume was 1.22 ml. Ethylene was measured at 6 min (data taken from Reference 10).

† Control formation of ethylene was 1.30±0.08 nmol (n=86). Except for urea ($p > 0.4$), all other results differed significantly from controls (0.5 mM ethanol, $p < 0.02$; all others, $p < 0.01$).

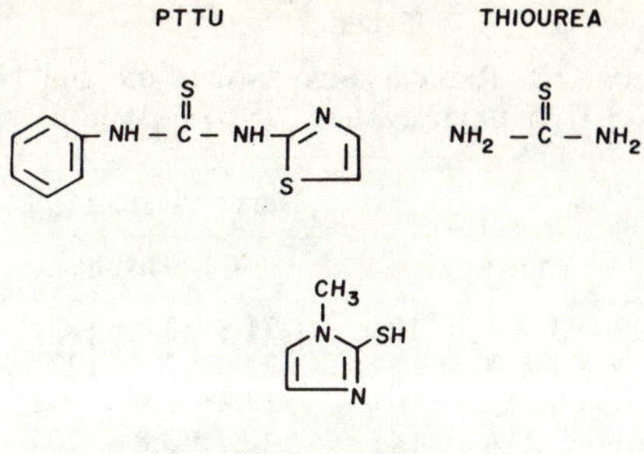

FIGURE 3. The structures of thiourea and two thiourea derivatives, namely, methimazole and PTTU [1-phenyl-3-(2-thiazolyl)-2-thiourea].

mented by the addition of extra hydrogen peroxide. In TABLE 1, very little ethylene was formed when 6-ADA was omitted. Ethylene production was suppressed partially by superoxide dismutase and more strongly by catalase. These results were in accord with the Haber-Weiss reaction, in which hydroxyl radicals arise from an interaction of superoxide and hydrogen peroxide (FIGURE 2).

Ethanol is a good scavenger for hydroxyl radicals; the estimated second-order rate constant is 10^9 M^{-1} sec^{-1}.[27] Thiourea is a more powerful scavenger; its rate constant is estimated at 5×10^9 M^{-1} sec^{-1}.[27] In TABLE 1, thiourea was more effective than ethanol as an inhibitor of ethylene production; this result was in accord with the relative rate constants for the two. TABLE 1 also shows results with two thiourea derivatives, PTTU and methimazole (cf. FIGURE 3); both agents appeared as effective as the unsubstituted thiourea in scavenging hydroxyl radicals. The last line of TABLE 1 shows that urea at a comparable concentration did not suppress ethylene production. Urea reacts relatively poorly with hydroxyl radicals ($k < 7 \times 10^5$ M^{-1} sec^{-1}),[27] and it served as a negative control.

In studies *in vivo,* various scavengers were tested as protective agents against the neurotoxicity of 6-OHDA or 6-ADA. Swiss-Webster mice received intraperitoneal injections of the scavengers, followed 1 hr later by intravenous injection of the neurotoxic phenylethylamine. The test organ was the left atrium, with its rich sympathetic nerve plexus. Jonsson and Sachs [28] had shown earlier that the destructive action of 6-OHDA was complete after 4 hr or less. In our studies,[10] left atria were studied at 4, 24, and 72 hr after injection. Some results are shown in TABLE 2.

PTTU provided strong protection against 6-OHDA or 6-ADA. In more detailed studies with 6-OHDA, protection fell off with higher injected doses of 6-OHDA (20 and 50 mg/kg) or with lower doses of PTTU (50 and 100 mg/kg). Protection was also less when 6-OHDA was injected at 3 hr after PTTU (200 mg/kg), and it was absent after 5 hr. TABLE 2 shows that several other hydroxyl radical scavengers, namely, ethanol, *n*-butanol, and methimazole (a ring-closed thiourea analog; cf. FIGURE 3), also provided lesser but distinct protection against 6-OHDA. In other experiments,[10] the protective scavengers failed to block the accumulation of injected [^{3}H]NE by the left atrium *in vivo.* The latter result signified that the agents did not block the axonal membrane transport system for amines and, therefore, probably did not block uptake of

TABLE 2

EFFECT OF VARIOUS HYDROXYL RADICAL SCAVENGERS ON THE NEURODEGENERATIVE ACTION OF 6-HYDROXYDOPAMINE AND 6-AMINODOPAMINE *

Radical Scavenger (g/kg)		Neurotoxin (mg/kg)†		Nerve Plexus (% control ± SEM) (n)			
				(−) Scavenger		(+) Scavenger	
PTTU	(0.2)	6–ADA	(10)	31.8±3.0	(22)	95.8±5.6	(19)
PTTU	(0.2)	6–OHDA	(7.5)	21.6±2.0	(22)	97.0±3.3	(22)
PTTU	(0.2)	6–OHDA	(10)	17.3±0.8	(109)	68.3±4.5	(62)
Ethanol	(4.0)	6–OHDA	(4)	40.7±3.8	(28)	58.7±5.6	(26)
n-Butanol	(0.8)	6–OHDA	(4)	32.8±2.9	(26)	49.7±3.9	(27)
Methimazole	(0.4)	6–OHDA	(4)	28.4±3.1	(18)	63.1±6.2	(21)

* Drugs were administered intraperitoneally to male Swiss-Webster mice 1 hr prior to intravenous 6-OHDA or 6-ADA. Then, after either 24 hr (PTTU experiments) or 4 hr (all other drugs), the left atria were removed, and the degree of neuronal damage was assessed. Measurement of the uptake and accumulation of [^{3}H]norepinephrine (compared to untreated controls) served as an index of the integrity of the adrenergic nerve plexus. All p values with scavengers were less than 0.01 compared to the corresponding groups not pretreated with scavengers (data taken from Reference 10).

† The doses of neurotoxin are expressed in terms of their acid salts, 6–OHDA·HBr and 6-ADA·2HCl, respectively.

6-OHDA. Recently, Jonsson [29] confirmed that the accumulation of [^{3}H]-6-OHDA by peripheral sympathetic nerves in PTTU-treated mice was essentially normal.

In Vivo STUDIES WITH 5,7-DHT

The ability of MAO inhibitors to block the action of 5,7-DHT on noradrenergic nerves indicates that a product derived from MAO is required for the neurotoxic action. To test the possibility that hydroxyl radicals might be involved, we evaluated the effects of two hydroxyl radical scavengers, PTTU and ethanol, on the neurotoxic action of 5,7-DHT in the mouse left atrium. The results (TABLE 3) showed that both PTTU and ethanol exerted protective actions. The decrease in protection by PTTU at higher injected doses of 5,7-DHT paralleled similar effects observed with 6-OHDA.[10] Nialamide was included as an internal control in these experiments. Nialamide, as had been reported for other MAO inhibitors,[3, 4] provided good protection.

In Vitro STUDIES WITH 5,6-DHT AND 5,7-DHT

Isolated rat brain mitochondria stimulated oxygen consumption by either 5,6-DHT or 5,7-DHT (FIGURE 4). Neither dopamine nor serotonin produced comparable results. Thus, the effect appeared relatively specific for the serotonin neurotoxins. Boiling the mitochondrial preparation virtually eliminated the stimulatory action. Thus, a heat-labile factor was responsible for the enhanced oxygen consumption.

TABLE 3

EFFECT OF PTTU AND ETHANOL (HYDROXYL RADICAL SCAVENGERS) AND NIALAMIDE (MONOAMINE OXIDASE INHIBITOR) ON THE NEURODEGENERATIVE ACTION OF 5,7-DIHYDROXYTRYPTAMINE *

Protective Agent (dose)		5,7-DHT (mg/kg) †	Nerve Plexus Integrity (% control ± SEM) (n) (−) Protective Agent	(+) Protective Agent
PTTU	(200 mg/kg)	40	62.2±6.9 (16)	102.8±5.3 (12)
Nialamide	(50 mg/kg)	40	same as above	95.6±5.7 (5)
PTTU	(200 mg/kg)	60	34.9±3.4 (35)	60.3±5.4 (23)
Ethanol	(4 g/kg)	60	same as above	72.5±4.9 (13)
Nialamide	(50 mg/kg)	60	same as above	81.4±6.8 (9)

* PTTU and ethanol were administered intraperitoneally to male Swiss-Webster mice 1 hr prior to intravenous 5,7-DHT. Nialamide was given intraperitoneally 2 hr prior to intravenous 5,7–DHT. Twenty-four hours later, the left atria were removed, and the degree of neuronal damage was assessed as in TABLE 2. All p values with protective agents were less than 0.005 compared to control mice not receiving protective agents (data taken from Reference 14).

† Doses of 5,7-DHT (administered as the creatinine sulfate salt) are expressed in terms of the free base.

Oxygen consumption with mitochondria and 5,6-DHT or 5,7-DHT was markedly inhibited by cyanide (FIGURE 4), an inhibitor of cytochrome oxidase and other metalloenzymes. Inhibition was also observed with azide (data not shown), but azide was not as effective as cyanide. Pargyline, on the other hand, did not have any apparent effect on the consumption of oxygen. For example, in an experiment with 5,7-DHT (0.5 mM), oxygen consumption was 52 ± 5

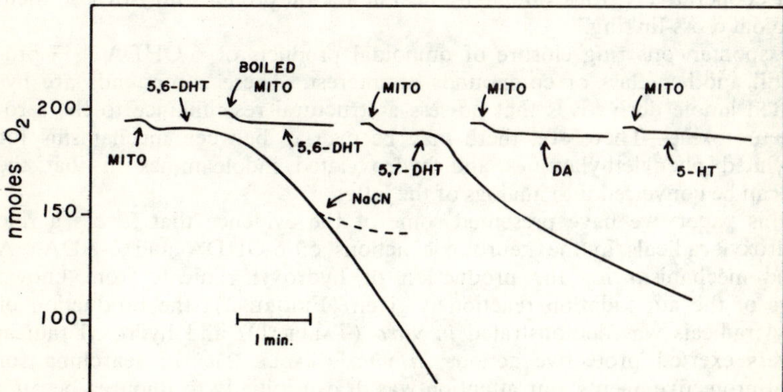

FIGURE 4. Oxygen electrode tracings that show the stimulation of oxygen consumption on adding 5,6- or 5,7-dihydroxytryptamine (0.5 mM) to rat brain mitochondria and the effect of cyanide. Controls consist of equivalent concentrations of dopamine (DA) and serotonin (5-hydroxytryptamine, 5-HT) and a boiled mitochondrial preparation. In the absence of mitochondria, the oxygen consumption by 5,6-DHT or 5,7–DHT was very small (less than 2 nmol/min).

nmol/min (mean ± SD; $n = 3$) without pargyline and 50 ± 4 nmol/min with pargyline (1.0 mM). Thus, the enhanced oxygen consumption in the presence of mitochondria did not appear to be dependent on MAO activity.

In other experiments, when catalase was added to the system after oxygen consumption had virtually ceased, very little oxygen was evolved. In comparison, 6-OHDA consumed oxygen rapidly without apparent catalysis by mitochondria; when catalase was added, roughly one half of the consumed oxygen was returned to solution (cf. Reference 16). These results indicated that in the presence of mitochondria, oxygen consumption by 5,6-DHT and 5,7-DHT did not result in a corresponding accumulation of hydrogen peroxide.

Discussion

With regard to the mechanisms of action of 6-OHDA and 6-ADA, attention has been focused on two possible forms of toxic products. On the one hand, these agents produce quinoidal products that become bound to proteins in either model *in vitro* systems or intraneuronally *in vivo*.[24, 29, 30] It has been suggested that the "covalent" binding of the quinones to tissue proteins constitutes part of the toxic action on neurons. On the other hand, the spontaneous autoxidation of the neurotoxic phenylethylamines produces hydrogen peroxide as a product and certain reactive oxygen radicals as either intermediates (superoxide radical) or by-products (hydroxyl radical).[16, 26] It has been suggested that hydrogen peroxide and the oxygen radicals damage neurons by oxidizing cellular constituents. There is experimental evidence with superoxide scavengers and hydroxyl radical scavengers to support the hypothesis that oxygen radicals play a critical role in evoking neuronal damage.[10, 19, 20] In addition, a recent study by Jonsson[29] has shown that the degree of covalent binding of [^{3}H]-6-OHDA correlates with the degree of cellular damage in the presence of protective agents (e.g., PTTU, desmethylimipramine) or potentiating agents (e.g., nialamide). Thus, both quinoidal and free radical mechanisms may be operative simultaneously or cooperatively; for example, radical attack on proteins may expose them to quinoidal cross-linking.[30]

The spontaneous ring closure of quinoidal products of 6-OHDA[31, 32] produces still another class of compounds of interest. These compounds are hydroxylated indole derivatives that possess a structural resemblance to the serotonin neurotoxins. Therefore, there may be overlap between mechanisms for hydroxylated phenylethylamines and hydroxylated indoleamines in that the former can be converted into analogs of the latter.

In this paper, we have presented some of the evidence that favors a role for hydroxyl radicals in the neurotoxic actions of 6-OHDA and 6-ADA. A proposed mechanism for the production of hydroxyl radicals from known products of the autoxidation reaction is given (Figure 2); the production of hydroxyl radicals was demonstrated *in vitro* (Table 1); and hydroxyl radical scavengers exerted protective actions *in vivo* (Table 2). In searching for potential protective agents, our attention was drawn initially to thiourea because of its high rate constant for reaction with hydroxyl radicals. Thiourea, however, proved ineffective against 6-OHDA,[10] although it did protect the pancreas against alloxan,[33] another cell toxin whose mechanism of action may involve hydroxyl radicals. PTTU and methimazole were subsequently selected for study because they represented thiourea analogs (Figure 3) and because they were known to be accumulated by peripheral sympathetic nerves by virtue of the fact that they inhibited dopamine β-hydroxylase *in vivo*.[34, 35] As seen in TABLE 1, both PTTU and methimazole proved equally effective as thiourea itself in scavenging hydroxyl radicals *in vitro*. Moreover, they both provided protection against 6-OHDA (TABLE 2).

The more recent studies have centered on the neurotoxic properties of 5,7-DHT on the sympathetic nerve plexus of the mouse left atrium. Two hydroxyl radical scavengers, PTTU and ethanol, both proved effective as antagonists (TABLE 3). Because the neurotoxicity of 5,7-DHT can also be blocked by inhibitors of MAO, these observations imply that some special interaction of 5,7-DHT with MAO may result in the generation of hydroxyl radicals, perhaps from hydrogen peroxide which is a known product of the MAO reaction. 5,7-DHT also reacts spontaneously with oxygen to form hydrogen peroxide (and a quinone); however, it does so at a considerably slower rate than does 6-OHDA. The rate for 5,7-DHT is comparable to that of such agents as 5-OHDA or α-methyl-5,7-DHT,[4] which do not exhibit toxicity in noradrenergic systems; therefore, the spontaneous reaction with oxygen does not appear significant.

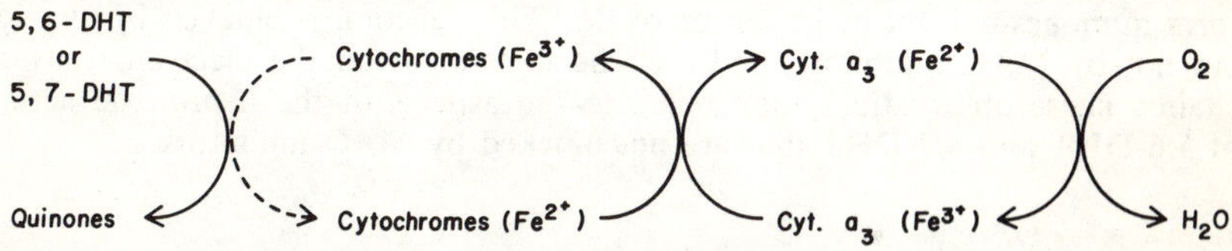

FIGURE 5. Proposed pathway for the interaction of 5,6-dihydroxytryptamine and 5,7-dihydroxytryptamine with rat brain mitochondria.

The experiments with rat brain mitochondria (FIGURE 4) were prompted by an interest in the potential interaction of 5,7-DHT and 5,6-DHT with mitochondrial MAO. The results were striking and clearly of interest, but they probably do not relate to the special action of 5,7-DHT on noradrenergic nerves.

The mitochondrial preparation spurred oxygen consumption by both 5,6-DHT and 5,7-DHT. This phenomenon was observed on the oxygen electrode (FIGURE 4) and also by visual observation of the formation of colored products, probably quinones (pink with 5,7-DHT, black with 5,6-DHT). On the other hand, pargyline did not inhibit oxygen consumption, and, therefore, MAO did not appear to be involved. The addition of catalase to the system did not produce any remarkable return of oxygen to the system (data not shown), and, therefore, hydrogen peroxide was not a major product of the oxygen-consuming reaction. The oxygen-consuming reaction was inhibited by cyanide (FIGURE 4) and less effectively by azide (data not shown). It was also inhibited by boiling (FIGURE 4) and by Triton® X-100 (data not shown), a detergent. On the other hand, it was not inhibited by antimycin A or rotenone (data not shown), two inhibitors of electron transport. It may be that the neurotoxic indoles react with a portion of the cytochrome system of the mitochondrial electron transport chain (FIGURE 5); if so, the indoles would donate electrons beyond the position of blockade by antimycin A. Such a mechanism, producing quinoidal products, would support hypotheses concerning quinoidal mechanisms of cell toxicity. These observations may reflect on the mechanisms of cell toxicity induced by 5,6-DHT on monoaminergic systems (namely, dopamine, serotonin, and noradrenaline neurons) and on the mechanism of action of 5,7-DHT on serotonergic neurons. On the other hand, these observations do not appear to relate to the special actions of 5,7-DHT on noradrenergic systems, where neurotoxicity can be prevented by inhibitors of MAO.

Conclusions

Hydroxyl radical scavengers, such as PTTU and ethanol, protect peripheral sympathetic neurons against destruction by 6-OHDA or 5,7-DHT. These results imply a similarity in mechanism of action for these two neurotoxins. During the autoxidation of 6-OHDA, hydroxyl radicals are generated as a by-product from hydrogen peroxide. Although direct experimental evidence is lacking, hydroxyl radicals may also be generated from hydrogen peroxide during an interaction of 5,7-DHT with MAO in noradrenergic nerves. This hypothesis would explain the protection by both MAO inhibitors and hydroxyl radical scavengers.

Rat brain mitochondria catalyze the oxidation of 5,6-DHT and 5,7-DHT to form quinones but not hydrogen peroxide. This reaction is blocked by cyanide, but not by MAO inhibitors, and may be dependent on the electron transport chain. These observations may relate to the aspects of the neurotoxic actions of 5,6-DHT and 5,7-DHT that are not blocked by MAO inhibitors.

References

1. Baumgarten, H. G., A. Björklund, A. Nobin, E. Rosengren & H. G. Schlossberger. 1975. Neurotoxicity of hydroxylated tryptamines: structure activity relationships. 1. Long-term effects on monoamine content and fluorescence morphology of central monoamine neurons. Acta Physiol. Scand. Suppl. **429:** 4–27.
2. Jacoby, J. H., L. D. Lytle & M. F. Nelson. 1974. Long-term effects of 5,7-dihydroxytryptamine on brain monoamines. Life Sci. **14:** 909–919.
3. Breese, G. R. & B. R. Cooper. 1975. Behavioral and biochemical interactions of 5,7-dihydroxytryptamine with various drugs when administered intracisternally to adult and developing rats. Brain Res. **98:** 517–527.
4. Creveling, C. R., J. Lundstrom, E. T. McNeal, L. Tice & J. W. Daly. 1975. Dihydroxytryptamines: effects on noradrenergic function in mouse heart *in vivo*. Mol. Pharmacol. **11:** 211–222.
5. Baumgarten, H. G., A. Björklund & D. F. Bogdanski. 1975. Similarities and differences in the mode of action of 6-hydroxydopamine and neurotoxic indoleamines. *In* Chemical Tools in Catecholamine Research. G. Jonsson, T. Malmfors & C. Sachs, Eds. Vol. **I:** 59–66. North Holland Publishing Company. Amsterdam, The Netherlands.
6. Breese, G. R. & T. D. Traylor. 1970. Effect of 6-hydroxydopamine on brain norepinephrine and dopamine: evidence for selective degeneration of catecholamine neurons. J. Pharmacol. Exp. Ther. **174:** 413–420.
7. Sachs, C. & G. Jonsson. 1975. Mechanisms of action of 6-hydroxydopamine. Biochem. Pharmacol. **24:** 1–8.
8. Tipton, K. F. 1968. The prosthetic groups of pig brain mitochondrial monoamine oxidase. Biochim. Biophys. Acta **159:** 451–459.
9. Heikkila, R. E. & G. Cohen. 1971. Inhibition of biogenic amine uptake by

hydrogen peroxide: a mechanism for the toxic effects of 6-hydroxydopamine. Science **172:** 1257–1258.

10. Cohen, G., R. E. Heikkila, B. Allis, F. Cabbat, D. Dembiec, D. MacNamee, C. Mytilineou & B. Winston. 1976. Destruction of sympathetic nerve terminals by 6-hydroxydopamine: protection by 1-phenyl-3-(2-thiazolyl)-2-thiourea, diethyldithiocarbamate, methimazole, cysteamine, ethanol and n-butanol. J. Pharmacol. Exp. Ther. **199:** 336–352.
11. Cohen, G. & P. Hochstein. 1963. Enzymatic mechanisms of drug sensitivity in brain. Diseases Nervous System **24**(Suppl.): 44–48.
12. Prohaska, J. R. & H. E. Ganther. 1976. Selenium and glutathione peroxidase in developing rat brain. J. Neurochem. **27:** 1379–1387.
13. Gaunt, G. L. & C. De Duve. 1976. Subcellular distribution of d-amino acid oxidase and catalase in rat brain. J. Neurochem. **26:** 749–759.
14. Allis, B. & G. Cohen. 1977. The neurotoxicity of 5,7-dihydroxytryptamine in the mouse atrium: protection by 1-phenyl-3-(2-thiazolyl)-2-thiourea and by ethanol. Eur. J. Pharmacol. **43:** 269–272.
15. Beauchamp, C. & I. Fridovich. 1970. A mechanism for the production of ethylene from methional. J. Biol. Chem. **245:** 4641–4646.
16. Cohen, G. & R. E. Heikkila. 1974. The generation of hydrogen peroxide, superoxide radical, and hydroxyl radical by 6-hydroxydopamine, dialuric acid and related cytotoxic agents. J. Biol. Chem. **249:** 2447–2452.
17. Clark, J. B. & W. J. Nicklas. 1970. The metabolism of rat brain mitochondria: preparation and characterization. J. Biol. Chem. **245:** 4724–4731.
18. Fridovich, I. 1972. Superoxide radical and superoxide dismutase. Accounts Chem. Res. **5:** 321–326.
19. Sachs, C., G. Jonsson, R. Heikkila & G. Cohen. 1975. Control of the neurotoxicity of 6-hydroxydopamine by intraneuronal noradrenaline in rat iris. Acta Physiol. Scand. **93:** 345–351.
20. Cohen, G. & R. E. Heikkila. 1977. In vivo scavenging of superoxide radicals by catecholamines. *In* Superoxide and Superoxide Dismutases. A. M. Michelson, J. McCord & I. Fridovich, Eds. Academic Press, Inc. New York, N.Y. In press.
21. Harber, F. & J. Weiss. 1934. The catalytic decomposition of hydrogen peroxide by iron salts. Proc. Soc. A **147:** 332–351.
22. Cohen, G. 1977. In defense of Haber-Weiss. *In* Superoxide and Superoxide Dismutases. A. M. Michelson, J. McCord & I. Fridovich, Eds. Academic Press, Inc. New York, N.Y. In press.
23. Myers, L. S., Jr. 1973. Free radical damage of nucleic acids and their components by ionizing radiation. Fed. Proc. **32:** 1882–1894.
24. Saner, A. & H. Thoenen. 1971. Model experiments on the molecular mechanism of action of 6-hydroxydopamine. Mol. Pharmacol. **7:** 147–154.
25. Lundstrom, J., H. Ong, J. Daly & C. R. Creveling. 1973. Isomers of 2,4,5-trihydroxyphenylethylamine (6-hydroxydopamine): long term effects on the accumulation of [^{3}H]-norepinephrine in mouse heart in vivo. Mol. Pharmacol. **9:** 505–513.
26. Heikkila, R. E. & G. Cohen. 1973. 6-Hydroxydopamine: evidence for superoxide radical as an oxidative intermediate. Science **181:** 456–457.
27. Anbar, M. & P. Neta. 1967. A compilation of specific bimolecular rate constants for the reactions of hydrated electrons, hydrogen atoms and hydroxyl radicals with inorganic and organic compounds in aqueous solution. J. Appl. Rad. Isotopes. **18:** 493–523.
28. Jonsson, G. & C. Sachs. 1972. Degenerative and nondegenerative effects of 6-hydroxydopamine on adrenergic nerves. J. Pharmacol. Exp. Ther. **180:** 625–635.
29. Jonsson, G. 1976. Studies on the mechanisms of 6-hydroxydopamine cytotoxicity. Med. Biol. **54:** 406–420.
30. Creveling, C. R., A. Rotman & J. W. Daly. 1975. Interactions of 6-hydroxy-

dopamine and related compounds with proteins: a model for the mechanism of cytotoxicity. *In* Chemical Tools in Catecholamine Research. G. Jonsson, T. Malmfors & C. Sachs, Eds. Vol. **I:** 23–32. North Holland Publishing Company. Amsterdam, The Netherlands.
31. Blank, C. L., R. L. McCreery, R. M. Wightman, W. Chey, R. N. Adams, J. R. Reid & E. E. Smissman. 1976. Intracyclization rates of 6-hydroxydopamine and 6-aminodopamine analogs under physiological conditions. J. Med. Chem. **19:** 178–180.
32. Borchardt, R. T. 1974. Affinity labeling of catechol O-methyltransferase by the oxidation products of 6-hydroxydopamine. Mol. Pharmacol. **11:** 436–449.
33. Heikkila, R. E., B. Winston, G. Cohen & H. Barden. 1976. Alloxan-induced diabetes: evidence of the hydroxyl radical as a cytotoxic intermediate. Biochem. Pharmacol. **25:** 1085–1092.
34. Stolk, J. M. & D. P. Hanlon. 1973. Inhibition of brain dopamine β-hydroxylase activity by methimazole. Life Sci. **12:** 417–423.
35. Johnson, G. A., S. J. Boukma & E. G. Kim. 1970. In vivo inhibition of dopamine β-hydroxylase by 1-phenyl-3-(2-thiazolyl)-2-thiourea (U-14,624). J. Pharmacol. Exp. Ther. **171:** 80–87.

Discussion

Dr. H. G. Baumgarten: Have you checked whether the scavengers you used inhibit uptake of 6-hydroxydopamine or 5,7-DHT into terminals?

Dr. Cohen: They do not.

Dr. E. M. Gál: Two other sites of action of superoxides produced by neurotoxins should be considered. One is the enzyme in brain, a superoxide-dependent dioxygenase, that destroys tryptophan. Superoxides may thus enhance the enzymatic destruction of tryptophan. The second is any process that is dependent on sulfhydryl compounds, such as coenzyme A or glutathione, since superoxides react rapidly with such compounds.

Dr. C. K. Meshul (*Downstate Medical Center, Brooklyn, N.Y.*): PTTU is a copper chelator. Could it not be acting by chelating copper or alternatively as a nucleophile inactivating a quinone?

Dr. Cohen: Yes and no. We don't exclude those possibilities, but they would not apply to ethanol, which is a good scavenger.

EFFECTS OF NEUROTOXINS *IN VITRO* ON THE BINDING OF SEROTONIN TO SEROTONIN-BINDING PROTEIN*

Hadassah Tamir and Maurice M. Rapport

Division of Neuroscience
New York State Psychiatric Institute
and Departments of Psychiatry and Biochemistry
Columbia University
College of Physicians and Surgeons
New York, New York 10032

The introduction of 5,6-dihydroxytryptamine (5,6-DHT) and 5,7-dihydroxytryptamine (5,7-DHT) as specific neurotoxins for serotonergic axon terminals supplied a powerful tool for study of the role of serotonin in chemical transmission.[1-3] Their specificity stems from the specificity of their active uptake at the nerve terminal membrane. Further studies, however, indicate that 5,6-DHT is selective for sertonin neurons only in a restricted very-low-dose range; it can cause nonspecific damage to other types of neurons, even at low doses.[4]

Concerning the molecular mechanism of the cytotoxicity, there are two schools of thought, both invoking products of autoxidation. One mechanism suggests superoxide and hydroxyl radicals as the major cytotoxic agents,[5] because removal of the radicals greatly reduces the effect of 6-hydroxydopamine, a neurotoxin that also causes degeneration of nerve endings.[6] The other suggests that quinone formed in the autoxidation of 5,6-DHT reacts with sulfhydryl groups of proteins in the nerve ending, causing irreversible polymerization.[7] While one or both mechanisms may serve for 5,6-DHT, which is readily oxidized and can form quinone, neither is suitable for 5,7-DHT. The neurotoxin 5,7-DHT binds poorly to several proteins and does not polymerize them. It has a low tendency to autoxidize and does not form quinones and, yet, is a very effective neurotoxin, showing more specificity than the other neurotoxins for serotonin neurons, particularly in the central nervous system.[7]

These observations imply that 5,6-DHT and 5,7-DHT act in similar ways, that they both exert inhibitory effect *before* their interaction with oxygen by acting on a protein(s) that binds them with high affinity.

Our studies provide evidence that both neurotoxins are potent inhibitors of the binding of serotonin to a specific protein, serotonin-binding protein (SBP), found in serotonergic tracts, and that this inhibition can be demonstrated both *in vivo* and *in vitro*.

The presence of a soluble high-affinity binding protein for serotonin that might function as a storage and/or transport form was predicted for several reasons. First, it had been observed that after induction of serotonin synthesis by injection of tryptophan, much of the serotonin was localized in the cytosol fraction.[8] Second, physiologic experiments indicated that newly synthesized or newly injected biogenic amines were utilized preferentially to the endogenous amines.[9-11] Third, heterogeneity of the response to reserpine of storage

* Supported by United States Public Health Service Grant NS–12506.

forms of biogenic amines had been observed in several organs.[12, 13] All of these observations strongly suggested that there exists more than one compartment for storage of biogenic amines. Since synaptic vesicles are the generally accepted storage form, we looked in the cytosol fraction for a soluble protein with high affinity for serotonin as another, physiologically more active, compartment. It should be stressed, however, that "soluble protein" is only an operational definition, and, in fact, we cannot exclude the possibility that the soluble protein originates in a very fragile particulate structure that is easily disrupted during fragmentation of tissue.

SBP is a soluble protein with high affinity for serotonin ($K_d = 10^{-8}$ M). It has been found in synaptosomal cytoplasm and cytosol of rat brain [14, 15] and in the cytosol of peripheral nervous tissue [16] (myenteric plexus) of rabbit and guinea pig. Studies of brain regions showed that hypothalamus, thalamus, and the raphe nuclei, regions rich in serotonin, have a higher binding capacity than does either whole brain or cerebellum. Association of the protein with serotonergic tracts was also demonstrated by the decrease of binding capacity in the forebrain of rats after placement of electrolytic lesions in the raphe nuclei.[17]

The binding capacity of the partially purified protein is enhanced 10- to 20-fold by Fe^{2+}.[15] Oxidation of Fe^{2+} to Fe^{3+} is not involved either before or after serotonin binding, as judged from the absence of an electron paramagnetic resonance signal near $g = 4.3$.[18] The binding of serotonin to SBP occurs via sulfhydryl groups, because sulfhydryl blocking and reducing agents inhibit the binding of both serotonin and $^{59}Fe^{2+}$. The requirement for sulfhydryl groups is a feature in common with the model protein systems studied by Creveling and Rotman.[7]

The specificity of the binding of serotonin to SBP was tested by determining the capacity of other indole derivatives to inhibit the binding of labeled serotonin to SBP.[15] The most effective inhibitors were 5,6-DHT, 5,7-DHT, *N*-methyl serotonin, *N,N*-dimethyl serotonin, 6-hydroxytryptamine, and CZ-74, an analog of 4-hydroxytryptamine. Tryptophan and tryptamine did not affect the binding of serotonin to SBP. At least one hydroxyl group on the indole ring is essential for binding to SBP, and the amino group may be blocked.

Several experiments imply that the physiologic role of the protein may involve storage and/or transport of the biogenic amine: much of the newly synthesized serotonin is bound to SBP,[19] serotonin bound to SBP is not attacked by mitochondrial monoamine oxidase,[15] and reserpine, which blocks storage, inhibits the binding.[15]

It is generally assumed that reserpine depletes only the storage of vesicles. However, Tranzer presented evidence for the existence of a new amine storage or transport compartment, a "tubular reticulum-like, electron dense, fine ultrastructure" that appears in peripheral nervous tissue and in the central nervous system and that is not seen in reserpinized animals.[20] SBP may be derived from this structure. The inhibition of binding by vinblastine, a drug that interacts with contractile proteins, in particular tubulin, and blocks axonal transport, supports the suggestion that SBP may be involved in translocation of the biogenic amine in synaptic contacts. Supportive evidence for this notion comes from the morphologic studies by Jonsson *et al.*,[21] who pointed out the similarity of the effect of neurotoxins and vinblastine: both vinblastine and 5,6-DHT caused an accumulation of Falck-Hillarp fluorescence in serotonergic tracts. We suggest, therefore, that both types of drugs exert their effect by inhibiting the binding of serotonin to SBP and thereby interfere with storage and/or transport of the amine. The very low concentrations of neurotoxins that inhibit the binding of serotonin to SBP, both *in vivo* and *in vitro,* suggest that SBP is a primary target for these toxins.

Materials and Methods

Preparation of Serotonin-Binding Protein

Partially purified SBP was prepared from rat brain (minus cerebellum). After centrifugation (100,000*g*, 60 min) of the 10% homogenate, the supernatant was fractionated with ammonium sulfate.[15]

Partially purified SBP was also isolated from a myenteric plexus-longitudinal muscle preparation of guinea pig by use of a slight modification of the procedure described for brain.[16]

Characteristics of the Binding of Serotonin to Serotonin-Binding Protein

The assay of serotonin binding to SBP is based on molecular sieve chromatography on Sephadex® G-50, which separates bound and free serotonin.[14] The binding of [^{3}H]-5-hydroxytryptamine is measured in the presence of Fe^{2+} ions. Specific binding is determined as the total binding minus binding in the presence of excess cold serotonin. In inhibition studies, the drug is added just prior to the labeled serotonin, at the indicated concentration.

In Vivo *Studies*

Rats (~150 g) were anesthetized with ether and injected with vinblastine sulfate intraventricularly (1 μg in 4 μl per 150-g rat), according to the procedure of Noble *et al.*,[22] by use of stereotaxic rat atlas (Pellegrino and Cushman) coordinates: Bregma, O; AP, O; H, 3; L, 2. Control animals received saline. Animals were sacrificed 20 or 26 hr after injection.

Guinea pigs were injected with 10, 25, or 50 mg/kg of 5,6-DHT (free base) intraperitoneally 48 or 24 hr before sacrifice. The solution of drug contained ascorbic acid (1 mg/ml) in the saline vehicle. Control animals were injected with vehicle.

Rats (~250 g) were injected with saline solutions of L-tryptophan methyl ester (350 mg/kg) and pargyline (100 mg/kg) intraperitoneally. Control animals received vehicle. Animals were sacrificed 3 hr after injection.

In all cases, brain was removed and frozen on dry ice after discarding the cerebellum. The myenteric plexus-longitudinal muscle of guinea pig was isolated and assayed as previously described.

Serotonin Determination

The serotonin content of brain was measured according to the method of Saavedra *et al.*[23] The tissue was incubated with serotonin *N*-acetyltransferase and hydroxyindole *O*-methyltransferase in the presence of [^{3}H-*methyl*]-*S*-adenosylmethionine, and the melatonin produced from serotonin was isolated and counted.

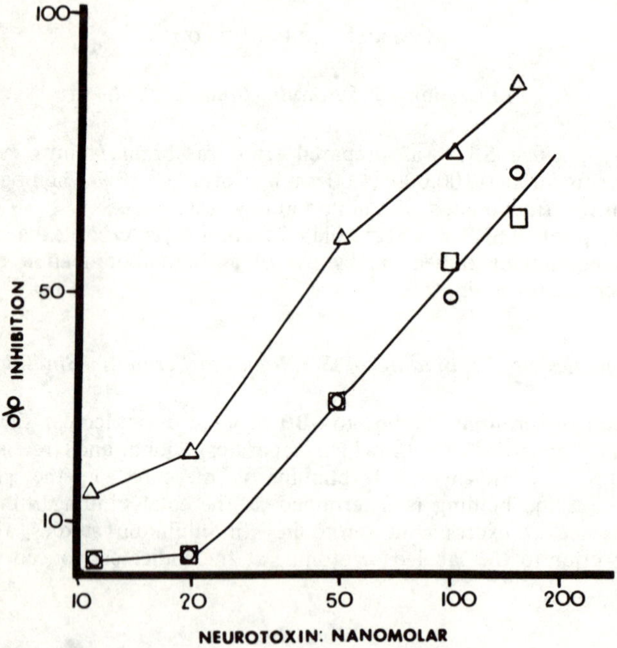

FIGURE 1. Inhibition of serotonin binding to SBP by neurotoxins. Incubation of protein (0.1 mg) with 10^{-4} M $FeSO_4$ for 8 min at room temperature was followed by addition of [^{3}H]serotonin (10^{-7} M), and binding was assayed as described in the text. Each point represents the average of two experiments. △, 5,6-DHT; ○, 5,7-DHT; □, 6-hydroxytryptamine.

RESULTS AND DISCUSSION

Effects of Serotonergic Neurotoxins and Related Compounds in Vitro

The interaction between the neurotoxins 5,6-DHT and 5,7-DHT and SBP is apparent from their inhibition of serotonin binding to SBP from brain; 50% inhibition occurred at 0.05 and 0.1 μM, respectively (FIGURE 1). The inhibitions produced by 6-hydroxytryptamine, bufotenine, and CZ-74 were similar (50% inhibition at 0.2 μM). 5-Hydroxytryptamine was less efficient in inhibiting the binding of serotonin, whereas tryptophan and tryptamine were not effective at all (up to 1 mM). Both neurotoxins had a very strong effect on partially purified SBP isolated from peripheral nervous tissue (myenteric plexus; 50% inhibition at 0.1 μM). The dependence of inhibition on concentration for the two neurotoxins was different. Whereas 5,7-DHT and 6-hydroxytryptamine had no effect below 3×10^{-8} M, 5,6-DHT affected the binding appreciably at this and even lower concentrations (25% inhibition at 10^{-8} M). This difference may be due to the operation of two different mechanisms for inhibition caused by 5,6-DHT, one predominating at very low concentrations and the other at somewhat higher concentrations. By measuring the effect of 5,6-DHT on serotonin binding to SBP under anaerobic conditions or in the presence of a thiourea derivative, it may be possible to determine whether the oxidation products of 5,6-DHT are responsible for its effectiveness at very low concentrations. Studies with serotonin analogs showed that compounds that contained a hydroxyl group in position 4,5, or 6 of the indole ring of tryptamine were as effective as 5,7-DHT in inhibiting the binding of serotonin to brain SBP; 5,6-DHT was twice as effective. This small difference in the effectiveness of the neurotoxins was not observed in the study of SBP from myenteric plexus.

Substituting the amino group with methyl groups did not diminish the inhibitory capacity. The inhibition by methyl derivatives, such as bufotenine, is intriguing because of recent reports on the existence of transmethylation reactions in the central nervous system that utilize serotonin and tryptamine as substrates.[24]

For SBP-indoleamine interaction, several observations tend to rule out the mechanism suggested by Rotman *et al.*,[7] according to which oxidation products of the neurotoxins cause inactivation of model proteins by irreversible polymerization.

SBP binds serotonin, even under anaerobic conditions.[18] Serotonin does not bind to the "model proteins" (and therefore cannot polymerize them).

The inhibition of serotonin binding to SBP by DHT can be demonstrated 15 min after addition. The binding of 5,6-DHT observed with the "model proteins" (and the polymerization that follows) is rather slow (3.5 hr for half saturation).

Both 5,6-DHT and 5,7-DHT are comparable in their effectiveness in inhibiting serotonin binding to SBP at concentrations of 0.5 and 0.1 μM, respectively. However, the concentrations at which they bind and polymerize "model proteins" are very different.

The rate and extent of autoxidation of the several hydroxytryptamine analogs studied are very different, and, yet, their inhibitory capacities for SBP-serotonin binding are similar.

These observations all suggest that both 5,6-DHT and 5,7-DHT exert their inhibitory effects by acting directly on SBP, and the very low concentrations at which inhibition can be demonstrated indicate that SBP may be the primary site of action.

Effect of 5,6-DHT in Vivo

The effects of 5,6-DHT *in vivo* are more complex than those *in vitro*: when injected intraperitoneally into guinea pigs, a biphasic effect was observed.[16] Very low concentrations of 5,6-DHT unexpectedly caused an increase in the binding capacity of serotonin to SBP isolated from myenteric plexus. However, as the concentration of 5,6-DHT increased, binding of serotonin was reduced. Considerable variability was encountered in the effectiveness of different batches of 5,6-DHT. However, the biphasic effect (illustrated for one

TABLE 1

EFFECT OF 5,6-DIHYDROXYTRYPTAMINE *in Vivo* ON THE SEROTONIN-BINDING CAPACITY OF SEROTONIN-BINDING PROTEIN FROM MYENTERIC PLEXUS

Treatment *	Binding Capacity (cpm/mg of protein)
Control	9400
10 mg/kg	21,500
25 mg/kg	32,150
50 mg/kg	1030

* Drug injected intraperitoneally 24 or 48 hr before sacrifice.

batch in TABLE 1) was seen in four separate experiments. These results may be explained by an initial decrease in endogenous serotonin (reducing competition with labeled serotonin used in the assay and thus enhancing binding capacity), followed by degeneration of the tracts and consequent reduction in SBP.

Effect of Vinblastine on Binding of Serotonin to Serotonin-Binding Protein

Morphologic studies by Jonsson *et al.*[21] point to similarities in the toxic effects of intraventricular injection of neurotoxins and vinblastine. Both caused accumulation of serotonin in serotonergic axons, although vinblastine was less specific, also causing accumulation of catecholamines in adrenergic axons. When both drugs were given simultaneously, an increased serotonin content was observed in B7–9 cell bodies in caudal raphe nuclei.

We have recently reported that vincristine and vinblastine inhibited binding of serotonin to SBP (50% inhibition at 1.5 and 7.5 μM, respectively).[15] Vinblastine (2 mM) in the presence of Mg^{2+} precipitated SBP together with tubulin and several other proteins, as judged by sodium dodecyl sulfate gel electrophoresis. Binding capacity was regained by dialysis of the suspended pellet to remove vinblastine. The inhibitory effect of the drug was also demonstrated *in vivo*. When rats were injected intraventricularly with vinblastine, the serotonin-binding capacity of brain cytosol was inhibited 42% after 20 hr and 60% after 26 hr. Serotonin levels were not elevated (TABLE 2), possibly because monoamine oxidase was not inhibited. Prolonged dialysis to remove both vinblastine and serotonin fully restored binding capacity.

Serotonin binding protein may well be a site at which DHT and vinblastine exert their effects and inhibit transfer of biogenic amines. Since tubulin is a protein implicated in transport, and because both tubulin[25, 26] and SBP[15] interact with vinblastine and nucleotides, one might think that tubulin and SBP are identical. However, there are several observations that rule out this possibility.

Colchicine does not inhibit the binding of serotonin either when added directly to the assay system or when added to the brain homogenate prior to partial purification of SBP, whereas tubulin has a specific binding site for colchicine.[27]

SBP has very poor binding capacity for labeled colchicine.

Purified brain tubulin has only a very low specific binding capacity for serotonin.

The nucleotides guanosine triphosphate and ATP have the same K_i for SBP,[15] whereas ATP does not bind to tubulin.[28]

However, SBP may be related to a "tubulin-like" material that is reported to be present in nerve endings. Differences in structure between the "tubulin" in nerve endings and that in cell bodies and axons is suggested by morphologic studies that indicate the greater fragility of the nerve ending material.[29]

Effects of Drugs Other Than Neurotoxins

Before we consider the role serotonergic neurotoxins play in serotonin metabolism, aside from their induction of degeneration of serotonergic terminals, let us review the effects of drugs other than the neurotoxins on the binding of serotonin to SBP. This information is particularly relevant for dopaminergic systems, because dopamine competes with serotonin in binding to SBP, and because the serotonergic neurotoxins appear to also affect dopaminergic systems.[33] A study of the effects of drugs with established sites of action may also assist in evaluating the functional role of SBP.

Antagonists like D-lysergic acid diethylamide, methysergide, and cyproheptadine did not affect the binding of serotonin to SBP. Thus, SBP is not a solubilized form of the postsynaptic membrane receptor for serotonin. Trazodone, a drug that blocks serotonin release induced by fenfluramine but does not affect serotonin storage,[30] did not affect the binding of serotonin to SBP, and fenfluramine itself did not affect binding.[15] These results suggest that SBP is not involved in release, but further experiments are required to establish this conclusion with certainty.

TABLE 2

EFFECT OF VINBLASTINE ON SEROTONIN-BINDING CAPACITY OF SEROTONIN-BINDING PROTEIN IN BRAIN

Treatment *	Time (hr)	Specific Binding (cpm/mg of protein)	Inhibition (%)	Serotonin † (μg/g of brain)
Expt I				
Saline (5)	20	9750±660 ‡		0.41
Vinblastine (5)	20	5700±390	42 §	0.40
Expt II				
Saline (5)	26	10,350±700		
Vinblastine (5)	26	4160±280	60 §	0.42

* Drug was injected intraventricularly (1 μg in 4 μl per 150-g rat), as described in the text.

† Measured by converting serotonin to melatonin, as described in the text.

‡ Values represent the mean±SEM.

§ $p < 0.01$.

The drug Lilly 110,140, a specific inhibitor of serotonin reuptake,[31] was found to be a moderate inhibitor of binding (50% inhibition at 10 μM). Because other drugs that inhibit reuptake, such as amitriptyline, imipramine, and desipramine, did not affect the binding of serotonin to SBP (up to 10 μM), this observation is puzzling. However, since uptake mechanisms comprise several steps that may be affected differently by different drugs, the mechanism of action of Lilly 110,140 may involve inhibition of the binding to a storage site (SBP) of the newly taken up serotonin.

Clozapine is also moderately effective in inhibiting the binding of serotonin to SBP (50% inhibition at 20 μM). This drug causes an increase in the concentration of tryptophan, serotonin, and 5-hydroxyindoleacetic acid in brain but does not directly affect release or degradation of serotonin.[32] Clozapine produces changes associated with serotonin, such as sedation and sleep induction, that are not found with other drugs, for example, amphetamines and salicylates, that cause an increase in the serotonin concentration in brain, and it has been suggested that clozapine may affect a special storage compartment of serotonin.[32] We suggest that SBP may function as this special storage compartment. Consistent with this suggestion is the ineffectiveness of amphetamines in inhibiting serotonin binding to SBP.

A very interesting similarity between the actions of clozapine and the neurotoxins is that both affect both serotonergic and dopaminergic systems. Clozapine increases dopamine concentrations in the striatum and serotonin levels in brain,[32] whereas 5,6-DHT is very effective in decreasing serotonin and dopamine in the diencephalon. Uptake of dopamine or serotonin by synaptosomes prepared from this region was also markedly reduced.[33]

We have previously demonstrated that dopamine very effectively inhibited the binding of serotonin to SBP, but neither Dopa nor other biogenic amines showed such inhibition.[14] It is suggested therefore that SBP or a very similar protein may be localized in dopaminergic tracts. This hypothesis may explain why serotonergic and dopaminergic systems show similar sensitivities to several drugs.

Effects of Neurotoxins on Physiologic Processes

It has been suggested that one mechanism that may account for the formation of stumplike swellings of axons after administration of neurotoxins is through interference with axonal transport.[3] Conceivably, this interference can be attributed to direct interaction of the neurotoxins with the proteins involved, such as neurotubular protein. Our results with SBP suggest that neurotoxins also inhibit storage functions, because newly synthesized serotonin is bound to SBP as the following experiment shows.

When the concentration of endogenous serotonin was increased (from 0.37 μg/g of brain to 3.5 μg/g of brain) by injecting rats with tryptophan (as the methyl ester) plus pargyline, the binding capacity of SBP from brain decreased 34% (TABLE 3). No significant decrease was obtained when either tryptophan or pargyline was injected. Full binding capacity was recovered after prolonged dialysis, indicating that no alteration in SBP had occurred. These results suggest strongly that much of the newly synthesized serotonin is bound to SBP.

Conclusions

The neurotoxins 5,6-DHT and 5,7-DHT are very effective inhibitors of the binding of serotonin to SBP. This soluble protein is associated with serotonergic tracts in both central and peripheral nervous systems, is found in synaptosomes, and has a high affinity for serotonin. The effectiveness of inhibition was very similar to the effectiveness of several other indole analogs, such as 6-hydroxytryptamine, the drug CZ-74, and bufotenine. The inhibition of binding of serotonin to SBP by neurotoxins can be demonstrated both *in vivo* and *in vitro*. We suggest that the neurotoxins may affect serotonergic tracts (and to a lesser degree dopaminergic tracts) not only through oxidation products but directly by high-affinity interaction with SBP. This protein is involved in storage and, possibly, transfer of serotonin and also binds much of the newly synthesized serotonin. We would attribute the specificity of the neurotoxins both to specific uptake into serotonergic terminals and to interaction with SBP in

Table 3

Binding of Newly Synthesized Serotonin by Serotonin-Binding Protein

Treatment *	Serotonin † (μg/g of brain)	Specific Binding (cpm/mg of protein)	Inhibition (%)
Saline (10)	0.37	12,240±830	—
Pargyline plus tryptophan (5)	3.50	8100±984	34 ‡
Pargyline (5)	1.75	11,400±860	7 §
Tryptophan (5)	0.40	9750±850	20 §

* Drug injected intraperitoneally 3 hr before sacrifice.
† Measured by converting serotonin to melatonin, as described in the text. Values represent mean±SEM.
‡ $p < 0.01$.
§ $p < 0.1$ (ns).

serotonergic tracts. This hypothesis is in accord with the conclusion of Baumgarten *et al.*,[34] who believe that uptake site affinity is not the only determinant for specificity and potency of neurotoxic indole amines, a conclusion based on the discrepancy between affinity for serotonin and catecholamine uptake sites. The interaction with SBP of other drugs, such as vinblastine and clozapine, is described, and the possibility that they and the neurotoxins have SBP as a common site of interaction is discussed.

References

1. Baumgarten, H. G., A. Björklund, L. Lachenmayer, A. Nobin & U. Stenevi. 1971. Acta Physiol. Scand. Suppl. **373:** 1–15.
2. Baumgarten, H. G., K. D. Evetts, R. B. Holman, L. L. Iversen, M. Vogt & G. Wilson. 1972. J. Neurochem. **19:** 1587–1597.
3. Baumgarten, H. G. & L. Lachenmayer. 1972. Z. Zellforsch. **135:** 399–414.
4. Baumgarten, H. G. & L. Lachenmayer. 1972. Brain Res. **38:** 228–232.
5. Cohen, G. & R. E. Heikkila. 1974. J. Biol. Chem. **249:** 2447–2452.

6. Cohen, G. R., R. E. Heikkila, B. Allis, F. Cabbat, D. Dembiec, D. McNamee, C. Mytilineou & B. Winston. 1976. J. Pharmacol. Exp. Ther. **199:** 336–352.
7. Rotman, A., J. W. Daly & C. R. Creveling. 1976. Mol. Pharmacol. **12:** 887–899.
8. Tissari, A. H. 1975. Med. Biol. **53:** 1–14.
9. Boullin, D. J., E. Costa & B. B. Brodie. 1966. J. Neurochem. **5:** 293–298.
10. Chakrin, L. W. & V. P. Whittaker. 1969. Biochem. J. **113:** 97–107.
11. Macon, J., L. Sokoloff & J. Glowinski. 1971. J. Neurochem. **18:** 323–331.
12. Glowinski, J. & R. J. Baldessarini. 1966. Pharmacol. Rev. **18:** 1201–1238.
13. Cattabeni, F., S. H. Koslow & E. Costa. 1973. Advan. Biochem. Psychopharmacol. **6:** 37–59.
14. Tamir, H. & Y. L. Huang. 1974. Life Sci. **14:** 83–93.
15. Tamir, H., A. Klein & M. M. Rapport. 1976. J. Neurochem. **26:** 871–878.
16. Jonakait, G. M., H. Tamir, M. M. Rapport & M. D. Gershon. 1977. J. Neurochem. **28:** 277–284.
17. Tamir, H. & M. Kuhar. 1975. Brain Res. **83:** 169–172.
18. Tamir, H., J. Peisach & M. M. Rapport. 1976. Neurosci. Abstr. **2:** 617.
19. Tamir, H. 1976. Trans. Amer. Soc. Neurochem. **7:** 189.
20. Tranzer, J. P. 1972. Nature (London) **237:** 57–58
21. Jonsson, G., P. Einarsson, K. Fuxe & H. Hallman. 1975. Med. Biol. **53:** 25–39.
22. Noble, E. P., R. J. Wurtman & J. Axelrod. 1967. Sciences **6:** 281–291.
23. Saavedra, J. M., M. Brownstein & J. Axelrod. 1973. J. Pharmacol. Exp. Ther. **186:** 508–515.
24. Snyder, S., S. P. Banerjee, H. I. Yamamura & D. Greenberg. 1974. Science **184:** 1243–1253.
25. Weisenberg, R. C. 1972. Science **177:** 1104–1105.
26. Berry, R. W. & M. L. Shelanski. 1972. J. Mol. Biol. **71:** 71–80.
27. Weisenberg, R. C., G. G. Borisy & E. W. Taylor. 1968. Biochemistry **7:** 4466–4478.
28. Jacobs, M. 1975. Ann. N.Y. Acad. Sci. **253:** 562–572.
29. Gray, E. G. 1975. Proc. Roy. Soc. London Ser. B **190:** 369–372.
30. Garattini, S., G. De Gaetano, R. Samanin, M. Bernasconi & M. C. Roncaglioni. 1976. Biochem. Pharmacol. **25:** 13–16.
31. Wong, D. T., J. S. Horning, F. P. Bymaster, K. L. Hauser & B. B. Molloy. 1974. Life Sci. **15:** 471–479.
32. Bürki, H. R., W. Ruch, H. Asper, M. Baggiolini & G. Stiller. 1974. Eur. J. Pharmacol. **27:** 180–190.
33. Saner, A., L. Pieri, J. Moran, M. Da Prada & A. Pletscher. 1974. Brain Res. **76:** 109–117.
34. Baumgarten, H. G., A. Björklund, A. Nobin, E. Rosengren & H G. Schlossberger. 1975. Acta Physiol. Scand. Suppl. **429:** 5–27.

Discussion

Dr. H. G. Baumgarten: Is there any evidence that a similar type of binding protein could exist in dopaminergic fibers?

Dr. Tamir: We believe that a similar or identical protein is present in dopaminergic systems. Its presence will explain the effect of 5,7-DHT on both serotonergic and dopaminergic neurons but not on noradrenergic neurons.

Dr. Baumgarten: The best explanation for the high content of dopamine in dopaminergic fibers is that there must be a high cytoplasmic pool.

DR. E. K. SILBERGELD (*NINCDS, Bethesda, Md.*): What other drugs interact with this binding protein? What would be the functional significance of competing for serotonin binding to this component?

DR. TAMIR: Reserpine affects this storage form (10^{-6} M gives 50% inhibition), whereas drugs that affect reuptake do not (neither chlorimipramine nor desipramine). Lilly 110,140 is a moderate inhibitor. Drugs that affect receptor sites or serotonin synthesis do not interfere with serotonin binding. The bound serotonin is protected from degradation by MAO, and so the serotonin-binding protein will be involved in regulating the physiologically active pool. The physiologic or behavioral effects that result from specific interference with this pool are not clear.

DR. SANKAR: Is the binding specific for 5-hydroxyindole derivatives?

DR. TAMIR: No. Hydroxyindoles substituted in the four, five, or six position are all effective inhibitors.

SITE OF ACTION OF NEUROTOXINS ON PROTEIN SYNTHESIS

Michael A. Moskowitz,* Deborah Rubin,†
Thaddeus S. Nowak, Jr.,‡ B. Suren Baliga,‡ and
Hamish N. Munro ‡

Laboratories of † Neuroendocrine Regulation and
‡ Physiological Chemistry
Department of Nutrition and Food Science
Massachusetts Institute of Technology
Cambridge, Massachusetts 02139
and Section of Neurology
Department of Medicine
Peter Bent Brigham Hospital-Harvard Medical School
Boston, Massachusetts 02115

Autoxidizable phenylethylamines, such as 6-hydroxydopamine (6-OHDA) or the dihydroxytryptamines, appear to selectively destroy catecholamine- and serotonin-containing neurons, respectively, when administered to experimental animals.[1, 2] Biochemical and physiologic evidence suggests that these drugs are taken up and concentrated within nerve terminals by a transport mechanism that is specific for certain catecholamine and indoleamine compounds.[3] Cytotoxicity is achieved when a critical concentration of drug is reached inside nerve terminals and other cellular structures, including the perikarya.[4] Two major molecular theories have been proposed to explain these toxic effects: hydrogen peroxide, hydroxyl radical, and superoxide ion, formed during oxidation, function as oxidizing agents to alter cytoplasmic and membrane proteins [5]; formation of oxidation products of dihydroxyindoles and 6-OHDA, which commonly bind covalently to protein and other macromolecules and thereby alter physiologic function.[6, 7]

We now report the interactions between these neurotoxic compounds and protein-synthesizing systems by use of reticulocyte lysate. Our results show that both dihydroxytryptamines and 6-OHDA are capable of profoundly inhibiting protein synthesis *in vitro* and of stabilizing polyribosomes onto their mRNA by blocking peptide chain elongation at concentrations that are comparable to those achieved *in vivo*.

Materials and Methods

L-[4,5-^{3}H]Leucine (sp act 2–10 Ci/mmol) was purchased from New England Nuclear, Boston, Massachusetts. 5,7-Dihydroxytryptamine creatinine sulfate, 6,7-dihydroxytryptamine creatinine sulfate monohydrate, 5,6-dihydroxytryptamine creatinine sulfate dihydrate, DL-octopamine hydrochloride, and 6-hydroxydopamine hydrobromide were obtained from Regis Chemical Com-

* Recipient of a Teacher-Investigator Award 11081 from the NINCDS and a Sloan Foundation Fellowship in Neuroscience.

0077–8923/78/0305–0096 $01.75/0

pany, Chicago, Illinois. D-Amphetamine sulfate, 5-hydroxydopamine hydrochloride, and DL-*p*-chloroamphetamine hydrochloride were obtained from Sigma Chemical Company, St. Louis, Missouri. Tyramine hydrochloride was purchased from Calbiochem, Los Angeles, California. *o*-Quinone and *p*-quinone derivatives of 6-OHDA were obtained as a gift from Dr. Richard Heikkila. All other chemicals were reagent grade.

Rabbit reticulocyte lysate was prepared according to the method of Gilbert and Anderson.[8]

For preparation of lysate to be used directly in protein synthesis incubations, washed reticulocytes were lysed in 1 vol of 1 mM dithiothreitol (DTT). Lysates to be used in preparation of fractionated protein-synthesizing systems were obtained by lysing cells in 4 vol of 2 mM $MgCl_2$, 1 mM DTT, and 0.1 mM EDTA (neutralized with sodium hydroxide).

Reticulocyte Ribosomes

Dilute lysate was centrifuged at 40,000 rpm for 2.5 hr in a Beckman Ti 50 rotor, and the supernatant was used for the preparation of the pH 5 enzyme fraction and reticulocyte tRNA (see below). The ribosomal pellet was the source of rabbit reticulocyte ribosomes and was used for the preparation of potassium chloride wash factors and globin mRNA, as described below. For use in protein synthesis, the ribosomal pellet was resuspended in 0.25 M sucrose, 1 mM DTT, and 0.1 mM EDTA to a final concentration of about 200 $A_{260\ nm}$/ml.

Potassium Chloride Wash Factors

Potassium chloride wash factors were prepared according to the method of Gilbert and Anderson.[8] Briefly, the ribosomes were adjusted to 0.5 M KCl and centrifuged as described above. The ribosomal pellet was resuspended in standard sucrose and used for preparation of mRNA. The supernatant from the potassium chloride wash was dialyzed in the cold overnight against 100 mM KCl, 10 mM Tris HCl (pH 7.4), 2 mM $MgCl_2$, 0.1 mM EDTA, and 1 mM DTT.

Globin Messenger RNA

Unwashed or salt-washed ribosomes were resuspended in 0.1 M NaCl, 0.2 M Tris HCl (pH 9.0), and 10 mM EDTA (pellet buffer). The suspension was brought to 0.5% sodium dodecyl sulfate (SDS) and was then extracted at room temperature with an equal volume of water-saturated phenol that contained 1 mg/ml of 8-hydroxyquinoline. The aqueous phase was decanted; the phenol phase was reextracted with an equal volume of pellet buffer. The aqueous extracts were combined, and RNA was precipitated by the addition of 2 vol of ethanol. The precipitate was collected, pellets were redissolved in buffer, and they were reprecipitated with ethanol, as described above. The precipitate was then dissolved in a small volume of water and dialyzed against 1 mM Tris HCl (pH 7.4).

Poly(A)-rich globin mRNA was obtained from the polysome-derived RNA fraction by chromatography on oligo(dT)-cellulose, according to the procedure of Aviv and Leder.[9] This fraction was approximately 2% of the total applied to the column.

pH 5 Enzyme Fraction

The supernatant obtained after ribosomes were pelleted from the lysate was adjusted to pH 5.1 with 1 M acetic acid and the precipitate was collected. The pellet was dissolved in 0.25 M sucrose, 25 mM Tris HCl (pH 7.4), 2 mM $MgCl_2$, 1 mM DTT, and 0.1 mM EDTA.

Reticulocyte Transfer RNA

Supernatant obtained after pelleting ribosomes was adjusted to 0.1 M NaCl, 0.2 M Tris HCl (pH 9.0), and 10 mM EDTA and brought to 0.5% in SDS. RNA was extracted and precipitated as described above for the preparation of mRNA. Transfer RNA was separated from other species according to the procedure of Anderson.[10]

Results

Inhibition of Protein Synthesis in Vitro

The capacity to inhibit protein synthesis in general was examined for two groups of toxins, namely, 6-OHDA and related compounds; this capacity was then examined for the corresponding serotonin neurotoxins. Utilizing the crude (unresolved) cell-free reticulocyte lysate system, we examined the extent to which 6-OHDA, 6-hydroxydopa, *o*-quinone, and *p*-quinone derivatives of 6-OHDA inhibit protein synthesis activity *in vitro.* FIGURE 1 illustrates the dose-response curves for various concentrations of 6-OHDA and for differing incubation periods. As early as 4 min after the start of incubation, inhibition of amino acid incorporation into protein was detectable at concentrations of 5 mM. At concentrations of 0.5 or 1.0 mM, inhibition was not observed at 4 min but was detectable at 8 min and gradually increased at 12 and 16 min. These observations, namely, the absence of inhibition at 4 min and the presence of increasing inhibition with time, raise the possibility that inhibition was mediated by products formed *in vitro* (e.g., oxidation products of 6-OHDA). We have also noted that prior neutralization of 6-OHDA and related compounds (pH 7.6) was necessary for them to most effectively inhibit protein synthesis in the reticulocyte lysate system. Unbuffered 6-OHDA (1 mM) did not inhibit protein synthesis when added to the reaction mixture.[10a] Various analogs of 6-OHDA also interfered with the incorporation of amino acids into growing peptide chains (FIGURE 2). 6-Hydroxydopa appeared to suppress protein synthesis as effectively as did 6-OHDA, whereas the *o*-quinone and *p*-quinone derivatives were only slightly less active than was 6-OHDA itself. To inhibit the incorporation of radio-labeled amino acid into nascent peptide chains, the addition of 1 mM 6-OHDA, 1 mM 6-hydroxydopa, 1.5 mM *o*-quinone, and 1.4 mM *p*-quinone was required.

When 6,7-, 5,6-, or 5,7-dihydroxytryptamine was added to the reticulocyte lysate system, significant inhibition of protein synthesis activity was also detected (FIGURE 2). 6,7-Dihydroxytryptamine (6,7-DHT), at an approximate concentration of 1 mM, caused 50% inhibition, whereas concentrations of 2 mM 5,7-dihydroxytryptamine (5,7-DHT) or 1 mM 5,6-dihydroxytryptamine (5,6-DHT) were needed to achieve this level of inhibition. The dihydroxy-

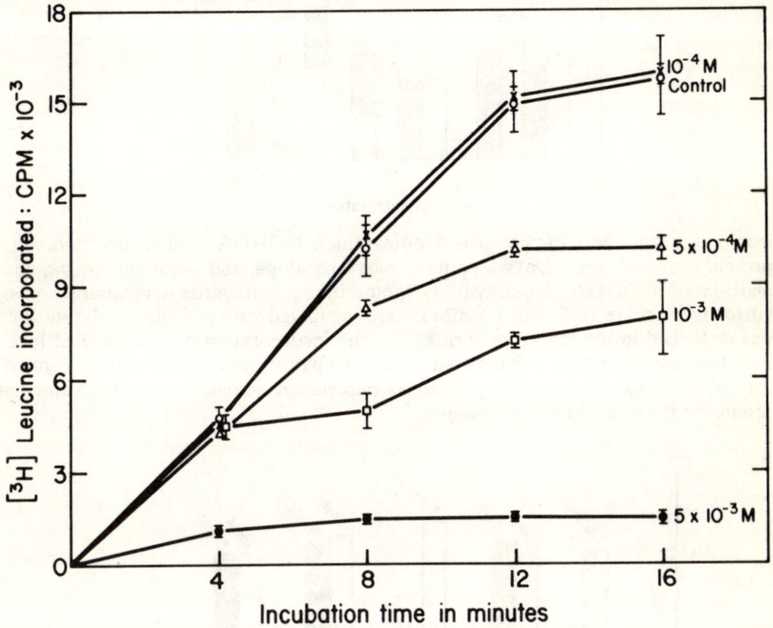

FIGURE 1. Effect of 6-hydroxydopamine on protein synthesis in a reticulocyte lysate. All incubations were performed at 30°C in a final volume of 50 μl. Reaction mixtures contained the following components at the final concentrations indicated: KCl, 60 mM; Tris HCl, 20 mM (pH 7.4); $MgCl_2$, 2 mM; DTT, 3 mM; ATP, 1 mM; guanosine triphosphate, 1.6 mM; phosphoenolpyruvate, 3 mM; pyruvate kinase 30–40 IU/ml; $[^{14}C]$leucine, 3 μCi/ml; unlabeled leucine, 40 μM each; 19 unlabeled amino acids (not including leucine), 60 μM each; hemin chloride 40 μM; and concentrated lysate, 50% of the reaction volume. 6-OHDA (pH 7.6) was added in various concentrations (0.10, 0.5, 1.0, 1.5 mM) immediately prior to the initiation of incubation. The amounts of radiolabeled leucine incorporated into trichloracetic acid-precipitable protein was measured after 4, 8, 12, and 16 min of incubation, according to the method of Mans and Novelli.[20] Data are expressed as the mean±SE.

indoleamines were also neutralized to pH 7.6 prior to their addition to the incubation mixture.

Analogs of 6-OHDA that autoxidize at a considerably slower rate than 6-OHDA (namely, dopamine, tyramine, octopamine, 5-OHDA) were next added to the rabbit reticulocyte lysate system to determine their effects on cell-free protein synthesis. At concentrations similar to those used for 6-OHDA

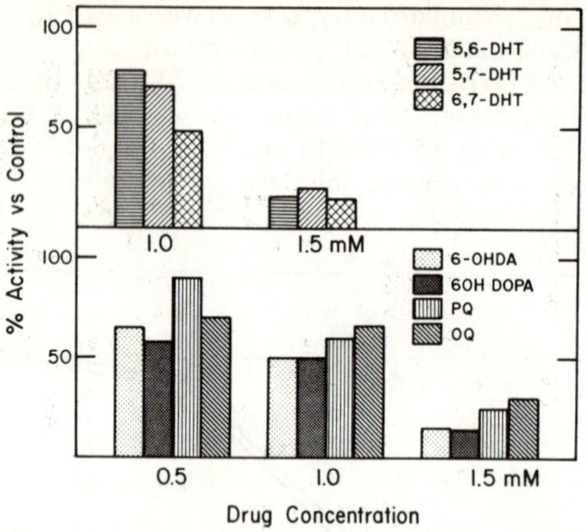

FIGURE 2. The effect of dihydroxyindoleamines, 6-OHDA, and related compounds on protein synthesis. 6-Hydroxydopamine, 6-hydroxydopa, and *p*-quinone or *o*-quinone derivatives of 6-OHDA, 5,6-dihydroxytryptamine, 5,7-dihydroxytryptamine, or 6,7-dihydroxytryptamine (0.5, 1.0, 1.5 mM) were incubated with reticulocyte lysate for 16 min as described in the legend to FIGURE 1. The incorporation or radiolabeled leucine into protein was determined for drug-treated incubation mixtures and then compared to control samples. Data are expressed as percentage activity vs control values and represent the mean of three experiments.

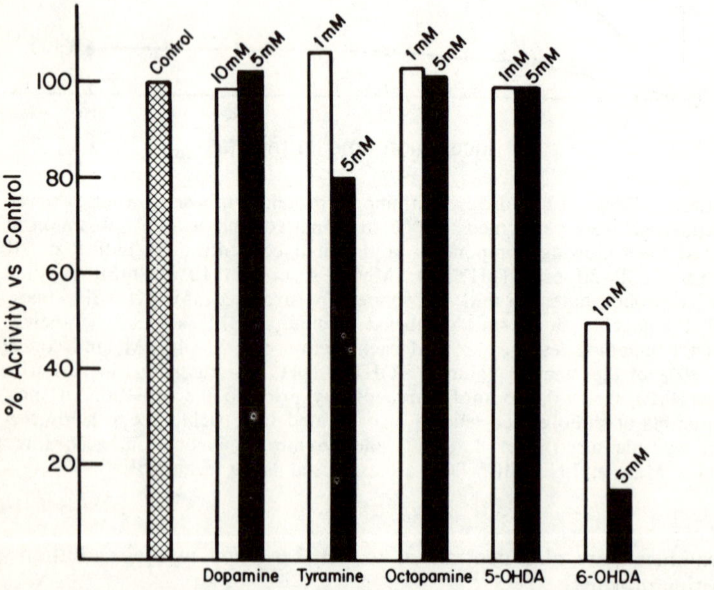

FIGURE 3. The effect of some slowly oxidizable phenylethylamine compounds on *in vitro* protein synthesis. Dopamine, 5-hydroxydopamine, tyramine, or octopamine was added to the reticulocyte lysate at the concentrations indicated. Protein synthesis was performed as described in the legend to FIGURE 1. Data are expressed as percentage activity vs control values.

and the dihydroxyindoleamines, these compounds had little, if any, inhibitory activity on amino acid incorporation into trichloroacetic acid precipitable protein (FIGURE 3). Taken together, these findings point to autoxidation of 6-OHDA as a necessary step in the blocking of protein synthesis activity *in vitro*.

Action of 6-OHDA on Initiation and Elongation in Vitro

With two cell-free systems prepared from the rabbit reticulocyte lysate, we examined the extent to which 6-OHDA inhibits peptide chain elongation and initiation in separate experiments; the results are shown in FIGURE 4. To test

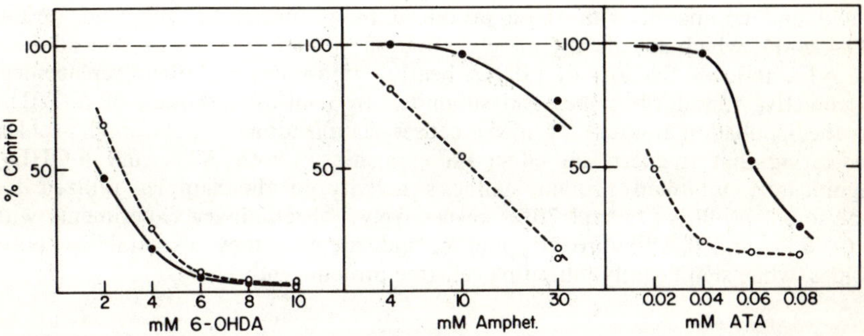

FIGURE 4. Effects of 6-hydroxydopamine, *d*-amphetamine sulfate, and aurintricarboxylic acid on initiating and elongating reticulocyte systems for *in vitro* protein synthesis. Incubations were performed at 30° C for 20 min in a final volume of 50 μl. The elongating system (indicated by one solid line) contained 0.34 $A_{260\ nm}$/ml reticulocyte polysomes supplemented with saturating amounts of the pH 5 enzyme fraction. The initiating system (indicated by the dashed line) contained 0.13 $A_{260\ nm}$/ml reticulocyte polysomes, the pH 5 enzyme fraction as in the elongating system, and saturating amounts of potassium chloride wash factors, globin mRNA, and reticulocyte tRNA. Both systems included other reaction mixture components, as indicated for the lysate system in FIGURE 1, but hemin and unlabeled leucine were omitted, potassium chloride was at 80 mM, magnesium chloride was at 3.5 mM, and guanosine triphosphate was at 0.4 mM. 6-OHDA (2–10 mM), *d*-amphetamine sulfate (4–30 mM), and aurintricarboxylic acid (0.02–0.08 mM) were added immediately prior to incubation. The amount of incorporation of radiolabeled amino acid into protein was then determined. Note that 6-OHDA has similar inhibitory action in both systems, whereas the initiating system is much more selectively inhibited by *d*-amphetamine sulfate.

the effect of 6-OHDA (2–10 mM) on the initiation of protein synthesis, the drug was added to an mRNA and potassium chloride wash-dependent fractionated reticulocyte system; the amount of radiolabeled amino acid incorporation into trichloroacetic acid-precipitable protein was then determined. At virtually all concentrations tested, 6-OHDA significantly inhibited incorporation of radiolabeled amino acids into proteins in this mRNA-dependent system (FIGURE 4). When 6-OHDA (2–10 mM) was incubated with polyribosomes and a pH 5 enzyme fraction prepared from reticulocytes (a cell-free system that primarily elongates nascent peptide chains), significant reductions in the amount of radiolabeled amino acid incorporated into trichloroacetic acid-precipitable protein was also evident. Since protein synthesis was inhibited in both the mRNA-dependent system (which initiates and then elongates nascent peptide chains) and the polysome pH 5 system (which specifically elongates preexisting chains), we can only be certain from our data that these compounds interfere with cellular elongation. Aurintricarboxylic acid (0.04 mM) and amphetamine (1–10 mM) (both specific inhibitors of protein synthesis initiation[11, 12] significantly reduced protein synthesis in the mRNA-dependent system only (FIGURE 4).

Effects of 6-OHDA on Reticulocyte Polysomes

The effects of 6-OHDA on reticulocyte polyribosomal profiles were investigated and compared with those produced by aurintricarboxylic acid (ATA, 0.04 mM), which specifically affects initiation.[12] FIGURE 5 shows that addition of ATA induces the loss of mRNA-bearing ribosomes and their replacement by inactive runoff ribosomes and subunits. By contrast, addition of 6-OHDA to the incubation mixture (1 mM) causes stabilization of polysomal profiles, indicating that it exerts an effect on elongation. Both ATA and 6-OHDA significantly inhibited protein synthesis activity in the samples utilized for polysomal profiles (75 and 70%, respectively). Preliminary experiments with 5,6-, 5,7-, and 6,7-dihydroxytryptamine indicate that they also stabilize polysomes, while significantly inhibiting cell-free protein synthesis.

DISCUSSION

Our studies have shown that addition of 5,6-, 6,7-, or 5,7-dihydroxytryptamine, 6-OHDA, *p*-quinone, or *o*-quinone significantly suppresses translation of mRNA in a reticulocyte lysate system by a mechanism that apparently interferes with nascent peptide chain elongation. The autoxidizable phenylethylamine compounds appear to be more effective on a molar basis as protein synthesis inhibitors than do the serotonin analogs (FIGURE 5). Molecular oxygen seems to be required to produce these effects because inhibition of protein synthesis cannot be induced by drugs that autoxidize very slowly nor can it be induced without prior neutralization of these compounds to pH 7.6, at which point they oxidize more rapidly.

The effective drug concentrations (0.5–1.5 mM) that we utilized for inhibition *in vitro* are actually much lower than the levels required *in vivo*. For example, it has been estimated that 50 mM 6-OHDA is the threshold intracellular concentration above which neuronal degeneration begins.[4, 13] To our knowledge, similar intracellular measurements have not yet been reported for the hydroxytryptamines. However, the concentrations of hydroxytryptamines required to deplete brain serotonin in rats are only slightly less than the minimal dose of 6-OHDA that produces the same effect. Both 6-OHDA and the hydroxytryptamines have been studied *in vitro* for their ability to inhibit membrane uptake of biogenic amines.[13, 14] For 6-OHDA, inhibitory concentrations were in the millimolar range, whereas the effective concentrations were in the micromolar range for the hydroxyindoleamines. In the crude reticulocyte lysate system, both drugs were effective in inhibiting peptide chain elongation when added in millimolar concentrations.

It is not unexpected that there are major differences between *in vivo* and *in vitro* activities with regard to effects on membrane functions. Toxicity in a cell-free system depends to a large extent on the intrinsic properties of the molecule. Toxicity in such organs as the brain is due in part to various factors, including the ability of the drug to penetrate the brain substance to reach synaptic terminals, to be taken up by the nerve ending and be selectively concentrated, and to be transported in a retrograde manner to reach the axon and cell body. Other variables that influence toxicity *in vivo* include the route of administration, the speed of injection, and the type of anesthesia.[15] It is therefore not surprising that the potency of these compounds in a cell-free system does not always coincide with the extent to which they deplete brain serotonin and catecholamines within the whole brain and spinal cord.[14, 16]

We have previously reported that the addition of 1–5 mM *d*-amphetamine sulfate and related compounds (notably *p*-chloroamphetamine) inhibits protein synthesis by a mechanism that is specific for initiation but that does not affect

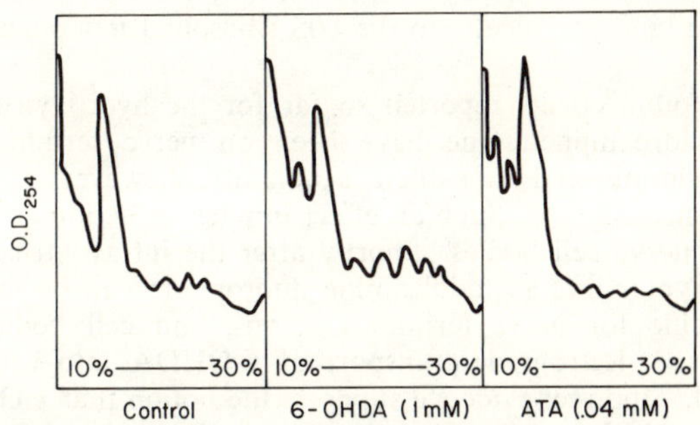

FIGURE 5. Effect of 6-hydroxydopamine and aurintricarboxylic acid on polysomal profiles in the reticulocyte lysate. The reaction mixture contained reticulocyte lysate and 6-OHDA (1 mM) or aurintricarboxylic acid (0.04 mM) and was prepared as described in the text and allowed to incubate for 16 min. Polyribosomes were then analyzed on sucrose density gradients (10–30%). Absorption profiles were recorded at 260 nm on a Gilford spectrophotometer.

peptide chain elongation.[11] By contrast, we have shown that the hydroxytryptamines and 6-OHDA do not interfere with peptide chain elongation. Therefore, the specific molecular events that result in toxicity apparently differ for the amphetamines and the autoxidizable compounds. 6-OHDA and 5,6-DHT reportedly complex with proteins and other macromolecules by covalently binding with sulfhydryl groups to form large polymers.[17] This covalent binding may also be the mechanism by which these compounds interact with ribosomes and other constituents of the cell system to inhibit protein synthesis. Preliminary experiments have shown that 6-OHDA, *p*-quinone, and *o*-quinone reduce the number of sulfhydryl groups that can bind to [^{3}H]iodoacetic acid after preincubation of brain ribosomes with these toxins.[17a] The formation of such ribosomal complexes would also explain the "freezing" effect of the same neurotoxins on polyribosomes, insofar as these toxins failed to disrupt reticulocytic polyribosomes (FIGURE 5) but profoundly inhibited incorporation of amino acids into nascent peptide chains (FIGURE 2). These neurotoxins may also complex with sulfhydryl-containing elongating factors and thereby limit the insertion of aminoacyl-tRNA into growing peptide chains.[18] They may also form complexes with the nascent peptide chain, as has been shown in the case of sulfhydryl groups on growing globin chains.[17] The relative contributions of hydrogen peroxide, superoxide radical, or hydroxyl ion generated during the oxidation of these amines must also be considered.[5]

In contrast, the site of action of amphetamine on protein synthesis systems differs from that of the other compounds studied, because the effect of amphetamine appears to be confined to an action on protein synthesis initiation.[11] This ability of amphetamine to inhibit initiation may account for the disaggregation of brain polysomes induced by the drug both *in vivo* and *in vitro*.[19, 20] When *d*-amphetamine or *p*-chloroamphetamine is added to a cell-free system prepared from wheat germ or reticulocytes that is made initiation factor dependent, thus requiring the addition of mRNA for activity, inhibition occurs. If amphetamine is added to a system prepared from polyribosomes that bear nascent peptide chains and an added pH 5 enzyme fraction, little reduction in activity is observed.[11, 20] Such a selective effect on initiation may correlate with the capacity of amphetamine to bind specifically to the 80S ribosomal monomer from wheat germ.[11]

The most profound effects reported so far for the hydroxytryptamines, 6-OHDA, and *p*-chloroamphetamine have been on nerve terminals. If these compounds are administered in sufficient quantity, however, they may also destroy the entire neuron.[20, 21] Such an effect implies an action (either primary or secondary) on nerve cell bodies. Shortly after the intraventricular injection of 5,6-dihydroxytryptamine, an indoleamine fluorophore can be visualized that is reportedly specific for nerve terminals, axons, and cell bodies of monoaminergic neurons.[21] Retrograde transport of 6-OHDA into cell bodies has also been reported.[1] Indirect evidence supports the notion that *p*-chloroamphetamine may also reach the perikarya, because subcellular fraction studies have shown that radiolabeled drugs can localize within microsomes of the cell body.[22] Possible consequences of toxin accumulation within the perikarya include complete disruption of protein synthesis, partial disruption that might interfere with the reparative phase of accelerated protein synthesis, or selective inhibition of synthesis of specific proteins, such as tryptophan hydroxylase.[23] Neurons may also be destroyed as an expression of a degenerative process that only secondarily affects the cell body. It should also be noted that a significant effect on protein synthesis activity has been reported previously within synaptosomes and microsomal fractions after the administration of 6-OHDA to adult rats.[24] Taken together, these observations indicate that the autoxidizable phenylethylamines, dihydroxytryptamines, and amphetamines and related compounds are potent inhibitors of protein synthesis *in vitro* at concentrations similar to those believed to be effective *in vivo*. Accordingly, the mechanism of *in vitro* inhibition of protein synthesis offers promise for elucidating the nature of their biologic actions on the central nervous system.

References

1. Kostrzewa, R. M. & D. M. Jacobowitz. 1974. Pharmacol. Rev. **26:** 199–286.
2. Baumgarten, H. G., A. Björklund, L. Lachenmayer & A. Nobin. 1973. Acta Physiol. Scand. Suppl. **391:** 1–19.

3. LUNDSTROM, J., H. ONG, J. DALY & C. R. CREVELING. 1973. Mol. Pharmacol. **9:** 505–513.
4. JONSSON, G. & CH. SACHS. 1971. Eur. J. Pharmacol. **16:** 55–62.
5. COHEN, G. & R. E. HEIKKILA. 1974. J. Biol. Chem. **249:** 2447–2452.
6. SANER, A. & H. THOENEN. 1971. Mol. Pharmacol. **7:** 147–154.
7. ROTMAN, A., J. W. DALY, C. R. CREVELING & X. O. BREAKFIELD. 1976. Biochem. Pharmacol. **25:** 383–388.
8. GILBERT, J. M. & W. F. ANDERSON. 1971. Methods Enzymol. **XX:** 542–549.
9. AVIV, H. & P. LEDER. 1972. Proc. Nat. Acad. Sci. USA. **69:** 1408–1412.
10. ANDERSON, W. F. 1969. Biochemistry **8:** 3687–3695.
10a. MOSKOWITZ, M. A. & D. RUBIN. 1977. Unpublished observations.
11. BALIGA, B. S., J. ZAHRINGER, M. TRACHTENBERG, M. A. MOSKOWITZ & H. N. MUNRO. 1976. Biochim. Biophys. Acta **442:** 239–250.
12. HOERZ, W. & K. S. MCCARTY. 1971. Biochem. Biophys. Acta **228:** 526–535.
13. JONSSON, G. & CH. SACHS. 1975 *In* Chemical Tools in Catecholamine Research. G. Jonnsson, T. Malmfors & Ch. Sachs, Eds. : 41–50. North Holland Publishing Company. Amsterdam, The Netherlands.
14. BAUMGARTEN, H. G., A. BJÖRKLUND, A. NOBIN, E. ROSENBREN & H. G. SCHLOSSBERGER. 1975. Acta Physiol. Scand. Suppl. **429:** 1–27.
15. BAUMGARTEN, H. G., K. D. EVETTS, R. B. HOLMAN, L. L. IVERSEN, M. VOGT & G. WILSON. 1972. J. Neurochem. **19:** 1587–1597.
16. BJÖRKLUND, A., A. S. HORN, H. G. BAUMGARTEN, A. NOBIN & H. G. SCHLOSSBERGER. 1975. Acta Physiol. Scand. Suppl. **429:** 30–60.
17. ROTMAN, A., J .W. DALY & C. R. CREVELING. 1976. Mol. Pharmacol. **12:** 887–899.
17a. MOSKOWITZ, M. A. & B. S. BALIGA. 1977. Unpublished observations.
18. STEINERT, P., B. S. BALIGA & H. N. MUNRO. 1974. Mol. Biol. **88:** 895–912.
19. MOSKOWITZ, M. A., B. W. WEISS, L. LYTLE, H. N. MUNRO & R. J. WURTMAN. 1975. Proc. Nat. Acad. Sci. USA **72:** 834–836.
20. MOSKOWITZ, M. A., B. S. BALIGA, J. ZAHRINGER & H. N. MUNRO. 1975. Neuroscience **5:** 288.
21. BAUMGARTEN, H. G., A. BJÖRKLUND, L. LACHENMAYER, A. RENSCH & E. ROSENGREN. 1974. Cell Tissue Res. **152:** 271–281.
22. NOBIN, A., H .G. BAUMGARTEN, A. BJÖRKLUND, A. LACHENMAYER & U. STENEVI. 1973. Brain Res. **56:** 1–24.
23. SHERMAN, S. D. & M. GAL. 1975. Psychopharmacol. Commun. **1**(3): 261–273.
24. FULLER, R. W., J. BAKER, K. W. PERRY & B. B. MOLLOY. 1975. Neuropharmacology **14:** 739–746.
25. LOH, H., S. STOLMAN & C. Y. LEE. 1971. Life Sci. **10:** 1171–1180.
26. MANS, R. J. & G. D. NOVELLI. 1961. Arch. Biochem. Biophys. **94:** 48–53.

DISCUSSION

DR. E. M. GÁL: When we inject 10 or 20 mg/kg of *p*-CA into animals, we don't see any disaggregation in brain ribosomes or early impairment of amino acid incorporation. Did you try the effect of *p*-CA on germinating wheat seedlings?

DR. MOSKOWITZ: With amphetamine in doses of 10 mg, and *p*-chloroamphetamine at 10 mg/kg, we observe disaggregation of ribosomes. We can block this effect by pretreatment with Lilly 110,140. The effects of these compounds on *in vitro* protein synthesis were first reported in 1971 by Rowe. Pro tein synthesis was inhibited not only in cell bodies but also in synaptosomal regions.

The effects of *p*-chloroamphetamine on the wheat germ system are more profound than those of amphetamine. Concentrations that inhibit protein synthesis are between 1 and 5 mM.

EFFECT OF DRUGS ON THE DISTRIBUTION OF [^{14}C]-5,6-DIHYDROXYTRYPTAMINE AND [^{14}C]-5,7-DIHYDROXYTRYPTAMINE IN RAT BRAIN *

H. G. Baumgarten, H. P. Klemm, and L. Lachenmayer

Department of Functional Anatomy
University of Hamburg
2000 Hamburg, Federal Republic of Germany

H. G. Schlossberger

Department of Organic Chemistry and Spectroscopy
Max Planck Institute for Biochemistry
D-8033 Martinsried, Munich, Federal Republic of Germany

It is generally accepted that the intraneuronal accumulation of amine neurotoxins by the amine-concentrating membrane transport systems of noradrenaline (NA), dopamine (DA), and 5-hydroxytryptamine (5-HT) neurons is one of the essential prerequisites for their selectivity of action. Thus, blockade of the NA transport sites in brain by, for example, desmethylimipramine (DMI) or related compounds has been shown to counteract the degeneration of noradrenergic axons and axon terminals produced by intraventricular injection of 6-hydroxydopamine (6-OHDA) or 5,7-dihydroxytryptamine (5,7-DHT).[1-4] The requirement of drug pretreatment for improvement of the selectivity of action of the amine neurotoxins indicates that these compounds have an affinity for more than one of the amine uptake mechanisms, and the results of studies in which IC_{50} values for such neurotoxins as inhibitors of [^{3}H]NA, [^{3}H]DA, and [^{3}H]-5-HT uptake have been determined [3, 5] tend to support this idea. Attempts to explain the *in vivo* behavior of some of these neurotoxins (e.g., the degree of specificity and extent of long-lasting depletion of NA, DA, and 5-HT) by the IC_{50} values of some of these drugs have not provided meaningful results (e.g., for the specificity and neurotoxic potency of 5,6-DHT and 5,7-DHT [3,6]), which suggests that the IC_{50} data may not necessarily reflect the true affinity of some of the indole neurotoxins to the NA, 5-HT, and DA uptake sites. Furthermore, the IC_{50} values do not reveal whether the unexplained *in vivo* specificity is determined by differences in the *rate* of incorporation of the false neurochemicals into the NA, DA, and 5-HT neurons, to what extent they are retained intraneuronally, and whether they are releasers of endogenous monoamines. To obtain at least an approximate idea of the extent of accumulation of the serotonin neurotoxins 5,6-DHT and 5,7-DHT in the different types of central monoaminergic neurons, the distribution of ^{14}C-labeled 5,6-DHT and 5,7-DHT was studied in the adult rat central nervous system by means of autoradiography.

Since it has been claimed that only the fraction of amine neurotoxin that is irreversibly bound to brain protein may correlate with and reflect the extent of damage inflicted on the neuron,[7, 8] and because metabolism of nonirreversibly

* Supported by grants from the Deutsche Forschungsgemeinschaft.

0077-8923/78/0305-0107 $01.75/0

bound neurotoxin may obscure the degree of irreversible binding soon after injection, the distribution of radioactivity was evaluated 24 hr after lateral ventricular injection of [^{14}C]-5,6-DHT and [^{14}C]-5,7-DHT. To obtain a semi-quantitative measure of the protein-bound radioactivity derived from [^{14}C]-5,6-DHT and [^{14}C]-5,7-DHT, the radioactivity distribution was also analyzed by liquid scintillation spectrometry in 13 central nervous system regions after homogenization of tissue in perchloric acid (PCA).

The data presented show that some of the questions regarding the selectivity of action of 5,6-DHT and 5,7-DHT can be resolved by such a combined approach and that drugs with known actions on the central nervous system alter to a remarkable extent the distribution of [^{14}C]-5,6-DHT and [^{14}C]-5,7-DHT in brain.

Distribution of Radioactivity in Pellet and Supernatant Fractions of Perchloric Acid-Homogenized Brain Regions

In studies on the distribution of radioactivity in pellet and supernatant fractions of brain regions homogenized in PCA at various time intervals (30 min to 8 days) after left lateral intraventricular injection of [^{14}C]-5,6-DHT and [^{14}C]-5,7-DHT, it was found that during the first few hours after injection, a fraction of the PCA-precipitable radioactivity derived from [^{14}C]-5,7-DHT was not truly protein bound and was extractable from the pellet by use of chloroform-methanol.[11] Therefore, a comparison of the proportions of protein-bound radioactivity derived from [^{14}C]-5,6-DHT and [^{14}C]-5,7-DHT was considered meaningful only after the disappearance from the brain of the lipid-bound fraction of [^{14}C]-5,7-DHT-derived radioactivity (i.e., about 24–48 hr after drug injection).

The radioactivity distributions in pellet and supernatant fractions of 13 central nervous system regions after injection of [^{14}C]-5,6-DHT and [^{14}C]-5,7-DHT (94 μg of free base [1-^{14}C]-5,7-DHT, 1.3 $\times$ 10^6 dpm, in 20 μl of saline, and 75 μg of free base [1-^{14}C]-5,6-DHT, 1.3 $\times$ 10^6 dpm, in 20 μl of saline, both drugs given as the creatinine sulfate-water complex) are given in Tables 1 and 2. Some differences in the radioactivity distribution patterns after injection of either drug are obvious. The amount of protein-bound radioactivity is higher in most central nervous system regions of the [^{14}C]-5,6-DHT-injected animals. However, in regions located at an increasing distance from the injection site (spinal cord), the reverse occurs, indicating that 5,7-DHT more readily distributes in the ventricular and subarachnoidal cerebrospinal fluid than does 5,6-DHT and/or that it more easily equilibrates with brain matter. A craniocaudal gradient in the amount of radioactivity retained in the pellet fractions is apparent with [^{14}C]-5,6-DHT and [^{14}C]-5,7-DHT. It is also evident that the ventricle-distant regions are less radioactive (e.g., the forebrain-rest sample) than are the ventricle-near ones (e.g., the hippocampus). When the amounts of radioactivity (disintegrations per minute per milligram of tissue) recovered from the 48-hr supernatants of all PCA-homogenized brain regions in the [^{14}C]-5,7-DHT- and [^{14}C]-5,6-DHT-injected animals are expressed as percentages of the pellet-bound radioactivity, it becomes apparent that the proportion of PCA-soluble activity is higher in [^{14}C]-5,7-DHT- than in [^{14}C]-5,6-DHT-injected animals (Tables 1 & 2).

Benztropine, DMI, and pargyline all had more pronounced effects on reten-

TABLE 1

REGIONAL PATTERNS OF ACID-SOLUBLE (SUPERNATANT) AND ACID-INSOLUBLE (PELLET) RADIOACTIVITY 48 HOURS AFTER INTRAVENTRICULAR INJECTION OF $[1\text{-}^{14}C]$-5,6-DHT *

	Supernatant	Pellet
Septum	15.6±6.5 †	214.6±66.2
Left striatum	3.2±1.7	152±78
Right striatum	2.5±0.5	108±21
Left hippocampus	7.5±1.6	130±19
Right hippocampus	4.8±0.9	69±11
Hypothalamus	8.6±2.9	187±45
Tectum	9.6±0.7	211±59
Tegmentum	6.1±1.4	99±17
Pons-medulla oblongata	3.5±1.3	87±29
Cervical spinal cord	2.4±1.3	53±0.8
Thoracic spinal cord	1.7±0.5	13.2±10.2
Lumber spinal cord	1.0±0.2	6.2±0.1
Forebrain rest	2.2±0.2	38.1±2.1

* Injected dose was 1.3×10^6 dpm (75 μg of free base).

† All results are expressed as disintegrations per minute per milligram of fresh tissue

TABLE 2

REGIONAL PATTERNS OF ACID-SOLUBLE (SUPERNATANT) AND ACID-INSOLUBLE (PELLET) RADIOACTIVITY 48 HOURS AFTER INTRAVENTRICULAR INJECTION OF $[1\text{-}^{14}C]$-5,7-DHT *

	Supernatant	Pellet
Septum	19.1±2.8 †	100.6±19.2
Left striatum	12.2±3.4	28.3±13.2
Right striatum	9.0±2.2	23.9±5.6
Left hippocampus	8.1±1.1	52.5±3.9
Right hippocampus	5.7±0.6	37.8±8.2
Hypothalamus	13.8±1.6	129.1±12.8
Tectum	10.4±0.7	86.3±11.7
Tegmentum	10.8±0.6	69.5±6.0
Pons-medulla oblongata	5.9±0.6	66.4±9.6
Cervical spinal cord	4.3±0.5	39.8±6.9
Thoracic spinal cord	3.7±0.6	31.9±3.6
Lumber spinal cord	2.3±0.7	11.0±5.3
Forebrain rest	4.1±0.9	30.9±1.0

* Injected dose was 1.3×10^6 dpm (94 μg of free base).

† All results are expressed as disintegrations per minute per milligram of fresh tissue.

tion of radioactivity in the brain after intraventricular injection of [^{14}C]-5,7-DHT than after [^{14}C]-5,6-DHT, and the same was true when the labeled drugs were administered into the brains of animals subjected to degeneration of their central catecholamine terminals (by intraventricular and intracisternal injections of 6-OHDA, combined with nialamide) or to destruction of their serotonin projections (by DMI-5,7-DHT). DMI (25 mg/kg ip, given 45 min before 5,6-DHT or 5,7-DHT) affected regional radioactivity content only marginally after administration of [^{14}C]-5,6-DHT (slight reduction in pellet-bound radioactivity in the septum, hypothalamus, and forebrain-rest sample) but moderately reduced soluble and insoluble radioactivity in nearly all central nervous system regions after intraventricular injection of [^{14}C]-5,7-DHT, suggesting that the percentage of radioactivity accumulated by the central NA neuron system is comparatively small. Surprisingly, when benztropine (60 mg/kg, ip) was administered prior to [^{14}C]-5,6-DHT (45 min before the labeled drug), it consistently increased the radioactivity in the pellet and supernatant fractions of most central nervous system regions (see FIGURES 1 & 2); the larger percentage increases in soluble versus PCA-insoluble radioactivity in some regions may indicate that benztropine somehow retards the metabolic clearance of 5,6-DHT in the brain. The changes produced by benztropine in the pattern and extent of binding of radioactivity in the brain after [^{14}C]-5,7-DHT administration were much more selective: benztropine decreased pellet-bound and soluble radioactivity in the striatum and hypothalamus but increased it in the hippocampus, which suggests that the drug caused a shift of activity from the ventricle-close dopaminergic to the ventricle-near noradrenergic forebrain regions. The most dramatic changes were noted in the pargyline (60 mg/kg, ip) -pretreated animals injected intraventricularly with [^{14}C]-5,7-DHT (FIGURES 3 & 4). While pellet-bound radioactivity was moderately increased in the striatum, hippocampus, and in the three spinal cord sections (FIGURE 4), soluble activity was found to be

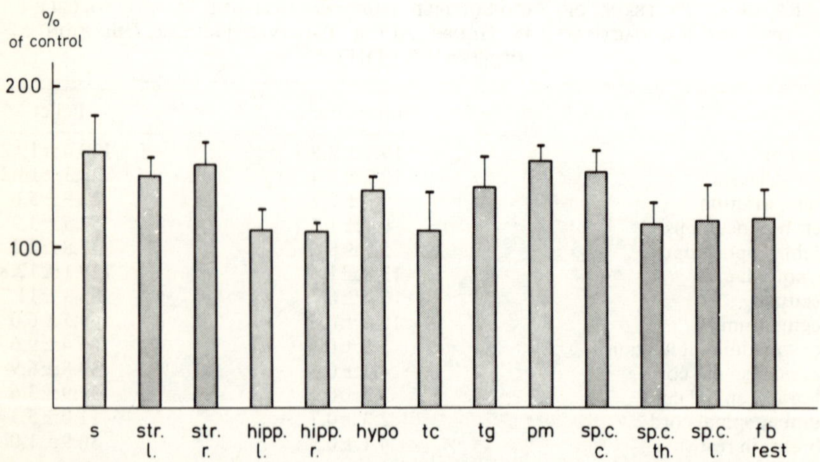

FIGURE 1. Regional pattern of acid-soluble (supernatant) radioactivity 48 hr after intraventricular injection of [1-^{14}C]-5,6-DHT in rats treated with benztropine (60 mg/kg, ip, 45 min before 5,6-DHT).

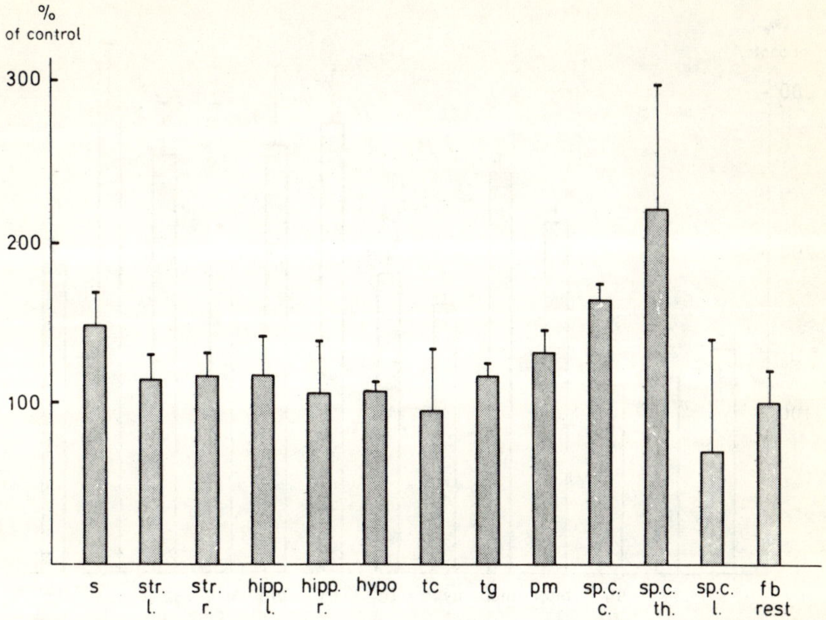

FIGURE 2. Regional pattern of acid-insoluble (pellet) radioactivity 48 hr after intraventricular injection of [1-^{14}C]-5,6-DHT in rats treated with benztropine (cf. legend to FIGURE 1).

highly significantly elevated in nearly all brain regions (FIGURE 3). The supernatant fractions of some regions contained more than twice the activity in the corresponding 48-hr controls. The ratio of soluble to bound radioactivity in individual regions varied considerably.

As expected, PCA-insoluble radioactivity was only moderately decreased in regions of the rat central nervous system known to be rich in catecholamine terminals after administration of [^{14}C]-5,6-DHT in rats that had previously undergone destruction of most of their NA and DA projections.

Similar changes were apparent after injection of [^{14}C]-5,7-DHT, but the decrease in insoluble radioactivity was significant in almost all central nervous system regions, except the septum, striatum, and the lumbar spinal cord (see FIGURE 7). The overwhelming importance of the central serotonin system for uptake of 5,7-DHT was confirmed by the substantial reduction in both PCA-soluble and protein-bound radioactivity in most central nervous systems that occurred after intraventricular injection in animals having much of their serotonin projections destroyed by treatment with DMI-5,7-DHT (cf. FIGURES 5 & 6).

By contrast, the same treatment had less effect on the retention of radioactivity in pellet and supernatant fractions from brain regions of [^{14}C]-5,6-DHT-injected animals.

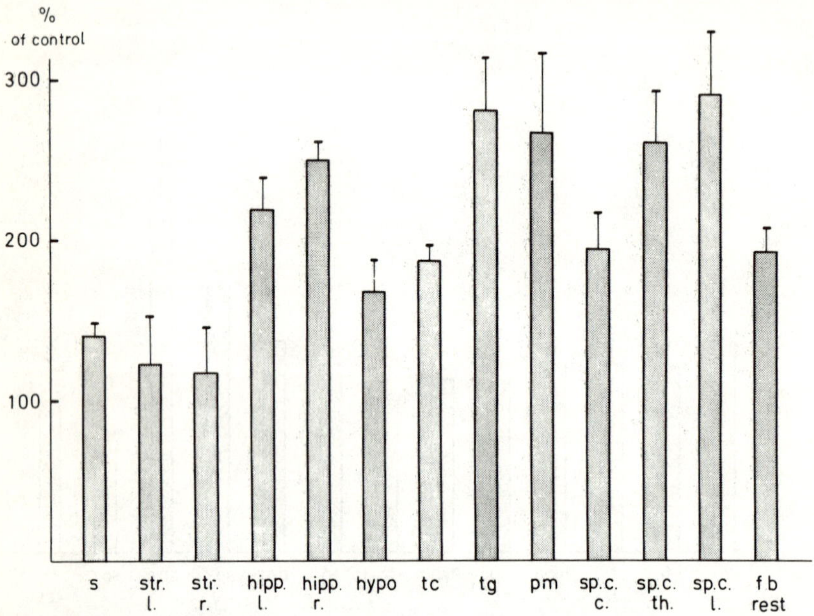

FIGURE 3. Regional pattern of acid-soluble (supernatant) radioactivity 48 hr after intraventricular injection of [1-^{14}C]-5,7-DHT in rats treated with pargyline (60 mg/kg, ip, 45 min before 5,7-DHT).

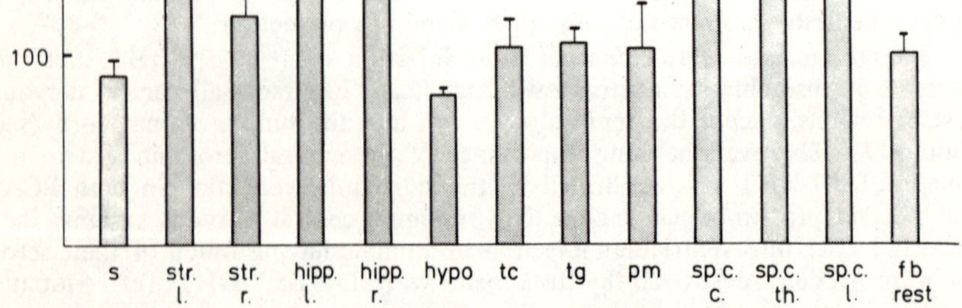

FIGURE 4. Regional pattern of acid-insoluble (pellet) radioactivity 48 hr after intraventricular injection of [1-^{14}C]-5,7-DHT in rats treated with pargyline (cf. legend to FIGURE 3).

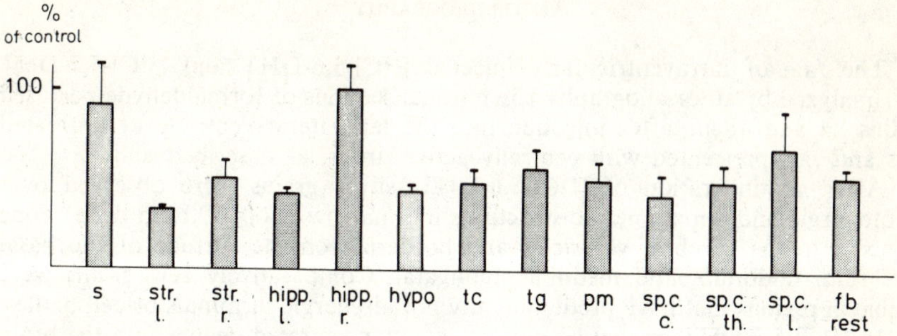

FIGURE 5. Regional pattern of acid-soluble (supernatant) radioactivity after intraventricular injection of [1-^{14}C]-5,7-DHT in rats treated with DMI (25 mg/kg, ip) and 5,7-DHT (120 μg, given intraventricularly, 30 min after DMI), 10 days before administration of the labeled drug.

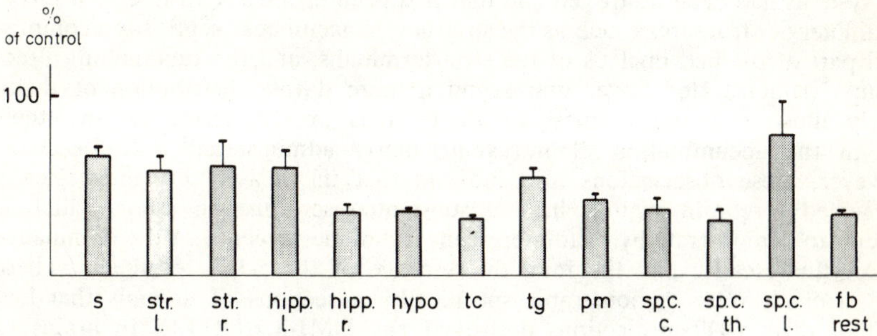

FIGURE 6. Regional pattern of acid-insoluble (pellet) radioactivity after intraventricular injection of [1-^{14}C]-5,7-DHT in rats treated with DMI and 5,7-DHT (cf. legend to FIGURE 5).

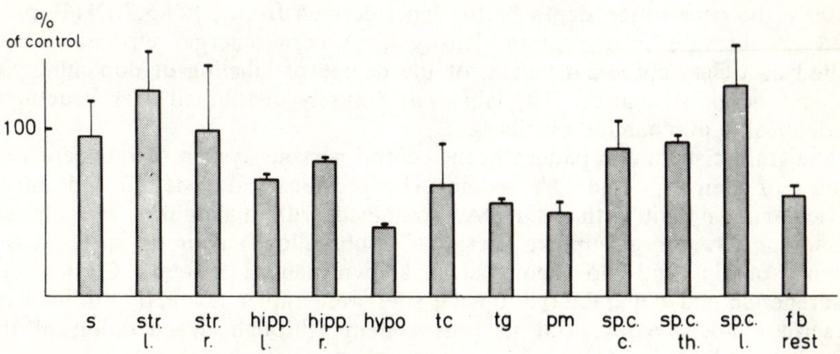

FIGURE 7. Regional pattern of acid-insoluble radioactivity after intraventricular injection of [1-^{14}C]-5,7-DHT in rats treated with 250 μg of 6-OHDA (given intraventricularly and intracisternally, 2 days apart, combined with 150 mg/kg of nialamide, ip), 10 days before [^{14}C]-5,7-DHT.

Autoradiography

The fate of intraventricularly injected [^{14}C]-5,6-DHT and [^{14}C]-5,7-DHT was analyzed by autoradiography † in coronal sections of formaldehyde-perfused brains 24 and 48 hr after injection into the left lateral ventricle of untreated rats and rats pretreated with centrally active drugs, as described above.

After administration of [^{14}C]-5,6-DHT, silver grains were observed over serotonergic and dopaminergic structures in a narrow (400–700 μm deep) zone adjacent to the cerebral ventricles and bordering on the surface of the basal forebrain, midbrain, and medulla oblongata. Comparatively few grains were visible in regions that have predominantly noradrenergic terminals or cell bodies.

In [^{14}C]-5,7-DHT-treated specimens, grains penetrated deeper into the brain parenchyma, both from the ventricles and from the basal cisternae, and there was a much more pronounced labeling of serotonergic than of dopaminergic and of noradrenergic structures throughout the rat central nervous system.

Neither benztropine nor DMI significantly altered the distribution of grains in the rat central nervous system, but in animals in which most central NA and DA systems had been destroyed and that had been injected with [^{14}C]-5,6-DHT, dopaminergic structures (such as the striatum, n. accumbens septi, supracommissural part of the bed nucleus of the stria terminalis, and the tuberculum olfactorium) demonstrated fewer grains and a more diffuse distribution of them, clearly illustrating the richness of these areas in DA terminals and their role in the accumulation of intraventricularly administered [^{14}C]-5,6-DHT. However, these observations also indicate that the heavily dopaminergically innervated forebrain regions have a substantial serotonergic input, which is difficult to demonstrate by histofluorescent techniques because of the dominance of DA fluorescence and the faint fluorescence of the 5-HT terminals. These DA terminals were demonstrable specifically in sections of animals that had most of their 5-HT projections destroyed (by DMI-5,7-DHT). In brains of [^{14}C]-5,7-DHT-injected rats, benztropine and DMI slightly depressed, but did not abolish, the labeling of ventricle-near DA- and NA-containing structures, respectively, but had no effect on the number of grains located over the remaining monoaminergic terminal systems.

The effects of pargyline pretreatment were remarkably different from those of benztropine or DMI. Pargyline caused a highly statistically significant increase in the penetration depth of the label derived from [^{14}C]-5,7-DHT, produced no increase in the grain density over noradrenergic terminals, and resulted in a clear-cut enhancement of the degree of labeling of dopaminergic and serotonergic structures. The label was diffusely distributed over structures not identical to monoaminergic fibers.

The grain distribution pattern in the central nervous system of rats deprived of most of their NA and DA systems (by previous intracisternal and intraventricular treatment with 6-OHDA, combined with nialamide) was almost specific for serotonergic uptake sites; the results allow an almost quantitative mapping of this hitherto incompletely known neuron system. Conversely, noradrenergic and dopaminergic uptake sites were rather distinctly outlined in the serotonin-deprived central nervous system (through degeneration of its projections by DMI-5,7-DHT).

† Details on the distribution of labeled 5,6-DHT and 5,7-DHT in the rat central nervous system will be published elsewhere.

Discussion

The doses of [^{14}C]-5,6-DHT and [^{14}C]-5,7-DHT (75 and 94 μg of free base, respectively) used in the present study are commonly administered to experimental animals for inducing lesions in the central serotonergic fiber systems. Surprisingly, the three psychotropic drugs employed all altered the amounts and regional patterns of radioactivity and the proportion of soluble versus bound radioactivity after injection into the cerebral ventricles, though not to the same extent. Generally, the percentage changes in pellet-bound and PCA-soluble radioactivity caused by these drugs were smaller in [^{14}C]-5,6-DHT-injected than in [^{14}C]-5,7-DHT-injected brains, even though the amount of protein-bound radioactivity after injection of [^{14}C]-5,6-DHT was twice (or more) that retained in most regions after [^{14}C]-5,7-DHT injection. This finding indicates that a major proportion of the [^{14}C]-5,6-DHT-derived, firmly bound radioactivity in the rat central nervous system is resistant to treatment with drugs known to affect the accumulation, half-life, regional distribution, and metabolism of exogenously applied monoamines; this radioactivity is therefore probably unspecifically bound to nonmonoaminergic brain structures. In most rat central nervous system regions subjected to prior destruction of serotonergic fibers (by combined treatment with DMI and 5,7-DHT [9]), the percentage reduction of bound radioactivity after [^{14}C]-5,6-DHT treatment was about 50% of control values, a finding that, again, supports our hypothesis above. Destruction of most of the central NA and DA fibers in the rat central nervous system (by intracisternal and intraventricular injection of 6-OHDA) resulted in smaller decreases of protein-bound radioactivity (20–30% in some brain regions) than did destruction of 5-HT projections, indicating that only a fraction of the total amount of pellet-bound radioactivity is accounted for by catecholamine neurons. This fraction of [^{14}C]-5,6-DHT-derived pellet-bound radioactivity is mainly confined to the dopaminergic forebrain terminal systems, as demonstrated by autoradiograms and as suggested by an earlier observation [3] that unspecific binding of radioactivity from [^{14}C]-5,6-DHT to slices of rat spinal cord was not influenced by prior destruction of noradrenergic fibers. Therefore, about 40% (or more) of the amount of radioactivity retained *in vivo* in the PCA-insoluble (pellet) fractions must be bound to nonmonoaminergic structures. The mechanism responsible for the elevation of bound and soluble radioactivity in the central nervous system of benztropine-pretreated [^{14}C]-5,6-DHT-injected animals is not known. Autoradiography did not provide a clear-cut explanation for this observation, because the distribution of silver grains was not strikingly altered by benztropine.

In agreement with the results obtained by autoradiography, benztropine decreased pellet-bound radioactivity in the striatum and hypothalamus of [^{14}C]-5,7-DHT-injected animals, a finding consistent with the fact that this drug is a moderately potent blocker of DA transport into dopaminergic synaptosomes.[10] Interestingly, this drug appeared to cause a shift of radioactivity to non-dopaminergically innervated forebrain regions (e.g., the hippocampus), thus changing the regional radioactivity distribution pattern. DMI decreased the availability of [^{14}C]-5,7-DHT to the noradrenergic terminal systems, as judged by autoradiography; however, this effect of DMI was not confirmed by quantitative data on radioactivity retention in individual brain regions, indicating that the proportion of radioactivity that, in the untreated rat brain, is incorporated into NA systems is negligible when compared to the amount accumulated by the 5-HT neurons. DMI attenuated but did not completely prevent the labeling of rat central nervous system noradrenergic terminals injected intraventricularly with [^{14}C]-5,7-DHT. This finding suggests that 5,7-DHT must reach a critical intraneuronal concentration to become toxic.

The effects of pargyline pretreatment on brain regional radioactivity distribution patterns after intraventricular injection of 5,7-DHT, which consisted of a dramatic increase in pellet-bound and in soluble radioactivity, suggest that monoamine oxidase is of considerable importance in the clearance of 5,7-DHT from the brain. The unproportional increases in soluble radioactivity when compared to the amount of bound radioactivity may be due to an increased half-life of unmetabolized 5,7-DHT, which cannot be excreted from the brain because of the existence for this amine of barrier mechanisms. Monoamine oxidase activity in the brain appears to restrict the penetration of this *m*-DHT into the brain parenchyma, because, in pargyline-pretreated animals, silver grains were found over ventricle- and surface-distant structures that were devoid of label in nonpretreated rats. If, as proposed here, monoamine oxidase inhibition increases the half-life and availability of unmetabolized 5,7-DHT in the brain, this drug may become toxic in the forebrain DA terminal systems. Studies with α-methyl-5,7-DHT support the latter contention.[11]

Conclusions

After [^{14}C]-5,6-DHT injection, the radioactive label is confined to monoaminergic structures in a narrow zone (400–700 μm) of the ventricle-near and surface-near brain parenchyma (mainly those parts that border on the basal forebrain, midbrain, and hindbrain); structures identifiable as dopaminergic and serotonergic have significantly more grains than do noradrenergic ones. In [^{14}C]-5,7-DHT-treated rats, the radioactive label is visible over serotonergic and, to a lesser degree, noradrenergic and (forebrain) dopaminergic structures. While the regional distribution pattern of the grains is similar after administration of either drug, penetration of [^{14}C]-5,7-DHT into brain matter from the ventricular and subarachnoidal cerebrospinal fluid occurs preferentially. DMI pretreatment slightly suppresses the accumulation of [^{14}C]-5,7-DHT in noradrenergic structures, and benztropine decreases the labeling of forebrain dopaminergic structures, without much effect on the extent of labeling of serotonergic fibers.

After paragyline pretreatment, penetration of label from [^{14}C]-5,7-DHT into the brain parenchyma is greatly enhanced, but the degree of labeling of individual 5-HT, NA, and DA fibers is similar to that in controls. In animals with most of their central 5-HT projections destroyed, grains were found to be more numerous over central catecholaminergic fiber systems, whereas intraventricular injection of [^{14}C]-5,7-DHT results in an almost selective labeling of 5-HT axons and terminals in rats that had been subjected to measures that cause degeneration of NA and DA fibers. Other than a slight reduction of the labeling intensity of forebrain DA terminals by benztropine, the drugs utilized had little effect on the distribution of radioactive label derived from [^{14}C]-5,6-DHT in the brain. Measurements of PCA-soluble and pellet-bound radioactivity in 13 central nervous system regions of rats injected intraventricularly with 1.3×10^6 dpm of [^{14}C]-5,6-DHT or [^{14}C]-5,7-DHT, and subjected to pretreatment with benztropine, pargyline, DMI, 6-OHDA (intraventricular or intracisternal administration, combined with nialamide), or DMI-5,7-DHT, confirmed the autoradiographic data.

References

1. Evetts, K. D. & L. L. Iversen. 1970. Effects of protryptiline on the depletion of catecholamines induced by 6-hydroxydopamine in the brain of the rat. J. Pharm. Pharmacol. **22:** 540–542.
2. Breese, G. R. & T. D. Traylor. 1971. Depletion of brain noradrenaline and dopamine by 6-hydroxydopamine. Brit. J. Pharmacol. **42:** 88–99.
3. Björklund, A., A. S. Horn, H. G. Baumgarten, A. Nobin & H. G. Schlossberger. 1975. Neurotoxicity of hydroxylated tryptamines: structure-activity relationships. 2. In vitro studies on monamine uptake inhibition and uptake impairment. Acta. Physiol. Scand. Suppl. **429:** 29–60.
4. Breese, G. R. & B. R. Cooper. 1975. Behavioral and biochemical interactions of 5,7-dihydroxytryptamine with various drugs when administered intracisternally to adult and developing rats. Brain Res. **98:** 517–527.
5. Horn, A. S., H. G. Baumgarten & H. G. Schlossberger. 1973. Inhibition of the uptake of 5-hydroxytryptamine, noradrenaline and dopamine into rat brain homogenates by various hydroxylated tryptamines. J. Neurochem. **21:** 233–236.
6. Baumgarten, H. G., A. Björklund, A. Nobin, E. Rosengren & H. G. Schlossberger. 1975. Neurotoxicity of hydroxylated tryptamines: structure-activity relationships. 1. Long-term effects on monamine content and fluorescence morphology of central monamine neurons. Acta Physiol. Scand. Suppl. **429:** 1–27.
7. Baumgarten, H. G., A. Björklund & D. F. Bogdanski. 1975. Similarities and differences in the mode of action of 6-hydroxydopamine and neurotoxic indoleamines. *In* Chemical Tools in Catecholamine Research. G. Jonsson, T. Malmfors & Ch. Sachs, Eds. : 59–66. North Holland Publishing Company. Amsterdam, The Netherlands.
8. Rotman, A., J. W. Daly & C. R. Creveling. 1976. Oxygen-dependent reaction of 6-hydroxydopamine, 5,6-dihydroxytryptamine, and related compounds with proteins in vitro: a model for cytotoxicity. Mol. Pharmacol. **12:** 887–899.
9. Björklund, A., H. G. Baumgarten & A. Rensch. 1975. 5,7-Dihydroxytryptamine: improvement of its selectivity for serotonin neurons in the CNS by pretreatment with desipramine. J. Neurochem. **24:** 833–835.
10. Horn, A. S. 1976. Characteristics of transport in dopaminergic neurons. *In* The Mechanism of Neuronal and Extraneuronal Transport of Catecholamines. D. M. Paton, Ed. : 195–214. The Raven Press. New York, N.Y.
11. Baumgarten, H. G., H. P. Klemm, L. Lachenmayer, A. Björklund, W. Lovenberg & H. G. Schlossberger. This monograph.

Discussion

Question: Why is there an increase in the amount of radioactively labeled 5,7-DHT in the pellet and supernatant fractions prepared from the caudal spinal cord region?

Dr. Baumgarten: I believe that it depends on the treatment used. One possibility is that the radioactively labeled 5,7-DHT molecules diffuse rather readily away from the injection site, and this alone may cause a time-dependent increase in pellet-bound radioactivity in the lower spinal cord. Another possibility is that part of the compound is actively transported down to the lower spinal cord region via an intact, axoplasmic transport system or via the caudal flow of cerebrospinal fluid. This transported material might be 5,7-DHT or a metabolite (possibly 5,7-dihydroxyindoleacetic acid) that is formed enzymatically via the action of monoamine oxidase in more rostrally located brain regions. Finally, it is well known that the caudal portions of the spinal cord have many more serotonin-containing nerve terminals than do more rostral portions. Perhaps the increased accumulation of 5,7-DHT in the lumbar spinal cord merely reflects differences in the density of serotonin (and 5,7-DHT) -binding sites.

DR. M. DA PRADA: We have seen similar differences in the accumulation of 5,7-DHT in the caudate nucleus. Do you think that this compound also lesions dopaminergic nerve terminals in this brain region?

DR. BAUMGARTEN: When we first started our work with 5,7-DHT, we thought that it might be neurotoxically specific for serotonin- and norepinephrine-containing nerves. However, the specificity of the drug appears to be determined at least in part by the amount that is administered. When intraventricular doses greater than 100 μg are administered, we see discrete lesions in some dopaminergic nerve tracts; even greater damage to these neurons occurs when the dose of 5,7-DHT exceeds 150 μg. Dopaminergic nerve terminal destruction then occurs in the caudate nucleus, nucleus accumbens, and in the olfactory tubercles.

In this regard, we find that if we use a congener of 5,7-DHT that is monoamine oxidase resistant, or if we inject animals with 5,7-DHT after pretreating them with a monoamine oxidase inhibitor, the neurotoxic effect of the compound on dopamine neurons is increased, whereas its cytotoxic effect on norepinephrine-containing neurons is decreased. On the basis of this evidence, we find that Dr. Breese's method of combining 5,7-DHT injections with a monoamine oxidase inhibitor pretreatment does not really fulfill the criteria for making 5,7-DHT any more specific than it might otherwise be when it is administered alone.

It should be noted that we and our colleagues in Lund find that animals pretreated with desmethylimipramine and then given 5,7-DHT appear to have less damage to caudate nucleus dopaminergic nerves than do animals that receive 5,7-DHT alone. Hence, I would issue a word of caution to investigators who use doses of 5,7-DHT greater than 100 μg of free base with or without various drug pretreatments: they should always check for the specificity of the drug treatment by using neuroanatomic methods. Otherwise, one has no way to be absolutely certain that physiologic or behavioral changes associated with the drug treatment result from a specific action of 5,7-DHT on serotonergic neurons alone.

DR. DA PRADA: We also have evidence that injections of 75 μg of 5,7-DHT lesion dopaminergic nerves. Although dopamine levels appear normal after this drug treatment, homovanillic acid concentrations are reduced by approximately 30%.

DR. BAUMGARTEN: Our neuroanatomic evidence indicates that much less than 1% of the dopaminergic nerve terminals in the caudate nucleus, nucleus accumbens, and olfactory tubercles is destroyed after treatment with 75 μg of 5,7-DHT. However, once the dose is increased above 100 μg, greater nerve terminal damage occurs.

CEREBRAL METABOLISM OF SOME SEROTONIN DEPLETORS *

E. M. Gál and A. D. Sherman

Department of Psychiatry
College of Medicine
University of Iowa
Iowa City, Iowa 52242

A single injection of *p*-chloro-α-methylphenylethylamine [*p*-chloroamphetamine (*p*-CA)] or of *N*-ethyl-α-methyl-*m*-(trifluoromethyl)phenylethylamine [fenfluramine (FE)] brings about a long-term reduction in the levels of 5-hydroxytryptamine (5-HT) and 5-hydroxyindoleacetic acid (5-HIAA) in the brain.[1-3] This action is related to cellular damage[4] and extensive impairment of synaptosomal uptake of 5-HT.[5] The change in the level of cerebral 5-hydroxyindoles as a consequence of the administration of *p*-CA is a reflection of a decrease in the synthesis of 5-hydroxytryptophan (5-HTP). This decrease is due to an almost 40% reduction in the enzyme tryptophan 5-hydroxylase (E.C. 1.14.16.4), as reflected by a time-dependent decrement in its V_{max} and turnover rate.[5] These changes are consequences of the neurotoxic nature of *p*-CA. The toxic levels of several analogs and homologs of *p*-CA and FE[6] complicated a single rational explanation, such as epoxide intermediates, for the neurotoxic effects of *p*-CA. In their long-term effects, *p*-CA and FE, however, resemble 5,6-dihydroxytryptamine,[7] a cytotoxic agent. Therefore, a hypothetical scheme that involves indoloid intermediates for the synthesis of toxic metabolites from *p*-CA was proposed.[5] To explore the rationale of this scheme, a study of the cerebral metabolism of *p*-CA and FE was undertaken.[8, 9] The scope of this paper, of necessity, is confined to the description of the nature of some of the cerebral metabolites of *p*-CA and FE and their implication as to the neurotoxicity of their parent drugs.

Materials and Methods

Male albino rats (Sprague-Dawley strain) maintained on a Purina Chow diet, with water ad lib., and weighing 250–400 g, were used throughout these studies. Materials and relevant methodology were described in detail in previous reports.[8-11]

Results

After intraventricular injection of various amounts of [^{3}H]-*p*-CA and [^{3}H]FE, it was observed that 0.5–0.6 mg of either drug was the minimal amount required to produce neurotoxic effects beginning at 4 hr, that is, a decrease in cerebral tryptophan 5-hydroxylase activity and 5-HT levels.

* Supported by National Institute of Mental Health Grant MH25711.

Inspection of the plots of the cerebral efflux of [^{3}H]-*p*-CA or [^{3}H]FE revealed them to be biphasic, irrespective of the administration route of the drugs. The rate of efflux, however, was dependent on the mode of injection. A comparison of the two administration routes indicates (TABLE 1) that only intraventricular injection of [^{3}H]-*p*-CA or [^{3}H]FE yielded efficient recovery of its cerebral metabolites. The values from intraperitoneal injection of the drugs confirmed that the appearance of the metabolites in brain was not due to "one-pass" recycling from the periphery.

Recovery of *p*-CA, FE, two major metabolites of *p*-CA, and three of FE by high-pressure liquid chromatography (HPLC) permitted their ultimate quantitation by electron capture-gas chromatography (EC-GC) and their identification by mass spectrometry.[8, 9, 11] The metabolites of *p*-CA that were recoverable from brain at 4 hr after its intraventricular injection were *p*-chloronorephedrine [PCNE, *p*-chloro-β-hydroxyamphetamine (P_2)] and 3,4-dimethoxyamphetamine [3,4-DMA (P_3)]. Their elimination was also biphasic. Between 0 and 12 hr after intraventricular injection, cerebral *p*-CA had a $t_{1/2}$ of 1.5 hr; the values for P_2 and P_3 were 1.9 and 2.8 hr, respectively, while beyond 12 hr, the $t_{1/2}$ of *p*-CA was 3.5 days, and that of each of its metabolites was 3.6 hr. During the same period, FE and norfenfluramine (NFE) had a $t_{1/2}$ of 2 hr; the $t_{1/2}$ of *N*-acetylnorfenfluramine (FE_3) was 5.9 hr. The $t_{1/2}$ of FE and NFE was 4.5 days during the slow phase of efflux, while that of FE_3 was 2.0 days. *m*-Trifluoromethylbenzoylglycine (FE_2) was only detectable at 6 hr. It was not present either at 4 or 24 hr.

After intraperitoneal administration of *p*-CA (2.5 mg/kg), its cerebral $t_{1/2}$ was 13 hr for the fast and 3.0 days for the slow phases, while FE (1.75 mg/kg) had a $t_{1/2}$ of 11 hr for the fast and 16.4 hr for the slow phases of elimination.

It was noted that following intraventricular injection of the parent compounds, the maximal accumulation of the cerebral metabolites from *p*-CA was at 4 hr and from FE at 6 hr (TABLE 2). No cerebral metabolites were detectable at any time from 4 hr on after intraperitoneal administration of *p*-CA (2.5 mg/kg) or FE (1.75 mg/kg). Intraventricular injection of *p*-CA and FE resulted in the excretion of the above metabolites in the urine. In the liver, the metabolites of *p*-CA, but not those of FE, were detectable. Additionally, the appearance of 4-hydroxy-3-chloroamphetamine in the urine[12] was confirmed.

TABLE 1

INFLUENCE OF ADMINISTRATION ROUTE ON CEREBRAL LEVELS OF *p*-CA AND FE

	p-CA Recovered (μg/g)		FE Recovered * (μg/g)	
Time (hr)	Intraperitoneal (2.5 mg/kg)	Intraventricular (600 μg)	Intraperitoneal (1.75 mg/kg)	Intraventricular (500 μg)
4	0.08	36.0	0.21	1.38
6	0.06	1.45	0.18	0.71
12	0.04	0.34	0.01	0.58

* Represents a mixture of recovered FE and NFE.

TABLE 2

ACCUMULATION OF METABOLITES AFTER INTRAVENTRICULAR INJECTION OF *p*-CA OR FE

Time (hr)	From *p*-CA (600 μg)		From FE (500 μg)		
	P_2	P_3	NFE	FE_2	FE_3
4	11.8±1.3 *	1.0±0.17	1.30±0.15	ND †	ND
6	0.22±0.02	0.20±0.03	0.70±0.05	0.05±0.005	0.005±0.001
24	0.15±0.01	0.15±0.02	0.45±0.05	ND	0.01±0.001
72	0.02±0.004	0.03±0.006	0.33±0.05	ND	0.02±0.002

* All values are expressed in micrograms per gram (wet weight) of brain tissue ± SE and are averages from four rats.

† Not detectable.

Iprindole®[13, 14] and Lilly 110,140[15] were demonstrated to be effective agents in modifying the metabolism and uptake of amphetamines. In our experiments, Iprindole (10 mg/kg) or Lilly 110,140 (10 mg/kg) was administered intraperitoneally 1 hr before injection of *p*-CA or FE. Pretreatment with Iprindole led to the disappearance of cerebral and hepatic P_2 and to lowered levels of *p*-CA. Iprindole did not affect dealkylation of FE to NFE; it enhanced the *N*-acetylation of NFE and the excretion of *m*-CF_3-benzoylglycine (FE_2). As expected, pretreatment of the rats with Lilly 110,140 resulted in a 90% reduction in the cerebral levels of *p*-CA and its metabolites. However, it had no effect on the metabolism of *p*-CA in liver nor did it influence the excretion rate of *p*-CA and its metabolites. In animals that were given Lilly 110,140 and FE, FE_2 and FE_3 were not detectable in brain or liver. The rate of excretion of FE_2 and FE_3 was also diminished during the first 24 hr.

Subcellular fractionation and density gradient separation of the crude mitochondria from brains of rats given [^{3}H]-*p*-CA or [^{3}H]FE one day earlier revealed marked differences in their distribution patterns (TABLES 3 & 4). Interestingly, the bulk of *p*-CA and its metabolites, except for 3,4-DMA, were in the nucleus, while most of the NFE and FE_3 appeared in the supernatant fraction. Again, on density gradient fractionation, the metabolites of *p*-CA and of NFE were mostly in the mitochondria and the synaptosomal layer, respectively.

It has frequently been suggested that an epoxide of *p*-CA might be the primary agent responsible for neurotoxicity. It was reasoned that simultaneous administration of reduced glutathione (GSH) with *p*-CA to rats might diminish, or even counteract, the toxicity of the drug. Glutathione is known to bind covalently to an epoxide, thereby minimizing the availability of epoxide for attachment to cellular structures. Our results indicate (TABLE 5) that neither intraventricular nor intraperitoneal injection of GSH had a statistically significant effect on 5-HT levels in brain, which was the index in assessing the neurotoxicity of *p*-CA. No experiments were performed with FE in combination with GSH, because there is as yet no evidence that would allow for a route to phenolic derivatives of FE *in vivo*. As indicated above, the detoxication pathway for FE is substantially different from that for *p*-CA.

It was noted that on intraventricular injection of about 2.5 μmol of [^{3}H]FE, 1 nmol of [^{3}H]trifluoromethylbenzoylglycine (FE_2) could be recovered from

TABLE 3

SUBCELLULAR DISTRIBUTION OF METABOLITES 24 HOURS AFTER INTRAVENTRICULAR INJECTION OF [^{3}H]-*p*-CA *

Fraction	*p*-CA Recovery (%)	PCNE Recovery (%)	3,4-DMA Recovery (%)
Homogenate †	100.0	100.0	100.0
Nuclear †	42.9	62.0	17.9
Microsomal †	3.4	0.0	29.8
Supernatant †	29.3	0.0	17.9
Myelin ‡	0.0	0.0	0.0
Synaptosomal ‡	2.3	3.1	2.4
Mitochondrial ‡	1.1	7.0	6.0
Total	79.0	72.1	73.8

* Six hundred micrograms of *p*-CA that contained 27.5 μCi of [^{3}H]-*p*-CA were administered intraventricularly 24 hr before sacrifice.

† Obtained directly.

‡ Obtained by density gradient centrifugation of the crude mitochrondrial pellet.

one rat brain. Intraperitoneal administration of FE did not lead to the formation of FE_2 in brain nor did synthetic hippurates reach this organ when administered intraperitoneally. Confirmation of cerebral synthesis of FE_2 and benzoylglycine (hippuric acid) was undertaken by assaying for their enzymatic synthesis *in vitro*. Incubation of [^{14}C]benzoic acid or *m*-CF_3-benzoic acid and [^{14}C]glycine with a 1-ml aliquot of a postmitochondrial supernatant for 1 hr resulted in the synthesis of FE_2 or hippuric acid [9] (TABLE 6). The identity of FE_2 was also established by its mass spectrum.

TABLE 4

SUBCELLULAR DISTRIBUTION OF METABOLITES 24 HOURS AFTER INTRAVENTRICULAR INJECTION OF [^{3}H]FE *

Fraction	NFE Recovery (%)	FE_2 Recovery (%)	FE_3 Recovery (%)
Homogenate †	100.0	ND ‡	100.0
Nuclear †	8.4	ND	ND
Microsomal †	6.0	ND	ND
Supernatant †	66.3	ND	60.0
Myelin §	ND	ND	ND
Synaptosomal §	10.8	ND	20.0
Mitochondrial §	3.6	ND	6.0
Total	95.1		86.0

* Five hundred micrograms of FE that contained 27.5 μCi of [^{3}H]FE were administered intraventricularly 24 hr before sacrifice.

† Obtained directly.

‡ Not detectable.

§ Obtained by density gradient centrifugation of the crude mitochondrial pellet.

TABLE 5

EFFECT OF REDUCED GLUTATHIONE (GSH) ON CEREBRAL 5-HT IN *p*-CA-TREATED RATS *

	Brain Stem 5-HT	Telencephalon 5-HT
Intraventricular		
GSH	6.11±0.29 †	2.63±0.08
GSH+*p*-CA	4.76±0.11	1.80±0.16
p-CA	4.38±0.31	1.26±0.33
Intraperitoneal		
GSH	6.14±0.24	2.77±0.13
GSH+*p*-CA	4.16±0.19	1.77±0.14
p-CA	4.08±0.07	1.57±0.11

* Animals received the same amount of *p*-CA, GSH, or both: 600 μg intraventricularly or 5 mg/kg intraperitoneally. 5-Hydroxytryptamine was assayed by optical reaction 24 hr after injection.

† Nanomoles per gram, average of four rats.

DISCUSSION

In general, *p*-CA and FE are administered peripherally; their intraventricular injection, nevertheless, allowed for demonstration of an active cerebral metabolism of these compounds, along with recoveries of the labeled metabolites of greater specific activities. By use of improved methodology, namely, a combination of HPLC and gas chromatography-mass spectrometry (GC-MS) the detection and isolation of several hitherto unidentified metabolites of *p*-CA or FE from rat brain have been achieved.

Two cerebral metabolites that appeared after intraventricular administration of *p*-CA were P_2 and P_3. P-chloronorephedrine from the brain was identical

TABLE 6

ENZYMATIC SYNTHESIS OF HIPPURATES IN SUPERNATANTS (12,000*g*) FROM RAT BRAIN HOMOGENATES

Substrates	Amount: Micrograms	Amount: Microcuries	Hippurates Recovered: Disintegrations per Minute	Hippurates Recovered: Nanograms	Hippurates Recovered: Conversion (%)
Benzoic acid and [14C]glycine (2)*	0.200 6.550 †	0.1	783	49.0	0.36
F_3C-benzoic acid and [14C]glycine (4)	0.150 6.550 †	0.1	762	98.0	0.35
[14C]Benzoic acid (2)	0.274 5.80 †	0.13	5540	7.5	1.84

* Number of experiments.

† The glycine pool in the supernatant was 5.8 μg/ml.

to a synthetic sample [9] (obtained from Dr. H. E. Smith, Vanderbilt University, Nashville, Tennessee) and was devoid of neurotoxic effect.[16] The other cerebral metabolite, 3,4-DMA, was identical to a synthetic sample (from Dr. A. T. Shulgin, Lafayette, California). Their trimethylsilyl derivatives were compared by GC-MS. Furthermore, their underivatized or derivatized samples from brain were isographic by EC-GC [9] with the synthetic 3,4-DMA but not with 2,3-, 2,4-, 2,5-, or 3,5-dimethoxyamphetamine.

The absolute level of 3,4-DMA in that brain attained 2 μg in 4 hr, with about 10% of it still persisting after 24 hr. At present, we have no information on the neurotoxicity of intraventricularly administered 3,4-DMA. As a psychotoxic agent, 3,4-DMA is relatively weak in rats and inactive in man.[17]

Both metabolites of *p*-CA appeared in the brain at the onset of detectable reduction of cerebral 5-HT and persisted in the central nervous system for several days. During the rapid phase of efflux, *p*-CA and its metabolites left the central nervous system at about the same rate; thereafter, the metabolites were eliminated at a much slower rate than was *p*-CA. This difference in the $t_{1/2}$ values between *p*-CA and its metabolites may be interpreted as a phenomenon due to either damage to specialized cells responsible for synthesis of these metabolites, which were therefore no longer metabolizing *p*-CA, or to the fact that *p*-CA was more structure bound during the first day. The recovery of metabolites from different cerebral regions agrees somewhat with the idea of cellular specialization, although transport of the metabolites from the site of synthesis to other regions cannot be ignored.

After administration of Iprindole, there is no P_2 detectable in either brain or liver, whereas 3,4-DMA disappears from the liver but not from the brain. This observation also suggests that after intraventricular injection of *p*-CA, the cerebral synthesis of 3,4-DMA continued independent of the peripheral organs. Conversely, the action of Lilly 110,140 led to the disappearance of 3,4-DMA from brain but not from the liver. The above observations are supportive of the cerebral synthesis of 3,4-DMA, in that the appearance of this compound in the brain occurred much earlier than in other organs of rats given Iprindole. This finding becomes particularly significant when one considers that by 4 hr, 90% of the intraventricularly injected *p*-CA was already in the peripheral organs; yet, the synthesis of 3,4-DMA was only detectable in the brain. In the brains of rats given Lilly 110,140, traces of the metabolites were detected only during the first 4 hr. This observation was indicative of the presence of intraventricular *p*-CA. Between 12 and 24 hr, no metabolites were present, a finding in agreement with the observation that unless *p*-CA is metabolized during the first 6 hr, no neurotoxicity will occur. Under similar experimental conditions, the hepatic and urinary levels of 3,4-DMA were not affected by Lilly 110,140. Interestingly, simultaneous injection of Iprindole and Lilly 110,140 increased the cerebral level of 3,4-DMA beyond that produced by either agent separately.

The metabolic conversion of *p*-CA to 3,4-DMA in rat brain is, to our knowledge, a hitherto unreported mode of methylation for this organ. At this time, we do not know the mechanism of this reaction. It is likely that the precursors of 3,4-DMA from the metabolism of *p*-CA could appear during the first 4 hr after injection of *p*-CA. Because of our primary emphasis on the course of the metabolic pattern of *p*-CA between the time of onset of its effect on cerebral 5-hydroxytryptamine levels and the full development of neurohistologic changes, there is an information gap concerning the metabolic changes of *p*-CA during the first 4 hr. The appearance of 3,4-DMA as a cerebral metabolite lends some support to the hypothesis [5] of the biosynthesis of a neurotoxic metabolite.

This hypothesis postulated the synthesis of a neurotoxic 5,6-dihydroxyindole or 2,3-dihydro-5,6-dihydroxyindole analogous to that proposed for the oxidation of L-Dopa by tyrosinase [21] or controversially invoked in the synthesis of melanins [22] or envisaged for the nonenzymatic formation from 2,5-dimethoxy-amphetamines.[23]

This hypothesis was advanced because of the observation that the long-term neurotoxicity produced by a single intraventricular or intraperitoneal injection of *p*-CA could not be altered either by administration of SKF 525A [19, 20] or by GSH.[9] Reduced glutathione, through the enzyme GSH transferase, is a biochemical trap for epoxides. This evidence, along with the lack of neurotoxic effects after administration of analogs or homologs of *p*-CA,[6] makes an epoxide intermediate of *p*-CA an unlikely neurotoxic agent.

The neurotoxicity produced by *p*-CA and FE is similar in many respects; yet, the duration of their effects and the degree of the changes induced in the various brain nuclei [24] might be an expression of the difference in their metabolism.

Fenfluramine, unlike *p*-CA, displayed a rapid efflux after intraventricular administration, with 99% of it recoverable from the periphery within 4 hr. Fenfluramine rapidly underwent dealkylation to NFE in the brain.[10] The cerebral metabolism of FE was rather slow; only 7% of it was metabolized to FE_2 and FE_3. The cerebral content of FE_3 increased over the first 3 days, suggesting either its continuous storage or synthesis. This increase of FE_3 was not noted in the liver either after intraperitoneal or intraventricular injection of NFE nor was it present in the brain when FE or NFE (1.75 mg/kg) was given intraperitoneally. Again, intraperitoneal pretreatment of the rats with Iprindole increased cerebral *N*-acetylation of NFE, whereas Lilly 110,140 decreased it. Iprindole appeared to be stimulatory to *N*-acetylation, because there was no increase in cerebral NFE levels. Of all of the cerebral regions, the neocortex had the highest level of FE_3, and it was also most affected by FE, as reflected by changes of its 5-HT content.[6]

Recovery of the other cerebral metabolite of FE or NFE, FE_2, from the brain was unexpected, because the prevailing concept indicates that the liver and kidney are the sites of synthesis of hippuric acid. Our experiments *in vitro*, with a postmitochondrial supernatant of rat brain homogenates, confirmed the enzymatic nature of this synthesis in the central nervous system.

The various degradative steps that convert FE or NFE to *m*-CF_3-benzoic acid in the brain are as yet not demonstrated. There is evidence, however, that FE, a secondary amine, can undergo oxidation to its *N*-oxide *in vivo* and *in vitro*.[25] In the presence of a supernatant (10,000*g*) of guinea pig liver, FE was metabolized to several distinct products, as identified by GC.[25] There is some question as to whether some of the products resulted from nonenzymatic degradation of FE. Nonetheless, the cerebral conversion of FE to *m*-CF_3-benzoic acid *in vivo* via its *N*-oxide and *m*-CF_3-benzylmethylketone is a distinct possibility, particularly since aromatic ketones will oxidize to benzoic acid if the side chain contains an odd number of carbon atoms. We noted that about 0.5% of the label, other than ^3H_2O, remained in the protein fraction after intraventricular administration of [^{3}H]FE or [^{3}H]NFE. It is conceivable that

R—CH_2COCH_3 could have formed a Schiff base with the amino groups of the protein.

In conclusion, a brief comparison of the differences noted in the distribution and metabolism of *p*-CA and FE might be instructive. Based on our data,[6, 7, 9-11] the differences can be summarized as follows:

The elimination rate of FE was 0.48 hr; for *p*-CA, it was 1.2 hr.

[^{3}H]Fenfluramine exhibited no major regional localization in the brain, whereas *p*-CA was concentrated in the subcortical area for several days.

The rate of metabolic conversion of FE was about 25% of that of *p*-CA.

[^{3}H]Fenfluramine converts into NFE, which partially metabolizes to *N*-acetylnorfenfluramine (FE_3) and *m*-CF_3-hippurate (FE_2), while *p*-CA is metabolized to *p*-chloro-β-hydroxyamphetamine (P_2) and 3,4-dimethoxyamphetamine (P_3).

The most significant difference between *p*-CA and FE is the subcellular distribution patterns of their metabolites. The microsomal fraction, for example, was devoid of detectable metabolites of FE, whereas 30% of a metabolite of *p*-CA was present in this fraction. After density gradient fractionation, 20% of *N*-acetylnorfenfluramine (FE_3) appeared in the synaptosomal fraction, compared to 3% of 3,4-dimethoxyamphetamine from *p*-CA.

[^{3}H]Fenfluramine did not affect tryptophan 5-hydroxylase activity in the B-9 nucleus, whereas *p*-CA had a marked effect.[24]

Finally, it is to be emphasized that there is as yet no evidence that any of the metabolites here reported is responsible for the neurotoxicity of FE.

References

1. Costa, E., A. Groppetti & A. Revuelta. 1971. Brit. J. Pharmacol. **41:** 57–64.
2. Sanders-Bush, E., J. A. Bushing & F. Sulser. 1972. Eur. J. Pharmacol. **20:** 385–388.
3. Sanders-Bush, E., J. A. Bushing & F. Sulser. 1975. J. Pharmacol. Exp. Ther. **192:** 33–41.
4. Harvey, J. A., S. E. McMaster & L. M. Yunger. 1975. Science **187:** 841–843.
5. Gál, E. M., P. A. Christiansen & L. M. Yunger. 1975. Neuropharmacology **14:** 31–39.
6. Sherman, A. D., E. M. Gál, R. W. Fuller & B. B. Molloy. 1975. Neuropharmacology **14:** 733–737.
7. Baumgarten, H. G. & L. Lachenmayer. 1972. Brain Res. **38:** 228–232.
8. Sherman, A. D., W. C. Hsiao & E. M. Gál. 1977. Neuropharmacology **16:** 17–24.
9. Sherman, A. D. & E. M. Gál. 1977. Neuropharmacology **16:** 309–315.
10. Sherman, A. D. & E. M. Gál. 1975. Psychopharmacol. Commun. **1:** 261–273.
11. Sherman, A. D. & E. M. Gál. 1977. Psychopharmacol. Commun. **2:** 409–427.
12. Parli, C. J. & B. Schmidt. 1975. Res. Commun. Chem. Pathol. Pharmacol. **10:** 601–605.
13. Freeman, J. J. & F. Sulser. 1972. J. Pharmacol. Exp. Ther. **183:** 307–315.
14. Fuller, R. W. & J. C. Baker. 1974. J. Pharm. Pharmacol. **26:** 912–914.
15. Wong, D. T., F. P. Bymaster, J. S. Horng & B. B. Molloy. 1974. Proc. Fed. Amer. Soc. Exp. Biol. **33:** 255.
16. Smith, H. E., E. P. Burrows, J. D. Miano, C. D. Mount, E. Sanders-Bush & F. Sulser. 1974. J. Med. Chem. **17:** 416–420.

17. SHULGIN, A. T., T. SARGENT & C. NARANJO. 1969. Nature (London) **221:** 537–540.
18. FULLER, R. W., K. W. PERRY & B. B. MOLLOY. 1975. Eur. J. Pharmacol. **33:** 119–124.
19. SANDERS-BUSH, E., J. A. BUSHING & F. SULSER. 1972. Biochem. Pharmacol. **21:** 1501–1510.
20. FULLER, R. W., H. D. SNODDY, B. ROUSH & B. B. MOLLOY. 1973. Neuropharmacology **12:** 33–42.
21. MASON, H. S. 1948. J. Biol. Chem. **172:** 83–99.
22. ZWEIG, J. S. & N. CASTAGNOLI, JR. 1974. J. Med. Chem. **17:** 747–749.
23. ROGERS, A. D. & G. CURZON. 1975. J. Neurochem. **24:** 1123–1129.
24. NECKERS, L. M., L. BERTILSSON & E. COSTA. 1976. Neurochem. Res. **1:** 29–35.
25. BECKETT, A. H., R. T. COUTTS & F. A. OGUNBONA. 1973. J. Pharm. Pharmacol. **25:** 190–192.

DISCUSSION

DR. E. SANDERS-BUSH: As I think I understand your argument, Dr. Gál, you feel that the neurotoxicity of *p*-chloroamphetamine is not mediated by an action of its metabolites. Along these lines, were you able to account for all of the radioactivity in the brain after injections of labeled *p*-chloroamphetamine as due to the presence of this compound and its two major metabolites?

DR. GÁL: We do not account for all of the radioactivity in brain as due to *p*-chloroamphetamine and its metabolites. However, we do not yet know about all of the possible metabolites of the parent compound, nor do we know if they are at all neurotoxic. However, I doubt that many of them are cytotoxic. The relative recoveries for *p*-chloroamphetamine were published in our recent article in *Neuropharmacology*.

DR. SANDERS-BUSH: Do you find any evidence of covalent binding of *p*-chloroamphetamine or fenfluramine to brain tissue?

DR. GÁL: You will notice that we started our analysis of the effects of these compounds at 4 hr, at a time when the neurotoxic effects of the drugs are first manifested. This time point is too late to determine clearly if there is any covalent binding of the compounds. However, we have recently found that there is binding of some ketone-like metabolites of fenfluramine to brain proteins. We also see some protein-bound radioactivity after treatment with radiolabeled *p*-chloroamphetamine, but we know relatively little about the nature of this binding.

QUESTION: Do you have any evidence of gross neurotoxic effects in animals treated with *p*-chloroamphetamine or fenfluramine?

DR. GÁL: I am sure that Dr. Harvey will cover this aspect in his paper; however, I might say that he and others have found histologic and neurochemical evidence of neurotoxicity in various brain regions after administration of these compounds. I would be wary of using fenfluramine as a human anorectic agent until it is proven from postmortem human biopsy material that these drugs do not cause brain lesions or neurotoxic changes.

STRUCTURE-ACTIVITY RELATIONSHIPS OF SEROTONIN NEUROTOXINS: EFFECTS ON SEROTONIN UPTAKE

A. S. Horn

Department of Pharmacy
University of Groningen
Groningen, The Netherlands

The selectivity of action of the neurotoxic transmitter analogs is thought to depend on their active uptake by and interaction with the different monoamine-containing neurons.[1-4] Previous experiments with a restricted number of toxic and nontoxic serotonin analogs have shown that their relative specificity of action is in part determined by their affinity for the 5-hydroxytryptamine (5-HT, serotonin), noradrenaline (NA), and dopamine (DA) uptake sites in the brain.[5,6] To gain a greater understanding of the structural requirements for 5-HT uptake site affinity and hence information of possible value in designing more specific serotonin-like neurotoxins, we have investigated a large number of other hydroxylated analogs as inhibitors of [^{3}H]-5-HT, [^{3}H]NA, and [^{3}H]DA transport into crude synaptosomal fractions from rat brain hypothalamus and corpus striatum.[7]

Materials and Methods

The drugs used in this study were synthesized by Dr. H. G. Schlossberger, Max Planck Institute for Biochemistry, Martinsried, Federal Republic of Germany. The numbering system for the tryptamine derivatives is shown in Figure 1. [G-^{3}H]-5-HT (5-hydroxy[G-^{3}H]tryptamine creatinine sulfate) and [7-^{3}H]NA (*d,l*-[7-^{3}H]noradrenaline hydrochloride) were both obtained from the Radiochemical Centre, Amersham, and had specific activities of 500 and 11.6 Ci/mmol, respectively. [2-^{3}H]DA (3,4-dihydroxyphenylethylamine [*ethyl*-2-^{3}H(N)]) was purchased from NEN-Chemical, Dreieichenhain, Federal Republic of Germany, and had a specific activity of 3.48 Ci/mmol.

A modification of the method of Horn and Snyder[8] was used to study acute interference of the tryptamine analogs with the uptake of ^{3}H-labeled amines into synaptosome-rich homogenates of rat hypothalamus (for [^{3}H]-5-HT and [^{3}H]NA uptake) and striatum (for [^{3}H]DA uptake). (Details of the procedure are reported in Horn *et al.*[5])

All drugs were dissolved immediately before use in distilled water that contained 0.25 mg of ascorbic acid per milliliter. The final concentration of the ^{3}H-labeled amines was 10^{-7} M in all cases, whereas the concentration of the tryptamines varied between 0.1 and 100 μM (cf. Table 1 & 2). For the determination of the IC_{50} value (the concentration of inhibitor required to produce a 50% inhibition of the uptake of [^{3}H]-5-HT), each of the tryptamine analogs was tested at three different concentrations, and a mean value $\pm$ SEM was obtained for three to five determinations at each cencentration. The IC_{50} values for the tryptamine analogs were obtained from these results by means of a graphic method.[5]

0077–8923/78/0305–0128 $01.75/0

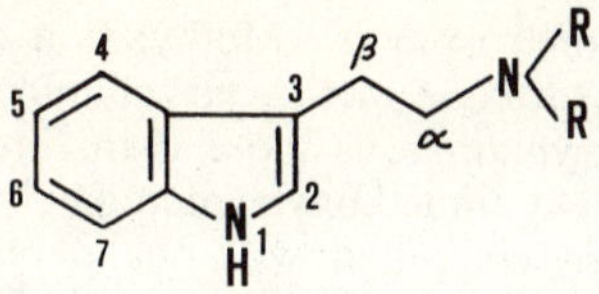

FIGURE 1. Numbering system for tryptamine derivatives.

RESULTS

The results are summarized in TABLES 1 and 2. In interpreting these results, it should be considered that for the most toxic compounds [e.g., 5,6-dihydroxytryptamine (5,6-DHT) and 5,7-dihydroxytryptamine (5,7-DHT)], the concentrations used in the affinity measurements are partly within the concentration range known to cause irreversible damage to amine uptake mechanisms.[7]

Of the compounds tested, *N*-methyl-5,6-DHT is, with an IC_{50} of 3.7×10^{-7} M, the most potent and selective inhibitor of 5-HT uptake into homogenates of rat hypothalamus, closely followed by 5,6-DHT. 6-Hydroxytryptamine is also a potent serotonin uptake inhibitor. 4,5-Dihydroxytryptamine has quite a good affinity for 5-HT uptake sites, whereas 5,7-DHT and 6,7-DHT are both more than six times less potent than 5,6-DHT. Of the other monohydroxylated compounds tested, 7-HT is more potent that 4-HT; the latter is the least potent

TABLE 1

EFFECT OF HYDROXYLATED TRYPTAMINES ON 5-HT UPTAKE INTO HYPOTHALAMIC HOMOGENATES *

Tryptamine Derivative	IC_{50}† (M)
N-methyl-5,6-DHT	3.7×10^{-7}
5,6-DHT	6.0×10^{-7}
6-HT ‡	8.0×10^{-7}
4,5-DHT	1.7×10^{-6}
7-HT	2.8×10^{-6}
5,7-DHT	4.0×10^{-6}
α-Methyl-5,7-DHT	4.4×10^{-6}
6,7-DHT	4.4×10^{-6}
4-HT	4.5×10^{-6}
4-Hydroxy-5-methoxytryptamine	5.6×10^{-6}
5-Hydroxy-6-methoxytryptamine	7.0×10^{-6}
5,6-Dibutyroxytryptamine	1.5×10^{-5}
N-Dimethyl-5,6-DHT	2.4×10^{-5}
5,6-Diacetoxytryptamine	7.4×10^{-5}
5,6,7-Trihydroxytryptamine	§

* Some of the values in TABLE 1 are from Horn *et al.*[5]

† The concentration of inhibitor required to produce a 50% inhibition of the uptake of [³H]-5-HT. Each drug was tested at three concentrations, and a mean value±SEM was obtained from three to five determinations at each concentration. The IC_{50} values were obtained from log-probit plots.

‡ Value for hypothalamic slices from Baumgarten *et al.*[9]

§ No inhibition of uptake at 1 or 10 μM.

serotonin uptake inhibitor of the serotonin positional isomers. Methylation of the 6-hydroxyl group in 5,6-DHT (to give 5-hydroxy-6-methoxytryptamine) leads to a fall in activity. 4-Hydroxy-5-methoxytryptamine is more than three times less potent than 4,5-DHT, whereas 5-hydroxy-6-methoxytryptamine has 11 times less affinity for 5-HT uptake sites when compared with the parent compounds. The most striking decrease in potency is obtained by acetylating both hydroxyl groups in 5,6-DHT (to give 5,6-diacetoxytryptamine), resulting in a greater than 120-fold reduction in affinity, similar results were obtained for the 5,6-*o*, *o*-dibutyrylated derivative of 5,6-DHT.

Methylation of the side chain nitrogen in 5,6-DHT (to give *N*-methyl-5,6-DHT), on the contrary, enhances uptake site affinity. The latter compound

TABLE 2

EFFECT OF HYDROXYLATED TRYPTAMINES ON UPTAKE OF 5-HT, NA, AND DA *

Tryptamine Derivative and Concentration	Inhibition of Uptake (%) †		
	5-HT (hypothalamus)	NA (hypothalamus)	DA (striatum)
N-methyl-5,6DHT, 1 μM	72.3±2.8(4)	39.1±4.8(4)	37.4±3.7(4)
5,6-DHT, 1 μM	59.0±2.3(4)	43.3±2.7(3)	44.7±0.7(3)
4,5-DHT, 10 μM	74.4±2.7(3)	32.0±3.6(3)	34.3±1.0(4)
7-HT, 1 μM	32.5±5.5(4)	60.8±2.2(3)	51.5±4.3(3)
5,7-DHT, 1 μM	25.7±1.2(3)	32.5±1.3(4)	15.4±1.7(4)
6,7-DHT, 10 μM	67.5±2.8(3)	30.1±0.7(3)	15.6±2.6(3)
4-HT, 10 μM	71.7±0.3(4)	43.0±1.7(4)	15.3±2.3(3)
4-Hydroxy-5-methoxytryptamine, 10 μM	54.1±3.5(5)	36.6±4.1(4)	8.3±0.8(3)
5-Hydroxy-6-methoxytryptamine, 10 μM	58.2±2.4(4)	25.9±1.2(3)	8.7±1.2(3)
5,6-Dibutyroxytryptamine, 50 μM	53.6±2.0(3)	62.9±2.0(4)	67.5±1.6(4)
N-Dimethyl-5,6-DHT, 10 μM	28.7±4.5(3)	57.7±4.2(3)	28.8±4.6(3)‡

* Some of the values in Table 2 are from Horn *et al.*[5]

† Homogenates from hypothalamus were used in studying the effect of the inhibitors on 5-HT and NA uptake, while homogenates of the corpus striatum were employed for DA uptake investigations. The results are expressed as the mean value±SEM with the number of experiments given in parentheses.

‡ The concentration for measuring DA uptake was 20 μM.

thus has an affinity for the 5-HT uptake sites that is only slightly lower than that of 5-HT itself (2.0×10^{-7} M). N^{ω}-dimethylation, on the other hand, produces a dramatic fall in affinity, similar to that which occurs after acylation of the hydroxyl groups. α-Methylation of the side chain in 5,7-DHT (to give α-methyl-5,7-DHT) does not, however, alter appreciably the affinity for the 5-HT uptake sites.

Surprisingly, the trihydroxy derivative, 5,6,7-trihydroxytryptamine, which has hydroxyl groups in positions compatible with reasonable affinity for the serotonin uptake sites (as judged from the behavior of the mono- and dihydroxylated positional isomers), fails to show any uptake site affinity at 1 or 10 μM.

TABLE 2 shows that, in general, with few exceptions, the hydroxylated tryptamines are more specific inhibitors of 5-HT than of NA or DA uptake. It should be noted, however, that the difference between the affinity of a given analog for the serotonin and catecholamine uptake sites is sometimes very small (e.g., for 5,6-DHT and 5,6-dibutyroxytryptamine). 7-Hydroxytryptamine differs from the remaining positional isomers of 5-HT in that it exhibits an inverse selectivity pattern, because it inhibits catecholamine and, in particular, NA uptake more efficiently than 5-HT uptake. The effect of the hydroxyl group in this position would appear to contribute to the unselectivity of 5,7-DHT. The high affinity of 7-HT for the NA and, especially, the DA uptake sites is, however, clearly attenuated in 5,7-DHT.

DISCUSSION

The data presented in this paper indicate that *N*-methyl-5,6-DHT has an affinity for 5-HT uptake sites that approximates that of 5-HT itself. Its potency as an inhibitor of 5-HT uptake is even greater than that of 5,6-DHT. The IC_{50} value for 5,6-DHT (6×10^{-7} M), measured in homogenates of rat hypothalamus, is in agreement with prior results in slices of rat hypothalamus.[9–11] When compared to 4- and 6-hydroxytryptamine, the parent *o*-dihydroxytryptamines, 5,6-DHT and 4,5-DHT have a slightly higher uptake site affinity. This difference may be due to the presence of one of the hydroxyl groups in the optimum position for 5-HT uptake, as suggested by the progressive fall in affinity noted with 5-HT, 6-HT, and 7-HT, with increasing distance of the hydroxyl group from the optimal 5-position. 4-Hydroxytryptamine is anomalous in this respect, because its hydroxyl group is closer to the 5-position than that of 7-HT but is a weaker uptake inhibitor. The *meta* position of the hydroxyl groups in 5,7-DHT decreases uptake site affinity despite the presence of one of the hydroxyl groups in the optimal 5-position. Methylation of the 5-hydroxyl group in 4,5-DHT or the 6-hydroxyl group in 5,6-DHT leads to a decrease in binding affinity (by threefold for 4-hydroxy-5-methoxytryptamine and by 11-fold for 5-hydroxy-6-methoxytryptamine). *O*-Methylation is a well-established principle of inactivation of catecholamines,[12, 13] and a corresponding decrease in uptake site affinity has been demonstrated for *O*-methylated catecholamine metabolites. An even more striking fall in affinity is obtained with 5,6-DHT by *O,O*-diacylation (to give 5,6-diacetoxytryptamine and 5,6-dibutyroxytryptamine) or by *N*ω-dimethylation.

The relative affinity of a given tryptamine for the 5-HT uptake site, on the one hand, and for the NA and DA uptake sites, on the other (measured at a fixed concentration of the compound; cf. TABLE 2), should be an important factor in determining the selectivity of action of the toxic indoleamines. Thus, neurotoxic compounds provided with only a small difference in their relative affinities for the serotonin and catecholamine uptake sites would be expected to have similar toxic potencies on both types of neurons. This is, however, not always the case. For instance, the difference in affinity of 5,6-DHT for the serotonin and catecholamine transport systems is rather small (TABLE 2), suggesting poor discrimination between both types of monoamine neurons. Nevertheless, our other studies[7] show that there is no indication of damage to the [^{3}H]NA uptake system by 5,6-DHT at either 10^{-6} or 10^{-5} M, that is, at concentrations that cause profound impairment of [^{3}H]-5-HT uptake. This observation is in agreement with previously reported findings *in vivo* that neither intraventricular [4] nor direct intracerebral injection of 5,6-DHT causes any significant damage to NA axons in rat brain.[15]

A similar situation occurs for 5,7-DHT: despite the fact that 5,7-DHT has a higher affinity for the NA than for the 5-HT uptake sites, the impairment of [^{3}H]-5-HT uptake is greater than that of [^{3}H]NA uptake.[7] The preferential toxic action of 5,7-DHT on serotonin neurons is also evident from observations *in vivo* on the effects of intracerebral injection of 5,7-DHT.[15] Intraventricular injection 5,7-DHT causes a dose-dependent long-term reduction in central nervous system 5-HT levels, whereas NA depletion is (when more than 50 μg of 5,7-DHT is administered to adult rats) generally of lesser magnitude and dose unrelated.

These observations demonstrate very clearly that uptake site affinity is certainly not the only determinant for the specificity and potency of a neurotoxic indoleamine. The possible importance of other factors, such as breakdown and metabolism of the drugs and the ability of their products to undergo tissue binding, has been discussed elsewhere.[7]

References

1. Thoenen, H. & J. P. Tranzer. 1973. The pharmacology of 6-hydroxydopamine. Annu. Rev. Pharmacol. **13:** 169–180.
2. Kostrzewa, R. M. & D. M. Jacobowitz. 1974. Pharmacological actions of 6-hydroxydopamine. Pharmacol. Rev. **26:** 199–287.
3. Jonsson, G. & Ch. Sachs. 1975. The mechanism of action of 6-hydroxydopamine. Biochem. Pharmacol. **24:** 1–8.
4. Baumgarten, H. G. & A. Björklund. 1976. Neurotoxic indoleamines and monoamine neurons. Annu. Rev. Pharmacol. **16:** 101–111.
5. Horn, A. S., H. G. Baumgarten & H. G. Schlossberger. 1973. Inhibition of the uptake of 5-hydroxytryptamine, noradrenaline and dopamine into rat brain homogenates by various hydroxylated tryptamines. J. Neurochem. **21:** 233–236.
6. Baumgarten, H. G., A. Björklund, A. S. Horn & H. G. Schlossberger. 1974. Studies on the neurotoxic properties of hydroxylated tryptamines. *In* Dynamics of Degeneration and Growth in Neurons. F. Fuxe, L. Ollson & Y. Zotterman, Eds. : 153–167. Pergamon Press. Oxford and New York.
7. Björklund, A., A. S. Horn, H. G. Baumgarten, A. Nobin & H. G. Schlossberger. 1975. Neurotoxicity of hydroxylated tryptamines: structure-activity relationships. Acta Physiol. Scand. Suppl. **429:** 29–60.
8. Horn, A. S. & S. H. Snyder. 1972. Steric requirements for catecholamine uptake by rat brain synaptosomes: studies with rigid analogs of amphetamine. J. Pharmacol. Exp. Ther. **180:** 523–530.
9. Baumgarten, H. G., K. D. Evetts, R. B. Holman, L. L. Iversen, M. Vogt & G. Wilson. 1972. Effects of 5,6-dihydroxytryptamine on monoaminergic neurons in the central nervous system of the rat. J. Neurochem. **19:** 1587–1597.
10. Heikkila, R. E. & G. Cohen. 1973. The inhibition of ^{3}H-biogenic amine uptake by 5,6-dihydroxytryptamine: a comparison with the effects of 6-hydroxydopamine. Eur. J. Pharmacol. **21:** 66–69.
11. Heikkila, R. E. & G. Cohen. 1974. The inhibition of ^{3}H-biogenic amine accumulation into rat brain tissue slices by various tryptamine derivatives. Res. Commun. Chem. Pathol. Pharmacol. **7:** 539–547.
12. Axelrod, J. 1959. The metabolism of catechol amines in vivo and in vitro. Pharmacol. Rev. **11:** 402–408.
13. Burgen, A. S. V. & L. L. Iversen. 1965. The inhibition of noradrenaline up-

take by sympathomimetic amines in the rat isolated heart. Brit. J. Pharmacol. **25:** 34–49.

14. BAUMGARTEN, H. G., A. BJÖRKLUND, L. LACHENMAYER, A. NOBIN & U. STENEVI. 1971. Longlasting selective depletion of brain serotonin by 5,6-dihydroxytryptamine. Acta Physiol. Scand. Suppl. **373:** 1–15.
15. BAUMGARTEN, H. G., A. BJÖRKLUND, L. LACHENMAYER & A. NOBIN. 1973. Evaluation of the effects of 5,7-dihydroxytryptamine on serotonin and catecholamine neurons in the rat CNS. Acta Physiol. Scand. Suppl. **391:** 1–9.

DISCUSSION

DR. H. G. BAUMGARTEN: What would be the IC_{50} of serotonin in inhibiting serotonin uptake? Why didn't you compare the IC_{50}s of the compounds rather than presenting their K_M values?

DR. HORN: The K_m value for 5-HT uptake was determined from a Lineweaver-Burk plot, and I have not arrived at an "IC_{50}" value for 5-HT uptake. The two values should be very similar, however.

DR. R. E. HEIKKILA: We have performed the examinations suggested by Dr. Baumgarten. The K_M values for transport, and the IC_{50} values are basically the same. Hence, there is really no difference between these values when you are working at concentrations of the drug that are under their K_Ms.

STRUCTURE-ACTIVITY RELATIONSHIPS IN COMPOUNDS ANALOGOUS TO 5,6-DIHYDROXYTRYPTAMINE *

Roger P. Maickel, Talmage R. Bosin, and Alan C. Donelson

Section on Pharmacology
Medical Sciences Program
Indiana University School of Medicine
Bloomington, Indiana 47401

Ernest Campaigne and Richard B. Rogers

Department of Chemistry
Indiana University
Bloomington, Indiana 47401

The disruption of normal neuronal transmission by interference with some aspect of neurotransmitter synthesis, uptake, storage, release, or metabolism is considered an important mode of drug action. Indeed, the use of drugs that block peripheral nervous transmission was indispensable in first establishing, then proving, the very nature of neuromuscular transmission,[1-3] and in investigations of the autonomic nervous system.[2,4] The method of blocking transmitter utilization has also been used in attempts to clarify the specific functions of putative neurotransmitters in the brain and to substantiate hypotheses that link the effects of psychoactive drugs to their interactions with neurochemical systems.[5,6]

Of greatest interest centrally have been the monoamines serotonin (5-HT) and norepinephrine (NE). Literally thousands of publications have covered their occurrence in nature, their distribution in various systems of vertebrate and invertebrate species, the nature of their nervous function, and the subcellular aspects of their utilization by nerves.[7-9] Their role, individually or together, in the neural processes responsible for behavior remains controversial and largely unresolved.[10] Present-day monoamine research seeks to provide the final, direct evidence that would confirm the status of NE and 5-HT as brain neurotransmitters, a goal heavily dependent on technologic advances, and to link directly the central function of monoamine-containing neurons to behavior.

Hypotheses that emphasize the role of NE and 5-HT as neurotransmitters in the normal function of the central nervous system are of two varieties: those that attempt to show evidence of this role, that is, to fulfill criteria necessary for the complete recognition of the compounds as neurotransmitters, and those that, assuming NE and 5-HT to be neurotransmitters, postulate aberrations in their metabolism or neuronal release mechanisms as causes for mental disorders or drug effects.

The technique of "selective depletion" of central monoamines as a means by which their functional role in behavior may be studied presumably has its origin in traditional biology. The surgical destruction or extirpation of an organ as a method of investigating its physiologic significance in a body has

* Supported by National Aeronautics & Space Administration Grant NGL 15–003–117 and by a grant from the Pennwalt Corporation.

been a mainstay in biologic research. The practical laboratory difficulties associated with surgical denervation in an organ as complex as the central nervous system encouraged a shift to the use of chemical means of differentiating the roles of the neurotransmitters.[11]

In 1959, Senoh *et al.*[12-14] reported that the administration of dopamine to rabbits resulted in the autoxidative and enzymatic formation of 6-hydroxydopamine (6-HDA). The peripheral administration of 6-HDA in the mouse led initially to the discovery that the compound produced a long-lasting depletion of NE in the heart.[15] Further studies showed that while 6-HDA had a lower ED_{50} than did either reserpine or metaraminol, its NE-depleting effects lasted considerably longer than did those of the other compounds tested.[16] It was originally suggested that 6-HDA damaged amine storage sites; [15, 17] electron microscopic studies indicated that the long-term loss of NE resulted from the destruction of adrenergic nerve terminals,[18] and concomitant loss of activity of NE-synthesizing enzymes was also found.[19] The effects of 6-HDA have been termed "chemical denervation"; [20] the compound has been widely used in studies concerned with the synthesis, uptake, storage, and functional aspects of both peripheral and central catecholamines.[21]

After central administration of 6-HDA, long-lasting selective decreases in catecholamine levels were observed without significant effect on 5-HT or γ-aminobutyric acid levels.[19, 22, 23] Generally, the behavior of animals given 6-HDA did not differ significantly from that of controls.[19, 22, 23] However, injections of 6-HDA into localized brain regions have been used to induce specific lesions of central monoamine neurons.[24] Spontaneous rotational behavior was observed in rats with unilateral 6-HDA lesions in the nigrostriatal DA system.[25, 26] Bilateral injection of 6-HDA into the lateral hypothalamus caused adipsia and aphagia, while animals lesioned in the substantia nigra showed marked hyperactivity, with behavioral reversal after 1 week.[27]

In 1971, the apparent discovery of a "serotonergic 6-HDA" was reported; intraventricular administration of 5,6-dihydroxytryptamine (5,6-DHT) resulted in the long-lasting depletion of brain 5-HT, with only transient effects on the levels of DA and NE.[28, 29] The percentage loss of 5-HT and the dose dependency of the effect varied in different brain regions. Subsequent fluorescent histochemical and electron microscopic studies revealed selective degeneration of indoleamine terminals and axons similar to the cytotoxic effects of 6-HDA on adrenergic neurons.[30, 31] The tendency of 5-HT levels to recover after the initial phase of depletion led to the discovery of axonal regeneration by the damaged serotonergic neurons.[32, 33]

In contrast to the results with 6-HDA, intraventricular injection of 5,6-HDT caused dramatic behavioral changes.[34] The acute disturbances have been described as "sedation during the first two hours after the injection (ether anesthesia), an increased irritability and hyper-reactivity against any type of sensory stimulation and bizarre social behavior (fighting in a stereotyped position and manner), and increased copulatory activity when kept in the same cage between the 1st and 3rd day after 5,6-DHT." [35] Both frequent fighting episodes and increased muricidal behavior have been reported, and decreased food consumption and loss of weight have also been observed. Intraventricular injection of 5,6-DHT produced significant hypothermia 30 min after administration.[36]

The advent of a selective neurotoxic agent, such as 5,6-DHT, opens fresh avenues for the functional differentiation of central monoamines and for possible discernment of their role in mediating behavior. The availability of such chemicals offers an alternative to the nonselective destruction of brain regions with heterogeneous cell populations, for example, by electrocoagulation or surgical lesioning. In addition, the functional significance of the depression of monoamine levels by selective synthesis inhibitors may be resolved through the use of cytotoxic agents. The use of chemicals to destroy monoamine nerve terminals may supplement brain amine concentrations as criteria for relating monoamine function to behavior.

Synthesis and pharmacologic testing of structural analogs are commonly employed to identify structural features responsible for a primary effect, to elucidate mechanisms of action, and/or to establish a rational basis for the development of more specific agents. The recent synthesis[37] of 3-(β-aminoethyl)-5,6-dihydroxybenzo[*b*]thiophene (5,6-DHT-S), the sulfur isostere of 5,6-DHT, offered the opportunity to investigate the relative importance of the indolic nitrogen in the action of 5,6-DHT. Benzo[*b*]thiophene isosteres of biologically active indoles have been studied previously by this laboratory in an attempt to estimate the pharmacologic contribution of the indolealkylamine ring nitrogen.[38-41] The availability of benzo[*b*]thiophene and 5,6-disubstituted tryptamines permitted a study of the contributions of the chemical subgroups present in 5,6-DHT to its monoaminergic activity and general effects on behavior. The parallel neurochemical ond behavioral structure-activity relationships make it possible to speculate on the relation of monoaminergic activity to behavior in the rat.

Contrasting Biochemical Effects of 5,6-DHT and 5,6-DHT-S in the Rat

Norepinephrine levels in the heart and spleen were decreased significantly by intraperitoneal injection of 30 mg/kg of each compound to adult male rats. Minimum levels in the heart were reached by 2 hr; each compound produced a 55–60% depletion of NE. Values returned to control levels by 16 hr. In the spleen, NE levels were similarly reduced, differing only in that 5,6-DHT caused a longer-lasting reduction (up to 16 hr) in the heart. 5-Hydroxytryptamine levels in the spleen were reduced to 50% by 5,6-DHT at 16 hr and were still decreased at 24 hr. 5,6-DHT-S, however, actually increased levels of 5-HT by 50% 2 hr after injection. No decrease in 5-HT levels was observed in the 24-hr period after administration of 5,6-DHT-S. The data on spleen 5-HT and NE levels are presented in Figure 1. In contrast to these effects, the intraperitoneal dosages had absolutely no effect on the levels of 5-HT and NE in the brains of the animals.

The results obtained after intraventricular injection of 80 μg of the compounds were most dramatic. 5,6-Dihydroxytryptamine significantly decreased the 5-HT content of the three brain regions: cerebral hemispheres, cerebellum, and brain stem. The decline differed only in the time course of the effect. A rapid drop of 25–35% was observed in the cerebral hemispheres by 4 hr after dosage; this fall persisted for at least 16 days. Brain stem 5-HT was lowered by 20–30% in a similar fashion, while cerebellar 5-HT showed a 25% decrease only between 16 hr and 4 days after dosage. The data for the effects of 5,6-DHT and 5,6-DHT-S on brain stem 5-HT and NE are shown in Figure 2.

In contrast to their effects on 5-HT, *both* compounds decreased levels of NE in the three brain regions. The decline in the cerebral hemispheres of approximately 30% was less sharp and more prolonged than that in the brain

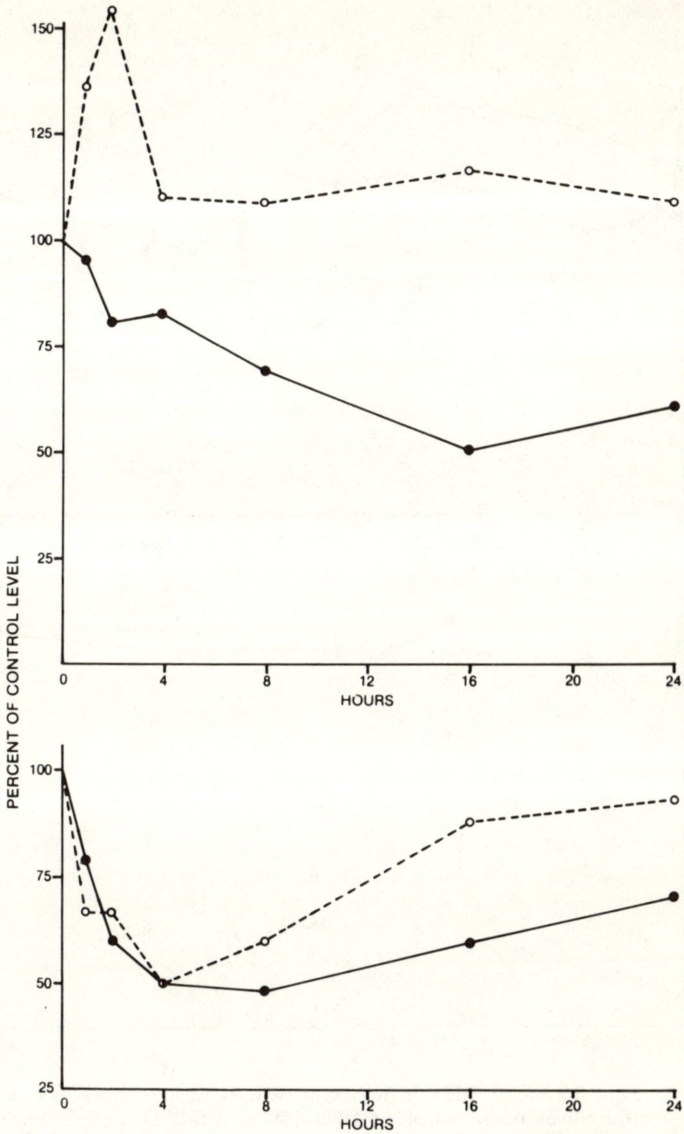

FIGURE 1. *Top:* Levels of 5-HT in rat spleen at various times after a single dose (30 mg/kg, ip) of 5,6-DHT (●) or 5,6-DHT-S (○). Each value is the mean of four animals; control values ($n=8$) were 1.40 ± 0.09 (SD) μg/g. *Bottom*: Levels of NE in rat spleen at various times after a single dose (30 mg/kg, ip) of 5,6-DHT(●) or 5,6-DHT-S (○). Each value is the mean of four animals; control values ($n=8$) were 0.42 ± 0.07 (SD) μg/g.

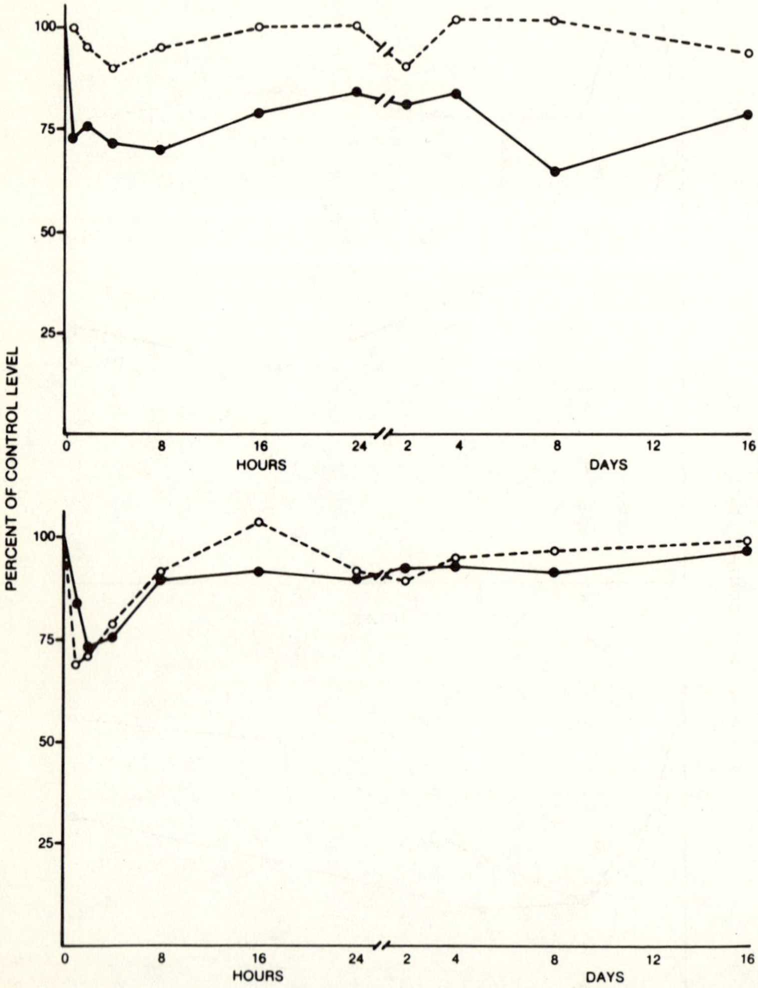

FIGURE 2. *Top:* Levels of 5-HT in rat brain stem at various times after a single dose (80 μg/rat, intraventricularly) of 5,6-DHT (●) or 5,6-DHT-S (○). Each value is the mean of four animals; control values ($n=8$) were 1.13 ± 0.17 (SD) μg/g. *Bottom:* Levels of NE in rat brain stem at various times after a single dose (80 μg/rat, intraventricularly) of 5,6-DHT (●) or 5,6-DHT-S (○). Each value is the mean of four animals; control values ($n=8$) were 1.01 ± 0.11 (SD) μg/g.

stem, where similar decreases were last observed at 4 hr. Decreased levels of NE in the cerebellum reached 75% of control values briefly, at 1–2 hr post-injection (FIGURE 2).

Both 5,6-DHT and 5,6-DHT-S produced a transient increase in 5-hydroxyindoleacetic acid (5-HIAA) levels. By 8 hr, 5,6-DHT had reduced 5-HIAA levels to 20–30% of control values in all three brain regions, paralleling similar reductions of 5-HT. 5,6-DHT-S, on the other hand, demonstrated an ability to increase 5-HIAA levels, which returned to control values by 24 hr. The effects of 5,6-DHT on 5-HIAA paralleled its decrease of 5-HT in the brain stem and cerebral hemispheres, and likewise in the cerebellum, where 5-HIAA levels returned to control levels after the fourth day. In contrast, 5,6-DHT-S caused a significant elevation in 5-HIAA levels 2–4 days after injection. Maximum increases of 40–50% were observed in the brain stem and cerebral hemispheres; a 12–20% decrease was seen in the cerebellum. 5-Hydroxyindoleacetic acid levels returned to nearly normal values by the 16th day. Lower doses (20 or 40 μg per rat) of both 5,6-DHT and 5,6-DHT-S produced similar effects, although they were lesser in magnitude and duration.

BEHAVIORAL EFFECTS OF 5,6-DHT AND 5,6-DHT-S IN THE RAT

Peripheral injections of 5,6-DHT and 5,6-DHT-S in the rat elicited similar behavior. When administered intraperitoneally (30 mg/kg), both compounds produced peripheral signs, such as exophthalmos, piloerection, rigid alert stance, and increased irritability, that indicated sympathetic autonomic stimulation. No increase in motor activity, indicative of central activity, was observed.

When administered centrally, 5,6-DHT and 5,6-DHT-S differed in their behavioral effects. Injection of 80 μg of 5,6-DHT-S into the lateral ventricle produced a behaviorally depressed animal; the sedated state could be described as a semiconscious stupor that lasted 30–45 min. After recovery, animals could not be distinguished from controls. Central administration of 5,6-DHT in doses of 80 μg caused an initial behavioral depression that, after 40–120 min, was usually succeeded by a period of sudden, intense activity.

The distinctive behavior that occurred after injection of 5,6-DHT was characterized by explosive jumping (rats at times leapt through covered cages to heights of 18 in.); erratic running behavior, in which the back legs, in a flipping motion, propelled the rat in large circles in open space or along the walls of an enclosure (the animal appeared to ignore barriers in its frantic movement and, at times, bloodied its head and cut its feet); frequent vocalization by high-pitched squeals or shrieks; convulsions during which the animal literally behaved as a "fish out of water," unable to breathe or make a sound; death occurred in 25–35% of the rats that entered into the latter type of behavior. Each such outburst lasted about 30–90 sec in confined spaces but could continue much longer in open spaces. Milder episodes were interspersed among the more intense ones and consisted of compulsive running and leaping at barriers, apparently a form of escape behavior. After each outburst, the animal lay still, breathing deeply, but was alert and hyperresponsive to external stimulation. As many as 30 mild and intense outbursts have been recorded in the 2 hr following injection of 5,6-DHT; the period of 45–75 min after injection contained the greatest frequency of such episodes. It was observed that rats that did not enter into the second phase of behavior but that remained behaviorally depressed and stuporous were less likely to recover than those that underwent the violent behavioral spasms and lived.

Long-term behavioral effects were observed after central administration of 5,6-DHT, but not 5,6-DHT-S. After the period of intense activity, rats given 5,6-DHT either recovered sufficiently to feed and water themselves or remained adipsic and aphagic, dying within 2–4 days postinjection. Those that recovered tended to weigh less than control animals and were often irritable and defensive when approached or handled. The long-term behavioral effects of 5,6-DHT were characterized by more frequent vocalization, unusual social interactions between cage-mates, and fighting episodes. One particular form of behavior was especially striking: two rats, standing on hind legs facing each other, would squeal, "hiss," or "spit" at each other but would not make physical contact. At times, this behavior lasted as long as 1 hr or more.

Biochemical and Behavioral Effects of 5-Hydroxy-6-methoxytryptamine (5-H-6-MT) and 5-Methoxy-6-hydroxytryptamine (5-M-6-HT) in the Rat

Each of these analogs was administered into the lateral ventricle of rats at a dosage of 100 μg per rat. No significant changes in the levels of 5-HT or NE were seen in any of the brain regions over the period 1–24 hr after dosage. 5-Hydroxy-6-methoxytryptamine caused a significant increase (20–25%) in 5-HIAA levels at 2 hr after injection, whereas 5-H-6-MT produced a significant decrease (25–35%) in 5-HIAA levels 1 hr postdosage.

Within 15 min after intraventricular injection of 5-H-6-MT, a behavioral syndrome identical to that caused by 5,6-DHT was manifested. Brief, periodic (3–8 min), convulsive episodes were combined with erratic running behavior and vocalization. The behavioral effects differed from those of 5,6-DHT in that they were quicker in onset and lasted less than 1 hr. There was evident an increased alertness over that of control animals following this syndrome; at this time, the animals could not be differentiated behaviorally.

In contrast, 5-H-6-MT produced no dramatic behavioral effects. Some increased exploratory activity soon after injection was observed, as was increased motor activity and alertness. However, these effects were slight, and the animals could not otherwise be differentiated from controls.

Biochemical and Behavioral Effects of 5,6-Dimethoxytryptamine (5,6-DMT) and 3-(β-Aminoethyl)-5,6-isopropylindendioxybenzo-[*b*]thiophene (IPST) in the Rat

Each compound was injected intraventricularly at a dose of 100 μg per rat. 5,6-Dimethoxytryptamine apparently elevated 5-HT levels, with no significant effect on 5-HIAA or NE. However, since 5,6-DMT plays a role in the method for the determination of 5-HT,[42] the increased fluorescence, which persisted for 24 hr, suggests a continued presence of 5,6-DMT in significant amounts. IPST, which does not interact with the assay methods, elevated 5-HT levels at 2 hr (33%) and 24 hr (44%) in the cerebral hemispheres; by 48 hr, however, these levels had declined to 59% of control values. Brain stem levels of 5-HT were increased by 20–30% at 24 hr postinjection, with a concomitant decrease in

5-HIAA. In the cerebral hemispheres, 5-HIAA levels were depressed for 48 hr after injection. Levels of NE remained essentially unchanged throughout the 48-hr period after administration of IPST.

Both IPST and 5,6-DMT-S were behaviorally active. IPST, in contrast to its parent compound 5,6-DHT-S, increased motor activity; the most prominent feature of the behavioral effect was compulsive cage circling interrupted by brief pauses. The effect commenced shortly after injection and lasted nearly 2 hr. Paradoxically, when administered peripherally (40 mg/kg, ip), IPST depressed all behavior. The rat given IPST would lay prone and stuporous but would lick a water bottle spout brought near to its mouth.

Rats given 100 μg of 5,6-DMT intraventricularly exhibited a disoriented, groping behavior, while rolling over and over. The latter behavioral effect could be elicited by moving the animal's cage. This behavior lasted up to 2.5 hr after injection, after which time the animals could not be distinguished from controls. No convulsive behavior similar to that produced by 5,6-DHT was observed.

Comparative Effects of 5,6-DHT and Analogs in the Rat

A summary of the biochemical and behavioral effects of the various compounds tested is presented in Table 1.

Although 5,6-DHT is generally considered as a selective depletor of brain 5-HT, the results obtained from both peripheral and brain tissues demonstrate an ability of the compound to lower the levels of both NE and 5-HT. That 5,6-DHT might share properties of both monoamines, and therefore have the potential to interact with either catecholaminergic or serotonergic systems, may be deduced readily from its structure. 5,6-Dihydroxytryptamine itself is the 6-hydroxy derivative of serotonin; the *o*-dihydroxyphenyl portion of the 5,6-DHT molecule is isomeric with catechol, a subgroup shared by the catecholamines NE and DA. In fact, prior to the discovery of its neurotoxic properties, 5,6-DHT was observed to be as active at 5-HT in the contraction of smooth muscle preparations [43] and to be a potent releaser of heart NE.[44] The diploid nature of the monoaminergic activity of 5,6-DHT *in vivo* has therefore been confirmed by us for peripheral and brain tissues.

If the structure of 5,6-DHT is used as the basis for making comparisons, the relation of chemical structure to monoaminergic activity may be summarized as follows. Replacement of the indolic nitrogen by sulfur abolishes 5-HT-depleting activity in spleen and in brain but does not alter NE-depleting activity in the heart, spleen, or brain. Methylation of either the 5- or 6-hydroxyl group of 5,6-DHT removes monoamine depletion activity. However, the increased and decreased levels of 5-HIAA that are observed soon after injection indicate that serotonergic interactions are not completely absent. Methylation of both hydroxyl groups on 5,6-DHT results in a complete abolition of monoamine depletion activity, although considerable behavioral activity is retained. In contrast, conversion of 5,6-DHT-S to its isopropylidene derivative produces a complex action on brain 5-HT and NE systems and causes a generalized stimulatory behavioral effect.

TABLE 1
EFFECTS OF 5,6-DHT AND ITS ANALOGS ON BRAIN BIOCHEMISTRY AND BEHAVIOR IN THE RAT

Compound	Biochemical Effects	Behavioral Results
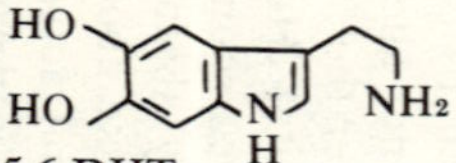 5,6-DHT	prolonged (20–30%) decrease in 5-HT; initial increase, then prolonged decrease in 5-HIAA; brief initial decrease in NE	initial behavioral depression, 30–45 min, then explosive jumping and erratic running behavior, convulsive episodes, and vocalization
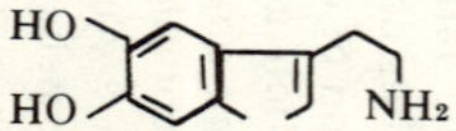 5,6-DHT-S	no effect on 5-HT; increased 5-HIAA levels 2–4 days postinjection; brief initial decrease in NE	behavioral depression lasting up to 30 min postinjection; sedation, animal lying prone
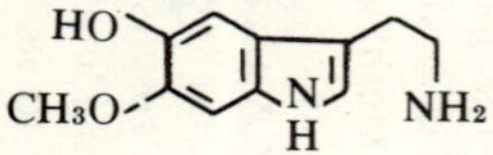 5-H-6-MT	no effect on 5-HT; brief initial increase in 5-HIAA; no effect on NE	brief, periodic (3–8 min) convulsive episodes within 15 min of injection, associated with erratic running behavior and vocalization, followed by period of increased alertness and responsivity
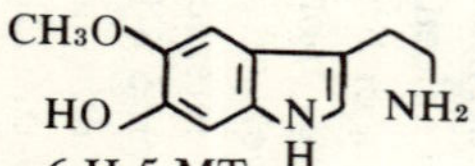 6-H-5-MT	no effect on 5-HT; brief initial decrease in 5-HIAA; no effect on NE	some increased exploratory activity within 15 min of injection but otherwise no difference from control animals
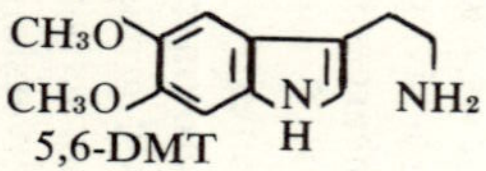 5,6-DMT	no significant effects on 5-HT, 5-HIAA, or NE	disoriented, groping behavior; animals exhibit much rotational behavior and hyperreactivity to sensory stimulation
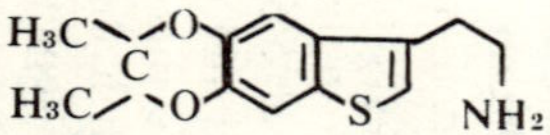 IPST	initial increase, then prolonged decrease in 5-HT; generally decreased 5-HIAA; brief initial decrease in NE	increased motor activity; cage circling with periodic pauses, lasting 2 hr after injection

Conclusions

The indolic nitrigen of 5,6-DHT appears to be an absolute requirement for the 5-HT neurotoxic effects of the compound, at least with respect to substitution of a sulfur atom. Similar specificity appears to reside in the phenolic state of the 5- and 6-hydroxyl groups. In contrast, the similar ability of 5,6-DHT-S to lower stored NE levels for a transient period appears to be related only to the catechol characteristics associated with an arylethylamine. The overall behavioral actions of the compounds are complex and suggest a multiplicity of effects including amine release/depletion and direct receptor interactions.

Acknowledgment

The 5-hydroxy-6-methoxytryptamine used in this study was supplied by Dr. A. A. Manian of the National Institute of Mental Health.

References

1. Eccles, J. C. 1937. Synaptic and neuromuscular transmission. Physiol. Rev. **17:** 538–557.
2. Dale, H. H. 1937. Transmission of nervous effects by acetylcholine. Harvey Lect. **32:** 229–247.
3. Brown, G. L. 1937. Transmission at nerve endings by acetylcholine. Physiol. Rev. **17:** 485–514.
4. Nickerson, M. 1949. The pharmacology of adrenergic blockade. Pharmacol. Rev. **1:** 27–56.
5. Gyermek, L. 1966. Drugs which antagonize 5-hydroxytryptamine and related indolealkylamines. *In* Handbook of Experimental Pharmacology. V. Erspamer, Ed. Vol. xix: 471–494. Springer Verlag. Berlin, Federal Republic of Germany.
6. Carlsson, A. & M. Lindqvist. 1961. DOPA analogues as tools for the study of dopamine and noradrenaline in brain. *In* Monoamines et Systeme Nerveux Central. J. De Ajuriaguerra, Ed. : 89–103. Georg & Company. Geneva, Switzerland.
7. Barchas, J. & E. Usdin. (Eds.) 1973. Serotonin and Behavior. Academic Press, Inc. New York, N.Y.
8. Blaschko, H. & E. Muscholl. (Eds.) 1972. Handbook of Experimental Pharmacology. Vol. XXXIII. Springer Verlag. Berlin, Federal Republic of Germany.
9. Erspamer, V. (Ed.) 1966. Handbook of Experimental Pharmacology. Vol. XIX. Springer Verlag. Berlin, Federal Republic of Germany.
10. Carlsson, A. 1964. Functional significance of drug-induced changes in brain monoamine levels. Progr. Brain Res. **8:** 9–20.
11. Brodie, B. B. & W. D. Reid. 1968. Serotonin in brain: functional considerations. Advan. Pharmacol. **6A:** 97–119.
12. Senoh, S., C. R. Creveling, S. Udenfriend & B. Witkop. 1959. Chemical, enzymatic and metabolic studies on the mechanisms of oxidation of dopamine. J. Amer. Chem. Soc. **81:** 6236–6243.
13. Senoh, S. & B. Witkop. 1959. Non-enzymatic conversions of dopamine to norepinephrine and trihydroxyphenethylamines. J. Amer. Chem. Soc. **81:** 6222–6230.
14. Senoh, S., B. Witkop, C. R. Creveling & S Udenfriend. 1959. 2,4,5-Trihydroxyphenethylamine, a new metabolite of 3,4-dihydroxyphenethylamine. J. Amer. Chem. Soc. **81:** 1768–1771.

15. PORTER, C. C., J. A. TOTARA & C. A. STONE. 1963. Effect of 6-hydroxydopamine and some other compounds on the concentration of norepinephrine in the hearts of mice. J. Pharmacol. Exp. Ther. **140:** 308–321.
16. STONE, C. A., J. M. STAVORSKI, C. T. LUDDEN, H. C. WENGER, C. A. ROSS, J. A. TOTARO & C. C. PORTER. 1963. Comparison of some pharmacologic effects of certain 6-substituted dopamine derivatives with reserpine, guanethidine, and metaramirol. J. Pharmacol. Exp. Ther. **142:** 147–158.
17. LAVERTY, R., D. F. SHARMAN & M. VOGT. 1965. Action of 2,4,5-trihydroxyphenylethylamine on the storage and release of noradrenaline. Brit. J. Pharmacol. **24:** 549–557.
18. TRANZER, J. P. & H. THOENEN. 1968. An electronmicroscopic study of selective, acute degeneration of sympathetic nerve terminals after administration of 6-hydroxydopamine. Experientia **24:** 155–157.
19. URETSKY, N. J. & L. L. IVERSEN. 1969. Effects of 6-hydroxydopamine on noradrenaline-containing neurones in rat brain. Nature (London) **221:** 557–558.
20. TRANZER, J. P. & H. THOENEN. 1967. Ultramorphologische Veranderingen der sympathischen Nervedenigungen der Katze nach Vorbehandlung mit 5- and 6-Hydroxy-Dopamin. Arch. Pharmakol. Exp. Pathol. **257:** 343–351.
21. SACHS, C. & G. JONSSON. 1975. Mechanisms of action of 6-hydroxydopamine. Biochem. Pharmacol. **24:** 1–8.
22. HAEUSLER, G., W. HAEFELY & H. THOENEN. 1969. Chemical sympathectomy of the cat with 6-hydroxydopamine. J. Pharmacol. Exp. Ther. **170:** 50–62.
23. BURKARD, W. P., M. JALFRE & J. BLUM. 1969. Effect of 6-hydroxydopamine on behaviour and cerebral amines in rats. Experientia **25:** 1295–1296.
24. UNGERSTEDT, U. 1971. Histochemical studies on the effect of intracerebral and intraventricular injections of 6-hydroxydopamine on monoamine neurons in the rat brain. *In* 6-Hydroxydopamine and Catecholamine Neurons. T. Malmfors & H. Thoenen, Eds. : 101–120. North Holland Publishing Company. Amsterdam, The Netherlands.
25. UNGERSTEDT, U. 1971. Postsynaptic supersensitivity after 6-hydroxydopamine induced degeneration of the nigro-striatal dopamine system. Acta Physiol. Scand. Suppl. **367:** 95–102.
26. UNGERSTEDT, U. 1971. Striatal dopamine release after amphetamine or nerve degeneration revealed by rotational behavior. Acta Physiol. Scand. Suppl. **367:** 49–66.
27. UNGERSTEDT, U. 1971. Adipsia and aphagia after 6-hydroxydopamine induced degeneration of the nigro-striatal dopamine system. Acta Physiol. Scand. Suppl. **367:** 95–110.
28. BAUMGARTEN, H. G., A. BJÖRKLUND, L. LACHENMAYER, A. NOBIN & U. STENEVI. 1971. Long-lasting selective depletion of brain serotonin by 5,6-dihydroxytryptamine. Acta Physiol. Scand. Suppl. **373:** 7–15.
29. BAUMGARTEN, H. G., K. D. EVETTS, R. B. HOLMAN, L. L. IVERSEN, M. VOGT & G. WILSON. 1972. Effects of 5,6-dihydroxytryptamine on monoaminergic neurones in the central nervous system of the rat. J. Neurochem. **19:** 1587–1597.
30. BAUMGARTEN, H. G., L. LACHENMAYER & H. G. SCHLOSSBERGER. 1972. Evidence for a degeneration of indoleamine containing nerve terminals in rat brain, induced by 5,6-dihydroxytryptamine. Z. Zellforsch. **125:** 553–569.
31. BAUMGARTEN, H. G., & A. BJÖRKLUND, A. F. HOLSTEIN & A. NOBIN. 1972. Chemical degeneration of indoleamine axons in rat brain by 5,6-dihydroxytryptamine: an ultrastructural study. Z. Zellforsch. **129:** 256–271.
32. BAUMGARTEN, H. G., L. LACHENMAYER, A. BJÖRKLUND, A. NOBIN & E. ROSENGREN. 1973. Long-term recovery of serotonin concentrations in the rat CNS following 5,6-dihydroxytryptamine. Life Sci. **12**(1): 357–364.
33. BJÖRKLUND, A., A. NOBIN & U. STENEVI. 1973. Regeneration of central

serotonin neurons after axonal degeneration induced by 5,6-dihydroxytryptamine. Brain Res. **50:** 214–220.
34. Massotti, M., A. Scotti de Carolis & G. Longo. 1974. Effects of three dihydroxylated derivatives of tryptamine on the behavior and on brain amine content in mice. Pharmacol. Biochem. Behav. **2:** 769-775.
35. Baumgarten, H. G. & H. G. Schlossberger. 1973. Effects of 5,6-dihydroxytryptamine on brain monoamine neurons in the rat. *In* Serotonin and Behavior. J. Barchas & E. Usdin, Eds. : 220–221. Academic Press, Inc. New York, N.Y.
36. Breese, G. R., B. R. Cooper, L. D. Grant & R. D. Smith. 1974. Biochemical and behavioural alterations following 5,6-dihydroxytryptamine administration into brain. Neuropharmacology **13:** 177–187.
37. Campaigne, E., R. B. Rogers, A. Donelson & T. Bosin. 1973. Benzo[b]thiophene derivatives. XX. The sulfur isostere of 5,6-dihydroxytryptamine. J. Heterocyclic Chem. **10:** 979–980.
38. Bosin, T. R., A. R. Buckpitt & R. P. Maickel. 1974. Substrate specificity of aromatic-L-amino acid decarboxylase. Life Sci. **14:** 899–908.
39. Harrison, S. D., Jr., T. R. Bosin & R. P. Maickel. 1974. Metabolism of the benzo[b]thiophene analog of N,N-dimethyltryptamine *in vivo* and *in vitro* in the rat: application of organic chemical and spectroscopic techniques to problems of drug metabolism. Drug Metab. Disposition **2:** 228–236.
40. Bosin, T. R., D. R. Hathaway & R. P. Maickel. 1974. Intestinal absorption of tryptophan isomers and analogs. Arch. Int. Pharmacodyn. **212:** 32–35.
41. Bosin, T. R., D. R. Hathaway & R. P. Maickel. 1975. Intestinal transport of tryptophan and its analogs. Amer. J. Physiol. **228:** 496–500.
42. Maickel, R. P. & F. P. Miller. 1966. Fluorescent products formed by reaction of indole derivatives and o-phthalaldehyde. Anal. Chem. **38:** 1937–1938.
43. Erspamer, V. 1966. Bioassay of indolealkylamines. *In* Handbook of Experimental Pharmacology. V. Erspamer, Ed. Vol. XIX: 113–139. Springer Verlag. Berlin, Federal Republic of Germany.
44. Daly, J. W., C. R. Creveling & B. Witkop. 1966. The chemorelease of norepinephrine from mouse hearts. Structure-activity relationships. I. Sympathomimetic and related amines. J. Med. Chem. **9:** 273–277.

Discussion

Dr. C. R. Creveling: Do you have any evidence that 5,6-DHT-S undergoes a spontaneous decarboxylation to generate 5,6-dihydroxy-S?

Dr. Maickel: No, if anything, this compound seems to be very stable under most biologic conditions.

Dr. Creveling: Is this compound sequestered in fatty tissues and stored there? What is its tissue half-life?

Dr. Maickel: We do not know the answers to your questions. We have only administered the compound intraventricularly and have not given it parenterally.

Dr. H. G. Baumgarten: Would you agree that if you had a *p*-chlorophenylalanine-pretreated animal, and then gave that animal 5,6-DHT, the multiple behavioral effects that you might see, apart from a possible slight tryptaminergic receptor stimulation caused by 5,6-DHT itself, would be caused mainly by an action of 5,6-DHT on dopaminergic neurons?

Dr. Maickel: I would agree with you; however, at this point, I am not sure that we could differentiate between the effects of 5,6-DHT on brain dopaminergic versus noradrenergic neurons. For example, we do not see signs of behavioral hyperactivity 24 hr after a dose of *p*-chlorophenylalanine, when both brain serotonin and norepinephrine concentrations are reduced. However, only in the 6–10-day period after the treatment, when only brain serotonin but not dopamine or norepinephrine concentrations are reduced, do we see signs of locomotor stimulation.

DR. M. DA PRADA: We do not see changes in brain norepinephrine after intraventricular injections of 5,6-DHT. How much of the drug do you use?

DR. MAICKEL: The minimum effective dose for these effects is about 40–80 μg, given intraventricularly to the rat.

DR. DA PRADA: You have also injected 5,6-DHT directly into the medial forebrain bundle, but you see no apparent lesions of the norepinephrine system.

DR. MAICKEL: We do not see great reductions with this route of administration (there is approximately a 25% decrease in the concentration of norepinephrine).

DR. M. RADULOVACKI (*University of Illinois, Chicago, Ill.*): I noticed that one of the compounds you studied, 5-hydroxy-6-methoxytryptamine, produced convulsive episodes. Is there ony evidence that this compound might alter dopaminergic neurotransmission?

DR. MAICKEL: That is possible, although we don't have any direct evidence to support that hypothesis. I don't know whether anyone has yet looked at the possible effects of this compound on brain dopaminergic neurons.

DR. J. H. JACOBY: I find that some of the actions of 5,6-DHT-S that you describe are very interesting. Some of these results might be interpreted as indicating that 5,6-DHT-S has a direct serotonin-receptor antagonistic action. Since you find that there are large increases in the levels of brain 5-hydroxyindoleacetic acid, even though serotonin levels do not change, I was wondering if you have looked at possible changes in brain tryptophan concentrations under these conditions?

DR. MAICKEL: We have only looked at them in one experiment. Although there were no apparent changes in this amino acid, the standard errors were quite large and may have masked an effect. In addition, we did not perform a detailed time study of possible changes in tryptophan concentrations in these experiments.

STRUCTURE-ACTIVITY RELATIONSHIPS AMONG THE HALOGENATED AMPHETAMINES

Ray W. Fuller

The Lilly Research Laboratories
Eli Lilly and Company
Indianapolis, Indiana 46206

Lowering of brain serotonin by halogenated amphetamines was initially observed by Pletscher *et al.*,[1-3] who primarily studied *p*-chloro-*N*-methylamphetamine (*p*-CMA). A series of halogenated amphetamines had been prepared in the Lilly Research Laboratories by Dr. Jack Mills as potential appetite suppressant drugs;[4, 5] we examined structure-activity relationships among those compounds and reported that *p*-chloroamphetamine (*p*-CA) was somewhat more active than *p*-CMA in lowering brain serotonin 16 hr after drug injection in rats.[6, 7] Later studies revealed that *p*-CMA is extensively converted to *p*-CA in rats and that most of its *in vivo* effects may be due to *p*-CA (see below). The mechanism of action of *p*-CA might initially have been described as obscure and now must be described as complex. Our studies of *p*-CA have emphasized the use of structural analogs as tools in elucidating and dissociating the multiple actions that *p*-CA apparently exerts on serotonin neurons.

Multiple Actions of *p*-CA on Serotonin Neurons

The initial observations that the concentrations of both serotonin and 5-hydroxyindoleacetic acid (5-HIAA) in the brain were depressed by *p*-CMA and *p*-CA[1, 2, 8] suggested that 5-hydroxyindole synthesis was inhibited (agents like reserpine that deplete serotonin by release increase 5-HIAA as serotonin is exposed to attack by monoamine oxidase). Brain tryptophan hydroxylase was not measured directly because methods were then unavailable, but the halogenated amphetamines were shown not to inhibit tryptophan hydroxylation by rat liver[2, 7] or by *Chromobacterium violaceum*[7] and to have no effect on 5-hydroxytryptophan decarboxylation.[2] Since the halogenated amphetamines released serotonin from platelets *in vitro* but prevented its conversion to 5-HIAA,[3] the proposal was made that the compounds lowered serotonin *in vivo* by releasing it from storage granules and lowered 5-HIAA by inhibition of monoamine oxidase. This mechanism left unanswered the fate of the released serotonin and was insufficient for explaining the effects produced by the halogenated amphetamines, though *p*-CA probably does release serotonin and inhibit monoamine oxidase *in vivo*.[9-11] Sanders-Bush *et al.*[12] made the important observation that *p*-CA does inhibit tryptophan hydroxylation *in vivo* in brain, and this inhibition of 5-hydroxyindole synthesis is probably the *major* mechanism that accounts for the lowering of serotonin and 5-HIAA levels. The molecular basis for this effect on tryptophan hydroxylase *in vivo* is unknown, because *p*-CA does not inhibit the enzyme *in vitro*, except at very high concentrations. *p*-CA and *p*-CMA are potent inhibitors of serotonin uptake by brain slices[13] and synaptosomes[14] *in vitro*, and inhibition of serotonin reuptake probably occurs *in vivo,* though such inhibition would not result in depletion of serotonin levels.

The actions of *p*-CA on serotonin neurons are summarized in TABLE 1. The initial effects probably include inhibition of tryptophan hydroxylation, release of serotonin from intraneuronal storage granules, inhibition of serotonin reuptake by the neuronal membrane pump, and inhibition of serotonin oxidation by monoamine oxidase. Lowering of serotonin by *p*-CA may occur largely through inhibition of tryptophan hydroxylation, but these actions seem to vary in prominence among various *p*-CA analogs, as will be discussed. Lowering of 5-HIAA may occur via inhibition of tryptophan hydroxylation, inhibition of monoamine oxidase, and inhibition of serotonin reuptake.

Not only does *p*-CA produce rapid depletion of brain serotonin, but the effects are also of extremely long duration. Sanders-Bush *et al.*[15] first recognized that serotonin levels were still depressed for as long as 4 months after a single injection of *p*-CA in rats. All parameters of serotonin neural function (tryptophan hydroxylase,[15] serotonin concentration[15, 16] and turnover,[16] 5-HIAA,[15, 16] and serotonin uptake[17]) are depressed at these long time intervals, which suggests that some serotonin neurons have been destroyed, and Harvey *et al.* have published histologic evidence for selective neurotoxic effects of *p*-CA.[18, 19] The initial effects of *p*-CA are not permanent, however, and can be readily reversed for a few hours after *p*-CA administration[20,21] (TABLE 2). Thus, the rapid initial effects and the long-lasting destructive effects of *p*-CA on serotonin neurons can be readily distinguished.

TABLE 1

ACTIONS OF *p*-CA THAT MAY INFLUENCE 5-HYDROXYINDOLE CONCENTRATION IN BRAIN AND THAT ARE POTENTIALLY DISSOCIABLE

Action	Influence on Concentration of	
	Serotonin	5-HIAA
Inhibition of tryptophan hydroxylation	decrease	decrease
Release of serotonin	decrease	increase
Inhibition of serotonin reuptake	no change	decrease
Inhibition of monoamine oxidase	increase	decrease
Irreversible damage to serotonin neuronal function *	decrease	decrease

* All parameters specifically associated with serotonin neurons are decreased: tryptophan hydroxylase, high-affinity uptake of serotonin, serotonin and 5-HIAA concentrations, and serotonin turnover.

By studying structural analogs of *p*-CA, we have been able to dissociate these initial and long-term effects and have further been able to distinguish the four multiple initial actions listed above. The structures of several of these analogs are shown in FIGURE 1.

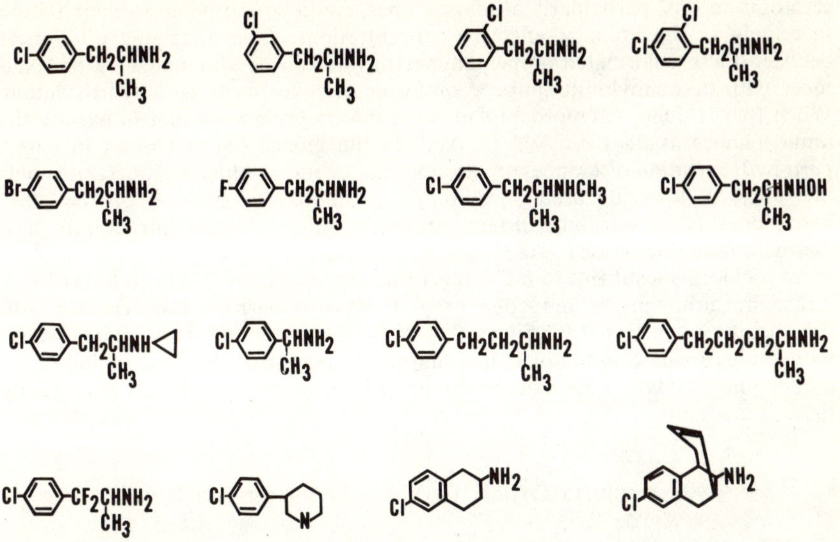

FIGURE 1. Structures of *p*-CA and some of its analogs. All of the compounds used were racemic mixtures of optical isomers.

INFLUENCE OF CHLORINE POSITION ON THE RING

In our initial studies, we observed no significant depletion of brain serotonin 16 hr after administration of either *o*-chloro- or *m*-chloroamphetamines, in contrast to the marked depletion that occurred after injection of *p*-CA.[7] A more careful analysis revealed that these compounds are, like amphetamine, rapidly metabolized in the rat.[22] The rat metabolizes amphetamine mainly by *para*-hydroxylation, and amphetamines that contain a substituent in this position persist in the brain and other tissues much longer than does amphetamine itself.[23] Thus, the inability of the *o*-chloro and *m*-chloro compounds to affect

TABLE 2

DURATION OF REVERSIBILITY OF SEROTONIN DEPLETION BY *p*-CA AND *p*-BROMOAMPHETAMINE (*p*-BA)

Hours After Injection of *p*-CA or *p*-BA *	Reversal (%)	
	p-CA	*p*-BA
4	100	92
8	87	79
24	31	20
48	0	0

*Uptake inhibitor (fluoxetine hydrochloride) was injected at this time, and serotonin concentration was measured 24 hr later. *p*-CA data were calculated from Reference 21, and *p*-BA data were from similar unpublished studies.[55]

serotonin in rats, particularly at longer times, could be attributed to their failure to remain in the brain at effective concentrations. The metabolic differences between these chlorinated amphetamines in rats can be eliminated by pretreatment with desmethylimipramine[22] or iprindole[24] to block *para*-hydroxylation. When that is done, *m*-chloroamphetamine lowers brain serotonin in exactly the same manner as does *p*-CA.[22, 24] And, in the guinea pig, a species in which *para*-hydroxylation of amphetamines does not occur, *m*-chloroamphetamine acts like *p*-CA in lowering brain serotonin.[25] *o*-Chloroamphetamine, on the other hand, does not lower but, instead, increases serotonin concentration in desmethylimipramine-treated rats[22] and guinea pigs.[26] Furthermore, the addition of an *o*-chloro substituent to *p*-CA markedly diminishes its ability to lower brain serotonin, although the reduction of 5-HIAA was perhaps even greater with the dichloro compound.[26] Since the dichloro compound is a more potent monoamine oxidase inhibitor,[26] this finding supports the idea that inhibition of monoamine oxidase does contribute to the decrease of 5-HIAA caused by these compounds.

Substituents Other Than Chlorine on the Ring

p-Bromoamphetamine[27] and *p*-bromomethamphetamine[28, 29] lower brain serotonin to an extent comparable to the reduction caused by the corresponding *p*-chloro compounds. *p*-Fluoroamphetamine decreases serotonin levels, but its effects do not persist to the same degree as those of *p*-CA.[27, 30] Other *para*-substituted amphetamines have less activity (trifluoromethyl, phenoxy) or no activity (methyl, methoxy) as serotonin depletors.[26]

Norfenfluramine is an analog of *p*-CA that bears an *m*-trifluoromethyl substituent instead of a *p*-chloro substituent on the ring. Norfenfluramine and fenfluramine (the *N*-ethyl compound) lower serotonin in rat brain for a very long time, though they are slightly less active than *p*-CA.[31–34] Although Costa and Revuelta[32] have reported that norfenfluramine increases serotonin turnover instead of decreasing it, as *p*-CA does, we find that norfenfluramine causes a rapid reduction of tryptophan hydroxylase accompanying the decline in serotonin, just as *p*-CA does (TABLE 3). Sanders-Bush *et al.* have reported short- and long-term depletion of serotonin and tryptophan hydroxylase in rat brain after injection of fenfluramine.[17] In general, the actions of norfenfluramine and fenfluramine on brain serotonin neurons may be very similar to those of *p*-CA and other chlorinated amphetamines.

Influence of Substituents on the Nitrogen

As mentioned previously, the initial observation of serotonin lowering by halogenated amphetamines was made with *p*-CMA,[1, 2] which is almost as effective as *p*-CA itself.[6, 7] In fact, more recent studies indicate that *p*-CMA is rapidly and extensively metabolized to *p*-CA in the rat (TABLE 4). We have compared other *N*-alkyl derivatives of *p*-CA and find that they all are less effective than *p*-CA in depleting brain serotonin in rats.[7, 55] *N*-Ethyl, *N*-isopropyl, and *N*,*N*-dimethyl derivatives lower brain serotonin and are metabolized by *N*-dealkylation to *p*-CA in the rat. The extent of their conversion to *p*-CA may be adequate to account for their lowering of serotonin; that is, they may

TABLE 3

RAPID DEPLETION OF BRAIN 5-HYDROXYINDOLES AND TRYPTOPHAN HYDROXYLASE BY *m*-TRIFLUOROMETHYLAMPHETAMINE (NORFENFLURAMINE) IN RATS *

Minutes	Brain 5-Hydroxyindoles (μg/g) Serotonin	5-HIAA	Tryptophan Hydroxylase
0	0.55±0.02	0.56±0.02	24.5±0.5
30	0.44±0.02 †	0.58±0.01	18.9±0.6 †
60	0.31±0.02 †	0.52±0.02	19.9±0.7 †
120	0.21±0.005 †	0.47±0.01 †	16.6±0.2 †
240	0.19±0.01 †	0.37±0.01 †	15.3±0.6 †

* Norfenfluramine hydrochloride (a gift from the A. H. Robins Company) was injected intraperitoneally at 15 mg/kg. Brain 5-hydroxyindoles [35] and tryptophan hydroxylase [36] were measured fluorometrically. Tryptophan hydroxylase units are expressed in micrograms of 5-hydroxytryptophan formed per gram of brain per hour. Mean values and standard errors for five rats per group are shown.

† $p < 0.01$, lower than time zero.

all act via conversion to *p*-CA. We have been unable to block this metabolic conversion by any drug to determine whether the *N*-alkyl compounds are capable of depleting serotonin on their own.

N-Hydroxyl-*p*-CA was an analog of interest because hydroxylamines are unstable and chemically reactive. We thought that *N*-hydroxyl-*p*-CA might be a (the) neurotoxic metabolite that has been considered as a possible mediator of the long-term effects of *p*-CA. However, we found that *N*-hydroxy-*p*-CA was, in fact, rapidly and quantitatively converted to *p*-CA, so that its effects on brain serotonin were indistinguishable from those of *p*-CA.[37]

N-Cyclopropylarylalkylamines are irreversible inhibitors of monoamine oxidase,[38, 39] in general, and *N*-cyclopropyl-*p*-CA is therefore a unique analog of *p*-CA in that the monoamine oxidase-inhibiting component of its activity plays

TABLE 4

SEROTONIN DEPLETION AND *p*-CA CONCENTRATION IN RAT BRAIN AFTER INJECTION OF *p*-CA OR *p*-CMA *

Drug	Treatment Interval/(hr)	5-Hydroxytryptamine (μg/g)	*p*-CA (nmol/g)
None		0.55±0.02	—
p-CA	2	0.32±0.01 †	91±3
	4	0.28±0.01 †	89±5
	8	0.19±0.01 †	86±4
p-CMA	2	0.37±0.01 †	58±3
	4	0.29±0.01 †	64±4
	8	0.19±0.01 †	77±7

* Drugs were injected intraperitoneally at 0.1 mmol/kg. Means and standard errors for five rats per group are shown.

† $p < 0.01$, different from control group.

a greater part in its action on serotonin. *N*-Cyclopropyl-*p*-CA does not lower serotonin initially but, instead, produces a delayed and long-lasting depletion.[40] 5-Hydroxyindoleacetic acid is lowered immediately, probably due largely to inhibition of monoamine oxidase.[40] *N*-Cyclopropyl-*p*-CA does cause a rapid reduction of tryptophan hydroxylase in brain, as does *p*-CA.[41] Thus, there are initially two opposing influences on serotonin concentration by this agent: inhibition of serotonin synthesis and inhibition of serotonin degradation, the net effect being little change in serotonin concentration at first. The intraneuronal effects of *N*-cyclopropyl-*p*-CA on serotonin metabolism can be prevented by prior administration of an uptake inhibitor to block its entry via the membrane pump.[41] Uptake inhibition has no detectable influence on overall inhibition of brain monoamine oxidase, probably because the monoamine oxidase inside serotonin neurons contributes an undetectably small portion to total brain activity.[41] Thus, an uptake inhibitor removes the inhibition of serotonin synthesis but not the inhibition of serotonin oxidation, so the result is a rapid increase in serotonin concentration.[41] Although serotonin oxidation may normally occur inside serotonin neurons, in the presence of an uptake inhibitor, serotonin presumably is oxidized by monoamine oxidase outside serotonin neurons.

Influence of Side Chain Length

The shorter homolog of *p*-CA, *p*-chloro-α-methylbenzylamine, is only a weak inhibitor of serotonin uptake and of monoamine oxidase *in vitro* and does not affect serotonin or 5-HIAA concentration in the brain *in vivo*.[42] Longer homologs of *p*-CA, on the other hand, selectively retain some of the pharmacologic actions of *p*-CA. The first longer homolog (one additional methylene unit connecting the phenyl ring and the amine-bearing carbon) is slightly more active than *p*-CA as an inhibitor of monoamine oxidase *in vitro* and slightly increases serotonin levels while decreasing 5-HIAA levels in rat brain, presumably due to weak inhibition of monoamine oxidase *in vivo*.[42] The next longer homolog, with two additional methylene units, is a weaker inhibitor of monoamine oxidase *in vitro* and does not significantly change serotonin levels *in vivo*.[42] It slightly lowers 5-HIAA levels *in vivo*.[42] Both of these two longer homologs of *p*-CA retain essentially full activity as a serotonin uptake inhibitor *in vitro;* this activity probably accounts fully for the 5-HIAA reduction by the longest homolog and partly for the 5-HIAA lowering by the intermediate homolog. Both of these longer homologs can block the depletion of serotonin by *p*-CA, demonstrating their ability to inhibit uptake into serotonin neurons *in vivo*.[42] Thus, lengthening the side chain of *p*-CA abolishes the serotonin-depleting action but retains other actions of *p*-CA.

β,β-Difluoro-*p*-CA

When the β-carbon of *p*-CA contains two fluorine atoms in place of hydrogen atoms, the basicity of the amine group is greatly reduced due to the strong electron-withdrawing property of fluorine.[43] This unique *p*-CA analog has a pK_a of 6.8 and so exists at physiologic pH mainly in the neutral form instead of being virtually completely ionized, as is *p*-CA itself (pK_a 9.3). Presumably, as a consequence of the existence of most molecules of β,β-difluoro-*p*-CA in the uncharged form and of nearly all molecules of *p*-CA as cations, these two substances differ strikingly in their biologic disposition. The difluoro compound is localized very heavily in adipose tissue, the tissue in which *p*-CA itself is present in lowest concentration.[43] Because of the localization of β,β-difluoro-*p*-CA in fat, its concentrations in brain are lower than those of *p*-CA after equimolar doses; it must be given at about four times the dose of *p*-CA to achieve initial concentrations in brain that are equivalent. Even then, the concentration of β,β-difluoro-*p*-CA is not maintained in the brain, because it is metabolized much more rapidly and extensively than is *p*-CA itself in the rat.[43] Although β,β-difluoro-*p*-CA initially depletes brain serotonin and 5-HIAA to the same extent as does *p*-CA itself when brain levels of the two drugs are equivalent, the effects of the difluoro compound on brain 5-hydroxyindoles in the rat are only transient and disappear within 24 hr.[43] Prior induction of liver microsomal enzymes with phenobarbital accelerates the already-rapid metabolism of β,β-difluoro-*p*-CA and prevents the transient depletion of 5-hydroxyindoles.[44] This effect of phenobarbital apparently occurs because the β,β-difluoro substitution on amphetamines makes the drugs more susceptible to deamination by hepatic enzymes.[45]

p-CA Analogs with Rigid Conformation

Chlorinated 3-phenylpiperidines are analogs of chloroamphetamines (though they lack the α-methyl group) with rigid conformations in which the nitrogen is part of a second ring structure that involves the β-carbon. The *p*-chloro, *m*-chloro, and *m,p*-dichloro compounds in this series lower serotonin and 5-HIAA levels in rat brain, but their effect is smaller than that of *p*-CA and is short-lasting.[46] In other respects, such as tissue distribution and influence of desmethylimipramine on brain levels, the chlorinated 3-phenylpiperidines resemble the chloroamphetamines.[46]

6-Chloro-2-aminotetralin is another analog of *p*-CA that has a rigid conformation. This compound also depletes brain serotonin and 5-HIAA, but, again, its effect is short-lasting and smaller than that of *p*-CA itself.[47] A particularly interesting compound that can be considered as a rigid *p*-CA analog derived from 6-chloro-2-aminotetralin by the addition of a third ring structure is a compound known as Org 6582. This compound was reported recently by Sugrue *et al.*[48] to be a potent and specific inhibitor of serotonin uptake. We have compared it to 6-chloro-2-aminotetralin.[49] Org 6582 has lost the ability of *p*-CA and 6-chloro-2-aminotetralin to inhibit tryptophan hydroxylase, to deplete serotonin, and to inhibit monoamine oxidase *in vivo* but retains the ability to inhibit serotonin uptake. It thus is a potent and specific inhibitor of serotonin uptake with a long duration of action.

Possible Mechanisms of Long-Term Effects of *p*-CA

The long-term depletion of 5-hydroxyindole levels and of tryptophan hydroxylase and serotonin reuptake capacity suggests that the function of some serotonin neurons has been permanently impaired by *p*-CA, and histologic changes consistent with neurotoxicity have been observed.[18, 19] The mechanism by which *p*-CA exerts these long-lasting effects is not known, but attention has focused on the possibility that a neurotoxic metabolite is formed from *p*-CA.[50-52] Despite this attention, no compelling evidence that a metabolite of *p*-CA plays any part in its actions on serotonin has emerged. If such a metabolite is involved, it most likely is formed inside the serotonin neuron, because *p*-CA itself is selectively taken up into serotonin neurons, but there is no *a priori* reason that a metabolite would be, and because a metabolite so highly reactive that it would destroy a neuron is unlikely to reach the site of action in brain if it were formed in a peripheral tissue, such as the liver. Since no evidence in favor of a toxic metabolite of *p*-CA has been uncovered, perhaps the idea that none is involved deserves more consideration. *p*-CA itself may cause irreversible damage to the serotonin neuron by virtue of being concentrated extensively inside the neuron over a prolonged time. Baliga *et al.*[53] have shown that amphetamines can inhibit protein synthesis by interfering with initiation; *p*-CA was more active than amphetamine itself in this regard. Perhaps *p*-CA attains concentrations inside the serotonin neuron high enough to inhibit the formation of key cell proteins and persists at these high concentrations for a long duration as a consequence of continual reuptake by the amine pump on the serotonin neuronal membrane. This mechanism for the long-lasting effects of *p*-CA would be compatible with the observations that an uptake inhibitor can reverse serotonin depletion (and tryptophan hydroxylase lowering) by *p*-CA initially and that the reversibility gradually declines to zero within 32–48 hr after *p*-CA injection. This period may be the time for which the serotonin neuron can survive without the resynthesis of essential proteins required for general cell function. The data summarized in TABLE 2 may be instructive in considering mechanisms by which serotonin depletion by *p*-CA and analogs becomes irreversible. One might anticipate that *p*-CA and *p*-bromoamphetamine would be metabolized at different rates, so that their effects would become irreversible at different rates if the irreversible effects were due to a toxic metabolite. This possibility would be particularly applicable to metabolism on the phenyl ring, for instance, the formation of an arene oxide intermediate due to attack by a mono-oxygenase.[54] However, the rates at which the effects of these two compounds become irreversible were virtually identical. This finding would be compatible with a theory that interference with some essential cellular process for a critical period is involved; that period would be independent of the inhibitory agent. The ability of some analogs of *p*-CA (e.g., 6-chloro-2-aminotetralin, *p*-fluoroamphetamine, *β*-*β*-difluoro-*p*-CA) to lower serotonin initially but not for a long duration may reflect a relatively poorer affinity for the uptake pump, so that these analogs are not maintained inside serotonin neurons at adequate concentrations for sufficiently long periods to produce irreversible damage.

SUMMARY

p-Chloroamphetamine and various analogs influence brain serotonin neurons through multiple actions. Comparison of these compounds has permitted the distinction between short-term and long-term depletion of serotonin and among inhibition of tryptophan hydroxylation, release of serotonin, inhibition of serotonin reuptake, and inhibition of monoamine oxidase as mechanisms involved in the actions of these agents on serotonin neurons.

ACKNOWLEDGMENTS

I am grateful to Drs. Jack Mills and Bryan B. Molloy and their associates for the synthesis of the *p*-CA analogs; to Drs. David T. Wong, C. John Parli, and E. Martin Gál for collaborative assistance in various phases of our studies; and to Harold D. Snoddy, Kenneth W. Perry, John C. Baker, and Susan Hemrick for their experimental work.

REFERENCES

1. PLETSCHER, A., W. P. BURKARD, H. BRUDERER & K. F. GEY. 1963. Decrease of cerebral 5-HT and 5-hydroxyindoleacetic acid by an aryl alkyl amine. Life Sci. **2:** 828–833.
2. PLETSCHER, A., G. BARTHOLINI, H. BRUDERER, W. P. BURKARD & K. F. GEY. 1964. Chlorinated arylalkylamines affecting the cerebral metabolism of 5-hydroxytryptamine. J. Pharmacol. Exp. Ther. **145:** 344–350.
3. PLETSCHER, A., M. DA PRADA, G. BARTHOLINI, W. P. BURKARD & H. BRUDERER. 1965. Two types of monoamine liberation by chlorinated aralkylamines. Life Sci. **4:** 2301–2308.
4. OWEN, J. E., JR. 1963. Psychopharmacological studies of some 1-(chlorophenyl)-2-aminopropanes. I. Effects on appetite-controlled behavior. J. Pharm. Sci. **52:** 679–683.
5. OWEN, J. E., JR. 1963. Psychopharmacological studies of some 1-(chlorophenyl)-2-aminopropanes. II. Effects on avoidance and discrimination behavior. J. Pharm. Sci. **52:** 684–688.
6. FULLER, R. W., C. W. HINES & J. MILLS. 1964. Lowering of rat brain serotonin levels by arylalkylamines. Fed. Proc. **23:** 146.
7. FULLER, R. W., C. W. HINES & J. MILLS. 1965. Lowering of brain serotonin level by chloroamphetamines. Biochem. Pharmacol. **14:** 483–488.
8. SANDERS-BUSH, E. & F. SULSER. 1970. p-Chloroamphetamine: *in vivo* investigations on the mechanism of action of the selective depletion of cerebral serotonin. J. Pharmacol. Exp. Ther. **175:** 419–426.
9. FULLER, R. W 1966. Serotonin oxidation by rat brain monoamine oxidase: inhibition by 4-chloroamphetamine. Life Sci. **5:** 2247–2252.
10. GALLAGER, D. W. & E. SANDERS-BUSH. 1973. *In vivo* measurement of the release of 5-hydroxytryptamine (5HT) from the hippocampus of the rat: effect of R04–1284, pargyline and p-chloroamphetamine (PCA). Fed. Proc. **32:** 303.
11. TRULSON, M. E. & B. L. JACOBS. 1976. Behavioral evidence for the rapid release of CNS serotonin by PCA and fenfluramine. Eur. J. Pharmacol. **36:** 149–154.
12. SANDERS-BUSH, E., J. A. BUSHING & F. SULSER. 1972. p-Chloroamphetamine—inhibition of cerebral tryptophan hydroxylase. Biochem. Pharmacol. **21:** 1501–1510.
13. CARLSSON, A. 1970. Structural specificity for inhibition of [^{14}C]-5-hydroxytryptamine uptake by cerebral slices. J. Pharm. Pharmcol. **22:** 729–732.
14. WONG, D. T., J.-S. HORNG & R. W. FULLER. 1973. Kinetics of serotonin accumulation into synaptosomes of rat brain—effects of amphetamine and chloroamphetamines. Biochem. Pharmacol. **22:** 311–322.
15. SANDERS-BUSH, E., J. A. BUSHING & F. SULSER. 1972. Long-term effects of p-chloroamphetamine on tryptophan hydroxylase activity and on the levels of 5-hydroxytryptamine and 5-hydroxyindole acetic acid in brain. Eur. J. Pharmacol. **20:** 385–388.
16. FULLER, R. W. & H. D. SNODDY. 1974. Long-term effects of 4-chloroamphet-

amine on brain 5-hydroxyindole metabolism in rats. Neuropharmacology **13:** 85–90.
17. Sanders-Bush, E., J. A. Bushing & F. Sulser. 1975. Long-term effects of p-chloroamphetamine and related drugs on central serotonergic mechanisms. J. Pharmacol. Exp. Ther. **192:** 33–41.
18. Harvey, J. A., S. E. McMaster & L. M. Yunger. 1975. p-Chloroamphetamine: selective neurotoxic action in brain. Science **187:** 841–843.
19. Harvey, J. A. & S. E. McMaster. 1976. Neurotoxic action of para-chloroamphetamine in the rat as revealed by Nissl and silver stains. Psychopharmacol. Bull. **12:** 62–64.
20. Fuller, R. W., K. W. Perry & B. B. Molloy. 1975. Effect of 3-(p-trifluoromethylphenoxy)-N-methyl-3-phenylpropylamine on the depletion of brain serotonin by 4-chloroamphetamine. J. Pharmacol. Exp. Ther. **193:** 796–803.
21. Fuller, R. W., K. W. Perry & B. B. Molloy. 1975. Reversible and irreversible phases of serotonin depletion by 4-chloroamphetamine. Eur. J. Pharmacol. **33:** 119–124.
22. Fuller, R. W., R. J. Schaffer, B. W. Roush & B. B. Molloy. 1972. Drug disposition as a factor in the lowering of brain serotonin by chloroamphetamines in the rat. Biochem. Pharmacol. **21:** 1413–1417.
23. Fuller, R. W. & C. W. Hines. 1967. Tissue levels of chloroamphetamines in rats and mice. J. Pharm. Sci. **56:** 302–303.
24. Fuller, R. W. & J. C. Baker. 1974. Long-lasting reduction of brain 5-hydroxytryptamine levels by 3-chloroamphetamine and 4-chloroamphetamine in iprindole-treated rats. J. Pharm. Pharmacol. **26:** 912–914.
25. Fuller, R. W. 1972. Species difference in the lowering of brain 5-hydroxytryptamine by meta-chloroamphetamine. J. Pharm. Pharmacol. **24:** 88.
26. Fuller, R. W., H. D. Snoddy, B. W. Roush & B. B. Molloy. 1973. Further structure-activity studies on the lowering of brain 5-hydroxyindoles by 4-chloroamphetamine. Neuropharmacology **12:** 33–42.
27. Fuller, R. W., J. C. Baker, K. W. Perry & B. B. Molloy. 1975. Comparison of 4-chloro-, 4-bromo- and 4-fluoroamphetamine in rats: drugs levels in brain and effects on brain serotonin metabolism. Neuropharmacology **14:** 739–746.
28. Leonard, B. E. & S. A. Shallice. 1971. Some neurochemical effects of amphetamine, methylamphetamine and p-bromomethylamphetamine in the rat. Brit. J. Pharmacol. **41:** 193–212.
29. Magyar, K. & J. Knoll. 1975. Para-substituted amphetamines and brain serotonin. Pol. J. Pharmacol. Pharm. **27**(Suppl.): 139–143.
30. Harvey, J. A., S. E. McMaster & R. W. Fuller. 1977. Comparison between the neurotoxic and serotonin depleting effects of various halogenated derivatives of amphetamine in the rat. J. Pharmacol. Exp. Ther. In press.
31. Costa, E., A. Groppetti & A. Revuelta. 1971. Action of fenfluramine on monoamine stores of rat tissues. Brit. J. Pharmacol. **41:** 57–64.
32. Costa, E. & A. Revuelta. 1972. Norfenfluramine and serotonin turnover rate in the rat brain. Biochem. Pharmacol. **21:** 2385–2393.
33. Clineschmidt, B. V., J. A. Totaro, J. C. McGuffin & A. B. Pflueger. 1976. Fenfluramine: long-term reduction in brain serotonin (5-hydroxytryptamine). Eur. J. Pharmacol. **35:** 211–214.
34. Harvey, J. A. & S. E. McMaster. 1975. Fenfluramine: evidence for a neurotoxic action on midbrain and a long-term depletion of serotonin. Psychopharmacol. Commun. **1:** 217–228.
35. Miller, F. P., R. H. Cox, Jr., W. R. Snodgrass & R. P. Maickel. 1970. Comparative effects of p-chloro-N-methylamphetamine on rat brain norepinephrine, serotonin and 5-hydroxyindole-3-acetic acid. Biochem. Pharmacol. **19:** 435–442.
36. Gál, E. M. & K. Patterson. 1973. Rapid nonisotopic assay of tryptophan-5-hydroxylase activity in tissues. Anal. Biochem. **52:** 625–628.

37. Fuller, R. W., K. W. Perry, J. C. Baker, C. J. Parli, N. Lee, W. A. Day & B. B. Molloy. 1974. Comparison of the oxime and the hydroxylamine derivatives of 4-chloroamphetamine as depletors of brain 5-hydroxyindoles. Biochem. Pharmacol. **23:** 3267–3272.
38. Mills, J., R. Kattau, I. H. Slater & R. W. Fuller. 1968. N-Substituted cyclopropylamines as monoamine oxidase inhibitors. Structure-activity relationships. Dopa potentiation in mice and *in vitro* inhibition of kynuramine oxidation. J. Med. Chem. **11:** 95–97.
39. Fuller, R. W. 1968. Kinetic studies and effects *in vivo* of a new monoamine oxidase inhibitor, N-[2-(o-chlorophenoxy)-ethyl]-cyclopropylamine. Biochem. Pharmacol. **17:** 2097–2106.
40. Fuller, R. W. & B. B. Molloy. 1973. Effects of N-cyclopropyl-4-chloroamphetamne on brain serotonin metabolism in rats. Res. Commun. Chem. Pathol. Pharmacol. **6:** 407–418.
41. Fuller, R. W. & K. W. Perry. 1977. Further studies on the effects of N-cyclopropyl-p-chloroamphetamine in rat brain. Neuropharmacology **16:** 495–497.
42. Fuller, R. W., K. W. Perry, D. T. Wong & B. B. Molloy. 1974. Effects of some homologues of 4-chloroamphetamine on brain serotonin metabolism. Neuropharmacology **13:** 609–614.
43. Fuller, R. W., H. D. Snoddy & B. B. Molloy. 1973. Effect of β,β-difluorosubstitution on the disposition and pharmacological effects of 4-chloroamphetamine in rats. J. Pharmacol. Exp. Ther. **184:** 278–284.
44. Fuller, R. W., J. C. Baker, K. W. Perry & B. B. Molloy. 1974. Effect of phenobarbital on the metabolism and serotonin-depleting action of 4-chloro-β,β-difluoroamphetamine in rats. Arch. Int. Pharmacodyn. **208:** 274–278.
45. Fuller, R. W. & B. B. Molloy. 1976. The effect of aliphatic fluorine on amine drugs. *In* Biochemistry Involving Carbon-Fluorine Bonds. R. Filler, Ed. : 77–98. American Chemical Society. Washington, D.C.
46. Fuller, R. W., B. B. Molloy, J. C. Baker, K. W. Perry & W. A. Day. 1975. Biochemical pharmacology of some chlorinated 3-phenyl-piperidines in rats. Arch. Int. Pharmacodyn. **214:** 263–270.
47. Fuller, R. W., K. W. Perry, J. C. Baker & B. B. Molloy. 1974. 6-Chloro-2-aminotetralin, a rigid conformational analog of 4-chloroamphetamine: pharmacologic properties of it and related compounds in rats. Arch. Int. Pharmacodyn. **212:** 141–153.
48. Sugrue, M. F., I. Goodlet & S. E. Mireylees. 1976. On the selective inhibition of serotonin uptake in vivo by Org 6582. Eur. J. Pharmacol. **40:** 121–130.
49. Fuller, R. W., D. T. Wong, H. D. Snoddy & F. P. Bymaster. 1977. Comparison of the effects of 6-chloro-2-aminotetralin and of Org 6582, a related chloroamphetamine analog, on brain serotonin metabolism in rats. Biochem. Pharmacol. **26:** 1333–1337.
50. Gál, E. M. 1976. The effects of intraventricularly administered para-chloroamphetamine and its analogs. Psychopharmacol. Bull. **12:** 52–54.
51. Sherman, A., E. M. Gál, R. W. Fuller & B. B. Molloy. 1975. Effects of intraventricular p-chloroamphetamine and its analogues on cerebral 5-HT. Neuropharmacology **14:** 733–737.
52. Sanders-Bush, E. 1976. Long-term effects of para-chloroamphetamine on central serotonergic mechanisms. Psychopharmacol. Bull. **12:** 57–59.
53. Baliga, B. S., J. Zahringer, M. Trachtenberg, M. A. Moskowitz & H. N. Munro. 1976. Mechanism of d-amphetamine inhibition of protein synthesis. Biochim. Biophys. Acta **442:** 239–250.
54. Parli, C. J. & B. Schmidt. 1975. Metabolism of 4-chloroamphetamine to 3-chloro-4-hydroxyamphetamine in rat: evidence for an *in vivo* "NIH shift" of chlorine. Res. Commun. Chem. Pathol. Pharmacol. **10:** 601–604.
55. Fuller, R. W., H. D. Snoddy & J. C. Baker. 1977. Unpublished results.

DISCUSSION

DR. A. S. HORN: Have you studied the various isomers of these compounds? If so, do they have differential actions?

DR. FULLER: All of the compounds that we have studied were administered as mixtures of the D- and the L-isomers. It would be interesting to know if they have differential effects; however, we haven't given this a high priority, because several years ago we found that the two isomers of *p*-chloroamphetamine appeared to be identical with respect to their effects on brain serotonergic neurons.

DR. E. SANDERS-BUSH: Have you looked at other possible antagonist effects of fluoxetine on *p*-chloroamphetamine or its derivatives? I ask this question because there is some debate about the mechanisms underlying the initial decrease in tryptophan hydroxylase activity seen after *p*-chloroamphetamine injections. Can this effect be reversed if animals are pretreated with fluoxetine?

DR. FULLER: In fact, all parameters that are depressed after *p*-chloroamphetamine treatment can be reversed by the administration of fluoxetine. For example, brain serotonin levels and tryptophan hydroxylase activity are normal within 4 hr after fluoxetine treatment. If the reversal in tryptophan hydroxylase activity is dependent on the synthesis of new enzyme protein, that implies that there is a more rapid rate of synthesis than one would anticipate on the basis of values for tryptophan hydroxylase enzyme turnover described previously in the literature. However, there is also no evidence that the reversal of *p*-chloroamphetamine-induced inhibition of tryptophan hydroxylase is related to differential inhibition by *p*-chloroamphetamine or its metabolites. In short, then, I really don't know what mechanisms might account for the ability of fluoxetine to rapidly reverse the effects of *p*-chloroamphetamine.

Some indirect evidence indicates to us that *p*-chloroamphetamine might act directly on serotonin-containing neurons, rather than via a neurotoxic metabolite of the compound. For example, in one study, we reasoned that if the metabolite was formed via a change in the ring structure of *p*-chloroamphetamine, a compound like *p*-bromoamphetamine might be converted to a neurotoxic metabolite at a rate different from that of *p*-chloroamphetamine. Since the rate of development of irreversibility of *p*-chloroamphetamine and *p*-bromoamphetamine were identical, we now think that *p*-chloroamphetamine exerts its effects directly on the neuron.

DR. E. M. GÁL: We have some evidence that *p*-chloroamphetamine produces a strong inhibition of tryptophan hydroxylase activity either *in vivo* or *in vitro,* if you use the natural cofactor for the enzyme in the assay. Have you or Dr. Sanders-Bush tried this cofactor?

DR. FULLER: The fact that *p*-chloroamphetamine-induced inhibition of tryptophan hydroxylase activity cannot be reversed by dialyzing tryptophan hydroxylase obtained from brain homogenates would argue against the possibility that *p*-chloroamphetamine's effects only occur in the presence of a natural cofactor but not an unnatural one. Hence, it seems unlikely that there is any direct inhibition of tryptophan hydroxylase by either *p*-chloroamphetamine or a metabolite that would be present in our preparation.

DR. SANDERS-BUSH: We have also found that *p*-chloroamphetamine decreases tryptophan hydroxylase activity *in vivo* whether we use Dr. Fuller's assay or the procedure developed by Dr. Gál.

Dr. Gál: There is some evidence that the more active form of the tryptophan hydroxylase enzyme molecule must be phosphorylated. Is there any possibility that this process might be interfered with shortly after *p*-chloroamphetamine administration?

Dr. Fuller: I think that something like your hypothesis might be possible.

Dr. J. H. Growdon: Immediately after injection of *p*-chloroamphetamine in normal animals, one sees a rather complex set of changes in motor behavior that can be reversed up to 30–45 min later by injecting these animals with fluoxetine.

ALTERATIONS IN THE NEUROCYTOTOXICITY OF 5,7-DIHYDROXYTRYPTAMINE BY PHARMACOLOGIC AGENTS IN ADULT AND DEVELOPING RATS *

G. R. Breese † and R. A. Mueller ‡

Departments of ‡ Pharmacology, † Psychiatry, and ‡ Anesthesiology
Biological Sciences Research Center
University of North Carolina School of Medicine
Chapel Hill, North Carolina 27514

In 1971, Baumgarten *et al.*[1] demonstrated that 5,6-dihydroxytryptamine (5,6-DHT) damages serotonin-containing fibers, an action comparable to that of 6-hydroxydopamine, which destroys catecholaminergic neurons.[2] Because systemic toxicity and nonspecificity of neural damage limited the usefulness of 5,6-DHT,[3, 4] analogs that might exhibit properties similar to those of 5,6-DHT were examined. One of these derivatives, 5,7-dihydroxytryptamine (5,7-DHT), was demonstrated to reduce brain serotonin and tryptophan hydroxylase and to induce axonal degeneration in areas that contain high concentrations of serotonin.[5-9] Unfortunately, 5,7-DHT also reduced brain norepinephrine content, indicating that noradrenergic fibers were affected.[5, 7, 8] Since some drugs had previously been found to reduce the cytotoxic actions of 6-hydroxydopamine on noradrenergic fibers,[2, 10] experiments were initiated to determine whether these drugs could antagonize the neurocytotoxic actions of 5,7-DHT on noradrenergic neurons.[8]

Another area of interest was the use of various dosage regimens or adjunctive drug treatments to magnify the destructive effect of 5,7-DHT on serotonin-containing neurons. Because administration of a monoamine oxidase (MAO) inhibitor had been shown to enhance the effect of 6-hydroxydopamine on dopaminergic fibers,[2, 11] parallel experiments were performed to determine whether MAO inhibitors would potentiate the effects of 5,7-DHT.[8] It has also been shown that administration of 6-hydroxydopamine produces a more dramatic neurocytotoxic effect on catecholamine-containing fibers when administered to immature rats.[12, 13] Therefore, the magnitude of the cytotoxic effect of 5,7-DHT was examined as a function of age. Finally, the various drug protocols developed to enhance the specificity of 5,7-DHT in adult rats were applied to animals of various ages with the hope of achieving the maximum biologic response with the greatest specificity. We will review those findings, discussing procedures that reduce the action of 5,7-DHT on noradrenergic fibers and maximize the destruction of serotonergic fibers.

* Supported by United States Public Health Service Grants HD-03110 and HD-10570.

0077–8923/78/0305–0160 $01.75/0

Materials and Methods

The first series of experiments was performed in adult male Sprague-Dawley rats (170–190 g) obtained from Zivic Miller Laboratories. Treated animals were given 200 μg of 5,7-DHT into the fourth ventricle (intracisternally) while under ether anesthesia. The 5,7-DHT was dissolved in sterile saline that contained 0.5% of ascorbic acid. Control animals received intracisternal injections of the saline vehicle. Five to 10 min prior to treatment, adult animals received phenobarbital sodium (20–30 mg/kg) or pentobarbital sodium (30 mg/kg) intraperitoneally to prevent seizures.[1]

Some animals were pretreated with various drugs before 5,7-DHT was injected intracisternally to determine what effect they would have on the monoamine changes produced by 5,7-DHT. These drugs included pargyline (50 mg/kg), iproniazid (100 mg/kg), tranylcypromine (10 mg/kg), pheniprazine (15 mg/kg), chlorpromazine (25 mg/kg), benztropine (5 mg/kg), *p*-chlorophenylalanine (2 × 150 mg/kg), brompheniramine (25 mg/kg), reserpine (2 × 2.5 mg/kg), chlorimipramine (25 mg/kg), fenfluramine (10 mg/kg), desipramine (25 mg/kg), and fluoxetine [Lilly 110,140: 3-(*p*-trifluoromethylphenoxy) *N*-methyl-3-phenylpropylamine hydrochloride, 15 mg/kg]. Drugs were injected 15 min (fluoxetine, brompheniramine, benztropine), 30 min (fenfluramine, pargyline, pheniprazine, tranylcypromine, iproniazid), or 60 min (chlorpromazine, desipramine, chlorimipramine) prior to administration of 5,7-DHT. Reserpine and *p*-chlorophenylalanine were administered 48 and 24 hr, respectively, before the intracisternal injection of 5,7-DHT.

In other experiments, 1-day-old rats received various doses of 5,7-DHT intracisternally in 10 μl of vehicle. Some animals were pretreated with desipramine (20 mg/kg) or pargyline (40 mg/kg) when 3 days of age before receiving a 50-μg dose of 5,7-DHT. Animals were sacrificed at 30–40 days of age to assay brain monoamine content. The effects of these treatments on locomotor activity and body weight gain were assessed at weekly intervals from 7 days of age. For assessment of locomotor activity, three rat pups were placed in photocell activity cages.[14] Each cage had a circular runway with an outside diameter of 12 in. Pulses generated from the six photocell sensors were recorded by a microprocessor. Activity was recorded at 5-min intervals over a 50-min test period. Other groups of rats were injected with *p*-chloroamphetamine (25 mg/kg) or fenfluramine (15 mg/kg) at 1, 3, and 5 days of age. The effects of these treatments on brain monoamine content were assessed at 60 days.

Serotonin and catecholamines were analyzed in the brain as described previously.[11]

Results

Effect of Various Centrally Acting Drugs on the Actions of 5,7-DHT in Adult Rats

In accord with previous findings,[5-8] intracisternal administration of 5,7-DHT to adult rats was found to cause a significant reduction in serotonin and norepinephrine content in the brain, whereas dopamine content was unaltered (Table 1). The reduction of serotonin was found to persist for 270 days after

TABLE 1

DURATION OF THE EFFECTS OF 5,7-DHT ON BRAIN MONOAMINE CONTENT *

Treatment	Duration (days)	Brain Monoamine Content (ng/g) 5-Hydroxy-tryptamine	Norepinephrine	Dopamine
Control (saline)	—	415±25	351±17	716±33
5,7-DHT	14	132±4 †	296±16 †	711±38
5,7-DHT	30	157±24 †	237±18 †	658±76
5,7-DHT	90	135±10 †	300±10 †	700±39
5,7-DHT	270	137±12 †	321±15	747±38

* Adult rats received 200 μg of 5,7-DHT intracisternally. Values represent the means±SEM of seven to 12 determinations.

† $p < 0.001$ when compared with control values.

treatment, but the effect on noradrenergic fibers diminished with time. The 5,7-DHT-induced alteration of serotonin content in the brain was most marked in the striatum, with progressively less effect in brain stem, hypothalamus, and the rest of the brain.[8]

Because MAO inhibitors had been found to enhance the effects of intracisternally administered 6-hydroxydopamine,[11] animals were pretreated with MAO inhibitors to determine how inhibition of MAO might influence the actions of 5,7-DHT. While none of the agents chosen affected the neurocytotoxic action of 5,7-DHT on serotonergic fibers, all were found to block the effects of 5,7-DHT on noradrenergic fibers (TABLE 2).

Another series of experiments was performed to determine possible effects of a variety of psychotropic drugs on the actions of 5,7-DHT (TABLE 3). In this work, agents like reserpine and *p*-chlorophenylalanine, which deplete brain

TABLE 2

EFFECT OF VARIOUS MONOAMINE OXIDASE INHIBITORS ON THE ACTIONS OF 5,7-DHT *

Treatment		Dose of Inhibitor (mg/kg)	Brain Monoamine Content (ng/g) 5-Hydroxy-tryptamine	Norepinephrine	Dopamine
Saline	+5,7-DHT	—	130±10 †	229±16 †	601±40
Iproniazid	+5,7-DHT	100	161±12 †	352±34	683±73
Pheniprazine	+5,7-DHT	15	170±30 †	414±25	676±81
Tranylcypromine	+5,7-DHT	10	132±10 †	402±22	625±59
Pargyline	+5,7-DHT	50	142±16 †	399±21	675±42
Saline		—	451±16	378±25	625±52

* Adult rats received 200 μg of 5,7-DHT intracisternally 30 min after injection of the various MAO inhibitors and were killed 21 days after treatment. Values represent the means±SEM of five to 20 determinations.

† $p < 0.001$ when compared with control values.

serotonin content, did not alter the ability of 5,7-DHT to reduce serotonin or norepinephrine in the brain. Desipramine and chlorimipramine, which reportedly decrease the uptake of norepinephrine and serotonin into monoaminergic neurons,[15, 16] antagonized the effect of 5,7-DHT on noradrenergic neurons without affecting 5,7-DHT-induced depletion of serotonin. Benztropine and brompheniramine had no effect on the neurocytotoxicity of 5,7-DHT. Furthermore, fluoxetine, a compound reported to inhibit the uptake of serotonin,[17, 18] did not alter the action of 5,7-DHT in reducing serotonin content (TABLE 3).

TABLE 3

EFFECT OF VARIOUS CENTRALLY ACTING DRUGS ON BRAIN MONOAMINE CONTENT AFTER 5,7-DHT TREATMENT *

Treatment Prior to 5,7-DHT	Dose (mg/kg)	Brain Monoamine Content (% of control)		
		5-Hydroxytryptamine	Norepinephrine	Dopamine
Saline	—	38.4±2.6 †	66.0± 2.6 †	102.1±4.4
Reserpine	(2×2.5)	29.2±3.8 †	59.6± 4.7 †	82.0±4.5
p-Chlorophenylalanine	(2×150)	30.6±1.8 †	66.4±10.9 †	108.3±8.6
Fenfluramine	10	36.2±1.3 †	86.0± 4.9 ‡	96.4±6.0
Desipramine	25	36.7±4.2 †	96.0± 7.6 ‡	98.3±4.6
Chlorimipramine	25	41.8±2.5 †	85.4± 4.8 ‡	105.3±4.8
Benztropine	25	32.4±2.5 †	67.0± 3.4 †	102.3±5.9
Chlorpromazine	25	34.3±4.5 †	89.7± 9.3 ‡	94.9±8.2
Brompheniramine	25	42.2±2.2 †	72.1± 4.4 †	91.4±2.0
Fluoxetine (Lilly 110,140)	15	30.7±4.0 †	70.1± 2.8 †	100.3±4.0

* All adult rats received 200 μg of 5,7-DHT intracisternally and were killed 30 days later. Control value for brain norepinephrine was 354±9 ng/kg, for dopamine was 686±31 ng/g, and for serotonin was 430±17 ng/g. Drug treatments are described in MATERIALS AND METHODS. Brain monoamine values for treatment groups that received drugs were not significantly different from controls. Values represent the means±SEM of 10–25 determinations.

† $p < 0.001$ when compared with saline-treated control values (no 5,7-DHT).

‡ $p < 0.01$ when compared with values from rats that received 5,7-DHT alone.

Effects of 5,7-DHT Administered to the Neonate

Previous studies have demonstrated that relatively high doses of 6-hydroxydopamine produce a marked destruction of catecholaminergic neurons when administered to the neonate,[12, 13] whereas administration of a comparable dose to an adult shows much less effect. To determine whether 5,7-DHT might induce more profound alterations in serotonergic neurons when administered to the neonate, several doses of 5,7-DHT were injected intracisternally to 1-day-old animals (TABLE 4). Because *p*-chloroamphetamine and fenfluramine have also been reported to induce neurocytotoxicity when administered to adult rats,[19–21] a group of neonates was also given these drugs (TABLE 5).

TABLE 4

EFFECT OF VARIOUS DOSES OF 5,7-DHT ADMINISTERED TO NEONATES ON BRAIN MONOAMINES *

Brain Amine	Dose of 5,7-DHT (μg/brain) 10	25	50	75
Serotonin	50.4±5.2 †	26.4±4.0 †	13.6±2.0 †	13.1± 1.9 †
Norepinephrine	97.9±3.6	74.3±2.0 †	51.9±6.3 †	54.9±10.6 †
Dopamine	94.7±5.7	94.3±5.2	77.8±6.5 †	76.4± 6.8 †

* Animals received various doses of 5,7-DHT when 1 day old. Animals were sacrificed when 28 days of age. Values represent the means±SEM (designated as percentage of control) of at least seven rats. Control values were comparable to those in TABLES 2 and 3.

† $p < 0.01$ when compared with control values.

Brain serotonin and norepinephrine were found to be drastically reduced by the 25-, 50-, and 75-μg doses of 5,7-DHT (TABLE 4). The amount of depletion was greater than that observed in adults, even though a lower dose was administered (cf. TABLES 1 & 4). It is of interest that the higher doses of 5,7-DHT also induced a significant reduction of dopamine, a response not apparent in adults. In contrast to the action of 5,7-DHT, neither *p*-chloroamphetamine nor fenfluramine significantly affected brain serotonin content at 60 days of age, suggesting that neurocytotoxicity did not occur or was reversible (TABLE 5). A single dose of 15 mg/kg of *p*-chloroamphetamine administered to an adult was found to reduce serotonin by 40 ± 4.2%.

After administration of low doses of 6-hydroxydopamine, norepinephrine content has been found to be elevated in the brain stem and the cerebellum, suggesting an enhanced growth of noradrenergic fibers.[22, 23] Since whole brain serotonin was reduced by all doses of 5,7-DHT injected, the amount of depletion found in both the brain stem and the forebrain was examined to determine whether one of these areas might demonstrate an increased content of serotonin. However, elevation of serotonin content was not observed in

TABLE 5

EFFECT OF NEONATALLY ADMINISTERED *p*-CHLOROAMPHETAMINE AND FENFLURAMINE ON BRAIN SEROTONIN CONTENT

Treatment *	Brain Serotonin Content (ng/g) Brain Stem	Forebrain
Control	870±23	449±16
Fenfluramine	824±24	396±13
p-Chloroamphetamine	880±19	401±13

* Rat pups received *p*-chloroamphetamine (25 mg/kg) or fenfluramine (15 mg/kg) on Days 1, 3, and 5 and were sacrificed when 60 days of age. Values represent the means±SEM of at least eight determinations.

these brain areas, suggesting that collateral sprouting of serotonin fibers did not occur. Sachs and Jonsson [24] have found elevated uptake of [^{3}H] serotonin in the brain stem after peripheral administration of 5,7-DHT. While it was concluded that outgrowth of serotonergic nerve terminals was stimulated, the content of serotonin was not reported. Therefore, additional investigations will be required to determine whether this failure to observe an elevated level of serotonin after intracisternal administration of 5,7-DHT is due to a special characteristic of serotonin fibers or to the need to choose even lower doses of 5,7-DHT than those already tested.

TABLE 6

EFFECT OF PARGYLINE AND DESIPRAMINE ON THE ACTIONS OF 5,7-DHT ADMINISTERED TO NEONATES

	Brain Monoamine Content (% of control)		
Treatment	5-Hydroxy-tryptamine	Norepinephrine	Dopamine
5,7-DHT	34.6±7.2 †	49.8±3.4 †	96.6±7.9
Pargyline+5,7-DHT	13.3±1.1 †	110.7±4.0	94.5±4.0
Desipramine+5,7-DHT	10.3±0.8 †	113.0±5.5	86.5±7.6

* Animals received 50 μg of 5,7-DHT intracisternally when 3 days of age. The 5,7-DHT was administered 30 min after pargyline (40 mg/kg) and 60 min after desipramine (20 mg/kg). Values represent the means±SEM of at least six determinations. Control values were comparable to those in TABLES 1 and 3.

† $p < 0.001$ when compared with control.

Effect of Pargyline and Desipramine on the Actions of 5,7-DHT Administered to the Neonate

Three-day-old rats were also given 5,7-DHT either alone or in combination with pargyline or desipramine. In these experiments, the monoamine changes were measured when the rats were 30–40 days old (TABLE 6). As described earlier for adults, both pargyline and desipramine were found to antagonize the effect of 5,7-DHT on noradrenergic fibers without affecting its action on serotonergic fibers (TABLE 6).

Effect of 5,7-DHT on Locomotor Activity

In previous work, the suggestion had been made that serotonergic pathways played an inhibitory role in locomotor activity.[25-27] For that reason, locomotor activity was tested in 5,7-DHT-treated rats with and without pargyline pretreatment, and the results were compared with activity measures in control animals (FIGURE 1). A marked elevation in locomotor activity was noted at 14 days in rats treated with 5,7-DHT alone and in those treated with pargyline prior to receiving 5,7-DHT. By 28 days of age, the activity of the 5,7-DHT-treated groups was significantly less than that of the control group.

Body weights of animals treated with 5,7-DHT were measured at weekly intervals. In accord with previously reported results,[8] animals treated with 5,7-DHT had significantly lower body weights than did control rats at 14 days of age (32.2 ± 0.5 g for control rats versus 21.9 ± 0.9 g for 5,7-DHT-treated rats; $p < 0.01$). Pretreatment with pargyline prevented this loss of body weight induced by 5,7-DHT (29.0 ± 0.8 g for rats treated with pargyline and 5,7-DHT). The mechanism of this weight loss has not been thoroughly examined, but one possible explanation may be that 5,7-DHT-treated rat pups appear to bite the mother's nipples during the period of suckling, possibly reducing the amount of nutrition received.

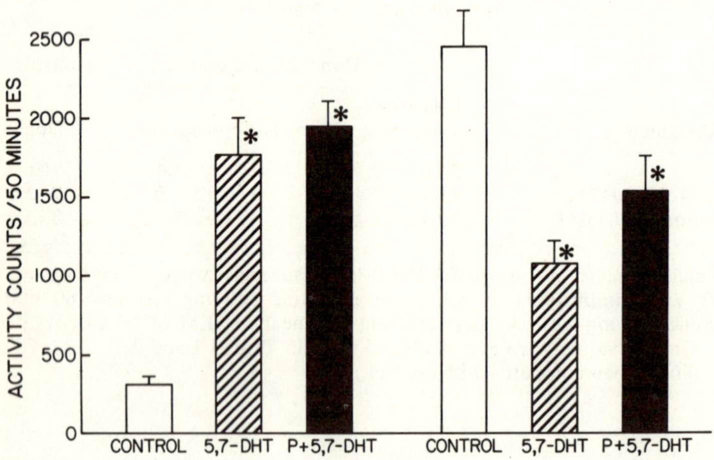

FIGURE 1. Effect of neonatally injected 5,7-DHT on locomotor activity at 14 (*left*) and 28 (*right*) days of age. Animals were given 50 μg of 5,7-DHT intracisternally on Day 3. Locomotor activity of three animals was determined for a period of 50 min. Each bar represents the mean$\pm$SEM of at least 10 determinations. The asterisks indicate that $p < 0.001$ when compared with control values.

Discussion

In confirmation of previous findings,[5, 8, 28, 29] administration of 5,7-DHT to adult and developing rats was shown to result in a reduction of brain serotonin and norepinephrine content. Since the action of 5,7-DHT on noradrenergic fibers is proposed to be due to its selective accumulation by specific uptake mechanisms,[31] the specificity of action of 5,7-DHT for serotonergic neurons should be improved if this uptake mechanism was blocked. In agreement with this hypothesis, drugs known to inhibit the uptake of norepinephrine were found to antagonize the depletion of norepinephrine by 5,7-DHT, just as they had been found to inhibit the actions of 6-hydroxydopamine on noradrenergic fibers.[2]

The protective effect of MAO inhibition against the action of 5,7-DHT on noradrenergic fibers concomitant to the lack of effect on the destruction of serotonergic fibers was not expected. Daly *et al.*[32] have reported that pargyline inhibits the action of 5,7-DHT on peripheral noradrenergic fibers. These workers have suggested that this action may be related to an ability of pargyline and other MAO inhibitors to prevent formation of a metabolite of 5,7-DHT. Presumably, the metabolite would be a good substrate for uptake into noradrenergic fibers and would retain its neurocytotoxicity. It follows that such a postulated metabolite is not essential for the neurocytotoxic effects of 5,7-DHT on serotonergic neurons, because destruction of serotonin fibers persisted in rats pretreated with MAO inhibitors. Regardless of this proposal, the block by pargyline of 5,7-DHT-induced norepinephrine neuronal destruction would seem to be distinct from that produced by inhibitors of norepinephrine uptake, because the antagonism caused by pargyline and the other MAO inhibitors does not correlate with any action they might have on the noradrenergic uptake system.[2, 30]

Because 5,7-DHT has been proposed to enter serotonergic fibers by way of a specific uptake system,[31] it was anticipated that drugs reported to inhibit this system would reduce the actions of 5,7-DHT.[15, 16, 18, 33] Surprisingly, none of the drugs tested opposed the ability of 5,7-DHT to reduce serotonin content. Not even fluoxetine (Lilly 110,140), a compound that inhibits uptake of serotonin into synaptosomes and prevents the depletion of serotonin produced by *p*-chloroamphetamine,[17, 18, 34] antagonized the serotonin depletion produced by 5,7-DHT. Whether this finding reflects an inability of fluoxetine to block comparatively high doses of 5,7-DHT, which might already saturate the specific uptake system, or is a consequence of cell accumulation of 5,7-DHT by a fluoxetine-insensitive process is yet to be determined. An observation that may have relevance to this discussion is the finding that *p*-chloroamphetamine produces no long-term depletion of brain serotonin when administered to the neonate.[35] If a system is required to concentrate *p*-chloroamphetamine, perhaps this uptake system has not yet developed or is inoperative in the neonate. Conversely, the lack of any diminution in the toxic effects of 5,7-DHT in the neonate suggests that such an uptake mechanism is not an obligatory requirement for the action of 5,7-DHT on serotonergic fibers. Because of the implications of the different responses to these two drugs in the neonate, the mechanism by which 5,7-DHT and *p*-chloroamphetamine enter serotonergic fibers and produce their toxic effects should receive further attention.

Administration of 5,7-DHT to the neonate was found to produce a greater destruction of serotonergic fibers than was possible in adult rats. This difference might be the result of an inherent increased susceptibility of differentiating cells to any toxin or a consequence of an immature mechanism for destruction of the toxin. For example, 5,7-DHT administered to 1-day-old rats produced depletions comparable to those in 3-day-old animals given pargyline pretreatment prior to injection of 5,7-DHT. Therefore, it could be proposed that MAO activity, which is low in the 1-day-old rat,[36] simulates treatment with a MAO inhibitor. Another possibility for the increased effectiveness of 5,7-DHT in the neonate that has received little attention is that free radical scavenging systems are not yet fully developed. Perhaps this alternative should be investigated, in view of the report that drugs that influence superoxide anion availability can alter the toxicity of 6-hydroxydopamine.[37]

The increased destruction of serotonergic neurons seen in the neonatal rats pretreated with desipramine is difficult to understand but may be a result of the presence of a higher concentration of 5,7-DHT when uptake into noradrenergic neurons is inhibited. However, this increased response to 5,7-DHT after desipramine treatment does not occur in adult rats.

Several studies in our laboratory have suggested that serotonergic pathways may have an inhibitory function on locomotor activity [25–27]; therefore, it was reasoned that the locomotor activity of developing rats might be significantly greater after neonatal administration of 5,7-DHT than that observed in control rats. As was predicted, developing rats treated with 5,7-DHT were found to be nearly five times as active as control rats at 14 days of age. Since the activity of the rats pretreated with pargyline before receiving 5,7-DHT was equivalent to that of those that were administered 5,7-DHT alone, this hyperactivity would not appear to require a reduction of noradrenergic fibers, a procedure known to enhance locomotor activity in adult rats.[13] Furthermore, the alteration in growth observed in the animals treated with 5,7-DHT would not seem to account for the increased locomotor activity observed at 14 days of age. This view is based on the observation that when 5,7-DHT is given to rats pretreated with pargyline, the growth is normal, yet the hyperactivity is unaltered. Curiously, at 28 days of age, the locomotor activity of the treated rats was significantly less than that observed in control animals. Although the reason for this change is unknown, it may reflect the activation of some compensatory neural mechanism(s). Based on the hyperactivity observed in neonatal rats given 6-hydroxydopamine, Shaywitz *et al.*[38] suggested that such changes in locomotion in immature rats may have relevance to the abnormal behavior observed in minimal brain dysfunction. Therefore, studies are under way to determine whether the hyperactivity observed at 14 days of age in 5,7-DHT-treated rats can be altered by drug treatment, such as with central nervous system stimulants or serotonin precursors.

There are reports that animals become hyperphagic after treatments that reduce only serotonin content.[39] If body weight is a measure of such alteration, this behavior was not evident in animals treated with 5,7-DHT during early development, because none of the treatments significantly elevated the weight gain of rats up to 35 days of age. This finding confirms an earlier report from our laboratory.[8]

References

1. Baumgarten, H. G., A. Björklund, L. Lachenmayer, A. Nobin & U. Stenevi. 1971. Long-lasting selective depletion of brain serotonin by 5,6-dihydroxytryptamine. Acta Physiol. Scand. Suppl. **373:** 1–15.
2. Breese, G. R. 1975. Chemical and immunochemical lesions by specific neurotoxic substances and antisera. *In* Handbook of Psychopharmacology. L. L. Iversen, S. D. Iversen & S. H. Snyder, Eds. Vol. **1:** 137–189. Plenum Publishing Corporation. New York, N.Y.
3. Baumgarten, H. G., A. Björklund, A. F. Holstein & A. Nobin. 1972. Chemical degeneration of indoleamine axons in rat brain by 5,6-dihydroxytryptamine; an ultrastructural study. Z. Zellforsch. **129:** 256–271.
4. Breese, G. R., B. R. Cooper, L. D. Grant & R. D. Smith. 1974. Biochemical and behavioral alterations following 5,6-dihydroxytryptamine administration into brain. Neuropharmacology **13:** 177–187.

5. BAUMGARTEN, H. G., A. BJÖRKLUND, L. LACHENMAYER & A. NOBIN. 1973. Evaluation of the effects of 5,7-dihydroxytryptamine on serotonin and catecholamine neurons in the rat CNS. Acta Physiol. Scand. Suppl. **391:** 1–19.
6. BAUMGARTEN, H. G., S. J. VICTOR & W. LOVENBERG. 1973. Effect of intraventricular injection of 5,7-dihydroxytryptamine on regional tryptophan hydroxylase of rat brain. J. Neurochem. **21:** 251–253.
7. JACOBY, J. G., L. D. LYTLE & M. F. NELSON. 1974. Long term effects of 5,7-dihydroxytryptamine on brain monoamines. Life Sci. **15:** 909–919.
8. BREESE, G. R. & B. R. COOPER. 1975. Behavioral and biochemical interactions of 5,7-dihydroxytryptamine with various drugs when administered intracisternally to adult and developing rats. Brain Res. **98:** 517–527.
9. BAUMGARTEN, H. G. & L. LACHENMAYER. 1972. 5,7-Dihydroxytryptamine: improvement in chemical lesioning of indoleamine neurons in the mammalian brain. Z. Zellforsch. **135:** 399–414.
10. EVETTS, K. D. & L. L. IVERSEN. 1970. Effects of protriptyline on the depletion of catecholamines induced by 6-hydroxydopamine in the brain of the rat. J. Pharmacol. **22:** 540–542.
11. BREESE, G. R. & T. D. TRAYLOR. 1970. Effect of 6-OHDA on brain NE and DA: evidence for selective degeneration of catecholamine neurons. J. Pharmacol. Exp. Ther. **174:** 413–420.
12. BREESE, G. R. & T. D. TRAYLOR. 1972. Developmental characteristics of brain catecholamines and tyrosine hydroxylase in the rat: effects of 6-OHDA. Brit. J. Pharmacol. **44:** 210–222.
13. SMITH, R. D., B. R. COOPER & G. R. BREESE. 1973. Growth and behavioral changes in developing rats treated intracisternally with 6-hydroxydopamine: evidence for involvement of brain dopamine. J. Pharmacol. Exp. Ther. **185:** 609–619.
14. COOPER, B. R., G. R. BREESE, L. D. GRANT & J. L. HOWARD. 1973. Effects of 6-hydroxydopamine treatments on active avoidance responding: evidence for involvement of brain dopamine. J. Pharmacol. Exp. Ther. **185:** 358–370.
15. CARLSSON, A. 1970. Structural specificity for inhibition of (C^{14}-5HT) uptake by cerebral slices. J. Pharm. Pharmacol. **22:** 729–732.
16. BUCZKO, W., D. DEGAETIANO & S. GARATTINI. 1974. Influence of some tricyclic antidepressant drugs on the uptake of 5-hydroxytryptamine by rat blood platelets. J. Pharm. Pharmacol. **26:** 814–815.
17. FULLER, R. W., K. W. PERRY & B. B. MOLLOY. 1974. Effect of an uptake inhibitor on serotonin metabolism in rat brain: studies with 3-(p-trifluoromethylphenoxy) N-methyl-3-phenylpropylamine (Lilly 110140). Life Sci. **15:** 1161–1171.
18. WONG, D. T., J. S. HORNG, F. P. BYMASTER, K. L. HAUSER & B. B. MOLLOY. 1974. A selective inhibitor of serotonin uptake: Lilly 110140, 3-(p-trifluoromethylphenoxy)-N-methyl-3-phenylpropylamine. Life Sci. **15:** 471–479.
19. HARVEY, J. A., S. E. MCMASTER & L. M. YUNGER. 1975. p-Chloroamphetamine: selective neurotoxic action in brain. Science **187:** 841–843.
20. SEKERKE, H. J., H. E. SMITH, J. A. BUSHING & E. SANDERS-BUSH. 1975. Correlation between brain levels and biochemical effects of the optical isomers of p-chloroamphetamine. J. Pharmacol. Exp. Ther. **193:** 835–844.
21. HARVEY, J. A. & S. E. MCMASTER. 1975. Fenfluramine: evidence for a neurotoxic action on midbrain and a long-term depletion of serotonin. Psychopharmacol. Commun. **1**(2): 217–228.
22. KONKOL, R. J., E. BENDEICH & G. R. BREESE. 1977. A morphological and biochemical study of the altered growth pattern of central catecholamine neurons following 6-hydroxydopamine. Brain Res. In press.
23. KOSTRZEWA, R. M. & J. W. HARPER. 1975. Comparison of the neonatal effects of 6-hydroxydopa and 6-hydroxydopamine on growth and development of noradrenergic neurons in the central nervous system. *In* Chemical Tools in Catecholamine Research. G. Jonsson, T. Malmfors & C. Sachs, Eds. Vol. I:

181–188. North Holland Publishing Company. Amsterdam, The Netherlands.

24. Sachs, C. & G. Jonsson. 1975. 5,7-Dihydroxytryptamine induced changes in the postnatal development of central 5-hydroxytryptamine neurons. Med. Biol. **53:** 156–164.

25. Breese, G. R., B. R. Cooper & A. S. Hollister. 1975. Involvement of brain monoamines in the stimulant and paradoxical inhibitory effects of methylphenidate. Psychopharmacologia **44:** 5–10.

26. Breese, G. R., B. R. Cooper & R. A. Mueller. 1974. Evidence for involvement of 5 hydroxytryptamine in the actions of amphetamine. Brit. J. Pharmacol. **52:** 307–314.

27. Hollister, A. S., G. R. Breese, C. Kuhn, B. R. Cooper & S. M. Schanberg. 1976. An inhibitory role for brain 5-HT-containing systems in the locomotor effects of d-amphetamine. J. Pharmacol. Exp. Ther. **198**(1): 12–22.

28. Baumgarten, H. G., S. J. Victor & W. Lovenberg. 1975. A developmental study of the effects of 5,7-dihydroxytryptamine on regional tryptophan hydroxylase in rat brain. Psychopharmacol. Commun. **1**(1): 75–88.

29. Lytle, L. D., J. H. Jacoby, M. F. Nelson & H. G. Baumgarten. 1974. Long-term effects of 5,7-dihydroxytryptamine administered at birth on the development of brain monoamines. Life Sci. **15:** 1203–1217.

30. Hendley, E. D. & S. H. Snyder. 1968. Relationship between the action of monoamine oxidase inhibitors on the noradrenaline uptake system and their antidepressant efficacy. Nature (London) **220:** 1330–1331.

31. Björklund, A., H. G. Baumgarten & A. Nobin. 1974. Chemical lesioning of central monoamine axons by means of 5,6-dihydroxytryptamine and 5,7-dihydroxytryptamine. *In* 5-Hydroxytryptamine and Other Indolealkylamines in Brain. E. Costa, M. Sandler & G. L. Gessa, Eds. Vol. 10: 13–33.

32. Daly, J., J. Lundstrom & C. R. Creveling. 1974. Structure activity correlations in tri-hydroxy-phenethylamines and di-hydroxytryptamine; relationship to cytotoxicity in adrenergic and serotonergic neurons. *In* Dynamics of Degeneration and Regeneration in Neurons. K. Fuxe, L. Olson & Y. Zotterman, Eds.: 29–42. Pergamon Press. Oxford, England.

33. Ross, S. B. & A. L. Renyi. 1969. Inhibition of the uptake of tritiated-6-hydroxytryptamine in brain tissue. Eur. J. Pharmacol. **7:** 270–277.

34. Fuller, R. W., K. W. Perry & B. B. Molloy. 1975. Reversible and irreversible phase of serotonin depletion by 4-chloroamphetamine. Eur. J. Pharmacol. **33:** 119–124.

35. Neckers, L. M., L. Bertilsson, S. H. Koslow & J. L. Meek. 1975. Reduction of tryptophan hydroxylase activity and 5-hydroxytryptamine concentration in certain rat brain nuclei after p-chloroamphetamine. J. Pharmacol. Exp. Ther. **196:** 333–338.

36. Baker, P. C., K. M. Hoff & M. D. Smith. 1974. The maturation of monoamine oxidase and 5-hydroxyindole acetic acid in regions of the mouse brain. Brain Res. **65:** 255–264.

37. Cohen, G., R. E. Heikkila, B. Allis, F. Cabbat, D. Dembiec, D. MacNamee, C. Mytilineou & B. Winston. 1976. Destruction of sympathetic nerve terminals by 6-hydroxydopamine: protection by 1-phenyl-3-(2 thiazolyl)-2-thiourea, diethyldithiocarbamate, methimazole, cysteamine, ethanol and n-butanol. J. Pharmacol. Exp. Ther. **199:** 336–352.

38. Shaywitz, B. A., J. H. Klopper, R. D. Yager & J. W. Gordon. 1976. Paradoxical response to amphetamine in developing rats treated with 6-hydroxydopamine. Nature (London) **261:** 153–155.

39. Saller, C. F. & E. M. Stricker. 1976. Hyperphagia and increased growth in rats after intraventricular injection of 5,7-dihydroxytryptamine. Science **192:** 285–287.

Discussion

Dr. H. Tamir: I notice that despite all of your various pharmacologic manipulations, you never seem to be able to reduce brain serotonin concentrations below 30–35% of control values. Would you care to speculate on this?

Dr. Breese: This observation encouraged us to try other pharmacologic manipulations to see if we could reduce brain serotonin concentrations below the 70–75% depletion that occurred after the administration of 5,7-DHT alone.

Dr. Tamir: Dr. Baumgarten warned us earlier that injection of 5,7-DHT greater than 100 μg of free base might induce lesions in brain dopaminergic neurons. However, dopamine concentrations in the brains of dogs are not altered apparently when they are given rather high doses of 5,7-DHT; if anything, concentrations of this neurotransmitter seem to be increased over control levels.

Dr. Breese: Some of these differences may be related to the route of administration used for 5,7-DHT. For example, 6-hydroxydopamine produces a much different pattern of catecholamine depletion when it is given intraventricularly compared to that seen after intracisternal injections. We prefer to give the drugs intracisternally, for no better reason than that, empirically, we feel this route works better.

Dr. H. G. Baumgarten: You are correct in that the intraventricular and intracisternal routes of administration produce different patterns of serotonin depletion, different magnitudes of depletion in different regions of the brain, and these effects have different time courses for both the rate of depletion and the extent of recovery of brain serotonin.

However, when you only measure total brain dopamine content, or total brain serotonin concentrations, you might overlook important regional changes that may occur but that might be lost because the tissues assayed are just too large. Additionally, your brain dopamine assay measurements may be misleading inasmuch as neurons that survive the drug treatment may somehow hyperfunction and compensate for the reduced levels reflected by the destruction of other cells. I believe that measurements of tissue neurotransmitter concentrations alone may thus be totally misleading under some conditions. I suggest that one should always perform careful morphologic evaluations and perform measurements on neurotransmitter concentrations and turnover rates at the same time in discrete, small regions of the brain or spinal cord whenever possible.

I would also like to make a point about your interpretations regarding the mechanisms underlying the enhanced specificity of 5,7-DHT in animals pretreated with a monoamine oxidase inhibitor. By using drugs that block monoamine oxidase prior to giving 5,7-DHT, you might impair the uptake of 5,7-DHT into noradrenergic neurons, thus making less of the neurotoxic drug available to these cells.

Dr. Breese: In a previous publication, it has been shown that regional levels of dopamine were not affected by 5,7-DHT administered to adult rats. The suggestion that MAO inhibitors may be inhibiting uptake into noradrenergic fibers seems unlikely, because pargyline does not affect the action of 6-hydroxydopamine on noradrenergic fibers. It is also interesting to note that 5,7-DHT is most toxic when it is given alone to immature rats. However, it is less toxic (actually, the mortality is less) when these animals are pretreated with desmethylimipramine prior to an injection of 5,7-DHT. The treatment that we found to be least toxic involves pretreating animals with a monoamine oxidase inhibitor prior to 5,7-DHT.

DR. M. O. CARRUBA: How do you interpret your findings that specific serotonin uptake inhibitors (e.g., fluoxetine) do not alter the neurotoxic effects of 5,7-DHT?

DR. BREESE: In these experiments, we gave a very large dose (i.e., 200 μg) of 5,7-DHT. It is possible that this dose of the compound might saturate the uptake system itself (especially if the uptake-inhibiting action of fluoxetine is a competitive one). Thus, under these conditions, the neurotoxic action of 5,7-DHT would not be blocked. However, another possibility is that 5,7-DHT is not taken up by a fluoxetine-sensitive uptake site. I think that the correct answer to this question can only be obtained by doing control dose-response relationship studies between the two drugs.

DR. CARRUBA: Aren't your hypotheses difficult to defend, in view of the fact that you only gave one dose of fluoxetine at only one time interval prior to the injection of 5,7-DHT?

DR. BREESE: Although your point is well taken, you must realize that the dose and time parameters that we used for fluoxetine are very effective in blocking the neurotoxic effects of *p*-chloroamphetamine on brain serotonergic nerves. Moreover, the goal of this experiment was to determine whether we could produce a destruction of noradrenergic nerves by 5,7-DHT while simultaneously blocking the effects of this drug on brain serotonin-containing neurons. Our results indicate that this did not happen with the combined treatment.

DR. R. W. FULLER: I do not believe that the results of Dr. Breese can be accounted for simply on the basis that fluoxetine fails to alter 5,7-DHT neurotoxicity because it does not have a sufficiently long enough half-life. Fluoxetine in the dose used blocks serotonin uptake for as long as 48 hr; moreover, the drug is effective in blocking the neurotoxic effects of *p*-chloroamphetamine for as long as 48 hr after a single injection.

I guess I tend to prefer Dr. Breese's explanation that the large doses of 5,7-DHT may competitively override the uptake-inhibiting properties of fluoxetine. The relative concentration of 5,7-DHT at the site of the neuronal uptake pump may be tremendous in comparison to the concentration of fluoxetine or of *p*-chloroamphetamine, when we are able to block its effects with fluoxetine.

DR. D. V. COSCINA: Dr. Breese, I share your interest in the behavioral effects of the neurotoxic drugs. Haven't you shown in the past that 5,7-DHT injections in neonatal animals produce a suppression in feeding behavior, and in subsequent body weight gain, that does not appear to recover with time? Do you see similar types of behavioral changes after injection of fenfluramine in neonatal animals?

DR. BREESE: We did not see any changes in body weight after injections of *p*-chloroamphetamine or fenfluramine in neonatal animals when they were killed at 60 days of age. However, there may have been an acute effect of the drug that we missed, since we only examined the animals at this one time point.

It may be that the reduced body weights of animals injected with 5,7-DHT during the neonatal period result from the fact that the drug-treated nursing offspring may bite the nipples of the lactating mother. As a result, the mothers may prematurely wean their offspring or, under some conditions, may actually kill them.

Dr. M. Da Prada: I think that in studies of this kind, explorng the possibility that various pharmacologic treatments might enhance the selectivity of the serotonin neurotoxins, one must measure not only monoaminergic neurotransmitter levels but also their metabolites. The situation appears to be very comparable to that which occurs when one looks at the effects of chlorpromazine treatment on brain dopamine concentrations. Dopamine levels appear to be normal, but there are tremendous decreases in homovanillic acid, and sometimes in MOPEG concentrations as well. Using a serotonin uptake inhibitor other than fluoxetine, we have found that animals pretreated with this compound and then given 75 μg of 5,6-DHT show reductions in brain homovanillic acid concentrations and increases in the levels of MOPEG sulfate. We do not yet know whether these changes result from a direct action of the neurotoxin on dopaminergic or noradrenergic nerves or whether they are caused by an imbalance that results from the serotonergic neurotoxic effects of the dihydroxytryptamine.

Dr. Breese: Were the changes in homovanillic acide and MOPEG sulfate long-term effects of the drug treatment or were they only transient?

Dr. Da Prada: We obtained these effects 10–12 days posttreatment.

Dr. Breese: We, too, have some animals that we will examine for possible changes in the metabolites of neurotransmitter compounds after injections of 5,7-DHT in developing animals. However, I think that I would agree with you that dihydroxytryptamine-induced changes in these other neurotransmitter metabolites most probably reflect an alteration in the balance between serotonin and these other neurotransmitter systems. I think that if you find changes in neurotransmitter metabolites or in the turnover rates for these compounds, but no changes in their levels, the changes most likely result from alterations in the interactions among various neuronal systems, rather than from a direct neurocytotoxic action of the drug on these other systems.

Dr. Da Prada: We think it surprising that our animals show decreases in homovanillic acid after 5,6-DHT injection, inasmuch as many investigators think that brain serotonergic nerves normally inhibit the dopaminergic system.

Dr. Breese: I agree with you. However, if we hypothesize that the dopaminergic system is inhibited by serotonergic nerves only after it is first activated, serotonergic systems in the brain may only come into play when the dopaminergic system is activated in some way. Under resting conditions, this serotonergic inhibitory system may not be very important for dopaminergic neurotransmission.

Dr. C. T. Giambalvo: We have data that show an interaction between brain dopaminergic and serotonergic nerves. Injections of 5,7-DHT increase brain dopamine turnover of the ipsilateral side of the brain when the drug is injected unilaterally into the raphe nuclei. These neurochemical effects are accompanied by marked rotational behavioral changes that can be accounted for by possible drug-induced alterations in dopaminergic nerve activity.

I have one question for Dr. Breese. Are the locomotor activity patterns of 28-day-old animals pretreated with 5,7-DHT depressed?

Dr. Breese: Yes, they are reduced significantly below those of control injected animals.

Dr. Giambalvo: Why do you think that the locomotor activity levels of these animals are reduced in 28-day-old animals but are actually increased in animals pretreated with 5,7-DHT but tested at 14 days of age? What is peculiar about 28 days of age?

DR. BREESE: I have no idea. We were not expecting these kinds of results. However, we have repeated these findings in 50–60 animals and have no doubt about their reliability.

DR. A.-L. BAROFSKY (*Howard University, Washington, D.C.*): Do you know if there are any sex-related differences in response to 5,7-DHT? For example, did you see any evidence of masculinization of female animals injected with the drug?

DR. BREESE: We are in the process of looking at this possibility by injecting 5-hydroxytryptophan in 5,7-DHT-pretreated animals and then looking for possible changes in prolactin levels. However, we have not seen any sex-related differences with respect to the effects of 5,7-DHT on shuttle box avoidance behavior, operant behavior, or in any of the other behavioral responses we have examined.

GENERAL DISCUSSION

Loy D. Lytle, *Moderator*

Department of Nutrition and Food Science
Massachusetts Institute of Technology
Cambridge, Massachusetts 02139

DR. E. M. GÁL: Dr. Baumgarten, have you looked at the possible neurotoxic effects of quinone compounds, such as the 2,3-dihydroxyindoles?

DR. H. G. BAUMGARTEN: No, we have not yet looked at these compounds.

DR. GÁL: Have you any data regarding the effects of chlorohydroxy-substituted indoles?

DR. BAUMGARTEN: The only chlorinated derivative that we have studied, 5-hydroxy-7-chlorotryptamine, appeared to be inactive.

DR. E. SANDERS-BUSH: Dr. Breese found that *p*-chloroamphetamine apparently had no neurotoxic effect when injected in newborn rats. Might it be possible that these age differences are due to the low drug-metabolizing capabilities of newborn animals?

DR. R. W. FULLER: With respect to developmental differences in response to the neurotoxins, we have also confirmed the observation that Dr. Breese mentioned in his paper, namely, that *p*-chloroamphetamine does not appear to have any neurotoxic effect in young rats. Interestingly, we can give 1-day-old rats as much as 50 mg/kg of *p*-chloroamphetamine, a dose that is lethal in adult rats. This dose of the drug not only did not kill these young rats but it also had no effect on their serotonin neurons. This lack of neurotoxic effect of the drug continues until rats are about 7 days of age; at this time, we begin to see initial short-term depletions of brain serotonin after drug administration. However, there were still no long-term neurotoxic effects of the drug. Thereafter, however, the neurotoxicity of p-chloroamphetamine progressively increased until adulthood was reached.

We feel that these age-related differences in response to *p*-chloroamphetamine may be associated with the ability of the neuronal uptake pump to accumulate sufficient quantities of the drug to produce a neurotoxic effect. Perhaps the uptake pump is less efficient for accumulating sufficient neurotoxic concentrations of *p*-chloroamphetamine early in development. In contrast, it may be more efficient in taking up the dihydroxytryptamines.

Interestingly, there appear to be at least two differences between the actions of halogenated amphetamine neurotoxins and the dihydroxytryptamines: the former compounds have no apparent toxic effects in young animals, and their neurotoxicity in mature animals can be blocked by drugs that inhibit their intraneuronal uptake. The neurotoxic effects of the dihydroxytryptamine compounds, on the other hand, are present in both mature and immature animals, and they do not appear to be blocked by drugs that inhibit the uptake pump.

DR. GÁL: It is well known that embryonic and neonatal animals lack fully mature detoxification mechanisms for most drugs. Isn't it still possible that *p*-chloroamphetamine cannot be converted to a neurotoxic metabolite in young animals? I think that the only way to resolve these differences would be to characterize and quantify the possible metabolites of *p*-chloroamphetamine in both young and old animals.

DR. FULLER: Although this is still a possibility, you have to remember that *p*-chloroamphetamine is not metabolized to any appreciable degree in adult animals. In rats, a large portion of an injected dose of the drug is simply excreted into the urine unchanged, although some of it is metabolized to 3,4-dimethoxyamphetamine prior to excretion. I don't know of any firm evidence that a metabolite is involved in the long-term neurotoxic effects of *p*-chloroamphetamine on brain serotonin. The initial short-term effects of the drug occur within 2 hr after its administration, and it would seem even more unlikely that a metabolite is involved in these effects.

DR. SANDERS-BUSH: Dr. Fuller, if you no longer favor the hypothesis that an active metabolite mediates the neurotoxic actions of *p*-chloroamphetamine, what are your current thoughts about why it takes so long for the drug to irreversibly exert its neurotoxic effects?

DR. FULLER: At the present time, I don't know the answer to your question. However, I like the idea that Dr. Moskowitz proposed earlier in the meeting. Dr. Moskowitz has shown that *p*-chloroamphetamine inhibits protein synthesis *in vitro* when administered in relatively high concentrations. Although at first glance one might be suspicious about the relevance of these findings to the *in vivo* condition, I think that it is possible that the concentrations of this neurotoxic drug might well be reached inside serotonin-containing neurons due to their active transport into the cells by the uptake pump. If the concentrations of the neurotoxins are maintained at these high levels for 24–32 hr, one might very well have a cytotoxic effect resulting from the long-term inhibition of cellular protein synthesis. This mechanism would be compatible with the idea that the rate of development of the irreversible neurotoxic action of *p*-chloroamphetamine is identical with the time that it takes other drugs, such as *p*-bromoamphetamine, to exert their neurotoxicity. According to this schema, the neurotoxic properties of some of these compounds might not depend so much on the precise structure of the neurotoxin or on how rapidly or slowly the compound might be converted to an active neurotoxic metabolite. Rather, their neurotoxic properties might depend simply on how long they inhibit protein synthesis inside the cell, thus leading to eventual cell death.

DR. BAUMGARTEN: Dr. Fuller, what is the approximate affinity of *p*-chloroamphetamine to brain serotonergic uptake sites, in comparison to the affinity of 5,7-DHT for these sites?

DR. FULLER: We have published this information previously, in 1973. As I recall, the K_I for *p*-chloroamphetamine is in the neighborhood of 10^{-6} M.

DR. BAUMGARTEN: So its affinity is similar to that seen for 5,7-DHT. Dr. Jonsson and his colleagues at the Karolinska Institute have shown that at birth, the V_{max} of the uptake system for 5,7-DHT is lower than in adult rats; however, the K_m for this compound doesn't change with age. Hence, these data suggest that the uptake system for these compounds in young animals is comparable to that seen in adult animals.

I, too, believe that the mechanisms underlying the neurotoxic effects of *p*-chloroamphetamine may be much different from those involved in the cytotoxicity of the dihydroxytryptamines. In the newborn animal, 5,7-DHT is even more neurotoxic on a molar basis than it is in the adult, whereas with the halogenated amphetamines, the opposite seems true. Maybe we should focus on the fact that in immature animals, serotonin neurons are still in the process of sprouting new terminals and increasing dendritic ramifications. Maybe your idea that these compounds inhibit protein synthesis becomes more important when neurons are growing in size early in development.

DR. FULLER: I agree that it seems likely that *p*-chloroamphetamine and the dihydroxytryptamines may have different mechanisms of action. However, I think that they also share at least one feature in common: each must be actively transported into serotonin neurons before it exerts its neurotoxic effects.

DR. BAUMGARTEN: Exactly right. Along these lines, we are presently studying the relative tissue distributions of *p*-chloromethoxyamphetamine with autoradiographic techniques. So far, our data show that uptake into individual serotonin neurons is highly comparable in both newborn and adult animals. However, that certainly is not the whole story.

DR. FULLER: A large amount of the *p*-chloromethoxyamphetamine that you administer will be *N*-demethylated very rapidly. Do you place the radioactive label for the compound on the methyl group or is it elsewhere?

DR. BAUMGARTEN: It is not on the methyl group.

DR. FULLER: I suppose that it really doesn't make too much difference, but I think that if you started with radioactively labeled *p*-chloroamphetamine, you might avoid the problems of having a lot of your drug immediately metabolized to other, possibly unidentifiable compounds.

NEUROCHEMICAL EFFECTS OF SEROTONIN NEUROTOXINS: AN INTRODUCTION

Ray W. Fuller

The Lilly Research Laboratories
Eli Lilly and Company
Indianapolis, Indiana 46206

The two major groups of serotonin neurotoxins currently known are the halogenated amphetamine derivatives, the most extensively studied being *p*-chloroamphetamine, and the hydroxy derivatives of serotonin, particularly 5,6- and 5,7-dihydroxytryptamines. Destruction of serotonin neurons by either type of agent results in neurochemical changes in all parameters specifically associated with brain serotonin neurons, namely, decreases in tryptophan hydroxylase, in serotonin and 5-hydroxyindoleacetic acid content, and in the high-affinity uptake of serotonin. The neurochemical effects of these two groups of serotonin neurotoxins can be compared briefly as follows.

Time Course of the Neurochemical Effects. Serotonin depletion occurs rapidly with halogenated amphetamines and dihydroxytryptamines. The halogenated amphetamines apparently have two separate actions on the serotonin neuron: an immediate effect, whereby inhibition of serotonin synthesis and release of bound serotonin rapidly deplete serotonin content, and the prolonged neurotoxic effect. Halogenated amphetamines competitively inhibit serotonin uptake shortly after they are given and rapidly lead to inhibition of tryptophan hydroxylase, but these effects and the depletion of 5-hydroxyindoles are not irreversible at first; that is, the halogenated amphetamines are not immediately neurotoxic.[1, 2] In contrast, the hydroxytryptamines probably exert neurotoxic effects on the serotonin neurons more rapidly; at least there has been no evidence that their effects can be reversed. The neurochemical manifestation of the neurotoxic effect depends partly on the turnover of tryptophan hydroxylase and on the uptake system. Though the halogenated amphetamines act at least as rapidly to deplete serotonin content, the dihydroxytryptamines are more rapidly neurotoxic.

Reversibility. As discussed above, the effects of the halogenated amphetamines are initially reversible. The effects of *p*-chloroamphetamine,[2] *p*-bromoamphetamine,[3] and fenfluramine[4] become partially irreversible within 6–8 hr and totally irreversible within 32–48 hr. The effects of the dihydroxytryptamines are apparently never reversible.

Route of Administration. The dihydroxytryptamines do not cross the blood-brain barrier and therefore must be injected intraventricularly, intracisternally, or otherwise directly into the central nervous system. The halogenated amphetamines, in contrast, cross the blood-brain barrier readily. In fact, the neurochemical effects of *p*-chloroamphetamine are more pronounced when it is given systemically than when it is given intraventricularly, apparently because in the latter case, it leaks out of the brain so rapidly. *p*-Chloroamphetamine is approximately as potent a serotonin neurotoxin on a milligram per kilogram dose basis when given intraperitoneally as when given intraventricu-

0077–8923/78/0305–0178 $01.75/0

larly.[5] *p*-Chloroamphetamine is also active when given orally. The systemic efficacy of halogenated amphetamines could be considered a technical advantage in the use of serotonin neurotoxins as pharmacologic tools. On the other hand, the dihydroxytryptamines can be injected directly into specific brain regions to produce local effects,[6] whereas this route of administration cannot be employed with the halogenated amphetamines because of their rapid diffusion.

Anatomic Specificity. The effects of halogenated amphetamines and dihydroxytryptamines on serotonin content of different brain regions are different. Sanders-Bush *et al.*[7] have pointed out that *p*-chloroamphetamine depletes serotonin most in the cerebral cortex and striatum, regions that contain serotonin mainly in nerve terminals. The hypothalamus and spinal cord, which are also rich in serotonin nerve terminals, and the medulla-pons regions, which contain some cell bodies of serotonin neurons, are relatively less depleted of serotonin after *p*-chloroamphetamine injection. Dihydroxytryptamines cause marked depletion of serotonin in the hypothalamus and pons-medulla and in the spinal cord.[8] Since dihydroxytryptamines are injected intraventricularly or intracisternally, diffusion to various brain regions is one factor that influences their regional specificity. Also, the regional specificity varies with time, because regeneration of serotonin neurons occurs more rapidly in some brain regions.[9]

Species Differences. Halogenated amphetamines depress serotonin levels initially in all species studied but are not neurotoxic in all species. For example, the serotonin depletion induced by *p*-chloroamphetamine in mice[10] and rabbits[11] is short-lasting, whereas in rats and guinea pigs,[12] *p*-chloroamphetamine is neurotoxic. The duration of the effect of *p*-chloroamphetamine in dogs[13] and cats[14] is still uncertain. Dihydroxytryptamines are probably neurotoxic in all species.

Ontogeny. Dihydroxytryptamines can produce neurotoxicity in rats of all ages, including neonates.[15] *p*-Chloroamphetamine, in contrast, is not neurotoxic in neonatal rats.[16, 22]

Modification of Effects by Other Drugs. Dihydroxytryptamines may be given with an inhibitor of uptake into catecholamine neurons so that they will more specifically destroy serotonin neurons[17] or with an inhibitor of monoamine oxidase to protect them from oxidative deamination.[18] Otherwise, dihydroxytryptamines cause some depletion of catecholamines.[8, 19] There is no need to give the halogenated amphetamines in combination with an inhibitor of uptake into catecholamine neurons, because they cause no neurotoxic effects on such neurons.[20] Although the depletion of serotonin caused by the halogenated amphetamines can be prevented by inhibitors of uptake into serotonin neurons,[1, 2, 4] attempts to prevent the effects of dihydroxytryptamines with an uptake inhibitor have so far been unsuccessful.[21, 23] The latter finding raises questions about the requirement for active uptake of the dihydroxytryptamines into serotonin neurons.

In summary, certain halogenated amphetamines and dihydroxytryptamines produce neurochemical changes in the brain related to neurotoxic effects on serotonin neurons. These agents are useful pharmacologic tools in studies of functional roles of serotonin neurons. These two groups of neurotoxic agents differ with respect to numerous pharmacologic properties, as revealed by neurochemical studies. These differences are important to consider when using the agents for behavioral, endocrine, neurologic, or other studies.

REFERENCES

1. MEEK, J. L., K. FUXE & A. CARLSSON. 1971. Blockade of p-chloromethamphetamine induced 5-hydroxytryptamine depletion by chlorimipramine, chlorpheniramine and meperidine. Biochem. Pharmacol. **20:** 707–709.
2. FULLER, R. W. & K. W. PERRY & B. B. MOLLOY. 1975. Reversible and irreversible phases of serotonin depletion by 4-chloroamphetamine. Eur. J. Pharmacol. **33:** 119–124.
3. FULLER, R. W. This monograph.
4. CLINESCHMIDT, B. V., A. G. ZACCHEI, J. A. TOTARO, A. B. PFLUEGER, J. C. MCGUFFIN & T. I. WISHOUSKY. This monograph.
5. SHERMAN, A., E. M. GÁL, R. W. FULLER & B. B. MOLLOY. 1975. Effects of intraventricular p-chloroamphetamine and its analogues on cerebral 5-HT. Neuropharmacology **14:** 733–737.
6. BAUMGARTEN, H. G. & A. BJÖRKLUND. 1976. Neurotoxic indoleamines and monoamine neurons. Annu. Rev. Pharmacol. **16:** 101–111.
7. SANDERS-BUSH, E., J. A. BUSHING & F. SULSER. 1975. Long-term effects of p-chloroamphetamine and related drugs on central serotonergic mechanisms. J. Pharmacol. Exp. Ther. **192:** 33–41.
8. BAUMGARTEN, H. G., A. BJÖRKLUND, L. LACHENMAYER & A. NOBIN. 1973. Evaluation of the effects of 5,7-dihydroxytryptamine on serotonin and catecholamine neurons in the rat CNS Acta Physiol. Scand. Suppl. **391:** 1–19.
9. BJÖRKLUND, A., A. NOBIN & U. STENEVI. 1973. Regeneration of central serotonin neurons after axonal degeneration induced by 5,6-dihydroxytryptamine. Brain Res. **50:** 214–220.
10. MILLER, K. W., E. SANDERS-BUSH & J. V. DINGELL. 1971. p-Chloroamphetamine —species differences in the rate of disappearance and the lowering of cerebral serotonin. Biochem. Pharmacol. **20:** 500–503.
11. PLETSCHER, A., G. BARTHOLINI, H. BRUDERER, W. P. BURKARD & K. F. GEY. 1964. Chlorinated arylalkylamines affecting the cerebral metabolism of 5-hydroxytryptamine. J. Pharmacol. Exp. Ther. **145:** 344–350.
12. FULLER, R. W. & K. W. PERRY. 1974. Long-lasting depletion of brain serotonin by 4-chloroamphetamine in guinea pigs. Brain Res. **82:** 383–385.
13. FULLER, R. W., C. W. HINES & J. MILLS. 1965. Lowering of brain serotonin level by chloroamphetamines. Biochem. Pharmacol. **14:** 483–488.
14. LIPPMANN, W. & M. WISHNICK. 1965. Effect of DL-p-chloro-N-methylamphetamine on the concentrations of monoamines in the cat and rat brain and rat heart. Life Sci. **4:** 849–857.
15. LYTLE, L. D., J. H. JACOBY, M. F. NELSON & H. G. BAUMGARTEN. 1974. Long-term effects of 5,7dihydroxytryptamine administered at birth on the development of brain monoamines. Life Sci. **15:** 1203–1217.
16. NECKERS, L. M., L. BERTILSSON, S. H. KOSLOW & J. L. MEEK. 1976. Reduction of tryptophan hydroxylase activity and 5-hydroxytryptamine concentration in certain rat brain nuclei after p-chloroamphetamine. J. Pharmacol. Exp. Ther. **196:** 333–338.
17. BJÖRKLUND, A., H. G. BAUMGARTEN & A. RENSCH. 1975. 5,7-Dihydroxytryptamine: improvement of its selectivity for serotonin neurons in the CNS by pretreatment with desipramine. J. Neurochem. **24:** 833–835.
18. BREESE, G. R., B. R. COOPER, L. D. GRANT & R. D. SMITH. 1974. Biochemical and behavioral alterations following 5,6-dihydroxytryptamine administration into brain. Neuropharmacology **13:** 177–187.
19. SANER, A., L. PIERI, J. MORAN, M. DA PRADA & A. PLETSCHER. 1974. Decrease of dopamine and 5-hydroxytryptamine after intracerebral application of 5,6-dihydroxytryptamine. Brain Res. **76:** 109–117.
20. MILLER, F. P., R. H. COX, JR., W. R. SNODGRASS & R. P. MAICKEL. 1970. Comparative effects of p-chlorophenylalanine, p-chloroamphetamine and p-chloro-

N-methylamphetamine on rat brain norepinephrine, serotonin and 5-hydroxyindole-3-acetic acid. Biochem. Pharmacol. **19:** 435–442.

21. Breese, G. R. & B. R. Cooper. 1975. Behavioral and biochemical interactions of 5,7-dihydroxytryptamine with various drugs when administered intracisternally to adult and developing rats. Brain Res. **98:** 517–527.
22. Fuller, R. W., K. W. Perry & J. A. Clemens. 1977. Unpublished results.
23. Fuller, R. W. & J. A. Clemens. 1976. Unpublished results.

ENZYME CHANGES AS AN INDEX OF NEUROTOXIN SPECIFICITY

Walter Lovenberg and Eleanor Bruckwick

Section on Biochemical Pharmacology
National Heart, Lung, and Blood Institute
Bethesda, Maryland 20014

Several compounds have been shown to cause destruction of serotonergic neurons. The specificity of the neurotoxins for a certain type of neuron apparently depends on the affinity of the neurotoxin for cellular uptake systems. Several indices of neuronal destruction have been used to monitor the interaction of the drug with specific neurons. These indices include histochemical techniques,[1] measurement of neurotransmitter content,[2] and measurement of enzymes associated with specific cell types.[3-5] These enzymes are generally uniquely associated with the synthesis of each transmitter. In the serotonergic neurons, the unique enzyme is tryptophan hydroxylase.

In this paper, we will review studies performed in our laboratory with tryptophan hydroxylase, tyrosine hydroxylase, and dopamine β-hydroxylase activities as indices of the viability of serotonergic, dopaminergic, and adrenergic (norepinephrine and epinephrine) neurons, respectively.

The amine neurotransmitters are formed in metabolic pathways that have certain common features (FIGURE 1). The initial step in the enzymatic pathway for either indoleamine or phenylethylamine transmitters is the hydroxylation of either tryptophan or tyrosine. The enzymes that catalyze these reactions are distinct but have many properties in common. The hydroxylated amino acids in either pathway are next decarboxylated by aromatic L-amino acid decarboxylase. In noradrenergic and adrenergic neurons, dopamine is converted to norepinephrine by the enzyme dopamine β-hydroxylase. Clearly, tryptophan hydroxylase is an unambiguous marker for serotonergic neurons in brain, as is dopamine β-hydroxylase for neurons that use either norepinephrine or epinephrine as a transmitter. Tyrosine hydroxylase is by definition present in all catecholaminergic cells, and its activity in tissue extracts must reflect contributions from all catecholamine cell types.

Based on work to be presented and on immunofluorescence studies,[6,7] it appears that far greater amounts of tyrosine hydroxylase are present in dopaminergic neurons than in noradrenergic or adrenergic neurons. Therefore, in most brain regions, tyrosine hydroxylase is a practical marker for dopaminergic neurons.

DETERMINATION OF TRYPTOPHAN, TYROSINE, AND DOPAMINE β-HYDROXYLASES IN A SINGLE HOMOGENATE

Rats are killed by decapitation, and the brain is removed quickly. The brain is dissected, as described previously,[3] by removing its various regions in the following order: septum, striatum, hypothalamus, remaining forebrain, pons-medulla, mesencephalic tectum, and mesencephalic tegmentum. Approximately one third of the spinal cord is removed from the upper part of the

0077-8923/78/0305-0182 $01.75/0

body. After removal, the tissues are placed on aluminum foil in dry ice. The dissected brains from three rats are pooled according to anatomic regions. The brain tissues are stored in liquid nitrogen until preparation for assay. For assay, pooled tissues from each region are weighed and placed in ground glass conical homogenizers that are immersed in ice. All of the following steps are performed at 4°C. An amount of 0.02 M Tris hydrochloride buffer (pH 6.8) that contains 0.2% Triton® X-100 equal to 4 vol of the tissue weight (milliliters/gram) is added to each homogenizer. The tissues are quickly homogenized, and aliquots to be subsequently used for dopamine β-hydroxylase (DBH) assay are removed and placed in centrifugation tubes. Before centrifugation, the aliquots for DBH analysis are diluted as follows:

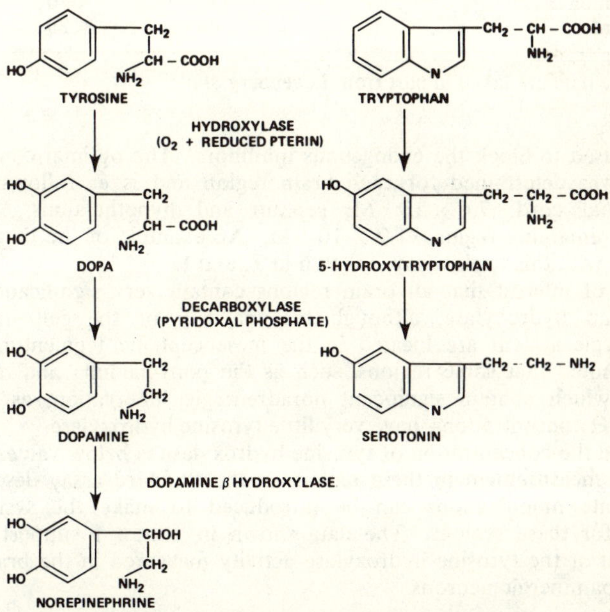

FIGURE 1. Synthetic routes for neurohumoral amines.

the septum, hypothalamus, and tegmentum, 1:10 with water; and the tectum, striatum, pons-medulla, spinal cord, and forebrain, 1:6 with water. Dithiothreitol (0.2 M) is added to the remainder of the homogenate to make the final concentration 0.002 M. This homogenate is used for analysis of tyrosine and tryptophan hydroxylases. All tissue homogenates are centrifuged for 20 min at 40,000*g*. From the supernatant fraction, 0.020-ml aliquots are taken for assay for tryptophan hydroxylase. The assay system used is a micromodification (1/10 vol of all components) of that described previously.[3] Aliquots of 0.04 ml are taken for assay of tyrosine hydroxylase in the system described by Cicero *et al.*[8] For both assays, 6-methyltetrahydropterin is used as the cofactor.

Dopamine β-hydroxylase is determined in the supernatant fraction by the procedure of Molinoff *et al.*[9] as modified by our laboratory.[10] In this procedure

TABLE 1

REGIONAL ACTIVITIES OF TRYPTOPHAN, TYROSINE, AND DOPAMINE β-HYDROXYLASES

Region	Hydroxylase Activity * (nmol/g/hr)		
	Tryptophan	Tyrosine	Dopamine
Striatum	46	724	35
Septum	100	166	270
Hypothalamus	183	84	188
Forebrain	46	26	77
Mesencephalic tegmentum	494	61	105
Mesencephalic tectum	150	<20	117
Pons-medulla	135	<20	131
Spinal cord	85	<20	55

* These data are taken in part from Lovenberg *et al.*[23]

Cu^{2+} is used to block the endogenous inhibitors. The optimal concentration of copper was determined for each brain region and is as follows: tegmentum and spinal cord, 7.6×10^{-5} M; septum and hypothalamus, 9.7×10^{-5} M; and the remaining regions, 1.2×10^{-4} M. An example of the data that can be obtained with this triple assay is shown in TABLE 1.

It is of interest that all brain regions contain very significant amounts of tryptophan hydroxylase, although the majority of the cell bodies of the serotonergic system are located in the mesencephalic tegmentum. It should also be noted that some regions, such as the pons-medulla and mesencephalic tectum, which contain significant noradrenergic innervation, as indicated by their DBH concentrations, have very little tyrosine hydroxylase.

While the concentration of tyrosine hydroxylase is below values required for accurate measurement in these regions in the standard assay described above, appropriate modifications can be introduced to make the system sensitive enough for these regions. The data shown in TABLE 1 support the concept that most of the tyrosine hydroxylase activity measured in the brain is derived from dopaminergic neurons.

5,6-DIHYDROXYTRYPTAMINE (5,6-DHT)

The discovery of the peripheral neurotoxic action of 6-hydroxydopamine (6-OHDA) by Thoenen and Tranzer in 1968[11] and the subsequent observation of Bloom *et al.*[12] that this compound also produced central noradrenergic neurotoxicity provided an impetus to the search for their specific neurotoxic agents. In 1971, Baumgarten *et al.*[2] reported that the 6-hydroxy derivative of serotonin apparently exhibited selective neurotoxicity for the serotonergic system. Subsequently, we examined the effect of an intraventricular injection of 75 μg of 5,6-DHT on the regional activity of tryptophan hydroxylase as a function of time after administration.[4] We found that 1 hr after injection of the drug, there was a 15–70% decrease in tryptophan hydroxylase activity in all regions of the brain and in the spinal cord. In separate studies, it was observed that 5,6-DHT inhibited tryptophan hydroxylase with an IC_{50} of less than 5×10^{-5} M. Thus, it is possible that this early loss of enzymatic activity is partly due to inhibition of the enzyme rather than to cytotoxicity. However, the mesencephalic tegmentum, an area that contains the serotonergic cell bodies of the Raphe nuclei, showed little or no loss of enzymatic activity, an observation that suggests that even this early event is dependent on the neurotoxin being concentrated within specific nerve endings. Two days after this single dose, the septal and hypothalamic areas recovered a portion of their tryptophan hydroxylase activity, whereas other regions showed further losses in activity. From 2 to 6 days after drug administration, there was a significant recovery of enzymatic activity in all brain regions, except the spinal cord, with near normal values observed in the septum, striatum, hypothalamus, pons-medulla, and mesencephalic tegmentum.

Twelve days after the injection, significant losses of tryptophan hydroxylase activity occurred in all brain regions. These depletions ranged from 10% in the striatum to more than 95% in the spinal cord. Even 60 days after the initial dose, there was only little recovery in any of the regions. From the above study,[3] it is concluded that 5,6-DHT induces total and permanent destruction of many serotonergic nerve endings and fibers. It appears that the brain regions that contain primarily nerve terminals are the most sensitive to the drug and that cell body areas undergo retrograde degeneration.

In separate experiments, we have examined the effects of a single dose of 5,6-DHT on the regional activity of tyrosine hydroxylase and DBH 10 days after administration. A typical experiment is presented in TABLE 2. In this experiment, rats received an intraventricular injection of either 75 μg of 5,6-DHT or vehicle. Almost without exception, the level of catecholaminergic enzymes appeared to be slightly higher in rats that received 5,6-DHT than in control animals. This finding suggests that 5,6-DHT is relatively specific for serotonin neurons and that such neurons may have a modulatory effect on catecholamine neurons, because, in general, the catecholamine enzymes are slightly elevated. It should be noted that the potential of neurotoxins for probing interneuronal regulation has not been fully utilized. In some preliminary experiments, we observed that an intraventricular dose (100 μg) of 6-OHDA resulted in an apparent increase in the tryptophan hydroxylase activity of the septum and mesencephalic tegmentum but had no significant effect on the enzyme in other regions.

TABLE 2

EFFECT OF 5,6-DHT ON REGIONAL TYROSINE AND DOPAMINE β-HYDROXYLASES

Region	Hydroxylase (% of control) Tyrosine	Dopamine
Striatum	108	121
Septum	61	102
Hypothalamus	146	147
Forebrain	142	97
Tegmentum	122	144
Spinal cord	—	146

5,7-Dihydroxytryptamine (5,7-DHT)

Many structural analogs of 5,6-DHT have been prepared, and 5,7-DHT was observed to show potential as a neurotoxin.[11] Because this compound has an apparently lower toxicity than does 5,6-DHT, it is possible to use a larger dose (150 μg) in rat studies. We found that at this dose, 5,7-DHT was more effective than was 75 μg of 5,6-DHT in causing a depletion of tryptophan hydroxylase.[3, 5] The activity of tryptophan hydroxylase was reduced to 50% or less in all brain regions, except the mesencephalic tegmentum, 2 days after a single dose. By 12 days, further depletion had occurred in all regions, with the mesencephalic tegmentum also having less than 50% of the control activity. Since independent studies [12] had indicated that 5,7-DHT effectively depleted norepinephrine and serotonin but did not deplete dopamine, it was of interest to find that little or no depletion of tyrosine hydroxylase activity occurred in any region. From these experiments and the histofluorescence data of others,[6, 7] it is clear that the majority of tyrosine hydroxylase activity in the central nervous system is located within dopamine-containing neurons. Other studies indicated that 5,7-DHT effectively interacted with the norepinephrine uptake system but that 5,7-DHT could be a powerful and selective neurotoxin for serotonergic systems [13] if animals were pretreated with desmethylimipramine.

In a separate series of experiments,[14] we administered 5,7-DHT to newborn rats and measured the normal developmental pattern for regional brain tryptophan hydroxylase. Although the content of this enzyme in the brain is low at birth, there is a relatively rapid increase in activity, with adult concentrations reached in most regions by 20 days of age. The development of the enzyme content was totally blocked if the animals were treated intracisternally on the second day after birth with 40 μg of 5,7-DHT. The enzyme content in treated animals was lower after 40 days in all regions, except the hypothalamus. Even in the hypothalamus, the hydroxylase content was only 20% of that of control animals at Day 40. At 7 months, there was a slight recovery of activity in all regions. A small transient depression of tyrosine hydroxylase activity was observed at Day 12 in the septum. Otherwise, the activity of tyrosine hydroxylase was not significantly different between experimental and control animals. The growth of animals in which the normal development of the serotonin systems had been blocked was significantly retarded. Preliminary studies, however, indicate that there are no significant differences in circulating growth hormone in the treated animals.[15]

p-Chloroamphetamine

It is known that *p*-chloroamphetamine (*p*-CA) can cause a long-lasting depletion of serotonin from the brain.[16] It was also observed that *p*-chlorophenylalanine (*p*-CP) can result in significant depletion of brain serotonin for up to 2 weeks after a single large dose.[17] In one of the first attempts to correlate enzymatic changes with transmitter depletion, we observed that a nearly total loss of tryptophan hydroxylase activity occurred after a single injection of *p*-CP.[18] After several days, the enzymatic activity of the brain stem began to return slowly; after 2 weeks, values near control levels were obtained. The mechanism of this enzymatic loss still remains a mystery, because total destruction of neurons does not appear to be involved.

There has been a continued interest in *p*-CA, because it shows evidence of neurotoxicity.[19, 20] Sanders-Bush *et al.*[21, 22] have conducted extensive studies on the biochemistry of this serotonin depletion and find significant long-term loss of tryptophan hydroxylase activity. Studies in our laboratory [23, 24] have confirmed the long-term loss of tryptophan hydroxylase activity that occurs after a single intraperitoneal dose of *p*-chloromethamphetamine (*p*-CMA). By a detailed regional analysis for both tryptophan and tyrosine hydroxylases, we make the following conclusions:

p-Chloromethamphetamine has little or no effect on catecholaminergic neurons.

The regions rich in nerve endings are most sensitive to the neurotoxic effects of *p*-CMA.

The mesencephalic tegmentum, which contains serotonergic cell bodies, exhibits a transient increase in tryptophan hydroxylase activity, a finding that probably reflects the primary interaction of the drug at nerve endings with a possible interruption of axoplasmic flow.

Except for the mesencephalic tectum, all serotonergic terminal regions show a partial recovery of tryptophan hydroxylase activity by the 25th day after injection.

Qualitatively similar results have been observed in other laboratories.[25]

It is clear that the neurotoxicity of the chlorinated amphetamine derivatives is dependent on their specific interaction with the neurotransmitter uptake system of the serotonergic neurons.[26] In this regard, these compounds are similar to the dihydroxyindole derivatives. However, the molecular mechanisms of action of these two classes of drugs are unknown. Much of the speculation concerning the mode of action of 6-OHDA and the DHTs relates to their highly reactive nature and their ability to generate potentially toxic oxygen radicals under aerobic conditions. Similar speculation is not applicable to the chlorinated amphetamines, although a mechanism for the conversion of *p*-CA to a dihydroxyindole derivative has been suggested.[27]

Conclusions

The measurement of neurotransmitter biosynthetic enzymes has provided an added dimension to the study of serotonergic neurotoxins. It appears that tryptophan hydroxylase activity in the brain is limited to the serotonergic cells. This enzyme provides a better index of serotonergic cellular damage than does neurotransmitter content. The potential nonspecificity of serotonin neurotoxins can also be monitored by measurement of other cell-specific enzymes.

References

1. Baumgarten, H. G., L. Lachenmayer & H. G. Schlossberger. 1972. Z. Zellforsch. **125:** 553.
2. Baumgarten, H. G., A. Björklund, L. Lachenmayer, A. Nobin & U. Stenevi. 1971. Acta Physiol. Scand. Suppl. **373**.
3. Baumgarten, H. G., S. J. Victor & W. Lovenberg. 1973. J. Neurochem. **21:** 251.

4. Victor, S. J., H. G. Baumgarten & W. Lovenberg. 1974. J. Neurochem. **22:** 541.
5. Lovenberg, W. & S. J. Victor. 1974. *In* Advances in Biochemical Psychopharmacology. Vol. **10:** 93. The Raven Press. New York, N.Y.
6. Reis, D. J., R. A. Ross, U. M. Pickel, T. Lewander & T. H. Joh. 1975. *In* Chemical Tools in Catecholamine Research. O. Almgren, A. Carlsson & J. Engel, Eds. Vol. **II:** 53. North Holland Publishing Company. Amsterdam, The Netherlands.
7. Fuxe, K. 1976. Paper presented at the American College of Neuropsychopharmacology, New Orleans, La.
8. Cicero, T. J., L. G. Sharpe, E. Robins & S. S. Grote. 1972. J. Neurochem. **19:** 2241.
9. Molinoff, P. B., R. Weinshilboum & J. Axelrod. 1971. J. Pharmacol. Exp. Ther. **178:** 425.
10. Nagaoka, A. & W. Lovenberg. 1977. Eur. J. Pharmacol. In press.
11. Baumgarten, H. G. & L. Lachenmayer. 1972. Z. Zellforsch. **135:** 399.
12. Baumgarten, H. G., A. Björklund, L. Lachenmayer & A. Nobin. 1973. Acta Physiol. Scand. Suppl. **391:** 1.
13. Björklund, A., H. G. Baumgarten & A. Rensch. 1975. J. Neurochem. **24:** 833.
14. Baumgarten, H. G., S. J. Victor & W. Lovenberg. 1975. Psychopharmacol. Commun. **1:** 75.
15. Skyler, J., H. G. Baumgarten & W. Lovenberg. 19xx Unpublished observations.
16. Pletscher, A., G. Bartholini, H. Bruderer, W. P. Burkard & K. Gey. 1964. J. Pharmacol. Exp. Ther. **145:** 334.
17. Koe, B. K. & A. Weissman. 1966. J. Pharmacol. Exp. Ther. **154:** 499.
18. Jequier, E., W. Lovenberg & A. Sjoerdsma. 1967. Mol. Pharmacol. **3:** 274.
19. Harvey, J. A. & S. E. McMaster. 1976. Psychopharmacol. Bull. **12:** 62.
20. Sanders-Bush, E. 1976. Psychopharmacol. Bull. **12:** 57.
21. Sanders-Bush, E., J. A. Bushing & F. Sulser. 1975. J. Pharmacol. Exp. Ther. **21:** 1501.
22. Sanders-Bush, E., J. A. Bushing & F. Sulser. 1975. J. Pharmacol. Exp. Ther. **192:** 33.
23. Lovenberg, W., M. N. Walker & H. G. Baumgarten. 1976. Monogr. Neural Sci. **3:** 109.
24. Walker, M. N., H. G. Baumgarten & W. Lovenberg. 1977. In preparation.
25. Neckers, L. M., L. Bertilsson, S. H. Koslow & J. L. Meek. 1976. J. Pharmacol. Exp. Ther. **196:** 333.
26. Fuller, R. W., K. W. Perry, H. D. Snoddy & B. B. Molloy. 1974. Eur. J. Pharmacol. **28:** 233.
27. Gál, E. M., P. A. Christiansen & Y. M. Yunger. 1975. Neuropharmacology **14:** 31.

Discussion

Dr. H. Tamir: The observation that 5,7-DHT affects the tegmentum indicates that cell bodies are altered by these neurotoxins. For awhile, we were led to believe that they affected only nerve endings and axons. However, you showed a remarkable effect on the tegmentum after 12 days.

Dr. Lovenberg: I'm sure there is a very significant effect on cell bodies. There is no way to determine whether this effect is due to loss of cell bodies or to a reduction in the amount of enzyme in each cell body. Certainly, however, there is an effect on cell body-containing regions by 5,7-DHT.

Dr. H. G. Baumgarten: There is no loss of cell bodies. There appears to be a retrograde temporary depression of enzyme synthesis that is reflected in decreased serotonin content in these cells. However, they recover progressively, beginning at 15–25 days, depending on the dose.

Dr. Lovenberg: I think the recovery is rather slow, because at 30 days there is still substantial loss of enzymatic activity.

Dr. C. T. Giambalvo: We injected 5,7-DHT directly into the medial raphe nucleus. Histologic sections revealed degeneration of cell bodies, so there was an effect on cell bodies in our studies.

STUDIES OF SEROTONIN NEUROTOXINS IN DISCRETE BRAIN NUCLEI

James L. Meek

Laboratory of Preclinical Pharmacology
National Institute of Mental Health
Saint Elizabeths Hospital
Washington, D.C. 20032

The study of the initial site of action of neurotoxins is a fundamental step in understanding their mechanism of action. The peculiar anatomic location of serotonergic neurons, with cell bodies in nuclei of pons and medulla, and with all other brain areas that contain only axonal projections, facilitates the study of the primary site of action of neurotoxins that affect serotonergic cells. This paper will summarize our studies on the mode of action of *p*-chloroamphetamine (*p*-CA) and *p*-chlorophenylalanine (*p*-CPA), which were conducted by taking advantage of this anatomic location of serotonergic neurons. We used the concentration of serotonin (5-hydroxytryptamine, 5-HT) and tryptophan hydroxylase in discrete brain nuclei as markers of the state of 5-HT neurons.

Methods have long been needed to selectively inactivate 5-HT neurons. The dihydroxytryptamines and *p*-CA both cause a long-lasting depletion of 5-HT that usually fails to be associated with a total and permanent destruction of 5-HT neurons. Another compound, *p*-CPA, causes a nearly complete but only temporary depletion of 5-HT. A potentially selective tool to study serotonergic neurons would be a false neurotransmitter. The use of α-methyl-5-HT to deplete 5-HT was therefore examined.

Materials and Methods

Tryptophan hydroxylase was measured by high-pressure liquid chromatography. 5-Hydroxytryptamine was measured by mass fragmentography and by high-pressure liquid chromatography (HPLC). The latter technique became the method of choice because of its simplicity and high sensitivity. High-pressure liquid chromatography permits simultaneous separation and quantitation of closely related indoles with a sensitivity in the picomole range; the simplicity of this method is unsurpassed.

The HPLC apparatus for measurement of 5-HT and tryptophan hydroxylase has been described in detail elsewhere.[1-3] Briefly, the 5-HT or α-methyl-5-HT in the samples was separated on 0.5 m × 2.1 mm column of Dupont® SCX resin with a mobile phase of 0.3 M sodium acetate (pH 6.0) pumped at 900 psi at room temperature. After elution from the column, the stream was mixed with concentrated perchloric acid and passed through a fluorometer with a flow cell. Under these conditions (Figure 1), α-methyl-5-HT eluted in the void volume (1.08); 5-HT and α-methyl-5-HT eluted at ratios of 2:35 and 5:15, respectively. For analysis of α-methyl-5-HT, a more acidic mobile phase was used: 0.1 M sodium formate plus perchloric acid to adjust the pH to 3.0. Under these conditions, 5-HT and α-methyl-5-HT are so strongly retarded that they do not interfere in the measurement of the latter. Tissues were extracted

0077–8923/78/0305–0190 $01.75/0

with 4–6 vol of 0.4 N $HClO_4$ and centrifuged. A 50-μl aliquot of the supernatant was injected into the column. FIGURE 1 (*middle*) shows the measurement of 5-HT in a 50-μl extract of whole rat brain (10 ml). In FIGURE 1 (*right*), the rat was pretreated intraperitoneally 6 hr before death with 200 mg/kg of α-methyl-5-hydroxytryptophan (a gift from The Upjohn Company). The 5-HT concentration was reduced, and a peak that represented α-methyl-5-HT was seen.

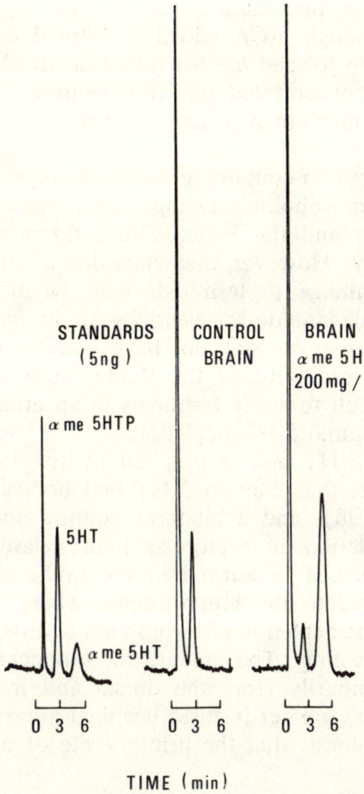

FIGURE 1. Measurement of 5-HT and α-methyl-5-HT by HPLC. The chromatograms show standards (5 ng each) and brain extracts of a normal rat and of one pretreated with α-methyl-5-HT phosphate (200 mg/kg, ip).

EFFECTS OF *p*-CA ON 5-HT IN DISCRETE BRAIN NUCLEI

Sanders-Bush *et al.*[4] noted that *p*-CA caused a 50–70% depletion of 5-HT and tryptophan hydroxylase in whole rat brain that lasted for 4 months or longer. Because they performed their measurements in whole brain, it was not clear whether certain 5-HT pathways were destroyed completely or whether a certain portion of 5-HT was lost in all axons. Though the local concentration of 5-HT in nuclei that contain cell bodies is high, the total 5-HT stored in Raphe nuclei represents only a small portion of the total 5-HT present in brain. Therefore, when a drug decreases the 5-HT content of whole brain by 70%, this drug must extensively deplete the 5-HT stored in terminal and preterminal axons. The depletion of 5-HT is not, however, a good index of neurotoxicity, because it can not be differentiated from depletion due to neuronal damage or from a simple displacement of 5-HT from binding sites. The measurement of tryptophan hydroxylase permits detection of neurotoxicity and allows differentiation from displacement or inhibition of binding. This enzyme has a half-life of 2–3 days,[5] which is longer than that of 5-HT by several orders of magnitude; moreover, synthesis of this enzyme occurs only in cell bodies, whereas 5-HT is synthesized in every part of the serotonergic neuron.

As previously reported,[6] tryptophan hydroxylase in the forebrain (the brain tissue anterior to the optic chiasm) is reduced within 1–4 hr after *p*-CA injection. One must conclude that *p*-CA or a metabolite inactivates the enzyme directly or increases its destruction. Although *p*-CA added *in vitro* does not affect tryptophan hydroxylase, a metabolite formed *in vivo* might do so. However, it is possible that *p*-CA, which is concentrated in 5-HT neurons,[7] may increase proteolysis by causing swelling, interfere with axonal transport of the enzyme, or disrupt lysosomal membranes.

Lorenz *et al.*[8] noted swelling in serotonin-containing axons 2 days after injection of *p*-CA but could not detect morphologic changes of terminals in various areas by use of light microscopy and the Falck-Hillarp fluorescence technique or by use of electron microscopy. However, their data do not contradict our suggestion that early severe damage to terminals does occur. As described below, *p*-CA has a considerable anatomic selectivity in its site of action, and the areas examined by Lorenz *et al.* may not have had extensive 5-HT depletion. In addition, the poor sensitivity of the fluorescence histochemical method for 5-HT makes it difficult to study terminals in an area like the hippocampus, where *p*-CA causes maximal 5-HT depletion.

The time course of the depletion of 5-HT and tryptophan hydroxylase is shown in FIGURE 2. We examined nuclei that contain 5-HT cell bodies [the dorsal raphe (B7)], the median raphe (B8), and a bilateral pontine nucleus known as B9. In these three nuclei, depletion of tryptophan hydroxylase was 50% or less, enzymatic activity had returned to normal in the raphe nuclei within 3 weeks after injection. In the caudate and hippocampus, where 5-HT is contained only in axons and terminals, the depletion of tryptophan hydroxylase and 5-HT was more complete and longer lasting. The caudate and hippocampus receive their serotonergic innervation primarily from the dorsal and median raphe, respectively. Since the depletion was greater in these terminal areas than in the raphe, these data support the hypothesis that the primary site of action of *p*-CA cannot be in the cell bodies.

ATTEMPTS TO INCREASE THE EXTENT OF 5-HT DEPLETION BY *p*-CA

Although 5-HT depletion produced by *p*-CA is long lasting, the extent is less than 70%. Since repeated doses do not appear to increase the extent of depletion,[9, 10] we have examined whether increasing the dose of *p*-CA leads to a greater depletion of 5-HT. Normally, the maximum tolerated dose of *p*-CA injected intraperitoneally is 20 mg/kg. We found that in rats pretreated with chlorpromazine (10 mg/kg) 1 hr before the injection, hyperthermia and hyperactivity were prevented, and a dose of 80 mg/kg of *p*-CA could be injected. One day after this combined treatment, whole brain 5-HT levels were less than 10% of those of untreated controls. However, 7 days after the injection of chlorpromazine plus *p*-CPA, 5-HT levels had increased to 20% of controls, indicating that complete destruction had not occurred. Similar results were obtained when a second dose of *p*-CA was given 2 days after the first injection and the rats killed 7 days after the first dose had been administered.

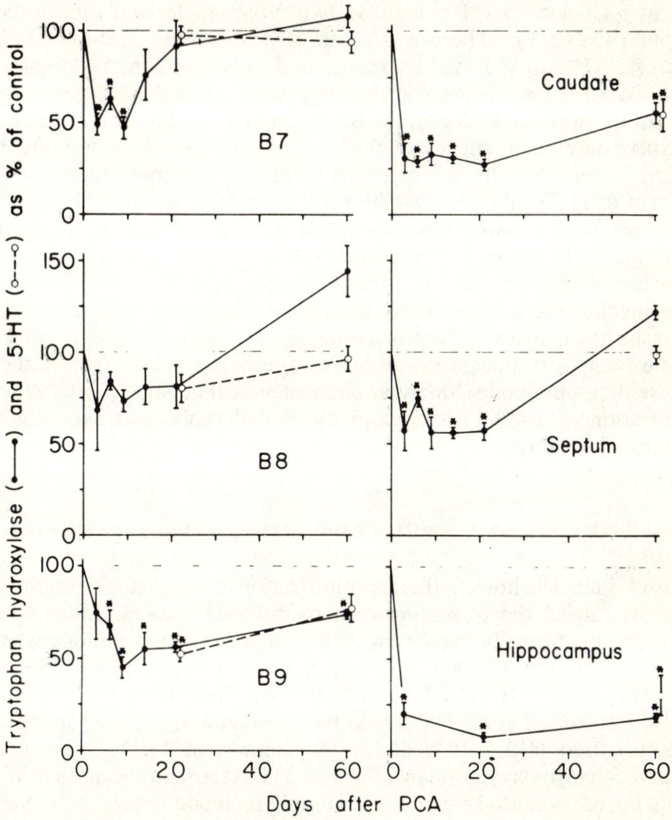

FIGURE 2. Time course of the changes in 5-HT (○) and tryptophan hydroxylase (●) after injection of *p*-CA (100 μmol/kg).

EFFECTS OF *p*-CPA ON 5-HT IN DISCRETE BRAIN NUCLEI

The mechanism of neurotoxicity of *p*-CPA remains to be established, but it is probably a direct inactivation of tryptophan hydroxylase by a *p*-CPA metabolite and not via inhibition of synthesis of tryptophan hydroxylase in the cell bodies. As with *p*-CA, anatomic considerations provide evidence for the location on the neuron of the initial site of action. *p*-Chlorophenylalanine causes a greater than 90% loss of tryptophan hydroxylase within 1 day after injection in brain and caudal and cranial portions of spinal cord.[5] The loss of tryptophan hydroxylase is therefore probably local and not a result of alterations in synthesis or transport due to the considerable distance of terminals in the cord from cell bodies in the raphe. Aghajanian *et al.*[11] noted that there was little change in the fluorescence of the raphe perikarya after administration of *p*-CPA. Their qualitative data support the hypothesis that the primary site of action is in the terminals, but this conclusion required quantitative confirmation. We measured the effects of *p*-CPA on 5-HT content in two raphe nuclei and the caudate and hippocampus (TABLE 1). There was about 80% depletion of 5-HT in the two terminal areas, 57% in the median raphe, and only 29% in the dorsal raphe. These data are direct evidence for the suggestion of Aghajanian *et al.*[11] that inhibition of tryptophan hydroxylase is incomplete in the cell bodies. As a possible explanation, they suggested that newly synthesized enzyme in the cell bodies might account for the incomplete inhibition, but a more likely hypothesis is nonuniform drug distribution. *p*-Chlorophenylalanine is probably transported by the same mechanism as is tryptophan; this transport differs in various nuclei. We have found[12] that after a tryptophan loading (200 mg/kg), the maximum tryptophan levels (percentages of control levels) are 450 in the dorsal raphe, 700 in the median raphe, and 1200–1500 in the superior colliculus or in the grisea centralis at the level of the substantia nigra. The tryptophan accumulated in the raphe nuclei disappears much more rapidly than does that in the other areas. These data obtained with tryptophan suggest that after a *p*-CPA loading, the concentration of *p*-CPA reached in the dorsal raphe was not sufficient to cause substantial toxicity.

α-METHYL-5-HT AS A SEROTONERGIC FALSE NEUROTRANSMITTER

Lahti and Platz[13] showed that administration of α-methyl-5-hydroxytryptophan to rats caused the formation of α-methyl-5-HT and a concomitant decrease in norepinephrine in the brain. They suggested that endogenous 5-HT concentrations were not affected, but lack of a suitable assay that differentiated α-methyl-5-HT from 5-HT made it difficult for them to study the question. Since a false transmitter for 5-HT would have potential usefulness in psychiatric treatments, we used HPLC to examine the kinetics of 5-HT, α-methyl-5-HT, and α-methyl-5-hydroxytryptophan (FIGURE 3). After intraperitoneal injection of 200 mg/kg of α-methyl-5-hydroxytryptophan, blood levels of α-methyl-5-hydroxytryptophan increased and then declined rapidly with a $t_{1/2}$ of about 1 hr. The penetration of α-methyl-5-hydroxytryptophan into brain was also rapid; the concentration then declined, with a $t_{1/2}$ of 4 hr. The concentration of α-methyl-5-HT formed in the brain reached a maximum level (1.3 μg/g) at 6 hr and then declined with a $t_{1/2}$ greater than 24 hr. The long half-life of α-methyl-5-HT probably reflects storage in neurons, its difficulty in passing the blood-brain barrier, and its metabolic stability.

The concentration of 5-HT in the brain decreased to about 35% of control levels 6 hr after injection of α-methyl-5-hydroxytryptophan but returned to near normal levels at 24 hr, when the concentration of α-methyl-5-HT was still high. The mechanism of 5-HT depletion by α-methyl-5-HT is probably one of displacement from binding sites, but the fact that high α-methyl-5-HT levels in the brain endure longer than does depletion of 5-HT indicates that α-methyl-5-HT is probably not retained in 5-HT neurons. Instead, α-methyl-5-HT probably redistributes to noradrenergic neurons, because depletion of norepinephrine is longer lasting.[13] The results with α-methyl-5-hydroxytryptophan do not en-

TABLE 1

EFFECT OF *p*-CPA ON 5-HT IN DISCRETE BRAIN NUCLEI *

	5-HT (ng/mg of protein)	
	Control	*p*-CPA
Dorsal raphe nucleus	25.5±1.1	18.1±1.6
Median raphe nucleus	20.4±1.4	8.9±1.9
Caudate nucleus	2.75±0.27	0.56±0.05
Hippocampus	2.43±0.07	0.43±0.06

* Rats were injected intraperitoneally with *p*-CPA methyl ester hydrochloride (400 mg/kg) 24 hr before sacrifice; $N=4$.

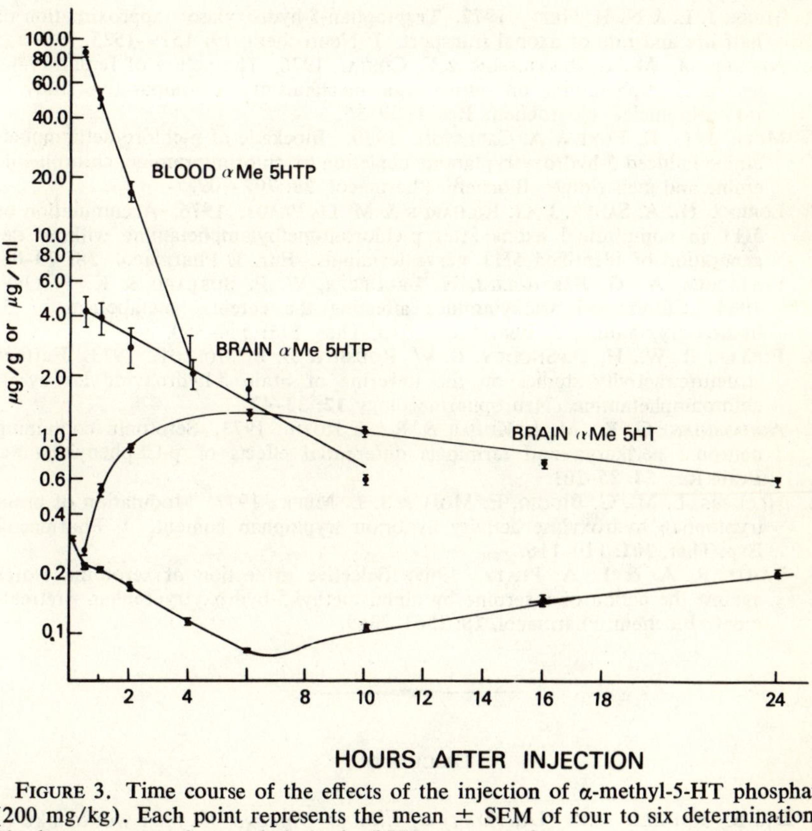

FIGURE 3. Time course of the effects of the injection of α-methyl-5-HT phosphate (200 mg/kg). Each point represents the mean ± SEM of four to six determinations. The bottom curve shows whole brain 5-HT concentrations.

courage its use for the selective depletion of serotonin, but they do suggest that examination of other compounds should reveal that some have greater potency in depleting 5-HT and that they might provide a selective functional blockade of serotonergic pathways.

References

1. MEEK, J. L. 1976. Applications of inexpensive equipment for high pressure liquid chromatography to assays for taurine, GABA and 5-hydroxytryptophan. Anal. Chem. **48:** 375–379.
2. NECKERS, L. M. & J. L. MEEK. 1976. Measurement of 5HT turnover rate in discrete nuclei of rat brain. Life Sci. **19:** 1579–1584.
3. NECKERS, L. M., L. BERTILSSON, S. KOSLOW & J. L. MEEK. 1976. Reduction of tryptophan hydroxylase activity and 5-hydroxytryptamine concentration in certain brain nuclei after p-chloroamphetamine. J. Pharmacol. Exp. Ther. **196:** 333–338.
4. SANDERS-BUSH, E., J. A. BUSHING & F. SULSER. 1972. Long term effects of p-Cl-amphetamine on the tryptophan hydroxylase activity and on the levels of 5-hydroxytryptamine and 5-hydroxyindoleacetic acid in brain. Eur. J. Pharmacol. **20:** 385–388.
5. MEEK, J. L. & N. H. NEFF. 1972. Tryptophan-5-hydroxylase: approximation of half life and rate of axonal transport. J. Neurochem. **19:** 1519–1525.
6. NECKERS, L. M., L. BERTILSSON & E. COSTA. 1976. The action of fenfluramine and p-Cl-amphetamine on serotonergic mechanisms: a comparative study in rat brain nuclei. Neurochem. Res. **1:** 29–35.
7. MEEK, J. L., K. FUXE & A. CARLSSON. 1970. Blockade of p-chloromethamphetamine induced 5-hydroxytryptamine depletion by chlorimipramine, chlorpheniramine and meperidine. Biochem. Pharmacol. **20:** 707–709.
8. LORENZ, H., A. SANER, J. G. RICHARDS & M. DA PRADA. 1976. Accumulation of 5HT in nonterminal axons after p-chloro-n-methylamphetamine without degeneration of identified 5HT nerve terminals. Eur. J. Pharmacol. **38:** 79–88.
9. PLETSCHER, A., G. BARTHOLINI, H. BRUDERER, W. P. BURKARD & K. F. GEY. 1964. Chlorinated aralkylamines affecting the cerebral metabolism of 5-hydroxytryptamine. J. Pharmacol. Exp. Ther. **145:** 344–350.
10. FULLER, R. W., H. D. SNODDY, B. W. ROUSH & B. B. MOLLOY. 1973. Further structure activity studies on the lowering of brain 5-hydroxyindoles by 4-chloroamphetamine. Neuropharmacology **12:** 33–42.
11. AGHAJANIAN, G. K., M. J. KUHAR & R. H. ROTH. 1973. Serotonin containing neuronal perikarya and terminals differential effects of p-Cl-phenylalanine. Brain Res. **54:** 85–101.
12. NECKERS, L. M., G. BIGGIO, E. MOJA & J. L. MEEK. 1977. Modulation of brain tryptophan hydroxylase activity by brain tryptophan content. J. Pharmacol. Exp. Ther. **201:** 110–116.
13. LAHTI, R. A. & P. A. PLATZ. 1969. Selective protection of serotonin stores against the action of reserpine by alpha methyl-5-hydroxytryptophan pretreatment. Biochem. Pharmacol. **18:** 2363–2369.

Discussion

DR. M. RADULOVACKI (*University of Illinois, Chicago, Ill.*): I would like to comment on the pronounced depletion of serotonin by *p*-chloroamphetamine in the hippocampus. The hippocampus is an interesting structure with respect to serotonin neurons. For example, loading with 5-hydroxytryptophan in cats results in the highest accumulation of serotonin in the hippocampus. Also, the highest rate of synthesis of serotonin in rats treated with 5-hydroxytryptophan is in the hippocampus. During hibernation, the highest accumulation of serotonin is in the hippocampus. Also, our studies show that during slow-wave sleep, serotonin and 5-hydroxyindoleacetic acid concentrations are highest in the hippocampus.

BIOCHEMICAL, BEHAVIORAL, AND PHARMACOLOGIC STUDIES OF THE EFFECTS OF DIHYDROXYTRYPTAMINES IN THE RODENT BRAIN *

R. Malcolm Stewart,† Sylvia C. Gerson, Günther Sperk, Alexander Campbell, and Ross J. Baldessarini

Departments of Psychiatry and † Neurology
Harvard Medical School
and Mailman Laboratories for Psychiatric Research
McLean Division of Massachusetts General Hospital
Belmont, Massachusetts 02178

A recent advance in the study of the physiology of the mammalian central nervous system (CNS) has been the ability to lesion specific classes of neurons by the use of selective neurotoxins. The dihydroxylated tryptamines, including 5,6-dihydroxytryptamine (5,6-DHT) and 5,7-dihydroxytryptamine (5,7-DHT), structural analogs of 5-hydroxytryptamine (5-HT, serotonin), are effective in lesioning serotonin neurons.[1] Lesioning by 5,6-DHT is largely restricted to the spinal cord and lower brain stem, and, due to its toxicity, 5,6-DHT can be administered only in a limited range of doses.[2, 3] In contrast, 5,7-DHT induces more extensive destruction of serotonin-containing neurons in both forebrain and brain stem-spinal cord but is evidently less selective, causing toxic damage to norepinephrine neurons as well.[4] Pretreatment with desmethylimipramine (DMI) prior to lesioning with 5,7-DHT protects norepinephrine neurons and allows selective destruction of indoleamine neurons in a more extensive distribution throughout the neuroaxis.[5] The systemic administration of the serotonin precursor 5-hydroxytryptophan (5-HTP) in rodents so lesioned with DMI plus 5,7-DHT produces a behavioral syndrome of tremor, jerking movements, and seizures. This response may be assessed by pharmacologic and behavioral techniques and permits an evaluation of the function of presumptive central serotonergic synapses *in vivo* after selective lesioning with serotonin neurotoxins. This myoclonic syndrome that occurs after administration of a serotonin precursor to rodents lesioned with DHT may reflect the development of supersensitivity related to central serotonergic synapses.[6, 7]

* Supported in part by National Institute of Mental Health Research Grants MH-16674 and MH-25515, Research Career Investigator Award MH-74370 (R. J. B.), an award from the Scottish Rite Research Fellowship Program, Supreme Council 33° A.A., Scottish Rite, Northern Masonic Jurisdiction of the United States (S. C. G.), and by a research grant from the Scottish Rite Benevolent Foundation. Dr. Stewart was supported by National Institute of Mental Health Postdoctoral Research Fellowship MH-01215, and by BRSG Research Grant B 75–39 to the Massachusetts General Hospital. Dr. Sperk was supported by the Max Kade Foundation for Austrian Scholars and by the Sandoz Foundation.

0077–8923/78/0305–0198 $01.75/0

Biochemical Studies

In our initial studies of the biochemical effects of dihydroxylated tryptamines on central indoleamine neurons, we found that intracisternally injected 5,6-DHT was more selective than was 5,7-DHT or 6,7-DHT in damaging indoleamine-containing neurons in the lower brain stem and spinal cord of the rat.[2-4] It reduced 5-HT levels by more than 80% in spinal cord but less in other regions; endogenous 5-hydroxyindoleacetic acid (5-HIAA) was also decreased, but tryptophan was unchanged. Unlike the other analogs, 5,6-DHT (up to 75 μg) did not affect the levels, uptake, or synthesis of catecholamines *in vitro*. Other indices of the impaired integrity of indoleamine-containing neurons after administration of 5,6-DHT included decreases by up to 75% of the high-affinity uptake of [^{3}H]-5-HT (0.01 μM) into crude synaptosomal (P_2) fractions at 1–4 weeks later. Uptake of 0.01 μM [^{3}H]norepinephrine or 10 μM 5-HT was not affected, and uptake of 5-HT was not altered by depletion of 5-HT with *p*-chlorophenylalanine. Release by 50 mM potassium of previously accumulated [^{3}H]-5-HT was also decreased after injection of 5,6-DHT but not after 6-hydroxydopamine administration, which, in turn, did diminish the uptake and release of [^{3}H]norepinephrine. *In vivo* accumulation of labeled 5-HT and 5-HIAA synthesized within 15 min after intracisternal injection of [^{3}H]tryptophan was markedly reduced in medulla-spinal cord; synthesis of 5-HT from [^{14}C]tryptophan in synaptosomal fractions *in vitro* was also decreased to an extent dependent on the dose of 5,6-DHT given a week earlier, and this effect persisted for at least 2 months. High-affinity uptake of tryptophan (0.1–1.0 μM but not at 1 mM) into spinal cord synaptosomes was also partially (30%) decreased after 5,6-DHT treatment. In general, these biochemical effects of 5,6-DHT appeared to be partially dose dependent and persisted for more than 1 month. These observations supported the conclusion that 5,6-DHT causes an apparently relatively selective degeneration of central indoleamine-containing axons, particularly those that descend from midbrain to spinal cord.

Although it seemed that the dihydroxylated tryptamines (DHTs) might be utilized as pharmacologic tools to produce selective degeneration of 5-HT-containing cells in the CNS, at least in the spinal cord, there were several factors that initially limited their use in functional studies of indoleamine-containing neurons.[2-4] Thus, due to its toxicity, 5,6-DHT can be administered only within a restricted range of doses. While a dose of 75 μg injected into the cerebrospinal fluid (CSF) induced a profound, prolonged, and selective depletion of 5-HT in the rat spinal cord ($>$80%), only moderate decreases were noted in the rest of the brain (about 30–50%).[4] Higher doses did not enhance the depletion of 5-HT and were poorly tolerated by the animals, because convulsions and deaths occurred frequently. Toxicity of the DHT compounds was decreased by DMI,[4] which selectively blocks uptake of amines into catecholamine-containing neurons, which therefore might contribute to the toxicity of high doses of DHTs. We attempted, without success, to increase the effectiveness of 5,6-DHT by pretreating with an inhibitor of monoamine oxidase (MAO), pargyline, or of catechol *O*-methyltransferase (COMT), pyrogallol, prior to injecting 5,6-DHT or by giving multiple doses of 5,6-DHT. The analog 6,7-DHT was found to be very toxic and to damage catecholamine-containing neurons in the rat brain.[5]

5,7-Dihydroxytryptamine was found to be a much less toxic substance. Intracranial doses of up to 200 μg could be administered, resulting in a more extensive destruction of indoleamine-containing neurons, particularly in the forebrain, where 5-HT was depleted by as much as 90%. However, 5,7-DHT was also found to have pronounced effects on central catecholamine-containing neurons; for example, concentrations of norepinephrine could be reduced by 40–50% in the brain and spinal cord by intracranial doses as low as 75 μg.[4, 5] Nevertheless, we found that the decreased levels of norepinephrine in the lower brain stem and spinal cord induced by even 100 μg of 5,7-DHT could be prevented by pretreatment of rats with DMI, a powerful and selective blocker of high-affinity uptake of norepinephrine that can also prevent the depletion of norepinephrine by 6-hydroxydopamine.[3, 4] Later, we found that pretreatment with DMI, followed by a high dose (200 μg) of 5,7-DHT, could be used to achieve a profound and selective destruction of 5-HT terminals in the forebrain (>75%), with an apparently high degree of selectivity, because both the levels of endogenous norepinephrine and the uptake of low concentrations of catecholamines (TABLE 1) were profoundly decreased, whereas high-affinity uptake of choline and of several amino acids proposed to act as neurotransmitters (TABLE 2) was unaffected; dopamine has also been reported to not be decreased by this dose of 5,7-DHT.[5] Moreover, the activity of tryptophan hydroxylase, but not of tyrosine hydroxylase, in isolated nerve endings of rat forebrain was markedly reduced by this treatment.[6] The efficacy of intracisternally administered 5,7-DHT after systemic DMI on the depletion of hypothalamic 5-HT (63%) was not quite as great as in cerebral cortex (79%), although administration of 5,7-DHT into the lateral cerebral ventricle was about as effective in depleting hypothalamic (78%) as cortical (79%) 5-HT.[9]

BEHAVIORAL STUDIES

The systemic administration of the serotonin precursor 5-hydroxytryptophan (5-HTP) produced a characteristic and reproducible neurologic syndrome,

TABLE 1

EFFECT OF INTRACISTERNAL INJECTION OF 5,7-DHT WITH OR WITHOUT PRETREATMENT WITH DMI ON LEVELS OF ENDOGENOUS AMINES AND UPTAKE OF LABELED AMINES IN RAT CEREBRAL CORTEX *

	Serotonin		Norepinephrine	
	Level	Uptake	Level	Uptake
Control	100±6	100±4	100±4	100±2
5,7-DHT	28±3 †	21±1 †	48±3 †	58±2 †
DMI + 5,7-DHT	21±5 †	25±1 †	101±3	99±3

* Male rats (200 g) were given DMI (25 mg/kg, ip) or saline 4 hr and again 1 hr prior to intracisternal injection of 5,7-DHT (200 μg of free base, in 25 μl) or its vehicle. Amine levels were determined 10 days later by fluorometric assay.[3] Uptake (5 min) was evaluated[5] with 50 nM [^{3}H]-5-HT or 30 nM [^{3}H]norepinephrine in a crude synaptosomal fraction; DMI pretreatment alone 10 days earlier had no effect. Values are expressed as mean percentage of control ± SEM; data were evaluated as nanograms of amine per gram of wet cortex, and as disintegrations per minute per milligram of synaptosomal protein, respectively, for at least 10 rats.

† $p < 0.001$ when compared to controls by Student's *t* test.

TABLE 2

EFFECT OF DMI AND 5,7-DHT ON *in Vitro* UPTAKE OF LOW CONCENTRATIONS OF VARIOUS SUBSTANCES IN A CRUDE SYNAPTOSOMAL FRACTION OF RAT BRAIN TISSUE *

	Control	DMI±5,7-DHT	Percentage of Control
Choline (70 nM)	2.0±0.1	1.9±0.1	95
Dopamine (20 nM) (striatum)	246.9±7.0	249.7±10.0	101
γ-Aminobutyric acid (10 nM)	32.7±1.5	31.4±0.9	96
Glutamic acid (10 nM)	107.3±3.1	102.9±4.5	96
Glycine (50 nM) (spinal cord)	20.8±1.0	22.2±0.5	107

* Results represent means for 10–20 determinations of dpm $\times$ 10^{-3}/mg of protein ± SEM taken up in 45 sec by cortical P_2 fractions (except glycine, for which spinal cord was used, and dopamine, for which striatal tissue and a 2.5-min incubation period were used). Rats received DMI (25 mg/kg, ip) 4 hr and again 1 hr prior to intracisternal injection of 5,7-DHT (200 μg of free base) or its vehicle and were sacrificed for uptake studies 10 days later.[5] None of these differences is statistically significant ($p > 0.1$ by Student's t test).

dependent on the dose of 5-HTP and of DHT, in rats with selective lesions of central serotonin neurons induced by previous treatment with systemic DMI and intracisternal 5,7-DHT.[7, 8] Approximately 5 min after intraperitoneal injection of 5-HTP (65 mg/kg), the rats became immobile and crouched. Soon thereafter, muscle twitches, at first intermittent and then becoming more continuous, were evident in the forelimbs and sometimes involved the whole body. The activity was typically most intense at 15 min, gradually lessened after 30 min, and usually ceased by 90 min. This motor activity was accompanied by flushing of the paws and by autonomic cholinergic responses, including salivation, piloerection, diarrhea, and ejaculation, that could be inhibited by pretreatment with atropine.[8]

Initially, this motor behavior was evaluated according to the following rating scale: 0, normal or quiet rat; 1, unsteady gait or head tremor; 2, occasional and intermittent myoclonus (jerking); 3, nearly continuous myoclonus; 4, generalized seizures or death.

The myoclonus in the rat could also be detected and recorded by the use of intramuscular electrodes.[8] In more recent studies, behavior has also been evaluated with an electronic activity monitor (Stoelting Company) in addition to this behavioral rating scale. Generally, the electronic activity monitor seems to be quite useful in quantifying the behavior, and the results obtained showed excellent agreement of activity monitor scores and behavior ratings (TABLES 3 & 4), although the activity monitor is not sensitive in detecting tremor.

Myoclonus was noted in rats lesioned with 5,6-DHT and challenged with 5-HTP, but this response was most striking in the hind limbs, possibly as a reflection of greatest lesioning effects in the lower spinal cord.[3] Tremor and autonomic responses were not as prominent in 5,6-DHT-treated rats as in 5,7-DHT-treated animals. We have also made lesions of serotonin neurons with DMI and 5,7-DHT in mice. After a challenge with 5-HTP, the mice also developed jerking movements but had appreciably less tremor than was noted in rats. In addition, mice usually did not develop obvious autonomic effects.

TABLE 3

MYOCLONIC RESPONSE INDUCED BY 5-HYDROXYTRYPTOPHAN 3 WEEKS AFTER INTRACISTERNAL INJECTION OF 5,7-DHT *

Dose of 5-Hydroxytryptophan (mg/kg, ip)	Total Behavioral Score	Total Counts per Session $\times 10^{-3}$
0	0	1.5±0.4
20	1.5±0.3	2.5±0.4
45	5.8±0.2	6.0±0.9
65	9.9±0.3	10.4±1.1
80	14.7±0.6	15.0±0.9
100	lethal †	lethal †

* Rats (N=4–6) were pretreated with DMI (25 μg/kg, ip), followed by 5,7-DHT (200 μg) and challenged 3 weeks later with 5-hydroxytryptophan (ip) at the doses stated.

† Doses above 80 mg/kg were lethal, so that a total behavioral score [8] cannot be stated. Correlation between behavioral rating and electronic monitoring scores was high (r >0.90).

The response to 5-HTP in rats was evident within 24 hr after intracranial injection of 5,7-DHT, became maximal by 10–14 days, and declined slightly thereafter to remain fairly stable after 3 weeks, persisting for several months. This complex time course for the development of myoclonus in response to an approximately half-maximally effective dose of 5-HTP (65 mg/kg) was first

TABLE 4

MYOCLONIC SYNDROME INDUCED IN RATS BY 5-HTP 10 DAYS AFTER INTRACISTERNAL INJECTION OF 5,7-DHT *

Dose of 5,7-DHT (μg)	Behavioral Score	Monitor Score (cpm)	Number of Rats	Number with Tremor	Number with Myoclonus	Number with Seizures
0	0.0	3.2	6	0	0	0
50	1.8	20.2	4	1	1	1
100	13.3	65.0	4	4	4	1
150	14.8	88.3	4	4	4	2
200	16.7	84.3	4	4	4	2
250	11.0	57.5	4	4	4	4

* Rats were pretreated with DMI (25 mg/kg, ip) 1 hr before receiving 5,7-DHT (intracisternally) at the doses stated and were challenged with 5-hydroxytryptophan (65 mg/kg, ip). The behavioral response was rated at 5 and 15 min and at 15-min intervals thereafter for a total of 90 min; reported behavioral scores are the mean total scores (sum of individual scores at each of the seven points of rating in the 90-min session).[8] Doses above 200 μg of DHT were sometimes lethal, and a total behavioral score cannot be accurately stated. Total activity counts recorded constantly for 90 min were converted to counts per minute, unless a rat died early; in such cases, the total counts were divided by the total minutes that the rat was alive. Correlation for behavioral ratings and electronic monitoring was high (r >0.90).

noted with the behavioral rating scale [7,8] and has since been confirmed with the activity monitor. A stable portion of the response curve at 3–4 weeks after 5,7-DHT administration was chosen for further pharmacologic studies.

Pharmacologic Studies

The behavioral response of the 5,7-DHT-lesioned rat to 5-HTP was dose dependent in a range of 10–100 mg of 5-HTP per kilogram of body weight, with the half-maximally effective dose being about 65 mg/kg (Table 3). Adult or newborn control (intracisternal placebo only) rats did not develop the behavior, even at doses of 5-HTP of up to 600 mg/kg. Doses above 80 mg/kg in DHT-pretreated rats were usually lethal in about 30 min. The myoclonic behavioral response to 5-HTP did not occur after systemic (intraperitoneal) administration of 5-HT.[7,8] Intracranial infusion of 5-HT in rats, however, also produced the myoclonic syndrome in 5,7-DHT-pretreated animals while depressing motor activity in placebo-pretreated controls.[10] The development of myoclonus in rodents appeared to be dependent on the intracranial dose of 5,7-DHT used for the initial lesioning, as is clearly demonstrated in the rat (Table 4), and was found in the mouse.

The development of the 5-HTP-induced myoclonus was diminished by several antiserotonin agents, including the proposed receptor blocking ergot alkaloid derivatives *d*-lysergic acid diethylamide (LSD), 2-bromo-LSD, and methysergide and by high, centrally active doses of the decarboxylase inhibitor RO4–4602 (Table 5).[7,8] The behavioral response to 5-HTP of DHT-pretreated rats was potentiated by the prior administration of chloroimipramine, a selective blocker of 5-HT uptake, and by pargyline, an MAO inhibitor. Pargyline also led to an excitatory and myoclonic response to tryptophan in intact rats, although tryptophan (up to 300 mg/kg, ip) had been ineffective by itself in DHT-pretreated rats (Table 5). Moreover, a powerful blocker of 5-HT uptake (Lilly 110,140) with 5-HTP provoked myoclonus in intact rats but only when high doses of both were used. The myoclonus syndrome could not be reproduced by L-dopa or apomorphine; nor could it be blocked by α-methyl-*p*-tyrosine, haloperidol, low doses of chlorpromazine (high doses of chlorpromazine produced sedation), phenoxybenzamine, phentolamine, low doses of propranolol (although high doses produced sedation and reduced the tremor), bicuculline, or atropine.

When rats were given a challenge dose of 5-HTP (65 mg/kg, ip) every 48 hr beginning on the second day of lesioning, the strength of the myoclonic response was considerably less than that in rats lesioned on the same day but subsequently challenged with only a single dose of 5-HTP alone (Table 6) or after a placebo injection every 48 hr. This finding suggests that increased agonism of 5-HT receptors may diminish their evolving supersensitivity. Moreover, in other experiments, the development of the myoclonic response after a low dose of 5,7-DHT in mice was also impeded with up to 70% inhibition of cerebral protein synthesis by cycloheximide for up to 10 days.[11] While this result suggests that protein synthesis may be required for the development of increased sensitivity to 5-HTP, it is not possible to draw a conclusion, because a small but important pool of new protein synthesis may have escaped the action of cycloheximide.

TABLE 5

EFFECTS OF DRUG TREATMENT ON 5-HYDROXYTRYPTOPHAN-INDUCED MYOCLONUS IN 5,7-DHT PLUS DMI-LESIONED RATS *

	Dose (mg/kg)	Score	Percentage of Control
Serotonin Precursors			
L-Tryptophan	10–300	0	0
L-5-Hydroxytryptophan	65	9.9±0.3	100
5-Hydroxytryptamine	65	0	0
Serotonin "Receptor Blockers"			
d-Lysergic acid diethylamide (LSD)	0.2	4.7±0.3	47
2-Bromo-LSD	0.5	5.4±1.8	54
Methysergide	1.0	6.1±0.6	61
Enzyme Inhibitors			
Pargyline (monoamine oxidase)	100	>30.0±9.0 †	>300 †
RO4–4602 (aromatic amino acid decarboxylase)	50	4.10±1.8	41
Serotonin Reuptake Blockers			
Chloroimipramine	15	11.7±2.5	119

* Rats were pretreated 1 month previously with 5,7-DHT plus DMI and challenged with 5-hydroxytryptophan (65 mg/kg, ip). The data are mean total behavioral scores [8] ± SEM (N=3–4 animals). All test drugs were given intraperitoneally prior to 5-hydroxytryptophan.

† Total score cannot be determined accurately due to intervention of generalized seizures and death.

DISCUSSION

The present work represents a considerable extension of the development of intracerebrally injected dihydroxylated indoleamines as selective lesioning agents of serotonin nerve terminals in the central nervous system of rat and mouse. We found that these agents produced striking losses of several biochemical characteristics associated with serotonin cells, including tryptophan hydroxylase activity, high-affinity uptake (and release) of serotonin, and a partial loss of high-affinity uptake of tryptophan. The relatively limited effects of 5,6-DHT on the brain, compared with its striking effects on the spinal cord, could be extended by the use of well-tolerated high doses of 5,7-DHT, provided that DMI pretreatment was given to avoid toxic effects on norepinephrine-containing neurons. This treatment produced strong depletions of serotonin, but not of catecholamines, in the diencephalon and forebrain and a selective loss of forebrain synaptosomal tryptophan hydroxylase activity. This lesioning method led to striking neurologic responses of rats or mice to systemic injections of the serotonin precursor 5-HTP, but not to serotonin itself, unless it was introduced directly into the lateral cerebral ventricles. The behavioral syndrome produced included myoclonus, which could be recorded by muscle electrodes or by an electronic activity monitoring technique. The response was greatly diminished by several serotonin antagonists but was very little affected by drugs that interact with other neurotransmitter systems.

This behavioral response may reflect changes in the sensitivity of central

TABLE 6

BEHAVIORAL RESPONSE TO 5-HYDROXYTRYPTOPHAN (65 mg/kg, ip) CHALLENGE AFTER ADMINISTRATION OF DMI PLUS 5,7-DHT *

Day After 5,7-DHT	Treatment: Controls	Repeated 5-Hydroxytryptophan	Once 5-Hydroxytryptophan
21	0.48±0.1 (6)	1.50±0.5 (6)	6.40±0.5 (8) †
41	0.70±0.1 (6)	2.75±0.2 (6)	7.50±1.0 (8) †
61	0.75±0.2 (6)	4.20±0.5 (6)	7.75±0.9 (6) †

* Rats pretreated with DMI (25 mg/kg) were lesioned with 5,7-DHT (200 μg). Behavior was evaluated in a Stoelting electronic activity monitor for 90 min. Data are mean monitor counts per 90-min session $\times 10^{-3} \pm$ SEM (*N*). Controls were prepared by placebo injection of CSF vehicle and given 5-hydroxytryptophan also every 2 days. Rats given repeated injections of 5-hydroxytryptophan received the amino acid every 2 days prior to the final challenge for scoring on the day stated. All rats were given 5-hydroxytryptophan and monitored on the day of challenge.

† Different from both comparison groups ($p < 0.01$) by Student's *t* test.

synapses to serotonin as a neurotransmitter. Several experiments suggested that presynaptic mechanisms, such as loss of reuptake of 5-HT, may partially explain the apparent supersensitivity to serotonin or its precursors after destruction of 5-HT terminals with DHT compounds. Thus, tryptophan or 5-HTP produced myoclonus in intact rats given an MAO inhibitor, suggesting that increased availability of 5-HT may be a sufficient condition. Moreover, at least a short-lasting myoclonic response could also be produced in normal rats by giving large doses of 5-HTP and of a specific inhibitor of presynaptic 5-HT uptake (Lilly 110,140). The myoclonic response to 5-HTP in the 5,7-DHT-treated mouse was not prevented by treatment up to a week after DHT treatment with doses of cycloheximide that effectively inhibited brain protein synthesis. Nevertheless, denervation supersensitivity, which involves increased effectiveness of postsynaptic serotonin receptor mechanisms, may contribute to the behavioral response to serotonin agonists observed after destruction of serotonin terminals with 5,7-DHT, although the precise mechanisms involved remain obscure and may be very complex. The appearance of myoclonus in response to 5-HTP after the restricted lesioning of serotonin neurons with 5,6-DHT and its presence with more extensive lesioning with 5,7-DHT suggest that the neural substrate for this response may be located in brain stem-spinal cord structures rather than in the cerebral cortex.

We conclude that the use of the dihydroxylated tryptamines represents an important and powerful strategy for the study of the functional importance of central indoleamine neuronal systems and for the investigation of still poorly understood pre- and postsynaptic regulatory mechanisms in these widespread and important central synaptic systems.

ACKNOWLEDGMENTS

Drug substances for these studies were kindly donated by Dr. Albert Manian and by Abbott Laboratories, Ayerst Laboratories, CIBA-Geigy Corporation,

Hoffmann-La Roche, Lakeside Laboratories, Eli Lilly and Company, McNeil Laboratories, and Smith Kline and French.

References

1. Baumgarten, H. G., A. Björklund, L. Lachenmayer, A. Nobin & U. Stenevi. 1971. Long-lasting selective depletion of brain serotonin by 5,6-dihydroxytryptamine. Acta Physiol. Scand. Suppl. **373:** 1–16.
2. Baldessarini, R. J. & S. Gerson. 1973. Effects of 5,6-dihydroxytryptamine on the metabolism of 5-hydroxytryptamine in the central nervous system of the rat. J. Pharm. Pharmacol. **25:** 647–648.
3. Gerson, S., R. J. Baldessarini & S. C. Wheeler. 1974. Biochemical effects of dihydroxylated tryptamines on central indoleamine neurons. Neuropharmacology **13:** 987–1004.
4. Gerson, S. & R. J. Baldessarini. 1974. Non-specific actions of dihydroxylated tryptamines in the central nervous system of the rat. J. Pharm. Pharmacol. **26:** 71–73.
5. Gerson, S. & R. J. Baldessarini. 1975. Selective destruction of serotonin terminals in rat forebrain by high doses of 5,7-dihydroxytryptamine. Brain Res. **85:** 140–145.
6. Harris, J. E., M. Karobath & R. J. Baldessarini. 1976. Characteristics of tyrosine hydroxylation in isolated nerve endings. Biochem. Pharmacol. **25:** 91–93.
7. Stewart, R. M., J. H. Growdon, D. Cancian & R. J. Baldessarini. 1976. Myoclonus after 5-hydroxytryptophan in rats with lesions of indoleamine neurons in the central nervous system. Neurology **26:** 690–692.
8. Stewart, R. M., J. H. Growdon, D. Cancian & R. J. Baldessarini. 1976. 5-Hydroxytryptophan-induced myoclonus: increased sensitivity to serotonin after intracranial 5,7-dihydroxytryptamine in the adult rat. Neuropharmacology **15:** 449–455.
9. Boyd, A. E., III, R. M. Stewart, S. C. Gerson, R. J. Baldessarini & S. Reichlin. 1977. In preparation.
10. Warbritton, J. D., III, R. M. Stewart & R. J. Baldessarini. 1977. Decreased locomotor activity with intraventricular infusion of serotonin in the rat. Brain Res. In press.
11. Sperk, G., R. M. Stewart, A. Campbell & R. J. Baldessarini. 1978. Inhibition of 5,7-dihydroxytryptamine-induced supersensitivity to 5-hydroxytryptophan in mice by repeated treatment with cycloheximide. Brain Res. In press.

Discussion

Dr. H. G. Baumgarten: Do you think there is any clinical pathologic correlate of such a syndrome in humans?

Dr. Stewart: Since I am a neurologist, I have thought about that question myself. Indeed, Drs. Adams and Lance have described myoclonus in people with postanoxic syndromes. In these conditions, administration of 5-hydroxytryptophan leads to improvement. On the other hand, there are syndromes in which giving 5-hydroxytryptophan will make the myoclonus worse.

Dr. D. V. Coscina: Can you produce this type of myoclonus with tryptophan instead of 5-hydroxytryptophan, either alone or with a monoamine oxidase inhibitor?

DR. STEWART: Yes. This response was not seen after systemic injections of serotonin, L-Dopa, apomorphine, or tryptophan alone but was produced by tryptophan plus an inhibitor of monoamine oxidase. 5-Hydroxytryptophan given to rats pretreated with 6-hydroxydopamine did not produce this myoclonus. We and others have found that in normal rats, either 5-hydroxytryptophan or tryptophan given with a monoamine oxidase inhibitor can produce myoclonus. Recently, we have given high doses of fluoxetine (Lilly 110,140) with 5-hydroxytryptophan and obtained a similar effect.

DR. B. L. JACOBS: Your LSD data are interesting because we saw supersensitivity to the LSD effect, whereas you seem to get a blocking effect.

DR. STEWART: Because agents like methysergide and LSD are mixed agonists-antagonists, different behavioral effects can be obtained, depending on the dose given.

DR. JACOBS: I am interested in the question of pre- versus postsynaptic actions. Our conclusion, as I will discuss tomorrow, is that the effect is probably exclusively presynaptic and not postsynaptic. Could your data possibly be attributed to a tolerance effect of continued dosing with 5-hydroxytryptophan rather than to a change in receptor sensitivity?

DR. STEWART: From the data available, assigning a specific mechanism is difficult. We have some recent evidence that cycloheximide, an inhibitor of protein synthesis, can prevent the effect. We are also now looking at binding to postsynaptic receptors. We are concerned about the same kinds of questions you are.

DR. J. H. GROWDON: In the 5,7-dihydroxytryptamine-pretreated rat, we have given the putative serotonin agonist quipazine and have reproduced the syndrome. That result would indicate that the syndrome is due to increased postsynaptic sensitivity.

IMMEDIATE AND LONG-TERM EFFECTS OF *p*-CHLOROAMPHETAMINE ON BRAIN AMINES *

Elaine Sanders-Bush and Larry R. Steranka

Department of Pharmacology
Vanderbilt University School of Medicine
and Tennessee Neuropsychiatric Institute
Nashville, Tennessee 37217

The administration of *p*-chloroamphetamine (*p*-CA) to rats induces a complex pattern of changes in brain amines and behavior. The immediate effects are relatively nonspecific, involving changes in serotonergic, noradrenergic, and dopaminergic neuronal systems that mediate a variety of behavioral effects. However, the persistent, long-term effects are specific to serotonin [5-hydroxytryptamine (5-HT)] neurons. Moreover, the mechanism that mediates the long-term reduction of brain 5-HT is different from that responsible for the immediate decrease in this amine. Therefore, these two phases of *p*-CA's action will be considered separately.

Immediate Biochemical Effects

The original studies of *p*-CA's action on brain 5-HT were entirely related to the mechanism of the early decrease, because the extremely long duration of this effect was not noted until 1972;[11] a significant decrease in brain 5-HT lasts for more than 4 months after a single dose. The initial decrease in the levels of 5-HT and its principal metabolite 5-hydroxyindoleacetic acid (5-HIAA) is due to many mechanisms, including release of 5-HT,[12–15] inhibition of its reuptake,[14, 16–18] and inhibition of its synthesis.[19, 20] Although *p*-CA administration reduces tryptophan hydroxylase activity at early times after injection,[20] the mechanism of this effect is not clear. The *in vitro* addition of *p*-CA, at concentrations below 1 mM, does not inhibit tryptophan hydroxylase,[20] although at higher concentrations (5–20 mM), a dose-related competitive inhibition has been reported.[21] Such a mechanism of competitive inhibition apparently is not responsible for the *in vivo* reduction in tryptophan hydroxylase activity after *p*-CA administration, because this effect is not reversed by dialysis.[20, 21] Moreover, the occurrence of such high concentrations *in vivo* seems unlikely, although the proposed active uptake of *p*-CA[22] may lead to locally high concentrations within 5-HT neurons. Another possible explanation for the discrepancy between the *in vivo* and *in vitro* data is the involvement of an active metabolite in the *in vivo* effects. However, pretreatment of rats with piperonyl butoxide, which inhibits the hepatic metabolism of many drugs and prolongs the half-life of *p*-CA, does not modify the onset or degree of reduction in enzymatic activity at early times after the injection of *p*-CA (Figure 1).

* Supported in part by United States Public Health Service Grant MH-26463, Research Scientist Development Award MH-70898 (E. S.-B.), and by the Tennessee Department of Mental Health and Mental Retardation.

0077–8923/78/0305–0208 $01.75/0

Thus, it would appear that a metabolite is not responsible for this effect of *p*-CA.

In addition to marked effects on 5-HT systems in the brain, administration of *p*-CA produces marked changes in the dynamics of catecholamine neurons. Although the levels of norepinephrine (NE) and dopamine (DA) are only slightly affected or not changed, every other biochemical parameter of catecholamine function that has been measured is substantially modified (TABLE 1). Moreover, in every case where temporal studies have been performed, the effects of *p*-CA on catecholamines are of relatively short duration and essentially disappear within 24 hr.

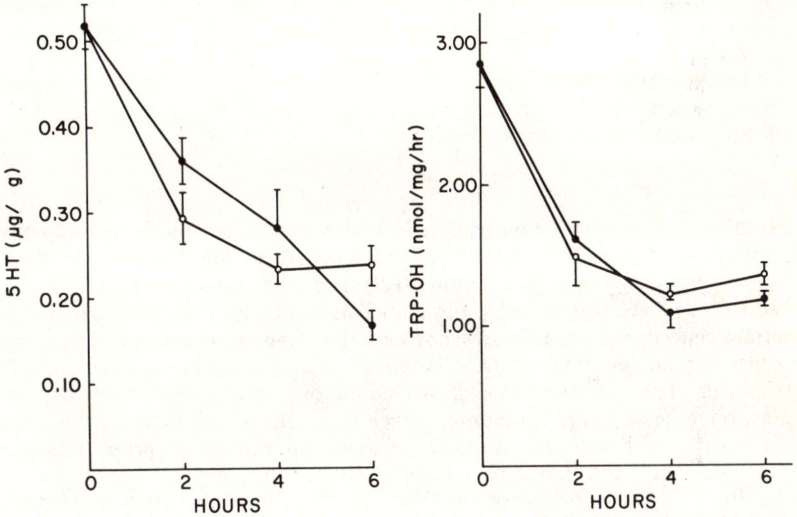

FIGURE 1. Effect of piperonyl butoxide (○) on *p*-chloroamphetamine's (*p*-CA) reduction of serotonin (5-HT) levels and tryptophan hydroxylase (TRP-OH) activity (●, saline). Rats were injected intraperitoneally with 800 mg/kg of piperonyl butoxide 1 hr prior to injection of 5 mg/kg of *p*-CA and killed at various times thereafter. The whole brain levels of 5-HT were assayed by the method of Bogdanski *et al.*,[62] and TRP-OH activity was assayed according to the method of Gál and Patterson.[63] The values presented are the means of five rats; vertical bars indicate the SEMs.

IMMEDIATE BEHAVIORAL EFFECTS

Perhaps because of the complex spectrum of biochemical effects that occurs immediately after *p*-CA administration, the initial behavioral effects are also quite complex. Thus, *p*-CA administration to rats induces stimulation of both ongoing behaviors (such as locomotion and learned responses) and stereotypy. These effects are similar to those caused by amphetamine and are presumably related to catecholaminergic mechanisms.[8–10, 23, 24] In addition, *p*-CA induces an abnormal motor response complex,[6, 9, 25] which is characteristic of increased 5-HT receptor stimulation[26] and is blocked by 5-HT synthesis inhibition.[25]

TABLE 1

EARLY CHANGES IN NOREPINEPHRINE (NE) AND DOPAMINE (DA) LEVELS AND DYNAMICS AFTER INJECTION OF *p*-CHLOROAMPHETAMINE

Biochemical Measure	Response *	Reference
NE levels	0, slight +	1, 2
DA levels	0, slight — or +	1, 3, 4
NE uptake	—	5, 6
DA uptake	—	5, 6
NE release	++	7
Deaminated metabolites	—	6, 8
Normetanephrine	++	6, 8
3-Methoxytyramine	+	6, 9
NE turnover	+	3, 4
DA turnover	+	3, 4, 10
In vivo tyrosine hydroxylase	++	4

* 0, No change; +, increase; —, decrease.

Results of drug interaction studies in which automated measures of motor activity are used to define operationally the concept of the "central stimulant" effect of *p*-CA suggest that both 5-HT and the catecholamines are involved.[7, 10, 23, 24] However, such automated measures may reflect partially the abnormal central 5-HT-mediated syndrome, thus leading to the conclusion that the central stimulant effect of *p*-CA is related to drug-induced changes in 5-HT metabolism. This problem was illustrated in our recent work in which we demonstrated that a rapid tolerance develops to the effect of *p*-CA on total motor activity with repeated doses.[6] As shown in FIGURE 2, prior treatment with one dose of *p*-CA markedly attenuates the increase in total activity counts caused by a subsequent dose of the drug. In these animals, the time between the two doses is only 24 hr, but equal reductions are found if the doses are given 2 weeks apart. It thus seems that at least a portion of this tolerance is mediated by the long-term reduction of 5-HT caused by the administration of *p*-CA. In agreement with this interpretation, prior depletion of 5-HT with *p*-chlorophenylalanine blocks the effects of *p*-CA on motor activity.[24] Many of the components of the 5-HT syndrome (i.e., head weaving, forepaw treading, tremor) are measured in activity cages, and, therefore, a reduction in these parameters may be partially responsible for the tolerance found when total activity counts are quantified. Indeed, by direct observation, we found that the induction of this syndrome is completely prevented by prior treatment with *p*-CA.[6]

Clearly, more well-defined measures of behavior are needed to determine the relative importance of 5-HT and the catecholamines in the behavioral effects of *p*-CA. Recently, we have successfully used a nondiscriminated, continuous bar-press (Sidman) avoidance paradigm for this purpose.[4] In this task, the administration of *p*-CA produces an amphetamine-like facilitation of avoidance responding that is apparently independent of its effects on 5-HT metabolism. Thus, this effect is not temporally related to the reduction of brain levels of 5-HT. Moreover, pretreatment with a dose of *p*-CA that maximally lowers the brain levels of 5-HT does not reduce the facilitatory effect of a subsequent dose of the drug, as occurs with a measure of gross motor activity.[6] In addition, 2.5 mg/kg of (S)-(+)-*p*-CA produces a marked facilitation of performance, whereas the same dose of (R)-(−)-*p*-CA has no effect on responding,[4] although the initial reductions in the brain levels of 5-HT and 5-HIAA, tryptophan hydroxylase activity, and the synaptosomal uptake of 5-HT are comparable after administration of these isomers.[27] Moreover, this dose of (R)-(−)-*p*-CA has no effect on the synthesis and turnover of catecholamines, whereas the (S)-(+)-isomer increases the conversion of [^{3}H]tyrosine to both [^{3}H]NE and [^{3}H]DA and increases *in vivo* tyrosine hydroxylase activity, which suggests that the facilitation of Sidman avoidance responding by *p*-CA is related to its effects

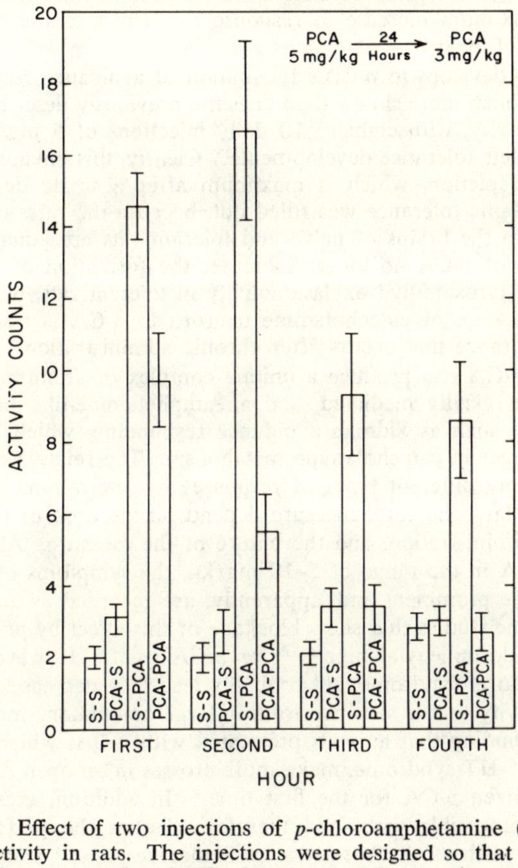

FIGURE 2. Effect of two injections of *p*-chloroamphetamine (*p*-CA) 24 hr apart on motor activity in rats. The injections were designed so that four different treatment schedules were employed. The first group (S-S) was given saline, followed 24 hr later by a second injection of saline; for the *p*-CA-S group, the first injection was 5 mg/kg of *p*-CA, and the second was saline; for S-*p*-CA, the first injection was saline, and the second was 3 mg/kg of *p*-CA; for *p*-CA-*p*-CA, the first injection was 5 mg/kg of *p*-CA, and the second was 3 mg/kg of *p*-CA.

Immediately after the second injection, the rats were placed in activity cages, and activity was monitored for 4 hr. The values presented are means for nine to 15 rats per group; Vertical bars indicate the SEMs. Based on data from El-Yousef *et al.*[6]

on catecholamines.[4] In contrast to the increase in total motor activity that is observed very soon after *p*-CA administration and that reaches a peak within the first hour,[6, 8] the facilitation of Sidman avoidance performance is not apparent in the first hour of testing and does not reach a maximum until 2 or 3 hr after *p*-CA administration.[4] In addition, this delay is associated with an initial disruptive effect, as evidenced by an increase in the number of shocks received during the first hour. The delays in the facilitatory effect and the immediate disruptive effects observed in Sidman avoidance are probably due to the presence of symptoms of the central serotonin syndrome, which are incompatible with the performance of the avoidance response. This conclusion is supported by the observation that the increase in the number of shocks that occurs after a single dose of *p*-CA is not observed after a second or third injection of the drug. Moreover, the maximum increase in response rate shifts to the first hour on repeated injection of the drug.

Tolerance also develops to *p*-CA's facilitation of avoidance responding, but this effect occurs much more slowly than does the previously described tolerance to total motor activity, with eight to 10 daily injections of 5 mg/kg of *p*-CA required for complete tolerance development.[28] Clearly, this phenomenon is not related to 5-HT depletion, which is maximum after a single dose of *p*-CA. Moreover, a metabolic tolerance was ruled out, because the rates of disappearance of *p*-CA from the brains of naive and tolerant rats are equal. However, the administration of *p*-CA no longer increases the formation of [^{3}H]DA and [^{3}H]NE or *in vivo* tyrosine hydroxylase activity in tolerant rats, which suggests that a reduced response of catecholamine neurons to *p*-CA is associated with the behavioral tolerance that occurs after chronic administration.

In summary, *p*-CA can produce a unique complex of abnormal behaviors, which are serotonergically mediated, and an amphetamine-like facilitation of ongoing behaviors, such as Sidman avoidance responding, which is related to drug-induced changes in catecholamine metabolism. The relative contributions of these qualitatively different types of responses in determining the effect of *p*-CA on a particular behavioral measure depend on the dose of the drug, the time after drug administration, and the nature of the measure. At early times after doses of *p*-CA in the range of 5–10 mg/kg, the symptoms of the central 5-HT syndrome are prominent and, apparently, are recorded as an increase in motor activity in the studies that show blockade of this effect by prior depletion of 5-HT with *p*-chlorophenylalanine[23, 24] or *p*-CA itself.[6] However, the presence of this behavioral syndrome will probably lead to a decrease in any measure of motor activity that is sensitive to only large ambulatory movements, as evidenced by the finding that animals pretreated with *p*-CA, which completely blocks the central 5-HT syndrome, make more crosses in an open field situation than do animals given *p*-CA for the first time.[6] In addition, these abnormal behaviors are incompatible with, and therefore disrupt the performance of, learned responses, such as the bar-press avoidance response[4] and intracranial self-stimulation.[45] However, the symptoms of the central 5-HT syndrome appear to be short-lived relative to the catecholamine-mediated, amphetamine-like effects of the drug, thus allowing the manifestation of amphetamine-like stereotyped behavior[9] and facilitation of the performance of learned behaviors[4, 64] at later times after drug administration. Moreover, development of tolerance to the central 5-HT syndrome and facilitation of avoidance performance have been demonstrated. However, tolerance develops at different rates, which presumably reflect adaptive changes in different neuronal systems to the continued presence of the drug.

Long-Term Biochemical Effects

In contrast to the complex neuronal effects of *p*-CA at early times after injection, the long-term biochemical effects are selective to serotonergic neurons and include a decrease in the levels of 5-HT and 5-HIAA, in the activity of the rate-limiting enzyme, tryptophan hydroxylase,[5, 11] in the high-affinity uptake of 5-HT (Table 2), and in the turnover of 5-HT.[29] All of these effects of *p*-CA can be explained by a single mechanism: cytotoxicity to 5-HT neurons. Indeed, direct morphologic evidence of degeneration of 5-HT cell bodies[30] and axon terminals[31] has recently appeared. In apparent conflict with the latter report, Lorez *et al.*[32] could not find electron microscopic evidence of degeneration of

Table 2

Synaptosomal Uptake of Serotonin at Various Times After a Single Dose of *p*-Chloroamphetamine (*p*-CA)*

Time After Injection (days)	Uptake (nmol/g/hr±SEM)	
	Control	*p*-CA
1	0.22±0.01	0.05±0.01 †
4	0.20±0.01	0.04±0.01 †
14	0.19±0.01	0.03±0.01 †
30	0.25±0.01	0.06±0.01 †
90	0.27±0.02	0.14±0.02 ‡

* Groups of four rats were injected intraperitoneal with a single dose of 10 mg/kg of *p-CA* or saline and killed at various times after injection. Data from Sanders-Bush *et al.*[5]

† $p < 0.001$

‡ $p < 0.01$

5-HT axon terminals after large doses of *p*-chloromethamphetamine, which exerts biochemical effects indistinguishable from those of *p*-CA.[1, 5] Instead, these authors demonstrated axonal swelling and accumulation of 5-HT fluorescence, an effect reminiscent of that produced by 5,6-dihydroxytryptamine. As a biochemical corrollary of this effect, we have recently measured the levels of 5-HT in axons and axon terminals after *p*-CA injection in a region of the central nervous system where these structures can be separated clearly, namely, the spinal cord.[33] These results showed that whereas *p*-CA causes a significant and long-lasting decrease in terminal 5-HT, 5-HT levels in axonal regions are simultaneously increased. Such differential effects may explain partially the regional differences in the effects of *p*-CA. Two regions that are particularly resistant to the depletion of 5-HT after administration of *p*-CA are the hypothalamus and the spinal cord. These regions contain prominent 5-HT axon bundles (e.g., the median forebrain bundle in the hypothalamus), and failure to find pronounced reductions in the overall levels of 5-HT may reflect the summation of changes in opposite directions in 5-HT axons and axon terminals. Another possible explanation for the regional differences observed may be related to the unusual selective effects of *p*-CA in 5-HT cell body areas, with long-term biochemical [46, 47] and histologic [30] changes occurring in neurons in the B_9 area but not in the main 5-HT cell body nuclei, B_7 and B_8. However, bilateral lesions of the B_9 cell group cause little or no changes in the levels of 5-HT, the high-affinity uptake of 5-HT,[34] or the activity of tryptophan hydroxylase [35] in several forebrain regions in which *p*-CA administration causes pronounced reductions. It thus appears that the regional differences in the prolonged biochemical effects of *p*-CA are not due to a selective cytotoxic action on cells in the B_9 area.

Both the immediate and the long-term effects of *p*-CA are prevented by the simultaneous injection of the 5-HT uptake inhibitor fluoxetine (Lilly 110,140, FIGURE 3). These results confirm those reported previously by Fuller *et al.*[22, 36]

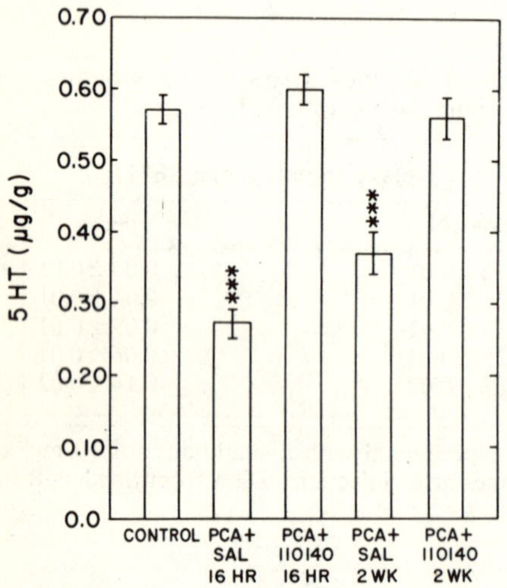

FIGURE 3. Effect of fluoxetine on immediate and long-term decreases in 5-HT after injection of *p*-chloroamphetamine (*p*-CA). Rats were injected simultaneously with 10 mg/kg of fluoxetine (Lilly 110,140) and 5 mg/kg of *p*-CA and either killed 16 hr later or injected again with 10 mg/kg of fluoxetine and killed 2 weeks later. Whole brain levels of 5-HT were assayed by the method of Bogdanski *et al.*[62] The values presented are the means of seven rats per group; vertical bars represent the SEMs.

and Ross.[37] Both of these effects are presumably due to a blockade of the active transport of *p*-CA into 5-HT nerve terminals. However, indirect data, such as those obtained in these drug interaction studies, are the only available evidence that *p*-CA, an extremely lipid-soluble drug, is actively accumulated in 5-HT neurons. In agreement with Ross,[37] we have not been able to demonstrate that [^{14}C]-*p*-CA is actively taken up into brain synaptosomes, under a variety of *in vitro* conditions. A significant amount of binding of [^{14}C]-*p*-CA does occur, but this binding is not energy or temperature dependent and apparently reflects the well-known affinity of apolar drugs for lipid membranes. Perhaps one reason for our inability to demonstrate an active uptake *in vitro* is that because *p*-CA is very lipid soluble, it passively diffuses back out of the tissue so rapidly that the active process is obscured. This hypothesis is analogous to Fuller's explanation for the reversal of *p*-CA's action by 5-HT uptake inhibitors; that is, *p*-CA continuously diffuses out of the nerve terminal, thus requiring a constantly active reuptake into the terminal.[22] However, attempts to demonstrate this phenomenon with rapid separation techniques, such as Millipore® filtration, have not been successful.

Interestingly, Ross[37] concluded that although the antagonism of the acute effects of *p*-CA by 5-HT uptake inhibitors is related to an inhibition of the 5-HT uptake system, the antagonism of the long-term irreversible effect of *p*-CA may also involve another mechanism. With regard to fluoxetine, at least, another striking effect is a profound inhibition of mixed-function oxidase activity in the liver.[38] We have also found a 70% reduction in the metabolism of aminopyrine and aniline by hepatic microsomes from rats pretreated with fluoxetine. Such an effect on *p*-CA metabolism could be important in fluoxetine's protection against the long-term irreversible depletion of 5-HT, particularly because earlier studies suggested that an active metabolite of *p*-CA may be responsible for this effect. Thus, the onset of the irreversible phase of *p*-CA's action is not immediate, as evidenced by the temporary recovery of tryptophan hydroxylase activity[5] and high-affinity 5-HT uptake capacity[27] and by the delay in the onset of the fluoxetine-resistent action.[36] Moreover, brain levels of *p*-CA disappear monoexponentially from 1 to 96 hr, with a half-life of 8.5 hr.[27] It thus appears that the long-term toxic effects of *p*-CA are unrelated to the presence of the unchanged drug, again leaving open the possibility of an active metabolite. Because of these considerations, we have recently examined the effects of classic inducers (phenobarbital and 3-methylcholanthrene) and an inhibitor (piperonyl butoxide) of hepatic mixed-function oxidase activity on the long-term toxic action of *p*-CA on brain 5-HT. These agents modify the toxic effect of *p*-CA in an unexpected manner. Pretreatment with phenobarbital has no effect on the long-term depletion of brain 5-HT by *p*-CA, nor does it change the half-life of the drug. Pretreatment with 3-methylcholanthrene provides nearly complete protection from this effect and increases the rate of disappearance of the drug. A significant protection is also provided by pretreatment with piperonyl butoxide, which inhibits hepatic mixed-function oxidase activity and prolongs the half-life of *p*-CA. These treatments themselves have no effect on the active uptake of 5-HT, thus ruling out a fluoxetine-like effect. Since a significant protection is afforded by treatments that either increase or decrease the half-life of *p*-CA, it appears that the long-term effect is independent of the brain levels of *p*-CA or its major metabolites. Instead, the long-term effect may be due to the formation of a minor, toxic metabolite, and such treatments may divert metabolism away from this route. One possible toxic route would involve the formation of a chemically active species, such as an epoxide or hydroxylamine, as an intermediate in ring hydroxylation or oxidative deamination. Such intermediates have been proposed to explain the hepatotoxicity of many drugs.[39] The former is the most likely candidate, because biochemical studies of *N*-hydroxy-*p*-CA suggest that this derivative is not itself toxic.[40] Moreover, the recent demonstration of ring hydroxylation of *p*-CA and a shift of chlorine to the *meta* position is clear-cut evidence for the occurrence of an epoxide intermediate.[38] To test indirectly for the possible involvement of a chemically active intermediate, we have attempted to modify the long-term effect of *p*-CA by manipulating the levels of glutathione. This sulfhydryl compound often plays a protective role by reacting with electrophilic intermediates, thus reducing their binding to cellular macromolecules.[42] Treatment of rats with large doses of cysteine, both a precursor of glutathione and an electrophil scavenger in its own right, fails to prevent or modify the long-term decrease in 5-HT caused by injection of *p*-CA (TABLE 3). Moreover, treatment with several doses of diethylmaleate, which reduces glutathione levels in the liver to less than one third of the normal value, does not potentiate the toxic effects of *p*-CA (TABLE 3). However, this treatment does not significantly reduce the brain levels of glutathione; therefore, if a reactive metabolite is formed within the brain, it may still be inactivated by reaction with glutathione. So far, we have not been able to reduce the brain levels of glutathione by either intraperitoneal or intraventricular administration of diethylmaleate. Attempts to demonstrate irreversible, "covalent" binding of [^{14}C]-*p*-CA to brain proteins, an indirect indication of the formation of a toxic intermediate,[43] have also been unsuccessful. Thus, although there is considerable indirect evidence to suggest that the long-term effect is due to a metabolite of *p*-CA rather than the parent compound, there is as yet no direct evidence for the involvement of a chemically active intermediate.

Species differences in the duration of the decrease in brain 5-HT that occurs after *p*-CA administration are consistent with the concept of an active metabolite mediating the long-term effects of *p*-CA. Mice, in particular, are unusually resistant to the toxic effect of *p*-CA. In earlier studies, it was found that 5-HT levels returned to control values within 1 day after moderate doses of *p*-CA.[1, 44] More recently, we have administered very large doses of *p*-CA to mice and found that even doses of 50 and 75 mg/kg (which kill 25–50% of the animals) failed to reduce 5-HT after 2 or 3 weeks. However, we have found evidence for an irreversible phase in this species also, with a duration apparently much less than in the rat. For example, administration of 45 mg/kg of *p*-CA to mice significantly reduces 5-HT levels at both 4 and 30 hr, but only the early decrease is reversed by fluoxetine (TABLE 4). Thus, as in the rat, *p*-CA administration to mice induces both a reversible and an irreversible phase of 5-HT depletion, but the duration of both phases is much shorter in mice. If *p*-CA itself mediates the reversible effects, and if a metabolite is responsible for the irreversible phase, such temporal differences between rats and mice can be explained by a more rapid disappearance of both the parent compound and the active metabolite in mice. Moreover, the relatively short duration of the irreversible phase in mice is difficult to reconcile with the hypothesis that the long-term irreversible effects of *p*-CA in rats are due to a cytotoxic effect on 5-HT neurons.

TABLE 3

EFFECT OF PRETREATMENT WITH CYSTEINE OR DIETHYLMALEATE (DEM) ON THE LONG-TERM EFFECT OF *p*-CHLOROAMPHETAMINE (*p*-CA)*

Pretreatment	*p*-CA Dose (mg/kg)	5-HT Levels (μg/g)	
		Control	*p*-CA
Vehicle	5	0.59±0.02	0.30±0.04 †
Cysteine	5	0.63±0.02	0.32±0.04 †
Vehicle	2	0.56±0.02	0.57±0.02
DEM	2	0.60±0.02	0.55±0.02

* Rats were pretreated with three doses of 225 mg/kg of cysteine, 300 mg/kg of diethylmaleate or vehicle prior to *p*-CA injection and then every 8–12 hr for 2 days. Rats were killed 2 weeks after *p*-CA treatment. The values presented are the mean of six animals±SEM.

† $p < 0.001$

TABLE 4

INTERACTION OF FLUOXETINE WITH *p*-CHLOROAMPHETAMINE (*p*-CA) IN MICE

Treatment	Hours	Brain Serotonin * (μg/g±SEM)	
		Saline †	Fluoxetine ‡
Control	—	0.52±0.02	0.48±0.02
p-CA	4	0.43±0.01 §	0.55±0.02
p-CA	30	0.34±0.03 ¶	0.34±0.02 ¶

* Groups of six mice were killed 4 or 30 hr after intraperitoneal administration of 45 mg/kg of *p*-CA, and whole brain levels of serotonin were assayed.

† Saline was injected 4 hr prior to sacrifice.

‡ Fluoxetine (10 mg/kg, ip) was injected 4 hr prior to sacrifice.

§ $p < 0.05$, *p*-CA versus control.

¶ $p < 0.01$, *p*-CA versus control.

LONG-TERM BEHAVIORAL EFFECTS

As reviewed earlier, the initial effects of *p*-CA on brain amines are extremely complex. However, after 2 or 3 days, the effects of the drug are limited to long-term decreases in the levels of 5-HT, 5-HIAA, and 5-HT uptake capacity and in tryptophan hydroxylase activity in brain. A few investigators have reported behavioral changes several days after administration of *p*-CA that are similar to those seen after other treatments whose primary actions are to decrease functional 5-HT in brain. Jouvet and Pujol [48, 49] have shown that reductions in functional brain 5-HT induced by either raphe lesions or synthesis inhibition with *p*-chlorophenylalanine result in marked reductions in sleep time in the cat. A similar effect has been observed after injection of *p*-chloromethamphetamine.[50] An increase in locomotor activity similar to that found after induction of raphe lesions [51–53] or synthesis inhibition by *p*-chlorophenylalanine [54–56] occurred several days after a single injection of *p*-CA.[57] In addition, facilitation of active avoidance behavior has been observed after induction of raphe lesions [58, 59] and after administration of *p*-chlorophenylalanine[60] and, more recently, *p*-CA.[61] These results corroborate the biochemical data reviewed above, which suggest that the long-term effects of *p*-CA result in a decrease in serotonergic transmission and support the view that *p*-CA may serve as a new pharmacologic tool in investigations of the role of 5-HT in the central nervous system.

REFERENCES

1. PLETSCHER, A., G. BARTHOLINI, H. BRUDERER, W. P. BURKARD & K. F. GEY. 1964. Chlorinated arylalkylamines affecting the cerebral metabolism of 5-hydroxytryptamine. J. Pharmacol. Exp. Ther. **145:** 344–350.
2. MILLER, F. P., R. H. COX, JR., W. R. SNODGRASS & R. P. MAICKEL. 1970. Comparative effects of p-chlorophenylalanine, p-chloroamphetamine and p-chloro-N-methylamphetamine on rat brain norepinephrine, serotonin and 5-hydroxyindole-3 acetic acid. Biochem. Pharmacol. **19:** 435–442.
3. GÖRLITZ, B. D. & H. H. FREY. 1972. Central monoamines and antinociceptive drug action. Eur. J. Pharmacol. **20:** 171–180.

4. STERANKA, L. R., R. J. BARRETT & E. SANDERS-BUSH. 1977. Facilitation of Sidman avoidance performance by p-chloroamphetamine: role of biogenic amines. Neuropharmacology. In press.
5. SANDERS-BUSH, E., J. BUSHING & F. SULSER. 1975. Long-term effects of p-chloroamphetamine and related drugs on central serotonergic mechanisms. J. Pharmacol. Exp. Ther. **192:** 33–41.
6. EL-YOUSEF, M. K., L. R. STERANKA & E. SANDERS-BUSH. 1977. Rapid tolerance to the motor effects of p-chloroamphetamine in rats. Psychopharmacology. In press.
7. STRADA, S. & F. SULSER. 1971. Comparative effects of p-chloroamphetamine and amphetamine on metabolism and *in vivo* release of ^{3}H-norepinephrine in the hypothalamus. Eur. J. Pharmacol. **15:** 45–51.
8. STRADA, S., E. SANDERS-BUSH & F. SULSER. 1970. p-Chloroamphetamine: temporal relationship between psychomotor stimulation and metabolism of brain norepinephrine. Biochem. Pharmacol. **19:** 2621–2629.
9. SCHEEL-KRUGER, J. 1972. Behavioral and biochemical comparison of amphetamine derivatives, cocaine, benztropine and tricyclic antidepressant drugs. Eur. J. Pharmacol. **18:** 63–73.
10. COSTA, E., K. NAIMZADA & A. REVUELTA. 1971. Effect of phenmetrazine, aminorex, and (±) p-chloroamphetamine on the motor activity and turnover rate of brain catecholamines. Brit. J. Pharmacol. **43:** 570–579.
11. SANDERS-BUSH, E., J. BUSHING & F. SULSER. 1972. Long-term effects of p-chloroamphetamine on tryptophan hydroxylase activity and on the levels of 5-hydroxytryptamine and 5-hydroxyindoleacetic acid in brain. Eur. J. Pharmacol. **20:** 385–388.
12. PLETSCHER, A., M. DA PRADA, W. BURKARD, G. BARTHOLINI, F. STEINER, H. BRUDERER & F. BIGLER. 1966. Arylalkylamines with different effects on the metabolism of aromatic monoamines. J. Pharmacol. Exp. Ther. **154:** 64–72.
13. PLETSCHER, A., M. DA PRADA & W. BURKARD. 1970. The effects of substituted phenylethylamines on the metabolism of biogenic monoamines. *In* Amphetamines and Related Compounds. E. Costa & S. Garattini, Eds. : 331–341. The Raven Press. New York, N.Y.
14. WONG, D., J. HORNG & R. FULLER. 1973. Kinetics of serotonin accumulation into synaptosomes of rat brain—effects of amphetamine and chloroamphetamine. Biochem. Pharmacol. **22:** 311–322.
15. SANDERS-BUSH, E., D. A. GALLAGER & F. SULSER. 1974. On the mechanism of brain 5-hydroxytryptamine depletion by p-chloroamphetamine and related drugs and the specificity of their action. Advan. Biochem. Psychopharmacol. **10:** 185–194.
16. CARLSSON, A. 1970. Structural specificity for inhibition of ^{14}C-5-hydroxytryptamine uptake by cerebral slices. J. Pharm. Pharmacol. **22:** 729–732.
17. MEEK, J. & K. FUXE. 1971. Serotonin accumulation after monoamine oxidase inhibition. Biochem. Pharmacol. **20:** 693–706
18. MEEK, J., K. FUXE & A. CARLSSON. 1971. Blockade of p-chloromethamphetamine induced 5-hydroxytryptamine depletion by chlorimipramine and meperidine. Biochem. Pharmacol. **20:** 707–709.
19. SANDERS-BUSH, E. & F. SULSER. 1970. p-Chloroamphetamine: *in vivo* investigations on the mechanism of action of the selective depletion of cerebral serotonin. J. Pharmacol. Exp. Ther. **175:** 419–426.
20. SANDERS-BUSH, E., J. A. BUSHING & F. SULSER. 1972. p-Chloroamphetamine inhibition of cerebral tryptophan hydroxylase. Biochem. Pharmacol. **21:** 1501–1510.
21. KOE, B. K. & R. F. CORKEY. 1976. Inhibition of rat brain tryptophan hydroxylase with p-chloroamphetamine. Biochem. Pharmacol. **25:** 31–36.
22. FULLER, R. W., K. W. PERRY & B. B. MOLLOY. 1975. Effect of 3-(p-trifluoromethylphenoxy)-N-methyl-3-phenylpropylamine on the depletion of brain serotonin by 4-chloroamphetamine. J. Pharmacol. Exp. Ther. **193:** 796–803.

23. FREY, H. & M. MAGNUSSEN. 1968. Different central mediation of the stimulant effects of amphetamine and its p-chloro analogue. Biochem. Pharmacol. **17:** 1299–1307.
24. LASSEN, J. B. 1974. The effect of p-chloroamphetamine on motility in rats after inhibition of monoamine synthesis, storage, uptake and receptor interaction. Psychopharmacology **34:** 243–254.
25. TRULSON, M. E. & B. L. JACOBS. 1976. Behavioral evidence for the rapid release of CNS serotonin by PCA and fenfluramine. Eur. J. Pharmacol. **36:** 149–154.
26. JACOBS, B. L. 1975. A behavioral syndrome for the study of CNS serotonin receptor activity: mediation by lower brainstem, evidence for denervation supersensitivity, and mechanism of action by p-chloroamphetamine. Neurosci. Abst. **1:** 295.
27. SEKERKE, H., H. SMITH, J. BUSHING & E. SANDERS-BUSH. 1975. Correlation between brain levels and biochemical effects of the optical isomers of p-chloroamphetamine. J. Pharmacol. Exp. Ther. **193:** 835–844.
28. STERANKA, L. R., E. SANDERS-BUSH & R. J. BARRETT. 1977. Tolerance to p-chloroamphetamine's effects on Sidman avoidance performance and catecholamine metabolism. Neuropharmacology. In press.
29. FULLER, R. W. & H. D. SNODDY. 1974. Long-term effects of 4-chloroamphetamine on brain 5-hydroxyindole metabolism in rats. Neuropharmacology **13:** 85–90.
30. HARVEY, J., S. MCMASTER & L. YUNGER. 1975. p-Chloroamphetamine: selective neurotoxic action in brain. Science **187:** 841–843.
31. HATTORI, T., P. L. MCGEER & E. G. MCGEER. 1976. Synaptic morphology in the ne ostriatum of the rat: possible serotonergic synapse. Neurochem. Res. **1:** 451–467.
32. LOREZ, H., A. SANER, J. G. RICHARDS & M. DA PRADA. 1976. Accumulation of 5HT in non-terminal axons after p-chloro-N-methyl-amphetamine without degeneration of identified 5HT nerve terminals. Eur. J. Pharmacol. **38:** 79–88.
33. MASSARI, V. J., Y. TIZABI & E. SANDERS-BUSH. 1977. Evaluation of the neurotoxic effects of p-chloroamphetamine: a histological and biochemical study. Neuropharmacology. In press.
34. MASSARI, V. J. & E. SANDERS-BUSH. 1975. Synaptosomal uptake and levels of serotonin in rat brain areas after p-chloroamphetamine or B-9 lesions. Eur. J. Pharmacol. **33:** 419–422.
35. MEEK, J. & L. BERTILSSON. 1975. Comparison of the effects of lesion in the B9 cell body group and p-chloroamphetamine on tryptophan hydroxylase and 5-hydroxytryptamine in rat brain nuclei. Brain Res. **100:** 140–144.
36. FULLER, R. W., K. W. PERRY & B. B. MOLLOY. 1975. Reversible and irreversible phases of serotonin depletion by 4-chloroamphetamine. Eur. J. Pharmacol. **33:** 119–124.
37. ROSS, S. B. 1976. Antagonism of the acute and long-term biochemical effects of 4-chloroamphetamine on the 5HT neurons in the rat brain by inhibitors of the 5-hydroxytryptamine uptake. Acta Pharmacol. Toxicol. **39:** 456–476.
38. PARLI, C. J. & P. A. KRIETER. 1976. Inhibition of drug metabolism by fluoxetine and norfluoxetine. Fed. Proc. **18:** 154.
39. GILLETTE, J. R., J. R. MITCHELL & B. B. BRODIE. 1974. Biochemical mechanisms of drug toxicity. Annu. Rev. Pharmacol. **14:** 271–288.
40. FULLER, R. W., K. W. PERRY, J. C. BAKER, C. J. PARLI, N. LEE, W. A. DAY & B. B. MOLLOY. 1974. Comparison of the oxime and the hydroxylamine derivatives of 4-chloroamphetamine as depletors of brain 5-hydroxy-indoles. Biochem. Pharmacol. **23:** 3267–3272.
41. PARLI, C. J. & B. SCHMIDT. 1975. Metabolism of 4-chloroamphetamine to 3-chloro-4-hydroxy amphetamine in rat: evidence for an *in vivo* "NIH shift" of chlorine. Res. Commun. Chem. Pathol. Pharmacol. **10:** 601–604.
42. MITCHELL, J. R., D. J. JOLLOW, W. Z. POTTER, J. R. GILLETTE & B. B. BRODIE.

1973. Acetaminophen-induced hepatic necrosis. IV. Protective role of glutathione. J. Pharmacol. Exp. Ther. **187:** 211–217.
43. JOLLOW, D. J., J. R. MITCHELL, W. Z. POTTER, D. C. DAVIS, J. R. GILLETTE & B. B. BRODIE. 1973. Acetaminophen-induced hepatic necrosis. II. Role of covalent binding *in vivo*. J. Pharmacol. Exp. Ther. **187:** 195–202.
44. MILLER, K. W., E. SANDERS-BUSH & J. V. DINGELL. 1971. p-Chloroamphetamine-species differences in the rate of disappearance and the lowering of cerebral serotonin. Biochem. Pharmacol. **20:** 500–503.
45. POSHEL, B. P. H. & F. W. NINTEMAN. 1971. Intracranial reward and the forebrain's serotonergic mechanism: studies employing p-chlorophenylalanine and p-chloroamphetamine. Physiol. Behav. **7:** 39–46.
46. BERTILSSON, L., S. H. KOSLOW & E. COSTA. 1975. 5-Hydroxytryptamine depletion in mesencephalic nuclei of rat brain following a single injection of p-chloroamphetamine. Brain Res. **91:** 348–350.
47. NECKERS, L. M., L. BERTILSSON, S. H. KOSLOW & J. L. MEEK. 1975. Reduction of tryptophan hydroxylase activity and 5-hydroxytryptamine concentration in certain rat brain nuclei after p-chloroamphetamine. J. Pharmacol. Exp. Ther. **196:** 333–338.
48. JOUVET, M. 1973. Serotonin and sleep in the cat. *In* Serotonin and Behavior. J. Barchas & E. Usdin, Eds. : 385–400. Academic Press, Inc. New York, N.Y.
49. JOUVET, M. & J. PUJOL. 1974. Effects of central alterations of serotonergic neurons upon the sleep-waking cycle. Advan. Biochem. Psychopharmacol. **11:** 211–216.
50. DELORME, F., J. L. FROMENT & M. JOUVET. 1966. Suppression du sommeil par la p-chloromethamphetamine et la p-chlorophenylalanine. C. R. Seances Soc. Biol. **160:** 2347–2351.
51. KOSTOWSKI, W., W. GIACALONE, S. GARATTINI & L. VALZELLI. 1968. Studies on behavioral and biochemical changes in rats after lesions in midbrain raphe. Eur. J. Pharmacol. **4:** 371–376.
52. NEILL, D. B., L. P. GRANT & S. P. GROSSMAN. 1972. Selective potentiation of locomotor effects of amphetamine by midbrain raphe lesions. Physiol. Behav. **9:** 655–657.
53. JACOBS, B. L., W. D. WISE & K. M. TAYLOR. 1974. Differential behavioral and neurochemical effects following lesions of the dorsal or median raphe nuclei in rats. Brain Res. **79:** 353–361.
54. CHRUSIEL, T. L. & Z. S. HERMAN. 1969. Effect of dopamine on behavior in mice depleted of norepinephrine or serotonin. Psychopharmacology **14:** 124–134.
55. FIBIGER, H. C. & B. A. CAMPBELL. 1971. The effect of parachlorophenylalanine on spontaneous locomotor activity in the rat. Neuropharmacology **10:** 25–32.
56. JACOBS, B. L., E. E. EUBANKS & W. D. WISE. 1974. Effect of indolealkylamine manipulations on locomotor activity in rats. Neuropharmacology **13:** 575–583.
57. MESSING, R. B., L. PHEBUS, L. A. FISHER & L. D. LYTLE. 1976. Effects of p-chloroamphetamine on locomotor activity and brain 5-hydroxyindoles. Neuropharmacology **15:** 157–163.
58. LORENS, S. A., J. P. SORENSEN & L. M. YUNGER. 1971. Behavioral and neurochemical effects of lesions in the raphe system of the rat. J. Comp. Physiol. Psychol. **77:** 48–52.
59. STERANKA, L. R. & R. J. BARRETT. 1974. Facilitation of avoidance acquisition by lesion of the median raphe nucleus: evidence for serotonin as a mediator of shock-induced suppression. Behav. Biol. **11:** 205–213.
60. TENEN, S. S. 1967. The effects of p-chlorophenylalanine, a serotonin depletor, on avoidance acquisition, pain sensitivity and related behavior in the rat. Psychopharmacologia **10:** 204–219.
61. VORHEES, C. V., G. J. SCHAEFER & R. J. BARRETT. 1975. p-Chloroamphetamine: behavioral effects of reduced cerebral serotonin in rats. Pharmacol. Biochem. Behav. **3:** 279–284.

62. Bogdanski, D., A. Pletscher, B. Brodie & S. Udenfriend. 1955. Identification and assay of serotonin in brain. J. Pharmacol. Exp. Ther. **117:** 82–88.
63. Gál, E. M. & K. Patterson. 1973. Rapid nonisotopic assay of tryptophan-5-hydroxylase activity in tissues. Anal. Biochem. **52:** 628.
64. Leith, N. 1977. Personal communication.

Discussion

Dr. W. Lovenberg: I was interested in your slide on Dopa accumulation. We have measured tyrosine hydroxylase extracted from the tissue and see no changes in the amount of this enzyme *in vitro.* What you are observing is probably an activation of the enzyme *in vivo.*

Dr. Sanders-Bush: Yes, I agree. We also found no changes in tyrosine hydroxylase assayed *in vitro.* These effects on tyrosine hydroxylase activity *in vivo* are reflected by an increase in synthesis of catecholamine from tritiated tyrosine.

Dr. C. K. Meshul (*Downstate Medical Center, Brooklyn, N.Y.*): Doesn't it seem strange that phenobarbital did not change the half-life of *p*-chloroamphetamine, because it induces cytochrome P-450? Are you assuming that a metabolite of *p*-chloroamphetamine is involved in the long-term effects on brain serotonin, and do you have any ideas as to what that metabolite might be?

Dr. Sanders-Bush: Dr. Fuller has also found that microsomal inducers have no effect on the rate of disappearance of *p*-chloroamphetamine. *p*-Chloroamphetamine may be an unusual substrate for the cytochrome P-450 system, the metabolism of which is not stimulated by phenobarbital. We showed that aminopyrine demethylation and aniline hydroxylation were increased three- to fourfold by phenobarbital pretreatment. It is a provocative hypothesis that a chemically active intermediate in the metabolism of *p*-chloroamphetamine mediates the toxic effects on serotonin neurons, but neither we nor anyone else has any direct evidence for the involvement of a chemically active intermediate. We have confirmed Dr. Gál's data that manipulation of glutathione levels does not affect the toxicity of *p*-chloroamphetamine, for example.

FENFLURAMINE AND BRAIN SEROTONIN

Bradley V. Clineschmidt, Anthony G. Zacchei, James A. Totaro,
A. Barbara Pflueger, Jodie C. McGuffin, and
Theodore I. Wishousky

Merck Institute for Therapeutic Research
West Point, Pennsylvania 19486

It has been 10 years since Duhalt and Verdavainne[1] and Opitz[2] found a decrease in serotonin [5-hydroxytryptamine (5-HT)] in brain a few hours after injecting animals with fenfluramine (*N*-ethyl-α-methyl-3-trifluoromethyl-β-phenylethylamine), a drug used mainly as an anorexigen in man. However, the long-lasting nature of the reduction in brain 5-HT caused by fenfluramine was only recently described.[3, 4] Brain 5-HT remains below the normal concentration for at least 30 days after a single injection of a high dose, for example, 15 mg/kg, of fenfluramine.[3, 4] A similar long-term decrease in brain 5-HT can also be achieved by repeated administration of fenfluramine in smaller amounts, such as 5 mg/kg/day for 5 days.[4]

According to Harvey and McMaster,[3] fenfluramine has a neurotoxic action on serotonergic cells of the B9 group, perhaps accounting for the compound's long-lasting effect on brain 5-HT. On the other hand, it would appear that fenfluramine also has prominent effects on serotonergic terminal processes. Pretreatment with chlorimipramine, an inhibitor of the uptake mechanism for 5-HT at the cell membrane,[5] prevents the short-term reduction of brain 5-HT caused by fenfluramine.[6] Garattini *et al.*[7] suggest that chlorimipramine prevents the access of fenfluramine into serotonergic terminals, thereby preventing the reduction in brain 5-HT. Fenfluramine inhibits the uptake and promotes the release of [^{3}H]-5-HT *in vitro,* using slices of cerebral cortex,[8] an area of the brain devoid of serotonergic cell bodies.

In this paper, evidence is presented that indicates that the long-term decrease in brain 5-HT results primarily from an action of fenfluramine on serotonergic brain terminals. We show also that the 5-HT uptake inhibitor fluoxetine given after fenfluramine can terminate its action, suggesting that continued uptake for several hours is required for fenfluramine to exert a long-lasting effect on the concentration of 5-HT in the brain. Studies are also included on the relationship of the concentration of fenfluramine and norfenfluramine in brain to the prolonged depletion of 5-HT.

Materials and Methods

After treatment, female Charles River rats weighing 160–210 g were sacrificed by decapitation; the whole brains were removed and either frozen on dry ice and stored at $-15°$ C for subsequent analysis or the brain was dissected into forebrain and a posterior midbrain block of tissue according to the method of Aghajanian *et al.*[9] The samples of posterior midbrain weighed from 26 to 52 mg. Whole brains were homogenized in 8 ml/g of tissue of 0.1 N HCl that contained 0.5% ascorbic acid. One half of the homogenate was used for determining 5-HT, and the other half was used for assaying 5-hydroxyindoleacetic acid (5-HIAA). 5-Hydroxytryptamine and 5-HIAA were determined, respectively, by the methods of Bogdanski *et al.*[10] and Giacalone and Valzelli.[11] The forebrain was ground in 10 ml/g of 0.1 N HCl, and a 10-ml aliquot of the homogenate was taken for assaying 5-HT. Midbrain samples were homogenized in a constant 2.5-ml volume of 0.1 N HCl, and 5-HT was determined in a 2-ml aliquot. In the assays of forebrain and midbrain 5-HT, 1 ml of the final aqueous extract was reacted with *o*-phthaldialdehyde,[12, 13] followed by extraction with chloroform.[14]

Newborn (1-day-old) or adult (170–200-g females) rats were treated at the same time and sacrificed 16 or 32 days later. After the newborns were several days old, the sex was ascertained, and the males were discarded. Injections were made subcutaneously, by use of a volume of 20 μl/10 g of body weight. 5-Hydroxytryptamine and 5-HIAA were ascertained as described above for whole brain.

The uptake of 5-HT was determined by use of a modification of the procedure of Carlsson.[5] A 200-mg portion of cerebral cortex was placed in a petri dish on ice and rinsed with cold Krebs-Henseleit solution that contained 0.1% glucose and equilibrated to pH 7.4 with 5% CO_2-95% O_2. The tissue was sliced into 0.2-mm cross sections with a McIlwain mechanical tissue chopper. The chopped tissue was transferred to a plastic centrifuge tube that contained 1 ml of ice-cold Krebs-Henseleit solution. After preparing all samples, 9 ml of cold Krebs-Henseleit solution was added to each tube, followed by 25 pmol of [^{3}H]-5-hydroxytryptamine creatinine sulfate (11.1 Ci/mmol) in 0.2 ml of solution. The tubes were gently agitated and incubated for 5 min at 37° C. Incubation was terminated by putting the tubes in ice water, followed by centrifugation at 1250*g* for 5 min. During each experiment, some samples were left in ice after the addition of [^{3}H]-5-HT (0° C controls). The supernatant was decanted, and the remaining tissue was washed by shaking for 10 min with 10 ml of cold 0.9% saline. After centrifugation at 1250*g* for 5 min, the aqueous layer was decanted, and the tissue residue was extracted with 2.5 ml of 0.4 N perchloric acid by mixing (Vortex) and then shaking for 15 min. Samples were centrifuged at 15,000*g* for 5 min at 4° C. Two-milliliter aliquots of the clear perchloric acid supernatant were added to counting vials that contained 20 ml of PCS solubilizer (Amersham/Searle) and shaken. The samples were counted on a Packard Tri-Carb liquid scintillation counter.

Dihydroxyphenylalanine (Dopa) and 5-hydroxytryptophan (5-HTP) were determined in whole brain after inhibition of decarboxylation with NSD-1015.[15] The tissue was ground in 6 ml/g of ice-cold 0.4 N perchloric acid that contained 0.5 mg/ml of sodium metabisulfite and 2 mg/ml of disodium EDTA. After grinding, 0.1 ml of potassium hydroxide-formate solution (10 N KOH added to concentrated formic acid in sufficient volume to adjust the latter to 5 N)[16] was added per milliliter of perchloric acid solution, precipitating excess chlorate and bringing the pH to about 2.8. After centrifugation for 15 min at 15,000*g* and 4° C, the clear supernatant was decanted and allowed to reach room temperature. Dihydroxyphenylalanine and 5-HTP were purified and separated by passing the supernatant through a Dowex® * 50W, X-4 column, 5 × 65 mm.[17]

* Dowex 200–400 mesh in the hydrogen form washed for 1 hr with 2 N NaOH that contained 1% disodium EDTA, followed by washing with distilled water until neutral, a 1-hr wash with 2 N NCl, and a final wash with distilled water until neutral. Sufficient 0.1 M phosphate buffer (pH 6.5) that contained 0.1% disodium EDTA was added to the resin until the pH stabilized at 6.5. The resin was stored as such until needed.

The column was washed with 5 ml of distilled water prior to applying the supernatant, and 6 ml of distilled water were passed through the column after the supernatant. Ten milliliters of 0.1 M sodium phosphate buffer (pH 6.5), which contained 0.1% disodium EDTA, was added to the column to elute Dopa and 5-HTP. The first 6 ml were collected for the assay of Dopa and the following 4 ml for 5-HTP. The fraction that contained Dopa was mixed with acid-washed alumina to adsorb Dopa, followed by elution with 0.05 N HCl and fluorometric determination.[18, 19] A 1-ml aliquot of the 5-HTP-containing eluate from the column was reacted with *o*-phthaldialdehyde,[12, 13] followed by chloroform extraction.[14]

Fenfluramine and norfenfluramine in plasma and brain were determined by use of a specific and sensitive gas chromatographic method, employing the internal standard *p*-chloro-*N*-methylamphetamine (*p*-CMA). The rats were decapitated after various times (2, 6, 18, or 24 hr; 2, 4, 8, 16 or 32 days) following per os injection of 15 mg/kg of fenfluramine. Blood was collected and centrifuged, and the plasma was separated. The whole brains were homogenized in 0.1 N HCl (8 ml/g of tissue). To a 50-ml glass-stoppered centrifuge tube that contained 1 ml of plasma (or an appropriate aliquot of brain homogenate, 1–4 ml), 0.5 ml of 1 N NaOH and 0.1 ml of *p*-CMA (5 μg/ml) were added. The contents of the tube were extracted twice with 10 ml of cyclohexane. After centrifugation, the pooled organic extract was added to a tube that contained 2 ml of 0.1 N HCl. After shaking and centrifugation, the organic phase was removed by aspiration. Five milliliters of fresh cyclohexane were added, and the aqueous phase was alkalinized (0.5 ml of 1 N NaOH). After extraction of the amines into cyclohexane, the organic phase (>4 ml) was transferred to a 13-ml tube and reacted with 100 μl of trichloroacetyl chloride[20] (0.1% in cyclohexane) for 30–60 min at 70° C. The tube was cooled, and the contents were shaken with 0.5 ml of water (Vortex, 1 min) and 0.5 ml of 5% NH_4OH. After centrifugation, the cyclohexane phase was transferred to a 13-ml tube and concentrated under nitrogen to about 1 ml. An appropriate aliquot (usually 5 μl) was injected directly into the gas chromatograph, Hewlett-Packard model 5830A equipped with a ^{63}Ni electron-capture detector and a 6 ft $\times$ 2 mm (i.d.) glass column packed with 10% OV-17 on 80–100 mesh Supelcoport®. The parameters for the temperature-programmed analysis were as follows:

oven temperature 1	= 190° C	carrier gas (argon/methane, 95/5)	= 42 ml/min
oven temperature 2	= 240° C	time 1 (initial hold)	= 4 min
ECD temperature	= 300° C	rate	= 5.0° C/min
injection port temperature	= 180° C	slope sensitivity	= 0.5

The retention times of norfenfluramine, fenfluramine, and *p*-CMA as the trichloroacetyl derivatives were, respectively, 4.4, 6.4, and 12.3 min. Standard plasma and brain curves were constructed by plotting the peak area ratios (norfenfluramine/*p*-CMA or fenfluramine/*p*-CMA) versus weight ratios. The peak area ratios obtained from the unknown were then used to determine the amount of norfenfluramine or fenfluramine present in the sample. The standard samples were run concurrently with the unknown samples.

Specificity of the method was confirmed by gas chromatographic-mass spectrometric analysis (LKB-9000S) of pooled brain and plasma samples that were obtained after gas chromatographic analysis. The individually pooled derivatized samples (brain, plasma, and standards) were concentrated to a small volume prior to injection on a 10% OV-17 column, 6 ft × 2 mm (i.d.). The gas chromatographic conditions were similar to those described above. The initial column, separator, and detector temperatures were 190, 260, and 270°C, respectively. Specificity of analysis was established on direct comparison of the mass spectra of fenfluramine and norfenfluramine obtained from the brain or plasma with authentic fenfluramine and norfenfluramine carried through the procedure. The spectra were identical. The base peak (*m/e* 188) in the mass spectrum of norfenfluramine as the trichloroacetyl derivative resulted from the ion $[CH_3CH{=}NHCCl_3]^{+}$. The base peak (*m/e* 216) in the mass spectrum of fenfluramine as the trichloroacetyl derivative resulted from the ion

$$\left[CH_3CH{=}N\begin{matrix} \diagup CCl_3 \\ \diagdown CH_2CH_3 \end{matrix} \right]^{+}.$$

DL-Fenfluramine hydrochloride, DL-norfenfluramine hydrochloride, fluoxetine hydrochloride, DL-*p*-chlorophenylalanine methyl ester hydrochloride, reserpine (Serpasil®), DL-5-hydroxy[G-^{3}H]tryptamine creatinine sulfate, NSD 1015 (3-hydroxybenzylhydrazine hydrochloride), and DL-*p*-chloromethamphetamine hydrochloride were obtained, respectively, from A. H. Robins Company, Richmond, Virginia; Les Laboratoires Servier, Neuilly-sur-Seine, France; Eli Lilly and Company, Indianapolis, Indiana; ICN Life Sciences Group (K & K), Plainview, New York; CIBA-Geigy Corporation, Summit, New Jersey; Amersham/Searle Corporation, Arlington Heights, Illinois; Sandev Ltd., Gilston Park, Harlow, Essex, Great Britain; and Regis Chemical Company, Morton Grove, Illinois. With the exception of *p*-chlorophenylalanine methyl ester hydrochloride, all doses were adjusted to account for the presence of salt.

Results

Time Course for the Decrease in 5-Hydroxytryptamine and 5-Hydroxyindoleacetic Acid Produced by Fenfluramine

Whole brain 5-HT and 5-HIAA were maximally reduced 5 hr after administration of 15 mg/kg of fenfluramine by gavage (Figure 1). Both 5-hydroxyindoles remained at this low level for 16 days without any apparent recovery. Thirty-two days after injection of fenfluramine, 5-HT was somewhat higher than at previous times, but there was no tendency toward a recovery in brain 5-HIAA. Between 32 and 64 days, there was an obvious return of both 5-HT and 5-HIAA toward concentrations found in controls, although recovery was still not complete, even 64 days after administration of fenfluramine.

Plasma and Brain Concentrations of Fenfluramine and Norfenfluramine

The levels of fenfluramine and norfenfluramine were determined 2, 6, 18, and 24 hr and 2, 4, 8, 16, and 32 days after oral administration of fenfluramine, 15 mg/kg. The plasma level of fenfluramine was highest at 2 hr and decreased relatively rapidly thereafter (Figure 2). Less than 10 ng/ml were present

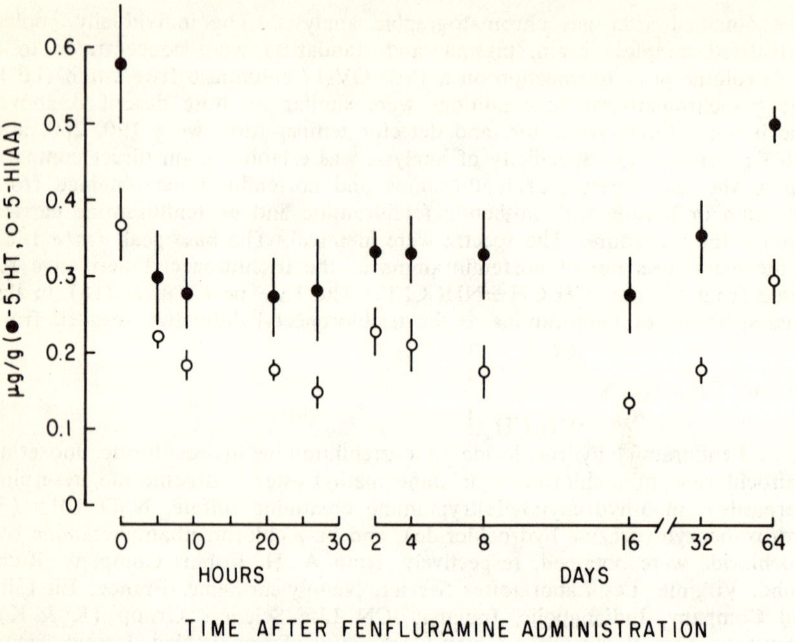

FIGURE 1. 5-Hydroxytryptamine and 5-HIAA concentrations in whole brain at various times after administeration of fenfluramine (15 mg/kg, po). Each point is the $\bar{X} \pm SD$ of 10 brains, except at 0 hr (control) $N=100$. At each of the 10 time intervals, 10 brains from vehicle-treated animals were assayed concurrently with 10 brains from fenfluramine-treated rats. For the purpose of illustation, the results from all vehicle-treated animals were pooled. Statistical evaluation was performed comparing brains from the 10 fenfluramine-treated animals with the 10 concurrently assayed vehicle-treated controls. At each time point, brain 5-HT and 5-HIAA were significantly ($p < 0.05$, two-tailed t test) lower in the group injected with fenfluramine.

48 hr after dosing. The plasma level of norfenfluramine rose slightly between 2 and 24 hr, then abruptly declined between 24 and 48 hr (FIGURE 2). Norfenfluramine was no longer detectable in the plasma 96 hr after administration of fenfluramine.

The appearance and disappearance of fenfluramine and norfenfluramine in brain were to a large extent parallel to the plasma levels, although the actual concentrations in brain were 10–20 times greater than in the plasma (FIGURE 3). Norfenfluramine exhibited a longer half-life (~14 hr) in the brain than did fenfluramine (~6 hr). Whereas fenfluramine was only detected up to 24 hr after dosing, norfenfluramine was still detectable in brain 96 hr after administration of fenfluramine.

Termination of Fenfluramine's Action by Fluoxetine

Rats treated with fenfluramine (15 mg/kg, po) were injected 2 hr later with fluoxetine or vehicle, and 3 hr later, the animals were sacrificed. As shown in FIGURE 4, the concentration of 5-HT in brains from animals that received a subsequent injection of fluoxetine was no different from control, while the concentration was markedly decreased in the brains of rats that received a subsequent injection of vehicle. When rats treated with fenfluramine and then fluoxetine 2 hr later were sacrificed 2 or 16 days after the second injection, the brain concentration of 5-HT was still no different from that in a control group (FIGURE 5). The reversal or termination of the 5-HT-lowering action of fenfluramine by fluoxetine was, therefore, complete and permanent. Fluoxetine also prevented the reduction in 5-HIAA caused by fenfluramine, as ascertained 16 days after treatment (FIGURE 6). Further investigation revealed that the time interval between injections of fenfluramine and fluoxetine is an important variable. In these studies, the rats were always sacrificed 3 hr after injection of fluoxetine, with the interval between the two drugs varying from 2 hr to 32 days (FIGURE 7). The shortest interval proved most effective, and, thereafter, the percentage recovery in brain 5-HT induced by the subsequent fluoxetine injection declined in an essentially linear fashion. When the interval between the two compounds was 2 or more days, fluoxetine was no longer effective. The small apparent recovery in brain 5-HT induced by fluoxetine at 2–32 days is not considered significant, because 3 hr after injection, fluoxetine alone caused an increase in 5-HT of approximately 6–7% (TABLE 1). Because fluoxetine itself caused a decrease in brain 5-HIAA (TABLE 1), studies

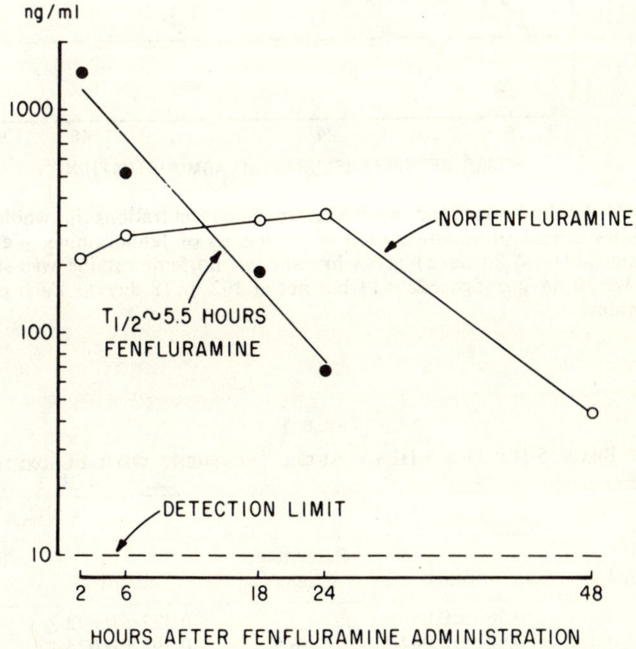

FIGURE 2. Plasma fenfluramine and norfenfluramine levels at various times after injection of fenfluramine (15 mg/kg, po). Fenfluramine and norfenfluramine were, respectively, not detectable (< 10 ng/ml) at 48 and 96 hr. Each point is the $\bar{X}$ of samples from five rats.

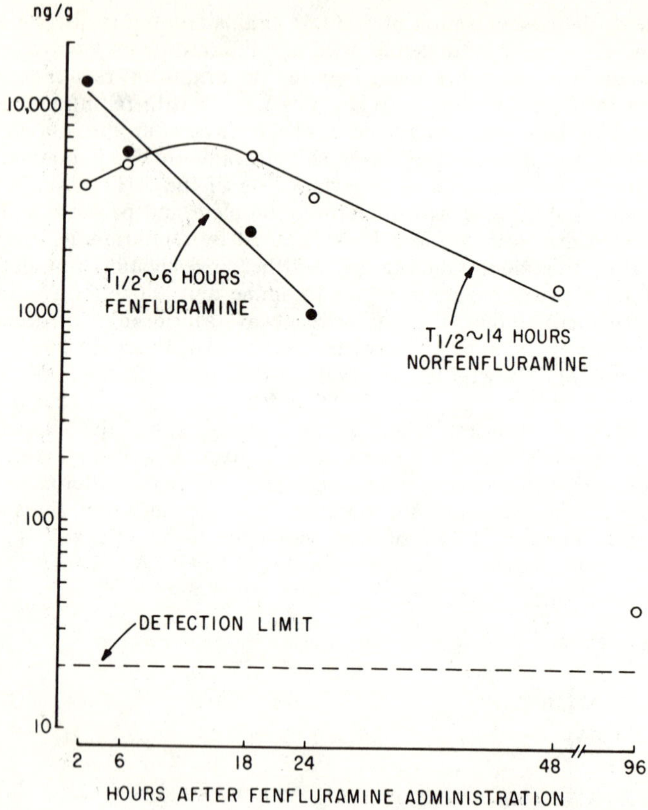

FIGURE 3. Fenfluramine and norfenfluramine concentrations in whole brain at different times after oral administration of 15 mg/kg of fenfluramine. Fenfluramine was not detectable (< 20 ng/g) at 48 hr, whereas norfenfluramine was still present at 96 hr ($\overline{X}$=39 ng/g; range=21–61) but not at 192 hr (8 days). Each point is the $\overline{X}$ of five brains.

TABLE 1

RAT BRAIN 5-HT AND 5-HIAA AFTER TREATMENT WITH FLUOXETINE

Treatment * (mg/kg, ip)	5-HT µg/g †	5-HT Percentage Change	5-HIAA µg/g †	5-HIAA Percentage Change
Vehicle	0.584±0.050		0.371±0.042	
Fluoxetine (10)	0.622±0.064 ‡	+6.5	0.293±0.026 ‡	−21.1

* Three hours prior to sacrifice.
† X±SD (N=30).
‡ $p < 0.05$ (two-tailed t test).

similar to those just presented for 5-HT were not conducted with respect to 5-HIAA.

Selectivity of Fluoxetine

When given 18 hr after fenfluramine, fluoxetine caused a 50% recovery in brain 5-HT (FIGURE 8, A). However, when administered subsequent to *p*-chlorophenylalanine (*p*-CPA) or reserpine, fluoxetine was without effect; that is, the 5-HT concentration in brain did not return toward the level found in brains from control rats (FIGURE 8, B & C).

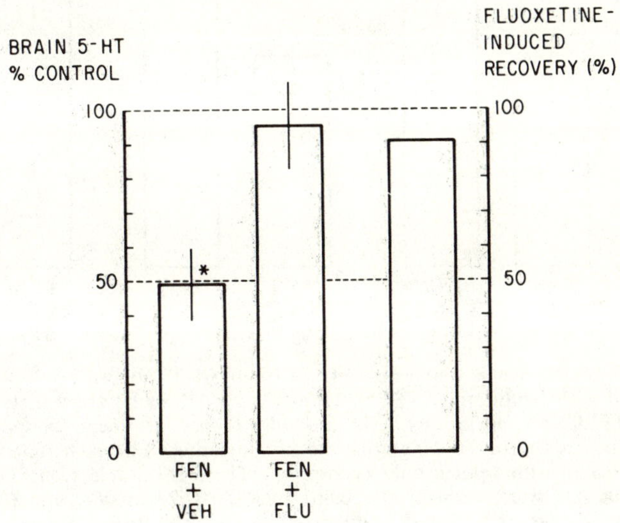

FIGURE 4. Fluoxetine-induced termination of the 5-HT lowering action of fenfluramine. Two hours after injection of fenfluramine (FEN, 15 mg/kg, po), the rats were injected with vehicle (VEH) or fluoxetine (FLU, 10 mg/kg, ip). Three hours later, the whole brains were removed and assayed for 5-HT. In this and all subsequent bar graphs, the top of the bar represents, the $\bar{X}$, and the vertical lines indicate the SD; $N=10$ per treatment group. Asterisk indicates that $p<0.05$ (two-tailed *t*-test) compared to group treated with fluoxetine. Concentration of 5-HT in brains from 10 control (VEH+VEH) animals was determined concurrently with the other groups, permitting calculation of percentage recovery induced by fluoxetine, as follows:

$$\frac{(\text{fenfluramine}+\text{fluoxetine})-(\text{fenfluramine}+\text{vehicle})}{(\text{vehicle}+\text{vehicle})-(\text{fenfluramine}+\text{vehicle})}\times 100.$$

The 5-HT level in the control group was 0.601 ± 0.057 ($\bar{X}\pm$SD) μg/g.

5-Hydroxytryptamine and 5-Hydroxyindoleacetic Acid After Norfenfluramine

A marked and long-lasting reduction in brain 5-hydroxyindoles occurred after the administration of norfenfluramine (15 mg/kg, po, FIGURE 9). Fluoxetine given shortly after norfenfluramine prevented the decrease in 5-HT normally produced by the compound (TABLE 2).

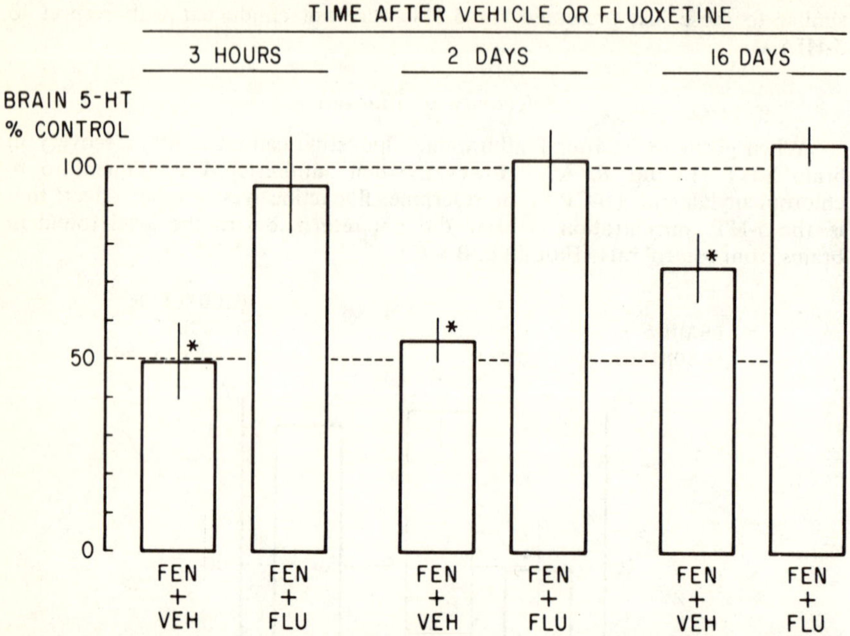

FIGURE 5. Permanence of the recovery of brain 5-HT induced by fluoxetine. Rats were treated with fenfluramine (FEN, 15 mg/kg, po) 2 hr before vehicle (VEH) or fluoxetine (FLU, 10 mg/kg, ip). The animals, 10 per treatment group, were sacrificed at 3 hr, 2 days, or 16 days after the last injection. For each time interval, 10 rats were treated with vehicle only (controls). The 5-HT levels ($\bar{X}\pm$SD) in brains from control rats were, respectively, 0.601 ± 0.057, 0.527 ± 0.064, and 0.604 ± 0.043 μg/g at 3 hr, 2 days, and 16 days. Other details are given in the legend to FIGURE 4.

TABLE 2

RECOVERY OF BRAIN 5-HT PRODUCED BY FLUOXETINE IN RATS TREATED WITH NORFENFLURAMINE

Treatment I * (mg/kg, po)	Treatment II † (mg/kg, ip)	5-HT μg/g ‡	5-HT Percentage Recovery §
Vehicle	vehicle	0.655 ± 0.022	
Norfenfluramine (15)	vehicle	0.201 ± 0.073 ¶	
Norfenfluramine (15)	fluoxetine (10)	0.641 ± 0.079	97.0

* Two hours prior to treatment II.

† Three hours before sacrifice.

‡ $\bar{X}\pm$SD of 10 brains per treatment group.

§ $$\frac{(\text{norfenfluramine}+\text{fluoxetine})-(\text{norfenfluramine}+\text{vehicle})}{(\text{vehicle}+\text{vehicle})-(\text{norfenfluramine}+\text{vehicle})}\times 100.$$

¶ $p < 0.05$ (two-tailed t test) versus both vehicle plus vehicle and norfenfluramine plus fluoxetine groups.

Newborn versus Adult

At subcutaneous dose levels of 7.5 and 15 mg/kg, fenfluramine caused a decrease in brain 5-HT in the adult rat but not in the newborn (FIGURE 10). With regard to 5-HIAA (FIGURE 11), fenfluramine clearly had a larger effect in the adult, although a significant decrease was also found in brains from newborn rats treated 16 or 32 days before sacrifice.

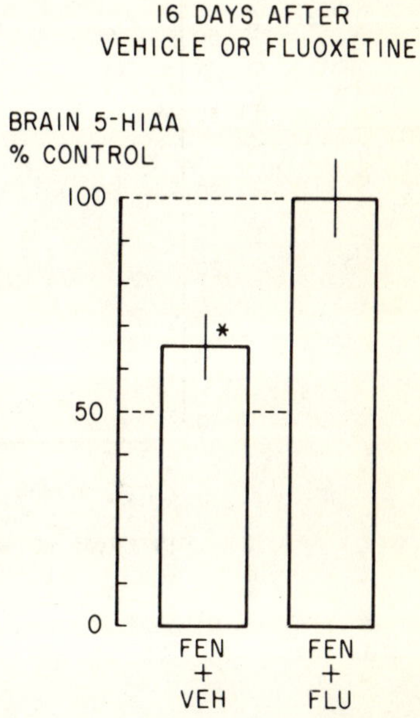

FIGURE 6. Permanence of the recovery in 5-HIAA induced by fluoxetine. Refer to the legend to FIGURE 5 for details. The 5-HIAA concentration in brains from the control group assayed concurrently was 0.339 ± 0.033 μg/g ($\overline{X} \pm$ SD).

Forebrain versus Midbrain

One or 4 days after injection of fenfluramine (15 mg/kg, po), 5-HT in forebrain was reduced by 60%, whereas 5-HT in posterior midbrain samples was decreased only about 30% (FIGURE 12). Sixteen days after administration of fenfluramine, 5-HT in both areas was still less than in vehicle-treated controls, but there was no difference in the magnitude of the reduction in the two regions.

[^{3}H]-5-Hydroxytryptamine Uptake

Sixteen days after the usual dose of fenfluramine, the uptake of [^{3}H]-5-HT into cerebral cortical tissue was inhibited by approximately 30% (TABLE 3).

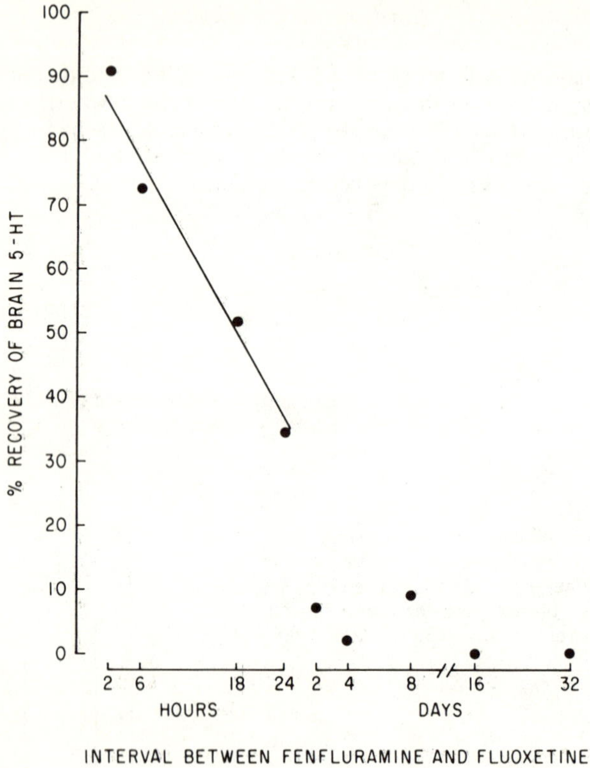

FIGURE 7. Importance of the interval between injection of fenfluramine and fluoxetine on the recovery of whole brain 5-HT. Ten rats per group were given 15 mg/kg (po) of fenfluramine, followed at various times thereafter by vehicle or 10 mg/kg(ip) of fluoxetine. The animals were sacrificed 3 hr after the second injection. At each time interval, a control group was treated (vehicle+vehicle), sacrificed, and the brains removed for concurrent assay with the other treatment groups. The percentage recovery in 5-HT was calculated from the $\bar{X}$ values obtained with the treatment groups (vehicle+vehicle, fenfluramine+vehicle, fenfluramine+fluoxetine), as described in the legend to FIGURE 4.

5-Hydroxytryptophan and Dopa Accumulation

Eight days after injection of fenfluramine, the rats were treated with NSD-1015 to inhibit aromatic amino acid decarboxylase and were sacrificed 30 min later. The level of 5-HTP was lower in the fenfluramine-treated group than in controls, while the concentration of Dopa was not different between the two groups (FIGURE 13). Brains from rats treated with vehicle instead of NSD-

1015 had similar levels of Dopa and 5-HTP, regardless of whether the animals had received fenfluramine or vehicle 8 days earlier. For Dopa, the values ($\bar{X} \pm$ SD) were, respectively, 0.006 ± 0.002 and 0.007 ± 0.004 μg/g for the vehicle- and fenfluramine-treated group. The corresponding values for 5-HTP were 0.135 ± 0.028 and 0.145 ± 0.067 μg/g.

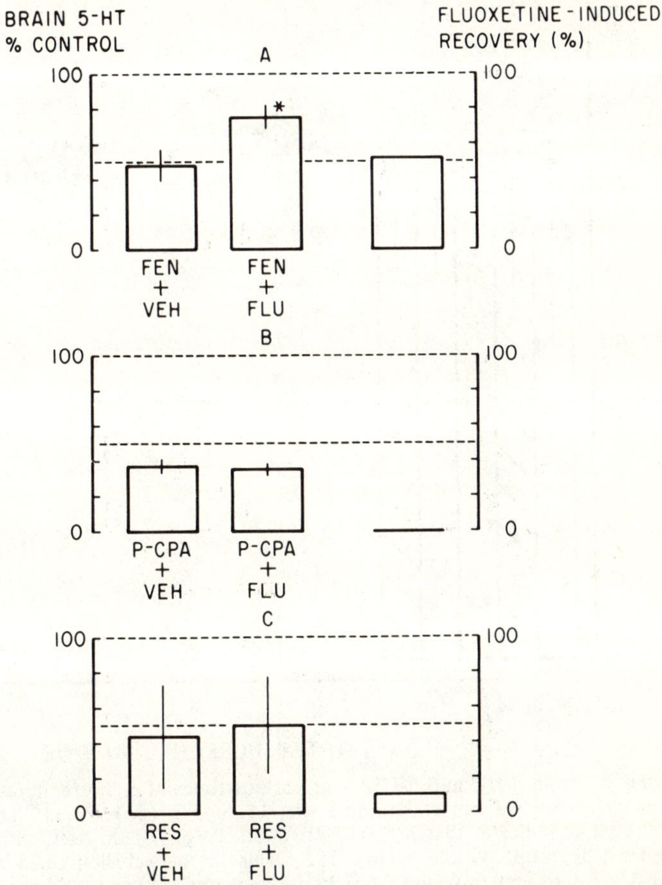

FIGURE 8. Selectivity of fluoxetine. A: Rats (10/group) treated with fenfluramine (FEN, 15 mg/kg, po) were injected 18 hr later with fluoxetine (FLU, 10 mg/kg, ip) or vehicle (VEH), and their brains were removed 3 hr later. Brains from a control (vehicle+vehicle) group of 10 animals assayed concurrently had 0.569±0.046 μg/g ($\bar{X}$±SD) of 5-HT. Fluoxetine-induced recovery was calculated according to the formula in FIGURE 4. Asterisk indicates that $p < 0.05$ (two-tailed *t* test) versus fenfluramine+vehicle. B: *p*-Chlorophenylalanine methyl ester hydrochloride (*p*-CPA, 300 mg/kg, ip) was injected 18 hr before fluoxetine. Other details are given in A. Concentration of 5-HT in brains from the concurrent control group was 0.498± 0.029 μg/g ($\bar{X}$±SD). C: Twenty rats were used for each group instead of 10, as in A and B. Reserpine (RES, 2.5 mg/kg, ip) was injected 18 hr prior to vehicle or fluoxetine. Brains from the control group of 20 animals had 0.598±0.071 μg/g ($\bar{X}$±SD) of 5-HT.

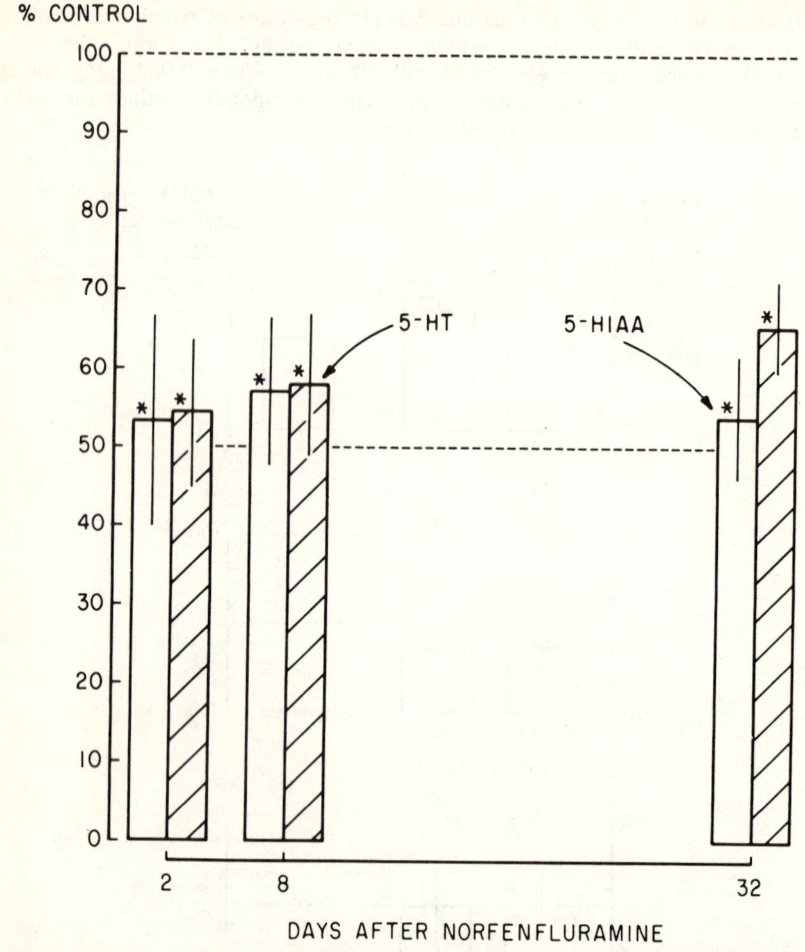

FIGURE 9. Brain 5-HT and 5-HIAA at various times after norfenfluramine administration. The dose of norfenfluramine was 15 mg/kg (po); $N=10$ per treatment group. Asterisk indicates that $p < 0.05$ (two-tailed t test) compared with concurrent vehicle-treated group. Whole brain 5-HT levels in the vehicle-treated groups were 0.501 ± 0.045, 0.616 ± 0.047, and 0.603 ± 0.052 μg/g at 2, 8, and 32 days, respectively; the corresponding 5-HIAA values were 0.432 ± 0.028, 0.353 ± 0.022, and 0.385 ± 0.051 μg/g. All data are expressed as $\bar{X} \pm$SD.

DISCUSSION

Eight or more days after the administration of fenfluramine, when fenfluramine and norfenfluramine were not detectable (< 20 ng/g) in the brain, a decrease was found in several indices of serotonergic function, including the uptake of 5-HT, synthesis rate (i.e., 5-HTP accumulation after injection of NSD-1015),[15] and the concentrations of 5-HT and 5-HIAA. This long-lasting effect of fenfluramine on serotonergic neurons does not extend to catecholamin-

TABLE 3

[³H]-5-HT UPTAKE IN CEREBRAL CORTEX 16 DAYS AFTER ADMINISTRATION OF FENFLURAMINE

Treatment (mg/kg, po)	[³H]-5-HT Uptake (pmol/g)*	Percentage Inhibition
Vehicle	6.50±0.75	
Fenfluramine (15)	4.45±0.90 †	31.6

* $\bar{X}$±SD of six (vehicle) or 16 (fenfluramine) determinations, comprised of duplicate determinations from three vehicle-treated and eight fenfluramine-treated animals.
† $p < 0.05$ (two-tailed *t*-test).

ergic ones, because dopamine and norepinephrine in brain are reduced to a lesser extent and return to control levels within 24 hr after administration of fenfluramine (15 mg/kg, po).[4] Furthermore, the accumulation of Dopa in brains of NSD-1015-treated rats was not affected by fenfluramine given 8 days earlier, whereas the accumulation of 5-HTP was diminished. The selectivity of fenfluramine clearly distinguishes it from such compounds as reserpine. In addition, reserpine caused an increase in brain 5-HIAA assessed 18 hr post-injection (data not presented), and the reduction in 5-HT elicited by reserpine was not reversed by fluoxetine. *p*-Chlorophenylalanine, an irreversible inhibitor

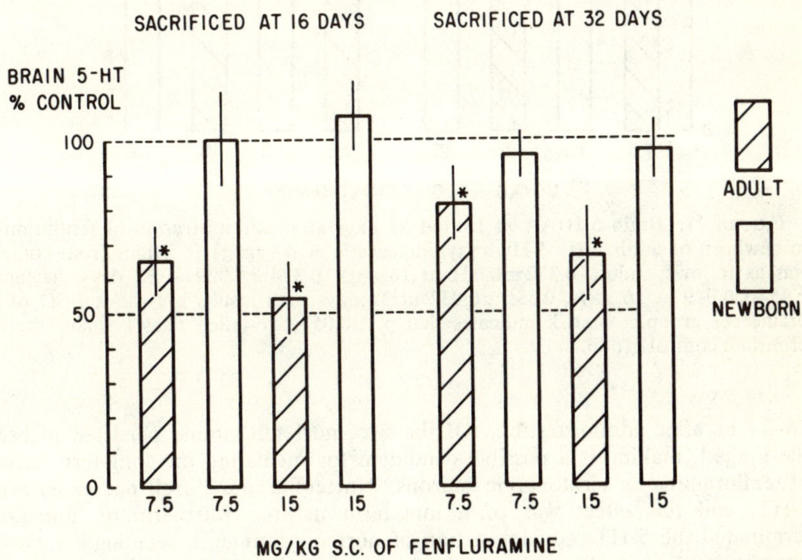

FIGURE 10. Brain 5-HT at 16 and 32 days after administration of fenfluramine to newborn or adult rats. For details, see MATERIALS AND METHODS. 5–Hydroxytryptamine (μg/g) in brains of vehicle-treated controls was as follows: adult, 0.541±0.075 at 16 days, 0.604±0.060 at 32 days; newborn, 0.305±0.025 at 16 days, 0.485±0.031 at 32 days. All results are the $\bar{X}$±SD of groups of 10 brains. Asterisk indicates that $p < 0.05$ (two-tailed *t* test) versus the appropriate control group.

of tryptophan hydroxylase,[21] resembled fenfluramine in reducing both 5-HT and 5-HIAA (data not shown). The duration of action of *p*-CPA is, however, shorter than that of fenfluramine, lasting only about 2 weeks.[22] The inability of fluoxetine to reverse 5-HT depletion induced by *p*-CPA also points to a difference in mode of action. Fenfluramine appears to be most similar to *p*-chloroamphetamine (*p*-CA) with regard to its effects on serotonergic neurons in the brain (*vide infra*).

In agreement with the findings of others,[23-25] norfenfluramine was present in plasma and brain as a major metabolite of fenfluramine. The plasma and brain concentrations of norfenfluramine actually exceeded those of fenfluramine

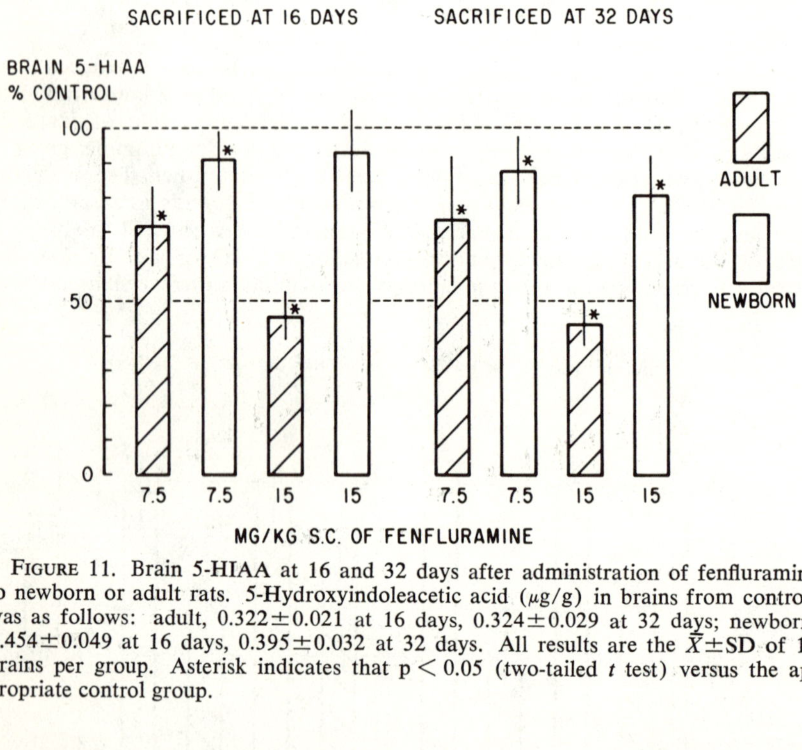

FIGURE 11. Brain 5-HIAA at 16 and 32 days after administration of fenfluramine to newborn or adult rats. 5-Hydroxyindoleacetic acid (μg/g) in brains from controls was as follows: adult, 0.322$\pm$0.021 at 16 days, 0.324$\pm$0.029 at 32 days; newborn, 0.454$\pm$0.049 at 16 days, 0.395$\pm$0.032 at 32 days. All results are the $\bar{X}\pm$SD of 10 brains per group. Asterisk indicates that $p < 0.05$ (two-tailed *t* test) versus the appropriate control group.

18–24 hr after administration. Of the two, norfenfluramine persisted in brain the longest, making it a possible candidate for mediating the long-term effects of fenfluramine on serotonergic neurons. Norfenfluramine itself decreased brain 5-HT, and this effect was of a long-term nature. Furthermore, fluoxetine terminated the 5-HT-reducing action of norfenfluramine, a seemingly requisite interaction if norfenfluramine is, in fact, the mediator of fenfluramine's long-term effect. This seems so because brain 5-HT was significantly restored by fluoxetine given 6–18 hr after fenfluramine, and by this time, norfenfluramine had already reached its peak level in brain. Thus, it seems quite unlikely that fluoxetine could have terminated the action of fenfluramine by blocking its conversion to norfenfluramine. It is unlikely, too, that fluoxetine might act by

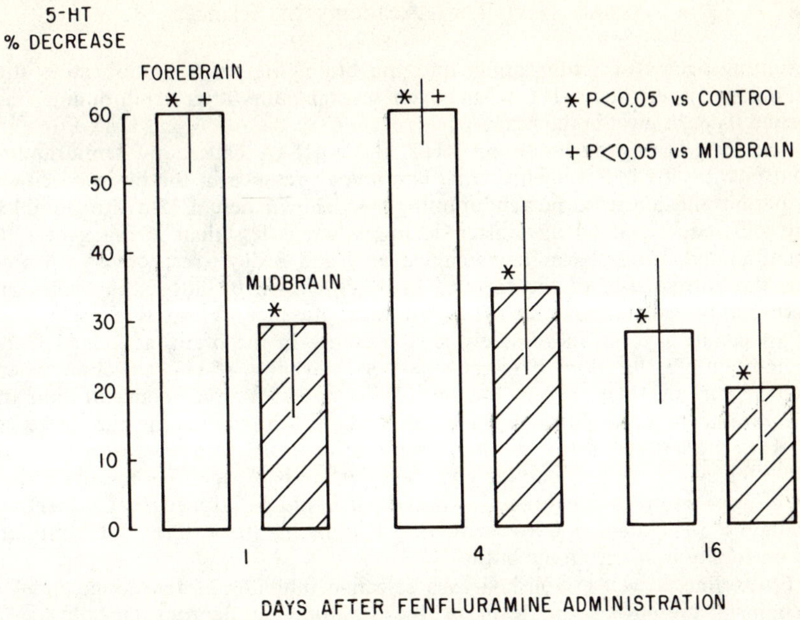

FIGURE 12. Forebrain and midbrain 5-HT after injection of fenfluramine. The rats were sacrificed 1, 4, or 16 days after administration of fenfluramine (15 mg/kg, po). The percentage decrease was calculated for each sample based on the mean value obtained with 10 vehicle-treated controls. 5-Hydroxytryptamine levels in forebrains from controls were, respectively, 0.677±0.038, 0.714±0.030, and 0.667±0.065 μg/g at 1, 4, and and 16 days; the corresponding values for midbrain were 1.148±0.127, 1.440±0.165, and 1.420±0.178 μg/g. All data are presented as $\bar{X}$±SD of 10 samples per group. Statistical significance was ascertained via a two-tailed *t* test.

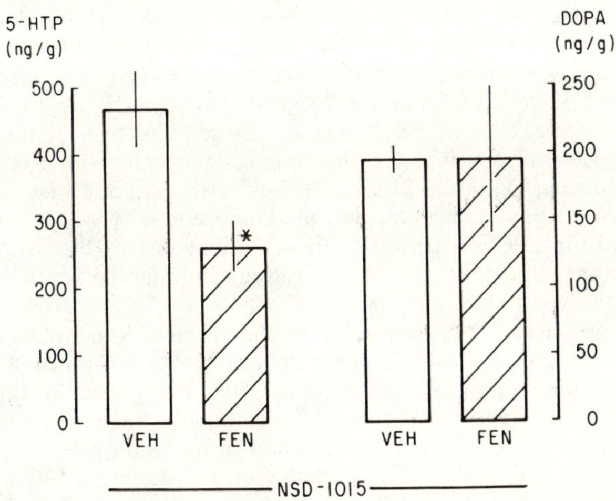

FIGURE 13. 5-HTP and Dopa accumulation. Eight days after administering fenfluramine (FEN, 15 mg/kg, po) or vehicle (VEH), NSD-1015 (150 mg/kg, ip) was given, and the rats were sacrificed 30 min later. Asterisk indicates that $p < 0.05$ (two-tailed *t* test); $N=6$ brains per treatment group.

preventing access of fenfluramine into the brain, inasmuch as fluoxetine still sigifinificantly restored 5-HT when given several hours after fenfluramine had reached its peak level in the brain.

The long-lasting depletion of 5-HT and 5-HIAA caused by fenfluramine could possibly be independent of the continued presence in the brain of either the parent substance or norfenfluramine. As shown herein, 5-hydroxyindoles were still reduced at 64 days after dosing, whereas less than 20 ng/g of fenfluramine and norfenfluramine remained at 2 and 8 days, respectively. Moreover, the permanence of the reversal in 5-HT caused by fluoxetine favors an interpretation other than that of a continued presence. Fluoxetine itself does not appear to have an inordinately long presence in the brain, as judged from the compound's duration of effectiveness in preventing 5-HT depletion when given at various times before *p*-CA.[26] If a persistent small concentration of fenfluramine or norfenfluramine was responsible for maintaining the decrease in 5-HT, the level of 5-HT in brain might be expected to increase for a short time after dosage with fluoxetine and then decline back again as the fluoxetine was eliminated from the brain. However, it could be argued that fluoxetine terminates 5-HT depletion by somehow facilitating the exit of fenfluramine and norfenfluramine from the brain.

Fluoxetine is well established as a selective inhibitor of the amine uptake mechanism present at the level of the serotonergic neuronal membrane,[26] accounting presumably for the compound's ability to reverse the action of fenfluramine. Pretreatment with 5-HT uptake inhibitors of the tricyclic type, for example, chlorimipramine, prevents the decrease in 5-HT caused by fenfluramine.[6] The structural dissimilarity of fluoxetine and chlorimipramine points to their common pharmacologic property of inhibiting the amine uptake system as the basis for antagonizing fenfluramine. The continued uptake of fenfluramine (or norfenfluramine) for several hours appears necessary for the drug to produce a long-term reduction in 5-HT. Fenfluramine resembles other analogs of amphetamine in this respect, because 5-HT was reported to return to normal levels when fluoxetine was given 3 hr after *p*-CA[26] or when chlorimipramine was injected 3 hr after *p*-CMA.[27] Fluoxetine induced a small but significant recovery in brain 5-HT when administered up to 24 hr after fenfluramine, whereas at 48 hr, it was no longer effective. Thus, the semipermanent process responsible for the long-term decrease in 5-HT seems to occur mainly during the initial 24 hr after administration and may depend on a sufficient intraneuronal concentration of fenfluramine (or norfenfluramine) being achieved for a critical period of time. Fluoxetine, by blocking the amine pump mechanism responsible for concentrating fenfluramine within the neuron, would interrupt this process.

Our findings favor 5-HT terminals as the primary site of action of fenfluramine, rather than the 5-HT-containing cell bodies.[3] One or 4 days after fenfluramine administration, the decrease in 5-HT was greater in the forebrain (terminal-rich area) than in the posterior midbrain section (cell body-rich region). Nevertheless, there was a diminution in midbrain 5-HT, perhaps indicating an effect on cell bodies as well. However, the samples of midbrain tissue undoubtedly contained some 5-HT terminals in addition to the cell bodies of the B7, B8, and medial portion of the B9 raphe nuclei. The relatively rapid decrease in forebrain 5-HT also points to serotonergic terminals as fenfluramine's main site of action. After extensive electrolytic lesioning of the midbrain cell body area, forebrain 5-HT and synaptosomal uptake activity remain at control values for 1–2 days, showing maximal reductions 8–14 days post-lesioning.[28, 29] In contrast, forebrain 5-HT was decreased maximally (60%) 1 day after administration of fenfluramine. In addition, fenfluramine failed to reduce brain 5-HT in the newborn rat, although a modest decrease in 5-HIAA did occur. Brains of newborn rats possess 5-HT cell bodies but have poorly differentiated terminal fields.[30] However, a plausible alternative interpretation is that the newborn rat might lack the enzymes required for the metabolic formation of an active metabolite of fenfluramine. Our findings with fenfluramine are consistent with previous studies with *p*-CA,[20, 31] indicating a primary action on serotonergic terminals as being responsible for the semipermanent decrease in 5-HT produced by *p*-CA.

The question arises as to whether fenfluramine destroys a fraction of 5-HT-containing terminals, acting similarly to the dihydroxylated tryptamines.[32] Although this possibility cannot be dismissed, it seems unlikely. The recovery of whole brain 5-HT after fenfluramine injection (nearly complete by 64 days) is more rapid than after administration of dihydroxylated tryptamines,[32] but regional analyses could reveal a longer-lasting effect of fenfluramine. Moreover, we did not find supersensitivity to 5-HTP-elicited forepaw clonus in rats treated 2–4 weeks earlier with fenfluramine, but such an effect was observed in animals 2 weeks after 5,6-dihydroxytryptamine administered intraventricularly.[33]

It is not known if the smaller but repeated dosage of fenfluramine used in man will cause a long-term reduction in brain 5-HT. Such an effect might explain, at least partially, tolerance, withdrawal depression, and perhaps other actions of this drug reported in humans.[34]

References

1. Duhalt, J. & C. Verdavainne. 1967. Modification du taux de serotonine cerebrale chez le rat par les trifluoromethylphenyl-2-ethyl-aminopropane (fenfluramine 768s). Arch. Int. Pharmacodyn. Ther. **170:** 276–286.
2. Opitz, K. 1967. Anorexigène phenylalkylamine und serotoninstoffwechsel. Arch. Exp. Pathol. Pharmakol. **259:** 56–65.
3. Harvey, J. A. & S. E. McMaster. 1975. Fenfluramine: evidence for a neurotoxic action on midbrain and a long-term depletion of serotonin. Psychopharmacol. Commun. **1:** 217–228.
4. Clineschmidt, B. V., J. A. Totaro, J. C. McGuffin & A. B. Pflueger. 1976. Fenfluramine: long-term reduction in brain serotonin (5-hydroxytryptamine). Eur. J. Pharmacol. **35:** 211–214.
5. Carlsson, A. 1970. Structural specificity for inhibition of [^{14}C]-5-hydroxytryptamine uptake by cerebral slices. J. Pharm. Pharmacol. **22:** 729–732.
6. Ghezzi, D., R. Samanin, S. Bernasconi, G. Tognoni, M. Genra & S. Garattini. 1973. Effect of thymoleptics on fenfluramine-induced depletion of brain serotonin in rats. Eur. J. Pharmacol. **24:** 205–210.
7. Garattini, S., W. Buczko, A. Jori & A. Samanin. 1975. The mechanism of action of fenfluramine. Postgrad. Med. J. **51**(Suppl. 1): 27–35.
8. Fuxe, K., L. O. Farnebo, B. Hamberger & S. O. Ogren. 1975. On the *in vivo* and *in vitro* actions of fenfluramine and its derivatives on central monoamine neurons, especially 5-hydroxytryptamine neurons, and their relation to the anorectic activity of fenfluramine. Postgrad. Med. J. **51**(Suppl. 1): 35–45.
9. Aghajanian, G. K., M. J. Kuhar & R. H. Roth. 1973. Serotonin-containing neuronal perikarya and terminals: differential effects of *p*-chlorophenylalanine. Brain Res. **54:** 85–101.

10. Bogdanski, D. F., A. Pletscher, B. B. Brodie & S. Udenfriend. 1956. Identification and assay of serotonin in brain. J. Pharmacol. Exp. Ther. **117:** 82–88.
11. Giacalone, E. & L. Valzelli. 1966. A method for the determination of 5-hydroxyindolyl-3-acetic acid in brain. J. Neurochem. **13:** 1265–1266.
12. Maickel, R. P. & F. P. Miller. 1966. Fluorescent products formed by reaction of indole derivatives and o-phthalaldehyde. Anal. Chem. **38:** 1937–1938.
13. Haubrich, D. R. & J. S. Denzer. 1973. Simultaneous extraction and fluorometric measurement of brain serotonin, catecholamines, 5-hydroxyindoleacetic acid and homovanillic acid. Anal. Biochem. **55:** 306–312.
14. Thompson, J. H., C. A. Spezia & M. Angulo. 1970. Fluorometric detection of serotonin using o-phthaldialdehyde: an improvement. Experientia **26:** 327–329.
15. Carlsson, A. & M. Lindqvist. 1973. In-vivo measurements of tryptophan and tyrosine hydroxylase activities in mouse brain. J. Neural Trans. **34:** 79–91.
16. Westerink, B. H. C. & J. Korf. 1975. Determination of nanogram amounts of homovanillic acid in the central nervous system with a rapid semiautomated fluorometric method. Biochem. Med. **12:** 106–115.
17. Atack, C. V. & T. Magnusson. 1970. Individual elution of noradrenaline (together with adrenaline), dopamine, 5-hydroxytryptamine, and histamine from a single, strong cation exchange column, by means of mineral acid-organic solvent mixtures. J. Pharm. Pharmacol. **22:** 625–627.
18. Anton, A. H. & D. F. Sayre. 1962. A study of the factors affecting the aluminum oxide-trihydroxyindole procedure for the analysis of catecholamines. J. Pharmacol. Exp. Ther. **138:** 360–375.
19. Porter, C. C., J. A. Totaro & A. Burcin. 1965. The relationship between radioactivity and norepinephrine concentrations in the brains and hearts of mice following administration of labeled methyldopa or 6-hydroxydopamine. J. Pharmacol. Exp. Ther. **150:** 17–22.
20. Sekerke, H. J., H. E. Smith, J. A. Bushing & E. Sanders-Bush. 1975 Correlation between brain levels and biochemical effects of the optical isomers of p-chloroamphetamine. J. Pharmacol. Exp. Ther. **193:** 835–844.
21. Jequier, E., W. Lovenberg & A. Sjoerdsma. 1967. Tryptophan hydroxylase inhibition: the mechanism by which p-chlorophenylalanine depletes rat brain serotonin. Mol. Pharmacol. **3:** 274–278.
22. Sanders-Bush, E., J. A. Bushing & F. Sulser. 1972. Long-term effects of p-chloroamphetamine on tryptophan hydroxylase activity and on the levels of 5-hydroxytryptamine and 5-hydroxyindole acetic acid in brain. Eur. J. Pharmacol. **20:** 385–388.
23. Campbell, D. B. 1970. Gas chromatographic measurement of levels of fenfluramine and norfenfluramine in human plasma, red cells and urine following therapeutic doses. J. Chromatogr. **49:** 442–447.
24. Bruce, R. B. & W. R. Maynard. 1968. Fenfluramine metabolism. J. Pharm. Sci. **57:** 1173–1176.
25. Morgan, C. D., F. Cattabeni & E. Costa. 1972. Methamphetamine, fenfluramine and their N-dealkylated metabolites: effect on monoamine concentrations in rat tissues. J. Pharmacol. Exp. Ther. **180:** 127–135.
26. Fuller, R. W., K. W. Perry & B. B. Molloy. 1975. Effect of 3-(p-trifluoromethylphenoxy)-N-methyl-3-phenylpropylamine on the depletion of brain serotonin by 4-chloroamphetamine. J. Pharmacol. Exp. Ther. **193:** 796–803.
27. Meek, J. A., K. Fuxe & A. Carlsson. 1971. Blockade of p-chloromethamphetamine induced 5-hydroxytryptamine depletion by chlorimipramine, chlorpheniramine and meperidine. Biochem. Pharmacol. **20:** 707–709.
28. Kuhar, M. J., R. H. Roth & G. K. Aghajanian. 1972. Synaptosomes from forebrains of rats with midbrain raphe lesions: selective reduction of serotonin uptake. J. Pharmacol. Exp. Ther. **181:** 36–45.
29. Kuhar, M. J., R. H. Roth & G. K. Aghajanian. 1971. Selective reduction of tryptophan hydroxylase activity in rat forebrain after midbrain raphe lesions. Brain Res. **35:** 167–176.

30. LOIZOU, L. A. 1972. The postnatal ontogeny of monoamine-containing neurones in the central nervous system of the albino rat. Brain Res. **40:** 395–418.
31. NECKERS, L. M., L. BERTILSSON, S. H. KOSLOW & J. L. MEEK. 1976. Reduction of tryptophan hydroxylase activity and 5-hydroxytryptamine concentration in certain rat brain nuclei after p-chloroamphetamine. J. Pharmacol. Exp. Ther. **196:** 333–338.
32. BAUMGARTEN, H. G., A. BJÖRKLUND, L. LACHENMAYER & A. NOBIN. 1973. Evaluation of the effects of 5,7-dihydroxytryptamine on serotonin and catecholamine neurons in the rat cns. Acta Physiol. Scand. **89**(Suppl. 391): 1–19.
33. CLINESCHMIDT, B. V. & J. C. MCGUFFIN. 1977. Unpublished results.
34. PINDER, R. M., R. N. BROGDEN, P. R. SAWYER, T. M. SPEIGHT & G. S. AVERY. 1975. Fenfluramine: a review of its pharmacological properties and therapeutic efficacy in obesity. Drugs **10:** 241–323.

DISCUSSION

DR. E. M. GÁL: What was the method of differentiating between norfenfluramine and fenfluramine?

DR. CLINESCHMIDT: The method involved gas chromatography, and specificity was confirmed by mass spectrometry.

DR. D. V. COSCINA: The lack of effect of fenfluramine in neonates might be due to the fact that serotonin nerve terminals are not developed, so that adequate uptake of the drug does not occur. Another possibility is that neonatal rats don't dealkylate fenfluramine to norfenfluramine.

DR. R. W. FULLER: *p*-Chloroamphetamine also doesn't deplete serotonin in the neonatal rat brain, and it doesn't require dealkylation.

DR. J. MCGINTY (*Downstate Medical Center, Brooklyn, N.Y.*): Have you tried giving norfenfluramine to the neonate?

DR. CLINESCHMIDT: No, but that would now be an obvious thing to try.

DR. G. R. BREESE: Have you determined whether fluoxetine (Lilly 110,140) blocks the anorexic effects of fenfluramine?

DR. CLINESCHMIDT: We haven't, but chlorimipramine has been reported to antagonize the anorexic activity of fenfluramine.

EFFECT OF CENTRAL NERVOUS SYSTEM-ACTING DRUGS AFTER SELECTIVE DESTRUCTION BY NEUROTOXINS OF 5-HYDROXYTRYPTAMINE FIBERS IN THE BRAIN

Michele O. Carruba

Department of Pharmacology
School of Medicine
University of Milan
20129 Milan, Italy

Giuseppe Nisticò and Giuliana Gargiulo

Department of Pharmacology
School of Medicine
University of Messina
98100 Messina, Italy

There have been many attempts to relate brain monoaminergic function to behavioral changes.

Several observations favor the hypothesis that 5-hydroxytryptamine (5-HT) may be involved in the central regulation of body temperature and sleep. In particular, Feldberg and Myers [1–3] reported that intraventricular injections of norepinephrine (NE) and 5-HT produced hypothermia and hyperthermia, respectively, in unanesthetized cats. On the basis of these results, they postulated that control of body temperature depends on the balanced release of these amines from the hypothalamus, which has been recently confirmed in rabbits by Shellemberger and Elder,[4] who also suggested a role for dopamine (DA) in thermoregulation. This neurotransmitter would act like 5-HT to increase body temperature, while NE would mediate the lowering of body temperature.

5-Hydroxytryptamine also elevated and catecholamines (CA) lowered body temperature in the chicken.[5,6] The sites of the effects of CA and 5-HT on body temperature have been localized in the hypothalamus in this animal species also.[7–10] In fowl, 5-HT given intraventricularly or infused into the hypothalamus, in addition to increasing body temperature, induced behavioral and electrocortical sleep.[9]

In recent years, evidence has been accumulated that interference with neurohumoral transmitters in the brain is at least partly responsible for the action of many of the modern psychotropic drugs.

The discovery of compounds able to produce a selective degeneration in monoaminergic pathways in the central nervous system (CNS) provides a new and useful tool for study of the functional role played by monoamines and the mechanism of action of many CNS-acting drugs.

Hydroxylated derivatives of tryptamine, such as 5,6-dihydroxytryptamine (5,6-DHT), when given intraventricularly, have been found to be capable of producing a rather selective degeneration of central indoleaminergic axons in rats [11–13] and fowl.[14] The experiments reported here have been designed to evaluate the involvement of the serotonergic system in mediating the hyperthermic response to some drugs. This goal has been accomplished either by

0077–8923/78/0305–0242 $01.75/0

blockade of 5-HT synthesis in the CNS with *p*-chlorophenylalanine (*p*-CPA) or by selective destruction of serotonergic nerve terminals with 5,6-DHT. In addition, evidence is also presented that supports a direct action of 5,6-DHT on central 5-HT receptors and the existence of a receptor supersensitivity to 5-HT at the time of the maximal degeneration of indoleaminergic neurons produced by 5,6-DHT.

Materials and Methods

Animals

Male albino New Zealand rabbits, 2.5–3.0 kg in body weight, and adult Rhode Island red chickens, 1.5–2.0 kg in body weight, were used. The animals were housed under controlled conditions (22 ± 2° C, 65% humidity, and 14 hr per day of artificial light, 6 AM to 8 PM).

Operative and Test Procedures in Rabbits

Intracerebroventricular injections of 5,6-DHT were made by a direct puncture technique under ether anesthesia; a small hole was drilled in the skull, 2 mm lateral to the midline and 1 mm rostral to the bregma, a 6.5 mm long needle mounted on a Hamilton microsyringe was inserted vertically, and a volume of 25 μl was injected into each ventricle. To prevent leakage of injected fluid, the needle was left in place for 1 min after the injection, and a drop of paste (Histoacryl-N-blau B®, Braun Melsungen) was placed around the needle immediately before its extraction.

Ten days before the experiment, the animals were pretreated with 5,6-DHT (50 μg of base into each cerebral ventricle, dissolved in 0.9% NaCl that contained 0.1% of ascorbic acid). *p*-Chlorophenylalanine dissolved in sesame oil was administered subcutaneously for three times, at 100 mg/kg on alternate days the last dose being administered 48 hr before the experiment.

All of the experiments were conducted in a temperature-controlled environment at 22.0 ± 1.0° C. Animals were restrained in open wooden stanchions, to which they had previously been conditioned.

Changes in rectal temperature were recorded continuously by a thermistor probe inserted 5 cm into the rectum and connected with a temperature recorder (Texas Instruments Inc.).

In some cases, temperature was recorded on coordinate paper by use of scales of 10 cm/° C on the ordinate and 8 cm/hr on the abscissa. The area between a mean control temperature and each fever curve over a 3-hr period after drug administration was determined by planimetric integration and expressed as the fever index (FI) in square centimeters. In this way, an FI was obtained for each animal, an expression of both the degree and duration of the fever.[15]

In cases in which there was no fever, the algebraic sum of the fluctuations above and below the baseline was calculated, giving a negative value in some instances.

Operative and Test Procedures in Fowl

Anesthesia with halothane, stereotactic implantation of infusion cannulae into both the third ventricle and the hypothalamus, brain coordinates for these cannulae, the procedure for implantation of a thermistor beneath the feathered skin of the interscapular region, and postoperative care were as previously described.[8] Methods for preparation and implantation of cortical electrodes for recording body temperature in a test chamber, with ambient temperature maintained near the middle of the thermoneutral range (17 and 26° C being the lower and upper thermoneutral limits, respectively, for adult fowl given by Barrot and Pringle[16] and King and Farner[17]), have also been described previously, as have the methods for infusions via the implanted cannulae (infusions into the hypothalamus of 1.0 μl or less and into the third cerebral ventricle of 10 μl or less) and for histologic preparations of the brains. Fowl were not tested until at least 1 week after recovery from operative procedures and thereafter at intervals of at least 1 week. They were placed in the test chamber between 9 and 10 AM, and control body temperature was recorded for 1 hr before drugs were given. Fowl were treated with a single injection of 5,6-DHT into the third cerebral ventricle. The dosage was 75 μg (calculated as free base).

Monoamine Determinations in Rabbits

On the day of the experiment, at 9 AM, some of the animals, either pretreated with 5,6-DHT or *p*-CPA or without pretreatment, were killed by decapitation, the brains were removed, and the hypothalami were dissected out. Catecholamines and 5-HT were estimated fluorometrically in the hypothalamus and residual brain by the method of Shellemberger and Gordon.[18]

Histofluorescence

Catecholamines and indoleamines were visualized by a histofluorescence technique.[19, 20]

Immediately after killing, the dissected brain areas were rapidly frozen in isopentane cooled by liquid nitrogen and dried *in vacuo* at −40° C for 5 days in an Edwards-Pearse apparatus. The specimens were then treated with formaldehyde gas (30% humidity) for 1 hr at 80° C and embedded in soft paraffin *in vacuo*. The sections were observed with a Leitz fluorescence microscope with an HBO 200-W high-pressure mercury lamp. Excitation (BG 38 and BG 3 + S405) and barrier (Leitz K 510 and 530) filters were used. Sections stained with cresylviolet and Bodian gold were used as controls.

Statistics

Data are presented as means ± standard errors (SE) and analyzed statistically by use of Student's one-tailed *t* test.

Drugs

The following drugs were used: 5,6-dihydroxytryptamine creatinine sulfate (Regis), *p*-chlorophenylalanine (synthesized by Dr. P. Di Vittorio, Istituto

Sieroterapico Milanese), *d*-amphetamine sulfate (Recordati), apomorphine (Sandoz), *dl*-fenfluramine (Farmos Group Medipolar Oy), phenylethylhydrazine (Angiolini), 5-hydroxytryptophan (Regis), 5-hydroxytryptamine hydrogen maleinate (Fluka), mebanazine oxalate (Actomol), and Pyrifer® VII (Aristopharm).

RESULTS

Effect of p-*CPA or 5,6-DHT Pretreatment on the Hyperthermia Induced in the Rabbit by* d-*Amphetamine, Apomorphine, and Bacterial Pyrogen*

p-Chlorophenylalanine or 5,6-DHT treatment per se did not alter the normal body temperature of the rabbits; in fact, on the day of the experiment, the basal

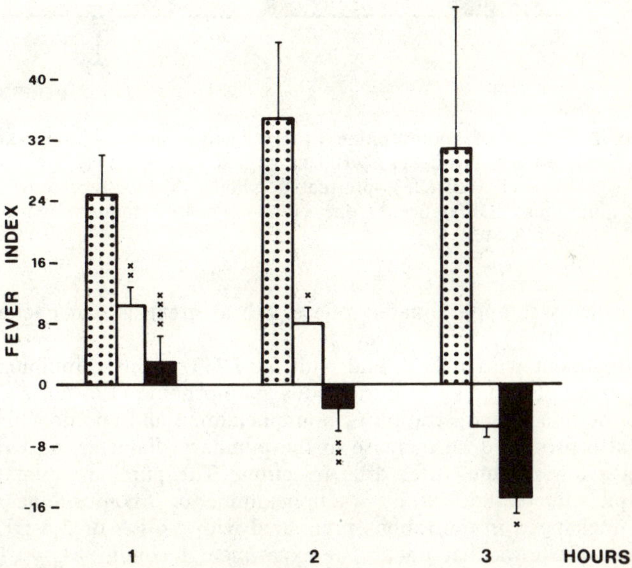

FIGURE 1. Effects of *d*-amphetamine (2 mg/kg; iv) on body temperature of saline (dotted bars), *p*-CPA (white bars, 100 mg/kg; sc, three times), or 5,6-DHT (black bars, 50 μg into each ventricle) pretreated rabbits. Values represent means±SE of 15 determinations. Differences from saline-pretreated rabbits: x, $p \leqslant 0.05$; xx, $p \leqslant 0.02$; xxx, $p \leqslant 0.01$; xxxx, $p \leqslant 0.001$; Student's *t* test.

temperature of the control animals was 40.2° C, while animals that received *p*-CPA or 5,6-DHT had mean values of 40.1 and 39.8° C, respectively ($p > 0.05$). As shown in FIGURE 1, in the animals pretreated with *p*-CPA, an intravenous dose of 2 mg/kg of *d*-amphetamine increased body temperature with a FI at 1 hr of approximately 10 cm², a significantly smaller effect than that observed in control animals. The animals pretreated with 5,6-DHT behaved like those pretreated with *p*-CPA in their hyperthermic response to *d*-amphet-

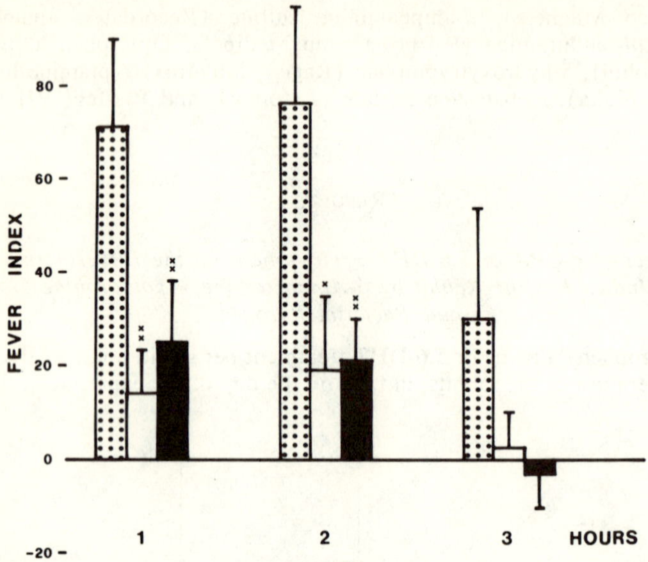

FIGURE 2. Effects of apomorphine (4 mg/kg, iv) on body temperature of saline (dotted bars), *p*-CPA (white bars, 100 mg/kg, sc, three times), or 5,6-DHT (black bars, 50 μg into each ventricle) -pretreated rabbits. Values represent means $\pm$ SE of six determinations. Differences from saline-pretreated rabbits: x, $p \leqslant 0.01$; xx, $p \leqslant 0.001$; Student's *t* test.

amine, which was approximately one eighth as great as that elicited in control animals (FIGURE 1).

Pretreatment with *p*-CPA and with 5,6-DHT strongly inhibited the hyperthermic effect of 4 mg/kg of intravenous apomorphine (FIGURE 2). In addition to hyperthermia, administration of *d*-amphetamine and apomorphine to control rabbits also produced an increase in the pupillary diameter, and rabbits began gnawing a few minutes after drug injection. The pupillary dilatation and the stereotypic effects induced by *d*-amphetamine or apomorphine appeared to remain unchanged in the rabbits pretreated with *p*-CPA or 5,6-DHT.

FIGURE 3 shows that under our experimental conditions, *p*-CPA and 5,6-DHT pretreatments were unable to modify the febrile response induced by intravenous injection of 0.0005 ml/kg of Pyrifer VII (which contains pyrogenic substances from nonpathogenic bacteria).

Effect of p-CPA *or 5,6-DHT Pretreatment on the Hyperthermia Induced in Rabbits by 5-HTP plus a MAO Inhibitor or by Fenfluramine*

In normal animals, an intravenous dose of 2 mg/kg of 5-hydroxytryptophan (5-HTP) given in combination with a monoamine oxidase (MAO) inhibitor (phenylethylhydrazine, 10 mg/kg, iv, 16 hr before) induced progressive and significant rises in body temperature; the animals always died at the peak of the temperature response, 232 ± 27 min after the injection (FIGURE 4). In 5,6-DHT pretreated animals, the same dose of 5-HTP plus the MAO inhibitor induced earlier and more rapid increases in body temperature; the rabbits died within 125 ± 18 min after the injection (FIGURE 4).

Rabbits pretreated with *p*-CPA showed at the time of 5-HTP injection a basal temperature 1° C lower than that of the control animals, and the hyperthermic action of 5-HTP (given 16 hr after the MAO inhibitor) was markedly attenuated (FIGURE 4). None of the rabbits died in hyperpyrexia. Thus, *p*-CPA completely protected the rabbits against the fatal hyperthermia induced by 5-HTP in combination with MAO inhibitor.

Pretreatment with either *p*-CPA or 5,6-DHT caused a significant decrease in the hyperthermic response to fenfluramine (5 mg/kg, iv; FIGURE 5).

Effect of p-*CPA and 5,6-DHT on Rabbit Brain Monoamine Concentrations*

The levels of the amines were determined in the hypothalamus and in the remaining brain; the results are shown in TABLE 1. 5,6-Dihydroxytryptamine reduced the 5-HT content in the hypothalamus and the remaining brain to about 60% of the control value, whereas norepinephrine and dopamine levels were only slightly modified.

The central 5-HT depletion induced by *p*-CPA treatment was also specific for 5-HT and even more pronounced (30–40% remaining).

Acute Effects of 5-HT and 5,6-DHT in Fowl

Infusion of 5-HT into the third cerebral ventricle induced immediate behavioral and electrocortical sedation and/or slow-wave sleep that lasted 6–8 hr

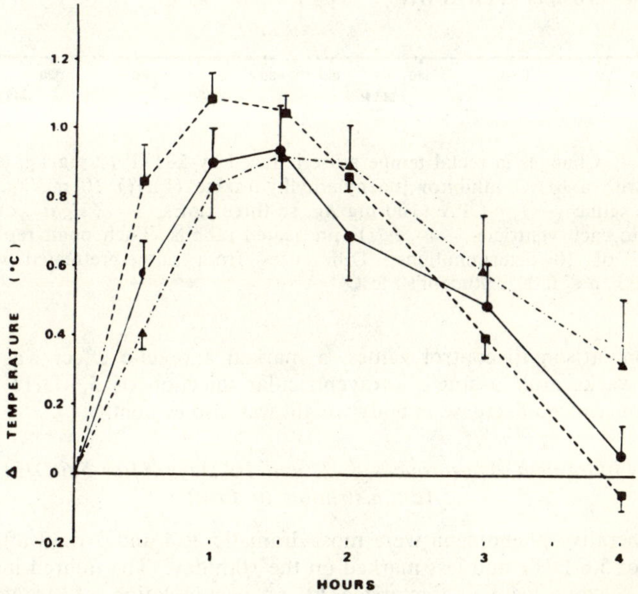

FIGURE 3. Changes in rectal temperature induced by Pyrifer VII (0.0005 ml/kg, iv) in saline (●), *p*-CPA (100 mg/kg, sc, three times, ■), or 5,6-DHT (50 μg into each ventricle, ▲), -pretreated rabbits. Each point represents the mean±SE of 12 determinations.

and that was accompanied by an increase in body temperature (maximum increase 0.8–1.3° C).[9]

Intraventricular infusion of a single dose (75 μg) of 5,6-DHT caused behavioral, electrocortical, and body temperature effects identical to those obtained in the same species after intraventricular infusion of 5-HT.

In some animals, an increase in locomotor activity with attempts to escape from the box occurred.

In addition, when two fowl were in the same box, aggressive behavior was elicited, culminating in several episodes of fighting during the first hour, after which it was necessary to keep the hens in separate boxes.

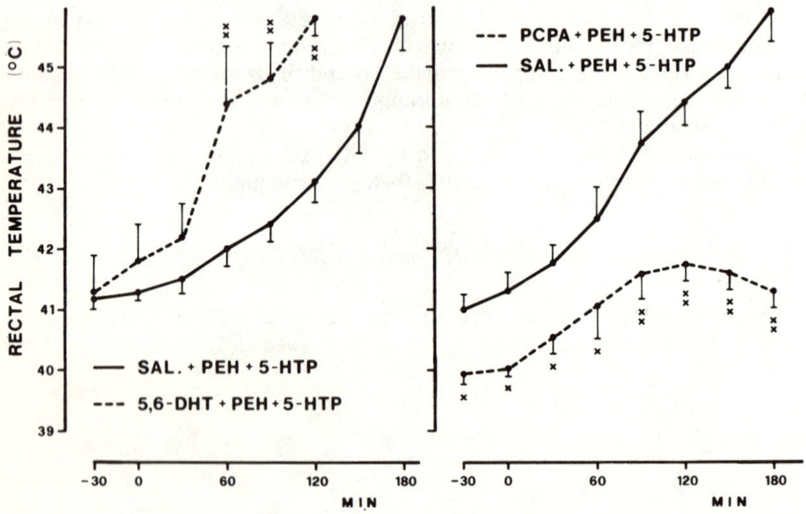

FIGURE 4. Changes in rectal temperature induced by 5-HTP (2 mg/kg, iv) in combination with a MAO inhibitor [phenylethylhydrazine (PEH) 10 mg/kg, iv, 16 hr before] in saline (—), *p*-CPA (100 mg/kg, sc three times, – – –, *right*), or 5,6-DHT (50 μg into each ventricle, – – –, *left*) -pretreated rabbits. Each point represents the mean±SE of 10 determinations. Differences from saline-pretreated rabbits: x, $p \leq 0.05$; xx, $p \leq 0.001$; Student's *t* test.

In comparison to control values, a marked anorectic effect was observed 1 and 2 weeks after a single intraventricular injection of 5,6-DHT (75 μg). At the same time, a decrease in body weight was also evident.

Fluorescent Appearances 4, 7, and 10 Days After 5,6-DHT Administration to Fowl

Degenerative phenomena were most dramatic at 4 and 7 days after administration of 5,6-DHT and less marked on the 10th day. The injured indoleaminergic (Ia) axons were characterized by an accumulation of strongly yellow fluorescent material that enabled them to be observed by histofluorescence, and the lesions were most conspicuous in the areas adjacent to the injection cannula.

TABLE 1

EFFECT OF *p*-CPA (100 MG/KG, SC, THREE TIMES) OR 5,6-DHT (50 μG INTO EACH VENTRICLE) ON MONOAMINE CONCENTRATIONS IN THE HYPOTHALAMUS AND RESIDUAL BRAIN OF THE RABBIT

	5-Hydroxytryptamine (μg/g±SE)		Norepinephrine (μg/g±SE)		Dopamine (μg/g±SE)	
Treatment	Hypothalamus	Rest of the Brain	Hypothalamus	Rest of the Brain	Hypothalamus	Rest of the Brain
Controls	0.97±0.048*	0.46±0.026	1.20±0.050	0.30±0.20	0.23±0.010	0.21±0.010
p-CPA	0.29±0.010†	0.19±0.006†	1.07±0.171	0.22±0.008‡	0.23±0.057	0.18±0.009
5,6-DHT	0.58±0.001†	0.30±0.020†	1.10±0.120	0.25±0.009	0.28±0.044	0.18±0.010

* Each value is the mean±SE of six determinations.
† $p \leqslant 0.05$ relative to controls.
‡ $p \leqslant 0.001$ relative to controls.

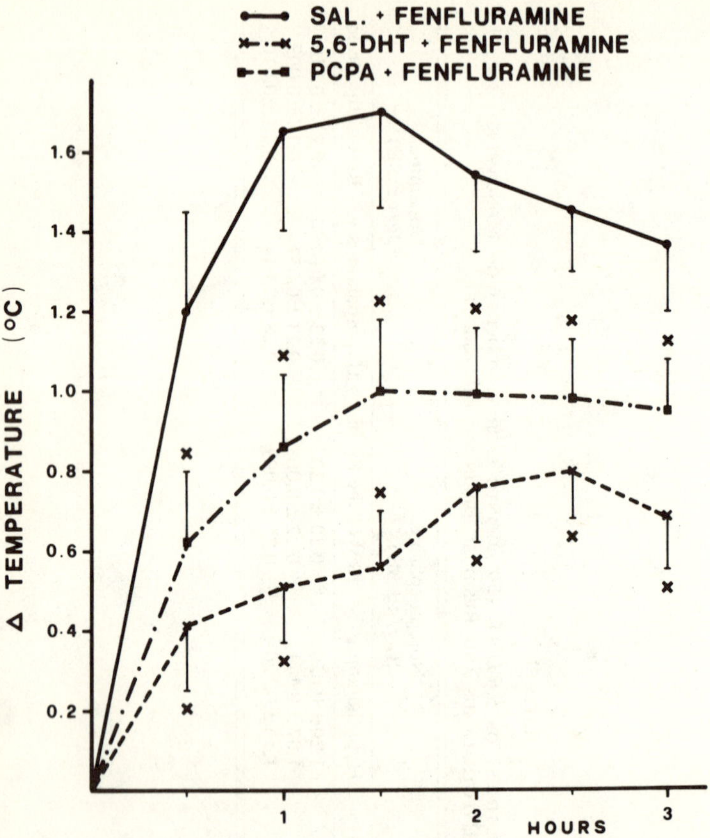

FIGURE 5. Changes in rectal temperature induced by fenfluramine (5 mg/kg, iv) in saline (●), *p*-CPA (100 mg/kg, sc, three times, x), or 5,6-DHT (50 μg into each ventricle, ■) -pretreated rabbits. Each point represents the mean±SE of eight to 10 determinations. Difference from saline-pretreated rabbits: x, $p \leqslant 0.01$; Student's *t* test.

In particular, 4 and 7 days after 5,6-DHT injection, highly distorted yellow fluorescent fibers, associated with some green fibers and terminals, were observed along the areas bordering the third ventricle, in the posterior, median, and anterior hypothalamus, the preoptic area (FIGURE 6) the median eminence, the nucleus tuberalis, and in the paleostriatum augmentatum. In addition, droplet-like swellings that contained intense yellow-orange material were observed in an area that comprised the paleostriatum augmentatum, preoptic area, infundibulum, hypothalamus (FIGURE 6), and mesencephalon, along the medial forebrain bundle and just beside the midline.

Control Bodian gold-impregnated preparations showed obvious signs of degenerated fibers in the same areas in which droplet-like swellings were seen. In addition to lesions in Ia neurons, 5,6-DHT produced a less intense involvement of CA neurons, as shown by the increase in green fluorescence in the areas adjacent to the third ventricle. An increase in yellow fluorescence more marked than that obtained after MAO inhibition was also demonstrated in terminals and cell bodies of the paraventricular organ, nucleus hypothalamicus posterior medialis (HPM), nucleus tuberalis, and preoptic region terminals (FIGURE 6).

Effects of 5-HT and Fenfluramine in Fowl After Pretreatment with 5,6-DHT

In animals pretreated with a single intraventricular dose of 5,6-DHT, the subsequent (1 week later) infusion into the third ventricle of 5-HT hydrogen maleinate (0.75 μmol) produced a more intense and longer-lasting body temperature increase than that evoked in control animals by the same doses of 5-HT given alone (TABLE 2). However, in some animals, a first phase of slight hypothermia (−0.5° C), which lasted 1–2 hr, preceded the longer-lasting (>4 hr) hyperthermic response.

The 5-HT effect on body temperature in 5,6-DHT-pretreated fowl was accompanied by changes in electrocortical activity; for example, during behavioral sleep or in the sedation state, higher-amplitude slower-frequency potentials were observed with increases in electrocortical integrals as compared to the effects of 5-HT in normal animals. These effects lasted more than 4 hr.

Fenfluramine given into the third cerebral ventricle (0.75 μmol) produced behavioral sedation that lasted about 60 min and an increase in body temperature that lasted about 90 min, accompanied by tachypnea (respiratory rate about 110/min). In comparison to these effects, infusion of the same dose of fenfluramine 1 week after a single intraventricular injection of 5,6-DHT (75 μg) had no effect on body temperature (TABLE 2) or behavior.

DISCUSSION

It has been established that various *d*-amphetamine effects, such as anorexia, hypermotility, compulsive gnawing, and other stereotypic behaviors, are mediated by the central dopaminergic system.[21–26]

The hyperthermia produced in rabbits by *d*-amphetamine has been postulated to be due to an action exerted on the CNS.[27] The indications for this central action include the findings that *d*-amphetamine-induced hyperthemia in this animal species is markedly reduced by prior curarization[28] or spinal section[29] but is not diminished by blockade of β- or peripheral α-adrenergic recep-

TABLE 2

EFFECTS OF INTRAVENTRICULAR 5-HT AND FENFLURAMINE ON BODY TEMPERATURE IN CONTROL AND 5,6-DHT-PRETREATED FOWL

Treatment	Mean Value of Maximum Increase in Temperature (±SE)	Mean Value of the Duration of the Increase (min±SE)	Statistical Analysis (p)
5-Hydroxytryptamine alone	1.13±0.11 (6)*	151.6±19.7 (6)	
One week after 5,6-DHT	1.96±0.12 (6)	350.0±23.9 (6)	0.001
Fenfluramine alone	0.73±0.08 (6)	93.3±13.8 (6)	
One week after 5,6-DHT	0.28±0.07 (6)	14.1± 3.7 (6)	⩽0.001

* Number of experiments is shown in parentheses.

tors.[29] Other evidence suggested that *d*-amphetamine may induce hyperthermia by influencing the central dopaminergic system. For instance, several of the neuroleptic drugs antagonize *d*-amphetamine-induced hyperthermia in rats.[30] These drugs were later shown to be potent inhibitors of central dopaminergic receptors.[31]

That low doses of *d*-amphetamine could elevate in the rat both body temperature and the turnover rate of brain DA without altering the turnover rate of brain NE was a further finding that implicated DA as the neurotransmitter involved in *d*-amphetamine-induced hyperthermia.[32] Finally, after the observation that pimozide, a selective dopaminergic antagonist,[31] counteracted the hyperthermic effect of *d*-amphetamine,[33] there was no doubt that *d*-amphetamine may enhance DA receptor activity by releasing DA from central neurons and consequently inducing hyperthermia.

Apomorphine is another drug known to stimulate central dopaminergic receptors.[34, 35] This drug, like *d*-amphetamine, is able to induce hypothermia or hyperthermia according to species,[36, 37] together with increases in motor activity,[38] compulsive gnawing, and other stereotypic behaviors.[39–41] The hyperthermic and other behavioral effects of apomorphine have been established as being dopaminergic in nature.[37]

The results reported herein provide evidence that, in addition to DA, an essential role is played by 5-HT in the mediation of the hyperthermic response of rabbits to *d*-amphetamine and apomorphine. This conclusion is based on the finding that after depletion of central 5-HT stores by either blockade of 5-HT synthesis through *p*-CPA or selective destruction of serotonergic nerve terminals by 5,6-DHT, the increase in rectal temperature induced by either *d*-amphetamine or apomorphine was inhibited.

A similar observation has been made by Quock and Horita,[42] who reported that prior treatment with *p*-CPA greatly impaired the hyperthermic effect of apomorphine.

Among the several possibilities, the requirement of the functional integrity of the serotonergic system for explication of the hyperthermic effect of the two dopaminergic agonists may be explained by the existence of a link in the CNS between the DA and 5-HT systems whereby stimulation of the dopaminergic neurons results ultimately in activation of 5-HT fibers involved in the hypothalamic regulation of body temperature.

It has been reported that depletion of 5-HT by *p*-CPA fails to inhibit *d*-

FIGURE 6. Degenerative effects of indoleaminergic neurons in the preoptic area-anterior hypothalamus and in areas bordering the third cerebral ventricle of adult fowl. Control experiments: **a.** Transverse section from the preoptic area after pretreatment with a MAO inhibitor, mebanazine (100 μmol/kg, 18 and 1 hr before). Intensely green-yellow fluorescent varicosities. **b.** Transverse section from the hypothalamic-periventricular area in an adult fowl pretreated with mebanazine. After 5,6-DHT: **c.** Transverse section from the preoptic area 7 days after a single intraventricular injection of 5,6-DHT (75 μg). Dramatic signs of degeneration of indoleaminergic axons shown by the conspicuous accumulation of yellow fluorescence in highly distorted and dilated axons. **d.** Similar degenerative phenomena in the periventricular area 7 days after a single intraventricular injection of 5,6-DHT. **e.** Transverse section from the anterior hypothalamus shows numerous intensely yellow fluorescent large-fiber swellings 7 days after a single intraventricular injection of 5,6-DHT. **f.** Highly distorted and dilated yellow fluorescent fiber in an area adjacent to the third cerebral ventricle 7 days after a single intraventricular injection of 5,6-DHT.

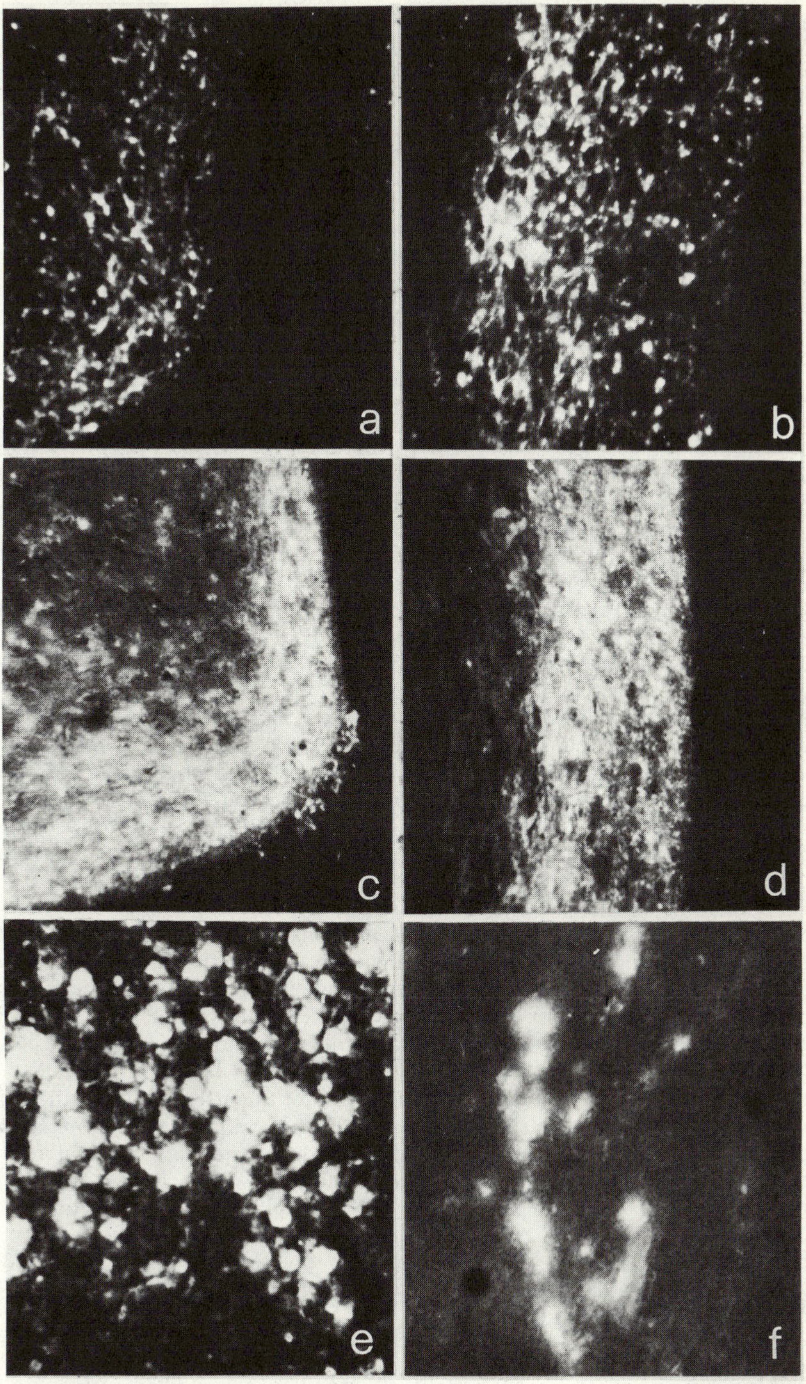
a
b
c
d
e
f

amphetamine-induced stereotypic and gnawing behavior in rats.[43] In agreement with these data, in our study, the gnawing behavior caused by either *d*-amphetamine or apomorphine in rabbits appeared to be independent of central 5-HT. In fact, selective reduction of brain 5-HT levels, induced by *p*-CPA or by 5,6-DHT pretreatment, did not modify significantly this behavioral effect of *d*-amphetamine and apomorphine.

Conflicting reports have appeared on the role played by 5-HT in the febrile response induced by pyrogens.

Giarman *et al.*[44] and Mašek *et al.*[45] have suggested that, in rabbits, the release of 5-HT may limit the fever induced by pyrogens. Other authors claimed that 5-HT is necessary for pyrogen-induced fever.[46–48]

Furthermore, Jacob *et al.*[49] and Peindaries and Jacob[50] reported that 5-HT may have opposite effects on the thermic response in rabbits and pointed out how this dualistic feature of the central action of 5-HT may solve the apparent discrepancy.

In the present work, we have investigated whether the functional impairment of central 5-HT activity affects the hyperthermic action of pyrogens. We found that, in contrast to the results obtained with *p*-CPA by Giarman *et al.*[44] and Mašek *et al.*,[45] either *p*-CPA or 5,6-DHT pretreatment was unable to modify the hyperthermic response induced by Pyrifer VII. This finding contradicts the hypothesis that 5-HT plays an important role in the mediation of pyrogen-induced fever.

The data here reported of an enhanced sensitivity to the hyperthermic effect of 5-HTP in combination with a MAO inhibitor in rabbits pretreated with 5,6-DHT may be taken as evidence for a denervation supersensitivity after 5-HT nerve ending degeneration.

5-Hydroxytryptophan administered intravenously reaches the brain and is decarboxylated not only in serotonergic but also in catecholaminergic neurons and, in addition, to a great extent also in capillaries,[51] so that brain 5-HT concentrations may be elevated after 5-HTP administration, even when intact serotonergic terminals are lacking.[52] In this condition, 5-HT formed from 5-HTP may be taken up by receptors,[53] so that the pharmacologic response observed is the result of an activation of postsynaptic 5-HT receptors.

Similar observations of increased sensitivity to 5-HTP after injection of 5-HT neurotoxins have been reported previously. For instance, in animals pretreated with 5,6-DHT or 5,7-DHT, there was a potentiation of the induction of tremor by 5-HTP in mice.[54] The inhibitory effect of 5-HTP on the extensor hind limb reflex in spinal rats,[55] the depressant effect of 5-HTP on a fixed-ratio bar press response,[56] and the myoclonus induced by 5-HTP[57] were all potentiated in animals previously treated with 5,6-DHT or 5,7-DHT.

Trulson *et al.*[58] hypothesized the existence of supersensitivity in the serotonergic system after destruction of central serotonergic nerve terminals on the basis of the observed shift to the left in a dose-response curve that related the dose of 5-methoxy-*N*,*N*-dimethyltryptamine (a direct acting 5-HT agonist[59]), 5-HTP, or tryptophan (plus a MAO inhibitor) to elicitation of a complex behavioral syndrome that is a specific reflection of the activity of central 5-HT receptors. These authors also pointed out a similarity and a difference between the behavior of the serotonergic system and the dopaminergic one in eliciting supersensitivity.

The similarity is the presence of supersensitivity in animals whose CA or 5-HT nerve terminals had been destroyed by 6-hydroxydopamine[60–63] or 5,6-

DHT, and the difference is the lack of supersensitivity after depletion of 5-HT by *p*-CPA-induced inhibition of synthesis, whereas depletion of CA produced by inhibition of tyrosine hydroxylase potentiates the behavioral response to CA precursors and agonists.[64, 65]

This difference between the two systems has been ascribed to the possibility that supersensitivity in the 5-HT system may be at least in part presynaptic in nature.[58] The failure of chronic pretreatment with *p*-CPA to enhance the 5-HTP plus MAO inhibitor effect on body temperature in rabbits confirms, by means of a different experimental approach, the findings of Trulson *et al.*[58] and may be due to some other nonspecific effect of *p*-CPA that may mask receptor supersensitivity.

However, the possibility that the supersensitivity observed after 5,6-DHT injection must be specifically ascribed to the degeneration of presynaptic neurons, which does not occur in the case of *p*-CPA, should also be taken into account. Support for receptor supersensitivity after 5,6-DHT-induced degeneration of 5-HT nerve endings is also provided by the observation that in fowl pretreated with 5,6-DHT, a subsequent infusion of 5-HT into the third cerebral ventricle caused more intense and longer-lasting hyperthermia and electrocortical effects than did the same dose of 5-HT given to control animals.

As opposed to the supersensitivity to 5-HTP plus a MAO inhibitor and 5-HT, the hyperthermic effect of fenfluramine, a drug that stimulates the 5-HT system by releasing stored amines from presynaptic neurons,[66] was partially or completely inhibited in rabbits and fowl whose brain 5-HT levels had been previously depleted by 5,6-DHT or *p*-CPA treatment.

It seems of interest to note that in rabbits, the neurotoxic effect of 5,6-DHT, as documented by the measurements of brain monoamine concentrations after drug administration, seems to be quite specific for 5-HT neurons, whereas in fowl, as shown by the increase in green fluorescence, the neurotoxin seems to also damage CA neurons, though to a less intense extent.

In previous work,[67, 68] by studying the effects of 5,6-DHT on isolated rat stomach fundus, duodenum, and ileum, on guinea pig bronchoconstriction, on cat blood pressure, and on rabbit and human platelets, it was possible to demonstrate that in peripheral tissues, 5,6-DHT not only acts on the transport, storage, and metabolism of 5-HT but is also a 5-HT receptor agonist.

The present experiments show that in fowl, intraventricular infusion of a single dose of 5,6-DHT produces behavioral, electrocortical, and body temperature effects identical to those obtained in the same species after intraventricular infusion of 5-HT, confirming the results of a previous investigation.[69]

This observation fits well with the results of Waller and Myers,[70] who also found that a dose-dependent hyperthermia similar to that evoked by 5-HT was present after acute 5,6-DHT injection into the hypothalamus of rhesus monkeys. It may be concluded on the basis of these findings that in the CNS also, the acute effects of 5,6-DHT might be attributed to a direct action on 5-HT receptors.

Summary

Some behavioral effects of CNS-acting drugs have been studied in two animal species after functional impairment of central serotonergic activity.

In rabbits, pretreatment with *p*-CPA or with 5,6-DHT counteracted the hyperthermia induced by *d*-amphetamine or apomorphine; the same pretreatments were ineffective in modifying pyrogen-induced fever. These data indicate a modulatory role of the 5-HT system in the hyperthermic response to dopaminergic agonists, such as *d*-amphetamine and apomorphine, and rule out 5-HT involvement in the hyperpyrexia induced by pyrogen.

A previous intraventricular injection of 5,6-DHT significantly potentiated the increase in body temperature induced by 5-HTP in combination with a MAO inhibitor. Pretreatment with *p*-CPA, instead, strongly inhibited the hyperthermic response to 5-HTP. In unanesthetized fowl, at the time of the most dramatic degenerative signs of indoleaminergic neurons induced by 5,6-DHT, as documented by histofluorescence pictures, intraventricular infusion of 5-HT induced more intense behavioral, electrocortical, and body temperature modifications than in control animals. These results suggest the existence of receptor supersensitivity after 5-HT nerve ending degeneration but not after brain 5-HT depletion.

The hyperthermic effect of a presynaptically acting drug, fenfluramine, was counteracted in rabbits and in fowl pretreated with either *p*-CPA or 5,6-DHT.

Since identical behavioral, electrocortical, and body temperature effects have been observed after 5-HT or 5,6-DHT infusion into the third ventricle of fowl, it may be concluded that 5,6-DHT behaves also as a central 5-HT receptor agonist.

References

1. Feldberg, W. & R. D. Myers. 1963. Nature (London) **200:** 1325.
2. Feldberg, W. & R. D. Myers. 1964. J. Physiol. (London) **173:** 25.
3. Feldberg, W. & R. D. Myers. 1965. J. Physiol. (London) **177:** 239.
4. Shellemberger, M. K. & J. T. Elder. 1967. J. Pharmacol. Exp. Ther. **158:** 219.
5. Allen, D. J. & E. Marley. 1967. Brit. J. Pharmacol. **31:** 290.
6. Nisticò, G., E. Marley & P. Preziosi. 1976. Pharmacol. Res. Commun. **8:** 53.
7. Marley, E. & G. Nisticò. 1971. J. Pharmacol. (Paris) **2:** 233.
8. Marley, E. & G. Nisticò. 1972. Brit. J. Pharmacol. **46:** 619.
9. Marley, E. & G. Nisticò. 1975. Brit. J. Pharmacol. **53:** 193.
10. Marley, E. & J. D. Stephenson. 1970. Brit. J. Pharmacol. **40:** 639.
11. Baumgarten, H. G., A. Björklund, A. F. Holstein & A. Nobin. 1972. Z. Zellforsch. **129:** 256.
12. Baumgarten, H G., L. Lachenmayer & H. G. Schlossberger. 1972. Z. Zellforsch. **125:** 553.
13. Daly, J., K. Fuxe & G. Jonsson. 1973. Brain Res. **49:** 476.
14. Gargiulo, G. & G. Nisticò. 1975. Comm. XXIX Conv. Soc. Ital. Sci. Vet. Elba.
15. Bornstein, D. L., C. Bredenberg & W. B. Wood. 1963. J. Exp. Med. **117:** 349.
16. Barrot, H. C. & E. M. Pringle. 1946. J. Nutrition **31:** 35.
17. King, J. R. & D. S. Farner. 1961. *In* Biology and Comparative Physiology of Birds. A. J. Marshal, Ed. Vol. **2:** 215. Academic Press, Inc. New York, N.Y.
18. Shellemberger, M. K. & J. H. Gordon. 1971. Arch. Biochem. **39:** 356.
19. Falck, B., N. A. Hillarp, G. Thieme & A. Torp. 1962. J. Histochem. Cytochem. **10:** 348.
20. Marsden, C. A. & G. A. Kerkut. 1970. *In* Experiments in Physiology and Biochemistry. G. A. Kerkut, Ed. Vol. 2: 327.
21. Fibiger, H. C., A. P. Zis & E. G. McGeer. 1973. Brain Res. **55:** 135.
22. Groppetti, A., F. Zambotti, A. Biazzi & P. Mantegazza. 1973. *In* Frontiers in Catecholamine Research. E. Usdin & H. S. Snyder, Eds. : 917. Pergamon Press. Oxford, England.

23. Beaz, L. A. 1974. Psychopharmacologia **35:** 91.
24. Clineschmidt, B. V., C. McGuffin & A. B. Werner. 1974. Eur. J. Pharmacol. **27:** 313.
25. Randrup, A. & J. Scheel-Krüger. 1966. J. Pharm. Pharmacol. **18:** 752.
26. Taylor, K. M. & S. H. Snyder. 1971. Brain Res. **28:** 295.
27. Hill, H. 1971. Fed. Proc. **30:** 206.
28. Belenky, M. L. & M. Vitolina. 1962. Int. J. Neuropharmacol. **1:** 1.
29. Hill, H. & A. Horita. 1970. Pharmacologist **12:** 197.
30. Morpurgo, C. & W. Theobald. 1967. Eur. J. Pharmacol. **2:** 287.
31. Andén, N. E., S. G. Butcher, H. Corrodi & K. Fuxe. 1970. Eur. J. Pharmacol. **11:** 303.
32. Costa, E. & A. Groppetti. 1970. *In* Amphetamines and Related Compounds. E. Costa & S. Garattini, Eds. : 231. The Raven Press. New York, N.Y.
33. Hill, H. & A. Horita. 1971. J. Pharm. Pharmacol. **23:** 715.
34. Ernst, A. M. 1967. Psychopharmacologia **10:** 316.
35. Andén, N. E., K. Rubenson, K. Fuxe & T. Hokfelt. 1967. J. Pharm. Pharmacol. **19:** 627.
36. Kruk, Z. L. 1972. Life Sci. **11:** 845.
37. Hill, H. & A. Horita. 1972. J. Pharm. Pharmacol. **24:** 490.
38. Randrup, A. & J. Munkvad. 1968. Pharmakopsychiat. Neuropsychopharmakol. **1:** 18.
39. Ernst, A. M. 1965. Psychopharmacologia **7:** 391.
40. Ernst, A. M. 1969. Acta Physiol. Pharmacol. Neerl. **15:** 144.
41. McKenzie, G. & K. Karpowiez. 1970. Pharmacologist **12:** 207.
42. Quock, R. M. & A. Horita. 1974. Science **181:** 539.
43. Ernst, A. M. 1972. Arch. Int. Pharmacodyn. **199:** 219.
44. Giarman, N. J., C. Tanaka, J. Mooney & E. Atkins. 1968. *In* Advances in Pharmacology. S. Garattini & P. A. Shore, Eds. : 307. Academic Press, Inc. New York, N.Y.
45. Mašek, K., H. Rašková & J. Rotta. 1972. Arch. Exp. Pathol. Pharmakol. **274:** 138.
46. Canal, N. & A. Ornesi. 1961. Atti Accad. Med. Lombarda **16:** 69.
47. Des Prez, R., R. Helman & J. A. Oates. 1966. Proc. Soc. Exp. Biol. Med. **122:** 746.
48. Mašek, K., H. Rašková & J. Rotta. 1968. J. Physiol. (London) **198:** 345.
49. Jacob, J., J. M. Girault & R. Peindaries. 1972. Neuropharmacology **11:** 1.
50. Peindaries, R. & J. Jacob. 1971. Eur. J. Pharmacol. **13:** 347.
51. Lichtensteiger, W., U. Mutzner & H. Langemann. 1967. J. Neurochem. **14:** 489.
52. Korf, J., K. Venema & F. Postema. 1974. J. Neurochem. **23:** 249.
53. Aghajanian, G. K. & I. M. Asher. 1971. Science **172:** 1159.
54. Longo, V. G., A. Scotti De Carolis, A. Liuzzi & M. Massotti. 1974. *In* Advances in Biochemical Pharmacology. Volume 10. Serotonin: New Vistas. E. Costa, G. L. Gessa & M. Sandler, Eds. : 109. The Raven Press. New York, N.Y.
55. Daly, J., K. Fuxe & G. Jonsson. 1974. Res. Commun. Chem. Pathol. Pharmacol. **7:** 175.
56. Breese, G. R. & B. R. Cooper. 1975. Brain Res. **98:** 517.
57. Stewart, R. M., J. H. Growdon, D. Cancian & R. J. Baldessarini. 1976. Neuropharmacology **15:** 449.
58. Trulson, E. M., E. E. Eubanks & B. L. Jacobs. 1976. J. Pharmacol. Exp. Ther. **198:** 23.
59. Fuxe, K., B. Holmstedt & G. Jonsson. 1972. Eur. J. Pharmacol. **19:** 25.
60. Ungerstedt, U. 1971. Acta Physiol. Scand. Suppl. **367:** 49.
61. Ungerstedt, U. 1971. Acta Physiol. Scand. Suppl. **367:** 69.
62. Schoenfeld, R. I. & N. Uretsky. 1972. Eur. J. Pharmacol. **19:** 115.
63. Schoenfeld, R. I. & N. Uretsky. 1973. J. Pharmacol. Exp. Ther. **186:** 616.

64. Thornburg, J. E. & K. E. Moore. 1973. Neuropharmacology **12:** 853.
65. Dominic, J. A. & K. E. Moore. 1969. Psychopharmacology **15:** 96.
66. Costa, E., A. Groppetti & A. Revuelta. 1971. Brit. J. Pharmacol. **41:** 57.
67. Barzaghi, F., H. R. Baumgartner, M. O. Carruba, P. Mantegazza & A. Pletscher. 1973. Brit. J. Pharmacol. **48:** 245.
68. Carruba, M. O. 1974. *In* Advances in Biochemical Pharmacology. Volume 10. Serotonin: New Vistas. E. Costa, G. L. Gessa & M. Sandler, Eds. : 103. The Raven Press. New York, N.Y.
69. Nisticò, G., G. Gargiulo, D. Rotiroti & P. Preziosi. 1976. Arch. Int. Pharmacodyn. **222:** 267.
70. Waller, M. B. & R. D. Myers. 1974. Res. Commun. Chem. Pathol. Pharmacol. **9:** 421.

Discussion

Dr. J. H. Jacoby: Why doesn't supersensitivity occur after injection of *p*-chlorophenylalanine? Is it possible that *p*-chlorophenylalanine is incorporated into a protein that somehow alters the development of postsynaptic receptor supersensitivity by impairment of protein synthesis?

Dr. Carruba: That is a possibility. Perhaps *p*-chlorophenylalanine masks in some way this supersensitivity of receptors.

Dr. H. G. Baumgarten: As long as the presynaptic component of the serotonin neuron is intact, minimum release of neurotransmitter may suppress the development of supersensitivity.

Dr. Jacoby: Determining the effect of *p*-chlorophenylalanine on the development of supersensitivity in a 5,7-dihydroxytryptamine-treated animal might be a definitive experiment.

GENERAL DISCUSSION

R. W. Fuller, *Moderator*
The Lilly Research Laboratories
Eli Lilly and Company
Indianapolis, Indiana 46206

DR. V. J. MASSARI (*Bethesda, Md.*): The relative differences in effectiveness of *p*-chloroamphetamine in depleting serotonin in different brain regions might be due to variations in the distribution of the drug or to some other mechanism. We have compared the relative effectiveness of *p*-chloroamphetamine on axons versus terminals by assaying serotonin in white versus gray matter in the spinal cord. Whereas *p*-chloroamphetamine causes a marked and long-lasting decrease in serotonin in the gray matter, it causes a simultaneous increase in serotonin in nonterminal axon areas (white matter). In areas of the brain that have prominent axon bundles, like the medial forebrain bundle, it is not surprising that one finds a less potent depletion of serotonin by *p*-chloroamphetamine due to a summation of effects in opposite directions.

DR. M. A. MOSKOWITZ: Dr. Meek, I would like to comment on your paper and disagree with some of your assumptions. First, your data showed that there are very significant effects of these neurotoxins on tryptophan hydroxylase activity in nerve terminals. But that doesn't necessarily reflect the synthesis of protein, because you are measuring enzymatic activity. Second, why was tryptophan hydroxylase in the B9 region reduced so markedly? Third, were you indicating that *p*-chloroamphetamine and *p*-chlorophenylalanine had identical mechanisms of action on serotonin neurons?

DR. J. L. MEEK: I am not suggesting that *p*-chloroamphetamine and *p*-chlorophenylalanine have similar mechanisms; I don't think there is any evidence that they do. I don't know why the B9 region, in particular, is affected; Dr. Harvey will talk about that tomorrow. There might be an effect on protein synthesis in cell bodies and a simultaneous acute effect on tryptophan hydroxylase in the forebrain, in which inhibition of protein synthesis is not involved; these effects would not be incompatible.

DR. J. A. HARVEY: There are differences in the relative neurotoxic effects of fenfluramine and *p*-chloroamphetamine. For example, 30 days after drug treatment, there is no effect of fenfluramine on the content of serotonin in the hippocampus, whereas *p*-chloroamphetamine causes a 75% decrease. In the amygdala, on the other hand, fenfluramine causes almost 50% depletion, whereas *p*-chloroamphetamine depletes by 75%. And there are similar differences throughout the brain. There are differences between the effects on terminals. Destroying a nerve terminal does not mean that the cell body will degenerate. Generally, there is no retrograde degeneration because of the vast branching. Hence, there is an independent effect on cell bodies, and the B9 cell group is destroyed. This happens with fenfluramine at a threshold dose of about 12.5 μmol/kg ($\sim$3.4 mg/kg). I don't think the absence of denervation supersensitivity under the conditions Dr. Clineschmidt mentioned reveals whether a nerve terminal is destroyed, because the system being examined for denervation supersensitivity may not be the system in which the terminals are destroyed.

Dr. J. Stein: Has anyone looked at other neurotransmitters, such as acetylcholine, to see if *p*-chloroamphetamine affects them?

Dr. E. Sanders-Bush: Dr. Schmidt at Vanderbilt has measured acetylcholine levels and choline acetylase activity and found essentially no effect with *p*-chloroamphetamine.

Dr. Stein: Could you comment on the lack of effect of *p*-chloroamphetamine on peripheral serotonin stores?

Dr. Sanders-Bush: Most of the studies on peripheral serotonin stores have been performed at relatively early times after administration of *p*-chloroamphetamine, at which time there is no decrease in such tissues as the small intestine. Presumably, most of the serotonin in the gut is not contained in neurons.

Dr. H. G. Baumgarten: Has anyone studied the effects of *p*-chloroamphetamine-like compounds on monoamine oxidase? The morphologic effects of *p*-chloroamphetamine look entirely different from those of 5,6- or 5,7-dihydroxytryptamine. *p*-Chloroamphetamine causes an accumulation of serotonin suggestive of less effective downward transport of serotonin along the axons. Is it possible that *p*-chloroamphetamine, apart from destroying some nerve terminals, has some monoamine oxidase-inhibiting activity?

Dr. R. W. Fuller: We reported in 1966 that *p*-chloroamphetamine is a rather potent inhibitor of serotonin oxidation by rat brain monoamine oxidase *in vitro* and, based on the measured concentrations of serotonin and *p*-chloroamphetamine in the brain, estimated that monoamine oxidase probably was inhibited substantially *in vivo*. Because *p*-chloroamphetamine is a reversible inhibitor, this point is rather difficult to prove directly. We have shown, however, that *p*-chloroamphetamine can protect against the irreversible inactivation of monoamine oxidase by an irreversible inhibitor, suggesting that it does occupy some active sites on monoamine oxidase in the brain. Monoamine oxidase inhibition doesn't seem to be a predominant action of *p*-chloroamphetamine, however, because the net effect is a reduction rather than an increase in serotonin content.

Dr. Sanders-Bush: What evidence is there that *p*-chloroamphetamine causes degeneration of serotonin nerve terminals? Even though *p*-chloroamphetamine causes toxic long-term decreases in function, perhaps it does not actually cause degeneration of the terminals.

Dr. M. Da Prada: Pertaining to the question of Dr. Baumgarten, we have found inhibition of monoamine oxidase activity in blood platelets by *p*-chloroamphetamine. I have a comment for Dr. Stewart. The results with LSD are difficult to interpret, because LSD can be a mixed agonist-antagonist, not only on serotonin receptors but also on dopamine receptors.

Dr. R. M. Stewart: We have been studying LSD binding to receptors and find that haloperidol can antagonize this binding, which would agree with some of your findings. We realize that LSD is not a pure serotonin receptor antagonist.

Dr. G. R. Breese: There is another method by which supersensitivity produced by 5,7-dihydroxytryptamine can be studied. Several years ago, Aprison described an operant behavior procedure whereby 5-hydroxytryptophan caused decreased responding in a fixed-ratio schedule. We reported that a 10 mg/kg dose of 5-hydroxytryptophan caused a 90% reduction of this bar-pressing behavior in 5,7-dihydroxytryptamine-pretreated animals but only a 5–10% reduction in controls.

Dr. Baumgarten: Dr. Stewart, do you have any evidence that the myoclonic convulsions you are studying might be mediated by disinhibited or disorganized cerebellar function?

DR. STEWART: We have not studied that, but certainly clinical evidence indicates that cerebellar dysfunction can lead to seizures.

DR. J. H. JACOBY: Dr. Meek, you tried to increase the neurotoxicity of *p*-chloroamphetamine by giving a high dose along with chlorpromazine to prevent the lethal hyperthermia. Chlorpromazine can act as an uptake inhibitor, which would tend to antagonize the serotonin depletion caused by *p*-chloroamphetamine. Perhaps if you subjected the rat to a lower ambient temperature to prevent the hyperthermia, the higher dose of *p*-chloroamphetamine might have induced a longer-lasting depletion of serotonin.

DR. MEEK: We essentially tried that experiment by putting the rats into a shallow water bath to facilitate heat transfer. Unfortunately, all of the animals died after doses of 30–40 mg/kg of *p*-chloroamphetamine. Apparently, animals treated with *p*-chloroamphetamine are very susceptible to acute stress. I think chlorpromazine is protecting the rats from *p*-chloroamphetamine in ways other than just preventing hyperthermia.

MORPHOLOGIC CONSEQUENCES OF SEROTONIN NEUROTOXIN ADMINISTRATION: NEURON-TARGET CELL INTERACTION IN THE RAT SUBCOMMISSURAL ORGAN

K. Møllgård and J. J. Lundberg

Anatomy Department A
University of Copenhagen
Universitetsparken 1, 2100 Denmark

L. Wiklund

Department of Histology
University of Lund
S-223 62 Lund, Sweden

L. Lachenmayer

Department of Neurology
University of Hamburg
2000 Hamburg, Federal Republic of Germany

H. G. Baumgarten

Department of Functional Anatomy
University of Hamburg
2000 Hamburg, Federal Republic of Germany

Neuron-Target Cell Interaction

Various neurotoxic compounds are now extensively used to obtain selective chemical neurodegeneration; the dihydroxylated tryptamines are employed for destruction of central serotonin fibers. The neurotoxic indoleamines could serve as tools for an analysis of serotonin neuron-target cell interaction, provided that the target cells are controlled mainly or exclusively by serotonergic [5-hydroxytryptamine (5-HT)] synapses, the time sequence of the specific destruction is known, and the cellular activity of the target cells can be easily defined and measured. The subcommissural organ (SCO) of the rat appears to fulfill the criteria for an ideal model of serotonin neuron-target cell interaction.

The subcommissural organ is a specialized area of the ependyma attached to the ventral surface of the posterior commissure (for references, cf. Discussion). It forms the lining of the roof of the cerebral aqueduct at its junction with the third ventricle. The function of the SCO remains obscure, but the pronounced secretory activity of the organ is well established. Secretory material of glycoprotein character is released apically into the ventricular fluid and forms the so-called Reissner's fiber. In addition to this apical secretion, a basal secretion into the blood vessels is postulated in some species. It has recently been demonstrated that the SCO of the rat is controlled mainly by a well-developed 5-HT input.[1]

Provided that 5-HT innervation is an important modulator of SCO cell function, it should be possible to correlate changes in the activity of these target

0077–8923/78/0305–0262 $01.75/0

cells with various stages of the degeneration cycle of 5-HT synapses. This goal is, however, possible only if the specific neurocytologic effects of the neurotoxic indoleamines can be clearly defined. Certain criteria have to be fulfilled before any specific change in a nerve terminal can be interpreted as degeneration (see, for example, the review by Raisman and Matthews[2]), and the following paragraphs summarize our present knowledge about selective destruction of presynaptic serotonergic elements.

Selective Destruction of Presynaptic Serotonergic Elements

From fluorescence microscopic studies, [^{3}H]serotonin uptake measurements, and biochemical investigations, 5,6-dihydroxytryptamine (5,6-DHT) is known to cause rather selective degeneration of serotonergic neurons.[3–5] Baumgarten *et al.*[6] found that intraventricular injection of this compound caused characteristic time-dependent ultrastructural changes in various subependymal regions of the rat brain. Early changes consisted of an enhancement of electron opacity in medium and large granular vesicles of unmyelinated axonal profiles, possibly reflecting accumulation of vesicle-bound 5,6-DHT, which is precipitated by glutaraldehyde and which reduces osmium tetroxide. Similar labeling of the storage compartment has been demonstrated[7,8] to occur in presumptive 5-HT fibers of the rat cerebellum, subependymal neuropil of the aqueduct, and supraependymal fibers of the lateral, third, and fourth ventricles. High concentrations of 5,6-DHT caused accumulation of electron-dense material in the axoplasm itself, in addition to enhancement of vesicular granularity.

Three to 6 hr after intraventricular injection of 5,6-DHT, axonal varicosities showed signs of damage, such as clumping of synaptic vesicles, often accompanied by an increase in their mean diameter, and rarefication and dissolution of the core material in granular vesicles.

Pronounced signs of degeneration were found 6–24 hr after injection. The most consistent feature of disintegration of axons was an impregnation of the axoplasm and other organelles with fine granular electron-dense material that corresponded to the classic type of dark degeneration. As pointed out earlier,[6] mitochondria revealed a conspicuous increase of electron density in their intermembraneous compartment, even in axons with relatively unimpaired ultrastructure. Disintegrated axonal and synaptic profiles were finally phagocytosed by neighboring glial elements.

Local signs of unspecific toxic damage to glial cells and myelinated axons were observed after left lateral ventricular injections of 5,6-DHT in the homolateral septum and caudate. These unspecific toxic side effects were undoubtedly due to the high concentrations of 5,6-DHT that accumulated in the surroundings of the injection cannula.

The sequence and features of neuropathologic alterations in central nervous system 5-HT fibers after administration of 5,7-DHT, a more recently introduced neurotoxic serotonin analog,[6,9] were similar to those documented after injection of 5,6-DHT, with the important advantage of having only minor cytotoxic side effects on nonmonoaminergic central nervous system structures.[1,10] Changes in the neuronal ultrastructure of serotonergic systems caused by intraventricular injection of 5,6-DHT or 5,7-DHT are strikingly similar to those induced in central catecholaminergic systems after intraventricular or intracisternal administration of 6-hydroxydopamine[11,12] reported previously.

Consequences at the Target Cell Level

At present, it seems almost impossible to study specific changes in the functional cytology of target neurons in the brain after selective chemical destruction of identified presynaptic elements mainly because of inadequate methods and the complex architecture involved. The SCO of the rat represents a unique model for the study of some important effects of serotonergic systems. The findings in this study of histochemical and ultrastructural changes in target cells due to deprivation of a nervous input can be correlated with the selective destruction of 5-HT terminals. The results thus indicate that the normal serotonergic input inhibits protein and glycoprotein synthesis and release of secretory material in the rat SCO.

Materials and Methods

Animals

Seventy-five adult female and male Sprague-Dawley rats, 180–220 g in body weight, were used.

Experimental Procedures

Under barbiturate anesthesia (Brietal®, 40 mg/kg) rats were injected stereotaxically in the lateral ventricle with one of the following drugs: 50 or 75 μg of 5,6-DHT (Regis), 50 μg of [^{14}C]-5,6-DHT (3 or 14 mCi/mmol), 150 μg of 5,7-dihydroxytryptamine (5,7-DHT, Regis), or 50 μg of 6-hydroxytryptamine (6-HT) (all calculated as free base). All drugs were injected in a volume of 20 μl of saline (which contained L-ascorbic acid, 0.2 mg/ml, as an antoxidant). Normal, untreated rats and rats injected in the lateral ventricle with 20 μl of saline were used as controls. At various time intervals after injection (see Results), the rats were processed for either fluorescence microscopy, autoradiography, electron microscopy, or histochemistry.

Fluorescence Histochemistry

All animals used for fluorescence histochemical detection of monoamines were pretreated according to the method of Aghajanian *et al.*[13] to increase intraneuronal serotonin levels. Injection of chloral hydrate (300 or 400 mg/kg, ip) was followed 10 min later by 300 mg/kg of nialamide (ip); 5 min after the second injection, 100 mg/kg of L-tryptophan (ip) were administered.

One to 3 hr after the last injection, the rats were killed by decapitation under light ether anesthesia. The brains were rapidly dissected, and blocks that contained the SCO were quickly frozen in liquid propane, cooled by liquid nitrogen, and subsequently freeze-dried (for technical details, see Björklund *et al.*[14]). To achieve a high fluorescence yield of serotonin, a two-step formaldehyde gas treatment was used.[15] The specimens were first treated for 1 hr at 80° C with paraformaldehyde equilibrated at about 50% relative humidity, followed by another treatment at 80° C for 1 hr with paraformaldehyde equilibrated at 70% relative humidity. The specimens were then embedded in paraffin, sectioned, and studied in a fluorescence microscope.[14]

Autoradiography

The rats used for autoradiography were fixed by vascular perfusion with a cold saline solution that contained 4% acid-free formalin and were embedded in paraffin. Serial 5-μm sections were mounted on glass slides, deparaffinized, and coated with undiluted Ilford L 4 emulsion (dipping technique). Exposure of the emulsion-coated sections was performed at 8° C in darkness for 6 weeks. The silver grains were developed with Kodak D 19 b and fixed with sodium thiosulfate. Subsequently, all sections were counterstained with gallocyanine chromalum (according to Einarson) for 24 hr and mounted in Caedax®.

Electron Microscopy

Twenty control rats and 12 experimental rats treated with 5,6-DHT or 5,7-DHT and processed as described earlier were fixed by perfusing a solution that contained 2.5% glutaraldehyde and 1% paraformaldehyde in 0.1 M cacodylate buffer (pH 7.4) at a constant pressure of 140 mm Hg through the left ventricle. After 15–20 min of perfusion fixation, the brains were removed and placed in fresh fixative for a total of 6 hr. Tissue blocks that contained the SCO were isolated and postfixed for 2 hr in 2% OsO_4 in 0.1 M cacodylate buffer. Specimens were block stained with uranyl acetate for 1 hr, quickly dehydrated in increasing concentrations of ethanol, and embedded in Epon®. The blocks were carefully oriented during embedding so that survey sections would be strictly in the sagittal or frontal plane. One-micron thick sections stained with toluidine blue were examined in the light microscope. Silver-to-gray thin sections were cut from selected areas and, after poststaining with uranyl acetate and lead citrate, viewed in a Hitachi HS8 microscope or in a Philips EM 300. From the SCO of two control and two experimental rats, blocks were oriented in the frontal plane, and thin sections were cut every 10 μm from the rostral to the caudal end of the organ. Altogether, more than 30 well-defined SCO areas were investigated in these rats.

Histochemical Demonstration of Nucleoproteins, Glycoproteins, and Neurosecretory Proteins

Three groups of rats received a saline or 5,6-DHT injection 1–4 days before sacrifice. At that stage, the animals had been assigned code numbers, so that the histochemists did not know the experimental condition of any individual rat. The brains were removed and fixed for 24–72 hr in Bouin's fixative.[17] Serial paraffin sections 3 μm thick were cut strictly in the frontal plane through the entire organ of each rat. The following histochemical methods were used: staining with 0.1% toluidine blue O (Merck) in 30% ethanol,[16] periodic acid-Schiff (PAS) used with de Tomasi's reagent[17]; [as a control, some sections were digested by Maltase® (Merck) before staining, and other sections were stained omitting the periodic acid oxidation], Bock's gallocyanine chromalum method,[18] and the reaction for cystine (PFA-AB)[17] was compared with the Bock reaction for "neurosecretory substance."

Results

Fluorescence Microscopic Observations

Control Rats

The fluorescence histochemical findings of the SCO in the untreated rat are illustrated in FIGURE 1. Ventral to the myelinated fiber bundles of the posterior

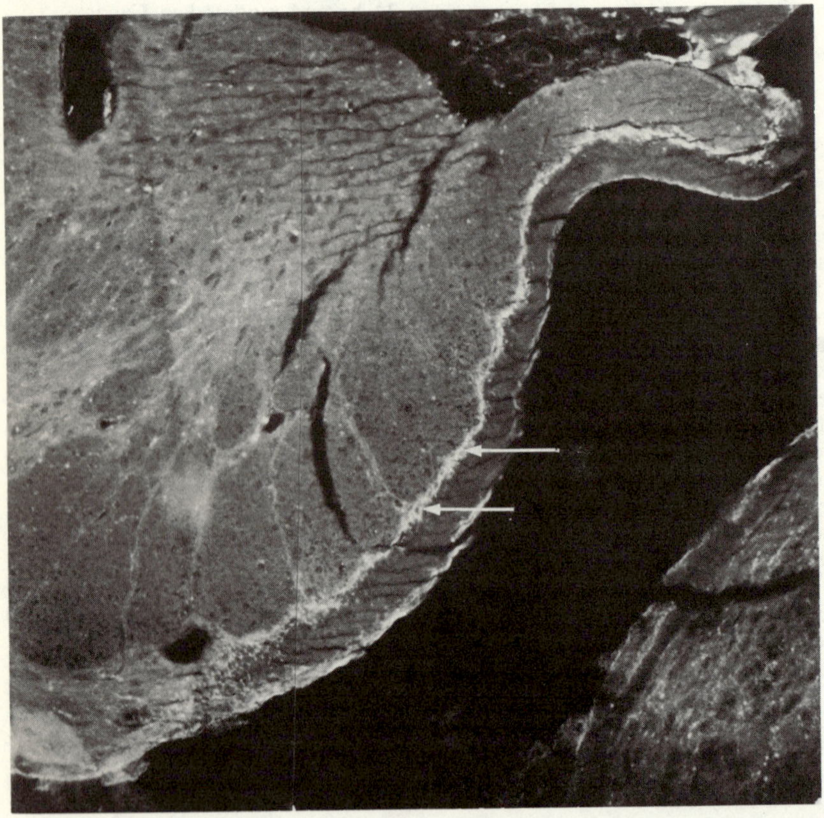

FIGURE 1. Fluorescence micrograph of a sagittal section through the entire SCO. Yellow fluorescent indoleamine-containing nerve fibers (probably serotonergic) are found in a densely packed layer (arrows) in the basal region of the organ. ×160.

commissure, a very dense plexus of yellow fluorescent nerve fibers is found in close relation to the basal parts of the ependymal cells of the SCO. This plexus attains such a high density that individual nerve fibers are difficult to distinguish. From this basal, very dense plexus, single varicose fibers extend apically between the SCO cells up to approximately the nuclear level of the pseudostratified ependyma, but no fibers are observed to reach beyond the nuclear level (i.e., the apical two thirds of the ependyma). These findings are in agreement with previous fluorescence histochemical observations of the rat SCO.[19-21] The fluorescent nerve fibers display typical characteristics of serotonin content, such as color, marked increase of fluorescence intensity by the pharmacologic pretreatment (see MATERIALS AND METHODS), and rapid fading by exposure to the short-wavelength light of the fluorescence microscope.[22]

Rats Treated with Neurotoxic Drugs

Intraventricular administration of either 5,6-DHT or 5,7-DHT caused at all doses used a virtually total destruction and disappearance of the yellow fluorescent fibers in relation to the SCO. Thus, in most rats, 1–3 weeks after injection of the neurotoxic drugs, no yellow fluorescent nerve fibers could be found. A few animals showed signs of a less efficient denervation, leaving single scattered serotonergic fibers in the basal SCO. To obtain a precise picture of the localization of 5-HT in the rat SCO, some animals were injected intraventricularly with a single dose of 6-HT. As seen in FIGURE 2, the 6-HT fluorophore is localized in distinct, finely varicose structures bordering the ventricular surface of the ependyma, distributed in between the basal processes of the SCO cells, and localized in the hypendymal neuropil and surrounding pretectum. In addition, some of the SCO cell nuclei exhibit a faintly yellow diffuse formaldehyde-induced fluorescence, suggesting uptake of 6-HT from the ventricle into the modified ependyma. After destruction of the indoleaminergic terminals by intraventricular 5,6-DHT or 5,7-DHT, there is no accumulation of 6-HT in varicose fiber profiles associated with the rat SCO or its surroundings.

Autoradiographic Localization of [^{14}C]-5,6-Dihydroxytryptamine

5,6-Dihydroxytryptamine has earlier been found to possess a remarkably high affinity for serotonergic uptake sites in the brain,[4] and it accumulates selectively in serotonergic neuronal cell bodies and axonal processes. This selective uptake and concentration of 5,6-DHT to high intracellular levels in serotonergic neurons accounts for the high specificity of its neurotoxic actions. In the present experiments, rats received an intraventricular injection of 75 or 50 μg of [^{14}C]-5,6-DHT 1, 24, or 48 hr before sacrifice. Subsequent autoradiographic studies revealed labeling (i.e., uptake of 5,6-DHT), with a distribution in the area of the SCO, that closely corresponds to the indoleaminergic innervation pattern in this part of the brain (FIGURE 3). Thus, silver deposits are sparse or lacking over the apical (supranuclear) portions of the SCO cells, but intense labeling is found over the basal parts of the organ. Rich labeling also occurs over the apical portions of the nondifferentiated, ordinary ependymal cells, which receive a rich plexus of supraependymal serotonergic nerve fibers.[23] In contrast, no labeling occurs over the apical portions of the SCO cells, indicating that they are not contacted apically by such a supraependymal 5-HT fiber system. Neither the pattern nor the intensity of labeling showed any significant time-dependent changes within the interval of the survival times used.

These autoradiographic experiments demonstrate that 5,6-DHT is specifically taken up in serotonergic nerve fibers in relation to the SCO, but no (or very little) uptake of the neurotoxin occurs into the specialized SCO ependymal or hypendymal cells. This observation probably excludes the possibility of direct cytotoxic effects on the SCO cells.

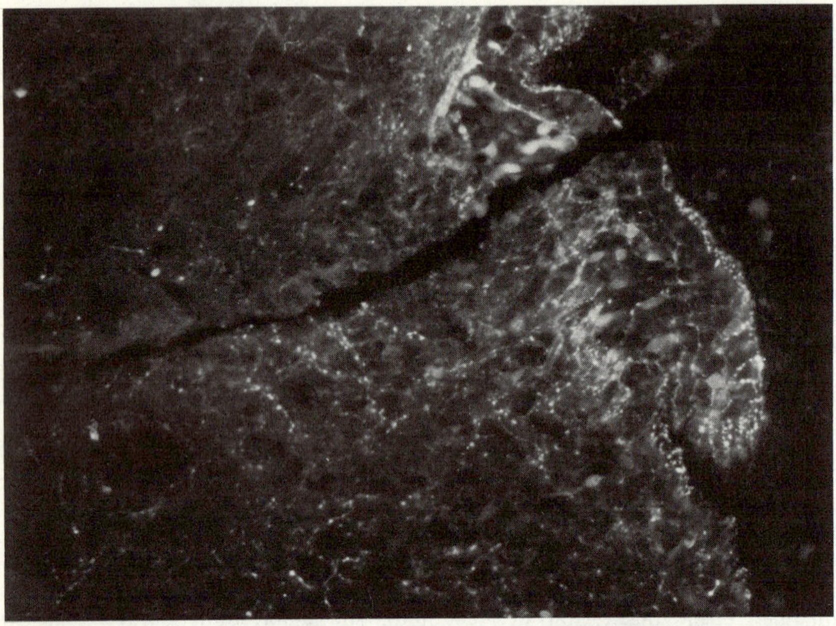

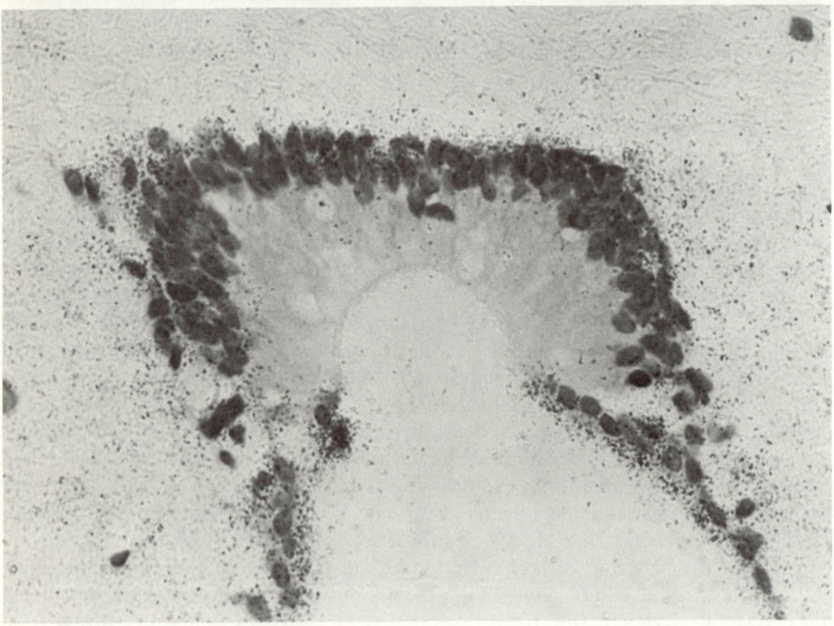

FIGURE 2. (*top*). Note the distinct varicose fluorescent fibers in the SCO area after intraventricular injection of 6-HT. ×420.

FIGURE 3 (*bottom*). Clusters of silver grains accumulate both in the basal region of the SCO and in connection with ordinary ependymal cells that flank the organ after injection of ^{14}C-labeled 5,6-DHT. ×2100.

Ultrastructural and Protein Histochemical Aspects

In the following description, some ultrastructural and histochemical features that may be relevant for an analysis of the secretory activity of the target cells (the rat SCO) will be emphasized. Two types of SCO cells can be recognized: one is an elongated ependymal cell in contact apically with the cerebrospinal fluid; the other is a basal or hypendymal cell that is probably detached from the ventricular surface. In addition to SCO cells, glial cells, nerve fibers, and some SCO-associated neurons [24] contribute to the hypendyma.

There are obvious regional differences between rostral and caudal parts of the SCO. We have circumvented these differences by focusing our attention mainly to the rather homogeneous midportion of the organ. However, even within the same territory, the SCO ependymal cells show varying degrees of cytoplasmic stainability, ranging from nearly no reaction (light cells in the electron microscope) to a very strong reaction (dark cells in the electron miscroscope). Both types are interpreted as being extremes of one and the same organ-specific cell type.

Electron Microscopic Observations

Based on several different experimental procedures, we concluded in a recent study [1] that the SCO in the rat is under the heavy influence of a serotonergic system. The same paper includes a detailed description of the ultrastructure of the synapses. Therefore, only a few observations relevant to the present study will be mentioned here.

Control Material

Both the SCO hypendymal cells and the perikarya and basal processes of the SCO ependymal cells have asymmetric (Gray type 1) synaptic contacts in the form of varicose vesicle-filled axon terminals (FIGURES 4–6). The presynaptic element contains mitochondria and a cluster of small clear vesicles adjacent to the presynaptic membrane. A few large granular vesicles are encountered in most terminals, usually located away from the synaptic junction. A dense particulate material is seen in the widened synaptic cleft, and a characteristic postsynaptic thickening of dense filamentous material is always found attached to the postsynaptic SCO cell membrane.

The SCO ependymal cells of the rat are characterized by a cylindrical apical process that extends to the ventricular lumen. Tight junctions and zonulae adherentia-like contacts are found between adjacent apical processes. The extracellular space is uniformly narrow in the apical zone, but at the level of the nuclei, and particularly in the basal portion of the organ, it has a labyrinthic arrangement. This appearance is due to a complex intermingling of the basal processes of SCO ependymal and processes of SCO hypendymal cells. This basal part of the SCO (the superficial hypendyma) borders on a neuropil (the deep hypendyma) that consists of dendrites and perikarya of the SCO-associated neurons and of axons, some of which penetrate into the superficial hypendyma.

The apical processes of SCO cells contain many cisternae of endoplasmic reticulum filled with a homogeneous finely flocculent material, numerous tubular

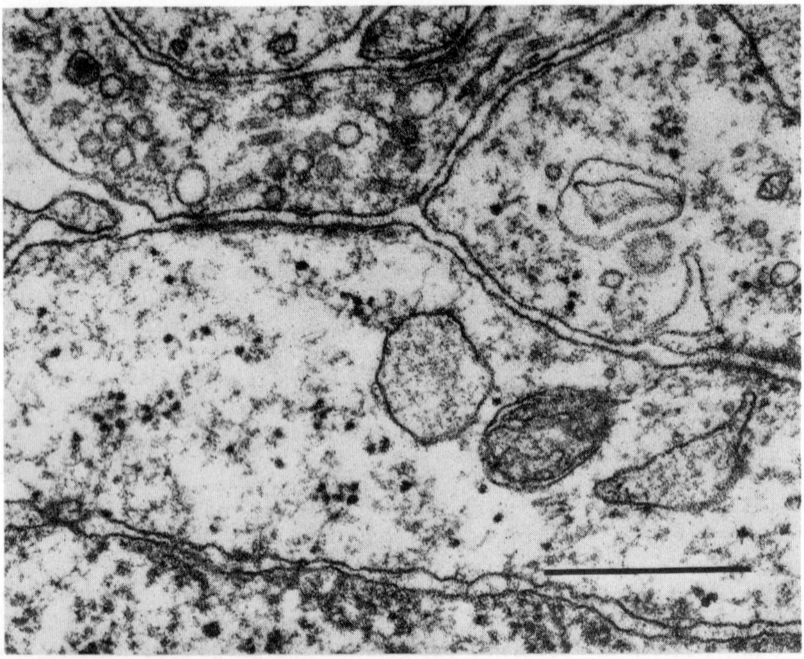

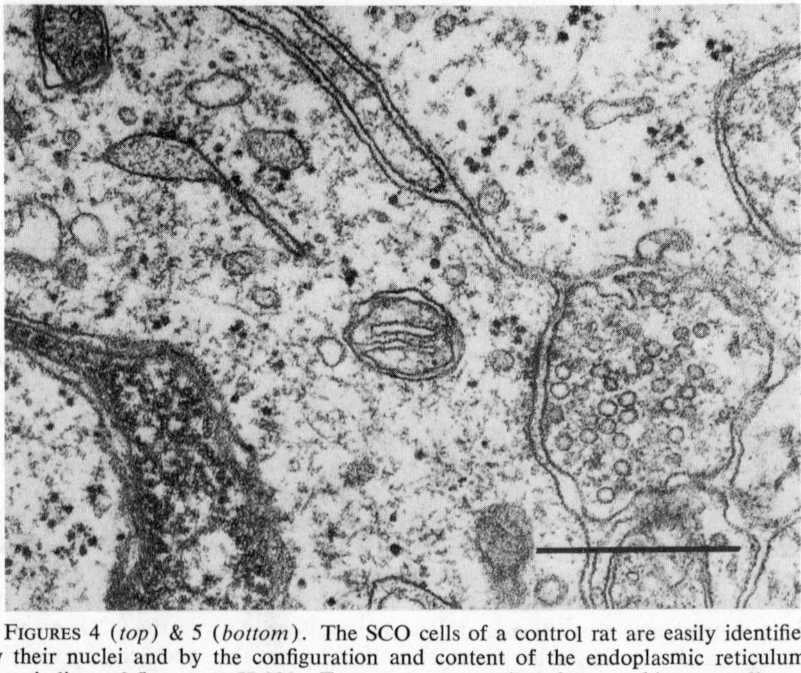

FIGURES 4 (*top*) & 5 (*bottom*). The SCO cells of a control rat are easily identified by their nuclei and by the configuration and content of the endoplasmic reticulum. Bars indicate 0.5 μm. ×57,000. FIGURE 4. A terminal from a thin unmyelinated axon forms a *bouton en passant*. The diversity of vesicular morphology is obvious. FIGURE 5. Note the small spherical vesicles with clear centers in the presynaptic element. This type of synapse is found very infrequently.

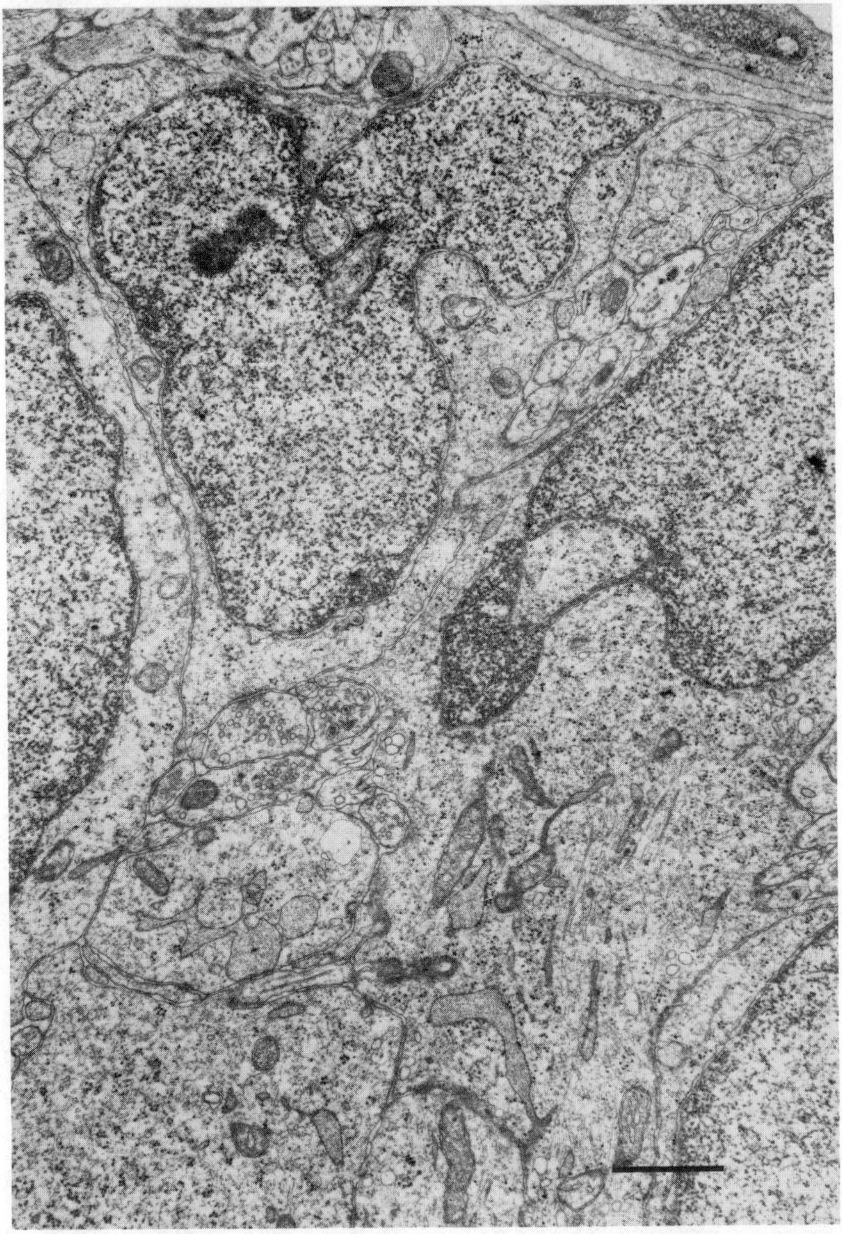

FIGURE 6. Basal part of the SCO of a control rat. Four axon terminals, one in synaptic contact with an apparent SCO cell, are seen near the center of the field. Observe the diversity of the synaptic vesicles. The SCO cell cytoplasm adjacent to the vessel in the upper right corner is devoid of secretory elements. Bar indicates 1 μm. ×15,000.

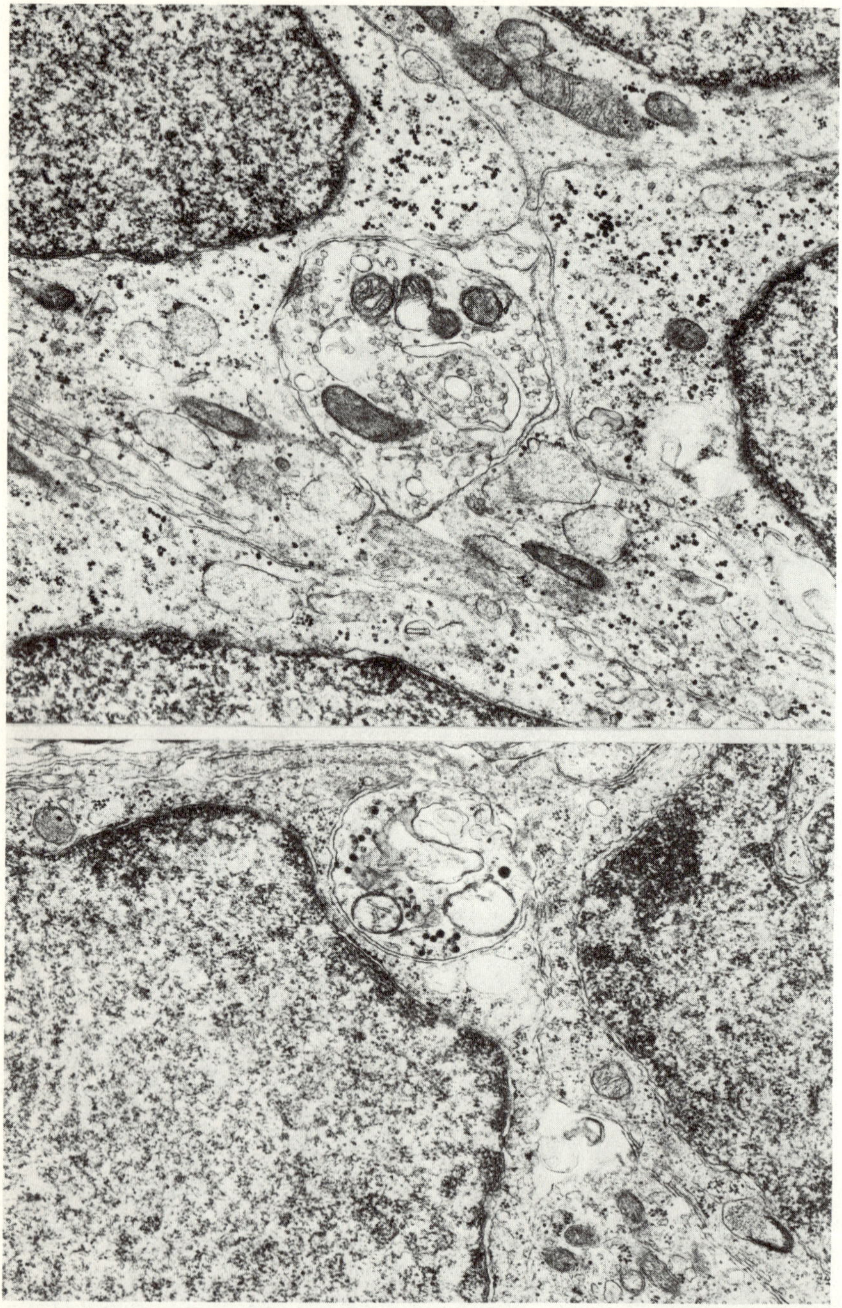

FIGURES 7a (*top*) & 7b (*bottom*). Rat SCO 24 hr after intraventricular injection of 5,7-DHT. FIGURE 7a. Discrete changes in the ultrastructure of a synaptic bouton innervating an SCO cell. Note swelling of a mitochondrion and enlargement of elec-

profiles, and dense granules of varying shape and density. Similar granules are infrequently observed in the perinuclear cytoplasm. The basal processes (FIGURE 6) are almost devoid of secretory elements, whereas they contain many polysomes and mitochondria. Small scattered Golgi complexes are found in both the apical and basal parts of the SCO cells, and within the cytoplasm, smaller dense granules with diameters of 200–400 Å appear to represent glycogen. The SCO cell nucleus is often very irregular in shape (FIGURE 6), and, except for a distinct condensation of chromatin clumped beneath the nuclear envelope, the karyoplasm seems rather homogeneous. Occasionally, a nucleolus is present.

Experimental Material

Administration of 5,6-DHT or 5,7-DHT causes damage and disintegration of small unmyelinated axons and presynaptic elements in the basal part of the SCO. These axons and presynaptic elements are indoleaminergic in nature, as shown by their capacity to accumulate serotonin neurotoxins. Disintegration of the indoleaminergic fiber system in the SCO occurs over a period of 1–4 days. The specific destruction of the presynaptic elements (FIGURE 7a & 7b) follows the pattern of degeneration described earlier. Virtually all presynaptic elements in contact with SCO cells are disintegrated 3 days after injection, but a few synapses with a distinct population of small spherical clear vesicles in the presynaptic elements (cf. FIGURE 5) apparently survive. These findings indicate that the rat SCO may not be entirely controlled by a monosynaptic serotonergic input.

Cytologic Consequences at the Target Cell Level of Removal of 5-HT Input

Distinct ultrastructural changes are found throughout the entire SCO 24 hr after injection of 5,6-DHT or 5,7-DHT. The endoplasmic reticulum forms a very complex network in the apical and perinuclear cytoplasm (FIGURE 8). Most of the cisternae are very distended and contain a rather dense flocculent material; in some cells, the dilated cisternae occupy almost the entire apical process, where it emerges from the nuclear area (FIGURE 8, *inset*).

Apically, the cells taper off and protrude into the ventricle with tonguelike processes filled with numerous electron-dense granules and/or the so-called light secretory sacs ("pale granules"), as demonstrated in FIGURE 9. Electron-dense granules are also observed in the perinuclear cytoplasm (FIGURE 8) and in the basal processes. Several Golgi complexes are found in apical, perinuclear, and basal portions of SCO cells. The extracellular space is increased and often contains fine flocculent material similar to the content of the distended endoplasmic reticulum cisternae.

Three days after the injection, the distension of the endoplasmic reticulum is even more pronounced. Dilated cisternae now often occupy the perinuclear

---

tron-lucent vesicles. ×20,600. FIGURE 7b. Early stage of degeneration in a presynaptic element in contact with SCO cell perikaryon. Note small and large dense core vesicles in the axoplasm and a swollen mitochondrion with disrupted cristae. ×20,600.

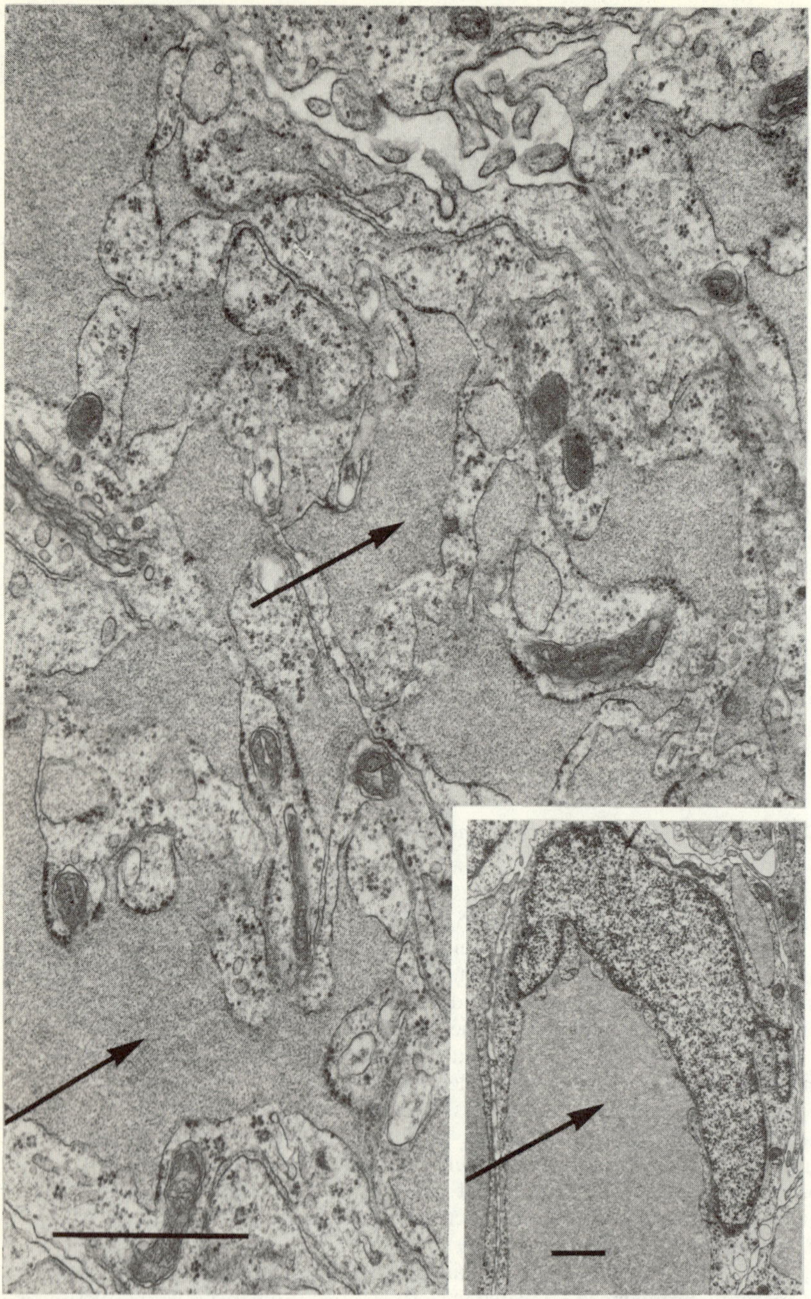

FIGURE 8. Perinuclear region of two rat SCO cells 24 hr after intraventricular injection of 5,6-DHT. Arrows point to some extremely distended cisternae of the endoplasmic reticulum loaded with dense flocculent material. Some electron-dense granules are also present. *Inset:* Apical part of the perinuclear cisterna. Bars indicate 1 μm. ×27,000; *Inset:*. ×7200.

part of the basal processes (FIGURE 10), whereas the distal part and the hypendymal cells contain numerous pale and dense secretory granules (FIGURE 11).

The extracellular space is particularly increased in the superficial hypendyma and often contains a flocculent material of the same density as that seen within the secretory sacs of the endoplasmic reticulum (FIGURE 12). Granules of

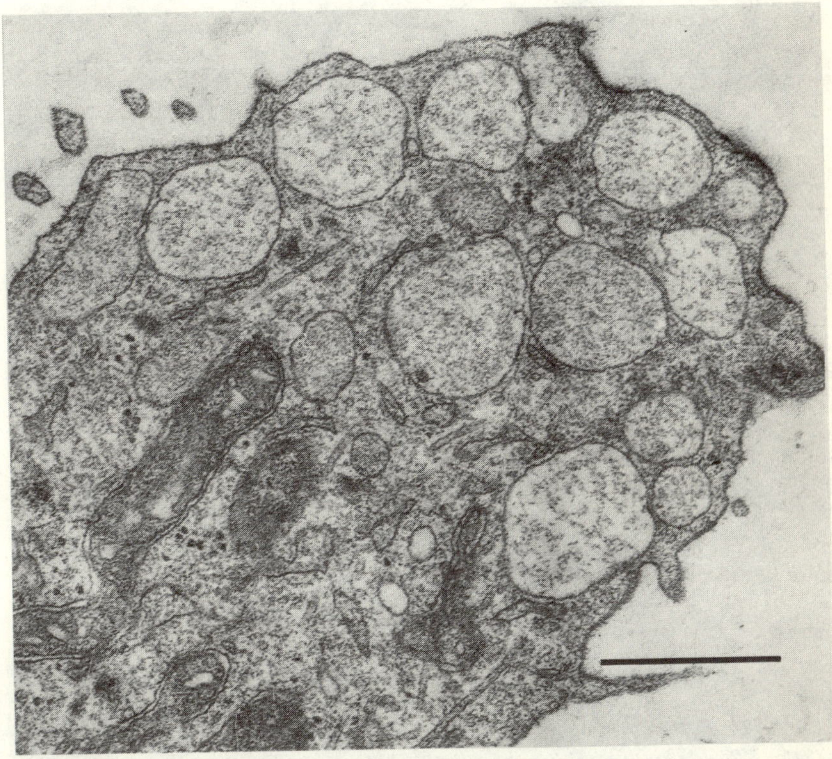

FIGURE 9. Apical process in contact with the cerebrospinal fluid. Rat SCO 24 hr after intraventricular injection of 5,7-DHT. Numerous "light secretory sacs" of different size and density are seen, probably in different phases of discharging secretory material. Bar indicates 0.5 μm. $\times$47,500.

varying density accumulate close to the cell membranes that border on the perivascular spaces. These membranes exhibit numerous pinocytotic and exocytotic profiles (FIGURE 12).

The ultrastructural and histochemical changes in nuclei, nucleoli, and glycogen content that occur after administration of 5,6-DHT or 5,7-DHT are similar and will be described in the histochemical result section.

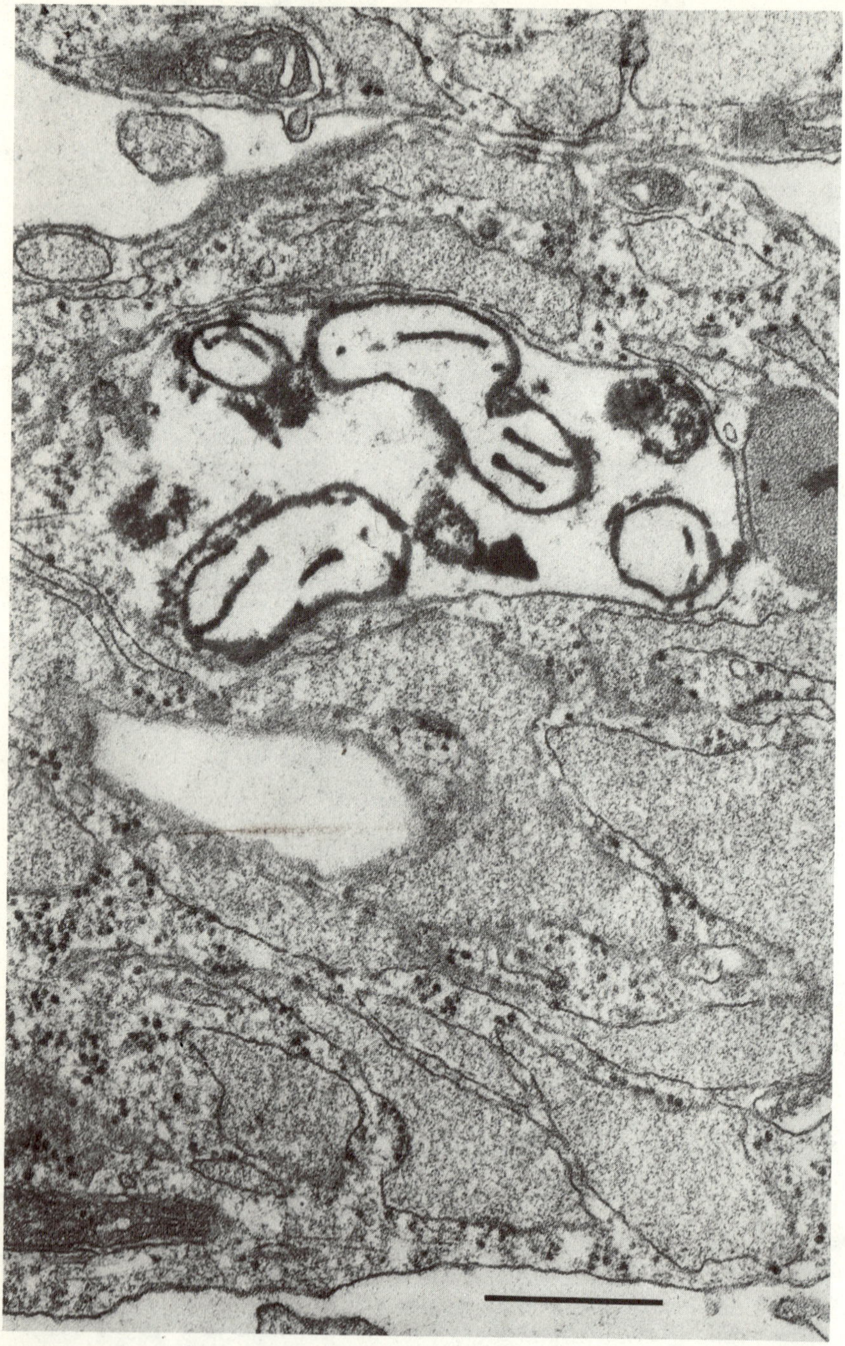

FIGURE 10. Rat SCO 3 days after intraventricular injection of 5,7-DHT. The perinuclear part of the basal cytoplasm demonstrates enlarged flocculent material-filled endoplasmic reticulum cisternae. In this tangential section, a characteristic degenerating presynaptic profile is seen surrounded by SCO cell cytoplasm. Bar indicates 0.5 μm. ×47,500.

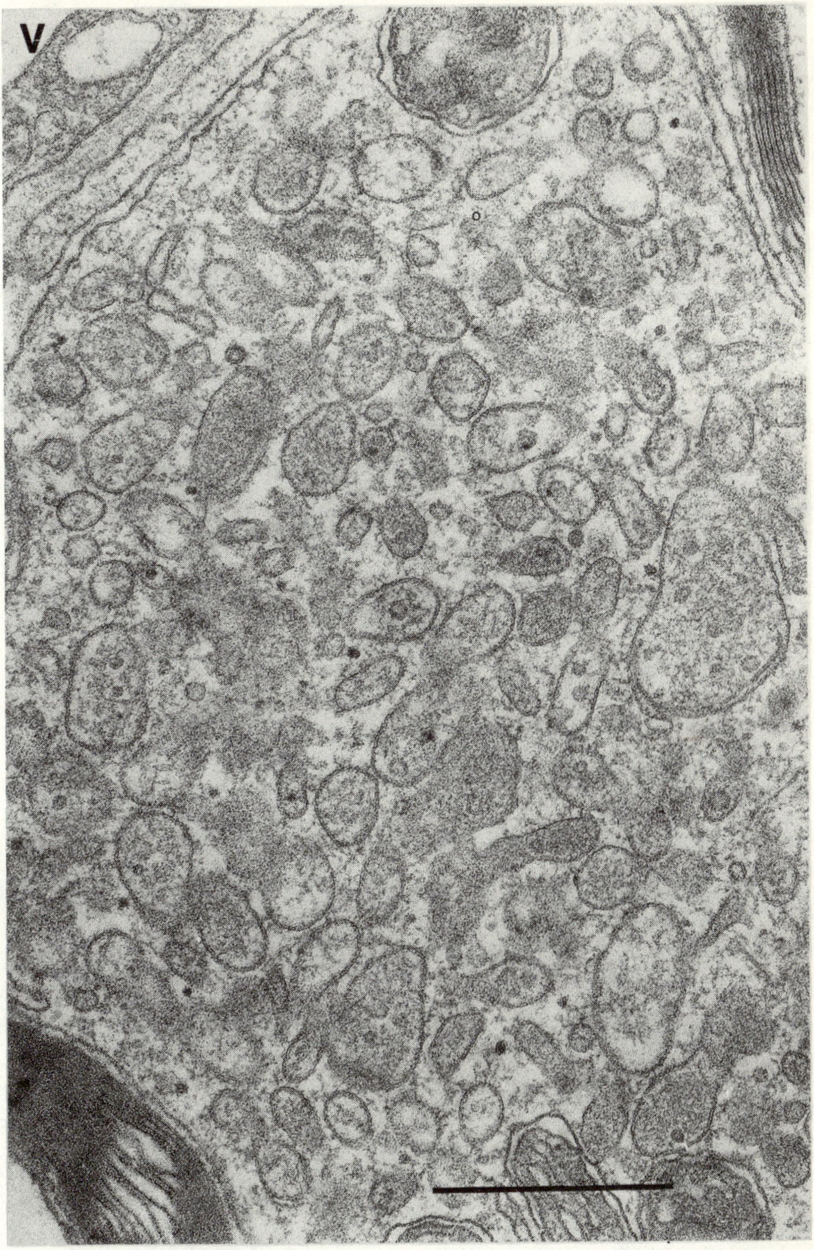

FIGURE 11. Basal process of SCO cell 72 hr after 5,7-DHT treatment. Numerous membrane-bound vesicles of varying shape and density are found adjacent to a vessel (V). Bar indicates 0.5 μm. ×65,000.

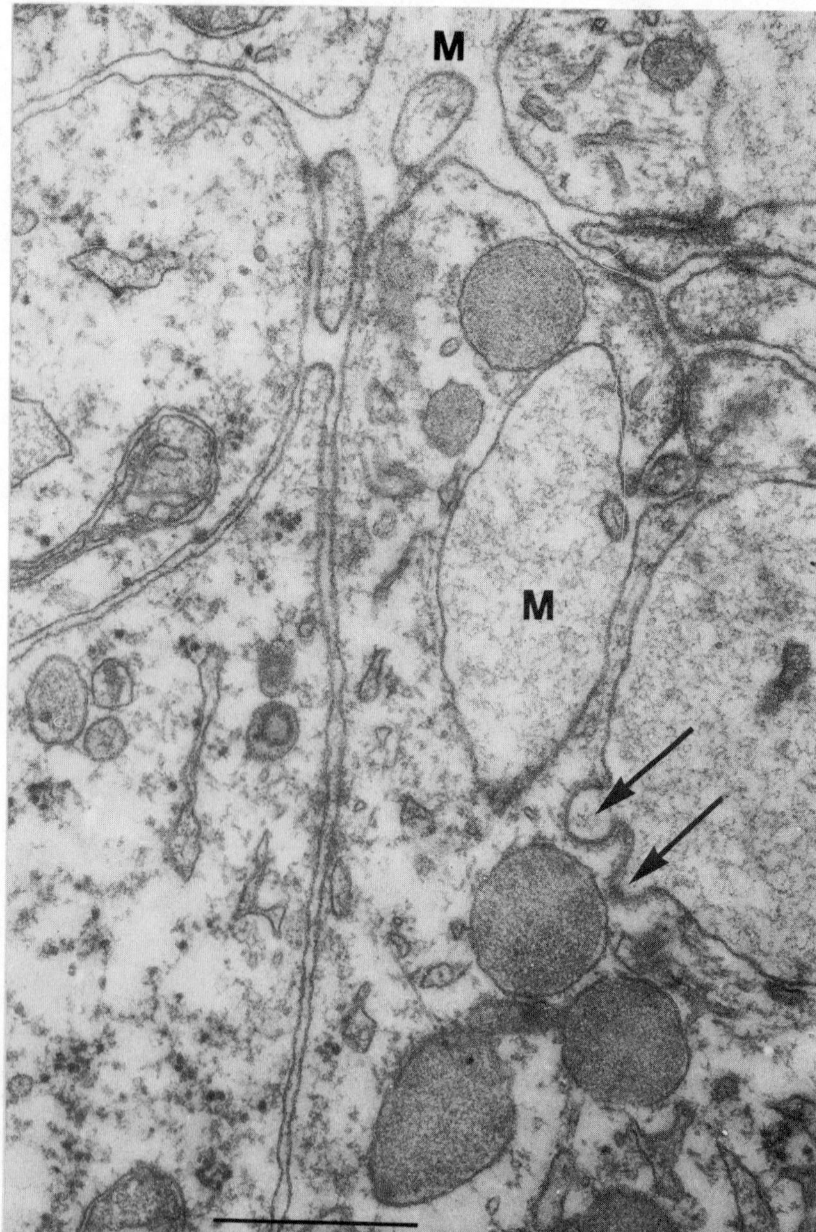

FIGURE 12. The increased extracellular space in the superficial hypendyma contains flocculent material (M) 72 hr after 5,6-DHT treatment. Note the large dense granules and the presumptive exocytotic profiles (arrows). Bar indicates 0.5 μm. $\times 57{,}000$.

Nucleoproteins, Glycoproteins, and Neurosecretory Proteins

Control Rats

The SCO cells demonstrate varying degrees of cytoplasmic Maltase-resistant PAS positivity (FIGURE 13) and of basophilia after toluidine blue staining. These features give the organ a striated appearance, the stria being perpendicular to the ventricular surface. Comparison of PAS positivity before and after digestion by Maltase demonstrates that all cells of the rat SCO contain glycogen. The cell nuclei exhibit inconspicuous nucleoli and are usually elongated, with condensed chromatin distributed beneath the nuclear envelope. They are often encircled by a thin light halo of nonreacting cytoplasm. In the perinuclear and basal cytoplasm, but not in the apical part of the SCO ependymal cells, an accumulation of fine granular material is consistently found. In contrast, coarse dense granules line the ventricular surface. This pattern of reactivity results in a characteristic stratification of the SCO, with the layers oriented parallel to the ventricular surface.

The histochemical reactions considered specific for the so-called neurosecretory substance (PFA-AB and Bock's reaction) revealed two kinds of Gomori-positive material in the cytoplasm: a fine granular material was especially pronounced at the level of the nuclei, and strongly reacting large granules were accumulated in the apical processes very close to the ventricular surface.

Experimental Conditions

Intraventricular injection of 5,6-DHT at doses known to cause destruction of 5-HT fibers induces some distinct histochemical changes in SCO cells. A more intensely stained fine granular substance seems to load many of the cells (FIGURE 14) by 24 hr after the injection. Similarly, an increased reactivity of the coarse granular material is observed at the ventricular border and sometimes in the basal processes. The same pattern of reactivity is observed after all of the histochemical reactions that have been performed. Because of this pattern, the organ appears more striated but less stratified. Morphologic changes include significant swelling of nuclei, which appear light, vesicular, and rounded, and the presence of enlarged nucleoli (cf. FIGURES 13 & 15). A pronounced accumulation of glycogen is apparent when Maltase-digested PAS-stained sections are compared with nondigested sections. The increased apical secretion apparently results in a partial loss of the apical cytoplasm, leading to distinct changes in the configuration of the SCO-aqueductal interface (cf. FIGURES 13–15). Four days after the injection (FIGURE 15), the histochemical reactivity of the SCO is further enhanced, and the light perinuclear halo has disappeared. At this stage of denervation, the basal parts of the SCO also exhibit strong reactivity compatible with the occurrence of basal secretion.

DISCUSSION

Ultrastructural and histochemical investigations have demonstrated that, in a wide range of species, the SCO has a secretory function (for thorough reviews of the literature, see References 25–28). Some basic questions concerning

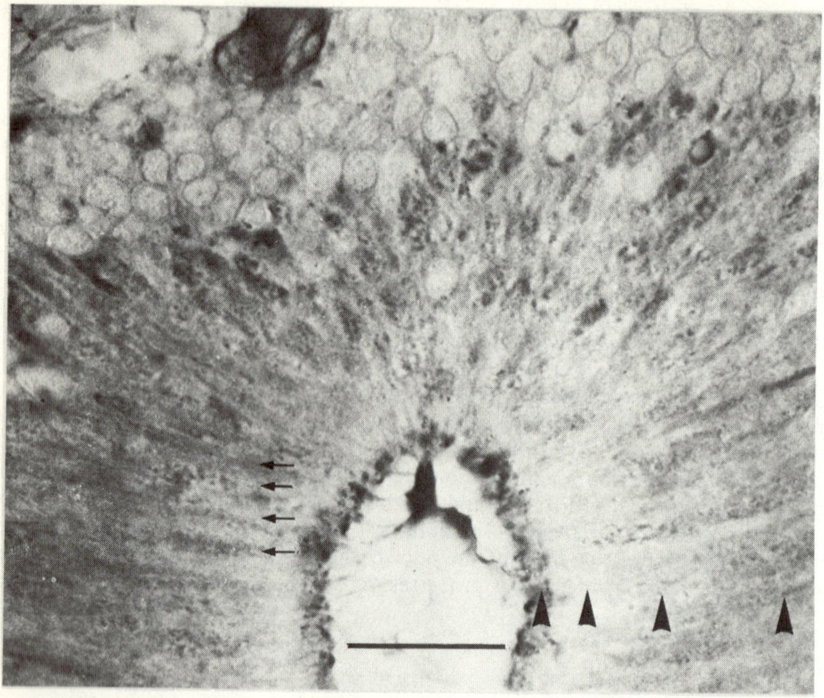

FIGURES 13–15. Frontal sections taken from the same area of the SCO all show PAS reactions after Maltase digestion. The two types of Maltase-resistent PAS-positive material are shown. Bars indicate 20 μm. FIGURE 13 (*above*). The SCO of a control rat injected with saline exhibits both striation (arrows) and stratification (arrowheads) due to the varying degrees of reactivity. FIGURE 14 (facing page, *top*). Note the increased PAS positivity of the SCO that occurs 24 hr after injection of 5,6-DHT. Compare the size of nuclei and nucleoli, the reactivity of the perinuclear cytoplasm, and the configuration of the SCO-aqueductal interface with the similar parameters in FIGURES 13 and 15. FIGURE 15 (facing page, *bottom*). The subventricular nonreacting layer has virtually disappeared, and, due to the very strong reactivity of the entire cytoplasm of many SCO cells, the organ appears very striated in this section obtained from a rat 4 days after injection of 5,6-DHT.

the synthetic machinery, the chemical nature of the secretory material, the packing in various types of granules, and the polarity of the secretion (apical and/or basal) are, however, subjects of much dispute. Part of this controversy is clearly a result of species differences, but the secretory activity is also a function of the developmental stage, of the individual secretory cycle of each cell, of the polarity of its secretion, and of the specific SCO region. Furthermore, the present study has demonstrated the crucial role played by the serotonergic system in the control of the secretory activity of the SCO of the adult rat.

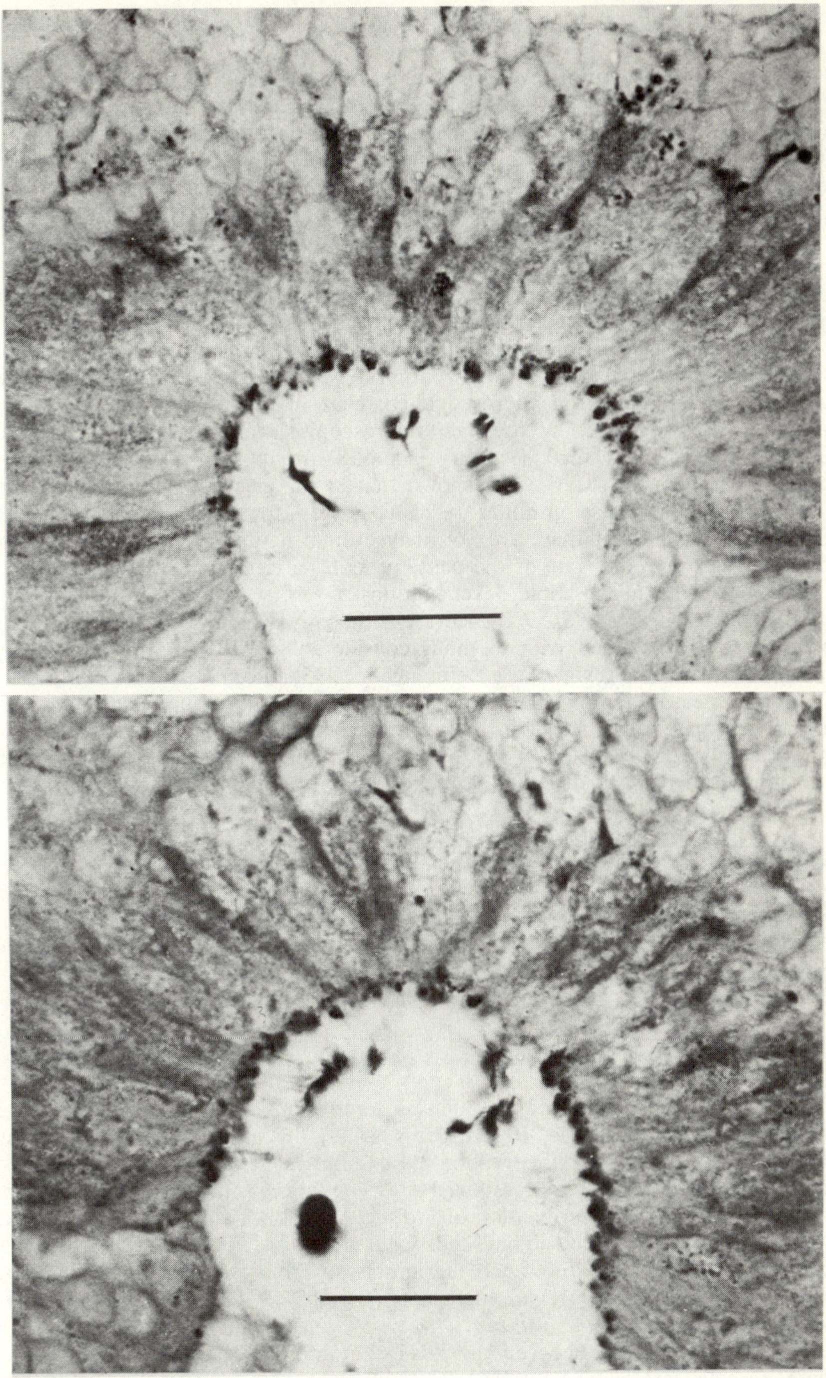

General Features of the Subcommissural Organ as a Secretory Organ

At the ultrastructural level, the secretory product of SCO cells is usually described as consisting of two populations of granules: light secretory sacs ("pale granules") and dense granules (see, for example, References 29–31). There is general agreement among most authors that the pale granules are derived from the endoplasmic reticulum, though their final fate and the role played by the Golgi complex are matters of controversy (see Reference 24). Although the secretory nature of the flocculent material has been questioned,[32] many observations support the idea that the light secretory sacs release their contents apically. Basal secretion into the blood vessels of the mammalian SCO has until now only been postulated in the dog [33] and the rabbit.[34]

The second population of granules, the dense granules, have also been the subject of much dispute. Some authors [28, 35] consider the electron-dense granules to be the electron microscopic correlate of the Gomori-positive neurosecretory substance seen at the light microscopic level, whereas Smith [32] has inferred that the so-called secretory granules are most likely lysosomes. A recent histochemical study [31] has shown that at the electron microscopic level, only few of the dense granules react positively for acid phosphatase and, therefore, that most of them are not of lysosomal nature. Not only the nature but also the origin and fate of the dense granules are matters of controversy.

At the light microscopic level, Stutinsky [36] first demonstrated Gomori-positive material in the SCO. Since then, the organ from various species has been extensively studied with methods considered selective for neurosecretory material (for critical reviews, see References 33, 35, and 37).

Some releasing factors (somatostatin,[38] luteinizing hormone-releasing factor, and thyrotropin-releasing factor [39]) and a hormone (arginine vasotocin [40]) have very recently been detected in the SCO with immunoperoxidase techniques and by radioimmunoassay. The significance of these findings is obscure at present.

Subcommissural Organ of the Rat as a Model for Neuron-Target Cell Interaction

The SCO of the adult rat is outstanding in several respects. Important in the present context is the heavy serotonergic innervation and the low secretory activity.

Innervation

Although there have been numerous electron microscopic studies of the rat SCO (see, for example, References 41–45), the ultrastructural equivalent of the dense serotonergic innervation, as demonstrated by fluorescence microscopy,[19-21] has not yet been established. The present investigation provides the first ultrastructural verification of a direct synaptic input to the mammalian SCO ependymal and hypendymal cells. Furthermore, we have established the serotonergic nature of the nerve terminals by fluorescence histochemistry, autoradiography, and electron microscopy, combined with selective chemical neurodegeneration.

The site of contact between the 5-HT-containing presynaptic elements and the SCO cells satisfies the ultrastructural criteria for a synapse, including the postsynaptic specialization characteristic for neurons. This feature clearly distinguishes the serotonergic synapse on SCO cells from the "synaptoid contacts" [46] between some unidentified nerve fibers and tanycytes of the median eminence. The serotonergically innervated SCO thus represents a unique model in which we can study various morphologic and functional effects of 5-HT terminals under normal and pathologic conditions.

Secretory Activity

Although the SCO of the normal rat at the ultrastructural level shows apical secretory granules, flocculent material in the endoplasmic reticulum, and some scattered Golgi complexes (cf. RESULTS), a comparison between the secretory activity of the rat and that of several other species indicates that the normal rat has a rather low secretory activity. Thus, in a thorough electron miscroscopic study, the complete absence of secretory elements in the basal processes was demonstrated.[42]

The secretory product of the rat SCO is known to be PAS positive, Gomori positive, and to contain protein.[25] These findings have been confirmed in the present study, and the results of the different histochemical reactions indicate the presence of glycoproteins (Maltase-resistant PAS positivity) combined with a neurophysin-like substance (the positive PFA-AB and Bock's reaction), which might well be the protein part of the glycoprotein, since the various reactions reveal parallel patterns of intensity and exhibit similar topographic distributions on serial sections.

Changes in Secretory Activity After Disruption of Serotonergic Input

Serotonergic synapses obviously exert a regulatory role in the rat SCO. As early as 24 hr after injection of 5,6-DHT or 5,7-DHT, we observed a tremendous engorgement of endoplasmic reticulum and an accumulation of pale and dense granules in apical, perinuclear, and basal parts of the organ. These phenomena might reflect either a stimulation of the synthetic machinery or an inhibition of release of secretory material or any combination of these possibilities. In a parallel light and electron microscopic study,[55] we found that the mean size (volume) of the SCO cell nucleus of 5,6-DHT-treated rats was significantly increased compared to that of control rats ($p < 0.001$). This dramatic change in nuclear volume (cf. FIGURES 13 & 14) and the presence of large nucleoli, together with indications of an increased release of secretory material (many exocytotic and pinocytotic profiles and presence of flocculent material in the extracellular space), suggest that the synthetic machinery is stimulated under our experimental conditions. Whether the increased release of secretory material is secondary to increased production of this material or whether it is directly dependent on the lack of serotonergic input is not known at present but open to experimental verification.

Possible Function of the Serotonergic System

Despite intense research efforts during the last two decades, little is known about the function of serotonergic neurons in the mammalian central nervous system.[47] Concepts based on pharmacologic interference with serotonergic transmission suffer from a lack of specificity. Thus, the behavioral changes that have been reported in raphe-lesioned animals (e.g., increased motor activity, enhanced irritability, and facilitated avoidance reactions) appear to result mainly from damage to nonserotonergic structures, because interruption of the ascending 5-HT projections by localized intracerebral injection of 5,7-DHT, which gives more selective and more profound reductions in telencephalic serotonin content that does raphe destruction, failed to induce similar abnormalities.[48, 49] Any prediction about the function of serotonergic transmission would require an identification of the 5-HT synapses and the postsynaptic effects mediated by them at their target cell level (cf. pp. 262–264).

That functional changes in central serotonergic neurons can cause region-specific changes in brain protein synthesis has, in fact, been demonstrated in mice treated with a rather selective serotonin-lowering [*p*-chlorophenylalanine (*p*-CPA)] or a rather selective serotonin-raising drug (6-methyltetrahydro-β-carboline).[50] Under conditions of reduced cerebral 5-HT content (*p*-CPA), brain microsomal protein synthesis was increased, whereas it was decreased when 5-HT levels were elevated (6-methyltetrahydro-β-carboline).

Recently published biochemical observations [51] indicate that 5-HT formed from systemically injected 5-hydroxytryptophan (5-HTP) by decarboxylation causes an immediate disruption of protein synthesis (polysomal disaggregation) in rat brain; this effect of 5-HTP appears to be mediated by 5-HT, because it can be prevented by pretreatment of animals with decarboxylase inhibitors or the known antiserotonergic drugs cyproheptadine or methysergide. Though this finding provides clear evidence for an inhibitory role of serotonin on protein synthesis in the adult rat brain, it remains to be established whether this effect of serotonin is mediated by the 5-HT receptors that are under physiologic influence of serotonergic synapses in the brain.

Whether the inhibitory effects of 5-HT on protein synthesis of SCO cells are mediated by hyperpolarization of the postsynaptic membrane or by a direct effect of 5-HT transported across the cell membrane into some serotoninoceptive sites in the SCO cell cytoplasm (analogous to the mechanism of steroid action on protein synthesis in target cells) remains to be elucidated. The proposed translocation of 5-HT to some receptive sites in the cytoplasm could well be mediated by the postulated serotonin-binding protein.[52] The former mechanism appears to support published findings by Bloom *et al.*,[53] who have found that iontophoretically administered 5-HT depresses electrical activity in many neurons of the rat central nervous system.

In the periphery of mammals, serotonin has been found coexistent with polypeptide secretion in the so-called amine precursor uptake decarboxylase (APUD) cells [54] of the gastric, intestinal, and bronchial mucosae, and the concept has been advanced that the synthesis and/or release of the secretory peptides is under the control of 5-HT. If so, serotonin, the suspected transmitter of 5-HT neurons in the mammalian central nervous system, might have some additional role as an internal modulator of release of as yet unidentified secretory peptide(s) that may affect the serotonergically innervated target neurons or, in case of the rat SCO, the modified ependymal cells.

Conclusions

Previous fluorescence microscopic observations of a very dense indoleaminergic innervation of the SCO are confirmed, and our ultrastructural studies have demonstrated that axonal boutons filled with both small, empty and large, granular vesicles establish synaptic contact with the basal processes and perikarya of the modified SCO ependymal cells. The indoleaminergic nerves in the region also take up and concentrate 6-HTP and [^{14}C]-5,6-DHT with remarkable specificity.

Both 5,6-DHT and 5,7-DHT given into the lateral cerebral ventricle as a single injection of 50 μg (free base) cause almost complete destruction of the synapses that innervate the SCO cells and also of varicose fibers of the hypendyma.

Degeneration of 5-HT synapses is associated with marked changes in the ultrastructural and light microscopic appearances of SCO cells. The dominant feature at the electron microscopic level is an enormous swelling of an extensive network of endoplasmic reticulum, the presence of scattered well-organized Golgi complexes, and accumulation of secretory granules in apical and basal processes. Increased release of secretory material was also observed. A moderate increase of the extracellular space in the SCO region and a dramatic increase in nuclear and nucleolar volumes are detected at both light and electron microscopic levels. Histochemically, an increased reaction for neurophysin-like material and for glycoproteins in both apical and basal processes is evident. Taken together, the reported changes indicate that the serotonergic input inhibits both synthesis and release of secretory material in the SCO of the adult rat.

The exact role of transmitters, in this case 5-HT, requires further clarification. It seems evident that the 5-HT system has effects other than those related to the transfer of electrical activity, at least in the rat SCO. Whether the influences on genetic expression, metabolism, and secretory activity are caused by serotonin per se or by some other substance(s) released concomitantly from the serotonergic system is not known at present.

Acknowledgments

We appreciate the skillful technical assistance received during this study from Lisbeth G. Jørgensen (histochemistry), Yael Balslev (thin sectioning), and K. Romer (photography). We also thank A. Björklund for ideas and suggestions.

References

1. Møllgård, K. & L. Wiklund. 1978. Neuro-ependymal serotonergic synapses in the rat subcommissural organ. J. Neurocytol. In press.
2. Raisman, G. & M. R. Matthews. 1972. Degeneration and regeneration of synapses. *In* The Structure and Function of Nervous Tissue. Volume IV. Physiology II and Biochemistry II. : 61–104. Academic Press. New York and London.
3. Baumgarten, H. G., A. Björklund, L. Lachenmayer, A. Nobin & U. Stenevi. 1971. Longlasting selective depletion of brain serotonin by 5,6-dihydroxytryptamine. Acta Physiol. Scand. Suppl. **373:** 1–16.
4. Baumgarten, H. G., K. D. Evetts, R. B. Holman, L. L. Iversen, M. Vogt & G. Wilson. 1972. Effects of 5,6-dihydroxytryptamine on monoaminergic neurons in the central nervous system of the rat. J. Neurochem. **19:** 1587–1597.

5. Björklund, A., A. Nobin & U. Stenevi. 1973. Effects of 5,6-dihydroxytryptamine on nerve terminal serotonin and serotonin uptake in the rat brain. Brain Res. **53:** 117–127.
6. Baumgarten, H. G., A. Björklund, A. F. Holstein & A. Nobin. 1972. Chemical degeneration of indolamine axons in the rat brain by 5,6-dihydroxytryptamine. Z. Zellforsch. **129:** 256–271.
7. Chan-Palay, V. 1975. Fine structure of labelled axons in the cerebellar cortex and nuclei of rodent and primates after intraventricular infusions with tritiated serotonin. J. Anat. Embryol. **148:** 235–265.
8. Chan-Palay, V. 1976. Serotonin axons in the supra- and subependymal plexuses and in the leptomeninges: their roles in local alterations of cerebrospinal fluid and vasomotor activity. Brain Res. **102:** 103–130.
9. Baumgarten, H. G. & L. Lachenmayer. 1972. 5,7-Dihydroxytryptamine. Improvement in chemical lesioning of indolamine neurons in the mammalian brain. Z. Zellforsch. **135:** 399–414.
10. Baumgarten, H. G. & L. Lachenmayer. 1974. Indoleamine-containing nerve terminals in the rat median eminence. Z. Zellforsch. **147:** 285–292.
11. Bloom, F. E. 1971. Fine structural changes in rat brain after intracisternal injection of 6-hydroxydopamine. *In* 6-Hydroxydopamine and Catecholamine Neurons. T. Malmfors & H. Thoenen, Eds. : 135–150. North Holland Publishing Company. Amsterdam, The Netherlands.
12. Richards, J. G. 1971. Ultrastructural effects of 6-hydroxydopamine on catecholamine neurons in the rat brain. *In* 6-Hydroxydopamine and Catecholamine Neurons. T. Malmfors & H. Thoenen, Eds. : 151–161. North Holland Publishing Company. Amsterdam, The Netherlands.
13. Aghajanian, G. K., M. J. Kuhar & R. H. Roth. 1973. Serotonin-containing neuronal perikarya and terminals: differential effects of p-chlorophenylalanine. Brain Res. **54:** 85–101.
14. Björklund, A., B. Falck & Ch. Owman. 1972. Fluorescence microscopic and microspectrofluorometric techniques for the cellular localization and characterization of biogenic amines. *In* Methods of Investigative and Diagnostic Endocrinology. Volume 1. The Thyroid and Biogenic Amines. S. A. Berson, Ed. : 318–368. North Holland Publishing Company. Amsterdam, The Netherlands.
15. Fuxe, K. & G. Jonsson. 1967. A modification of the histochemical fluorescence method for the improved localization of 5-hydroxytryptamine. Histochemie **11:** 161–166.
16. Kramer, H. & G. M. Windrum. 1955. The metachromatic staining reaction. J. Histochem. Cytochem. **3:** 227–237.
17. Pearse, A. G. E. 1968. Histochemistry, Theoretical and Applied. 3rd edit. Vol. 1. Churchill, Ltd. London, England.
18. Bock, R. 1966. Über die Darstellbarkeit neurosekretorischer Substanz mit Chromalaun-Gallocyanin im supraoptico-hypophysären System beim Hund. Histochemie **6:** 362–369.
19. Fuxe, K., T. Hökfelt & U. Ungerstedt. 1968. Localization of indolealkylamines in CNS. Advan. Pharmacol. **6A:** 235–251.
20. Björklund, A., Ch. Owman & K. A. West. 1972. Peripheral sympathetic innervation and serotonin cells in the habenular region of the rat brain. Z. Zellforsch. **127:** 570–579.
21. Wiklund, L. 1974. Development of serotonin-containing cells and the sympathetic innervation of the habenular region in the rat brain. Cell Tissue Res. **155:** 231–243.
22. Björklund, A., B. Falck & U. Stenevi. 1971. Microspectrofluorometric characterization of monoamines in the central nervous system: evidence for a new monoamine-like compound. *In* Histochemistry of Nervous Transmission. O. Erankö, Ed. Vol. 34: 63–73. Elsevier. Amsterdam, The Netherlands.
23. Richards, J. G. & J. P. Tranzer. 1974. Ultrastructural evidence for the localization of an indolealkylamine in the supra-ependymal nerve from combined cytochemistry and pharmacology. Experientia **30:** 287–289.

24. KIMBLE, J. E. & K. MØLLGÅRD. 1975. Subcomissural organ-associated neurons in fetal and neonatal rabbit. Cell Tissue Res. **159:** 195–204.
25. OLSSON, R. 1958. Studies on the subcommissural organ. Acta Zool. **39:** 71–102.
26. PALKOVITS, M. 1965. Morphology and function of the subcommissural organ. Stud. Biol. Hung. **4:** 1–105.
27. STERBA, G. 1969. Morphologie und Funktion des Subkommissuralorgans. *In* Zirkumventrikuläre Organe und Liquor. : 17–27. Gustav Fischer Verlag. Jena, German Democratic Republic.
28. HERRLINGER, H. 1970. Licht- und elektronenmikroskopische Untersuchungen am Subcommissuralorgan der Maus. Ergeb. Anat. Entwickl.-Gesch. **42:** 1–73.
29. PAPACHARALAMPOUS, N., A. SCHWINK & R. WETZSTEIN. 1968. Elektronenmikroskopische Untersuchungen am Subcommissuralorgan des Meerschweinchens. Z. Zellforsch. **90:** 202–229.
30. RODRÍGUEZ, E. 1970. Ependymal specializations. III. Ultrastructural aspects of the basal secretion of the toad subcommissural organ. Z. Zellforsch. **111:** 32–50.
31. MURAKAMI, M., T. SHIMADA, T. ORIBE & T. HIRAKI. 1972. An electron microscopic study on the subcommissural organ of the monkey. Macacus fuscatus. Arch. Histol. Jap. **34:** 61–72.
32. SMITH, U. 1970. Aspects of fine structure and function of the subcommissural organ of the embryonic chick. Tissue Cell **2:** 19–32.
33. OKSCHE, A. 1969. The subcommissural organ. J. Neuro-Visc. Relat. Suppl. **9:** 111–139.
34. KIMBLE, J. E. & K. MØLLGÅRD. 1973. Evidence for basal secretion in the subcommissural organ of the adult rabbit. Z. Zellforsch. **142:** 223–239.
35. VIGH, B., P. RÖHLICH, I. TEICHMANN & B. AROS. 1967. Ependymosecretion (ependymal neurosecretion). VI. Light and electron microscopic examination of the subcommissural organ of the guinea pig. Acta Biol. Acad. Sci. Hung. **18:** 53–66.
36. STUTINSKY, F. 1960. Colloide, corps de Herring et substance Gomori positive de la neurohypophyse. Compt. Rend. **144:** 1357–1360.
37. MØLLGÅRD, K., M. MØLLER & J. KIMBLE. 1973. Histochemical investigations on the human fetal subcommissural organ. II. The "large granules." Histochemie **37:** 61–74.
38. PELLETIER, G., R. LECLERC, D. DUBE, F. LABRIE, R. PUVIANI, A. ARIMURA & A. V. SCHALLY. 1975. Localization of growth hormone-release-inhibiting hormone (somatostatin) in the rat brain. Amer. J. Anat. **142:** 397–400.
39. KIZER, J. S., M. PALKOVITS & J. BROWNSTEIN. 1976. Releasing factors in the circumventricular organs of the rat brain. Endocrinology **98:** 311–317.
40. ROSENBLOOM, A. A. & D. A. FISCHER. 1975. Notes and comments. Arginine vasotocin in the rabbit subcommissural organ. Endocrinology **96:** 1038–1039.
41. WETZSTEIN, R., A. SCHWINK & P. STANKA. 1963. Die periodisch strukturierten Körper im Subcommissuralorgan der Ratte. Z. Zellforsch. **61:** 493–523.
42. STANKA, P., A. SCHWINK & R. WETZSTEIN. 1964. Elektronenmikroskopische Untersuchung des Subcommissuralorgans der Ratte. Z. Zellforsch. **63:** 277–301.
43. SCHWINK, A. & R. WETZSTEIN. 1966. Die Kapillaren im Subcommissuralorgan der Ratte. Elektrononmikroskopische Untersuchungen an Tieren verschiedenen Lebensalters. Z. Zellforsch. **73:** 56–88.
44. LIN, H.-S. & I.-L. CHEN. 1969. Development of the ciliary complex and microtubules in the cells of rat subcommissural organ. Z. Zellforsch. **96:** 186–205.
45. KRSTIĆ, R. 1973. Ultrastrukturelle Lokalisation von Mukosubstanzen der Zellhülle im Subcommissuralorgan der Ratte. Z. Zellforsch. **139:** 237–252.
46. GÜLDNER, F.-H. & J. R. WOLFF. 1973. Neurono-glial contacts in the median eminence of the rat: ultrastructure, staining properties and distribution on tanycytes. Brain Res. **61:** 217–234.
47. CHASE, T. N. & D. L. MURPHY. 1973. Serotonin and central nervous system function. Annu. Rev. Pharmacol. **13:** 181–197.

48. LORENS, S. A., C. KÖHLER, B. SREBRO & C. H. GULDBERG. 1975. Behavioral effects of central 5-HT depletion: comparison of pcpa, 5,7-DHT and electrolytic midbrain raphe lesions. Exp. Brain Res. Suppl. **23:** 130.
49. HOLE, K., K. FUXE & G. JONSSON. 1975. The effect of 5,7-dihydroxytryptamine lesions of ascending 5-hydroxytryptamine pathways on pain sensitivity and motor activity. Exp. Brain Res. Suppl. **23:** 92.
50. ESSMAN, W. B. 1975. Diurnal differences in altered brain 5-hydroxytryptamine-related regional protein synthesis. J. Pharmacol. (Paris) **6**(3): 313–322.
51. WEISS, B. F., J. L. LIEBSCHÜTZ, R. J. WURTMAN & H. N. MUNRO. 1975. Participation of dopamine- and serotonin-receptors in the disaggregation of brain polysomes by l-Dopa and l-5-HTP. J. Neurochem. **24:** 1191–1195.
52. TAMIR, H. & M. M. RAPPORT. 1976. Is the serotonin binding protein (SBP) a soluble storage form for serotonin? Res. Commun. Chem. Pathol. Pharmacol. **13:** 225–235.
53. BLOOM, F. E., B. J. HOFFER, C. NELSON, Y. SHEU & G. R. SIGGINS. 1973. The physiology and pharmacology of serotonin mediated synapses. *In* Serotonin and Behavior. J. Barchas & E. Usdin, Eds. : 249–262. Academic Press. New York and London.
54. PEARSE, A. G. E. 1969. The cytochemistry and ultrastructure of polypeptide hormone-producing cells of the APUD series and the embryologic, physiologic and pathologic implications of the concept. J. Histochem. Cytochem. **17:** 303–313.
55. LUNDBERG, J. J. & K. MØLLGARD. 1977. Unpublished results.

NEUROTOXIC ACTION OF HALOGENATED AMPHETAMINES *

John A. Harvey

Department of Psychology
The University of Iowa
Iowa City, Iowa 52242

The development of selective monoamine neurotoxins, beginning with the demonstration of peripheral [1, 2] and central [3, 4] actions of 6-hydroxydopamine (6-OHDA) on catecholamine neurons and the more recent demonstration of central neurotoxic actions of 5,6-dihydroxytryptamine and 5,7-dihydroxytryptamine (5,6-DHT and 5,7-DHT) on serotonergic neurons,[5–7] has provided valuable tools for functional and morphologic studies of central monoamine systems. More recently, it has been demonstrated that several halogenated amphetamines form a third class of compounds that produce a selective neurotoxic action, which appears to be restricted primarily to serotonergic neurons.[8, 9]

The first evidence for a possible neurotoxic action of halogenated amphetamines resulted from the finding that 4-chloroamphetamine produced decreases in both serotonin and 5-hydroxyindoleacetic acid (5-HIAA) content of brain that persisted for as long as 4 months after a single injection.[10] Subsequently, it was shown that 4-bromoamphetamine [11] and fenfluramine [9, 12] also produced long-term and apparently irreversible decreases in the 5-hydroxyindole content of brain. In contrast, decreases in serotonin and 5-HIAA contents of brain produced by other halogenated amphetamines,[13] including β,β-difluoro-4-chloroamphetamine [14] and 4-fluoroamphetamine,[11] persist for less than 24 hr. The ability of these drugs to produce either short- or long-term depletion of 5-hydroxyindoles is blocked by inhibitors of serotonin uptake, such as chlorimipramine [15] and fluoxetine,[16] suggesting a requirement for transport into serotonergic neurons. These findings led to the suggestion that 4-chloroamphetamine and related compounds or their metabolite(s) may produce a neurotoxic action after their uptake into serotonin neurons.[13, 17]

The present paper will describe some of the anatomic evidence that has been obtained concerning the neurotoxic effects of various halogenated amphetamines by the use of Nissl- and silver-stained sections of brain.

Materials and Methods

Male albino rats (Holtzman, Madison, Wisconsin) were housed two per cage with free access to Purina Laboratory Chow and tap water in an animal colony maintained at a temperature of 22° C and an alternating 12-hr light-dark cycle. Rats weighing 150–200 g were injected intraperitoneally with various drugs dissolved in sterile physiologic saline. Control animals received equivalent volumes of saline (2 ml/kg). 4-Chloroamphetamine was purchased from Regis

* Supported by United States Public Health Service Grant MH16841 from the National Institute of Mental Health.

0077–8923/78/0305–0289 $01.75/0

Chemical Company, while fenfluramine was a gift from A. H. Robins. Fluoxetine, β,β-difluoro-4-chloroamphetamine, 4-bromoamphetamine, and 4-fluoroamphetamine were synthesized in the Lilly Research Laboratories. All of these drugs were racemic mixtures of hydrochloride salts. 5,7-Dihydroxytryptamine creatinine sulfate (Regis Chemical Company) was dissolved in a 1% (w/v) solution of ascorbic acid, and 10 μl that contained 100 μg of the free base were injected into each lateral ventricle by the method of Noble *et al.*[18] At various times after injection, as indicated in the text, animals were perfused transcardially, under pentobarbital anesthesia, by a modification of the method of DeOlmos and Ingram.[19] Frozen, 30 μm thick, coronal sections, were stained by a modification[8] of the cresyl violet stain described by Fernstrom[20] or by the silver impregnation method (procedure I) of Fink and Heimer.[21] Separate groups of rats were decapitated at various times after injection, the brains, minus olfactory bulbs and cerebellum, were homogenized in 4 ml of 0.1 N HCl, and serotonin content was determined by the butanol extraction method of Bogdanski *et al.*[22] In some cases, an aliquot of the 0.1 N HCl homogenate was combined with an equal volume of 20% (w/v) trichloroacetic acid for the simultaneous determination of catecholamines. Catecholamines were extracted by the method of Crout *et al.*,[23] and dopamine and norepinephrine contents of the extract were determined by the method of Chang.[24] Serotonin creatinine sulfate (Regis Chemical Company), dopamine hydrochloride, and norepinephrine bitartrate (Nutritional Biochemical Corporation) served as standards. In some cases, selected brain regions were analyzed for serotonin content by the *o*-phthaldehyde condensation technique after butanol extraction.[25]

Results

Comparison of Nissl-stained material obtained from control rats injected with physiologic saline and from rats injected with 100 μmol/kg of 4-chloroamphetamine revealed clear signs of an intense neurotoxic reaction in a large number of cells (approximately 30 μm in diameter) located in the ventral midbrain tegmentum.[8] These cells demonstrated a shrunken, pyknotic appearance with a dense staining of Nissl substance throughout the cytoplasm and nucleus and a prominent perineuronal space. The distribution of these reacting cell bodies in the ventral midbrain tegmentum corresponded to the distribution of the serotonergic cell group designated B-9 by Dahlström and Fuxe[26] and Jonsson *et al.*[27] This neurotoxic reaction could be seen as early as 1 day after injection and was still present at 9, 14, and 30 days, though the intensity of Nissl staining and the presence of perineuronal spaces were greatly reduced. However, from 1 day after injection, there existed an increased staining of glial cells and evidence of glial reactions and neuronophagia.[8] The material obtained at 30 days after 100 μmol/kg of 4-chloroamphetamine indicated that at least 80% of the cell bodies in this region had undergone irreversible degeneration. Dosages of 50 (Figure 1c) or 100 μmol/kg (Figure 2a) of 4-chloroamphetamine produced comparable effects in the B-9 region. There was some slight reduction in reacting cells after the 25 μmol/kg dose and 12.5 μmol/kg appeared to be the threshold dose (Figure 1b; cf. control section in Figure 1a).

An identical neurotoxic action of 4-chloroamphetamine could be observed in sections stained by the silver impregnation method. Cells within the B-9 region demonstrated a shrunken appearance, a clumping of argyrophilic deposits in the cytoplasm, a dense argyrophilia of the nucleus, and a prominent perineuronal space. However, there was never any detectable evidence of degenerating axons or terminals in any region of the brain in silver-stained sections taken from animals at 1, 3, 9, 14, or 30 days after injection of 100 μmol/kg of 4-chloroamphetamine.[8]

The neurotoxic effects of 4-chloroamphetamine appeared to be highly specific for cell bodies in the B-9 region. There was no evidence of a similar reaction accompanied by evidence of neuronophagia in any other area of the brain, though there was an occasional and inconsistent occurrence of some abnormally staining cells (approximately 10–15 μm in diameter) in the raphe nuclei (B-7 and B-8). However, such reacting neurons were also seen in the B-7 region of control animals and hence were most likely due to an artifact of formalin fixation rather than a neurotoxic effect of the drug. Although the abnormally staining cells in the B-8 region occurred primarily after drug treatment, we could not detect any evidence of neuronophagia in the B-8 region after longer survival times.[8]

TABLE 1

EFFECT OF 100 μMOL/KG OF 4-CHLOROAMPHETAMINE ON BRAIN MONOAMINES AT 30 DAYS AFTER INJECTION

Monoamine	Monoamine Content (nmol/g±SEM) Control	4-Chloroamphetamine	Percent Change
Serotonin	3.71±0.17	2.08±0.06 *	−44
Norepinephrine	2.07±0.24	2.08±0.21	0
Dopamine	4.14±0.50	3.97±0.29	−4

* Significantly different from control value ($p < 0.001$, two-tailed t test).

The neurotoxic action of 4-chloroamphetamine on cells within the B-9 region was accompanied by a highly specific action on serotonin content of the brain. At 30 days after injection of 100 μmol/kg of 4-chloroamphetamine, there was a significant reduction in serotonin but no significant change in either norepinephrine or dopamine (TABLE 1).

An examination of several other halogenated amphetamines revealed that 100 μmol of either 4-bromoamphetamine[28] or fenfluramine[9] also produced signs of neurotoxic reactions in cell bodies located within the B-9 region, while 4-fluoroamphetamine (100 μmol/kg) or β,β-difluoro-4-chloroamphetamine (400 μmol/kg) had no detectable neurotoxic action.[28] 4-Chloroamphetamine and 4-bromoamphetamine had equivalent neurotoxic actions, while the number of reacting neurons was less after equimolar dosages of fenfluramine, so that a number of normally staining cell bodies could be observed in the B-9 region. Fenfluramine also differed in its effects within the B-9 region. In the caudal portions, the reacting cells were identical in appearance to those seen after 4-chloroamphetamine administration, while in the rostral two thirds of B-9, the cells differed in that they demonstrated a smaller perineuronal space. However, as in the case of 4-chloroamphetamine, the threshold dose for detecting neurotoxicity after fenfluramine dosing was also 12.5 μmol/kg.[29] This differen-

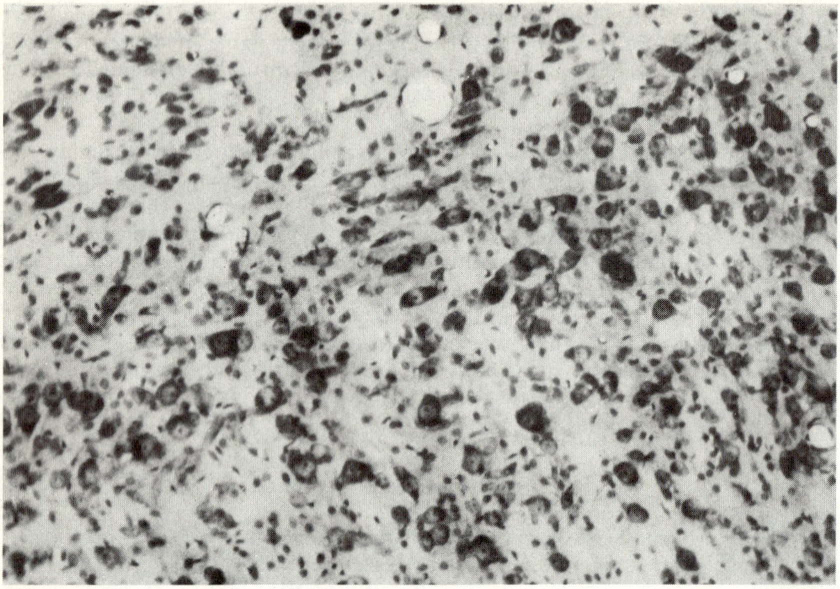

FIGURES 1a–1c. Appearance of cells in the B-9 region in cresyl violet-stained sections ($\times$100) at 3 days after (FIGURE 1a, *above*) saline injection. Note the normal appearance of cells in this region. FIGURE 1b (facing page, *top*). Appearance 3 days after 12.5 μmol/kg of 4-chloroamphetamine. Note the normal appearance of most cells but the occurrence of some darkly staining, shrunken cell bodies with a perineuronal space located especially in the lower left portion of this Figure. FIGURE 1c (facing page, *bottom*). Appearance 3 days after 50 μmol/kg of 4-chloroamphetamine. Note the large number of shrunken and darkly staining cells with prominent perineuronal spaces scattered throughout this region.

tial neurotoxic potency of 4-chloroamphetamine as compared to fenfluramine was also reflected in a differential effect on brain serotonin content at 30 days after injection of 100 μmol/kg (TABLE 2).[9] 4-Chloroamphetamine not only produced a greater decrease in serotonin content of the brain but the pattern of decrease also differed. For example, fenfluramine did not significantly decrease the hippocampal content of serotonin, whereas 4-chloroamphetamine produced a large (76%) decrease. In contrast, the decrease in serotonin content of the amygdala was significant after injection of fenfluramine (46%) and 4-chloroamphetamine (75%).

Both the neurotoxic and serotonin-depleting actions of fenfluramine and 4-chloroamphetamine at dosages of 100 μmol/kg could be completely blocked by giving an animal 10 mg/kg of fluoxetine[28] (dosage calculated as the salt) 10 min before the intraperitoneal injection of either drug (FIGURE 2b; cf. FIGURE 2a).

Since a single injection of fenfluramine that produces a just-detectable neurotoxic reaction in some cell bodies within the B-9 region is quite low (12.5 μmol/kg), we also investigated the effects of chronic treatment (10 daily injections) at this dosage and at a dosage of 6.25 μmol/kg, which has no detectable neurotoxic effects after a single injection.[29] Animals were either

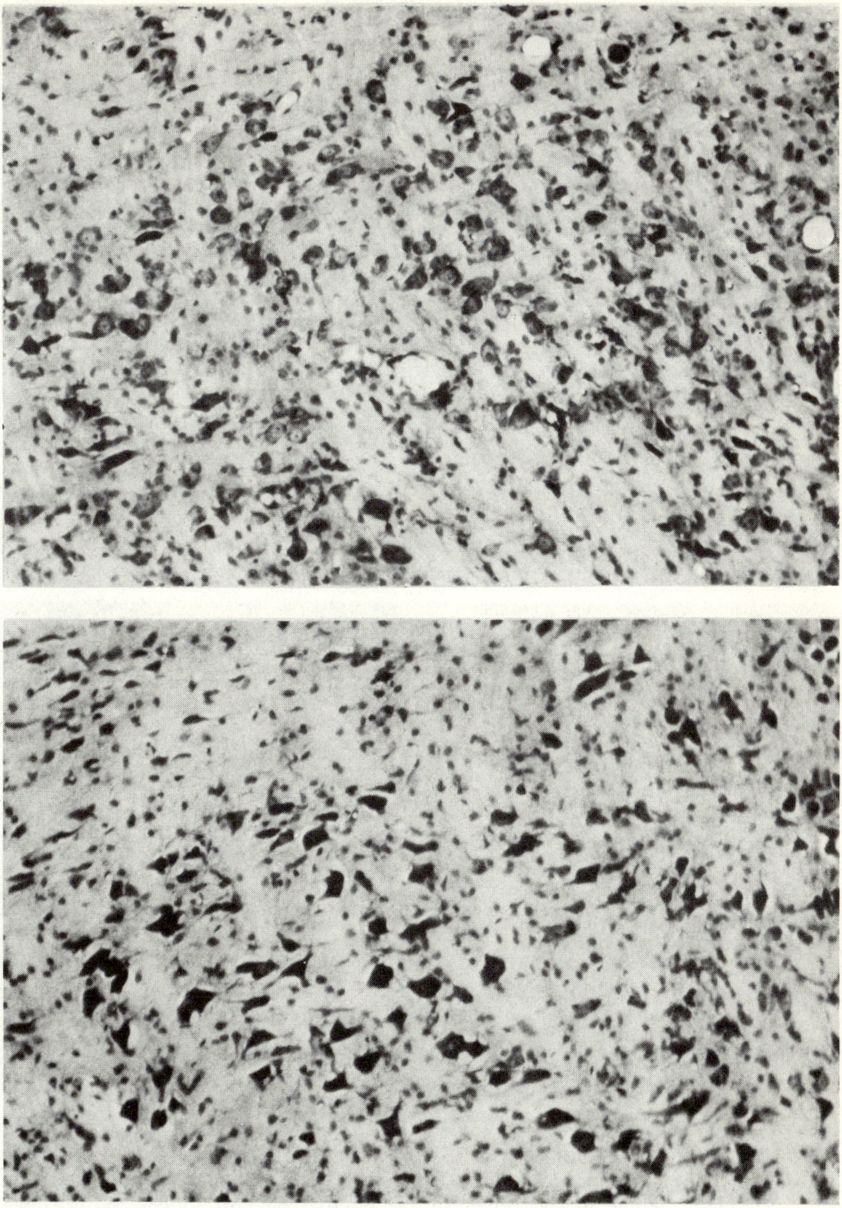

perfused for detection of neurotoxicity or decapitated for analysis of brain content of serotonin 3 days after the last injection. Ten daily injections of fenfluramine at 6.25 μmol/kg had no neurotoxic effect on cell bodies in the B-9 region (FIGURE 3a). However, 10 daily injections of 12.5 μmol/kg (FIGURE 3b) had a cumulative neurotoxic action, as indicated by the presence of a greater number of reacting neurons in the B-9 region and by a significantly

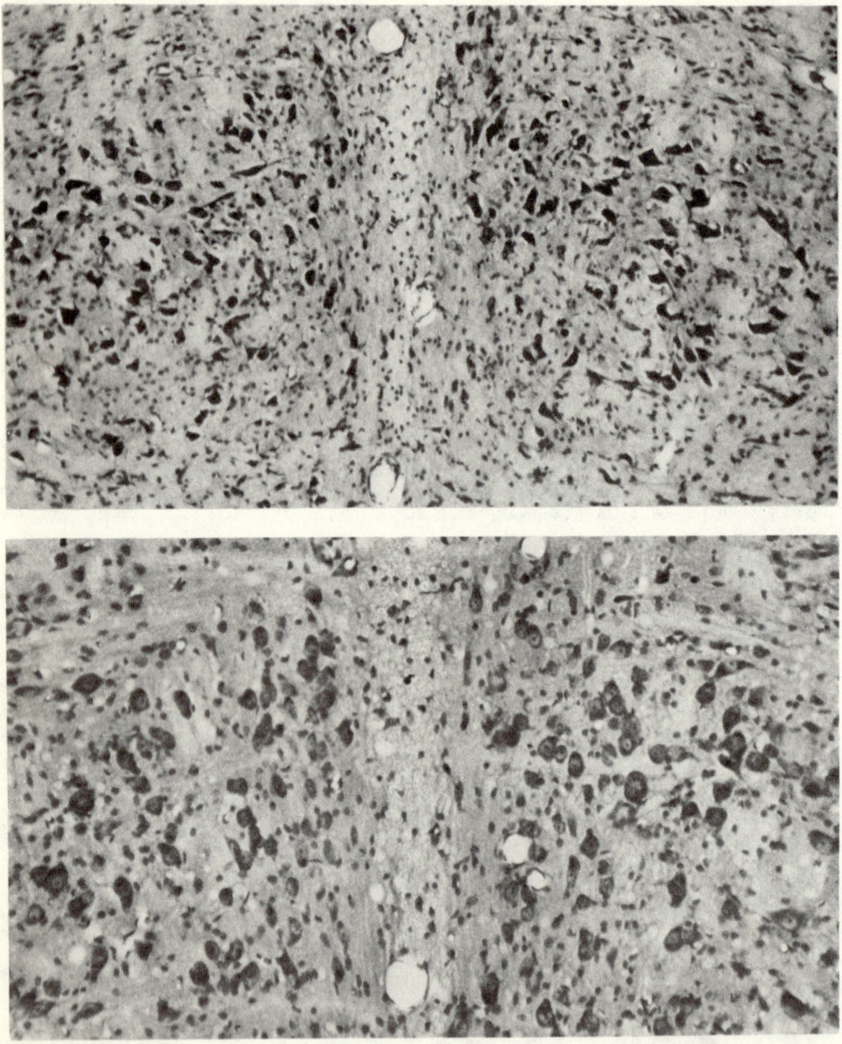

FIGURE 2. Appearance of cells in the B-9 region in cresyl violet-stained sections ($\times$100) at 3 days after treatment with 100 μmol/kg of 4-chloroamphetamine (FIGURE 2a, *top*). The midline at the center of the photograph corresponds to the three prominent blood vessels. Note the darkly staining, shrunken appearance and prominent perineuronal spaces surrounding the reacting neurons. FIGURE 2b (*bottom*). Fluoxetine (10 mg/kg) given 10 min prior to 100 μmol/kg of 4-chloroamphetamine. Note the normal appearance of cells.

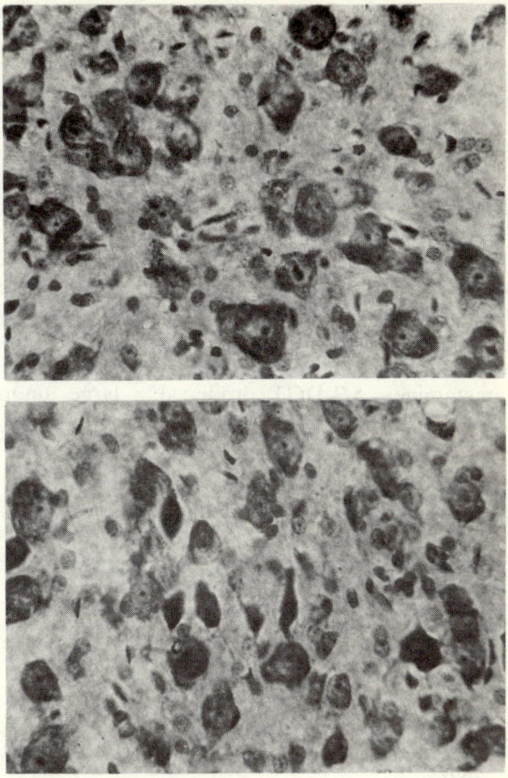

FIGURE 3. Appearance of cells in the B-9 region in cresyl violet-stained sections (×430) at 3 days after 10 daily injections of 6.25 μmol/kg of fenfluramine (FIGURE 3a, *top*). Note the normal appearance of all cells. FIGURE 3b (*bottom*). Appearance 3 days after 12.5 μmol/kg of fenfluramine. Note the dark staining, shrunken appearance and perineuronal spaces, demonstrated by several reacting cell bodies, and the normal appearance of many cells in the field.

TABLE 2

EFFECT OF 100 μMOL/KG OF 4-CHLOROAMPHETAMINE OR FENFLURAMINE ON SEROTONIN CONTENT OF VARIOUS BRAIN REGIONS AT 30 DAYS AFTER INJECTION

Brain Region	Serotonin Content (nmol/g±SEM) Saline Injection	Percentage Change in Serotonin Content: Fenfluramine	Percentage Change in Serotonin Content: 4-Chloroamphetamine
Hippocampus	2.43±0.31	−11	−76*
Amygdala	4.42±0.12	−46*	−75*
Telencephalon	2.75±0.08	−20*	−52*
Brain stem †	5.24±0.13	−12*	−24*

* Significant difference from control value ($p < 0.05$, two-tailed *t* test). Values are based on from three to six rats per group. The telencephalon was assayed after the hippocampus and amygdala had been removed.

† Includes the diencephalon.

greater reduction in brain serotonin (TABLE 3) than seen after a single injection of either 12.5 or 25 μmol/kg of fenfluramine.

It is of interest to compare the neurotoxic effects of the halogenated amphetamines after intraperitoneal injection with the neurotoxic effects of intraventricularly applied 5,7-DHT by use of the same conventional stains. 5,7-Dihydroxytryptamine (200 μg) produced a much greater neurotoxic reaction in the midbrain than did the halogenated amphetamines. In silver-stained sections, approximately 50% of the cells in the B-7, B-8, and B-9 regions demonstrated a shrunken appearance, intense argyrophilia, and, occasionally, a perineuronal space. The reacting cells in the B-8 region though greater in number looked similar to cells that reacted after treatment with 4-chloroamphetamine, 4-bromoamphetamine, or fenfluramine. The abnormally staining cells in the B-9 region appeared identical to cells that reacted after fenfluramine injection. However, in contrast to the neurotoxic effects of the halogenated amphetamines, 5,7-DHT produced a large amount of terminal degeneration confined to the midline of the midbrain and extending from the B-7 to the B-9 region. A great deal of the terminal degeneration surrounded the reacting cells in these regions (see FIGURES 4a, 4b & 4c).

TABLE 3

EFFECT OF ACUTE AND CHRONIC INJECTION OF FENFLURAMINE ON SEROTONIN CONTENT ON RAT BRAIN

Fenfluramine Dosage (μmol/kg)	Number of Injections	Percent Change in Serotonin Content 3 Days After Last Injection
6.25	1	0
6.25	10	−4
12.5	1	−6
12.5	10	−18*
25.0	1	−9*
50.0	1	−43*

* Significant change from saline controls ($p < 0.02$, two-tailed t test). The mean brain content of serotonin in saline controls ($n=9$) was (nmol/g±SEM) 3.89 ± 0.06.

DISCUSSION

Although it is now clear that several halogenated amphetamines produce a neurotoxic action within the B-9 region, it has not been definitely established that the reacting neurons are serotonergic. Such proof must await studies that employ histochemical fluorescence methods. However, the following lines of evidence suggest that we are, in fact, observing the irreversible degeneration of B-9 serotonergic cell bodies: 1) Only drugs that induce a long-term and apparently irreversible decrease in the brain content of serotonin (4-chloroamphetamine, 4-bromoamphetamine, and fenfluramine) produce signs of neurotoxicity in the B-9 region.[28] 2) The extent of the neurotoxic reaction in the B-9 region is related to the amount of serotonin depletion.[28, 29] 3) Both the neurotoxic effects of 4-chloroamphetamine and fenfluramine and the long-term depletion of serotonin are blocked by inhibitors of serotonin uptake, suggesting a requirement for transport into serotonergic neurons.[28] 4) The serotonin content and tryptophan 5-hydroxylase activity of the B-9 region remain significantly decreased for up to 2 months after 4-chloroamphetamine administration, whereas the serotonin content and tryptophan 5-hydroxylase activity in the B-7 and B-8 regions return to normal within this time period.[30] 5) The effect of *p*-chloroamphetamine on serotonin content of the brain appears to be due to a reduction in the V_{max} but not the K_m of tryptophan 5-hydroxylase.[17] 6) The reduction in the serotonin content of the telencephalon is associated with a decrease in the ability of synaptosomal fractions [12, 17] or brain slices [31] to take up serotonin.

It is well known that the conventional silver stains, including the one employed in this study, do not impregnate the very fine unmyelinated fibers of serotonergic and noradrenergic neurons.[32, 33] Thus, the absence of terminal degeneration that occurs after injection of the halogenated amphetamines does not preclude the presence of damage to serotonergic terminals. Indeed, there is ample data to suggest that terminals of the B-7 and B-8 neurons are being damaged, since 4-chloroamphetamine produces a decrease in the serotonin content of both the caudate and the hippocampus, both of which are innervated by cell bodies located in the B-7 and B-8 raphe nuclei [34, 35] but not by serotonergic cell bodies of the B-9 region.[36] It should be pointed out that such damage to serotonergic terminals could easily occur without any evidence of retrograde degeneration of serotonin cell bodies in the B-7 and B-8 regions, because it is well known that retrograde degeneration is not readily apparent in Nissl-stained sections, due presumably to the extensive collateral branching of their axons. Thus, we need to know a great deal more about the full extent of the neurotoxic action of these halogenated amphetamines. The same considerations apply to the dihydroxytryptamines. The occurrence of profuse terminal degeneration in the B-7, B-8, and B-9 regions after 5,7-DHT injection also suggests that this drug most likely has the ability to produce extensive damage to terminals other than those of serotonin or norepinephrine neurons. Damage to neuronal elements other than those that contain serotonin or norepinephrine, such as myelinated fibers, has been previously noted to occur after 5,6-DHT,[7] 5,7-DHT,[37] and 6-hydroxydopamine [33, 38] treatment, either intraventricularly or intracerebrally. As noted by Björklund *et al.*,[37] the effects of these neurotoxins are selective but not specific. To quote an old dictum, the dihydroxytryptamines and halogenated amphetamines have two neurotoxic actions, those we know about, and for which reason they are used, and those about which we are still ignorant.

The neurotoxic actions of 4-chloroamphetamine, 4-bromoamphetamine, and fenfluramine observed in the B-9 region are most likely due to a direct effect of these drugs or some metabolite(s) on the cell body, because the neuropathologic changes observed in both Nissl- and silver-stained sections are not the characteristic response seen in cell bodies undergoing retrograde degeneration. It is possible that these drugs are first taken up by axons and then transported to the cell body, where they then exert a direct neurotoxic effect. It is interesting to note that the neurotoxic reactions observed after intraperitoneal injection of haloamphetamines or after intraventricular dosing with 5,7-DHT in this study are identical to those reported to occur in catecholamine cell bodies after intraventricular injection of the neurotoxin 6-hydroxydopamine.[33] Based on the

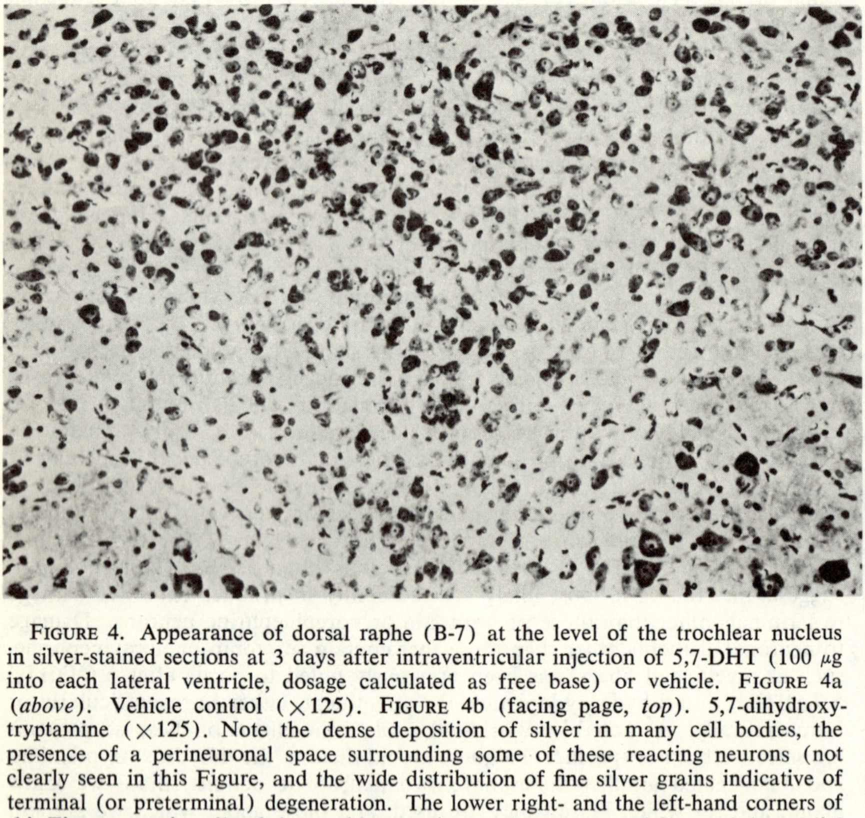

FIGURE 4. Appearance of dorsal raphe (B-7) at the level of the trochlear nucleus in silver-stained sections at 3 days after intraventricular injection of 5,7-DHT (100 μg into each lateral ventricle, dosage calculated as free base) or vehicle. FIGURE 4a (*above*). Vehicle control (×125). FIGURE 4b (facing page, *top*). 5,7-dihydroxytryptamine (×125). Note the dense deposition of silver in many cell bodies, the presence of a perineuronal space surrounding some of these reacting neurons (not clearly seen in this Figure, and the wide distribution of fine silver grains indicative of terminal (or preterminal) degeneration. The lower right- and the left-hand corners of this Figure contain cells of the trochlear nucleus and unsuppressed fibers of the medial longitudinal fasciculus. FIGURE 4c (facing page, *bottom*). 5,7-Dihydroxytryptamine (×538). An enlarged view of the same region shown in FIGURE 4b, somewhat overexposed with respect to cell bodies to improve visualization of terminal degeneration.

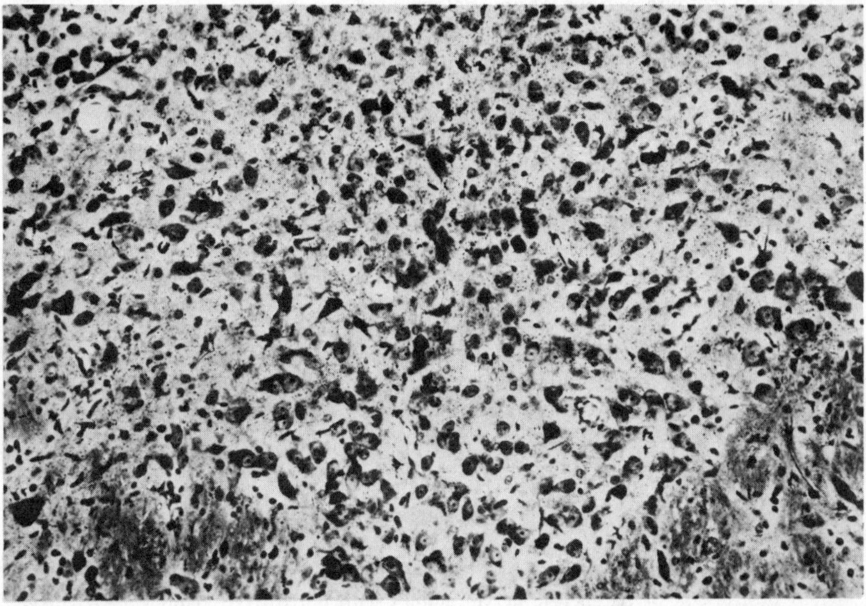

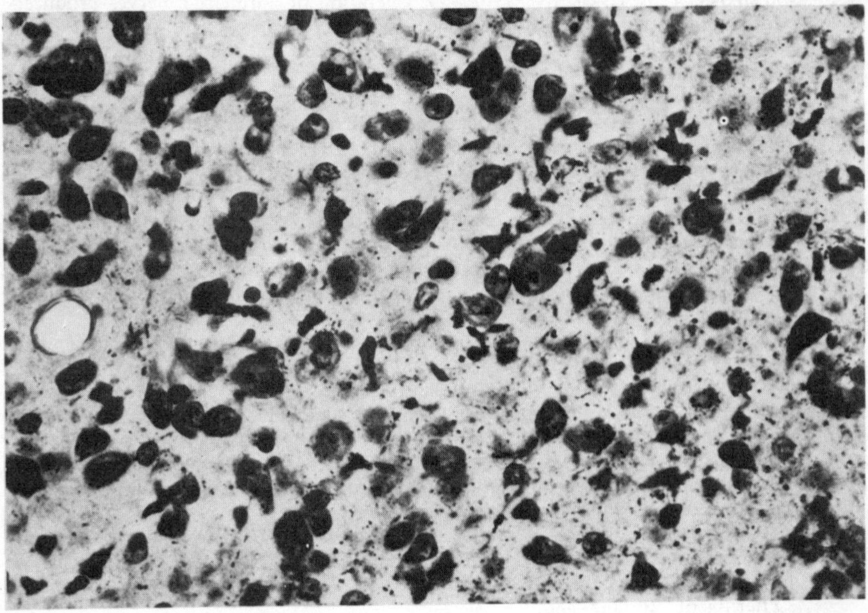

neurotoxic reactions observed in silver- or Nissl-stained sections, Hedreen and Chalmers[33] also concluded that 6-hydroxydopamine was producing a direct neurotoxic effect on the catecholamine cell bodies.

The neurotoxic actions of the halogenated amphetamines are of particular interest in that they appear to exert a more selective action on the B-9 serotonergic cell bodies as compared with the B-7 and B-8 cell bodies. The basis for such a selective neurotoxicity may reside in some basic biochemical differences among these various cell groups. The halogenated amphetamines also appear to produce a more specific neurotoxic action on the serotonin system of brain than does either 5,6-DHT or 5,7-DHT, because the decrease in serotonin content is not accompanied by a decrease in norepinephrine or dopamine. The neurotoxic actions of 4-chloroamphetamine and fenfluramine are also of practical concern, because both drugs have been employed in humans, and because fenfluramine is still prescribed as an anorexogenic agent. The cumulative neurotoxicity obtained with 12.5 μmol/kg of fenfluramine suggests a possible toxic hazard in man. Indeed, chronic treatment of nonobese humans with 7.5 μmol/kg of fenfluramine has been shown to produce a 66% decrease in the probenecid-induced accumulation of 5-HIAA in the cerebrospinal fluid.[39] Similar decreases in probenecid-induced accumulation of 5-HIAA in rat brain have also been noted after fenfluramine treatment.[40]

References

1. Thoenen, H. & J. P. Tranzer. 1968. Chemical sympathectomy by selective destruction of adrenergic nerve endings with 6-hydroxydopamine. Arch. Exp. Pathol. Pharmakol. **261:** 271–288.
2. Malmfors, T. & C. Sachs. 1968. Degeneration of adrenergic nerves produced by 6-hydroxydopamine. Eur. J. Pharmacol. **3:** 89–92.
3. Uretsky, N. J. & L. L. Iversen. 1969. Effects of 6-hydroxydopamine on noradrenaline-containing neurons in the rat brain. Nature (London) **221:** 557–559.
4. Bloom, F. E., S. Algeri, A. Groppetti, A. Revuelta & E. Costa. 1969. Lesions of central norepinephrine terminals with 6–OH–dopamine: biochemistry and fine structure. Science **166:** 1284–1286.
5. Baumgarten, H. G., A. Björklund, L. Lachenmayer, A. Nobin & U. Stenevi. 1971. Long-lasting selective depletion of brain serotonin by 5,6-dihydroxytryptamine. Acta Physiol. Scand. Suppl. **373:** 1–15.
6. Baumgarten, H. G. & L. Lachenmayer. 1972. 5,7-Dihydroxytryptamine: improvement in chemical lesioning of indoleamine neurons in the mammalian brain. Z. Zellforsch. **135:** 399–414.
7. Baumgarten, H. G., A. Björklund, A. F. Holstein & A. Nobin. 1972. Chemical degeneration of indoleamine axons in rat brain by 5,6-dihydroxytryptamine. An ultrastructural study. Z. Zellforsch. **129:** 256–271.
8. Harvey, J. A., S. E. McMaster & L. M. Yunger. 1975. p-Chloroamphetamine: selective neurotoxic action in brain. Science **187:** 841–843.
9. Harvey, J. A. & S. E. McMaster. 1975. Fenfluramine: evidence for a neurotoxic action on midbrain and a long-term depletion of serotonin. Psychopharmacol. Commun. **1:** 217–228.
10. Sanders-Bush, E., J. A. Bushing & F. Sulser. 1972. Long-term effects of p-chloroamphetamine on tryptophan hydroxylase activity and on the levels of 5-hydroxytryptamine and 5-hydroxyindole acetic acid in brain. Eur. J. Pharmacol. **20:** 385–388.
11. Fuller, R. W., J. C. Baker, K. W. Perry & B. B. Molloy. 1975. Comparison

of 4-chloro-, 4-bromo- and 4-fluoroamphetamine in rats: drug levels in brain and effects on brain serotonin metabolism. Neuropharmacology **14:** 739–746.

12. Sanders-Bush, E., J. A. Bushing & F. J. Sulser. 1975. Long term effects of p-chloroamphetamine and related drugs on central serotonergic mechanisms. J. Pharmacol. Exp. Ther. **192:** 33–41.
13. Fuller, R. W. & B. B. Molloy. 1974. Recent studies with 4-chloroamphetamine and some analogues. Advan. Biochem. Psychopharmacol. **10:** 195–205.
14. Fuller, R. W., H. D. Snoddy & B. B. Molloy. 1973. Effect of β,β-difluoro substitution on the disposition and pharmacological effects of 4-chloroamphetamine in rats. J. Pharmacol. Exp. Ther. **184:** 278–284.
15. Meek, J. L., K. Fuxe & A. Carlsson. 1971. Blockade of p-chloromethamphetamine induced 5-hydroxytryptamine depletion by chlorimipramine, chlorpheniramine and meperidine. Biochem. Pharmacol. **20:** 707–709.
16. Fuller, R. W., W. Perry & B. B. Molloy. 1975. Effect of 3-(p-trifluoromethylphenoxy)-N-methyl-3-phenylpropylamine on the depletion of brain serotonin by 4-chloroamphetamine. J. Pharmacol. Exp. Ther. **193:** 796–803.
17. Gál, E. M., P. A. Christiansen & L. M. Yunger. 1975. Effect of p-chloroamphetamine on cerebral tryptophan-5-hydroxylase *in vivo:* a reexamination. Neuropharmacology **14:** 31–39.
18. Noble, E. P., R. J. Wurtman & J. Axelrod. 1967. A simple and rapid method for injecting ^{3}H-norepinephrine into the lateral ventricle of the rat brain. Life Sci. **6:** 281–291.
19. De Olmos, J. S. & W. R. Ingram. 1971. An improved cupric-silver method for impregnation of axonal and terminal degeneration. Brain Res. **33:** 523–529.
20. Fernstrom, R. C. 1958. A durable Nissl stain for frozen and paraffin sections. Stain Technol. **33:** 175–176.
21. Fink, R. P. & L. Heimer. 1967. Two methods for selective silver impregnation of degenerating axons and their synaptic endings in the central nervous system. Brain Res. **4:** 369–374.
22. Bogdanski, D. F., A. Pletscher, B. B. Brodie & S. Udenfriend. 1956. Identification and assay of serotonin in brain. J. Pharmacol. Exp. Ther. **117:** 82–88.
23. Crout, J. R., C. R. Creveling & S. Udenfriend. 1961. Norepinephrine metabolism in rat brain and heart. J. Pharmacol. Exp. Ther. **132:** 269–277.
24. Chang, C. C. 1964. A sensitive method for spectrophotofluorometric assay of catecholamines. Int. J. Neuropharmacol. **3:** 643–649.
25. Maickel, R. P., R. H. Cox, J Saillant & F. P. Miller. 1968. A method for the determination of serotonin and norepinephrine in discrete areas of rat brain. Int. J. Neuropharmacol. **7:** 275–281.
26. Dahlström, A. & K. Fuxe. 1964. Evidence for the existence of monoamine containing neurons in the central nervous system. I. Demonstration of monoamines in the cell bodies of brain stem neurons. Acta Physiol. Scand. **62** (Suppl. 232): 1–55.
27. Jonsson, G., P. Einarsson, K. Fuxe & H. Hallman. 1974. Microspectrofluorimetric studies on central 5-hydroxytryptamine neurons. Advan. Biochem. Psychopharmacol. **10:** 55–65.
28. Harvey, J. A., S. E. McMaster & R. W. Fuller. 1977. Comparison between the neurotoxic and serotonin depleting effects of various halogenated derivatives of amphetamine in the rat. J. Pharmacol. Exp. Ther. **202:** 581–589.
29. Harvey, J. A. & S. E. McMaster. 1977. Fenfluramine: cumulative neurotoxicity after chronic treatment with low dosages in the rat. Commun. Psychopharmacol. **1:** 3–17.
30. Neckers, L. M., L. Bertilsson, S. H. Koslow & J. L. Meek. 1975. Reduction of tryptophan hydroxylase activity and 5-hydroxytryptamine concentration in certain rat brain nuclei after p-chloroamphetamine. J. Pharmacol. Exp. Ther. **196:** 333–338.
31. Carlsson, A. 1970. Structural specificity for inhibition of ^{14}C-5-hydroxytryptamine uptake by cerebral slices. J. Pharm. Pharmacol. **22:** 729–732.

32. Aghajanian, G. K., F. E. Bloom & M. H. Sheard. 1969. Electron microscopy of degeneration within the serotonin pathway of rat brain. Brain Res. **13:** 266–273.
33. Hedreen, J. C. & J. P. Chalmers. 1972. Neuronal degeneration in rat brain induced by 6-hydroxydopamine: a histological and biochemical study. Brain Res. **47:** 1–36.
34. Jacobs, B. L., W. D. Wise & K. M. Taylor. 1974. Differential behavioral and neurochemical effects following lesions of the dorsal or median raphe nuclei in rats. Brain Res. **79:** 353–361.
35. Lorens, S. A. & H. C. Guildberg. 1974. Regional 5-hydroxytryptamine followlowing selective midbrain raphe lesions in the rat. Brain Res. **78:** 45–56.
36. Massari, V. J. & E. Sanders-Bush. 1975. Synaptosomal uptake and levels of serotonin in rat brain areas after p-chloroamphetamine or B-9 lesions. Eur. J. Pharmacol. **33:** 419–422.
37. Björklund, A., H. G. Baumgarten & A. Nobin. 1974. Chemical lesioning of central monoamine axons by means of 5,6-dihydroxytryptamine and 5,7-dihydroxytryptamine. Advan. Biochem. Psychopharmacol. **10:** 13–33.
38. Poirier, L. J., P. Langelier, A. Roberge, R. Boucher & A. Kitsikis. 1972. Non-specific histopathological changes induced by the intracerebral injection of 6-hydroxydopamine. J. Neurol. Sci. **16:** 401–416.
39. Shoulson, I. & T. N. Chase. 1975. Fenfluramine in man: hypophagia associated with diminished serotonin turnover. Clin. Pharmacol. Ther. **17:** 616–621.
40. Sanders-Bush, E., D. A. Gallagher & F. Sulser. 1974. On the mechanism of brain 5-hydroxytryptamine depletion by p-chloroamphetamine and related drugs and the specificity of their action. Advan. Biochem. Psychopharmacol. **10**: 185–194.

Discussion

Dr. W. H. Funderburk (*A. H. Robins Company, Richmond, Va.*): Dr. Harvey notified me of his original findings with fenfluramine in July 1975, and we instigated extensive studies to confirm or contradict his results. Several drugs were studied. Forty rats were divided randomly into eight groups of five animals each. One group was given saline for a negative control, and *p*-chloroamphetamine (10 mg/kg) was given to another group for a positive control. Methamphetamine (10 mg/kg), reserpine (5 mg/kg), and four doses of fenfluramine, which ranged from 1 to 31.6 mg/kg, were given to the remaining six groups of animals. Animals were sacrificed 72 hr after single doses of the drugs for histologic examination of the midbrain. Histologic techniques were used according to the directions of our consulting neuropathologist, Dr. W. Rosenblum, of the Medical College of Virginia. When negative results were obtained in the first study, three other groups of five rats each were given saline, 10 mg/kg of *p*-chloroamphetamine, and 31.6 mg/kg of fenfluramine. These animals were sacrificed 30 days after single doses of the drugs. The results of this study were also negative; Drs. Harvey and McMaster then kindly provided details of their procedures, which were followed as precisely as possible. For this study, three groups of five animals each were given saline, 10 mg/kg of *p*-chloroamphetamine, and 31.6 mg/kg of fenfluramine. They were all sacrificed 72 hr after a single injection of the drugs.

A total of 2515 sections of midbrain from 70 rats were studied by Dr. G. T. Mann, State Pathologist of Florida, Dr. R. Jones, a clinical pathologist, and Dr. J. M. Powers, a neuropathologist. Hyperchromatic, misshapened (dark) neurons and perineural spaces were indeed found in the B7, B8, and B9 cell groups. However, these spaces were also found in the controls and in the mesencephalic nucleus of V, the third nerve nucleus, the lateral midbrain reticular formation, and the pontine nuclei. No selective involvement of the B7, B8, or B9 regions could be found in the experimental animals. In all of the pathologists' opinions, the changes found in all groups of animals were artifactual. Dr. J. Cammermeyer discussed similar findings (An evaluation of the significance of the "dark" neuron, Ergeb. Anat. Entwickl. **36:** 1–61, 1962). Drs. Mann, Jones, and Powers reported, "It is our considered opinion that there were no changes in any of the groups of animals studied which could be associated with any particular group and/or drug."

Evidence in support of a lack of neuroanatomic toxicity caused by fenfluramine continues to grow. Shoulson and Chase showed that, clinically, fenfluramine did not produce irreversible changes (Clin. Pharmacol. Ther. **17:** 616–621, 1975). Belin *et al.* found that 5,6-DHT, a known neurotoxin, released fenfluramine from nerve terminals and concluded that fenfluramine did not induce degenerative lesions of 5-HT neurons (Neuropharmacology **15:** 613–617, 1976). Henriksen studied the effects of fenfluramine on the B9 region in rats by use of fluorescent techniques. He used single and multiple doses and sacrificed animals 1, 3, and 30 days after drug administration. He found no evidence of serotonergic cell loss in the B9 group (personal communication). There is no behavioral evidence for lesions being produced in midbrain raphe nuclei by fenfluramine, because it is well known that such lesions lead to a period of total wakefulness that lasts for several days (Jouvet, M. 1973. *In* Serotonin and Behavior. J. Barcaus & E. Usdin, Eds. Academic Press, Inc. New York, N.Y.). It has been shown repeatedly that fenfluramine produces sedation (Practitioner **195:** 677–680, 1965; Arch. Int. Pharmacodyn. **161:** 206–232, 1966). There have been no reported side effects of fenfluarmine when given to man that could be associated with lesions of the midbrain. The above data refute Dr. Harvey's presentation concerning fenfluramine cytotoxicity.

Dr. Harvey: Dr. Funderburk questions our findings, and those of others, which demonstrate that fenfluramine and 4-chloroamphetamine produce a long-term depletion of brain serotonin and a neurotoxic action on B9 cell bodies. The neuroanatomic studies of Drs. Rosenblum, Mann, Jones, and Powers on 4-chloroamphetamine and fenfluramine quoted by Dr. Funderburk are impossible to evaluate because they are unpublished studies, and no photographs of their histologic material were presented. The same problem arises with the studies of Dr. Henriksen with fenfluramine that are cited as a personal communication. I could just as easily cite unpublished studies by Dr. Robert Y. Moore with fenfluramine and 4-chloroamphetamine and by Dr. David M. Jacobowitz with 4-chloroamphetamine, both of whom have informed me that they have seen clear evidence of neurotoxic actions of these compounds in histologic material. However, it is best to stay with published material.

The long-term and apparently irreversible depletion of serotonin by 4-chloroamphetamine and fenfluramine is well documented by published articles (for example, see the papers in this monograph by Gál and Sherman, Fuller, Lovenberg and Bruckwick, Meek, Sanders-Bush and Steranka, Clineschmidt *et al.,* and Harvey). Moreover, Massari and Gottesfeld [Morphological evidence for a toxic action of p-chloroamphetamine (PCA) on serotonergic neurons. Fed. Proc. **36**(3): 105 (Abst.), 1977] have reported that there is a marked decrease in the number of serotonergic cells in the B9 region by 53 days after injection of 10 mg/kg of 4-chloroamphetamine, as determined by histochemical fluorescence techniques. No effects were noted on cells in either the B7 or B8 area. In summary, it would appear that published data in support of neuroanatomic toxicity caused by 4-chloroamphetamine and fenfluramine and long-term depletion of serotonin continue to accumulate.

Dr. Funderburk is in error with respect to his interpretations of published articles. Shoulson and Chase (1975) did not study long-term effects of fenfluramine. In that study, they administered fenfluramine or a placebo to patients for 8 consecutive days and on the last day of treatment measured the probenecid-induced accumulation of 5-hydroxyindoleacetic acid (5-HIAA) in the cerebrospinal fluid. They report that fenfluramine treatment led to 53–93% reductions (average 66%) in probenecid-induced accumulation of 5-HIAA. Thus, these data say nothing about irreversible changes, but they do indicate that fenfluramine, at a dosage of 7.5 μmol/kg per day, may have a more pronounced effect on serotonergic neurons of humans than in rodents. Similarly, Belin *et al.* did not examine whether fenfluramine induces degenerative lesions. Finally, destruction of the raphe neurons does lead to insomnolence in the cat but not in the rat. However, even in the cat, this effect is due to destruction of the raphe system (B7 and B8 neurons) and not B9 neurons, so the relevance of this effect is difficult to understand.

DEGENERATIVE EFFECTS OF VARIOUS NEUROTOXIC INDOLEAMINES ON CENTRAL MONOAMINE NEURONS *

Anders Nobin and Anders Björklund

Department of Histology
University of Lund
S-223 62 Lund, Sweden

Inspired by Thoenen and Tranzer's discovery of the selective neurotoxic properties of 6-hydroxydopamine (6-OHDA) on adrenergic neurons,[1] Baumgarten and Schlossberger synthesized the analogous indoleamine derivative 5,6-dihydroxytryptamine (5,6-DHT; see References 2 and 3). In a series of biochemical, fluorescence histochemical, and electron microscopic studies, performed in collaboration with Baumgarten, Lachenmayer, and Schlossberger, we showed that 5,6-DHT caused profound and fairly selective degeneration of serotonin-containing axons and terminals after intraventricular injection in the rat brain.[4-7] The development of neurotoxic drugs with selective or preferential actions on the indoleaminergic systems is, of course, of great interest in functional studies due to the widespread distribution and relatively diffuse organization of these systems. It is obvious that chemical denervation offers possibilities for overcoming the drawbacks inherent in denervation through mechanical or electrolytic lesions of the serotonin systems, which in the brain inevitably involve several different neuronal systems and nonneuronal tissue elements. Furthermore, the serotonin neurotoxins have proved to be highly useful tools in neuroanatomic research, because the axonal lesions induced by these substances in indoleamine axons are much superior to mechanical or electrolytic lesions in producing extensive accumulation of fluorescence in the lesioned axonal pathways;[6, 8-10] this advantage facilitates greatly the fluorescence histochemical tracing of monoamine-containing fibers.

5,6-Dihydroxytryptamine and other neurotoxic tryptamines were soon applied to a broad field of indoleamine research, and, as reflected, for instance, in this monograph, a vast literature has appeared since 1971 dealing with the serotonin neurotoxins in relation to the morphology and function of serotonin neurons in the central nervous system. This review will summarize the *in vitro* and *in vivo* studies performed in our laboratory dealing with the potency of 17 substituted tryptamines in producing toxic damage to serotonergic and catecholaminergic axons in the central nervous system.

In Vitro Studies on Monoamine Uptake Impairment and Uptake Inhibition

Impairment of Uptake

To obtain a quantitative measure of damage to the membrane-bound monoamine uptake mechanism induced by neurotoxic indoleamines, slices of rat

* Supported by Swedish Medical Research Council Grants 04X–3874 and 04X–56.

0077–8923/78/0305–0305 $01.75/0

cerebral cortex were preincubated at 37° C with the various indoleamines at concentrations of 10^{-7}–10^{-5} M for up to 60 min. After a thorough washout, uptake of tritium-labeled 5-hydroxytryptamine (5-HT) and noradrenaline (NA) was measured in a subsequent 10-min incubation.[11] The experiments demonstrated that the neurotoxicity develops rapidly, because incubation of the cortical slices for as little as 5 min in 10^{-5} M 5,6-DHT was sufficient to produce a profound reduction of subsequent [^{3}H]-5-HT uptake. With prolonged exposure to the drug, there was only little further reduction of [^{3}H]-5-HT uptake (FIGURE 1).

When the various tryptamines were tested for their ability to impair [^{3}H]-5-HT uptake in brain slices, it was found that 5,6-DHT, 5,7-dihydroxytryptamine (5,7-DHT), and their *N*-methylated analogs were the most potent compounds (see TABLE 1). 4,5-Dihydroxytryptamine, 6,7-dihydroxytryptamine, and 5,6-diacetoxytryptamine (5,6-DAcOT) were also active, whereas monohydroxylated tryptamines, such as 4-, 5-, and 7-hydroxytryptamine, did not cause any uptake impairment, although, in some cases, they induced, rather, an increase in the capacity of the slices to accumulate [^{3}H]-5-HT. The damage caused to the amine pump of the 5-HT neurons was strongly attenuated or even abolished if the exposure to the drug was performed at 0° C, indicating that the neurotoxic compound has to be transported into the neuron to exert its actions.

With the most potent neurotoxins (5,6-DHT, 5,7-DHT, and their *N*-methylated analogs), maximum uptake impairment was clearly obtained within the dose range tested (10^{-5}–10^{-7} M). Moreover, at the lowest concentration where uptake impairment was detectable with these compounds (generally 10^{-6} M), it was already near maximal. This finding suggests a threshold phenomenon in the development of cellular damage, which appears to be a characteristic feature of all of the dihydroxytryptamines that have earlier been shown to be the most potent long-term serotonin depletors *in vivo* (see below). With the less potent compounds, maximal effect on [^{3}H]-5-HT uptake probably did not develop within the dose range studied. These compounds (e.g., 4,5-DHT, 6,7-DHT, and 5,6-DAcOT) show a more graded impairment response and are also less efficient long-term serotonin depletors *in vivo* (see below).

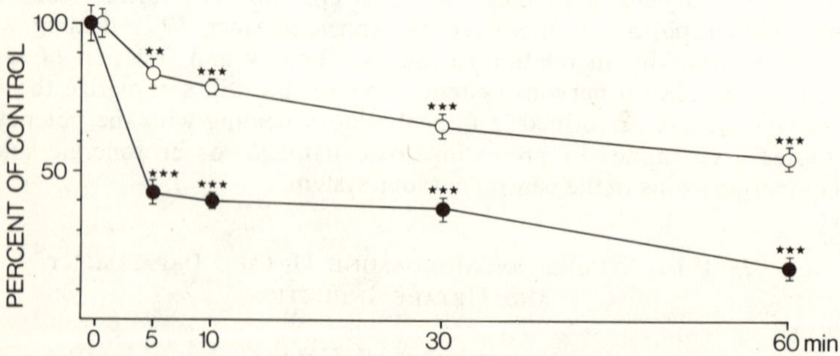

FIGURE 1. Impairment of [^{3}H]-5-HT uptake (○, total tritium; ●, unchanged 5-HT) in cortical slices induced by a preincubation in 10^{-5} M 5,6-DHT. The axis gives the time of preincubation in 5,6-DHT. Differences from zero-time values: ★★, $0.01 > p > 0.001$; ★★★; $p < 0.001$. Student's *t* test. (From Björkund *et al.*[11] By permission of *Acta Physiologica Scandinavica Supplementum.*)

TABLE 1

EFFECTS OF PREINCUBATION WITH VARIOUS NEUROTOXIC INDOLEAMINES ON SUBSEQUENT [^{3}H]-5-HT AND [^{3}H]-NA UPTAKE *in Vitro* *

Compound	Reduction of [^{3}H]-5-HT Uptake (%)			Reduction of [^{3}H]-NA Uptake (%)	
	10^{-7} M	10^{-6} M	10^{-5} M	10^{-6} M	10^{-5} M
5-HT	—	1	7	—	—
5,6-DHT	12	57 †	62 †	+9	+2
5,7-DHT	8	65 †	67 †	24 ‡	16
4,5-DHT (EDTA + ascorbic acid) §	—	+35	30 ¶	8	9
6,7-DHT (EDTA + ascorbic acid) §	—	29	42 ¶	11	13 ¶
5,6-DAcOT	1	48 †	64 †	12	20
α-Methyl-5,6-DHT	3	5	19	3	1
α-Methyl-5,7-DHT	0	17 ¶	43 †	+18 ¶	3
N-methyl-5,6-DHT	17	64 †	75 †	26 ‡	31 †
N-methyl-5,7-DHT	19 ¶	62 †	69 †	18	38 †

* Measured in thin cortical slices as the effect of a 30-min incubation at 37° C on the uptake of [^{3}H]-5-HT (0.5×10^{-7} M) or [^{3}H]-NA (10^{-7} M) during a subsequent incubation for 10 min. Data from Reference 11. Student's *t* test.

† Significant difference from control: $p < 0.001$.

‡ Significant difference from control: $0.01 > p > 0.001$.

§ EDTA (0.05 mg/ml) and ascorbic acid (0.2 mg/ml) were added during preincubation to protect the amines against oxidation.

¶ Significant difference from control: $0.05 > p > 0.01$.

α-Methylation of 5,6-DHT or 5,7-DHT counteracted the impairment of [^{3}H]-5-HT uptake seen with 5,6-DHT or 5,7-DHT, as did monoamine oxidase inhibition prior to, and during, preincubation with 5,6-DHT or 5,7-DHT. The α-methylated analogs are probably resistant to monoamine oxidase (MAO). These observations, therefore, suggest an important role for MAO in the development of the acute damage to the membrane-bound amine uptake mechanism (for discussion, see Reference 11).

Significant impairment of [^{3}H]-NA uptake was seen with 5,7-DHT (10^{-6} M), 6,7-DHT (10^{-5} M), *N*-methyl-5,6-DHT (10^{-6} M), and *N*-methyl-5,7-DHT (10^{-5} M). Preincubation with 10^{-5} M 5,6-DAcOT, α-methyl-5,6-DHT, or α-methyl-5,7-DHT or with 10^{-4} M 5,6-DHT or 4,5-DHT did not cause any impairment in the uptake of [^{3}H]-NA.[11]

Uptake Site Affinity

Our studies [11] (see also Reference 12) demonstrated that *N*-methyl-5,6-DHT and 5,6-DHT have an affinity for the 5-HT uptake sites close to that of 5-HT itself (2.0×10^{-7} M; see Reference 13). Also, 6-HT and 4,5-DHT have a good affinity for the 5-HT uptake sites, whereas 5,7-DHT and 6,7-DHT are both more than six times less potent than 5,6-DHT (see TABLE 2). Whereas methylation of the side chain nitrogen in 5,6-DHT (to give *N*-methyl-5,6-DHT) enhances uptake site affinity, methylation of the 6-hydroxyl group in 5,6-DHT (to give 5-hydroxyl-6-methoxytryptamine) leads to a fall in activity. The most striking decrease in potency is obtained by acetylating both hydroxyl groups in 5,6-DHT (to give 5,6-diacetoxytryptamine), resulting in a greater than 120-fold decline in affinity.[11]

The relative affinity of a given tryptamine for the 5-HT uptake site, on the one hand, and for the NA and DA uptake sites, on the other, should be an important factor in determining the selectivity of action of the toxic indoleamines. Thus, neurotoxic compounds provided with only a small difference in their relative affinities for the serotonin and catecholamine uptake sites would be expected to have similar toxic potencies on both types of neurons. This assumption seems, however, to not always be the case. For instance, the difference in the affinity of 5,6-DHT for the serotonin and catecholamine transport systems is rather small, suggesting poor discrimination between both types of monoamine neurons. Nevertheless, there is no indication of damage to [^{3}H]-NA uptake by 5,6-DHT *in vitro* at concentrations that cause profound impairment of [^{3}H]-5-HT uptake (see TABLE 2). This observation is in agreement with the findings *in vivo* that intraventricular[4] or direct intracerebral injection of 5,6-DHT[9, 14] causes only slight damage to NA axons in the rat brain. The situation is similar for 5,7-DHT, which, according to the data in TABLE 2, seems to have a higher affinity for NA than for 5-HT uptake sites. Despite this difference, impairment of [^{3}H]-5-HT uptake *in vitro* is greater than that of [^{3}H]-NA uptake. Intraventricular and intracerebral injections of 5,7-DHT also give a preferential toxic action on serotonin neurons.[8–10, 15]

These observations demonstrate clearly that uptake site affinity is certainly not the only determinant for the specificity and potency of a neurotoxic indoleamine. In considering other factors that might be involved, at least two possibilities seem obvious. It is conceivable that the catecholamine neurons, particularly the NA-containing ones, have a capacity to protect themselves against damage by 5,6-DHT or 5,7-DHT or to rapidly detoxify the drug by some enzymatic mechanism. Alternatively, it seems possible that the capacity of the uptake mechanism is lower in NA axons than in 5-HT axons, so that the amount of the drug that is transported into an NA fiber is markedly lower than that transported into an 5-HT fiber. If so, it would be more difficult for the drug to reach toxic concentrations within NA fibers.[11] Indirect support for this hypothesis comes from observation of [^{14}C]-5,6-DHT uptake into spinal cord slices, where prior removal of serotonin fibers caused a 40–50% reduction in [^{14}C]-5,6-DHT uptake, whereas prior removal of NA innervation had no effect (see FIGURE 3). This finding indicates that much more 5,6-DHT is accumulated in serotonin fibers than in NA fibers[11] (see also below).

Accumulation and Retention of [^{14}C]-5,6-DHT, [^{14}C]-5,7-DHT, and [^{3}H]-5-HT in Tissue Slices

Time course studies on the accumulation of [^{14}C]-5,6-DHT and [^{14}C]-5,7-DHT in tissue slices *in vitro* revealed interesting differences in the uptake and retention characteristics between these two amines and those of [^{3}H]-5-HT[11] (FIGURE 2). After an initial rapid uptake, the rate of accumulation of [^{3}H]-5-HT into the tissue slices rapidly levels off beyond 5 min. By contrast, [^{14}C]-5,6-DHT continues to accumulate at a high rate for at least 1 hr of incubation (FIGURE 2, A). During the first 5 min of incubation, [^{14}C]-5,6-DHT is taken

TABLE 2

UPTAKE SITE AFFINITY AND UPTAKE INHIBITION OF VARIOUS NEUROTOXIC INDOLEAMINES AS EVALUATED *in Vitro* *

Drug	Affinity for 5-HT Uptake Sites † (M)	Inhibition of Uptake ‡ (%) 5-HT	NA	DA	Concentration of Tryptamine Derivative (μM)
N-methyl-5,6 DHT	3.7×10^{-7}	72	39	37	1
5,6-DHT	6.0×10^{-7}	59	43	44	1
6-HT	8.0×10^{-7}	—	—	—	
4,5-DHT	1.7×10^{-6}	74	32	34	10
5,7-DHT	4.0×10^{-6}	25	32	15	1
α-Methyl-5,7-DHT	4.4×10^{-6}	—	28	10	1
6,7-DHT	4.4×10^{-6}	67	30	15	10
5-Hydroxy-6-methoxytryptamine	7.0×10^{-6}	58	25	8	10
5,6-Diacetoxytryptamine	7.4×10^{-5}	53	23	31	10

* Data from References 11 and 12.

† Measured as the concentration of the compound that caused 50% inhibition of [^{3}H]-5-HT uptake by synaptosomes prepared from rat hypothalamus.

‡ Measured in synaptosomes from rat hypothalamus ([^{3}H]-5-HT and [^{3}H]-NA) or striatum ([^{3}H]-DA). Inhibition was measured at a fixed concentration of the compound present during incubation with the tritiated amine (10^{-6} or 10^{-7} M).

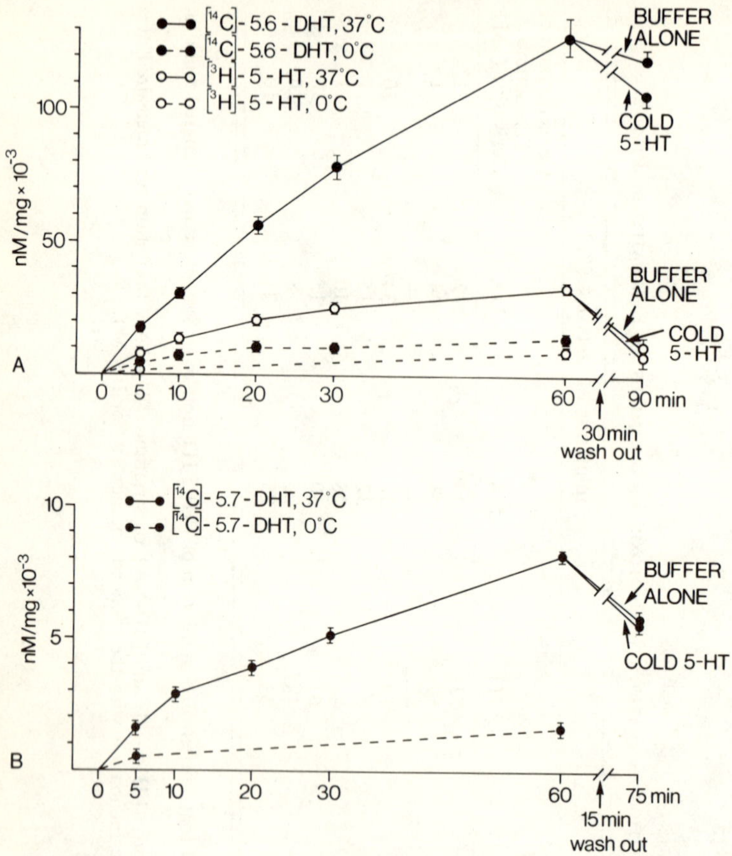

FIGURE 2. Time course of accumulation of radioactivity in cortical slices after incubation in 10^{-5} M [^{14}C]-5,6-DHT or [^{3}H]-5-HT at 37 or 0°C (A) and in 10^{-5} M [^{14}C]-5,7-DHT at 37 or 0°C (B). Part of the slices, incubated for 1 hr, were washed in fresh buffer or buffer contained 10^{-5} M unlabeled 5-HT for 15–30 min. Values are means±SEM from six to eight observations. (From Björklund *et al.*[11] By permission of *Acta Physiologica Scandinavica Supplementum.*)

up about twice as rapidly as is [^{3}H]-5-HT, whereas the initial rate of uptake of [^{14}C]-5,7-DHT is about five times slower than that of [^{3}H]-5-HT (FIGURE 2, B). There are pronounced differences in the uptake and mode of binding among 5,6-DHT, 5,7-DHT, and 5-HT in the tissue. While most of the previously accumulated [^{3}H]-5-HT can be displaced from the slices by buffer or buffer that contains cold 5-HT, only an insignificant amount of [^{14}C]-5,6-DHT and about 40% of the accumulated [^{14}C]-5,7-DHT are displaceable by the same procedure (FIGURE 2). This observation can be explained by the high tendency of the 5,6-*o*-quinone formed from 5,6-DHT to undergo covalent binding to tissue proteins.[39] The uptake of [^{14}C]-5,6-DHT into slices of rat spinal cord is reduced by 40–50% when the serotonergic innervation is removed (through a previous intraventricular injection of 5,6-DHT), whereas removal of catecholaminergic fibers is without effect (FIGURE 3). Similar denervations result in a 80–90% reduction in [^{14}C]-5,7-DHT or [^{3}H]-5-HT uptake.[11] From these observations, it can be concluded that 5,6-DHT becomes firmly bound in the tissue and that significant binding occurs also in nonmonoaminergic structures. Such nonspecific binding is much less pronounced for 5,7-DHT. This difference may explain at least some of the nonspecific effects noted after 5,6-DHT treatment.

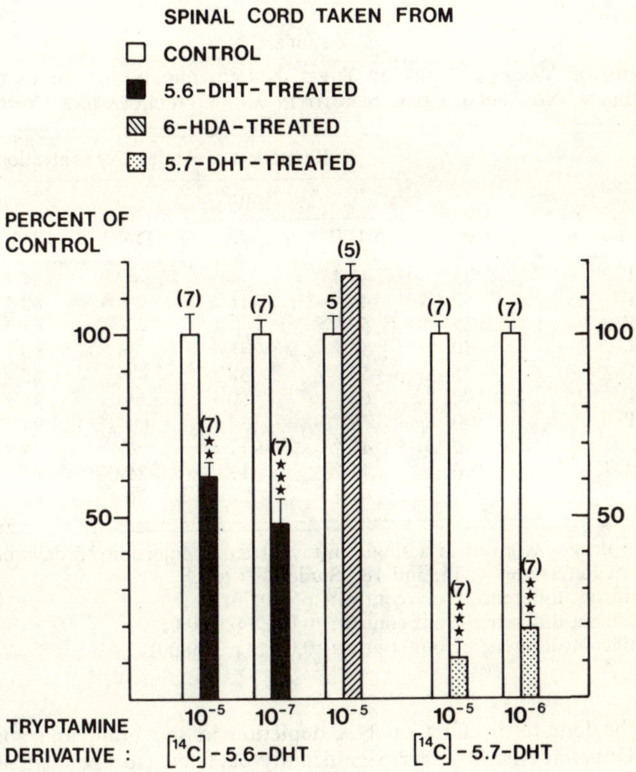

FIGURE 3. Accumulation of radioactivity (from [^{14}C]-5,6-DHT or [^{14}C]-5,7-DHT) in spinal cord slices from control, 5,6-DHT (75 μg), 5,7-DHT (150 μg), or 6-OH DA (250 μg) -treated animals during a 10-min incubation at 37°C. The drugs were injected intraventricularly 3–4 weeks before the incubation experiments. The values represent active uptake calculated as the uptake measured at 37°C minus the uptake at 0°C. Bars give means+SEM with number of determinations with parentheses. Differences from controls: ★★, $0.01 > p > 0.001$; ★★★, $p < 0.001$. Student's *t* test. (From Björklund *et al.*[11] By permission of *Acta Physiologica Scandinavica Supplementum*.)

Effects on Monoamine Content and Uptake of Central Monoamine Neurons

Effects on Monoamine Levels

One dose of 25–75 μg of 5,6-DHT injected into the cerebrospinal fluid causes a marked long-lasting depletion of rat brain serotonin[4, 6, 7, 10, 16–19] (Table 3). After a 75-μg injection, the long-term depletion is highest, about 85%, in the spinal cord and lowest, about 20%, in the pons and medulla oblongata, whereas the remaining brain regions show approximately 40–60% reductions in serotonin concentration.[4] After 50–75 μg of 5,6-DHT are injected, the long-term effect on NA depletion in the brain and spinal cord is slight. Dopamine is, however, significantly depleted (to about 70% of normal at 10 days). More clear-cut and permanent reductions of brain NA and DA are seen after an injection of 100 μg of 5,6-DHT.[4, 19]

Table 3

Effects of Various Doses of Three Neurotoxic Indoleamines on Whole Brain and Spinal Cord Serotonin and Catecholamine Content *

		Reduction of Amine Concentration (%)				
		Brain			Spinal Cord	
Compound	Dose (μg)	5-HT	NA	DA	5-HT	NA
5,6-DHT	25	41 †	—	—	84 †	+10
5,6-DHT	50	44 †	16 ‡	29 †	88 †	+21 §
5,6-DHT	75	50 †	—	—	84 †	+26 ‡
5,7-DHT	50	52 †	48 †	9	88 †	58 †
5,7-DHT	75	55 †	38 †	+1	85 †	50 †
5,7-DHT	100	62 †	50 †	5	91 †	65 †
5,7-DHT	200	78 †	38 †	+1	94 †	80 †
5,6-DAcOT	50	41 †	+17 ‡	7	79 †	+58 ‡
5,6-DAcOT	100	41 †	13	25	91 †	22
5,6-DAcOT	200	41 †	22	28	85 †	18

* Each drug was given in a single intraventricular injection 10 days before killing. Data from References 4, 15, and 18. Student's *t* test.

† Significant difference from control: $p < 0.001$.

‡ Significant difference from control: $0.05 > p > 0.01$.

§ Significant difference from control: $0.01 > p > 0.001$.

When administered intraventricularly, 5,6-DHT has definite limitations as a tool for total and selective destruction of the indoleamine-containing axons in the brain. Doses higher than 75 μg have thus turned out to be severely toxic to the rat. These drawbacks of 5,6-DHT seem, at least partly, to be related to its high tendency to become oxidized at physiologic pH. The greater chemical stability of 5,7-DHT confers better diffusion properties and less nonspecific toxicity. Thus, 200 μg of 5,7-DHT cause a 75–95% reduction in brain and spinal cord serotonin (Table 3), but, in contrast to 5,6-DHT, 5,7-DHT also has pronounced and persistent effects on central NA neurons.[8, 15, 16, 18, 20] After 200 μg of 5,7-DHT are administered, the decrease in brain and spinal cord NA levels is 40–85%.[15] At the same dose level, for example, 50 μg, 5,6-DHT and 5,7-DHT have roughly the same depletory effect on whole brain and spinal cord 5-HT concentrations (TABLE 4). At this dose level, 5,7-DHT caused a 50% reduction in NA levels in the brain and spinal cord.[15, 18] This drawback can be overcome by blocking the uptake of 5,7-DHT into NA neurons with desipramine[21, 22] (see below).

We have also studied the 5-HT-depleting potency of several other tryptamines at the 50-μg dose level after intraventricular injection (TABLE 4).[18] Not only 5,6-DHT and 5,7-DHT but also 6,7-DHT, 5,6-diacetoxytryptamine, α-methyl-5,7-DHT, and *N*-methyl-5,7-DHT significantly reduced brain and spinal cord 5-HT. A 50-μg dose of 4,5-DHT, 5,6-dibutyroxytryptamine, and *N*-methyl-5,6-DHT caused significant decreases of 5-HT in the spinal cord only.

Acylation of the ring hydroxyls in 5,6-DHT attenuates its serotonin-depleting effects, as does *N*-methylation or α-methylation of the side chain. α-Methylation, which probably renders the compounds resistant to MAO, blocks the long-term depletory effect of 5,6-DHT on serotonin and of 5,7-DHT on noradrenaline, whereas the serotonin-depleting effect of 5,7-DHT is partly counteracted. Since α-methylation, at least with respect to 5,7-DHT, does not affect the uptake site affinity of the drug (see TABLE 2), it appears that MAO also plays a role *in vivo* in promoting the toxic effects of the dihydroxylated tryptamines. This notion finds further support in the observations of Breese and Cooper[22] and Daly *et al.*[23] that prior MAO inhibition counteracts the neurotoxic effect of 5,7-DHT on central and peripheral NA neurons.

The most potent depletors of brain and spinal cord NA at the 50-μg dose level are 5,7-DHT and *N*-methyl-5,7-DHT (TABLE 4). Minor decreases of NA have been observed after treatment with 5,6-DHT, 4,5-DHT, 6,7-DHT, and 4-hydroxy-5-methoxytryptamine. Dopamine, on the other hand, was reduced in whole brain after administration of 50 μg of 5,6-DHT and *N*-dimethyl-5,6-DHT but not with any of the other compounds tested (TABLE 4).

Effects on [^{3}H]-5-HT and [^{3}H]-NA Uptake

5,6-Dihydroxytryptamine (75 μg), 5,7-DHT (200 μg), and 5,6-diacetoxytryptamine (75 μg) were tested for their ability to reduce the accumulation of tritiated 5-HT and NA in brain and spinal cord slices during a short incubation period (10 min) *in vitro* 8–12 days after intraventricular administration of the drug.[18] These experiments were performed to substantiate the contention that the reduction in central nervous system monoamine content that occurred after treatment with the three compounds is due to destruction of monoaminergic axons and terminals and, in turn, to a reduction in the number of neuronal uptake sites for monoamines.[13]

At the doses tested, all three drugs cause a similar strong reduction (80–90%) in uptake and retention of [^{3}H]-5-HT in the spinal cord, but their potency varies in the hypothalamus and cortex.[18] 5,6-Diacetoxytryptamine is less efficient than 5,6-DHT in decreasing [^{3}H]-5-HT accumulation in these two brain regions (25–35 vs 45–55% reduction). 5,7-Dihydroxytryptamine caused a considerably higher reduction (80–90%), but this decrease should be attributed to the higher dose used (200 μg). The effect on [^{3}H]-NA uptake varies among the three compounds; whereas 5,6-DHT and 5,6-DAcOT have little effect, 5,7-DHT reduces [^{3}H]-NA uptake by 40–60% in all regions analyzed.[18]

Improvement of Selectivity of 5,7-DHT by Desipramine Pretreatment

For selective lesioning of serotonin neurons, 5,7-DHT has a major disadvantage in that it discriminates poorly between 5-HT-containing and NA-containing structures. Among the several possibilities for improving the specificity of action of 5,7-DHT against central nervous system indoleamine neurons, two methods have been found to be satisfactory: interference with the uptake of 5,7-DHT into CA neurons by uptake blockers, such as desipramine,[21, 24] and counteraction of the toxic effects of 5,7-DHT on central NA neurons by prior MAO inhibition.[22] Thus, a single intraperitoneal injection of 25 mg/kg of desipramine prevents the effect of a subsequent 5,7-DHT (150 μg intraventricularly) injection on NA neurons, as determined by NA and [^{3}H]-NA uptake measurements and by fluorescence histochemistry. The reduction in 5-HT and [^{3}H]-5-HT is, however, as great as in the animals given 5,7-DHT

TABLE 4

EFFECTS OF 50 μG OF NEUROTOXIC INDOLEAMINES ON WHOLE BRAIN AND SPINAL CORD SEROTONIN AND CATECHOLAMINE CONTENT *

Compound	Reduction of Amine Concentration (%)				
	Brain			Spinal Cord	
	5-HT	NA	DA	5-HT	NA
5,6-DHT	44 †	16 ‡	29 †	88 †	+21 §
5,7-DHT	52 †	48 †	9	88 †	58 †
4,5-DHT	12	48 †	16	58 †	—
6,7-DHT	21 ‡	33 †	+3	46 †	5
5,6,7-Trihydroxytryptamine	1	—	—	+4	—
5,6-DAcOT	41 †	+17 ‡	7	79 †	+58 ‡
5,6-Dibutyroxytryptamine	18	13	15	54 †	16 ‡
α Methyl-5,6-DHT	2	19	9	7	+1
α-Methyl-5,7-DHT	38 †	20	+3	52 †	6
N-methyl-5,6-DHT	24	+5	2	77 †	+16
N-dimethyl-5,6-DHT	—	12	31 §	+6	+16 ‡
N-methyl-5,7-DHT	37 †	35 †	10	84 †	31 §
4-Hydroxy-5-methoxytryptamine	15 §	14 ‡	11	21	+27 ‡
4-HT	+21 ‡	+5	+17	+46 §	+27
6-HT	+16	+25 §	+50 †	+25	+27 ‡
7-HT	+37	—	+25	+36 ‡	+27

* Each drug was given in a single intraventricular injection 10 days before killing at a dose of 50 μg (free base). Data from References 4, 10, and 18. Student's *t* test.
† Significant difference from control: $p < 0.001$.
‡ Significant difference from control: $0.05 > p > 0.01$.
§ Significant difference from control: $0.01 > p > 0.001$.

alone[21] (FIGURE 4). Desipramine thus greatly reduces the damage caused to the CA fiber systems, but it should be emphasized that this protection is not complete: some NA axons are still lesioned, for example, in the bulbospinal system,[21] and there are, fluorescence microscopically, minor signs of damage to DA axons in the caudate nucleus close to the third ventricle and in the basal forebrain.[38]

EFFECTS OF 5,6-DHT ON SEROTONIN SYNTHESIS AND METABOLISM

Cytotoxic activity has, in case of 5,6-DHT, also been studied with respect to [^{3}H]serotonin formation from [^{3}H]tryptophan and its further metabolism to [^{3}H]-5-hydroxyindoleacetic acid (5-HIAA).[25] Rats were given one injection of 75 μg of 5,6-DHT into the lateral ventricle. At 10–14 days after injection, slices were prepared from the hypothalamus and spinal cord and incubated with [^{3}H]tryptophan for 30 min. [^{3}H]-5-Hydroxytryptamine and [^{3}H]-5-HIAA formed were isolated from the tissue, and the [^{3}H]-5-HIAA released was isolated from the incubation medium, as described by Björklund *et al.*[25]

As shown in TABLE 5, 5,6-DHT treatment caused a 74–88% reduction in the total formation of hydroxylated products. This reduction in the spinal cord is similar to that obtained after a total surgical denervation (compression); the reduction in the hypothalamus is similar to that obtained after two injections of *p*-chlorophenylalanine (350 mg/kg, ip, 1 and 3 days before killing). The reductions affected both the ability of the tissue to synthesize [^{3}H]-5-HT and its ability to store the synthesized amine and release it into the medium. Similar results have been reported in *in vivo* experiments by Hyyppä *et al.*[26] and Gerson *et al.*[16] Gerson *et al.*[16] also reported a 29% reduction in high-affinity [^{3}H]-tryptophan uptake into synaptosomes prepared from the spinal cord of 5,6-DHT-treated rats. Endogenous tryptophan levels were, however, not affected. Interestingly, in our *in vitro* experiments,[25] 5,6-DHT treatment had no effect on, or caused even an increase in, the synthesis of nonhydroxylated indoleamines (tryptamine and 5-methoxytryptamine), suggesting that the elements that synthesize these amines in the central nervous system are unaffected by this serotonin neurotoxin.

FLUORESCENCE MORPHOLOGY AFTER INTRAVENTRICULAR INJECTIONS

The cellular events induced by 5,6-DHT treatment have been studied in some greater detail by fluorescence histochemical methods in the bulbospinal indoleamine-containing neuron system.[6] This system has its cell bodies in the medulla oblongata and innervates, by short local projections, nuclei of the medulla oblongata and, by long descending projections, the entire spinal cord. Signs of acute axonal damage were observed by 1 day after injection as abundant strongly fluorescent axonal dilatations in the descending axonal pathway, mainly within the white matter in the caudal medulla oblongata and at the junction between the medulla oblongata and the spinal cord. Further caudally, there was a concomitant loss of histochemically detectable, nonterminal axons in the white matter of the spinal cord. The disappearance of indoleamine-containing terminals in the grey matter of the spinal cord was protracted: by 2 days, about half of the terminals had disappeared; by 4 days, less than

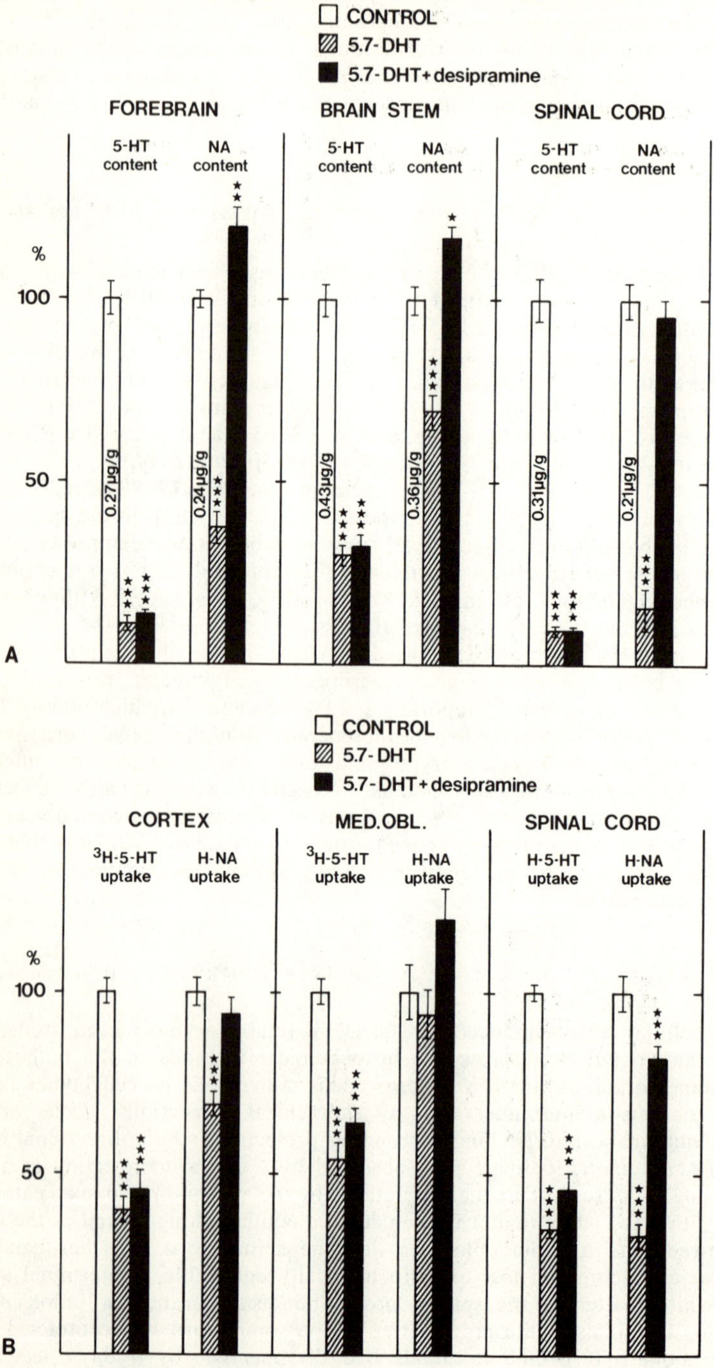
CONTROL
5.7-DHT
5.7-DHT+desipramine
FOREBRAIN
BRAIN STEM
SPINAL CORD
5-HT content
NA content
%
100
50
0.27µg/g
0.24µg/g
0.43µg/g
0.36µg/g
0.31µg/g
0.21µg/g
A
CONTROL
5.7-DHT
5.7-DHT+desipramine
CORTEX
MED.OBL.
SPINAL CORD
3H-5-HT uptake
H-NA uptake
H-5-HT uptake
B

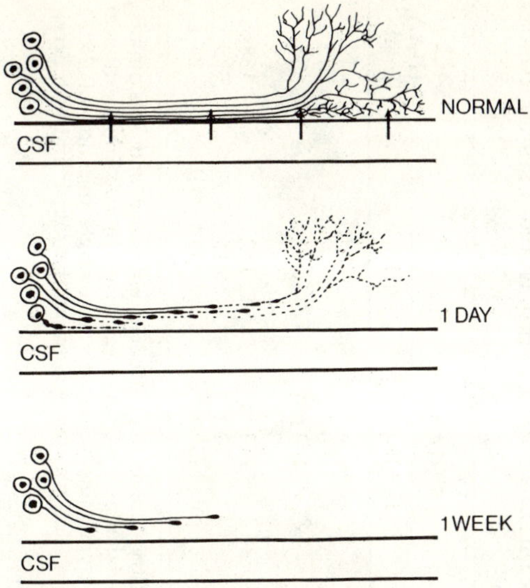

FIGURE 5. Schematic representation of the major events in the ascending and descending serotonin-containing neuron systems after 5,6-DHT treatment. The arrows indicate the diffusion from the cerebrospinal fluid (CSF). For further explanation, see text.

one third was left; and after 10 days or more, the thoracic and lumbar spinal cord appeared to be almost completely devoid of indoleamine innervation (FIGURE 5). With a similar time course, there was also a disappearance of the local indoleamine innervation of the inferior olive.[27] By contrast, the bulbospinal NA neuron system appeared to be entirely unaffected. No acute damage by 5,6-DHT was detected in the indoleamine-containing cell bodies in the medulla oblongata, but 30–90 days after injection, there was a partial disappearance of superficially located indoleamine cell bodies, probably as a result of

---

FIGURE 4. A: Effect of desipramine pretreatment (25 mg/kg, ip, 45–60 min before injection of 5,7-DHT) on long-term depletion of brain and spinal cord 5-HT and NA as seen 10 days after injection of 150 μg of 5,7-DHT. Forebrain comprises the diencephalon (excluding the pineal body) and telencephalon; brain stem comprises the medulla oblongata, pons (excluding the cerebellum), and mesencephalon. The bars give means±SEM of five determinations. B: Effects of 5,7-DHT (150 μg, 8–14 days) on the uptake of [^{3}H]-5-HT and [^{3}H]-NA, with and without desipramine pretreatment (25 mg/kg, ip 45–60 min before injection of 5,7-DHT). Thin slices were incubated in [^{3}H]-5-HT (0.5×10^{-7} M in the presence of 0.05 mg/ml of EDTA, 0.2 mg/ml of ascorbic acid and 0.005 mg/ml of nialamide or [^{3}H]-NA (10^{-7} M) for 10 min. After rinsing from 5 min in fresh buffer, the slices were dissolved in Soluene® 100 (Packard) and counted. The values represent active uptake, calculated as the uptake measured at 37°C minus the uptake at 0°C. Values are the means±SEM of 10–12 determinations. Differences from control values: ★, $0.05 > p > 0.01$; ★★, $0.01 > p > 0.001$; ★★★, $p < 0.001$. Student's *t* test. (From Björklund *et al.*[21] By permission of *Journal of Neurochemistry*.)

TABLE 5

FORMATION OF [^{3}H]-5-HT AND [^{3}H]-5-HIAA FROM [^{3}H]TRYPTOPHAN IN SLICES FROM DIFFERENT RAT CENTRAL NERVOUS SYSTEM REGIONS EXPRESSED AS PERCENT OF CONTROL *

Brain Region	Treatment	5-HT in Tissue	5-HIAA			5-HT+ 5-HIAA Total
			In Tissue	In Medium	Total	
Hypothalamus	*p*-CPA	14 †	19 †	4 †	14 †	14 †
Median eminence	5,6-DHT	17	17	0	7	12
Subcommissural organ	5,6-DHT	9	16	104	31	15
Spinal cord	Compression					
	Above lesion	154	85	144	112	138
	Below lesion	4 †	12 ‡	17 †	15 †	9 †
	5,6-DHT	24 †	32 †	33 ‡	33 ‡	26 †

* Incubation was performed for 30 min at 37°C in the presence of 5μCi of [^{3}H]tryptophan/ml. Median eminence and subcommissural organ values are the means of two independent determinations. *p*-Chlorophenylalanine (*p*-CPA) was given in two intraperitoneal injections (350 mg/kg each dose) 3 days and 1 day before sacrifice. 5,6-Dihydroxytryptamine was given in a single injection of 75 μg (free base) into the lateral cerebral ventricle 10 days before sacrifice. A total compression was made between the fourth and fifth cervical vertebrae 10–14 days before killing. Data from Reference 25. Student's *t* test.

† Significant difference from control: $p < 0.001$.

‡ Significant difference from control: $0.01 > p > 0.001$.

retrograde degeneration after severe axonal damage. The results indicate that in the bulbospinal system, 5,6-DHT has a direct neurotoxic action, primarily on the superficially located nonterminal parts of the indoleamine-containing axons, and that terminal degeneration largely occurs as an anterograde degeneration of the lesioned axons[6] (see FIGURE 5). The indoleamine cell bodies are much less susceptible to the direct action of 5,6-DHT.

In a subsequent study, we investigated axonal degeneration caused by other serotonin neurotoxins in the monoamine-containing neuron systems.[18] We found a rapid and long-lasting disappearance of fluorescent indoleamine axons situated near the ventricles or near the surface (e.g., in the suprachiasmatic nucleus, subcommissural organ, and neuropil underneath the fourth ventricle). This phenomenon occurred within 24 hr after injection with 50 μg of 5,7-DHT, *N*-methyl-5,7-DHT, 5,6-DHT, 5,6-diacetoxytryptamine, *N*-methyl-5,6-DHT, 6,7-DHT, 4,5-DHT, and, to a minor degree also, after α-methyl-5,7-DHT. At 1–2 days after treatment, the ventricle-distant indoleamine terminals (such as those in the amygdaloid complex, the ventral horn of the spinal cord, the interpeduncular nucleus, or the hypoglossal nucleus) still had a rather normal fluorescence pattern. These more deeply located terminal systems either remained largely unaffected up to 12 days (seen after 4,5-DHT, 6,7-DHT, α-methyl-5,6-DHT, and 4-hydroxy-5-methoxytryptamine) or disappeared partly during this postinjection period (seen after 5,7-DHT, 5,6-DHT, α-methyl-5,7-DHT, or 5,6-diacetoxytryptamine). None of the drugs studied caused any obvious acute damage to the indoleamine cell bodies at the 50-μg dose level. At a dose of 150 μg, 5,7-DHT caused, however, a degeneration of many indoleamine cell bodies in groups B-1 and B-3 of the medulla oblongata 1–3 months after injection.[10, 28]

4-Hydroxytryptamine, in addition to 6-HT and 7-HT, did not cause any sign of toxic damage to monoaminergic fiber systems. The accumulation of these substances within monoaminergic structures could be visualized in the fluorescence microscope 45–60 min after their intraventricular injection.[18] This accumulation was similar to that documented for 5,6-DHT and 5,7-DHT shortly after injection.[10]

The potential toxicity of the tryptamine derivatives toward CA neurons was evaluated by searching for signs of axonal and/or terminal damage in central NA and DA projections and terminals. Signs of lesioning in the NA pathways are observed after injection of 50 μg of 5,7-DHT, *N*-methyl-5,7-DHT, 4,5-DHT, or 6,7-DHT. In contrast, 5,6-DHT, *N*-methyl-5,6-DHT, α-methyl-5,7-DHT, and 4-hydroxy-5-methoxytryptamine caused only scattered dilatations in the NA fiber systems at this dose level. 5,7-Dihydroxytryptamine is particularly potent in damaging the central NA pathways and their terminal systems.[15] 5,6-Dihydroxytryptamine, *N*-methyl-5,6-DHT, 4,5-DHT, 5,6-diacetoxytryptamine, and 5,6-dibutyroxytryptamine, in decreasing order of potency, depleted the most ventricle-near DA terminals in the caudate nucleus. The same drugs also produced some minor damage to the nigroneostriatal DA axons, as did also 5,7-DHT and α-methyl-5,7-DHT.[38]

Intraventricular injection of 5-hydroxy-6-methoxytryptamine, 5-methoxy-6-hydroxytryptamine, *N*-acetyl-5-hydroxy-6-methoxytryptamine, and 5-hydroxy-7-chlorotryptamine (50 μg of free base) did not cause any significant damage to central catecholamine or indoleamine neurons observable in the fluorescence microscope.[18]

Lesioning of Axonal Pathways by Local Intracerebral Injections

The efficacy of intracerebral injections of 4,5-DHT, 5,6-DHT, and 5,7-DHT in inducing regional denervations in the central nervous system was evaluated after administration into the ventromedial tegmentum and in the grey matter of the cervical spinal cord, close to the ascending and descending indoleamine-containing fiber systems [9] (Figure 6). For the intracerebral injections, a volume of 2–4 μl was found most suitable, and a concentration of 1 μg/μl was chosen to produce the least possible unspecific damage. 5,7-Dihydroxytryptamine, which was the most efficient compound, caused a 85% reduction in forebrain 5-HT concentration and a 90–95% reduction in [^{3}H]-5-HT uptake in slices from the hypothalamus and cortex 10 days after its injection into the ventromedial tegmentum [9] (Figure 6). With surgical lesions, reductions of this magnitude are obtained only after large lesions destroying almost the entire midbrain raphe region.[29, 30] The efficacy of 5,7-DHT administration is also illustrated by spinal cord experiments in which the effect of two intraspinal injections of 2 μg of 5,7-DHT on the descending serotonin neurons was comparable to that obtained after a total transection of the cord.[9]

With respect to the effects of 4,5-DHT, 5,6-DHT, and 5,7-DHT on catecholamine neurons, it is obvious that the degenerative action of the dihydroxytryptamines is selective for serotonin neurons only within a certain dose range. With the doses used, the general tissue damage, particularly with 5,7-DHT, was very limited, and the denervating effects of the intracerebral injections were due primarily to a selective damage to the monoamine axonal bundles within seemingly intact (partly pigmented) tissue. The area of efficient, and probably complete, destruction of the indoleamine-containing axons extended far beyond the area of general tissue damage. In the locations tested, and with the doses used in the present study, efficient selective damage to indoleamine axons could be estimated to cover an area that reached approximately 1.0–1.5 mm outside the area of general damage. Within this area, the indoleamine cell bodies in the mesencephalic raphe nuclei appeared to be much less susceptible to the neurotoxic action of the dihydroxytryptamines.

When a dose of 2–4 μl (1 μg/μl) was used for intracerebral injections, all three compounds caused partial damage to the ascending DA systems, and 5,7-DHT induced significant damage to the NA fiber systems as well (Figure 6). Lorens *et al.*[31] and Hole *et al.*[32] have shown that injections of 5,7-DHT placed somewhat more caudally cause profound reductions in forebrain 5-HT without any significant damage to the ascending DA and NA fiber systems. It seems, therefore, that intracerebral injections can be used for fairly selective lesioning of 5-HT axonal pathways, provided that the dose and the position of the injection are carefully selected. It seems likely that the selectivity of action of local 5,7-DHT injections can also be improved by desipramine pretreatment.

When using the dihydroxytryptamines intracerebrally for functional studies, consideration must always be given to the degree of general tissue damage caused by the drug. These compounds are almost definitely generally cytotoxic, and selective damage is achieved only when the tissue concentration of the drug is so low that toxic levels are reached only within the cellular elements (i.e., the monoamine neurons) that actively concentrate the drug by means of the so-called membrane pump. This statement applies also to the effects caused by 6-OHDA. Poirier *et al.*[33] have clearly shown that when 6-OHDA is given in too high dosages or concentrations, it can produce almost unlimited tissue

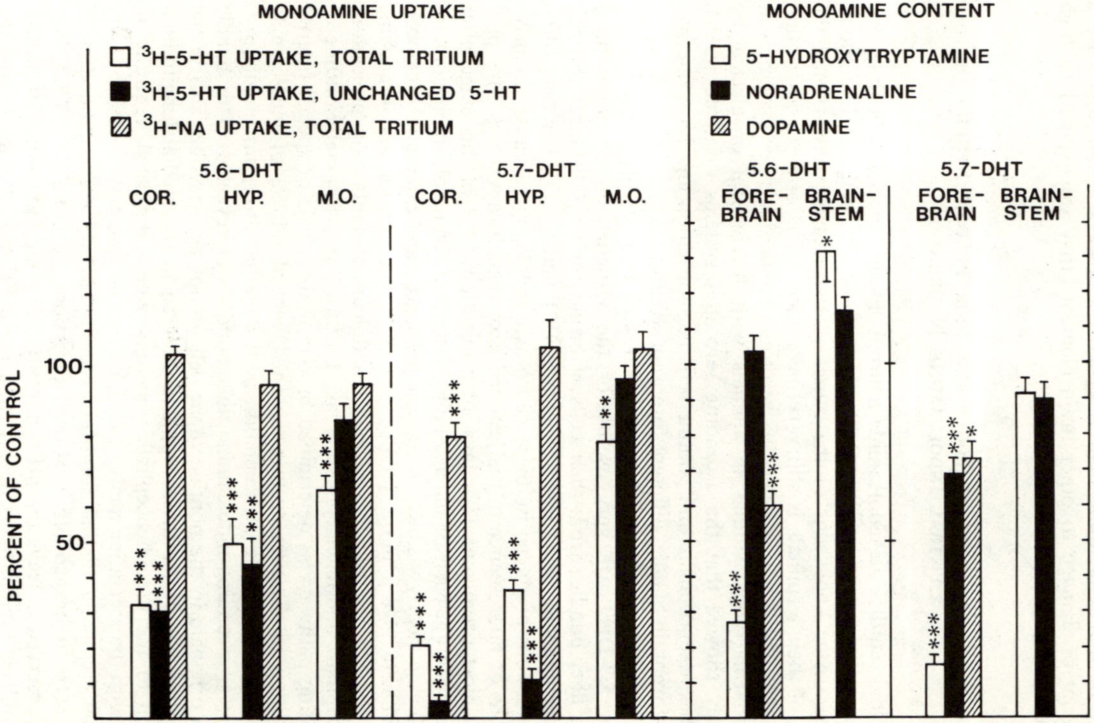

FIGURE 6. Effects of intracerebral injections of 4 μg (1 μg/μl) of 5,6-DHT or 5,7-DHT into the ventromedial tegmentum on monoamine uptake and monoamine content 9–15 days after injection. [^{3}H]-5-HT (0.5×10^{-7} M, 10 min) and [^{3}H]-NA (10^{-7} M, 10 min) uptake was measured in brain slices from cortex (COR), hypothalamus (HYP), and medulla oblongata (MO), and the values are expressed as percentages of control. Concentrations of 5-HT, NA, and DA were measured rostral and caudal to the injection site. Values are the means±SEM of from four to 15 determinations. Student's *t* test. (From Björklund *et al.*[9] By permission of *Zeitschrift für Zellforschung.*)

damage. Thus, as with surgical brain lesions, the nature, localization, and extent of the chemical lesions must always to be evaluated microscopically; biochemical controls alone can be obviously entirely misleading. Since the neurotoxic amines have no absolute specificity for monoamine neurons, it seems likely that extensive damage of the monoamine neuron systems is always accompanied by a certain amount of damage to other brain elements that are exposed to high concentrations of the drug.[5]

Dihydroxytryptamines as Tools for Morphologic Studies of Central Indoleamine Neurons

Morphologic studies on indoleamine neurons involve special difficulties due to insufficient sensitivity of the available fluorescence histochemical methods. For this reason, the knowledge of the causes of the axonal pathways and the distribution of their terminals is still very incomplete. The original description of Fuxe *et al.*[34] of the major ascending serotonin-containing fiber systems was based on pharmacologic studies in animals with mechanical or electrolytic lesions. They showed that the ascending serotonin systems originate in cell groups of the mesencephalic and pontine reticular formation and innervate, via the medical forebrain bundle, different forebrain structures.

We have found that both intraventricular and intracerebral injections of 5,6-DHT or 5,7-DHT are very useful for the visualization of nonterminal indoleamine fiber bundles from their cells of origin for long distances through the main fiber bundles. The usefulness of 5,6-DHT and 5,7-DHT is due to the fact that the lesions produced by these neurotoxic indoleamines cause extensive accumulations of fluorescence in the lesioned axonal pathways.[6, 8–10] This phenomenon is most prominent 1–3 days after injection;[6, 8, 9] at this stage after the injection, different parts of the ascending system of the indoleamine-containing fibers became clearly visible. In such animals, we[9] and Fuxe and Jonsson[35] have been able to give a more detailed picture of the courses and branching patterns of the pontine and mesencephalic indoleamine-containing projections to the diencephalic and telencephalic areas. Figure 7 is a schematic outline of these ascending pathways as determined by our study.[9] It is of special interest that, in addition to the previously established ventral bundle of indoleamine fibers ascending from the mesencephalon in the medial forebrain bundle,[34, 37] a smaller dorsal ascending indoleamine fiber bundle was revealed. This bundle appeared to originate in the pontine raphe cell groups and possibly also in the dorsal raphe nucleus of the mesencephalon and ran in the medial longitudinal fasciculus up to the mesodiencephalic junction, where it turned ventrally to intermingle with the ascending medial forebrain bundle system.

From the ascending medial forebrain bundle (MFB) system, bundles of fibers were seen to branch off in several directions (Figure 7). Thus, one branch was observed to leave the MFB dorsally along the fasciculus retroflexus; in the mammillary region, fibers depart from the MFB medially just behind the mamillothalamic tract toward the mammillary bodies. In the rostral hypothalamus, indoleamine-containing fibers leave the MFB laterally to run through the ansa lenticularis and the most ventral portion of the internal capsule toward the basal ganglia and the amygdala (Figure 7). Part of these fibers fan out into, and probably also through, the globus pallidus. Other fibers course laterally and caudally in the dorsal supraoptic commissure, above the optic tract.[9]

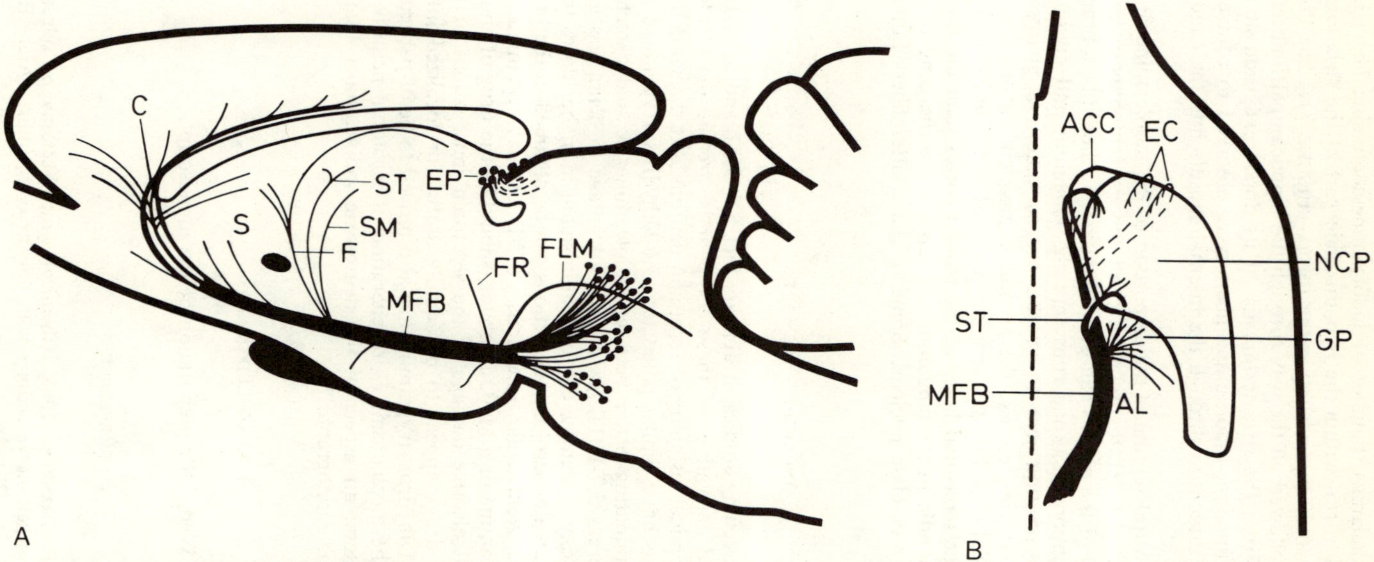

FIGURE 7. A: Schematic representation of the major ascending indoleamine axonal pathways from the mesencephalic raphe region in the rat brain, as observed after intraventricular or intracerebral injections of 5,6-DHT or 5,7-DHT. Indoleamine systems caudal to the mesencephalic reticular formation are not included in the Figure. B: Indoleamine pathways to the basal ganglia, as observed 2 days after two injections of 5,7-DHT (4 or 8 μg in 4 μl each) into the rostral and caudal portions of the nucleus caudatus putamen. The axonal pathways are represented schematically in a horizontal projection. ACC, nucleus accumbens; AL, ansa lenticularis; C, cingulum; EC, external capsule; EP, epithalamic indoleamine cell bodies; F, fornix; FLM, medial longitudinal fasciculus; FR, fasciculus retroflexus; GP, globus pallidus; MFB, medial forebrain bundle; NCP, nucleus caudatus putamen; S, septum; SM, stria medullaris; ST, stria terminalis. (From Björklund *et al.*[9] *By permission of Zeitschrift für Zellforschung.*)

In the rostral hypothalamus, bundles of indoleamine-containing fibers also leave the MFB dorsally to run within the hypothalamic part of the stria terminalis, within the stria medullaris and within the fornix (FIGURE 7). The fibers in the fornix could be followed all the way through the hippocampal fimbria to the hippocampus.[9, 35] From the stria terminalis, at its dorsal extension at the level of the fornix commissure, indoleamine fibers were observed to course in rostral and rostrolateral directions through the internal capsule into the caudate nucleus (FIGURE 7, B).

At the level of the rostral septum, a major portion of the indoleamine fibers in the MFB turn dorsally through the diagonal band and the septohypothalamic tract. At the rostral end of the corpus callosum (FIGURE 7), these fibers branch off into four major components: one runs into the cingulum and could be followed caudally in a position just above the corpus callosum, giving off branches toward the overlying cortex; the second component turns caudally underneath the corpus callosum into the septum; the third one courses rostrally into the lower part of the molecular layer of the frontal cortex; and the fourth component leaves the bundle in the cingulum to course above the genu of the corpus callosum into the caudate nucleus. Some of these latter fibers fan out into the external capsule.[9]

Concluding Remarks

From the present screening studies, it appears that 5,6-DHT and 5,7-DHT, and some closely related derivatives of these compounds, are the most potent and useful drugs for chemical lesioning of serotonin neurons. At the same dose level, the two compounds have similar potencies. 5,6-Dihydroxytryptamine has the advantage of discriminating better between indoleaminergic and catecholaminergic neurons, whereas 5,7-DHT has a great advantage in being less generally toxic. 5,7-Dihydroxytryptamine can be administered in higher doses than can 5,6-DHT, and thus greater serotonin depletions can be obtained in the brain with 5,7-DHT after intraventricular injections. 5,7-Dihydroxytryptamine, in combination with desipramine or a MAO inhibitor, should be the drug of choice for extensive and fairly selective damage of the serotonin innervations of the brain, whereas the two drugs appear to be equally useful for localized intracerebral injections. 5,6-Dihydroxytryptamine has also proved to be of particular interest for studies on the mechanisms of regeneration of serotonin neurons in the central nervous system, an aspect of the neurotoxic indoleamines that is discussed elsewhere in this monograph.[36]

Acknowledgment

We thank Christina Wedin for preparation of the manuscript.

References

1. THOENEN, H. & J. P. TRANZER. 1968. Chemical sympathectomy by selective destruction of adrenergic nerve endings with 6-hydroxydopamine. Arch. Exp. Pathol. Pharmakol. **261:** 271–288.

2. Schlossberger, H. G. & H. Kuch. 1963. Synthese von 5,6-Dihydroxytryptophan und verwanten Verbindungen. Liebigs Ann. Chem. **662:** 132–138.
3. Baumgarten, H. G. & H. G. Schlossberger. 1973. Effects of 5,6 dihydroxytryptamine on brain monoamine neurons in the rat. *In* Serotonin and Behavior. J. D. Barchas & E. Usdin, Eds. : 209–224. Academic Press, Inc. New York, N.Y.
4. Baumgarten, H. G., A. Björklund, L. Lachenmayer, A. Nobin & U. Stenevi. 1971. Long-lasting selective depletion of brain serotonin by 5,6-dihydroxytryptamine. Acta Physiol. Scand. Suppl. **373:** 1–16.
5. Baumgarten, H. G., A. Björklund, A. F. Holstein & A. Nobin. 1972. Chemical degeneration of indolamine axons in rat brain by 5,6-dihydroxytryptamine. An ultrastructural study. Z. Zellforsch. **129:** 256–271.
6. Nobin, A., H. G. Baumgarten, A. Björklund, L. Lachenmayer & U. Stenevi. 1973. Axonal degeneration and regeneration of the bulbospinal indolamine neurons after 5,6-dihydroxytryptamine treatment. Brain. Res. **56:** 1–24.
7. Björklund, A., A. Nobin & U. Stenevi. 1973. Effects of 5,6-dihydroxytryptamine on nerve terminal serotonin and serotonin uptake in the rat brain. Brain Res. **53:** 117–127.
8. Baumgarten, H. G. & L. Lachenmayer. 1972. 5,7-Dihydroxytryptamine: improvement in chemical lesioning of indoleamine neurons in the mammalian brain. Z. Zellforsch. **135:** 399–414.
9. Björklund, A., A. Nobin & U. Stenevi. 1973. The use of neurotoxic dihydroxytryptamines as tools for morphological studies and localized lesioning of central indolamine neurons. Z. Zellforsch. **145:** 479–501.
10. Björklund, A., H. G. Baumgarten & A. Nobin. 1974. Chemical lesioning of central monoamine axons by means of 5,6-dihydroxytryptamine and 5,7-dihydroxytryptamine. Advan. Biochem. Psychopharmacol. **10:** 13–33.
11. Björklund, A., A. S. Horn, H. G. Baumgarten, A. Nobin & H. G. Schlossberger. 1975. Neurotoxicity of hydroxylated tryptamines: structure-activity relationships. 2. In vitro studies on monoamine uptake inhibition and uptake impairment. Acta Physiol. Scand. Suppl. **429:** 29–60.
12. Horn, A. S., H. G. Baumgarten & H. G. Schlossberger. 1973. Inhibition of the uptake of 5-hydroxytryptamine, noradrenaline and dopamine into rat brain homogenates by various hydroxylated tryptamines. J. Neurochem. **21:** 233–236.
13. Shaskan, E. G. & S. H. Snyder. 1970. Kinetics of serotonin accumulation into slices from rat brain: relationships to catecholamine uptake. J. Pharmacol. Exp. Ther. **178:** 404–418.
14. Daly, J., K. Fuxe & G. Jonsson. 1973. Effects of intracerebral injections of 5,6-dihydroxytryptamine on central monoamine neurons: evidence for selective degeneration of central 5-hydroxytryptamine neurons. Brain Res. **49:** 476–482.
15. Baumgarten, H. G., A. Björklund, L. Lachenmayer & A. Nobin. 1973. Evaluation of the effects of 5,7-dihydroxytryptamine on serotonin and catecholamine neurons in the rat CNS. Acta Physiol. Scand. Suppl. **391:** 1–20.
16. Gerson, S., R. J. Baldessarini & S. C. Wheeler. 1974. Biochemical effects of dihydroxylated tryptamines on central indoleamine neurons. Neuropharmacology **13:** 987–1004.
17. Baumgarten, H. G., A. Björklund, A. S. Horn & H. G. Schlossberger. 1974. Studies on the neurotoxic properties of hydroxylated tryptamines. *In* Dynamics of Degeneration and Growth in Neurons. K. Fuxe, L. Olson & Y. Zotterman, Eds. : 153–167. Pergamon Press. Oxford and New York.
18. Baumgarten, H. G., A. Björklund, A. Nobin, E. Rosengren & H. G. Schlossberger. 1975. Neurotoxicity of hydroxylated tryptamines: structure-activity relationships. 1. Long-term effects on monoamine content and fluorescence morphology of central monoamine neurons. Acta Physiol. Scand. Suppl. **429:** 1–28.
19. Baumgarten, H. G., K. D. Evetts, R. B. Holman, L. L. Iversen, M. Vogt &

G. Wilson. 1972. Effects of 5,6-dihydroxytryptamine on monoaminergic neurons in the central nervous system of the rat. J. Neurochem. **19:** 1587–1597.

20. Jacoby, J. H., L. D. Lytle & M. F. Nelson. 1974. Long-term effects of 5,7-dihydroxytryptamine on brain monoamines. Life Sci. **14:** 909–919.
21. Björklund, A., H. G. Baumgarten & A. Rensch. 1975. 5,7-Dihydroxytryptamine: improvement of its selectivity for serotonin neurons in the CNS by pretreatment with desipramine. J. Neurochem. **24:** 833–835.
22. Breese, R. B. & B. R. Cooper. 1975. Behavioral and biochemical interactions of 5,7-dihydroxytryptamine with various drugs when administered intracisternally to adult and developing rats. Brain Res. **98:** 517–527.
23. Daly, J., J. Lundström & C. R. Creveling. 1974. Structure-activity correlations in trihydroxyphenylethylamines and dihydroxytryptamines. Relationship to cytotoxicity in adrenergic and serotonergic neurons. *In* Dynamics of Degeneration and Growth in Neurons. K. Fuxe, L. Olson & Y. Zotterman, Eds. : 29–42. Pergamon Press. Oxford and New York.
24. Gerson, S. & R. J. Baldessarini. 1975. Selective destruction of serotonin terminals in rat forebrain by high doses of 5,7-dihydroxytryptamine. Brain Res. **85:** 140–145.
25. Björklund, A., S. Axelsson & B. Falck. 1976. Intraneuronal indolamines in the central nervous system. Advan. Biochem. Psychopharmacol. **25:** 87–94.
26. Hyyppä, M. T., D. P. Cardinali, H. G. Baumgarten & R. J. Wurtman. 1973. Rapid accumulation of H^3-serotonin in brains of rats receiving intraperitoneal H^3-tryptophan: effects of 5,6-dihydroxytryptamine or female sex hormones. J. Neural Transmission **34:** 111–124.
27. Wiklund, L. & A. Björklund. 1977. Mechanism of regrowth of the bulbospinal serotonin system after 5,6-DHT-induced axotomy. 2. Fluorescence microscopical observations. In preparation.
28. Baumgarten, H. G., A. Björklund, L. Lachenmayer, A. Rensch & E. Rosengren. 1974. De- and regeneration of the bulbospinal serotonin neurons in the rat following 5,6- or 5,7-dihydroxytryptamine treatment. Cell Tissue Res. **152:** 271–281.
29. Kuhar, M. J., G. K. Aghajanian & R. H. Roth. 1972. Tryptophan hydroxylase activity and synaptosomal uptake of serotonin in discrete brain regions after midbrain raphe lesions: correlations with serotonin levels and histochemical fluorescence. Brain Res. **44:** 165–176.
30. Kuhar, M. J., R. H. Roth & G. K. Aghajanian. 1971. Selective reduction of tryptophan hydroxylase activity in rat forebrain after midbrain raphe lesions. Brain Res. **35:** 167–176.
31. Lorens, S. A., H. C. Guldberg, K. Hole, C. Köhler & B. Srebro. 1976. Activity, avoidance learning and regional 5-hydroxytryptamine following intra-brain stem, 5,7-dihydroxytryptamine and electrolytic midbrain raphe lesions in the rat. Brain Res. **108:** 97–113.
32. Hole, K., K. Fuxe & G. Jonsson. 1976. Behavioral effects of 5,7-dihydroxytryptamnie lesions of ascending 5-hydroxytryptamine pathways. Brain Res. **107:** 385–399.
33. Poirier, L. J., P. Langelier, A. Roberge, R. Boucher & A. Kitsikis. 1972. Non-specific histopathological changes induced by the intracerebral injection of 6-hydroxy-dopamine (6-OH-DA). J. Neurol. Sci. **16:** 401–416.
34. Fuxe, K., T. Hökfelt & U. Ungerstedt. 1968. Localization of indolalkylamines in CNS. Advan. Pharmacol. **6:** 235–251.
35. Fuxe, K. & G. Jonsson. 1974. Further mapping of central 5-hydroxytryptamine neurons: studies with the neurotoxic dihydroxytryptamines. Advan. Biochem. Psychopharmacol. **10:** 1–12.
36. Wiklund, L., A. Björklund & A. Nobin. This monograph.
37. Heller, A., J. A. Harvey & R. Y. Moore. 1962. A demonstration of a fall in brain serotonin following central nervous system lesions in the rat. Biochem. Pharmacol. **11:** 859–866.

38. WUTTKE, W., A. BJÖRKLUND, H. G. BAUMGARTEN, L. LACHENMAYER & M. FENSKE. 1977. De- and regeneration of brain serotonin neurons following 5,7-dihydroxytryptamine treatment: effects on serum LH, FSH and prolactin levels in male rats. Brain Res. **134:** 317–332.
39. BAUMGARTEN, H. G., H. P. KLEMM, L. LACHENMAYER, A. BJÖRKLUND, W. LOVENBERG & H. G. SCHLOSSBERGER. This monograph.

DEVELOPMENTAL PLASTICITY OF CENTRAL SEROTONIN NEURONS AFTER 5,7-DIHYDROXYTRYPTAMINE TREATMENT *

G. Jonsson, T. Pollare, H. Hallman, and Ch. Sachs

Department of Histology
Karolinska Institute
S-10401 Stockholm, Sweden

It is well known that the response to injury of the immature or developing nervous system and the mature nervous system is different in at least two ways. On the one hand, there is experimental evidence that the infant brain is more vulnerable than is the adult brain after certain damaging treatments. On the other hand, it has frequently been observed in various types of brain lesion experiments that there is more sparing or a greater extent of functional recovery when operations are performed in infants than in adults (see Reference 1). The latter findings have often been ascribed to the fact that the developing brain possesses a greater potential for neuronal regeneration and/or reorganization. Despite extensive studies, this contention is still open to debate. With this controversy in mind, we have focused our attention on an attempt to determine how the central catecholamine and 5-hydroxytryptamine (5-HT) neurons in the rat respond to early damage induced by the monoamine neurotoxins 6-hydroxydopamine (6-OHDA) and 5,7-dihydroxytryptamine (5,7-HT).[2-7] The neurotoxin-induced effects have mainly been analyzed by means of histochemical and neurochemical techniques. In these studies, we elected to administer the neurotoxins systemically in the neonatal stage, when the blood-brain barrier is not yet fully developed, allowing the neurotoxin to enter the brain, where it can elicit its neurodegenerative action on monoamine neurons.[8-10] This route of neurotoxin administration is advantageous with respect to specificity, as compared to, for example, intracisternal or local intracerebral injections; also, few nonspecific cytotoxic effects have yet been observed after systemic injection. Since the monoamine neurotoxins are generally cytotoxic and the specificity of the neurodegenerative action is solely dependent on their selective accumulation within monoamine neurons,[8-10] it is essential to avoid injection of high concentrations of neurotoxin locally to obtain a high degree of specificity in terms of neurotoxic effect. The latter requirement may be difficult to satisfy by intracisternal or intracerebral injection. A disadvantage with the systemic route of administration is that it may lead to rather heterogeneous effects with respect to extent of denervation of the various monoamine nerve terminal projections and thus to difficulties in defining the damage exactly from a neuroanatomic standpoint, which is of greatest importance for an adequate interpretation of the effects observed after neurotoxin administration. Preliminary studies of the effect of neonatal administration of the monoamine neurotoxins revealed that the immature monoamine neurons in the brain reacted in a rather complex manner, with

* Supported by Grant 04X–2295 from the Swedish Medical Research Council and by grants from the Karolinska Institute and from the Nathorsts and Bergvalls Foundations.

0077–8923/78/0305–0328 $01.75/0

marked regional differences, and quite differently from the response in adult animals. In this paper, we will review our studies on the altered development of central 5-HT neurons induced by 5,7-HT, with special reference to characterization of the changes, determination of the specificity of action of 5,7-HT, and the mechanisms involved in altered ontogenetic development.

Materials and Methods

Albino Sprague-Dawley rats were kept in a temperature-controlled room under a standardized dark-light (14–10 hr) schedule (light on at 6.00 AM, off at 8.00 PM). Each litter contained not more than eight pups, which were separated at 21 days of age. Compounds administered subcutaneously were

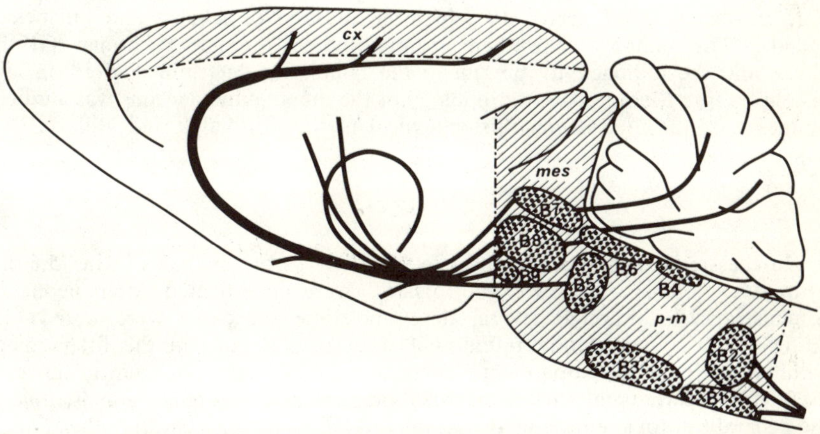

FIGURE 1. Schematic representation of rat brain (near midsagittal plane) with cutting lines indicated for the dissection of the neocortical (cx) and mesencephalon-pons-medulla (mes p-m) preparations. The 5-HT cell body groups (B1–B9), according to Dahlström and Fuxe,[18] with major pathways are also indicated.

injected in a volume of 0.25–0.1 ml in 0.9% NaCl. Intracisternal injections of 5,7-HT (10 or 50 μg in 5 μl of 0.9% NaCl) were made by means of a free-hand technique with a Hamilton syringe. In all experiments, the control animals received an equal volume of the solvent. The rats were killed by cervical dislocation, and the brains were rapidly removed and placed in beakers that contained cold Krebs-Ringer buffer. The various brain regions were thereafter dissected out for further analysis. Two central nervous system regions, namely, the neocortex and the mesencephalon-pons-medulla (which contained the 5-HT cell bodies), were of principal interest in these studies. The cutting lines used for dissection of these regions are indicated in FIGURE 1.

In vitro [^{3}H]monoamine uptake measurements were made on homogenates of brain tissue (which contained monoamine synaptosomes), as previously described.[7] The homogenates were incubated in Krebs-Ringer bicarbonate buffer (pH 7.4, which contained 1.1 mM ascorbic acid and 0.01 mM pargyline) with 0.05 μM [^{3}H]monoamine for 5 min at 37° C. The initial rate of [^{3}H]monoamine uptake can be considered to give a relative measure of the monoamine nerve terminal density (see Reference 7). Endogenous 5-HT in brain tissue was determined by means of a radioenzymatic assay,[11] and endogenous catecholamines were determined according to the method of Keller *et al.*[12] Release of [^{3}H]-5-HT from brain slices (1 $\times$ 2 $\times$ 0.3 mm produced with a McIlwain chopper) *in vitro* by field stimulation was performed as described by Farnebo and Hamberger.[13] The *in vitro* binding of [^{3}H]-5-HT and [^{3}H]-*d*-lysergic acid diethylamide (LSD) to homogenates of brain tissue for studies of 5-HT receptors was performed mainly according to the procedure of Bennett and Snyder.[14] The homogenates used for these binding studies were prepared by homogenization of the brain tissue in 0.05 M Tris HCl buffer (pH 7.4) with a Branson B-30 ultrasonic cell disruptor. After centrifugation at 50,000*g* for 10 min, the pellets were washed and resuspended in Tris HCl buffer that contained 0.1% ascorbic acid, 4 mM $CaCl_2$, and 10 μM pargyline. Aliquots of the suspensions were incubated at 37° C for 10 min after addition of labeled and unlabeled ligands. The samples were then vacuum-filtered through Whatman GF/B filters, and the radioactivity present in the filters was determined after rinsing in cold buffer. Fluorescence morphology of the monoamine neurons was studied by means of the fluorescence histochemical method of Falck and Hillarp.[15–17]

Results

In the initial experiments, the effects of the serotonin neurotoxins 5,6-dihydroxytryptamine (5,6-HT) and 5,7-HT were investigated after neonatal administration. Although several dose and time schedules were tested for 5,6-HT, it was consistently found that most of the pups in the litters died within a few days after treatment. Therefore, 5,6-HT does not seem to be very useful for experimental studies on 5-HT neurons in developing animals, unless the generalized toxic effects of this agent can be overcome. On the other hand, 5,7-HT was found to be well tolerated by young rats, even after subcutaneous doses of up to 200 mg/kg, which could be injected repeatedly without any lethal effects. The 5,7-HT-treated animals showed almost unchanged postnatal body weight gains as compared to those of saline-treated littermate controls, and the animals did not exhibit any obvious alteration of gross behavior in the adult stage. Furthermore, no consistent 5,7-HT-induced differences in the parameters studied were noted between male and female rats. When we investigated the effect of neonatal administration of 5,7-HT on the *in vitro* uptake of [^{3}H]-5-HT in the cerebral cortex and spinal cord to evaluate the neurotoxic potency of the drug, dose-dependent changes in [^{3}H]-5-HT uptake were found, with maximal effects observed after subcutaneous doses of 5,7-HT in the range of 50–100 mg/kg. Therefore, these doses were employed for the remainder of the experiments.

Analysis of the Effects of Neonatal 5,7-HT Treatment on 5-HT Neurons

In one series of experiments, the postnatal development of the blood-brain barrier in response to the neurotoxic action of 5,7-HT on 5-HT neurons in the cerebral cortex was investigated by administration of 5,7-HT at different developmental stages (FIGURE 2). The results showed that 5,7-HT injection on postnatal Days 1 and 5 led to very marked and permanent reductions in [^{3}H]-5-HT uptake (50–60% reductions), reflecting a pronounced 5,7-HT-induced denervation of 5-HT nerve terminals. No significant changes were observed when 5,7-HT was injected later during development. These data thus point to a rapid development of a blood-brain barrier for 5,7-HT between Days 5 and 7 after birth that prevents 5,7-HT from entering the brain in amounts necessary to induce a permanent degeneration of 5-HT nerve terminals. A similar time course for barrier development has been found for 6-OHDA with respect to neurotoxic action on central noradrenaline (NA) nerves,[19] although the results suggest a somewhat earlier formation of the barrier for 5,7-HT.

The acute and long-term effects of 5,7-HT treatment at birth on [^{3}H]-5-HT uptake are shown in FIGURE 3. The results from the acute experiments (FIGURE 3, A) showed that [^{3}H]-5-HT uptake was reduced considerably in both the

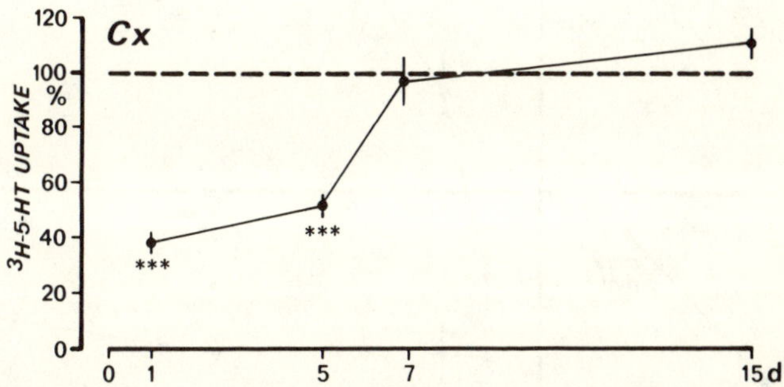

FIGURE 2. Effect of 5,7-HT (2 × 100 mg/kg, sc, 24-hr interval) administered to rats of various ages on the *in vitro* uptake of [^{3}H]-5-HT (0.05 μM, 5 min) in homogenates from the cerebral cortex (Cx). The first dose of 5,7-HT was injected on postnatal Days 1, 5, 7, and 15, respectively, and the rats were killed 12 weeks after the first injection. Each point represents the mean ±SEM of from five to eight determinations, expressed as a percentage of the uptake in homogenates from each littermate control group, respectively. Student's *t* test: ***, $p < 0.001$. Data from Reference 6.

cerebral cortex and the mesencephalon-pons-medulla within 1 hr, with maximal effects observed between 2 and 4 hr. The relative decrease in [^{3}H]-5-HT uptake in the cerebral cortex and the spinal cord remained low and approximately constant up to postnatal Day 28, after which time a tendency for a small recovery was observed. However, in the mesencephalon-pons-medulla, the 5-HT cell body-containing region, there was considerable recovery of [^{3}H]-5-HT uptake by 16 hr after 5,7-HT treatment. Between postnatal Days 3 and 14, a marked increase in [^{3}H]-5-HT uptake developed, which was maximal at Day 14: 75% over control levels. This increase continued throughout the study period, although the relative increase was somewhat lower at Day 56 than at Day 14 postnatally. Determinations of endogenous 5-HT demonstrated that, in agreement with the uptake results, 5-HT was reduced by about 40% (0.31 ± 0.05 μg/g; control, 0.55 ± 0.03 μg/g; $n = 5$) in the cerebral cortex, whereas there was a similar 5-HT increase in the mesencephalon-pons-medulla (1.32 ± 0.11 μg/g; control, 0.96 ± 0.10 μg/g; $n = 5$).

Regarding interpretation of these results, the reductions of [^{3}H]-5-HT uptake and endogenous 5-HT are definitely due a 5,7-HT-induced degeneration of 5-HT nerve terminals. This conclusion is also supported by fluorescence histochemical evidence that demonstrated that the number of 5-HT nerve terminals observable with the help of 6-hydroxytryptamine loading[20] was markedly reduced in both the cerebral cortex (FIGURE 4) and the spinal cord. The fluorescence histochemical analysis did not reveal any noticeable changes in the fluorescence morphology of the 5-HT cell bodies (B1–B9). Concerning the 5,7-HT-induced increases in [^{3}H]-5-HT uptake and endogenous 5-HT in the mesencephalon-pons-medulla, the most obvious explanation for these findings is that they are due an increased outgrowth of 5-HT nerve terminals in this

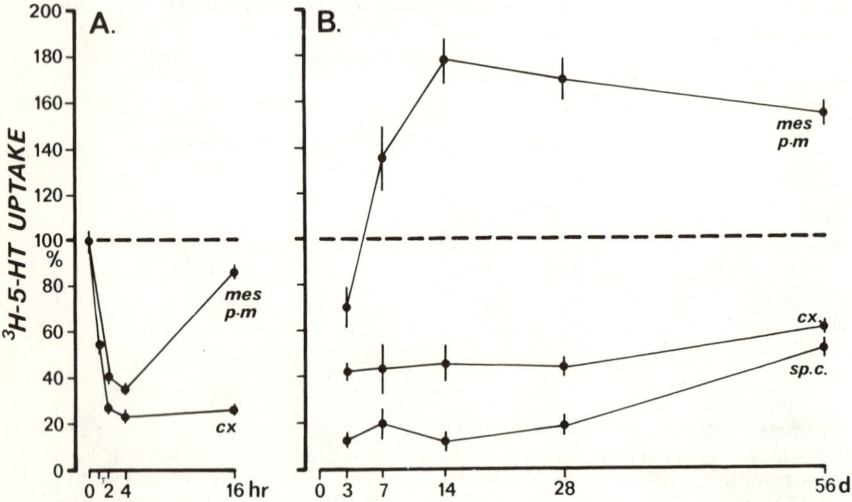

FIGURE 3. Effect of neonatal 5,7-HT administration on *in vitro* uptake of [^{3}H]-5-HT (0.05 μM, 5 min) in homogenates from the cerebral cortex (cx), mesencephalon-pons-medulla (mes p-m), and spinal cord (sp. c.). A: Acute effect of 5,7-HT (100 mg/kg, sc 1–16-hr interval). B: Long-term effect of 5,7-HT (2$\times$50 mg/kg, sc, 24-hr interval). Each point represents the mean $\pm$SEM of from five to eight determinations, expressed as a percentage of the uptake in saline-treated littermate control homogenates, respectively. Data from Reference 6.

region. This interpretation is strongly corroborated by the observation that the *in vitro* uptake of [^{3}H]-5-HT in the mesencephalon-pons-medulla after 5,7-HT treatment shows kinetic and pharmacologic properties identical to those of controls.[6] Although we have not been able to confirm the hypothesis of a 5,7-HT-induced increase in the outgrowth of 5-HT nerve terminals in an unambiguous way by fluorescence histochemical analysis, this lack of proof does not invalidate the present interpretation. The reason for the lack of firm evidence is most likely due to the fact that the formaldehyde fluorescence method is hampered by serious limitations with respect to demonstration of 5-HT nerve terminals in many areas of the brain (see Reference 20). Further evidence to support the hypothesis of an increased number of 5-HT nerve

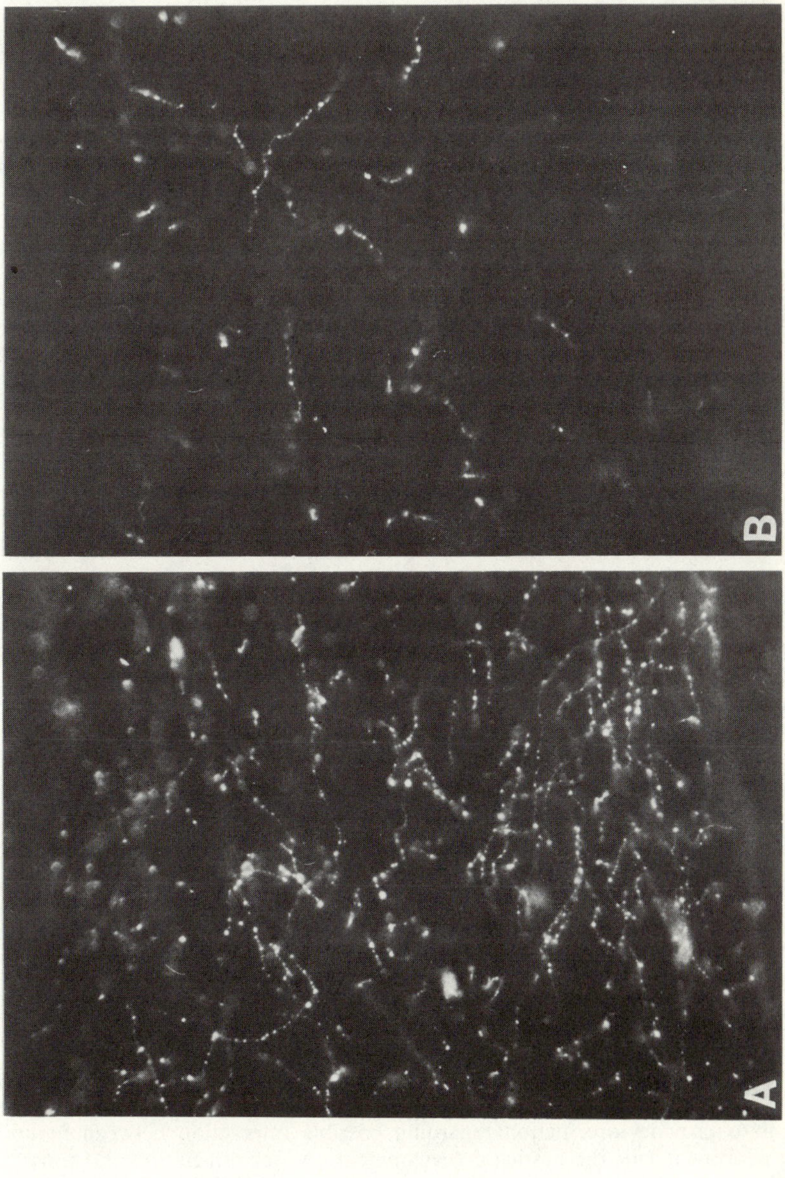

FIGURE 4. Effect of 5,7-HT treatment at birth on the fluorescence histochemical demonstration of 5-HT nerve terminals after incubation of slices from cerebral cortex *in vitro* with 1 μM 6-HT and 5 μM DMI. The 5,7-HT (100 mg/kg, sc) was injected at birth, and the animals were killed 6 weeks later. A, Vehicle-treated control; B, 5,7-HT. $\times$200.

terminals in the mesencephalon-pons-medulla after 5,7-HT treatment was obtained after studies on the field stimulation-induced release of [^{3}H]-5-HT previously taken up (FIGURE 5). It was thus observed that 5,7-HT treatment led to an increase in [^{3}H]-5-HT release that was quantitatively approximately proportional to the increase in [^{3}H]-5-HT uptake and endogenous 5-HT.

To determine whether the 5,7-HT-induced rise in [^{3}H]-5-HT uptake in the cell body-near regions might be related to the permanent 5-HT denervation in the cerebral cortex, partial brain hemisections were performed in 2-day-old rats to lesion the ascending 5-HT pathways (FIGURE 6). This operation was found to lead to a marked reduction in [^{3}H]-5-HT uptake in the cerebral cortex concomitant with increased uptake in the mesencephalon-pons-medulla. These results thus point to a close relationship between the 5-HT denervation in the cerebral cortex and the changes in the mesencephalon-pons-medulla region.

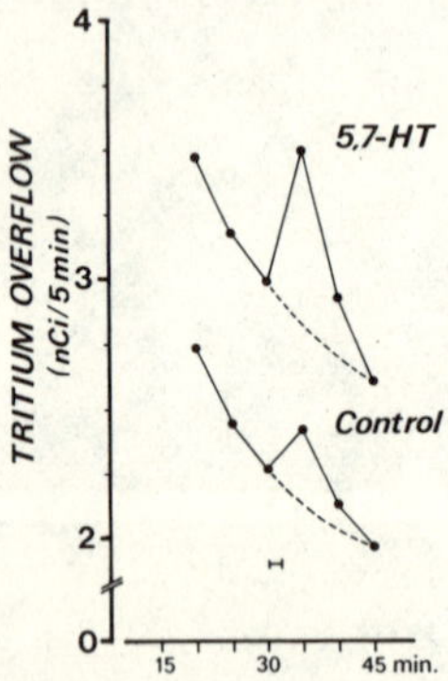

FIGURE 5. Tritium overflow during superfusion and stimulation of slices from the mesencephalon-pons-medulla of rats treated at birth with 5,7-HT (100 mg/kg, sc) or saline (control). The animals were killed at 6–8 weeks of age. The slices were first preincubated *in vitro* in Krebs-Ringer buffer that contained 0.1 μM [^{3}H]-5-HT for 30 min. Thereafter, the slices were superfused for 30 min and stimulated for 2 min. (|—|, biphasic pulses, 2 msec, 10 Hz, 12 mA) and then superfused for another 13 min according to reference 13. Tritium overflow per 5 min, ●—●; extrapolated spontaneous overflow, – – –. Each point represents the mean of six determinations.

Analysis of the Specificity of 5,7-HT

Since 5,7-HT has been shown to possess no absolute specificity in its neurotoxic action for 5-HT neurons [21–23] but to also affect catecholamine neurons to a certain extent, experiments were performed to analyze the specificity of 5,7-HT action after neonatal administration. These experiments were also thought to provide information regarding possible interaction between monoamine neurons during the postnatal development. As seen in FIGURE 7, neonatal 5,7-HT treatment led to qualitatively similar changes of [^{3}H]-5-HT uptake and [^{3}H]NA uptake in both the cerebral cortex and the mesencephalon-pons-medulla, although the effect on [^{3}H]NA uptake was quantitatively less pronounced. These results are thus in agreement with previous findings that indicated a preferential neurotoxic action of 5,7-HT on 5-HT neurons.[21–23] However, pre-

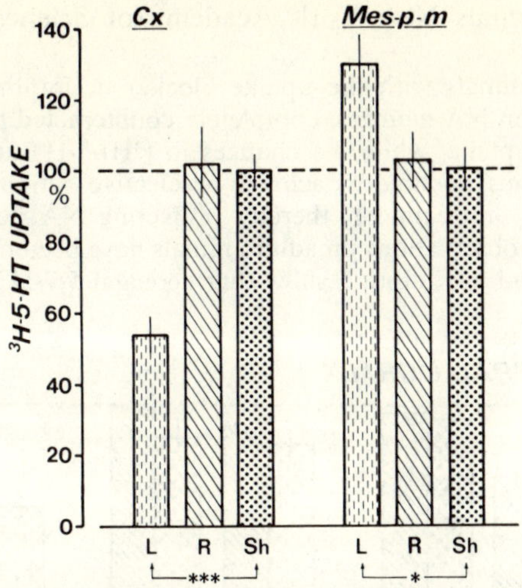

FIGURE 6. Effect of partial hemisection on the left (L) side at the level of the rostral mesencephalon in 2-day-old rats on *in vitro* uptake of [^{3}H]-5-HT (0.05 μM, 5 min) in homogenates from the cerebral cortex (Cx.) and the mesencephalon-pons-medulla (Mes-p-m). The animals were killed 8 weeks after the operation. Each column represents the mean ±SEM of from four to six determinations, expressed as a percentage of the uptake in sham-operated (Sh) controls. R, Right side. Student's *t* test. *, $0.001 < p < 0.05$; ***, $p < 0.001$.

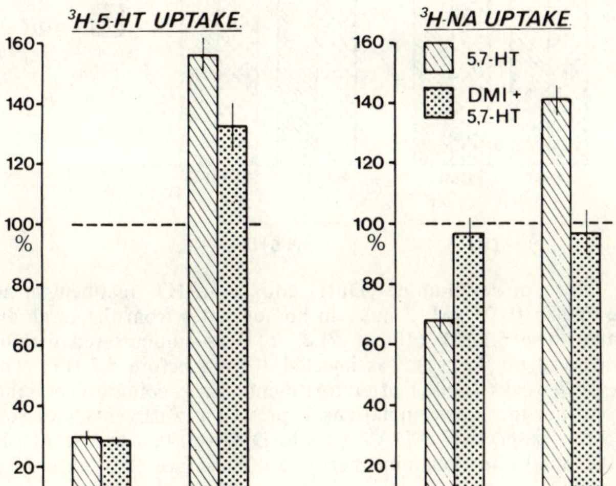

FIGURE 7. Effect of desipramine (DMI) on the 5,7-HT-induced changes in [^{3}H]-5-HT and [^{3}H]NA uptake (0.05 μM, 5 min) *in vitro* in homogenates from the cerebral cortex (Cx.) and the mesencephalon-pons-medulla (Mes-p-m). Desipramine (20 mg/kg, sc) was injected 30 min before each dose of 5,7-HT (2 × 100 mg/kg, sc, 24-hr interval). The first dose of 5,7-HT was injected within 4 hr after birth, and the animals were killed at the age of 8 weeks. Each column represents the mean ±SEM of from five to eight determinations, expressed as a percentage of the uptake in vehicle-treated littermate controls, respectively. Data frcm Reference 6.

treatment of the animals with the uptake blocker desipramine (DMI), which acts preferentially on NA neurons, completely counteracted the 5,7-HT-induced effect on [^{3}H]NA uptake, while the changes in [^{3}H]-5-HT uptake were largely unaffected. It is thus possible to achieve a selective action on 5-HT neurons by means of DMI pretreatment, thereby protecting NA neurons from attack by 5,7-HT. Similar observations on adult animals have been made by others.[23, 24] The results presented in FIGURE 5 show that neonatal 5,7-HT treatment did not

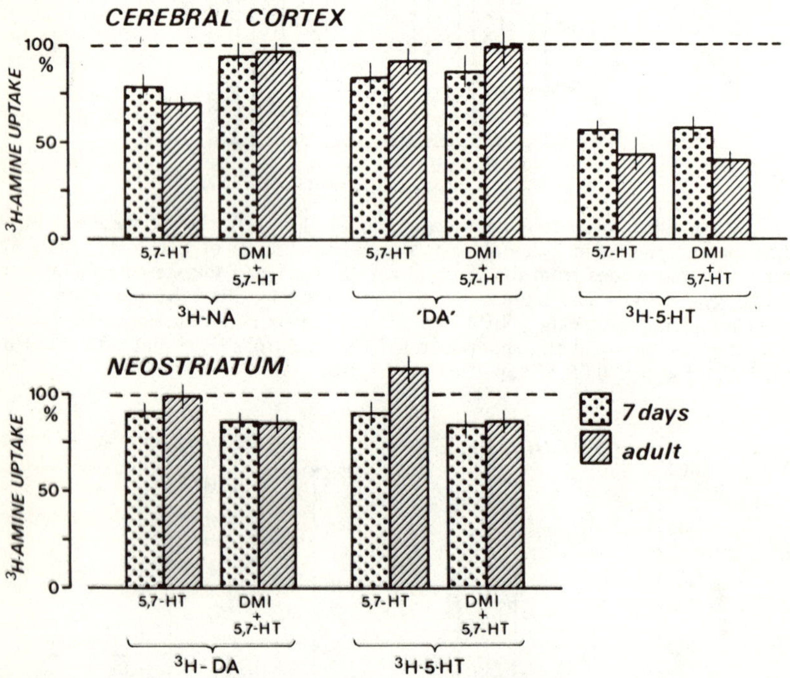

FIGURE 8. Effects of desipramine (DMI) and/or 5,7-HT treatment of newborn rats on [^{3}H]amine uptake (0.05 μM, 5 min) in homogenates from the cerebral cortex and the neostriatum. The 5,7-HT (100 mg/kg, sc) was administered within 4 hr after birth, and DMI (20 mg/kg, sc) was injected 30 min before 5,7-HT. The rats were killed 7 days or 7 weeks (adult) after treatment. Each column represents the mean $\pm$SEM of from four to six determinations, expressed as a percentage of vehicle-treated littermate controls, respectively. 'DA' refers to DMI-resistant uptake of [^{3}H]NA, most likely reflecting uptake in dopamine nerve terminals (see Reference 7). Data from Reference 7.

produce any significant neurodegenerative effects on dopamine (DA) nerve terminals in the cerebral cortex and the neostriatum, because [^{3}H]amine uptake in these structures was not significantly affected. This contention is also supported by data from chemical assays of DA in these regions, which revealed no change in endogenous DA concentration. It is of interest to note that the relative reductions in [^{3}H]NA and [^{3}H]-5-HT uptake were quantitatively the same in the cerebral cortex 7 days after 5,7-HT treatment, compared with the situation when the animals had reached adulthood (FIGURE 8). Since the tissue mass of the cerebral cortex increases three- to fourfold from postnatal Day 7 to the adult stage, the present data indicate that the nerves that are spared after 5,7-HT injection undergo an apparently normal postnatal development.

Effects of Neonatal 5,7-HT Administration on 5-HT and [³H]LSD Receptor Binding

Recent studies have shown that [^{3}H]-5-HT and [^{3}H]LSD bind to membrane preparations or brain homogenates with characteristics that resemble those of the physiologic postsynaptic 5-HT receptor.[14, 26, 27] It was therefore thought of interest to study the binding properties of these ligands to homogenates from the cerebral cortex and mesencephalon-pons-medulla after neonatal 5,7-HT treatment. Since these two regions represent the extremes of the altered ontogeny of 5-HT neurons, one markedly denervated region and one with hyperinnervation, it was considered that study of [^{3}H]-5-HT and [^{3}H]LSD binding in these regions could provide some information on the relationship between postsynaptic 5-HT receptor ontogeny and differentiation and maturation of 5-HT neurons.

The results presented in FIGURE 9 show a saturable binding of both [^{3}H]-5-HT and [^{3}H]LSD in both regions investigated, in agreement with findings reported by Bennett and Snyder.[14] Scatchard analyses of [^{3}H]-5-HT and [^{3}H]LSD binding suggest an apparent single population of binding sites for both ligands, both in the cerebral cortex and the mesencephalon-pons-medulla. The maximal number of [^{3}H]-5-HT-binding sites in the former region was about 15 pmol/g and about 7 pmol/g in the latter. The values for [^{3}H]LSD-binding sites were about 35 and 25 pmol/g, respectively. In all of the experiments, we consistently found that neonatal 5,7-HT treatment did not alter the binding characteristics or number of binding sites for [^{3}H]-5-HT and [^{3}H]LSD (FIGURES 9–11) to either homogenates from the cerebral cortex or the mesencephalon-pons-medulla. These results would imply, then, that the postsynaptic 5-HT receptor develops independent of the presence or absence of presynaptic 5-HT nerve terminals.

DISCUSSION

The present results show that 5,7-HT is the neurotoxin of choice for experimental work on 5-HT neurons during ontogenesis. The disadvantage of its neurodegenerative action on central NA neurons also can be circumvented by pretreatment with DMI, thus providing an experimental tool for inducing rather selective denervations of 5-HT neurons. However, it is also obvious from the data presented that 5-HT neurons may undergo rather complex changes after systemic 5,7-HT administration in the neonatal stage. Changes of this kind do not occur in adult animals, indicating that they must be associated with some special properties of the developing neuron. From the analyses performed, it can be concluded that this treatment produces marked and permanent denervations of the most distantly situated 5-HT nerve terminal projections, that is, in the cerebral cortex and the spinal cord, whereas 5-HT cell bodies are apparently more or less intact, according to the histochemical analysis. However, in the cell body-near regions, the results point to a stimulation of outgrowth, leading to an increased number of 5-HT nerve terminals. These 5-HT nerve terminals

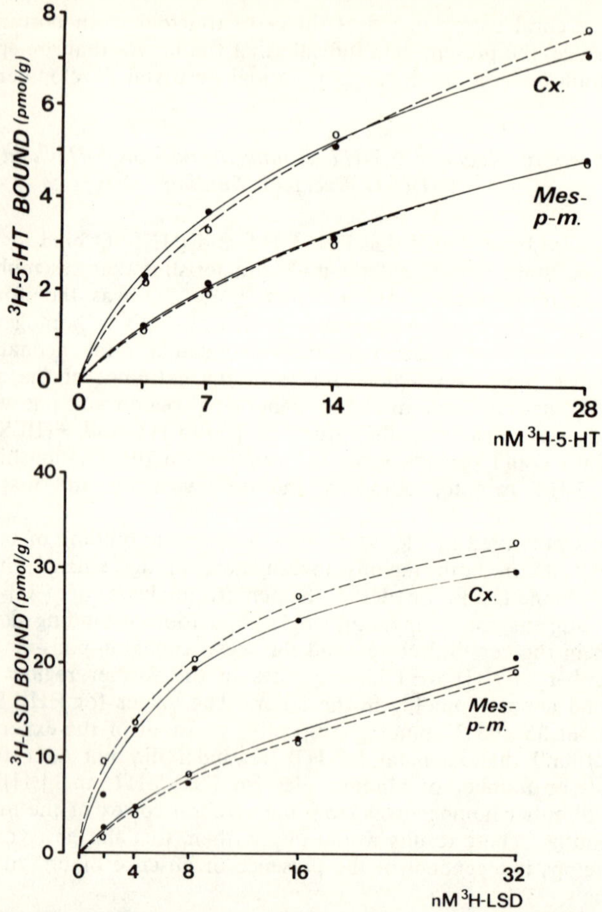

FIGURE 9. Binding of [^{3}H]-5-HT (*top*) and [^{3}H]LSD (*bottom*) to homogenates from the cerebral cortex (Cx.) and the mesencephalon-pons-medulla (Mes-p-m.) prepared from rats treated with 5,7-HT (○, 100 mg/kg, sc) at birth or saline (●, control). The animals were killed at the age of 5–8 weeks. The homogenates were incubated with various concentrations of [^{3}H]-5-HT (3.5–28 nM) and [^{3}H]-*d*-LSD (2–32 nM) in the presence (nonspecific binding) or absence (total binding) of 10 μM unlabeled 5-HT and 1 μM *d*-LSD, respectively. The values presented were obtained by subtracting nonspecific binding from total binding. Each point represents the mean of from six to nine determinations.

exhibit normal 5-HT uptake, storage, and release properties and pharmacologic characteristics. Therefore, it seems that the overall effect of neonatal 5,7-HT injection is redistribution of the terminal branchings of 5-HT neurons. The question then arises as to the mechanism responsible for this redistribution. In view of the fact that surgical lesion of the ascending 5-HT pathways caused a permanent 5-HT denervation in the cerebral cortex and also resulted in an increased [^{3}H]-5-HT uptake, most likely reflecting an increased number of nerve terminals in the cell body-near regions, it is conceivable that the effects in this region that occur after neonatal administration of 5,7-HT are also mainly related to the permanent 5,7-HT-induced denervation of distant 5-HT projections. The 5-HT neurons that have cell bodies in the brain stem are known to possess widespread projections that innervate many brain regions. Considering this fact, it seems likely that this phenomenon is a "pruning effect," or a conservation of total axonal arborization (FIGURE 12), which has previously been observed after the induction of neonatal lesions in the visual and olfactory systems of the hamster (see Reference 28). Thus, 5-HT neurons apparently have a developmental program that includes formation of a rather fixed quantity of nerve terminal branchings. When the most distant axons, that is, those that innervate the cerebral cortex and the spinal cord, are permanently damaged after treatment with 5,7-HT, their developmental program will be expressed in collateral systems closer to the cell bodies, leading to a redistribution of the nerve terminal projections. The present observations on 5-HT neurons after administration of 5,7-HT are quite analogous to the results seen in central NA neurons after neonatal 6-OHDA injection.[2-5] In fact, the present results show also that 5,7-HT is capable of inducing a "pruning effect" on NA neurons due to its neurotoxic action on this neuron type also. In view of this finding and the recent observations of Schneider,[28] it seems that the "pruning effect" is a general response mechanism for developing neurons and may very well account for most of the plasticity phenomena observed after early brain injuries in mammals.

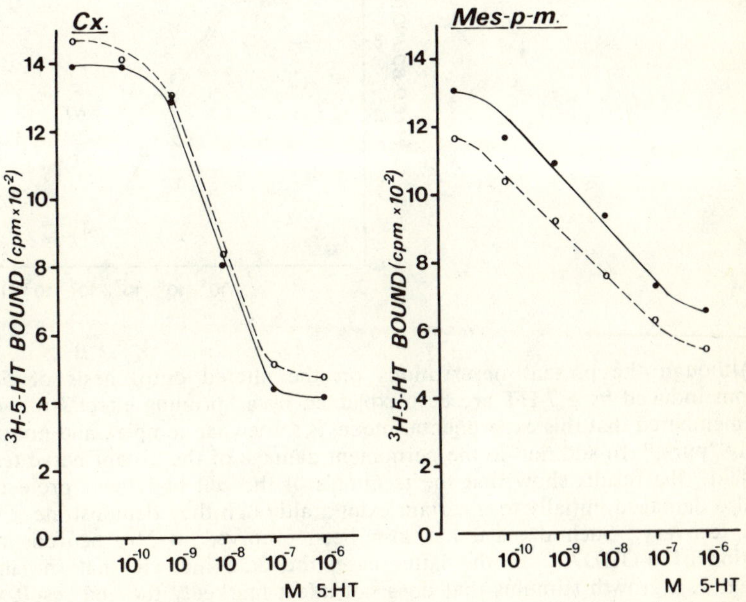

FIGURE 10. Competition of unlabeled 5-HT for [^{3}H]-5-HT binding to homogenates prepared from the cerebral cortex (Cx.) and the mesencephalon-pons-medulla (Mes-p-m.) from 5,7-HT-treated (○) and vehicle-treated (●) rats as in FIGURE 9. The homogenates were incubated in 6 nM [^{3}H]-5-HT with increasing concentrations of unlabeled 5-HT. Each point represents the mean of six determinations.

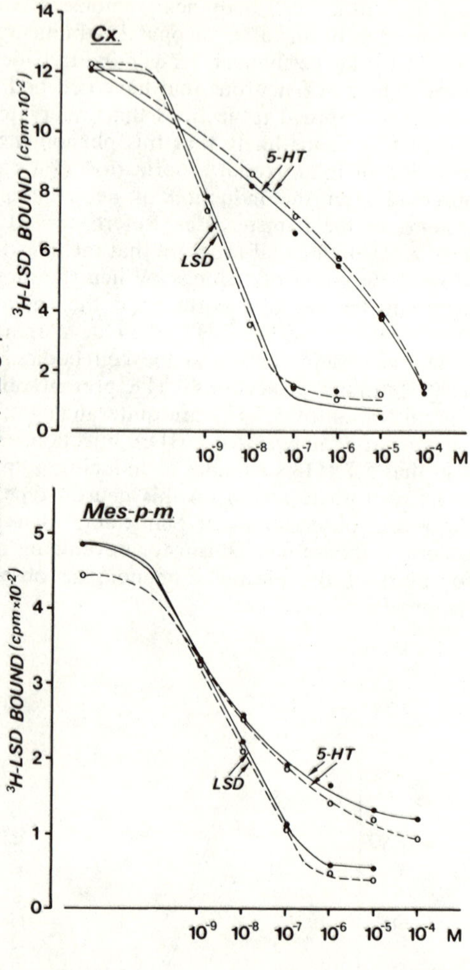

FIGURE 11. Competition of unlabeled 5-HT and LSD for [^{3}H]-*d*-LSD binding to homogenates of cerebral cortex (Cx.) and the mesencephalon-pons-medulla (Mes-p-m.) prepared from 5,7-HT-treated (○) and vehicle-treated (●) rats as in FIGURE 9. The homogenates were incubated in 3 nM [^{3}H]-*d*-LSD with increasing concentrations of unlabeled 5-HT and *d*-LSD. Each point represents the mean of six determinations.

Although the present observations on the altered ontogenesis of 5-HT neurons induced by 5,7-HT are best explained by a "pruning effect," it should be remembered that this experimental model is somewhat complex and not completely "pure." In addition to the permanent damage of the distant nerve terminal fields, the results show that the terminals of the cell body-near projections are also damaged initially to a certain extent, although they demonstrate a very rapid recovery. Such damage has also been observed in NA neurons after injection of 6-OHDA.[4] In the latter case, the data indicate that the initial damage is a growth stimulus that does not affect markedly the end result with respect to quantity of NA nerve terminals. Concerning 5-HT neurons after treatment with 5,7-HT, it is conceivable that the initial damage stimulates regrowth, but it is not possible on the basis of the available data to make a conclusion regarding whether such damage affects the end result with respect to quantity of nerve terminal arborizations in the cell body-near projections.

Further experiments are needed to determine whether this factor is significant for regulation of nerve terminal field size.

In this context, it may be relevant to comment on other studies that have employed 5,7-HT for studying the developmental aspects of 5,HT neurons.[29, 30] In these studies, which utilized the intracisternal route of 5,7-HT administration, the results point to permanent 5-HT denervations without any clear-cut signs of the "pruning effect" observed in this study. These findings are in good agreement with the results obtained from experiments performed in our laboratory by use of intracisternal injection of 5,7-HT and analysis of [^{3}H]-5-HT uptake *in vitro*.[33] Although it is difficult to provide any exact explanation for the apparent difference between the results observed after systemic as compared to intracisternal 5,7-HT injection, it is likely that the type of damage induced is quite different neuroanatomically and is thus responsible for the different end results. It is suggested that this difference might be due to the fact that intracisternal 5,7-HT injection causes more pronounced damage to collaterals or

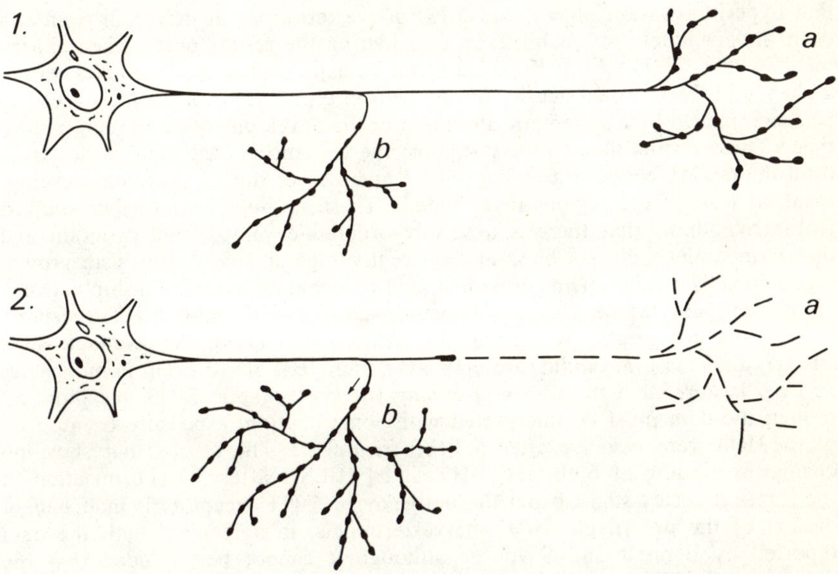

FIGURE 12. Schematic representation of the 5,7-HT-induced alterations of the postnatal development of central 5-HT neurons after systemic 5,7-HT administration at birth. The central 5-HT neurons have a marked collateralization that innervates many brain areas. *1.* The normal situation, where *a* represents distant nerve terminal projections, for example, in the cerebral cortex or spinal cord and *b* denotes projections closer to the cell bodies, for example, in the mesencephalon and pons-medulla. *2* After neonatal 5,7-HT treatment, there is a marked and permanent denervation of distant projections (*2a*) and a concomitant compensatory sprouting and increased outgrowth of 5-HT nerve terminals in cell body-near projections. The results obtained so far are compatible with the belief that the end results after systemic 5,7-HT administration in the neonatal stage are mainly due to a "pruning effect."[28] The developing 5-HT neurons thus seem to be programmed to produce a certain quantity of nerve terminal branchings, which they try to conserve after the 5,7-HT-induced injury.

even death of whole neurons, preventing the "pruning effect" to be observable. It is obvious also that the dose must be taken into consideration. This conclusion does not exclude, however, the belief that "pruning" cannot occur after intracisternal 5,7-HT injection in young animals. This phenomenon may very well be observable after an appropriate dose and/or a modification of the injection technique, in the same way as has been shown with NA neurons after intracisternal injection of 6-OHDA.[4]

In accord with the notion of the rather strict regulation of the development of 5-HT neurons are results that indicate that the neurons that are unaltered after injection of 5,7-HT undergo an apparently normal development with respect to quantity of nerve terminal arborizations. This conclusion can be deduced from results that show that the relative reductions in the number of nerve terminals in the spinal cord and cerebral cortex were approximately constant throughout the postnatal development period studied. These results would also imply that an early 5-HT denervation is not associated with any substantial "collateral sprouting" of nerves not damaged by this neurotoxin. Furthermore, it is of interest to note that NA and DA neurons apparently develop normally after selective 5-HT denervation (DMI plus 5,7-HT) and that hyperinnervation of NA and 5-HT nerve terminals in the brain stem can occur independently of each other. Considering the results observed in experiments with 6-OHDA,[6] this hypothesis can be expressed in more general terms: a change in the terminal density of one neuron type has never been observed to be accompanied by a systematic alteration of the development of another neuron type. These results thus strongly support the contention that there is no structural interaction between growing monoamine nerves during postnatal development, at least in the regions investigated. These findings lend further support to the hypothesis that there is a strictly ordered developmental program and that monoamine neurons have an apparently high degree of intrinsic growth regulation. The experiments presented here to elucidate the relationship between ontogenetic development of 5-HT neurons and postsynaptic 5-HT receptors merit some comments. Regarding the methodology used, the available data strongly indicate that binding of [^{3}H]-5-HT and [^{3}H]LSD to brain homogenates *in vitro* is a reliable means for assessing the postsynaptic 5-HT receptor, although the data must be interpreted with some caution, especially because all of the data were negative after 5,7-HT treatment. The results that show no change in binding of both [^{3}H]-5-HT and [^{3}H]LSD after 5,7-HT injection in the cerebral cortex suggest that the ontogeny of 5-HT receptors is independent of that of the presynaptic 5-HT nerve terminals, in agreement with the data reported by Bennett and Snyder,[14] although it cannot be excluded that the amount of 5-HT present in the cerebral cortex before 5,7-HT treatment (at birth) is sufficient to trigger receptor formation during postnatal development. The data obtained from the mesencephalon-pons-medulla region, which is hyperinnervated after 5,7-HT treatment, would imply that the receptors may not play a crucial role in guiding ingrowing 5-HT nerve terminals, again suggesting that the formation of receptors does not depend on 5-HT neurotransmission. This conclusion has been reached previously by other investigators studying noradrenergic and cholinergic systems.[31, 32] If the present interpretation concerning a developmental program for the receptors that is apparently independent of the presynaptic structure is correct, the question arises as to how specificity for connectivity and adequate neurotransmission is achieved. Although only speculation, it may be that local regulatory mechanisms (receptor supersensitivity and subsensitivity) and feedback systems that affect transmittor release and turnover are more important factors for the appropriate function of 5-HT systems than is a precise relationship between nerve terminals and receptors from a structural viewpoint.

Finally, regarding the functional significance of the rearrangement of 5-HT neurons after 5,7-HT administration in the neonatal stage, no definite conclusions can be made, because functional data are not available. While the consequences of denervations must to a large extent depend on the development of compensatory mechanisms, for example, receptor supersensitivity, increase in transmitter turnover of spared neurons, and/or readjustment of the functional balance between other systems, it is possible that the hyperinnervation in the brain stem may for certain functions represent a growth mechanism that leads to sparing and functional recovery. However, for other functions, it may very well have pathologic consequences and represent a mechanism that leads to developmental disturbances. The answers to these questions must await further studies.

Summary

Systemic administration of 5,7-HT to newborn rats produces an altered development of the 5-HT neurons in the central nervous system, with marked regional differences. 5,7-Hydroxytryptamine can enter the brain and elicit its neurotoxic actions after systemic administration in the neonatal stage due to an incompletely developed blood-brain barrier, which for 5,7-HT is elaborated between postnatal Days 5 and 7. Treatment with 5,7-HT at birth produces marked and permanent 5-HT denervation in the cerebral cortex and spinal cord, whereas hyperinnervation occurs in the 5-HT cell body-near regions (mesencephalon-pons-medulla). The latter effect is seen within the first week postnatally. Treatment with 5,7-HT also affects NA neurons in a similar manner, although the action is exerted preferentially on 5-HT neurons. A selective effect on 5-HT neurons can be achieved by DMI pretreatment, after which both NA and DA neurons develop normally. No signs of any interaction among growing 5-HT, NA, and DA neurons can be observed. Studies of the postsynaptic 5-HT receptor *in vitro* with [^{3}H]-5-HT and [^{3}H]LSD binding indicate that this receptor develops independently of presynaptic 5-HT nerve terminals. Neither 5-HT denervation nor 5-HT hyperinnervation was accompanied by any change in receptor-binding characteristics or receptor density. The results available are compatible with the view that the consequences for 5-HT neurons that occur after neonatal 5,7-HT administration are mainly due to a "pruning effect." The developing 5-HT neurons seem to be programmed to produce a certain quantity of nerve terminal arborizations, which they try to conserve after 5,7-HT-induced injury, leading to the observed rearrangement of 5-HT nerve terminals.

Acknowledgments

The skillful technical assistance of Miss B. Drevinger, Mrs. Bodil Käller, and Mrs. E. Lindqvist is gratefully acknowledged.

REFERENCES

1. STEIN, D. G., J. J. ROSEN & N. BUTTERS. (Eds.) 1974. Plasticity and Recovery of Function in the Central Nervous System. Academic Press, Inc. New York, N.Y.
2. JONSSON, G., CH. PYCOCK & CH. SACHS. 1973. Plastic changes of central noradrenaline neurons after 6-hydroxydopamine. *In* Frontiers in Catecholamine Research. E. Usdin & S. Snyder, Eds. : 459–461. Pergamon Press. New York, N.Y.
3. SACHS, CH., CH. PYCOCK & G. JONSSON. 1974. Altered development of central noradrenaline neurons during ontogeny by 6-hydroxydopamine. Med. Biol. **52:** 55–65.
4. SACHS, CH. & G. JONSSON. 1975. Changes in the development of central noradrenaline neurons following neonatal 6-OH-DA administration. *In* Chemical Tools in Catecholamine Research. G. Jonsson, T. Malmfors & Ch. Sachs, Eds. Vol. I: 163–171. North Holland Publishing Company. Amsterdam, The Netherlands.
5. JONSSON, G. & CH. SACHS. 1976. Regional changes in ^{3}H-noradrenaline uptake, catecholamines and catecholamine synthetic and catabolic enzymes in rat brain following neonatal 6-hydroxydopamine treatment. Med. Biol. **54:** 286–297.
6. SACHS, CH. & G. JONSSON. 1975. 5,7-Dihydroxytryptamine induced changes in the postnatal development of central 5-hydroxytryptamine neurons. Med. Biol. **53:** 156–164.
7. JONSSON, G. 1976. Developmental characteristics of central monoamine neurons and their reciprocal relations. Exp. Neurol. **53:** 801–814.
8. THOENEN, H. & J. P. TRANZER. 1968. Chemical sympathectomy by selective destruction of adrenergic nerve endings with 6-hydroxydopamine. Arch. Exp. Pathol. **261:** 271–288.
9. SACHS, CH. & G. JONSSON. 1975. Mechanisms of action of 6-hydroxydopamine. Biochem. Pharmacol. **24:** 1–8.
10. BAUMGARTEN, H. G. & A. BJÖRKLUND. 1976. Neurotoxic indoleamines and monoamine neurons. Annu. Rev. Pharmacol. Toxicol. **16:** 101–111.
11. SAAVEDRA, J. M., M. BROWNSTEIN & J. AXELROD. 1973. A specific and sensitive enzymatic isotopic microassay for serotonin in tissues. J. Pharmacol. Exp. Ther. **186:** 508–515.
12. KELLER, R., A. OKE, I. MEFFORD & R. N. ADAMS. 1976. Liquid chromatographic analysis of catecholamine routine assay for regional brain mapping. Life Sci. **19:** 995–1004.
13. FARNEBO, L.-O. & B. HAMBERGER. 1971. Drug-induced changes in the release of ^{3}H-monoamines from field stimulated rat brain slices. Acta Physiol. Scand. Suppl. **371:** 35–44.
14. BENNETT, J. P. & S. H. SNYDER. 1976. Serotonin and lysergic acid diethylamide binding in rat brain membranes: relationship to postsynaptic serotonin receptors. Mol. Pharmacol. **12:** 373–388.
15. FALCK, B., N.-Å. HILLARP, G. THIEME & A. TORP. 1962. Fluorescence of catecholamines and related compounds condensed with formaldehyde. J. Histochem. Cytochem. **10:** 348–354.
16. CORRODI, H. & G. JONSSON. 1967. The formaldehyde fluorescence method for the histochemical demonstration of biogenic monoamines.—A review on the methodology. J. Histochem. Cytochem. **15:** 65–78.
17. FUXE, K., T. HÖKFELT, G. JONSSON & U. UNGERSTEDT. 1970. Fluorescence microscopy in neuroanatomy. *In* Contemporary Research in Neuroanatomy. W. J. H. Nauta & S. O. E. Ebbesson, Eds. : 275–314. Springer Verlag. Berlin, Heidelberg and New York.
18. DAHLSTRÖM, A. & K. FUXE. 1964. Evidence for the existence of monoamine-containing neurons in the central nervous system. I. Demonstration of mono-

amines in the cell bodies of brain stem neurons. Acta Physiol. Scand. **62:** (Suppl. 232): 1–55.

19. SACHS, CH. 1973. Development of the blood-brain barrier for 6-hydroxydopamine. J. Neurochem. **20:** 1753–1760.
20. JONSSON, G., K. FUXE, B. HAMBERGER & T. HÖKFELT. 1969. 6-Hydroxytryptamine—a new tool in monoamine fluorescence histochemistry. Brain Res. **13:** 190–195.
21. BAUMGARTEN, H. G., A. BJÖRKLUND, L. LACHENMAYER & A. NOBIN. 1973. Evaluation of the effects of 5,7-dihydroxytryptamine on serotonin and catecholamine neurons in the rat CNS. Acta Physiol. Scand. Suppl. **391:**
22. DALY, J., K. FUXE & G. JONSSON. 1974. 5,7-Dihydroxytryptamine as a tool for the morphological and functional analysis of central 5-hydroxytryptamine neurons. Res. Commun. Chem. Pathol. Pharmacol. **1:** 175–187.
23. GERSON, S., R. J. BALDESSARINI & S. C. WHEELER. 1974. Biochemical effects of dihydroxylated tryptamines on central indoleamine neurones. Neuropharmacology **13:** 987–1004.
24. BJÖRKLUND, A., H. G. BAUMGARTEN & A. RENSCH. 1975. 5,7-Dihydroxytryptamine: improvement of its selectivity for serotonin neurons in the CNS by pretreatment with desipramine. J. Neurochem. **24:** 833–835.
25. GERSON, S. & R. J. BALDESSARINI. 1975. Selective destruction of serotonin terminals in rat forebrain by high doses of 5,7-dihydroxytryptamine. Brain Res. **85:** 140–145.
26. BENNETT, J. L. & G. K. AGHAJANIAN. 1975. D-LSD binding to brain homogenates: possible relationship to serotonin receptors. Life Sci. **15:** 1935–1944.
27. FILLION, G., M.-P. FILLION, C. SPIRAKIS, J.-M. BAKERS & J. JACOB. 1976. 5-Hydroxytryptamine binding to synaptic membranes from rat brain. Life Sci. **18:** 65–74.
28. SCHNEIDER, G. E. 1973. Early lesions of superior colliculus: factors affecting the formation of abnormal retinal projections. Brain Behav. Evol. **8:** 73–109.
29. LYTLE, L. D., J. H. JACOBY, M. F. NELSON & H. G. BAUMGARTEN. 1974. Long-term effects of 5,7-dihydroxytryptamine administered at birth on the development of monoamines. Life Sci. **15:** 1203–1217.
30. BAUMGARTEN, H. G., S. J. VICTOR & W. LOVENBERG. 1975. A developmental study of the effects of 5,7-dihydroxytryptamine on regional tryptophan hydroxylase in rat brain. Psychopharmacol. Commun. **1:** 75–88.
31. WOODWARD, D. J., B. J. HOFFER, G. R. SIGGINS & F. E. BLOOM. 1971. The ontogenetic development of synaptic functions, synaptic activation and responsiveness to neurotransmitter substances in rat cerebellar Purkinje cells. Brain Res. **31:** 73–97.
32. KOVELAS, E. D. & L. A. GREENE. 1976. The binding properties and regional ontogeny of receptors for α-bungarotoxin in chick brain. Brain Res. **113:** 111–126.
33. JONSSON, G. Unpublished observations.

5,7-DIHYDROXYTRYPTAMINE AS A TOOL TO STUDY THE FUNCTIONAL ROLE OF CENTRAL 5-HYDROXYTRYPTAMINE NEURONS *

Kjell Fuxe,† Sven-Ove Ögren,‡ Luigi F. Agnati,† Gösta Jonsson, and Jan-Åke Gustafsson §

† Department of Histology
and § Department of Chemistry
Karolinska Institute
S-104 01 Stockholm 60, Sweden
and ‡ Research Laboratories
AB Astra
Södertälje, Sweden

The first serotonin neurotoxin to be introduced was 5,6-dihydroxytryptamine (5,6-DHT).[1-5] 5,6-Dihydroxytryptamine has a rather selective neurotoxic action on 5-hydroxytryptamine (5-HT) neurons, with limited effects on catecholamine (CA) neurons. However, it has the disadvantage of causing a substantial degree of unspecific damage to nonmonoamine-containing nerve cells present within the area close to the injection site. In the search for better and more selective serotonin neurotoxins, several groups[6,7] found that 5,7-dihydroxytryptamine (5,7-DHT) was a better tool than 5,6-DHT for the functional mapping of central 5-HT neurons. The main advantage of 5,7-DHT versus 5,6-DHT was the relative lack of unspecific damage close to the injection site in the brain tissue. The unspecific damage was greater than that induced after injection with the solvent alone but was small compared with the large unspecific lesions that are produced by electrolytic lesions.[8] 5,7-Dihydroxytryptamine also causes substantial degeneration of noradrenaline (NA) neurons and, to some extent, of dopamine (DA) systems, which is a disadvantage. By using NA uptake blockers, such as protriptyline, however, the NA systems can be protected from degenerative effects, without interfering with the degeneration of the 5-HT systems.

In our laboratory, we used 5,7-DHT to study regeneration mechanisms within central 5-HT pathways. Thus, in our early studies with 5,6-DHT, we noticed that the degenerative action of this compound was not irreversible. We observed signs of sprouting and regeneration from lesioned 5-HT axonal bundles in the injection area about 2 weeks after induction of the lesion.[5] We also observed that this sprouting of 5-HT axons innervated not only nerve cells that were normally innervated by 5-HT terminals but also nonmonoamine-containing nerve cells not normally innervated by 5-HT nerve terminals. These early findings prompted an investigation of the possible regeneration of descending bulbospinal 5-HT systems to the spinal cord after degeneration of these pathways by intraventricular injection of 5,6-DHT (50 μg).[9,10] This degeneration is probably caused mainly by axotomy of the

* Supported by Grant 04X–715 from the Swedish Medical Research Council, Grant MH 25504–03 from the National Institutes of Health, and by grants from Magn. Bergvalls Stiftelse and Karolinska Institute Foundation.

0077–8923/78/0305–0346 $01.75/0

descending 5-HT axons at the level of the upper cervical spinal cord. Thus, the 5-HT axons in this area have a very superficial location and can be easily reached by the neurotoxin from the subarachnoidal space. We also observed a marked 5-HT denervation of the spinal cord. Tritiated 5-HT uptake was found to be 25% of control levels after these injections; after 3 months, [^{3}H]-5-HT uptake was about 75% of control values, suggesting that downgrowth of 5-HT terminal systems had occurred. A good correlation existed between [^{3}H]-5-HT uptake and the number of 5-HT nerve terminals observed within the gray matter. It is clear from recent work in our laboratory that higher doses of 5,6-DHT or 5,7-DHT should not be used in this type of study,[11] because large amounts of a neurotoxin will also damage 5-HT nerve cell bodies within the medulla oblongata; therefore, no regeneration of 5-HT terminals will occur. At least two mechanisms can mediate the regeneration of 5-HT terminals: either downgrowth of axons in the white matter of newly formed axons or continuous sprouting of terminals within the gray matter in a rostrocaudal direction. The latter possibility is favored by recent work in our laboratory by Nygren and Olsson [11] that revealed that after 6-hydroxydopamine (6-OHDA) -induced lesions of the descending bulbospinal NA systems, regeneration of NA systems will occur via downgrowth of NA nerve terminal systems within the ventral horn, which utilizes the longitudinally oriented dendrites of the motor neurons to reach the lumbosacral part of the spinal cord. In our study,[10] the interesting observation was also made that with the development of new 5-HT terminal plexa, the supersensitivity of the 5-HT receptors disappeared, as shown by reduced responsiveness to 5-HT receptor agonists, such as 5-methoxydimethyltryptamine (5-MeODMT), and to tryptophan after monoamine oxidase inhibition. These results strongly indicate that functional regeneration does occur, because the loss of supersensitivity suggests that the 5-HT boutons have in some way approached the 5-HT receptor sites in the spinal cord. The model used to demonstrate supersensitivity was the extensor reflex model, which in acutely spinalized animals is highly dependent on 5-HT receptor activity.[9, 10]

After this initial investigation of degeneration within 5-HT systems by use of 5,6-DHT, we began functional analysis of the ascending 5-HT systems to the telencephalon and diencephalon by use of injections of 5,7-DHT into the ascending 5-HT bundles within the caudal midbrain. Within the midbrain, there exists a large medial 5-HT pathway that runs close to the midline, which, *inter alia,* innervates large parts of the cortex cerebri, hypothalamus, and subcortical limbic regions. A main source of the cortical 5-HT systems appears to be the nucleus raphe dorsalis (group B7, according to Dahlström and Fuxe [12]), except for the hippocampal 5-HT terminals, which appear to arise from axons that originate in the nucleus raphe medianus (group B8, according to Dahlström and Fuxe [12]).[13–15] The neostriatal 5-HT terminals arise mainly from the nucleus raphe dorsalis.[13–15] The total amount of 5,7-DHT injected into the dorsomedial tegmentum was 4 μg in a volume of 4 μl. After induction of these lesions of the ascending 5-HT bundles, there was an 80–90% degeneration of 5-HT terminals within the hypothalamus and a 60–70% degeneration of cortical 5-HT terminal systems. By use of the lesions produced, the role of ascending 5-HT systems in behavior, sleep, and wakefulness mechanisms has been evaluated. In collaboration with Everitt,[16] we were able to obtain further support for the view that the ascending 5-HT systems inhibit sexual behavior in ovariectomized estrogen-primed female rats. Also, signs of increased aggression were observed in these rats, as evidenced by the fact that female rats attacked mounting males. In a subsequent study on castrated, testosterone proprionate-treated male rats, Larsson *et al.*[17] demonstrated that after a 5,7-DHT-induced lesion of the ascending 5-HT pathways, there was a marked increase in the proportion of rats that exhibited ejaculation behavior. It was also shown that pretreatment with the NA uptake blocker protriptyline, which protects NA pathways from the degeneration caused by 5,7-DHT, did not influence this increase in sexual behavior. It was found that the increase in sexual activity correlated with a reduction of [^{3}H]-5-HT uptake in the hypothalamus but not in the cortex cerebri, suggesting that 5-HT terminal systems within the hypothalamus were mainly involved in inhibiting sexual behavior. Also, the influence of 5,7-DHT-induced lesions of the ascending 5-HT pathways on feeding behavior were evaluated.[18] In this study, we demonstrated that the anorectic action of fenfluramine, a 5-HT-releasing agent, was reduced in rats with 5,7-DHT-induced degeneration of the ascending 5-HT pathways. These results suggest that the basic control of food intake regulated by glycostatic and other mechanisms is under the inhibitory control of hypothalamic 5-HT terminals. However, the decrease in lever-pressing activity for a food reward was potentiated after 5,7-DHT-induced lesions of the 5-HT ascending pathways. To explain these findings, we speculated that motivational control of food intake may involve mainly the limbic system. Since the degeneration of 5-HT nerve terminals was less in these regions than in the hypothalamus, sufficient amounts of 5-HT may be released from the remaining 5-HT terminals to reach the supersensitive 5-HT receptors, and an enhanced activation of these receptors may thus be obtained. In support of this hypothesis, we have observed that after a partial 5-HT denervation of the spinal cord, enhancement of the extensor reflex action of fenfluramine is not reduced but, rather, is increased. Taken together, these experiments suggest an inhibitory action of the hypothalamic and preoptic 5-HT terminals on both food drive and sexual drive in males and females.

In collaboration with Hole,[19] we also studied the effects of 5,7-DHT-induced lesions of the ascending 5-HT pathways on locomotor activity in an open field, on pain sensitivity, and on learning, as evaluated by the ability of such rats to acquire a one-way active avoidance response. We also attempted to determine whether such lesions impaired morphine analgesia and apomorphine-induced locomotor activity and stereotyped behavior. The conclusion drawn from this study was that the ascending 5-HT systems are involved in maintaining open-field ambulation. However, no support was obtained for the idea that they are involved in pain mechanisms, in morphine-induced analgesia, in apomorphine-induced locomotor behavior, or in one-way avoidance learning. These results clearly underline the importance of the use of serotonin neurotoxins to produce as specific lesions as possible of the 5-HT systems. Thus, electrolytic lesions of the raphe nuclei [20] have been found to produce a reduced analgesic effect of morphine and a defective acquisition of a one-way active avoidance response.[8] Finally, electrolytic lesions of the median raphe nucleus produced increases of locomotor activity in the open field,[8] a result opposite to what we observed in our analysis with 5,7-DHT-induced lesions of these nuclei. In these experiments, protriptyline was used to protect the NA pathways from degeneration by 5,7-DHT. The reduction of locomotor activity observed in these experiments by Hole *et al.*[19] could perhaps be related to an increased number of sensory stimuli reaching the cortex cerebri, which should lead to a decreased exploration of the environment by the animal.

Finally, we lesioned the ascending 5-HT pathways with 5,7-DHT to evaluate Jouvet's hypothesis that 5-HT neuron systems play an important role in maintaining slow-wave sleep II and paradoxic sleep events.[21] In this study,[22] we tried to obtain a control for the unspecific damage to nonmonoamine nerve cells produced by 5,7-DHT by pretreatment with chlorimipramine, which, according to several papers, should be a potent blocker of 5-HT uptake. Chlorimipramine was given in a dose of 25 mg/kg 10 min before the stereotaxic injection of 5,7-DHT. The results demonstrated that these fairly specific lesions of the ascending 5-HT systems produced a significant reduction of slow-wave sleep II and increase of wakefulness 3–4 days after induction of the lesion. It could be shown that these effects on the sleep-wakefulness cycle were related to a degeneration of cortical, but not of hypothalamic, 5-HT networks, because a significant correlation was observed between degeneration of cortical systems and reduction of slow-wave sleep II and paradoxic sleep and increased slow-wave sleep I. It should be emphasized that these results lend further strong support for the Jouvet's hypothesis that ascending 5-HT systems, particularly cortical 5-HT systems, are involved in the maintenance of slow-wave sleep II. In view of the lack of reduction of paradoxic sleep in these experiments, the ascending 5-HT systems are probably not involved in paradoxic sleep regulation, at least with regard to electrocortical activation. We were surprised to see that chlorimipramine in these experiments protected the NA terminals from degeneration but not the 5-HT terminals. Thus, the affinity for 5,7-DHT formed at the 5-HT uptake sites in the nerve cell membrane of the 5-HT axons may be too high to allow chlorimipramine to compete efficiently with 5,7-DHT. Therefore, new and more potent 5-HT uptake-blocking agents will be used to control for specificity. In view of the fact that the changes in slow-wave sleep II were related to the degeneration of cortical 5-HT systems, these changes in sleep-wakefulness regulation are probably not due to changes in sexual and aggressive behavior produced by the 5,7-DHT lesions, because these behavioral changes are related to the degeneration of the hypothalamic 5-HT systems and not of cortical 5-HT systems.

In the present paper, we have continued our analysis of the role of the ascending 5-HT pathways in behavior. In view of our failure to protect the 5-HT terminals from degeneration by pretreatment of the animals with chlorimipramine before the 5,7-DHT-induced lesion, we have in the present analysis injected the potent 5-HT uptake-blocking agent (*z*)-3-dimethylamino-1-(4-bromophenyl)-1-(3-pyridyl)propene (zimelidine; H 102/09)[23] in a dose of 20 mg/kg 1 hr before the 5,7-DHT injection into the dorsomedial tegmentum. The actions of such lesions on the acquisition, retention, and forced extinction of a one-way active avoidance response have been evaluated with or without pretreatment with zimelidine. Effects of pain sensitivity were also evaluated. Possible deficits in behavior were evaluated[4–5] 10 days after the lesion had been produced. Effects on the various monoamine systems were evaluated by measuring DA, NA, and 5-HT levels in the telencephalon and diencephalon. To understand the possible mechanisms for a recovery of a behavioral deficit observed soon after a lesion had been produced, the effects of the 5,7-DHT-induced lesions of the ascending 5-HT pathways on [^{3}H]LSD and [^{3}H]-5-HT binding were also evaluated in the cerebral cortex 2, 7, and 14 days after lesioning. Our previous results on the effects of 5,7-DHT-induced lesions of ascending 5-HT pathways mentioned above certainly indicate a recovery and restoration of the sleep-wakefulness cycle within 10–14 days after lesioning.

Materials and Methods

Adult male Sprague-Dawley rats (300–350 g in body weight) were used. They were housed one in a cage under standardized laboratory conditions. Food and water were available ad lib. The animals were divided randomly into two groups of eight animals each. They were treated as follows.

Group 1. The ascending 5-HT axons that project from the raphe nuclei were lesioned in the ascending 5-HT pathways by a bilateral injection of 5,7-DHT (4 μg/4 μl).[19, 22] 5,7-Dihydroxytryptamine was dissolved in isotonic sodium chloride together with 0.02% ascorbic acid; 5,7-DHT was injected by the use of a Kopf stereotaxic instrument in front of the descending 5-HT bundle in the mesencephalon. The coordinates were as follows: A, 5.6; L, ±0.6; V, 7.0 mm.

Group 2. These animals were treated in the same way as the animals described in group 1, except that zimelidine, a potent 5-HT uptake-blocking agent, was given in a dose of 20 mg/kg intraperitoneally 1 hr before the 5,7-DHT injection, as described above.

Group 3. These rats were sham operated for the ascending raphe projection by injection of the solvent for 5,7-DHT. The coordinates were the same as for group 1.

Four days after surgery, the animals were trained for the acquisition of an unsignaled one-way avoidance response in a modified shuttle box divided into two compartments of equal size (25 × 25 × 25 cm) by a wall.[24] A semicircular opening in the wall (7 × 8 cm) allowed the animals to escape a scrambled shock. The animals were placed in the "shock" compartment, and after a delay of 5 sec, the shock was delivered through the grid floor. Acquisition continued until a criterion of nine of 10 avoidance responses was achieved. Retention and forced extinction were tested on the following day. The retention experiment was performed in analogy with the acquisition trials. Forced extinction was performed by punishment of all escape responses to the previously safe compartment by a brief shock (1.0 sec). The animals were trained until they did not escape from the previously unsafe side within 20 sec for nine of 10 consecutive trials. The following parameters were calculated in all acquisition, retention, and forced extinction experiments: trials to second avoidance, trials to criterion, and number of shocks to criterion.

Five days after surgery, following the learning trials, pain sensitivity was measured in the hot-plate test. The surface temperature was 48°C. The time required for the animals to respond (by jumping, paw licking, and/or leg lifting) was recorded.

The extent of anterograde degeneration of the 5-HT neurons was evaluated by measuring the 5-HT levels in the cortex cerebri, including the hippocampal formation, and in the rest of the brain rostral to the 5,7-DHT-induced lesion of the ascending 5-HT pathways. To evaluate the neurotoxic actions on the DA and NA systems, DA and NA levels were also measured in the two brain regions mentioned above. The animals were killed by decapitation. The brain regions were immediately removed and homogenized in 0.4 N perchloric acid, and the extract was purified on a strong cation-exchange column (Dowex® 50, X-4), after which spectrophotofluorimetric analysis was performed.[25]

The lesion in the midbrain produced by 5,7-DHT was studied in the light microscope after formalin fixation and van Gieson staining.

In a separate experiment, we investigated whether pretreatment with zimelidine in a dose of 20 mg/kg given intraperitoneally 1 hr before the 5,7-DHT injection into the midbrain could antagonize the neurotoxic action of 5,7-DHT degeneration on the ascending 5-HT axonal bundles. In this experiment, the animals were killed 5 days after the operation, because results from the behavioral experiments indicated the presence of deficits in this time interval. The extent of degeneration of 5-HT nerve terminals was evaluated by measuring the reduction of the *in vitro* uptake of [^{3}H]-5-HT in hypothalamic synaptosomes and in slices of cerebral cortex.[26] Previous studies on central 5-HT nerve terminals have shown that measurement of 5-HT uptake is a good index of the relative number of 5-HT nerve terminals present.[27] [^{3}H]-5-Hydroxytryptamine (16–17 Ci/mmol) was obtained from New England Nuclear Corporation; the concentration used was 0.05 μM. Outside the DA-rich regions, this concentration of [^{3}H]-5-HT is preferentially taken up into the 5-HT nerve terminals.

Recently, Harvey *et al.*[28] reported that *p*-chloroamphetamine produces selective neurotoxic damage to the brain, particularly neuropathologic changes in the 5-HT cell bodies of group B9, according to Dahlström and Fuxe.[12] Therefore, we have compared the actions of 5,7-DHT and *p*-chloroamphetamine on the 5-HT systems in the brain by use of the methodology of Falck and Hillarp. *p*-Chloroamphetamine was given twice in a 24-hr interval at a dose of 10 mg/kg intraperitoneally. The animals were killed by decapitation 48 hr after the first injection of *p*-chloroamphetamine, and various parts of the brain and spinal cord were removed for fluorescence histochemical analysis of CA and 5-HT by the routine Falck-Hillarp procedure.[29, 30] This treatment schedule for *p*-chloroamphetamine has been shown by Ögren *et al.*[31] to cause a marked retardation of the acquisition of a two-way active avoidance response but not of the acquisition of escape responses.

Studies on [^{3}H]-5-HT Binding and on [^{3}H]-d-LSD Binding in the Dorsal Cortex Cerebri in Control Rats and Rats with 5,7-DHT-Induced Lesions of the Ascending 5-HT Pathways to the Telencephalon and Diencephalon

Adult male Sprague-Dawley rats were operated on 2, 7, and 14 days before killing. 5,7-Dihydroxytryptamine was injected into the dorsomedial part of the tegmentum of the midbrain, as described above. The animals were killed by decapitation. The routine binding procedure was almost identical to that published by Bennett and Snyder.[32] The liquid scintillation spectrometer used (Packard Intertechnique) gave counts and disintegrations per minute for each sample.

Hill and Scatchard plots with their 95% confidence intervals were calculated according to nonparametric procedures (Sen's test). The parallellism between the lines has been tested by means of nonparametric procedures (Hollander and Wolfe's test [33]).

Results

The controls acquired the one-way avoidance response to criterion within eight to 10 trials (FIGURE 1). Lesioning the ascending 5-HT neurons by 5,7-DHT (group 1) or zimelidine plus 5,7-DHT (group 2) did not significantly affect avoidance acquisition (trials to second avoidance, trials to criterion, and shocks to criterion). In the retention experiment, the zimelidine plus 5,7-DHT-treated animals (group 2) required significantly fewer trials ($p < 0.05$) to criterion than did the control group (FIGURE 2). The 5,7-DHT-treated animals were markedly impaired in the forced extinction experiment, as evidenced by significantly more trials required to obtain second avoidance,

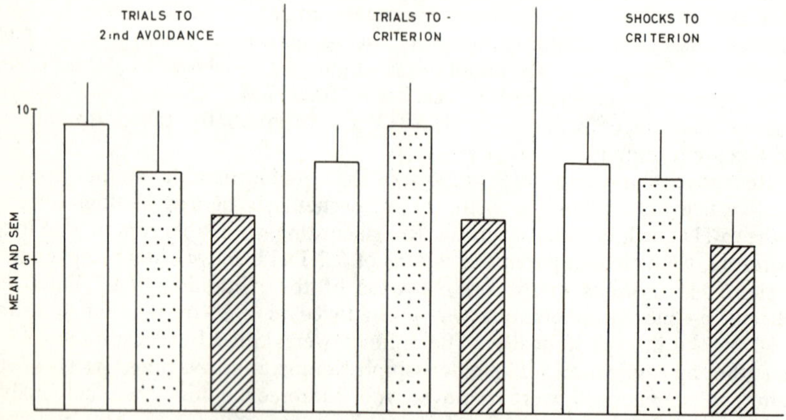

FIGURE 1. Effect of 5,7-DHT-induced lesions of the ascending 5-HT pathways in the midbrain on the acquisition of a one-way active avoidance response 4 days after the operation. Three groups were studied: a sham-operated group injected with the solvent alone (empty bar), a group injected with 5,7-DHT alone (dotted bar), and a group injected with 5,7-DHT after pretreatment with the 5-HT uptake blocker zimelidine (H 102/09) (striped bar). Zimelidine was injected intraperitoneally in a dose of 20 mg/kg 1 hr before killing. For further details, see MATERIALS AND METHODS. Means±SEM from eight rats are shown for each group. No statistically significant difference was observed when using a one-way analysis of variance.

criterion, and shocks to criterion compared with both the control group and the zimelidine plus 5,7-DHT group (FIGURE 3). In contrast, the zimelidine plus 5,7-DHT group was not different from the control group.

Pain sensitivity as measured by the hot-plate test 4 days after surgery was not changed after treatment with 5,7-DHT or zimelidine plus 5,7-DHT (TABLE 1). However, 50% of the animals were clearly hyperresponsive to the electric shock. They reacted by jumping, vocalization, and other avoidance behavior. This hyperresponsiveness was not observed 10 days after operation.

No signs of neurologic deficits were observed in the 5,7-DHT-treated rats. Furthermore, these animals did not differ from control animals in their responses in a locomotor box (data not shown).

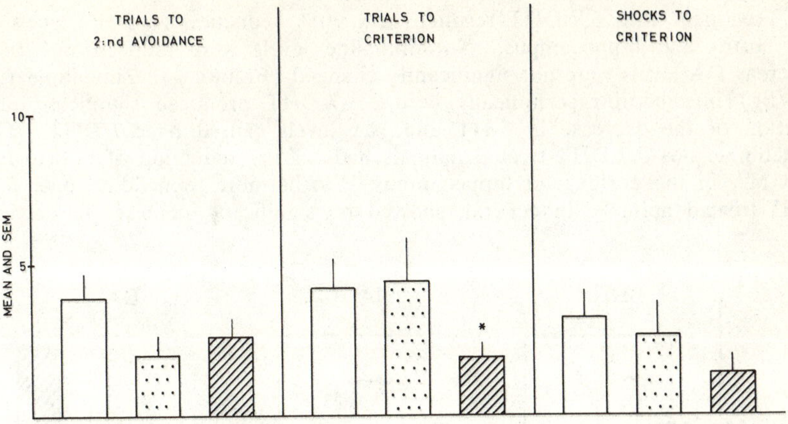

FIGURE 2. Effect of 5,7-DHT-induced lesions of the ascending 5-HT pathways in the midbrain on the retention of a one-way active avoidance response. For explanation of treatments and symbols, see the legend to FIGURE 1. Retention of the one-way active avoidance response was tested 5 days after the operation. Means±SEM from eight rats are shown for each group. One of the parameters used to study retention, namely, trials to criterion, revealed a significant reduction of trials to criterion within the group treated with 5,7-DHT and zimelidine (*, $p < 0.05$) when this group was compared with the sham-operated group. Statistical analysis was performed according to a one-way analysis of variance.

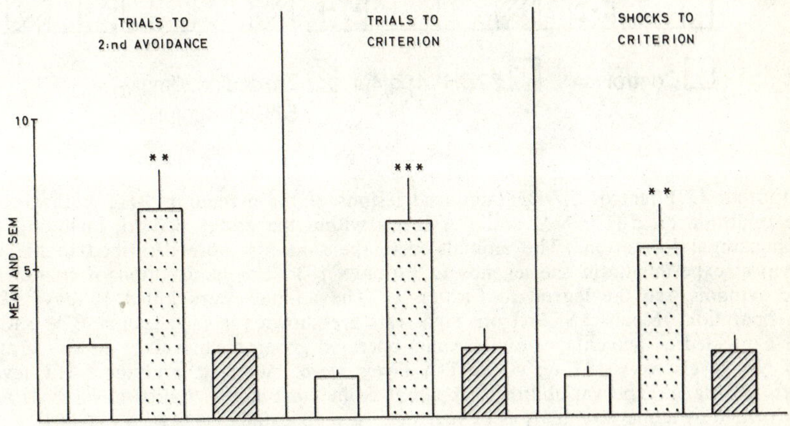

FIGURE 3. Effect of 5,7-DHT-induced lesions of the ascending 5-HT pathways in the midbrain on the forced extinction of a one-way active avoidance response. For explanations of treatments and symbols, see the legend to FIGURE 1. The tests were performed 5 days after the operation. Significant increases in the number of trials to second avoidance, trials to criterion, and shocks to criterion were observed in the 5,7-DHT-treated group. Means±SEM from eight rats are shown for each group. The 5,7-DHT group was significantly different from both the sham-operated control group and the group that received both zimelidine and 5,7-DHT (**, $p < 0.01$; ***, $p < 0.001$). Statistical analysis was performed according to a one-way analysis of variance.

Treatment with 5,7-DHT resulted in a 90% reduction of 5-HT levels in the cortex and hippocampus. Noradrenaline levels were reduced by 70%, whereas DA levels were not significantly changed (FIGURE 4). Zimelidine (20 mg/kg) injected intraperitoneally before 5,7-DHT produced significant prevention of the decrease in 5-HT and NA levels caused by 5,7-DHT. The zimelidine plus 5,7-DHT-treated animals had a 55% reduction of both 5-HT and NA in the cortex and hippocampus. Furthermore, zimelidine plus 5,7-DHT-treated animals, in general, showed nonsignificant increases in cortical

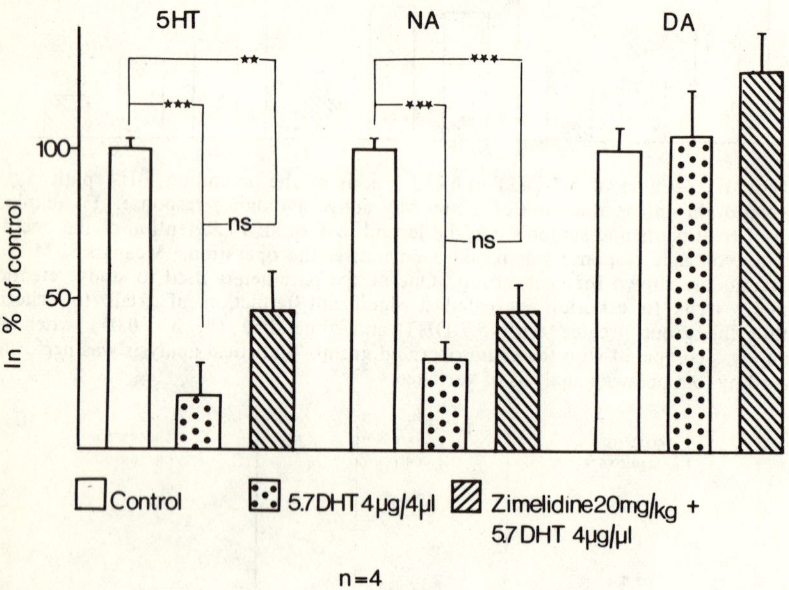

FIGURE 4. Effect of 5,7-DHT-induced lesions of the ascending 5-HT pathways in the midbrain on 5-HT, NA, and DA levels within the cortex cerebri, including the hippocampal formation. The animals were the same as those described in the behavioral experiments in the legends to FIGURES 1–3. For explanation of treatments and symbols, see the legend to FIGURE 1. The animals were killed 12 days after the operation. Means±SEM from eight rats are shown for each group. The values are expressed as percentages of the sham-operated group mean value. In this group, the NA levels were 187 ng/g, the DA levels were 250 ng/g, and the 5-HT levels were 150 ng/g. The variability is apparent from the Figure. Statistical analysis was performed by a one-way analysis of variance. ★★, $p < 0.05$; ★★★, $p < 0.001$.

DA levels, which ranged between 15 and 50%. In the rest of the brain in front of the lesion, 5-HT and NA levels were reduced in the 5,7-DHT group by 67 and 35%, respectively (FIGURE 5). The reduction of NA levels appeared to be counteracted by zimelidine, and the decrease in 5-HT levels was partly counteracted (48% reduction).

When zimelidine was given 1 hr before and 24 and 48 hr after 5,7-DHT (4 μg/4 μl), the depletion of 5-HT levels was significantly ($p < 0.001$) counteracted in the rest of the brain in front of the lesion [129 ± 15 ng/g (5,7-DHT) versus 305 ± 8 ng/g; means ± SEM for five rats] but not in the cerebral cortex

TABLE 1
EFFECT OF LESIONING THE ASCENDING 5-HT PATHWAYS BY USE OF 5,7-DHT ON PAIN SENSITIVITY

Group	Treatment	Hot-Plate Latency * (sec±SEM)
3	sham operation	8.2±0.7 †
1	5,7-DHT	8.1±1.4
2	zimelidine ‡ plus 5,7-DHT	8.4±0.8

* Time between placement on a 48°C hot plate and observation of a response.
† Means ±SEM of eight animals.
‡ Zimelidine (20 mg/kg, ip) was given 1 hr prior to surgery.

[26 ± 9 ng/g (5,7-DHT) versus 62 ± 12 ng/g; means ± SEM for five rats]. Similar results were obtained on the depletion of NA levels in the rest of the brain [413 ± 12 ng/g (5,7-DHT) versus 647 ± 25 ng/g; means ± SEM; $p < 0.001$] and in the cerebral cortex [152 ± 21 ng/g (5,7-DHT) versus 192 ± 16 ng/g; means ± SEM].

The variability observed was large, particularly in the group pretreated with

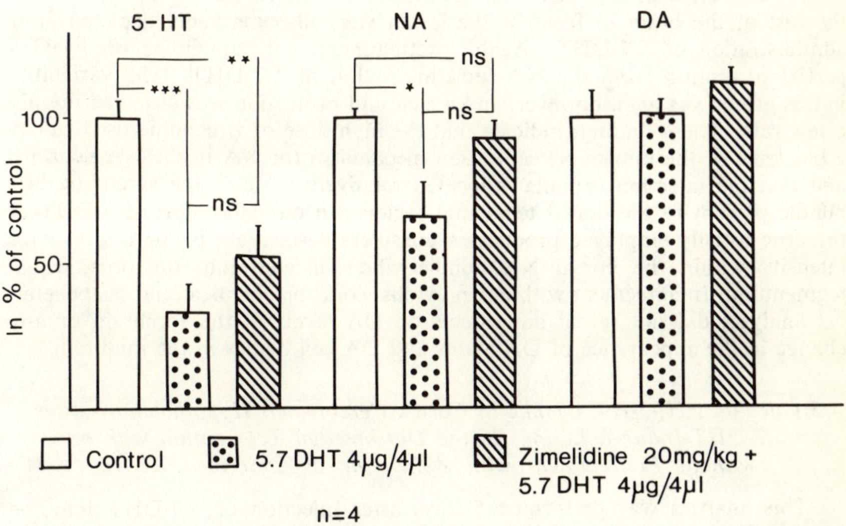

FIGURE 5. Effect of 5,7-DHT-induced lesions of the ascending 5-HT pathways in the midbrain on 5-HT, NA, and DA levels in the rest of the brain in front of the lesion (telencephalon plus diencephalon minus cortex cerebri). The animals were the same as those described in the legends to FIGURES 1–4. For explanation of treatments and symbols, see the legends to FIGURES 1–4. Means±SEM from eight rats are shown for each group. The values are expressed as percentages of the sham-operated group mean value. In this group, the NA levels were 600 ng/g, the DA levels were 1700 ng/g, and the 5-HT levels were 205 ng/g. The variability in this sham-operated group is apparent from the Figure and is on the order of 5–10%. Statistical analysis was performed by a one-way analysis of variance. ★, $p < 0.025$; ★★, $p < 0.005$; ★★★, $p < 0.001$.

zimelidine, suggesting that some animals have very poor protection against degeneration, whereas others show fairly good protection of 5-HT axons from the neurotoxic actions of 5,7-DHT. The reason for these differences among animals could be that the type of treatment schedule used for zimelidine only gives threshold amounts of the 5-HT uptake blocker at the site of the lesion. For example, 5,7-DHT may remain at the site of the lesion for a longer time period than does the 5-HT uptake-blocking agent, and/or sufficiently high concentrations of zimelidine may not be reached acutely at the site of the lesion, allowing sufficient amounts of 5,7-DHT to enter the axons via the 5-HT membrane pump to induce neurotoxic action. It must be underlined also that part of the failure of zimelidine to counteract the neurotoxic action of 5,7-DHT may be due to the somewhat larger unspecific damage found in the 5,7-DHT-treated groups versus the solvent-injected group. The maximal area of unspecific damage in the 5,7-DHT-treated group or the zimelidine plus 5,7-DHT-injected group occupied a volume around the tip of the cannula of about 6 mm^3, whereas the solvent-treated groups showed a somewhat smaller volume of damage ($\sim$1.4 mm^3). "Unspecific" damage indicates nerve cell loss, glial infiltration, and scattered vacuolization of the neuropil. It must be emphasized, however, that this unspecific damage was localized laterally and outside of the area of the large ascending medial 5-HT bundle. Within this region and within the raphe nuclei that contain the 5-HT nerve cell bodies, no unspecific damage could be detected.

As illustrated in FIGURES 4 and 5, NA levels within the cortex cerebri and the rest of the brain in front of the lesion were also markedly depleted after administration of 5,7-DHT. Again, pretreatment with zimelidine produced a partial protection from the NA-depleting action of 5,7-DHT. The variability in this group was high, however, and a clear-cut protection was observed in only a few rats. These findings indicate that the high dose of zimelidine used causes a blockade of the uptake concentration mechanism for NA in the NA neurons, and that the injection cannula is located relatively close to the ascending NA bundle present in the dorsal tegmental region. In our laboratory, 5,7-DHT at the dose usually employed produces a considerable increase in the fluorescence intensity within the dorsal NA bundle when injected into the dorsomedial tegmentum. In agreement with our previous work, biochemical and histochemical analyses did not reveal any effects on DA levels in the forebrain or any change in the appearance of DA axons and DA cell bodies in the midbrain.

Studies on [^{3}H]-5-HT Uptake in Cortex Cerebri and Hypothalamus After 5,7-DHT-Induced Lesions in the Dorsomedial Tegmentum with or without Pretreatment with Zimelidine as Described Above

This analysis was performed 5 days after injection of 5,7-DHT into the dorsomedial tegmentum. In the cerebral cortex, there was no difference in the extent of 5-HT uptake between the two groups. Thus, in the 5,7-DHT-injected group, [^{3}H]-5-HT uptake was 7403 $\pm$ 382 cpm/slice (means $\pm$ SEM for five rats); in the zimelidine-pretreated group, [^{3}H]-5-HT uptake was 7630 $\pm$ 457 cpm/slice (means $\pm$ SEM for five rats). Also, within the hypothalamus, there was no difference between the two groups. Thus, with the present treatment, zimelidine cannot protect 5-HT axonal bundles from the neurotoxic action of 5,7-DHT, at least not the bundles that project into the dorsal neocortex and the hypothalamus.

Fluorescence Histochemical Analysis of the Neurotoxic Action of p-*Chloroamphetamine*[47]

As shown in the FIGURE 6, treatment with *p*-chloroamphetamine induces the appearance of 5-HT fluorescence within many of the ascending 5-HT axons to the telencephalon and diencephalon. Strongly fluorescent 5-HT axons are observed to ascend within the medial midbrain tegmentum to reach the central

FIGURE 6. Fluorescence photograph from the medial forebrain bundle at the level of the retrochiasmatic region; routine Falck-Hillarp procedure. The animal was treated with *p*-chloroamphetamine (2×10 mg/kg, 24-hr interval, ip) and killed 48 hr after the first injection. With this procedure, the 5-HT pathways in the medial forebrain bundle are not visualized in normal untreated rats. However, as seen in the Figure, after treatment with *p*-chloroamphetamine there appear strongly yellowish fluorescent 5-HT axons (arrows) within the medial forebrain bundle. $\times 255$.

and dorsal parts of the medial forebrain bundle, in which they can be traced into its most rostral part; at this level, they pass dorsally within the tractus diagonalis region. The difference between treatment with *p*-chloroamphetamine and 5,7-DHT-induced lesions of the ascending 5-HT bundles is that many more 5-HT axons become visible after injection of 5,7-DHT. For example, we were able to observe immediately large numbers of 5-HT axons running in the ventrolateral part of the lateral hypothalamic area that did not fluoresce after treatment with *p*-chloroamphetamine. These 5-HT axons probably innervate the hypothalamus. It should be pointed that no fluorescence appeared within 5-HT axons descending into the spinal cord, nor did any 5-HT axons become fluorescent within the medulla and pons regions. The present observations confirm the conclusion of Köhler *et al.*[48] that there appears to occur a preferential degeneration of 5-HT terminal systems within the cortical regions, the neostriatum, thalamus, and superior collicle after treatment with *p*-chloroamphetamine. The present analysis reveals that this degeneration is due in part to a neurotoxic action of *p*-chloroamphetamine on the ascending 5-HT pathways to these regions. The *p*-chloroamphetamine approach to lesion 5-HT pathways to the forebrain has the obvious advantage over the use of 5,7-DHT-induced lesions that very few unspecific effects are produced by *p*-chloroamphetamine. Furthermore, its neurotoxic actions can be blocked completely by pretreatment with zimelidine, as shown by Köhler *et al.* In our opinion, *p*-chloroamphetamine should be the drug of choice for analysis of the functional role of the 5-HT systems to the forebrain, thalamus, and the superior collicle.

*Studies on [^{3}H]-5-HT and [^{3}H]-*d*-LSD Binding*

As demonstrated in FIGURE 7(A), specific binding of [^{3}H]-5-HT and of [^{3}H]-*d*-LSD occurs within the dorsal cerebral cortex of normal male Sprague-Dawley rats. From Scatchard plots, the dissociation constants and maximal number of binding sites can be determined. The number of maximal binding sites gives an indication of the total number of receptors present. In agreement with the observations of Bennett and Snyder,[32] a higher number of binding sites for *d*-LSD than for 5-HT was observed (FIGURE 7, B). Furthermore, *d*-LSD appears to have a higher affinity for its binding site than does 5-HT for its binding sites. Thus, the dissociation constant is clearly lower for *d*-LSD binding than for that of 5-HT. This conclusion is also in good agreement with the recent work of Bennett and Snyder.[32]

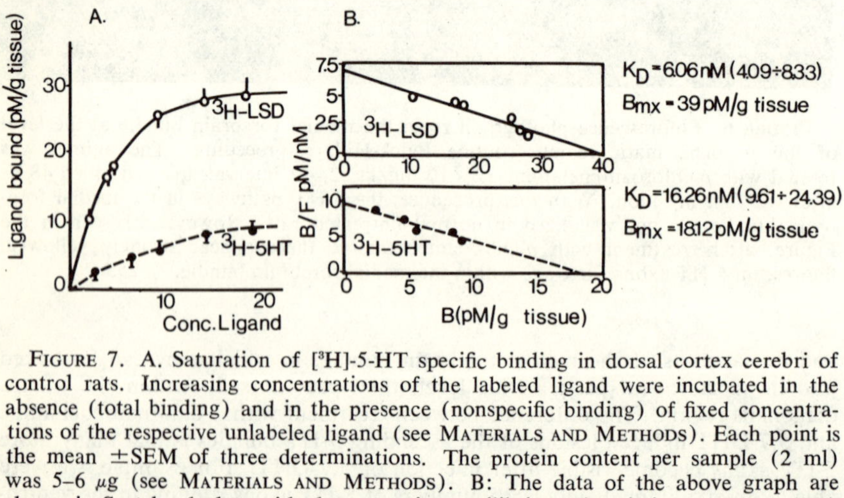

FIGURE 7. A. Saturation of [^{3}H]-5-HT specific binding in dorsal cortex cerebri of control rats. Increasing concentrations of the labeled ligand were incubated in the absence (total binding) and in the presence (nonspecific binding) of fixed concentrations of the respective unlabeled ligand (see MATERIALS AND METHODS). Each point is the mean ±SEM of three determinations. The protein content per sample (2 ml) was 5–6 μg (see MATERIALS AND METHODS). B: The data of the above graph are shown in Scatchard plots with the respective equilibrium dissociation constants (K_D) and the maximal number of binding sites (B_{max}). The K_D values are reported with their respective 95% confidence intervals (Sen's procedure; see MATERIALS AND METHODS).

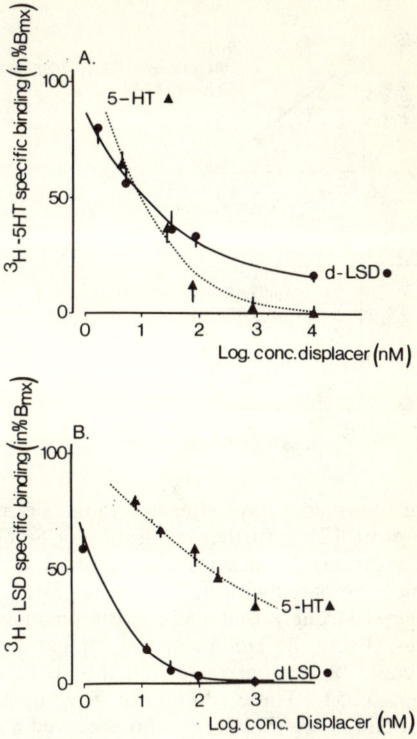

FIGURE 8. Displacement of [³H]-5-HT (A) and [³H]-*d*-LSD (B) specific binding by *d*-LSD and 5-HT. A: [³H]-5-HT (6.4 nmol) was incubated with increasing concentrations of unlabeled 5-HT and *d*-LSD. The study was performed on the dorsal cortex cerebri of normal male rats. The binding procedure was conducted as described by Bennett and Snyder.[32] The data shown represent specific [³H]-5-HT binding (total binding minus unspecific binding). Total binding was determined in the absence and unspecific binding in the presence of 10^{-5} M unlabeled 5-HT. Each point represents the mean ±SEM of three determinations. B: [³H]-*d*-LSD (2,3 nmol) was incubated with increasing amounts of unlabeled 5-HT and *d*-LSD. For the binding procedure, see Bennett and Snyder.[32] Specific *d*-LSD binding is shown. Unspecific binding is that found in the presence of 10^{-6} M unlabeled *d*-LSD. Each point is the mean±SEM of three observations.

FIGURES 8–10 show how 5-HT can displace [³H]-*d*-LSD binding and how *d*-LSD can displace [³H]-5-HT binding. These competition phenomena have been evaluated by Hill plots. From these plots, it can be observed that *d*-LSD has a 10-fold higher affinity for *d*-LSD-binding sites than for 5-HT-binding sites. Similarly, 5-HT appears to exhibit about a 16-fold higher affinity for 5-HT-binding sites than for *d*-LSD-binding sites.¶

FIGURES 11 and 12 show an analysis of the effects of 5-HT denervation of the forebrain on [³H]-5-HT and [³H]-*d*-LSD binding in the cerebral cortex. The 5,7-DHT-induced lesions of the ascending 5-HT pathways have already been described, and we have studied binding 2, 7, and 14 days after the operation. Two days after induction of the lesion, there was marked decrease in the number of binding sites for 5-HT. With regard to both specific 5-HT and *d*-LSD binding, there occurred a marked change in the pattern of specific binding. Thus, the saturation curves changed from a Langmuir type of curve into a sigmoid curve. The Hill plot for specific *d*-LSD binding clearly reveals interactions between binding sites. Assuming that the interacting sites are identical, a rough evaluation of the apparent interactional energy can be obtained from a consideration of the perpendicular distance between the parallel asymptotes of the tails (FIGURE 11, B). After 7 days, there is a tendency toward restoration of 5-HT binding sites. For *d*-LSD binding, there was no clear-cut change from what

¶ The Hill coefficients indicate that 5-HT may not affect all *d*-LSD-binding sites and that *d*-LSD may not affect all 5-HT-binding sites.

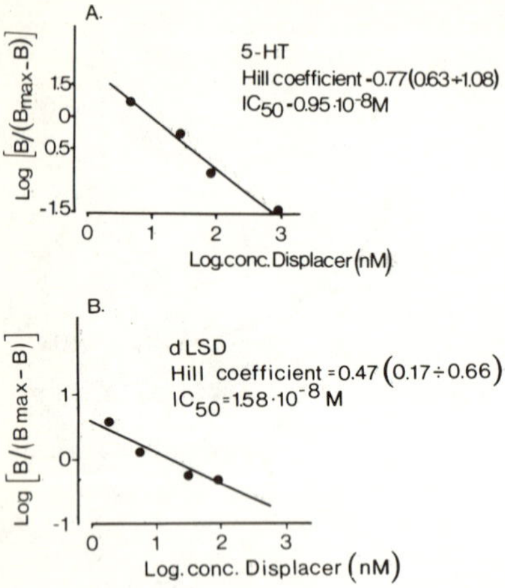

FIGURE 9. A: Hill plots of the data shown in FIGURE 8. Competition of 5-HT (A) and *d*-LSD (B) for specific [^{3}H]-5-HT binding. The maximum point was determined from the binding in the absence of added drug. The Hill coefficients are shown with their 95% confidence intervals (Sen's procedure; see MATERIALS AND METHODS). B: Competition of unlabeled *d*-LSD for specific [^{3}H]-5-HT binding. For further details, see the first part of this legend.

was observed 2 days after lesioning. Fourteen days after induction of the lesion (FIGURE 12), a further restoration of 5-HT-binding sites to normal has occurred; in addition, the dissociation constant for [^{3}H]-5-HT binding was reduced 2.5 times compared with that found for 5-HT binding in control rats. These findings suggest strongly that there exists an increased affinity for 5-HT at its binding sites. From the Hollander test for parallelism (see MATERIALS AND METHODS), it could be demonstrated that this increased affinity was statistically significant ($p < 0.05$). These results are in complete agreement with the recent findings of Bennett and Snyder,[32] who observed a similar change in affinity after electrocoagulations of the midbrain raphe region. In this experiment, we also noted a doubling of the number of binding sites for [^{3}H]-*d*-LSD, whereas the dissociation constant for [^{3}H]-*d*-LSD binding was not changed by 5-HT denervation. This increase in the number of binding sites for *d*-LSD was not observed by Bennett and Snyder.[32]

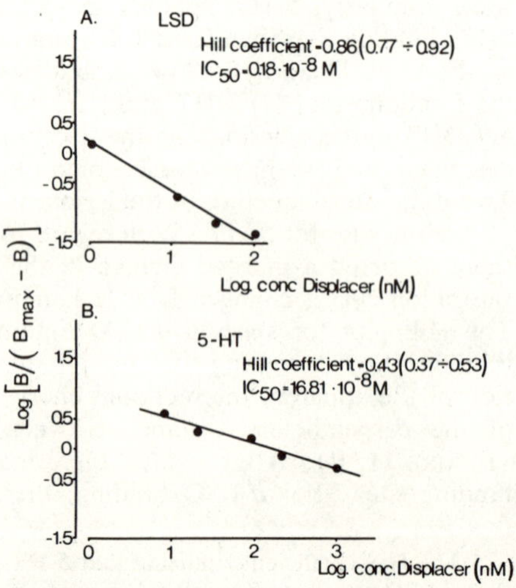

FIGURE 10. Hill plots for competition of *d*-LSD (A) and 5-HT (B) for specific [^{3}H]-*d*-LSD binding. For further details, see Figure 9 legend. The data used to calculate the Hill coefficients and the IC_{50} values were obtained in the experiment described in FIGURE 8.

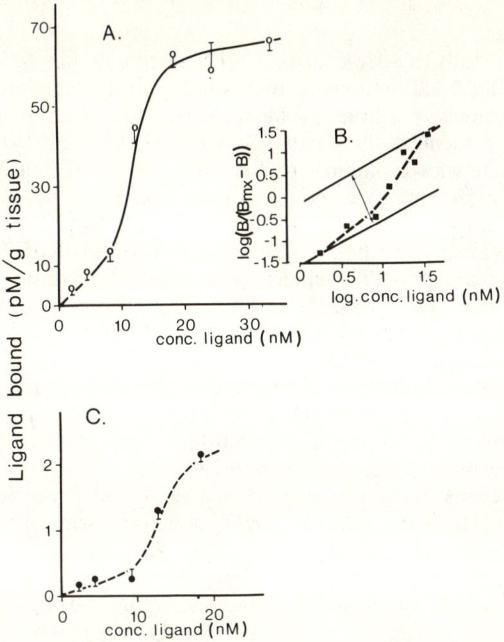

FIGURES 11 & 12. Saturation of [^{3}H]-*d*-LSD (A & B) and [^{3}H]-5-HT (C) specific binding in the dorsal cortex cerebri of rats with 5,7-DHT-induced lesions of the ascending 5-HT pathways in the midbrain 2 (FIGURE 11) and 14 (FIGURE 12) days after the operation. For further details, see the legend to FIGURE 7 and MATERIALS AND METHODS. In B, the Hill plot for specific [^{3}H]-*d*-LSD binding is shown. To evaluate the apparent interactional energy between sites, the asymptotes of the upper and lower tails of the plot have been drawn by eye. The perpendicular distance between the asymptotes can be used to make an evaluation of the interactional energy between sites.[46]

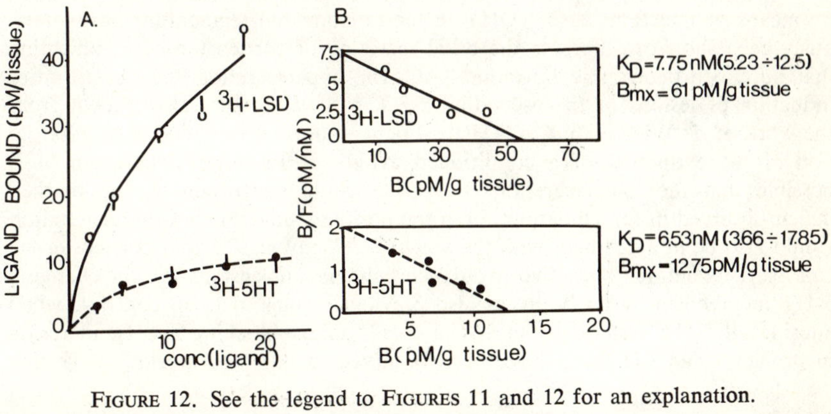

FIGURE 12. See the legend to FIGURES 11 and 12 for an explanation.

Discussion

In a previous study by Hole *et al.*,[19] it was found that 5,7-DHT-induced lesions of ascending 5-HT pathways in the midbrain did not interfere with the acquisition of a one-way active avoidance response. This finding has been confirmed in the present study, further supporting the view that the retarded acquisition of a one-way avoidance response observed after electrolytic lesions of the midbrain raphe area[8] is not due to a lesion of the ascending 5-HT systems. In the present analysis, we demonstrated that the retention of a one-way active avoidance response was also unimpaired by 5,7-DHT-induced lesions of the ascending 5-HT systems. The major behavioral finding of the present paper is that the forced extinction of the one-way active avoidance response was markedly retarded in rats with 5,7-DHT-induced lesions of the ascending 5-HT systems. This deficit in forced extinction was observed 5 days but not 10 days after induction of the lesion, illustrating the compensatory mechanisms that can occur in response to injury. The nature of these compensatory mechanisms is discussed in relation to our results obtained in the [^{3}H]-5-HT- and [^{3}H]-*d*-LSD-binding experiments.

The present results explain why no deficit in forced extinction in previous experiments with intrabrain stem 5,7-DHT injections was observed, because effects on behavior were studied 16–23 days after lesioning. The observed deficit in forced extinction of the early time interval studied after lesioning is probably related to an impairment of the ascending 5-HT terminal system, because the action on forced extinction by 5,7-DHT was markedly counteracted by pretreatment with the 5-HT uptake-blocking agent zimelidine, and because 6-OHDA-induced lesions of the dorsal NA bundle to the forebrain do not cause any change in forced extinction of this one-way active avoidance response.[34] It should be mentioned that a significant protection by zimelidine of the 5,7-DHT-induced depletion of 5-HT stores was only observed in the rest of the brain due to a high variability within the cerebral cortex and then only after repeated treatment with 5,7-DHT. It is therefore possible that the deficit in forced extinction observed could be related mainly to a degeneration of subcortical 5-HT systems. The point should be made that even a small protection of the 5-HT pathways, as revealed in higher 5-HT levels, seems to lead to a marked recovery of behavioral functions.

Lesions of the ascending 5-HT pathways as performed in the present study by means of injections of 5,7-DHT in the dorsomedial tegmentum have previously been shown to increase the basal secretion of corticosterone,[35] indicating that adrenocorticotrophic hormone (ACTH) hypersecretion may occur after induction of lesions of the ascending 5-HT pathways. It is also known from the work of de Wied[36] that ACTH treatment in normal rats will cause retardation of the extinction of a conditioned avoidance response. Therefore, it is possible that the hypersecretion of ACTH could contribute to the marked deficit observed in forced extinction in the present study. Such an interpretation would also be in agreement with the view of Telegdy *et al.*[37] that corticosterone can exert its actions on active avoidance behavior at least partly via changing 5-HT metabolism in the brain, as also previously suggested. In contrast, when altering all 5-HT systems by means of 5-HT uptake-blocking agents, increases in prolactin and LH levels have been observed in our laboratories, suggesting that some 5-HT systems may also have a stimulatory role in the control of LH and prolactin secretion. Growth hormone secretion, on the other hand, was reduced by 5-HT uptake-blocking agents, such as zimelidine.[49]

Of course, it cannot be excluded that the degeneration of the cortical NA systems could have contributed to the deficit in forced extinction observed. The deficit observed demonstrates that suppression of behavior is retarded despite punishment when the ascending 5-HT systems have been removed. These results are therefore in agreement with the conclusion of Stein *et al.*[38] that the 5-HT systems are implicated in negative reinforcement and punishment. Thus, Stein *et al.* have shown that 5,6-DHT-induced degeneration of 5-HT terminals will release suppressed behavior induced, in turn, by punishment. Similar results have recently been obtained by Tye *et al.*[39] after 5,7-DHT treatment. In agreement with this hypothesis, it has also been found that anxiety-reducing drugs, for example, the benzodiazepines, reduce 5-HT turnover.[38, 40] Stein *et al.* have emphasized that the reduction of 5-HT activity within the forebrain plays an important role in the anxiety-reducing actions of the benzodiazepines.[38] From the work of Stein *et al.*, it also seems that the benzodiazepines may act partly over an inhibitory γ-aminobutyric acidergic system, which, on activation by the benzodiazepines, can reduce activity within the ascending 5-HT systems.[38] In agreement with the discussion above, it has also recently been shown that *d*-LSD can attenuate the effect of punishment in the rat[41] in as low doses as 1 μg/kg, which will preferentially stimulate 5-HT autoreceptors and turn off the firing rate within the ascending 5-HT system.[42] In a previous study together with Kiianmaa,[22] we observed, in agreement with the pioneering work of Jouvet,[21] that 2–4 days after induction of 5,7-DHT lesions of the ascending pathways, there exists a reduction of slow-wave sleep II and an increase in wakefulness. This change in the electrocorticogram could, of course, be of importance in causing the observed behavioral deficit. If it is true that the increased wakefulness could be of importance for producing the behavioral deficit in forced extinction, even more clear-cut results would have been obtained had it been possible to avoid lesioning of the ascending NA systems to the cortex cerebri. This lesioning is, of course, easily performed by pretreatment of the animals with a relatively selective NA uptake-blocking agent, such as protriptyline or desipramine.

Early after lesioning, we also noted that when handled, the animals were hyperresponsive. They were also hyperresponsive to the electrical shock in the learning experiments and sometimes even jumped out of the cages. Thus, sensory stimuli appeared to have a greater influence on the 5,7-DHT-lesioned animals than on the solvent-treated animals. In support of this conclusion, it has been found that *d*-LSD in doses that act mainly on the autoreceptors for 5-HT that turn off activity in the 5-HT systems will sensitize the startle response in rats.[43] Thus, when turning off the 5-HT systems, an increased sensory stimulation may be obtained,[44] which could, for example, be the mechanism for the *d*-LSD-induced hallucinations, as proposed by Aghajanian and Haigler.[45] In such a state of increased sensory input and increased episodes of attentive wakefulness, it seems reasonable that a deficit in forced extinction can develop, because punishment stimuli now also have to compete with large numbers of other types of incoming stimuli, the transmission of which into the brain is no longer inhibited by 5-HT systems. Also, if the 5-HT systems indeed are involved in mediating negative reinforcement, the impact of punishment stimuli would be further reduced.

Specific [^3H]-5-HT and [^3H]-d-LSD Binding in Cerebral Cortex After 5-HT Denervation of the Forebrain

The results indicate that very complex events occur in the receptors after removal of their transmitter input. The findings also clearly indicate the importance of studying the time course of changes in receptor binding that occur after disruption of the transmitter pathway in question. First, we would like to emphasize that the lesions should be made as specific as possible by use of, for example, the serotonin neurotoxins and that lesions should be made by injection of the neurotoxins into the 5-HT pathways. In this way, the lesions will mimic the anterograde degeneration that occurs during the aging process. Furthermore, if 5,7-DHT is, instead, given intraventricularly or intracisternally, the neurotoxins will reach the receptor areas also and could, then, cause an unspecific destruction of the receptor sites.

It is of considerable interest to compare the specific 5-HT and *d*-LSD binding 2 and 14 days after lesioning. Thus, immediately after injection of the neurotoxin into the 5-HT axonal bundles, there occurs a disruption of impulse activity to the 5-HT terminal networks in the cerebral cortex. Two days after induction of the lesion, the 5-HT terminals still remain, but they have not released 5-HT for 2 days, and, at most, a slow leakage of 5-HT can occur. Fourteen days after induction of the lesion, on the other hand, the 5-HT boutons in the cerebral cortex have disappeared to a considerable degree, as evidenced by a substantial reduction (60%) of 5-HT uptake sites, as shown in synaptosomal fractions or in cerebral cortical slices, and a marked reduction also of cortical 5-HT levels. It should be realized, however, that the lesion is not complete and that some 5-HT terminal networks still remain within the cortical regions. The possibility should be considered that the deficit in 5-HT receptor activity may be more marked 2–5 days after lesioning than 14 days after this procedure. Thus, at the later time interval, there may occur increased activity within the remaining 5-HT terminals, and some 5-HT projections that were only partly lesioned by the neurotoxin may have recovered. With this background information, it is of particular interest that there occurs a marked reduction of 5-HT-binding sites 2 days after lesioning and a conformational change in the *d*-LSD- and 5-HT-binding sites in view of an altered shape of the saturation curve. Fourteen days after the operation, the picture is completely different. At this point, the number of 5-HT-binding sites is approaching normal, but they now probably show a conformational change in view of their marked increase in affinity. The *d*-LSD-binding site, on the other hand, no longer shows any sign of a conformational change, but the number of binding sites is still increased by 100%. These results illustrate well the dynamics of the receptor structures. Thus, when a transmitter deficit occurs, the receptors are capable of showing increases in the number of binding sites, in addition to conformational changes.. These two types of changes represent two different ways of increasing synaptic transmission at the receptor level to normalize a deficit in receptor activity. These mechanisms could at least partly be responsible for the normalization of the deficit in forced extinction responses that the

5,7-DHT-lesioned animals exhibited 5 but not 10 days after the operation. Thus, hyperactivity within the remaining 5-HT terminal systems in the cerebral cortex would result in increased 5-HT release into the extraneuronal space, allowing increased diffusion into neighboring receptor fields. In view of the data obtained on changes in 5-HT and *d*-LSD binding, these receptor fields may contain 5-HT receptors with a higher affinity for 5-HT. The increase in the number of *d*-LSD-binding sites could certainly potentiate the actions of *d*-LSD and related ergot drugs on postsynaptic 5-HT receptor sites. Thus, this study demonstrates that the *d*-LSD-binding sites are changed in relation to a specific removal of the 5-HT projections to the telencephalon and diencephalon, and therefore the *d*-LSD-binding site is probably in some way linked to the 5-HT receptor.

Perhaps we should also not forget to point out that the restoration of slow-wave sleep II 10–14 days after 5,7-DHT-induced lesions of the ascending 5-HT pathways could also be related to the phenomenon described above. Thus, 2–4 days after the 5,7-DHT-induced lesion of the ascending 5-HT pathways, there is a significant reduction of slow-wave sleep II and an increase in wakefulness. These effects have disappeared 10–14 days after lesioning. These results and those obtained on forced extinction clearly underline the fact that the behavioral and the electrocorticographic deficits occur mainly soon after lesioning (2–4 days). Thus, the disappearance in 5-HT-binding sites found 2 days after the 5,7-DHT-induced lesion may be to at least some extent responsible for the deficits observed in these parameters. Alternatively, the disappearance of 5-HT-binding sites could represent the disappearance of an inactivation site, which then gradually reappears as compensatory increases of 5-HT turnover occur. In agreement with this view, a conformational change toward a positive cooperativity occurs in the 5-HT-binding site 2 days after induction of the lesion, when the number of such sites is markedly reduced. Furthermore, 5-HT receptor supersensitivity to 5-HT agonists, such as 5-methoxydimethyltryptamine, has been demonstrated 4 days after induction of such a lesion.[9, 10] Of course, the major mechanism for the functional deficits is still the lack of release of 5-HT.

Conclusions

The ascending 5-HT pathways to the telencephalon and diencephalon were lesioned by way of stereotaxic injections of 5,7-DHT into the dorsomedial tegmentum of the midbrain, resulting in damage to the ascending 5-HT axons. This lesion was found to produce a marked deficit in the forced extinction of a one-way active avoidance response 5 days after the operation. Ten days after the operation, this deficit was no longer observed.

Measurements of DA, NA, and 5-HT levels in the cerebral cortex, including the hippocampal formation, and in the rest of the brain in front of the 5,7-DHT-induced lesion indicated that this behavioral deficit observed 5 days after the operation was related to a degeneration of 5-HT nerve terminals in the telencephalon and diencephalon, perhaps mainly those present within the subcortical region. By pretreatment with a 5-HT uptake-blocking agent, zimelidine, it was possible to partially protect the ascending 5-HT axons from the neurotoxic action of 5,7-DHT. This protection became significant in the 5-HT axons that project into subcortical regions. These results further link the behavioral deficits observed in forced extinction to the ascending 5-HT pathways, because zimelidine in the behavioral experiments was found to block the behavioral deficit produced by 5,7-DHT. Thus, it seems as if a partial protection of ascending 5-HT pathways can be sufficient to completely counteract the behavioral deficit produced by 5,7-DHT. The reason why it was impossible to fully protect the 5-HT axons from the neurotoxic actions of 5,7-DHT by pretreatment 1 hr earlier with zimelidine in a high dose is probably due to the high concentrations of 5,7-DHT present around the 5-HT axons, which will make it very difficult for the blocking agent to compete with 5,7-DHT for the uptake sites in the nerve cell membrane. Taken together, these results demonstrate that ascending 5-HT pathways are involved in the control of forced extinction of a one-way active avoidance response. Furthermore, these results support the view that 5-HT pathways are of importance in negative reinforcement mechanisms in the brain.

To further understand the mechanism responsible for the behavioral deficit observed and its recovery after 2 weeks, specific [^{3}H]-5-HT- and [^{3}H]-*d*-LSD-binding sites were analyzed 2, 7, and 14 days after 5,7-DHT-induced lesioning. The cerebral cortex was studied. Two days after lesioning, there was a marked reduction in the number of 5-HT-binding sites, whereas 2 weeks after the operation, they had partly recovered, and we could now observe an increase in the affinity of the 5-HT-binding sites, as evaluated from the dissociation constants. These results indicate that the lack of release of 5-HT may be due not only to the behavioral deficit observed 5 days after the operation but also to a marked initial reduction in the number of 5-HT-binding sites. Alternatively and more likely is the interpretation that the loss of 5-HT receptor sites represents the loss of an inactivation site to compensate for loss of 5-HT release. In agreement with this hypothesis, a conformational change toward positive cooperativity occurs within the few 5-HT-binding sites that remain 2 days after lesioning. Furthermore, the increased affinity for 5-HT at 5-HT-binding sites observed 14 days after lesioning may partly explain the functional recovery observed and also partly explain the supersensitivity phenomena observed by Hole *et al.*[19] after administration of 5-hydroxytryptophan to rats with 5,7-DHT-induced lesions of the 5-HT ascending axons, following pretreatment with a peripheral decarboxylase inhibitor. Marked changes in the *d*-LSD-binding sites were also observed after 5-HT denervation of the cortex cerebri. Fourteen days after lesioning, there existed a doubling of *d*-LSD-binding sites; 2 days after the operation, results were obtained that suggested a plasticity change within the *d*-LSD- and 5-HT-binding sites. Thus, the saturation curve changed from the Langmuir type into a sigmoid type, suggesting a change toward positive cooperative interactions within the binding sites for *d*-LSD and 5-HT, respectively.

Acknowledgments

We thank Ulla-Britt Finnman, Agneta Eliasson, and Beth Hagman for excellent technical assistance.

References

1. Baumgarten, H. G., A. Björklund, L. Lachenmayer, A. Nobin & U. Stenevi. 1971. Long-lasting selective depletion of brain serotonin by 5,6-dihydroxytryptamine. Acta Physiol. Scand. Suppl. **373**.

2. BAUMGARTEN, H. G., A. BJÖRKLUND, A. F. HOLSTEIN & A. NOBIN. 1972. Chemical degeneration of indolamine axons in rat brain by 5,6-dihydroxytryptamine. Z. Zellforsch. **129:** 256–271.
3. COSTA, E., J. DALY, H. LEFEVRE, J. MEEK, A. REVUELTA, F. SPARRO & S. STRADA. 1972. Serotonin and catecholamine concentrations in brain of rats injected intracerebrally with 5,6-dihydroxytryptamine. Brain Res. **44:** 304–308.
4. DALY, J., K. FUXE & G. JONSSON. 1972. 5,6-Dihydroxytryptamine: a new tool in mapping out of central 5-HT neurons. Abstr. 4th Int. Congr. Histochem. Cytochem., Kyoto, Japan.
5. DALY, J., K. FUXE & G. JONSSON. 1973. Effects of intracerebral injections of 5,6-dihydroxytryptamine on central monoamine neurons: evidence for selective degeneration of central 5-hydroxytryptamine neurons. Brain Res. **49:** 476–482.
6. BAUMGARTEN, H. G., A BJÖRKLUND, L. LACHENMAYER & A. NOBIN. 1973. Evaluation of the effects of 5,7-dihydroxytryptamine on serotonin and catecholamine neurons in the rat CNS. Acta Physiol. Scand. Suppl. **391**.
7. DALY, J., K. FUXE & G. JONSSON. 1974. 5,7-Dihydroxytryptamine as a tool for the morphological and functional analysis of central 5-hydroxytryptamine neurons. Res. Commun. Chem. Pathol. Pharmacol. **7:** 175–187.
8. LORENS, S. A., H. C. GULDBERG, K. HOLE, C. KÖHLER & B. SREBRO. 1976. Activity, avoidance learning and regional 5-hydroxytryptamine following intra-brain stem 5,7-dihydroxytryptamine and electrolytic midbrain raphe lesions in the rat. Brain Res. **108:** 97–113.
9. FUXE, K., G. JONSSON, L.-G. NYGREN & L. OLSSON. 1974. Studies on central 5-hydroxytryptamine neurons using dihydroxytryptamines: evidence for regeneration of bulbospinal 5-hydroxytryptamine axons and terminals. *In* Dynamics of Degeneration and Growth in Neurons. K. Fuxe, L. Olsson & Y. Zotterman, Eds. : 169–179. Pergamon Press. Oxford, England.
10. NYGREN, L.-G., K. FUXE, G. JONSSON & L. OLSSON. 1974. Functional regeneration of 5-hydroxytryptamine nerve terminals in the rat spinal cord following 5,6-dihydroxytryptamine induced degeneration. Brain Res. **78:** 377–394.
11. NYGREN, L.-G. & L. OLSON. 1977. Intracisternal neurotoxins and monoamine neurons innervating the spinal cord: acute and chronic effects on cell and axon counts and nerve terminal densities. Histochemistry **129:** 227–235.
12. DAHLSTRÖM, A. & K. FUXE. 1974. Evidence for the existence of monoamine-containing neurons in the central nervous system. Acta Physiol. Scand. **62** (Suppl. 232): 1–55.
13. LORENS, S. A. & H. C. GULDBERG. 1974. Regional 5-hydroxytryptamine following selective midbrain raphe lesions in the rat. Brain Res. **78:** 45–56.
14. BOBILLIER, P., F. PETITJEAN, D. SALVERT, M. LIGIER & S. SEGUIN. 1975. Differential projections of the nucleus raphe dorsalis and nucleus raphe centralis as revealed by autoradiography. Brain Res. **85:** 205–210.
15. GEYER, M. A., A. PUERTO, W. J. DAWSEY, S. KNAPP, W. P. BULLARD & A. J. MANDELL. 1976. Histologic and enzymatic studies of the mesolimbic and mesostriatal serotonergic pathways. Brain Res. **106:** 241–256.
16. EVERITT, B. J., K. FUXE & G. JONSSON. 1975. 5,7-Dihydroxytryptamine (5,7-DHT) lesions of ascending 5-hydroxytryptamine pathways and the sexual and aggressive behavior of female rats. J. Pharmacol. **6:** 25–32.
17. LARSSON, K., K. FUXE, B. J. EVERITT, M. HOLMSGREN & P. SÖDERSTEN. 1977. Sexual behavior in male rats after intracerebral injection of 5,7-dihydroxytryptamine. Brain Res. In press.
18. FUXE, K., L.-O. FARNEBO, B. HAMBERGER & S.-O. ÖGREN. 1975. On the *in vivo* and *in vitro* actions of fenfluramine and its derivatives on central monoamine neurons, especially 5-hydroxytryptamine neurons, and their relation to the anorectic activity of fenfluramine. Postgrad. Med. J. **51:** 35–45.
19. HOLE, K., K. FUXE & G. JONSSON. 1976. Behavioral effects of 5,7-dihydroxytrypt-

amine lesions of ascending 5-hydroxytryptamine pathways. Brain Res. **107:** 385–399.

20. SAMANIN, R., W. GUMULKA & L. VALZELLI. 1970. Reduced effect of morphine in midbrain raphe lesioned rats. Eur. J. Pharmacol. **10:** 339–343.
21. JOUVET, M. 1969. Biogenic amines and states of sleep. Science **163:** 32–41.
22. KIIANMAA, K. & K. FUXE. 1977. The effects of 5,7-dihydroxytryptamine-induced lesions of the ascending 5-hydroxytryptamine pathways on the sleep-wakefulness cycle. Brain Res. **131:** 287–301.
23. ROSS, S. B., S. O. ÖGREN & A. L. RENYI. 1976. (Z)-Dimethylamino-1-(4-bromophenyl)-1-(3-pyridyl) propene (H 102/09), a new selective inhibitor of the neuronal 5-hydroxytryptamine uptake. Acta Pharmacol. Toxicol. **39:** 152–166.
24. ÖGREN, S.-O. & K. FUXE. 1974. Learning, brain noradrenaline and the pituitary-adrenal axis. Med. Biol. **52:** 399–405.
25. CARLSSON, A., W. KEHR & M. LINDQVIST. 1976. The role of intraneuronal amine levels in the feedback control of dopamine, noradrenaline and 5-hydroxytryptamine synthesis in rat brain. J. Neural Transmission **39:** 1–19.
26. SACHS, C. & G. JONSSON. 1972. Degeneration of central and peripheral noradrenaline neurons produced by 6-hydroxy-DOPA. J. Neurochem. **19:** 1561–1575.
27. LIDBRINK, P., J. JONSSON & K. FUXE. 1971. The effect of imipramine-like drugs and antihistamine drugs on 5-hydroxytryptamine neurons. Neuropharmacology **10:** 521–536.
28. HARVEY, J. A., S. E. MCMASTER & L. M. YUNGER. 1975. *p*-Chloroamphetamine selective neurotoxic effects in brain. Science **187:** 841–843.
29. FALCK, B., N.-Å. HILLARP, G. THIEME & A. TORP. 1962. Fluorescence of catecholamines and related compounds condensed with formaldehyde. J. Histochem. Cytochem. **10:** 348–354.
30. FUXE, K. & G. JONSSON. 1973. The histochemical fluorescence method for the demonstration of catecholamines. Theory, practice and application. J. Histochem. Cytochem. **21:** 293–311.
31. ÖGREN, S. O., CH. KOHLER, S. B. ROSS & B. SREBRO. 1976. 5-Hydroxytryptamine depletion and avoidance acquisition in the rat. Antagonism of the long-term effects of *p*-chloroamphetamine with a selective inhibitor of 5-hydroxytryptamine uptake. Neurosci. Lett. **3:** 341–347.
32. BENNETT, J. P. & S. H. SNYDER. 1976. Serotonin and lysergic acid diethylamide binding in rat brain membranes: relationship to postsynaptic serotonin receptors. Mol. Pharmacol. **12:** 373–389.
33. HOLLANDER, M. & D. A. WOLFE. 1973. Nonparametric Statistical Methods. John Wiley & Sons, Inc. New York, N.Y.
34. ÖGREN, S.-O. & K. FUXE. 1977. On the role of brain noradrenaline and the pituitary adrenal axis in learning. I. Studies with corticosterone. Neurosci. Lett. **5:** 291–296.
35. FUXE, K., T. HÖKFELT, G. JONSSON, S. LEVINE, P. LIDBRINK & A. LÖFSTRÖM. 1973. Brain and pituitary-adrenal interactions studies on central monoamine neurons. : 239–269. S. Karger. Basel, Switzerland.
36. DE WIED, D. 1969. Opposite effect of ACTH and glucocorticosterone on extinction of conditioned avoidance behaviour. Excerpta Med. Int. Congr. Ser. **132**.
37. KOVACS, G. L., G. TELEGDY & K. LISSAK. 1976. 5-Hydroxytryptamine and the mediation of the pituitary-adrenocortical hormones in the extinction of active avoidance behaviour. Psychoneuroendocrinology **1:** 219–230.
38. STEIN, L., C. D. WISE & J. D. BELLUZZI. 1975. Effects of benzodiazepines on central serotonergic mechanisms. *In* Mechanism of Action of Benzodiazepines. E. Costa & P. Greengard, Eds. Vol. 14: 29–44. The Raven Press, New York, N.Y.
39. TYE, N. C., B. J. EVERITT & S. D. IVERSEN. 1977. 5-Hydroxytryptamine and punishment. Nature (London) **268:** 741–743.

40. Lidbrink, P., H. Corrodi & K. Fuxe. 1974. Benzodiazepines and barbituates: turnover changes in central 5-hydroxytryptamine pathways. Eur. J. Pharmacol. **26:** 35–40.
41. Schoenfeld, R. I. 1976. Lysergic acid diethylamide- and mescaline-induced attenuation of the effect of punishment in the rat. Science **192:** 801–803.
42. Aghajanian, G. K., H. J. Haigler & F. E. Bloom. 1972. Lysergic acid diethylamide and serotonin: direct actions on serotonin-containing neurons. Life Sci. **11:** 615–622.
43. Davis, M. & M. H. Sheard. 1974. Effects of lysergic acid diethylamide (LSD) on habituation and sensitization of the startle response in the rat. Pharmacol. Biochem. Behav. **2:** 675–683.
44. Bradley, P. B. & B. J. Key. 1958. The effect of drugs on arousal responses produced by electrical stimulation of the reticular formation of the brain. Electroencephalogr. Clin. Neurophysiol. **10:** 97–110.
45. Haigler, H. J. & G. K. Aghajanian. 1974. Lysergic acid diethylamide and serotonin: a comparison of effects on serotonergic neurons and neurons receiving a serotonergic input. J. Pharmacol. Exp. Ther. **188:** 688–699.
46. Triggle, D. J. 1971. Neurotransmitter-Receptor Interactions. Academic Press. London and New York.
47. Fuxe, K. & S.-O. Ögren. To be published.
48. Köhler, C., S. B. Ross, B. Srebro & S.-O. Ögren. This monograph.
49. Fuxe, K., S.-O. Ögren, B. J. Everitt, L. F. Agnati, P. Eneroth, J.-Å. Gustafsson, G. Jonsson, P. Skett & A. C. Hohn. 1978. The effect of antidepressant drugs of the imipramine type on various monoamine systems and their relation to changes in behaviour and neuroendocrine function. Symp. Med. Hoechst no. 13, Depression. In press.

REGENERATION OF SEROTONIN NEURONS IN THE RAT BRAIN AFTER 5,6-DIHYDROXYTRYPTAMINE-INDUCED AXOTOMY *

Leif Wiklund, Anders Björklund, and Anders Nobin

Department of Histology
University of Lund
S-223 62 Lund, Sweden

Regeneration is now well established in monoaminergic and other non-myelinated fiber systems in the adult mammalian central nervous system (see References 1 and 2 for reviews). Axonal regeneration has also been shown to occur after axotomy induced by the neurotoxic serotonin congeners. This phenomenon is in contrast to the very feeble or even abortive regenerative responses that have been registered in the catecholamine-containing neurons after 6-hydroxydopamine (6-OHDA) treatment.[3, 4] There is a risk, therefore, that experience from work with 6-OHDA might lead to neglect of the time-dependent changes related to regeneration in animals treated with serotonin neurotoxins. As will be discussed in the present review, the central serotonin neurons regenerate extensively after axotomy induced by 5,6-dihydroxytryptamine (5,6-DHT), and this is true also after 5,7-dihydroxytryptamine treatment. This information should be of profound importance for the interpretation of functional changes caused by the neurotoxic indoleamines.

Regeneration as Revealed by Biochemical Parameters

The axonal degeneration induced by an intraventricular injection of 5,6-DHT (75 μg) is reflected in a reduction in both endogenous 5-hydroxytryptamine (5-HT) and in [^{3}H]-5-HT uptake, as shown in Figures 1 and 2.[5, 6] The time course of 5-HT depletion is different in different regions. In areas close to the ventricles (such as the septum and hypothalamus), depletion is already maximal within 3–24 hr, and partial recovery occurs as early as after 4 days. In cortical areas, which are more remote from the ventricles, the reduction is more protracted, and maximal depletion is not reached until the 10th day after injection. This difference is probably due to the limited penetration of 5,6-DHT from the cerebrospinal fluid, which causes only terminal systems that are close to the ventricles to be reached directly by the drug.[7] In other regions (most apparently, the cerebral cortex and spinal cord), the slow 5-HT depletion is due to anterograde degeneration of axons lesioned at sites closer to the cell bodies.[8]

During the subsequent months, there is a recovery of both the 5-HT levels and the capacity of the tissue to take up [^{3}H]-5-HT *in vitro* (Figures 1 & 2).[6, 9] The recovery was least in areas most distant from the cell body regions, namely,

* Supported by Grants 04X–3874 and 04X–56 from the Swedish Medical Research Council and by grants from the Magnus Bergvall and Åke Wiberg Foundations and the Trygg-Hansa Research Fund.

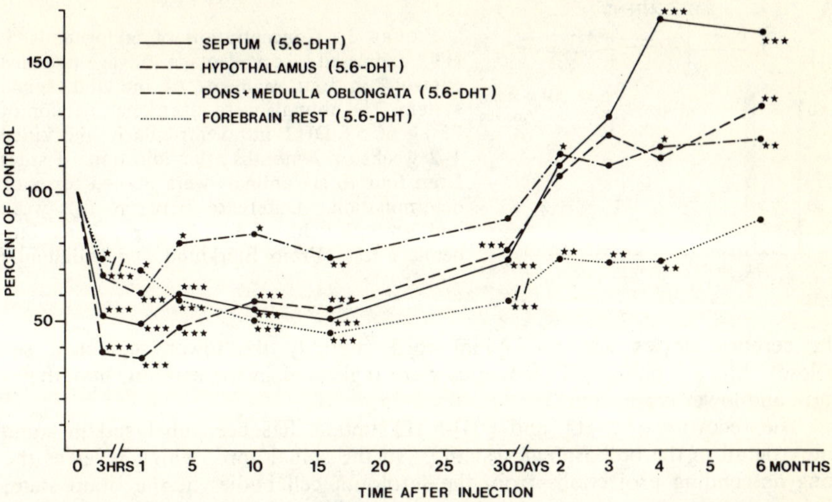

FIGURE 1. Time course of depletion and recovery of 5-HT in four brain regions after a single intraventricular injection of 75 μg (free base), of 5,6-DHT. The brain was dissected according to the method described in Reference 5. Differences from control values: ★★★, $p < 0.001$; ★★, $0.01 > p > 0.001$; ★, $0.05 > p > 0.01$. Student's *t* test. Data from Baumgarten et al.[5,9]

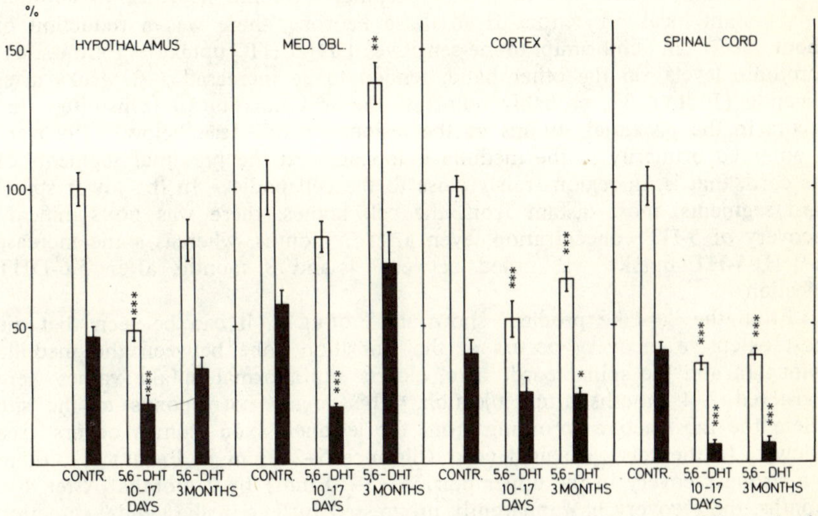

FIGURE 2. Uptake of total tritium (□) and unchanged [^{3}H]-5-HT (■) by brain slices and slices from thoracic segments of the spinal cord *in vitro* (at 0.5×10^{-7} M for 10 min). For each region, the values are expressed as the percentages of the total tritium uptake of the control slices and are given as means ±SEM of from five to 10 determinations. The values represent active uptake, calculated as the uptake measured at 37°C minus the uptake measured at 0°C. Differences from control values: ★★★, $p < 0.001$; ★★, $0.01 > p > 0.001$; ★, $0.05 > p > 0.01$. Student's *t* test. Data from Björklund *et al.*[6]

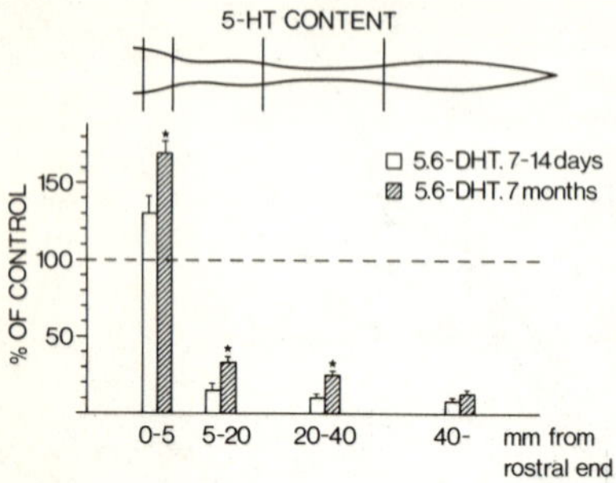

FIGURE 3. Concentration of endogenous 5-HT, expressed as percentage of age-matched controls, in four segments of the bulbospinal system. The animals were given one injection of 75 μg of 5,6-DHT intraventricularly and killed 1–2 weeks or 7 months after injection. Tissues from four to six animals were pooled for each determination. Difference between 1–2 week and 7-month values: ★, $0.05 > p > 0.01$. Student's *t* test. From Björklund and Wilklund.[10]

the cerebral cortex and the spinal cord (notably the lower segments; see below). More pronounced recoveries were registered in the septum, hypothalamus, and lower brain stem.[9]

The recovery of 5-HT and [^{3}H]-5-HT uptake has been analyzed in some more detail in the bulbospinal system.[10] In the spinal cord, which receives the long descending projections from the serotonin cell bodies in the brain stem, there was by 1–2 weeks postinjection a 90–95% reduction in 5-HT levels in all segments (FIGURE 3). There was also a uniform profound reduction in [^{3}H]-5-HT uptake throughout the spinal cord (FIGURE 4). The reduction amounted to 75–85% in terms of active uptake (i.e., uptake measured at 37°C minus uptake measured at 0°C) and to more than 90% in terms of chlorimipramine-sensitive uptake (see FIGURE 4). In the medulla oblongata, which contains cell bodies of the bulbospinal serotonin neurons, in addition to abundant local projections from those neurons, there was a reduction of about 50% in chlorimipramine-sensitive [^{3}H]-5-HT uptake (FIGURE 4). Serotonin levels, on the other hand, tended to be increased 1–2 weeks after injection (FIGURE 3), probably reflecting the accumulation of transmitter that occurs in the proximal stumps of the lesioned axons (see below). Recovery is observed primarily in the medulla oblongata and the proximal segments of the cord, that is, in regions fairly close to the cell bodies. In the lower spinal cord segments, most distant from the cell bodies, there was no significant recovery of 5-HT concentration, even after 7 months, whereas some increase in [^{3}H]-5-HT uptake was noted between 4 and 8 months after 5,6-DHT injection.

From the "uptake profiles" shown in FIGURE 4, it can be seen that the most extensive recovery occurs at the transition zone between the medulla oblongata and the spinal cord; here, clearly supranormal uptake values were registered 2–4 months after injection. This region corresponds to the site where the most active sprouting from the lesioned axon stumps occurs (see below). In the cervical spinal cord (distance 5–20 mm in FIGURE 4), there is a strong recovery in the upper portion (5–10 mm) by 2 months. After 4–8 months, the recovery has apparently progressed further in the caudal direction and now involves the entire cervical cord. In the thoracic and lumbar cord (distance 25 and 50 mm, respectively), there is no recovery up to 4 months after injection, but by 8 months, recovery is noted also at these distal levels.

These data thus provide support for the conclusion that the regrowth of new axonal sprouts starts in the medulla oblongata-upper spinal cord region and progresses slowly in the distal direction, resulting in a partial restitution of [^{3}H]-5-HT uptake in progressively more caudal segments of the cord. Generally, this observation is in accordance with the findings of Nygren *et al.*[11] These authors reported a more rapid recovery in the spinal cord, which was probably due to the lower doses of 5,6-DHT that they used.

FIGURE 5 illustrates the synthesis and metabolism of 5-HT studied *in vitro* in different segments of the bulbospinal system. In these experiments,[12] transverse slices were incubated in [^{3}H]tryptophan ([^{3}H]TP) for 30 min; the [^{3}H]-5-HT and ^{3}H-labeled hydroxyindoleacetic acid ([^{3}H]HIAA) formed were isolated from the tissue, and the [^{3}H]-5-HIAA released from the tissue was isolated from the incubation medium.

The 5,6-DHT-induced axotomy and denervation resulted acutely (by 2 weeks) in a 40% reduction in both [^{3}H]TP hydroxylation and [^{3}H]-5-HIAA formation and release in the medulla oblongata (region A). The reductions were greater, about 75–80%, in the cervical and lumbar spinal cord (regions B and C). With time, recovery occurred in all regions; at 6–8 months after injection, [^{3}H]TP hydroxylation reached 40–50% of control values in the spinal cord segments and 155% of those values in medulla oblongata. Tritium-labeled HIAA formation and release had recovered to between 50 and 60% of control values in the spinal cord. In the medulla oblongata, [^{3}H]-5-HIAA

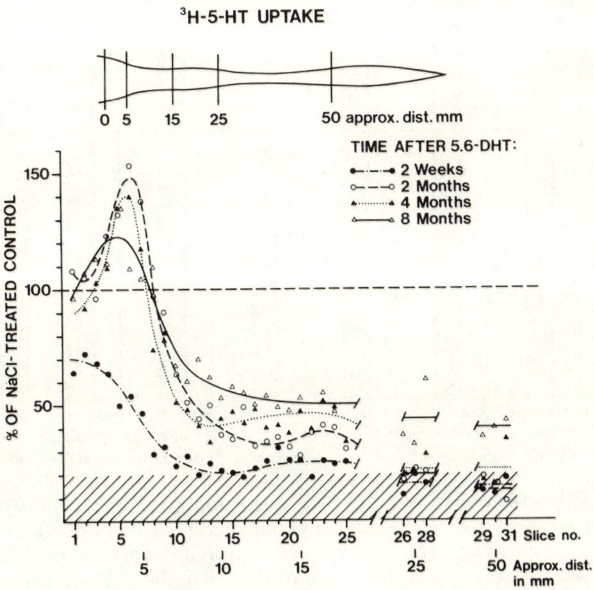

FIGURE 4. [^{3}H]-5-HT uptake in 31 consecutive transverse slices from the medulla oblongata and the spinal cord after 5,6-DHT treatment. Slice 1 was taken from the level of the area postrema, and the following 24 slices extend approximately 17 mm caudally, as indicated in the Figure. Slices 26–28 were taken from the thoracic cord (level 25 mm), and slices 29–31 were obtained from the upper lumbar or lower thoracic cord (level 50 mm). Each symbol represents the mean of three or four determinations as the percentages of age-matched controls. Hatched area denotes the level of the chlorimipramine-resistant part of uptake (chlorimipramine concentration 10^{-4} M). Data from Björklund and Wiklund.[10]

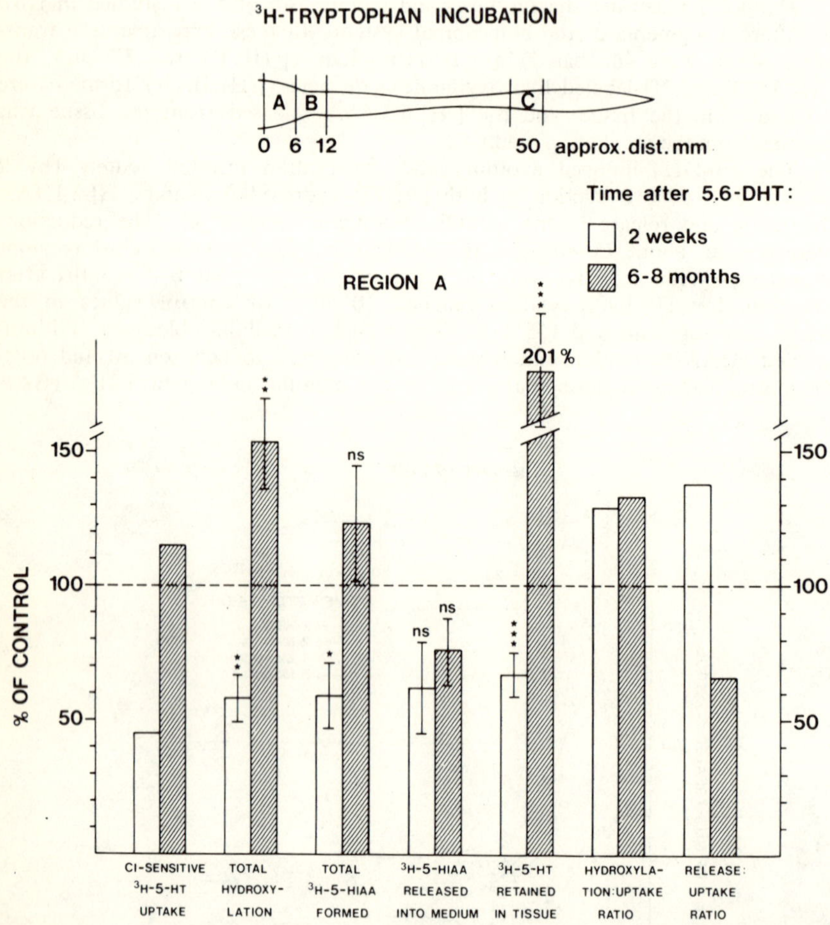

FIGURE 5. [^{3}H]-5-HT and [^{3}H]-5-HIAA formation from [^{3}H]TP during a 30-min incubation *in vitro* acutely (2 weeks) and long-term (6–8 months) after 5,6-DHT treatment (75 μg). Transverse slices were prepared from the medulla oblongata (region A), the upper cervical cord (region B), and the lumbar cord (region C). Means ± SEM are expressed as the percentages of age-matched controls. Total hydroxylation indicates the sum of [^{3}H]-5-HT and [^{3}H]-5-HIAA recovered from the tissue and the medium. Total [^{3}H]-5-HIAA represents the sum of [^{3}H]-5-HIAA recovered from the tissue and the medium. Chlorimipramine (Cl) -sensitive uptake is calculated from FIGURE 4. Differences from control values: ★★★, $p < 0.001$; ★★, $0.01 > p > 0.001$; ★, $0.05 > p > 0.01$. Student's paired *t* test. Data from Björklund and Wiklund.[10]

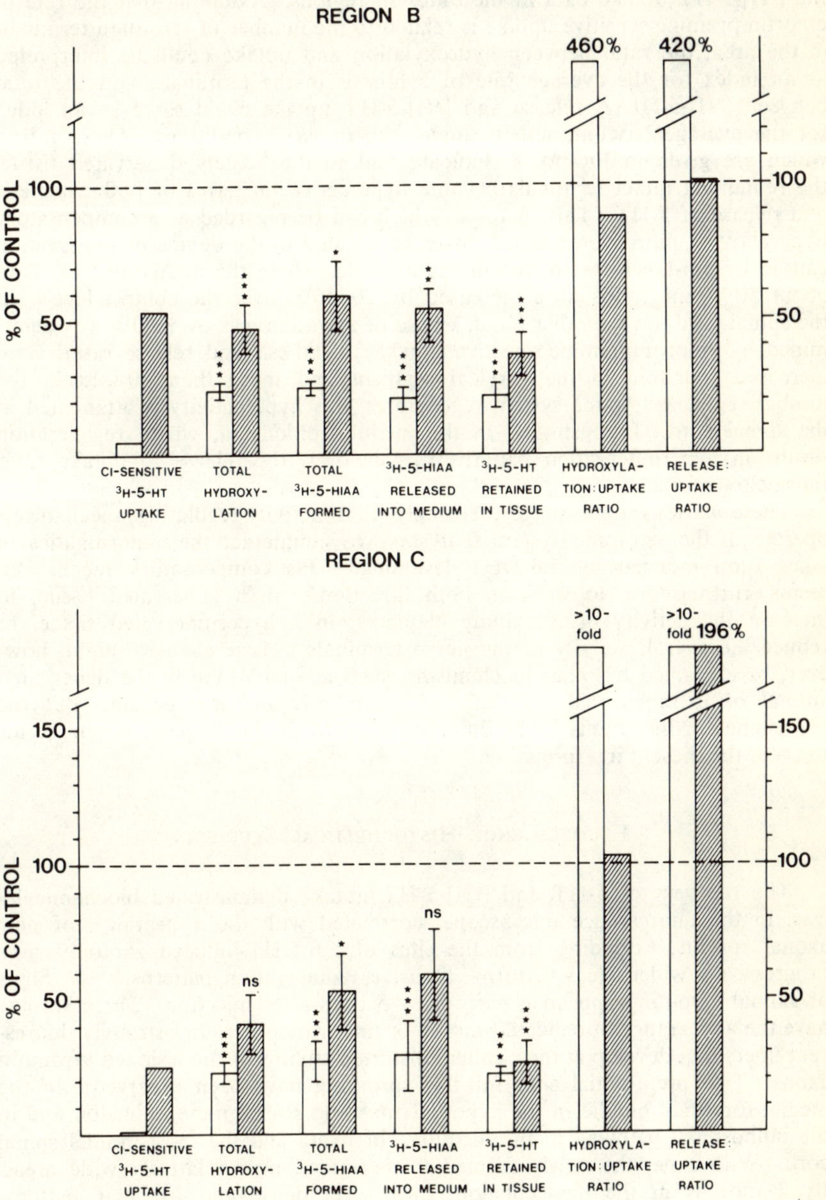
REGION B
460%
420%
% OF CONTROL
100
50
100
50
CI-SENSITIVE ³H-5-HT UPTAKE
TOTAL HYDROXY-LATION
TOTAL ³H-5-HIAA FORMED
³H-5-HIAA RELEASED INTO MEDIUM
³H-5-HT RETAINED IN TISSUE
HYDROXYLA-TION: UPTAKE RATIO
RELEASE: UPTAKE RATIO
REGION C
>10-fold
>10-fold
196%
150
100
50
ns
ns
CI-SENSITIVE ³H-5-HT UPTAKE
TOTAL HYDROXY-LATION
TOTAL ³H-5-HIAA FORMED
³H-5-HIAA RELEASED INTO MEDIUM
³H-5-HT RETAINED IN TISSUE
HYDROXYLA-TION: UPTAKE RATIO
RELEASE: UPTAKE RATIO

release tended to remain depressed even though [^{3}H]-5-HIAA formation had recovered.

It is of particular interest to compare the [^{3}H]TP conversion figures with the [^{3}H]-5-HT uptake data in the different regions. Assuming that the rate of chlorimipramine-sensitive uptake is related to the number of serotonin terminals in the area, the ratio between hydroxylation and uptake could be interpreted as an index for the average rate of synthesis in the terminals, and the ratio between [^{3}H]-5-HIAA release and [^{3}H]-5-HT uptake could serve as an index for the average baseline activity under the *in vitro* conditions. These ratios, which are given in FIGURE 5, indicate that in the acutely denervated tissue, the remaining intact terminals become hyperactive, in terms of both synthesis and release of 5-HT. This increase, which can be regarded as a compensatory hyperactivity, moreover, seemed, inversely related to the degree of denervation caused by 5,6-DHT treatment: in the medulla, where the uptake decreased to about 50%, the ratios were increased by 20–30% over the control levels. In the spinal cord, on the other hand, where denervation was over 90% (as determined by chlorimipramine-sensitive uptake), synthesis and release ratios were increased four-fold in the cervical segment and more than 10-fold in the lumbar segment. As the system regenerates, this hyperactivity is attenuated in the spinal cord. Interestingly, in the medulla oblongata, where regeneration results in hyperinnervation patterns (see below), the release to uptake ratio falls below normal.

These various data suggest strongly that activity-regulating mechanisms operate in the serotonin system that serve to counteract the abnormalities in innervation induced by 5,6-DHT lesioning. This compensatory mechanism seems, furthermore, to work in both directions: in a denervated tissue, to increase the activity of remaining elements; in a hyperinnervated tissue, to reduce the overall activity of the nerve terminals. These changes might, however, be explained by other mechanisms, such as unmasking in the denervated animal of extraneuronal sites of 5-HT synthesis and metabolism. But the consistency of the results at the different levels of the bulbospinal system seems to favor the present interpretation.

FLUORESCENCE HISTOCHEMICAL STUDIES

The recovery of 5-HT and [^{3}H]-5-HT uptake demonstrated biochemically was, in the fluorescence microscope, correlated with the appearance of new axonal sprouts, extending from the sites of 5,6-DHT-induced axotomy over progressively wider areas to form extensive reinnervation patterns.[6, 8, 12] Signs of axonal sprouting appear as early as 4–5 days after injection. These sprouts have the appearance of delicate, smooth or fine, varicose, rather strongly fluorescent fibers that develop at the swollen, distorted stumps of the lesioned serotonin axons. Two preferential sites for this sprouting have been observed: in the medial forebrain bundle in the hypothalamus and rostral mesencephalon and in the bulbospinal pathway in the medulla oblongata and the most cranial spinal cord. With time, the newly formed fibers extend over relatively wide areas. By 3 months in the diencephalon, they form dense and irregular patterns throughout the lateral hypothalamic area as far rostral as the suprachiasmatic region. In the diencephalic areas where indoleamine innervation is readily detectable in the normal animal, particularly in the suprachiasmatic nucleus and the subcommissural organ, an injection of 75 μg of 5,6-DHT causes an acute, complete denervation. By 3 months, no reinnervation was noted in the suprachiasmatic nucleus, whereas in some animals, a few fibers had reappeared in the subcommissural organ.[6]

The regrowth process has been analyzed in greater detail in the bulbospinal system.[12] This system consists of both short local projections to nuclei of the brain stem and of very long descending projections to the spinal cord gray matter. These short and long serotonergic projections presumably have common (or similar) origins from the raphe nuclei of the brain stem. Therefore, the system provides interesting possibilities for comparing the regenerative phenomena in terminal areas at different distances from the cell bodies of origin in the raphe nuclei.

In the spinal cord, the anterior horn and the intermediolateral cell column receive dense serotonergic innervation. Biochemical data on 5-HT uptake, content, and synthetic capacity indicate that administration of 75 μg of 5,6-DHT causes a virtually total denervation of the spinal cord (see above). In accordance with this observation, fluorescence histochemistry showed an almost total disappearance of yellow-fluorescent fibers in the gray matter. Only a few scattered fibers could sometimes be seen to remain. Axotomy of the serotonergic fibers occurs at a high, rather proximal level. Large numbers of sprouting proximal stumps of the lesioned axons are found in the upper cervical cord and medulla oblongata but not at more caudal levels of the spinal cord. Therefore, the regenerating fibers will have to grow for long distances before reaching their original terminal areas. Still, during the course of a protracted regrowth process, all levels of the spinal cord will eventually be partly reinnervated (cf. the biochemical data above). Fluorescence microscopically, this reinnervation was observed after long survival times (6–16 months) as a local reappearance of sparse serotonergic plexuses in the gray matter of the thoracic and lumbar spinal cord. At these levels, the reinnervation was markedly patchy; some areas of the anterior horn gray matter could thus be rather well supplied with branching terminal axons, while neighboring tissue lacked or showed only scattered fibers. However, not even locally did the reformed terminal plexus demonstrate a normal density. This irregular distribution of the regenerated terminals probably also explains why we did not find evidence of reinnervation of the lower spinal cord in all specimens analyzed, as would be expected from the biochemical data.

Many nuclei of the medulla oblongata receive characteristic serotonergic innervations. The regeneration of these short local projections occurs under quite different conditions than in the spinal cord, because the sprouting lesioned fibers are found in close proximity to, or even within, the denervated nuclei. Clearly, this finding favors reinnervation of the terminal areas in the medulla oblongata. The regenerative phenomena observed in the inferior olive will be described in more detail, but in all other brain stem nuclei investigated (including facial, hypoglossal, and trigeminal nuclei), similar observations were made.

The inferior olive is a nuclear complex that consists of the medial accessory (MAO), the dorsal accessory (DAO), and the principal nuclei (PO). The main features of the normal distribution of monoaminergic innervation in the rat inferior olive are illustrated in FIGURE 6. A rich serotonergic innervation is found in the lateral DAO, whereas other regions are devoid of innervation. In the dorsal lamella of the PO, a very dense catecholaminergic innervation is found (hatched area in FIGURE 6; for a more detailed description, see Reference 13). The proximity of the lateral DAO to the ventral brain surface provides for a good penetration of 5,6-DHT from the cerebrospinal fluid and ensures an efficient degeneration of the serotonergic terminals. Thus, the serotonergic denervation of the lateral DAO seemed complete by 5 days after injection. Reinnervation of the inferior olive will start almost immediately, however, because it is located most favorably in relation to the sprouting serotonergic fibers. Some lesioned serotonergic axonal stumps will even terminate within the inferior olive. Furthermore, the main descending serotonergic projection runs in the region immediately lateral to the inferior olive, and here a vigorous sprouting activity from a multitude of lesioned axons will ensue after the chemically induced axotomy. Within just 1 week after injection, the first

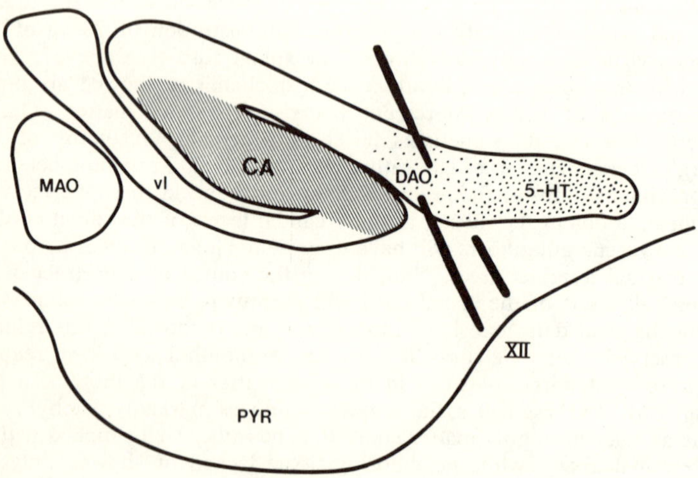

FIGURE 6. Schematic representation of the inferior olive of the rat. The main features of monoaminergic innervation are illustrated at a rostral coronal level. A dense serotonergic innervation (5-HT, represented by dots) is found in the lateral part of the dorsal accessory nucleus (DAO). The very dense catecholaminergic (CA) innervation of the dorsal lamella of the principal nucleus is indicated by hatching. Abbreviations: MAO, medial accessory nucleus; PYR, pyramidal tract; vl, ventral lamella of the principal nucleus; XII, roots of hypoglossal nerve.

thin sprouts are observed to extend into the denervated DAO (FIGURE 7, b). The regrowth process progresses rapidly, and after about 1 month, a terminal plexus of normal density has already been reformed (FIGURE 7, c). However, it is typical for the regeneration of the short local projections of the brain stem that growth does not stop when innervation of normal density has been reformed. Instead, it continues, forming hyperinnervations of increasing density. FIGURE 7(d) illustrates the situation at 7 months after 5,6-DHT administration, when a pronounced hyperinnervation has developed in the lateral DAO (and in other parts of the olivary complex). We have followed this regrowth for up to 16 months and found no sign of retraction of excess terminals; in fact, the formation of new terminals within the olive seems to continue for at least 1 year.

The explanation for this seemingly uncontrolled development of abnormally dense terminal plexuses in the short projections to the brain stem nuclei remains obscure, but it is obvious that this regrowth is guided by rather precise mechanisms of recognition. Thus, the correct terminal areas are found with high reproducibility, and surrounding areas are not invaded by the regrowing fibers. However, that there are limitations to these guiding mechanisms is illustrated in the inferior olive. Here, the regrowing fibers are apparently not able to distinguish among the various subnuclei. The regrowing fibers also invade parts of the olive that are not normally innervated by serotonergic fibers, such as the MAO, the dorsomedial DAO, and the ventral lamella of the PO (FIGURE 8). The plexus formed in these areas attains a density that is almost as high as that in the lateral DAO. An observation of particular interest is that the regrowing fibers do not invade the dorsal lamella of the PO. It seems possible that the presence of a dense catecholaminergic innervation in this area gives the regrowing fibers some additional cues.

To summarize, the regrowth of serotonergic innervation follows different patterns in distant projections, as in spinal innervation, and in short local projections, as in the brain stem. In distant terminal areas, regeneration only succeeds to partially reform the normal innervation, while abnormal hyperinnervations will occur in the brain stem nuclei. It seems as if the brain stem nuclei, due to their proximity to the sprouting fibers or raphe cell bodies, are favored in a competition for the regrowing fibers.

DISCUSSION

The capacity for regenerative growth of lesioned monoamine-containing axons has been demonstrated after mechanical and electrolytic lesions in rat mesencephalon and spinal cord.[14-16] After substantial mechanical lesions, the growing monoamine axon sprouts do not appear to reach across the necrosis into the original axonal pathway, and thus the growing fibers do not reestablish their original connections. In the chemically lesioned animal, the growth situation should be much more favorable, because the sprouting axons will grow in brain tissue that is morphologically grossly intact, and in the absence of necroses and scars. It seems that under these conditions, the regrowing serotonergic axonal sprouts will partly find their way back to areas normally innervated by serotonin fibers, enabling the reestablishment of connections disrupted by the chemically induced axotomy. From the observations in the spinal cord, it seems that this regrowth can occur over relatively long distances (probably over at least a few centimeters along the bulbospinal system).

Regeneration occurs both in serotonin and in noradrenaline neurons after 5,7-DHT treatment.[17, 18] After a high dose of 5,7-DHT (150 μg), the regeneration of the lesioned indoleamine neuron is, however, more restricted than after treatment with 75 μg of 5,6-DHT. This difference can be explained by the fact that the high dose of 5,7-DHT produces an axotomy that lies closer to the cell bodies, resulting in a higher incidence of retrograde cell loss. A most active and extensive regeneration after 5,7-DHT treatment occurs in the noradrenaline neuron systems, which are partially lesioned by this drug. Many diencephalic and telencephalic areas initially partly or completely denervated are thus efficiently reinnervated by the sprouting noradrenergic axons.

It is obvious that the regrowth phenomena observed in 5,6-DHT- or 5,7-

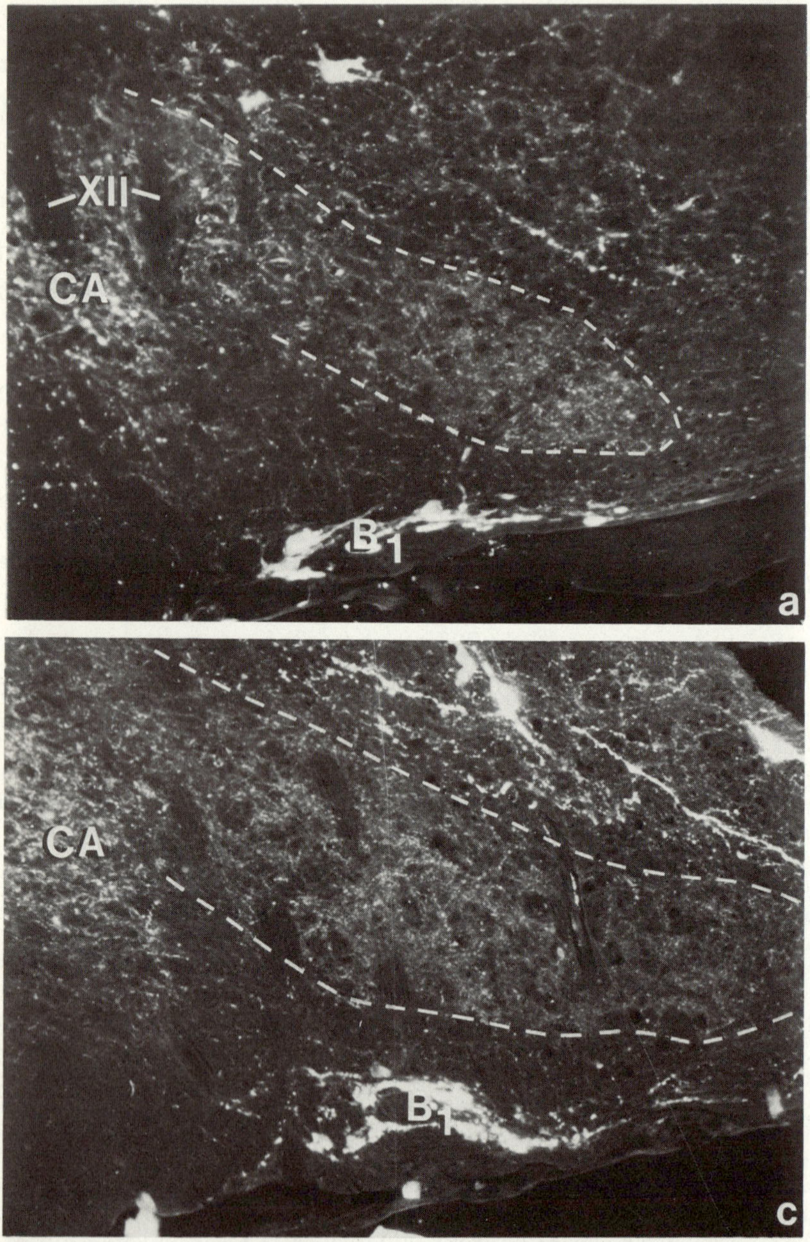

FIGURE 7. The lateral part of the dorsal accessory nucleus (DAO) of the inferior olive. a: Normal serotonergic innervation of the lateral DAO in a control animal. Note the occurrence of serotonergic B1 cell bodies and the proximity to the ventral brain surface. b: The appearance 8 days after an injection of 75 μg of 5,6-DHT in the lateral ventricle. The normal serotonergic innervation of the lateral DAO has degenerated, but newly formed sprouts are already extending into the denervated tissue from lesioned axons located within the olive (arrows). Note the rich distribution of

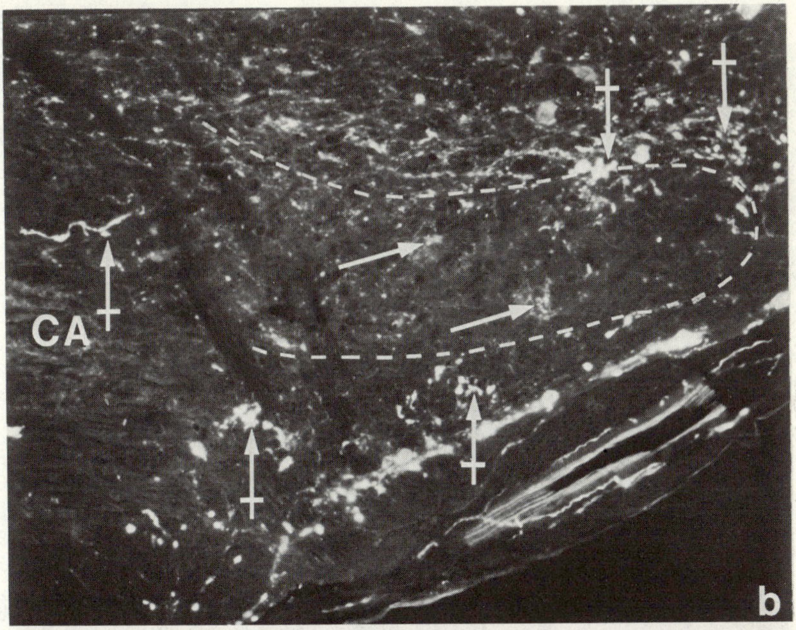

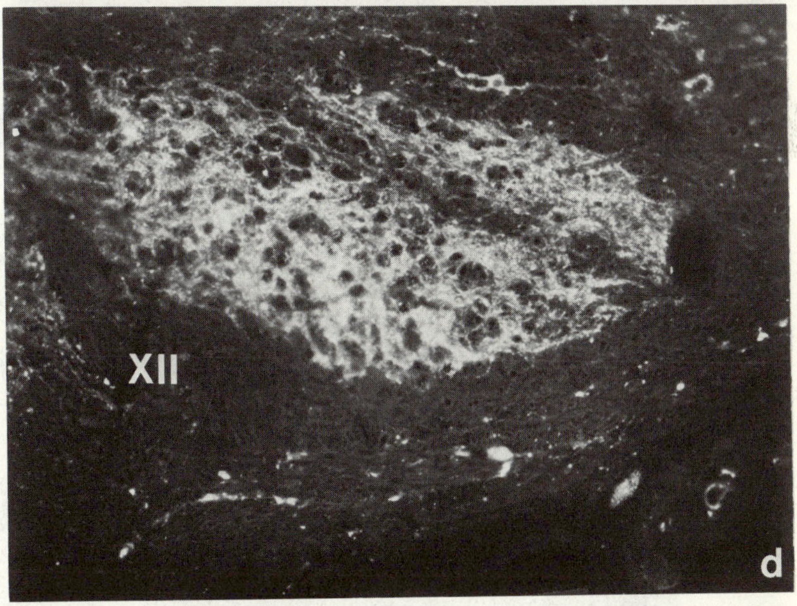

lesioned axonal stumps lateral and ventral to the olive and within the olivary complex (crossed arrows). c: The appearance 35 days after 5,6-DHT treatment. A serotonergic innervation of approximately normal density has been reformed. d: Striking hyperinnervation of the lateral DAO, which has formed 7 months after 5,6 DHT treatment. Abbreviations: CA, catecholaminergic innervation of the dorsal lamella of the principal olivary nucleus; XII, roots of the hypoglossal nerve. From Wiklund and Björklund.[12] ×140.

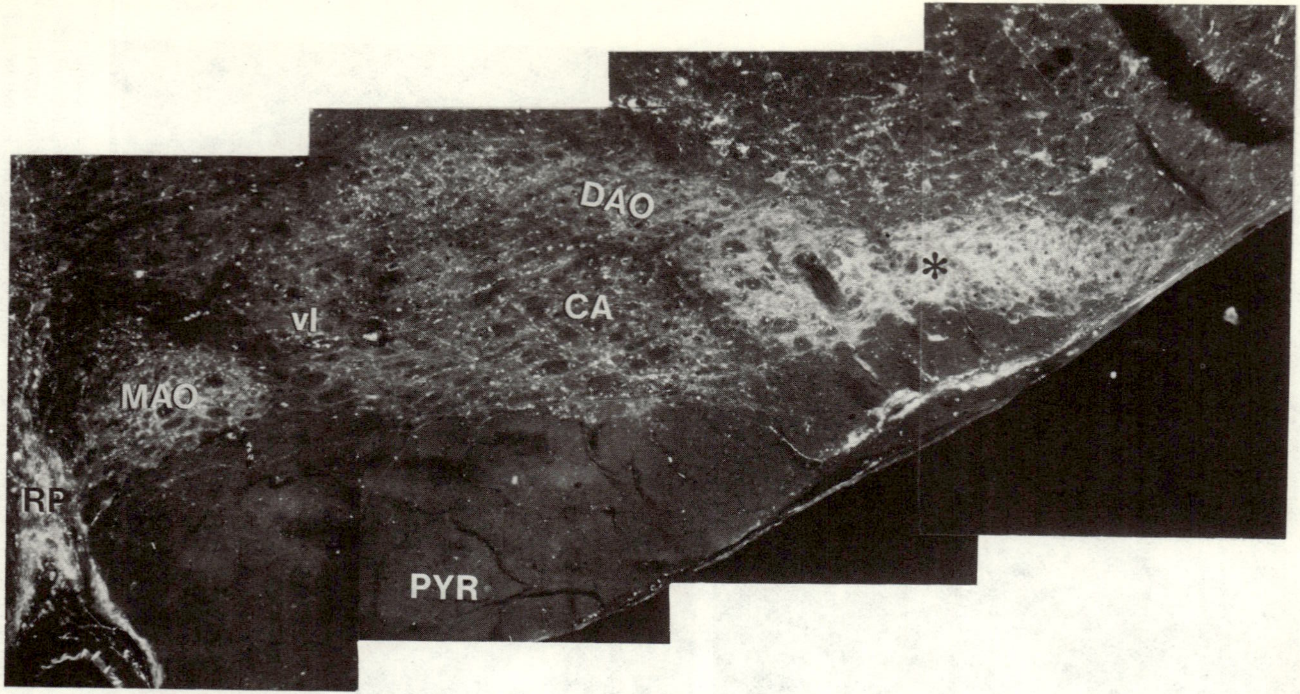

FIGURE 8. Photomontage of the inferior olive 7 months after administration of 5,6-DHT. The lateral dorsal accessory nucleus (DAO), which is the only part of the olive that receives serotonergic innervation in the normal rat (cf. FIGURE 6), displays hyperinnervation of greatly abnormal density (asterisk). Several other areas of the olive, which normally are devoid of serotonergic innervation, also exhibit very dense serotonergic plexuses. These plexuses include the dorsomedial part of the DAO, the ventral lamella of the principal olive (vl), and the medial accessory olive (MAO). The dorsal lamella, which receives a dense catecholaminergic (CA) innervation, is, however, not invaded by the regrowing serotonergic plexuses. Note the large number of serotonergic cell bodies in the nucleus raphe pallidus (RP). Abbreviations: PYR, pyramidal tract; XII, roots of hypoglossal nerve. From Wiklund and Björklund.[12]. ×130.

DHT-lesioned animals are of great importance for the interpretation of behavioral or functional data. In animals injected with 5,7-DHT in combination with desipramine (a treatment that results in a fairly selective serotonergic lesion), Wuttke *et al.*[19] have obtained evidence that regeneration of the drug-lesioned axons can, at least partly, result in functional recovery. In these experiments, the 5,7-DHT treatment resulted in a marked reduction in pituitary luteinizing hormone secretion, probably due to the removal of a stimulatory serotonergic input to the hypothalamus. Luteinizing hormone secretion recovered, however, within 2 months after injection, and this recovery was closely paralleled by a recovery of [^{3}H]-5-HT uptake and [^{3}H]-5-HT synthesis and release in the anterior hypothalamus. It seems likely that the regenerated serotonin axons reestablish a functional innervation in the initially denervated areas, and these axons could thus play a critical role in the recovery of many functional parameters in the lesioned animals. The growth phenomena that occur in the central nervous system of 5,6-DHT- or 5,7-DHT-lesioned animals are also highly interesting from a purely neurobiologic point of view and should provide unique possibilities for studies on the mechanisms of regrowth of lesioned axons in the central nervous system. Neurotoxic drugs have many potential uses for both morphologic and experimental studies, but perhaps the most intriguing applications of these drugs are as tools for studies of neuronal plasticity and growth within the complex neuronal circuitry of the brain.

Acknowledgment

We thank Christina Wedin for preparation of the manuscript.

References

1. Stenevi, U., A. Björklund & R. Y. Moore. 1973. Morphological plasticity of central adrenergic neurons. Brain Behav. Evol. **8:** 110–134.
2. Svendgaard, N., A. Björklund & U. Stenevi. 1975. Regenerative properties of central monoamine neurons. Ergeb. Anat. Entwicklungsgesch. **51**(4)**:** 1–77.
3. Uretsky, N. J. & L. L. Iversen. 1970. Effects of 6-hydroxydopamine on catecholamine containing neurons in the rat brain. J. Neurochem. **17:** 269–278.
4. Wiklund, L. & A. Björklund. 1975. Abortive regeneration of catecholamine neurons after intraspinal 6-hydroxydopamine treatment in the adult rat. *In* Chemical Tools in Catecholamine Research. G. Jonsson, T. Malmfors & Ch. Sachs, Eds. Vol. **1:** 133–139. North Holland Publishing Company. Amsterdam, The Netherlands.
5. Baumgarten, H. G., A. Björklund, L. Lachenmayer, A. Nobin & U. Stenevi. 1971. Long-lasting selective depletion of brain serotonin by 5,6-dihydroxytryptamine. Acta Physiol. Scand. Suppl. **373:** 1–16.
6. Björklund, A., A. Nobin & U. Stenevi. 1973. Regeneration of central serotonin neurons after axonal degeneration induced by 5,6-dihydroxytryptamine. Brain Res. **50:** 214–220.
7. Nobin, A. & A. Björklund. This monograph.
8. Nobin, A., H. G. Baumgarten, A. Björklund, L. Lachenmayer & U. Stenevi. 1973. Axonal degeneration and regeneration of the bulbospinal indolamine neurons after 5,6-dihydroxytryptamine treatment. Brain Res. **56:** 1–24.
9. Baumgarten, H. G., L. Lachenmayer, A. Björklund, A. Nobin & E. Rosengren. 1973. Long-term recovery of serotonin concentrations in the rat CNS following 5,6-dihydroxytryptamine. Life Sci. **12:** 357–364.

10. Björklund, A. & L. Wiklund. 1977. Mechanism of regrowth of the bulbospinal serotonin system following 5,6-dihydroxytryptamine induced axotomy. I. Biochemical findings. In preparation.
11. Nygren, L.-G., K. Fuxe, G. Jonsson & L. Olson. 1974. Functional regeneration of 5-hydroxytryptamine nerve terminals in the rat spinal cord following 5,6-dihydroxytryptamine induced regeneration. Brain Res. **78:** 377–394.
12. Wiklund, L. & A. Björklund. 1977. Mechanism of regrowth of the bulbospinal serotonin system following 5,6-dihydroxytryptamine induced axotomy. II. Fluorescence histochemical findings. In preparation.
13. Wiklund, L., A. Björklund & B. Sjölund. 1977. The indolaminergic innervation of the inferior olive. 1. Convergence with the direct spinal afferents in the areas projecting to the cerebellar anterior lobe. Brain Res. **131:** 1–21.
14. Björklund, A. & U. Stenevi. 1971. Growth of central catecholamine neurones into smooth muscle grafts in the rat mesencephalon. Brain Res. **31:** 1–20.
15. Björklund, A., R. Katzman, U. Stenevi & K. A. West. 1971. Development and growth of axonal sprouts from noradrenaline and 5-hydroxytryptamine neurones in the rat spinal cord. Brain Res. **31:** 21–33.
16. Katzman, R., A. Björklund, Ch. Owman, U. Stenevi & K. A. West. 1971. Evidence for regenerative axon sprouting of central catecholamine neurons in the rat mesencephalon following electrolytic lesions. Brain Res. **25:** 579–596.
17. Baumgarten, H. G., A. Björklund, L. Lachenmayer, A. Rensch & E. Rosengren. 1974. De- and regeneration of the bulbospinal serotonin neurons in the rat following 5,6- or 5,7-dihydroxytryptamine treatment. Cell Tissue Res. **152:** 271–281.
18. Björklund, A., H. G. Baumgarten, L. Lachenmayer & E. Rosengren. 1975. Recovery of brain noradrenaline after 5,7-dihydroxytryptamine-induced axonal lesions in the rat. Cell Tissue Res. **161:** 145–155.
19. Wuttke, W., A. Björklund, H. G. Baumgarten, L. Lachenmayer & M. Fenske. 1977. De- and regeneration of brain serotonin neurons following 5,7-dihydhroxytryptamine treatment: effects on serum LH, FSH and prolactin levels in male rats. Brain Res. **134:** 317–332.

GENERAL DISCUSSION

John A. Harvey, *Moderator*

Department of Psychology
The University of Iowa
Iowa City, Iowa 52242

DR. M. DA PRADA: Dr. Lachenmayer, do you have any comments about the physiologic role of the serotonergic terminals in the ependymal cells of the SCO?

DR. L. LACHENMAYER: The exact functional role of this innervation is not yet known, but these cells are specialized ependymal cells with a secretory activity.

DR. M. D. GERSHON (*Columbia University, New York, N.Y.*): Dr. Lachenmayer, I could not follow your evidence for increased secretion after 5,6-DHT or 5,7-DHT treatment, that is, for actual release of material.

DR. LACHENMAYER: We have evidence only for increased production of protein but not for actual release.

DR. R. D. MYERS: Dr. Harvey, some of us have been concerned about both peripheral and central serotonin. Do you or does anyone else know what effects *p*-CA has on mast cells or enterochromaffin cells?

DR. HARVEY: There does not seem to be any effect.

DR. C. KÖHLER: Dr. Harvey, do you have some information on the projections of the B9 cells to the forebrain, and do you see degeneration in other, nonserotonergic areas of the brain?

DR. HARVEY: The projections of the B9 cells have not been well established. Dr. Fuxe has reported projections to the caudate nucleus, and Massari and Sanders-Bush have reported decreases in the serotonin content of the thalamus after B9 lesions. Regarding your other question, we have not seen any reliable signs of degenerative changes produced by halogenated amphetamines outside of the B9 region.

DR. MYERS: Dr. Nobin, have you studied the recovery of dopaminergic and noradrenergic neurons after injection of 75 μg of 5,7-DHT?

DR. A. NOBIN: My colleague, Dr. Lindvall, has looked at such regeneration of norepinephrine and dopamine neurons after 5,7-DHT treatment by use of the glyoxylic acid fluorescence method. There are obvious signs of vigorous sprouting from the dopamine and norepinephrine neurons.

MS. J. MCGINTY (*Downstate Medical Center, Brooklyn, N.Y.*): Dr. Nobin, do your animals show any behavioral changes that could be correlated with denervation and regeneration?

DR. NOBIN: We have not studied behavior. Endocrinologic studies that Dr. Wuttke will present tomorrow demonstrate clear signs of functional recovery after 5,7-DHT injection.

DR. HARVEY: Dr. Fuxe, have you tried giving something like *p*-CPA in a very small dose, similar to giving AMPT to 6-hydroxydopamine-treated animals, to see if you can bring back the deficits in the 5,7-DHT-treated animals? Also, do these animals show an increased responsiveness to *d*-amphetamine?

Dr. K. Fuxe: No, not in these experiments. We performed experiments of that type previously but did not find any effects; however, we were looking at later time intervals. The time interval may be important. It now seems that it is very important to study the early actions of these neurotoxins because, otherwise, you can miss their effects. This is the reason why there are some discrepancies in the literature.

Part V. Physiologic Effects

NEUROENDOCRINE CONSEQUENCES OF CHEMICAL INDOLECTOMY *

Jacob H. Jacoby

Department of Pharmacology
College of Medicine and Dentistry of New Jersey
New Jersey Medical School
Newark, New Jersey 07103

Ronald F. Thomas

Department of Biology
Bronx Community College
The City University of New York
Bronx, New York 10453

This summary article will attempt to provide greater background information regarding the role of serotonin in neuroendocrine regulatory processes than can be attained by reading the ensuing research papers that deal with this sometimes complex subject. The authors hope that this short paper will lead to a greater appreciation of the research findings that relate to anterior pituitary function that were presented at this symposium. We have departed somewhat from the usual considerations of the actions of serotonin neurotoxins in that we have attributed to these drugs the ability to destroy neurons that contain indoleamines other than serotonin (i.e., tryptamine), a possible action of these drugs that has not as yet been studied but that, by a slight of pen, enables us to call the ensuant cytotoxic events a chemical indolectomy.

Regulation of Anterior Pituitary Secretion

Sufficient evidence exists to demonstrate a relationship of brain monoamines in the regulation of hormone secretion from the anterior pituitary. The final common pathway that links the hypothalamus to the pituitary is via neurohormones known as releasing and inhibiting factors. These peptides are secreted into the hypothalamohypophyseal portal capillaries in the median eminence and pass down the portal veins to alter the release of pituitary hormones. Presumably, neurotransmitters either act on hypothalamic neurons that synthesize and secrete these releasing factors or act directly on pituitary tissue to modulate the subsequent release of anterior pituitary hormones.

Luteinizing Hormone and Follicle-Stimulating Hormone

Abundant pharmacologic evidence is available to demonstrate a role of hypothalamic catecholamine-containing neurons that arrive from the mesen-

* Supported in part by NINCDS Grant 12876.

0077–8923/78/0305–0387 $01.75/0

cephalon (noradrenergic) or from within the hypothalamus itself (tuberoinfundibular dopamine-containing neurons) in the regulation of pituitary gonadotropin secretion. While strong evidence exists to support a facilitatory role for norepinephrine in luteinizing hormone (LH) release, the role of dopamine is less clear, and interpretations of results have been somewhat dependent on methods employed and brain site studied. Because this summary deals primarily with the effects of serotonin, a detailed discussion of these conflicting results is considered beyond the scope of this text; instead, the reader is referred to several recent review articles on catecholamine-pituitary interactions.[1-6] These reviews can also serve as reference sources for catecholamine regulatory effects on other hormone systems dealt with in proceeding sections.

A large body of data indicates that serotonin exerts a tonic inhibitory tone on gonadotropin release. Initial observations, obtained with peripheral serotonin administration, indicated that this amine blocked LH release[7] (as does more localized application to the medial basal hypothalamic-median eminence area[8]). Similar suppression of LH or follicle-stimulating hormone (FSH) is seen with intraventricular serotonin administration to intact[9, 10] or spayed rats.[11] A drawback to these and many similar studies to be subsequently cited is the large dose of transmitter employed; often, greater than total brain concentrations of the amine. Furthermore, it is to be doubted whether amine distribution and eventual receptor interaction is limited to sites normally acted on.

Elevating brain serotonin levels by administering precursor (5-hydroxytryptophan) systemically (intraperitoneally)[12] or increasing synaptic levels of serotonin by electrically stimulating midbrain raphe neurons also suppresses the preovulatory surge of LH.[13] Other indoleamines, such as melatonin or 5-methoxytryptophol, which are of pineal origin, also exert an inhibitory influence on pituitary gonadotropin secretion.[14] Both of these methoxyindoles, long known to have an antigonadotrophic effect when given in high systemic doses,[15, 16] are also able to delay the onset of puberty when given in small intraventricular doses;[14] melatonin can also block ovulation when given by this route.[14] Both these pineal indoles and serotonin also suppress the LH stimulatory effect of luteinizing hormone-releasing factor from rat neonatal pituitary in organ culture.[17] In further support of an inhibitory role of serotonin in gonadotrophic release is the observation, not supported by findings reported in this symposium,[18-20] that 5,6-dihydroxytryptamine (5,6-DHT) greatly enhances the release of these hormones.[21]

Although the studies mentioned above apparently attribute to serotonin and other indoles an inhibitory role in gonadotropin secretion, there is a growing body of evidence, including data presented at this symposium, that confounds this simplistic concept. Depletion of brain serotonin with 5,6-DHT given to prepubertal rats at 21–23 days of age did not lower plasma or pituitary FSH to a greater extent than seen in pair-fed animals that did not receive the neurotoxin, indicating that the reduced food intake [the hyperphagia seen with 5,7-dihydroxytryptamine (5,7-DHT), desmethylimipramine- (DMI) pretreated animals,[22] or after intraventricular *p*-chlorophenylalanine (*p*-CPA)[23] was not noted] and not serotonin depletion caused these hormonal effects.[19] Moreover, by 4 weeks postinjection, brain serotonin levels in drug-treated animals had recovered to the extent that they were similar to those of control animals allowed ad lib. access to food during the same interval.[19] Findings presented at this conference indicate that destruction of serotonin-containing neurons by administration of 5,7-DHT to animals whose catecholamine-containing neurons are protected from a similar fate (by prior treatment with a drug that blocks reuptake into such neurons, i.e., DMI) or in animals that receive *p*-chloroamphetamine (*p*-CA) does not result in long-term reduction of plasma LH or FSH levels in either intact, castrated, or estrogen-primed ovariectomized rats.[18-20] Where reductions of LH levels or alterations of hormonal cyclicity are noted,[20] they are of a transient nature and return to normal when brain serotonin levels are still reduced.[20] Similarly, reduction of brain serotonin concentrations with *p*-CPA does not result in reduced blood LH levels.[24] Indeed, it has been demonstrated that pretreatment of rats with 5,6-DHT does not alter the firing pattern of the medial preoptic area after electrical stimulation of the medial basal hypothalamus, amygdala, or mesencephalon.[25]

Recently, Hery *et al.*[26] demonstrated that the cyclicity of LH secretion observed in ovariectomized rats implanted with estrogen can be impaired by serotonin synthesis inhibition with *p*-CPA. 5-Hydroxytryptophan (5-HTP) restores the cyclicity, whereas methiothepin, a central serotonin receptor antagonist, gave results similar to synthesis inhibition. The authors speculate that serotonin may exert a permissive effect in the presence of estrogens and indicate that the timing of serotonin depletion and repletion may be critical in determining the pharmacologic effects of such manipulations on LH release. These interesting findings are certainly worthy of future considerations and recall earlier observations that indicated a delay in the onset of vaginal opening when rats are treated with *p*-CPA.[27]

Prolactin

Prolactin, unlike other anterior pituitary hormones, appears to be under tonic inhibitory control of the hypothalamus [prolactin inhibitory factor (PIF)]. A prolactin-releasing factor (PRF) is also thought to exist, and this neurohormone is probably different from thyroid-releasing hormone, which can also stimulate the release of prolactin.

Pharmacologic data indicate that such drugs as the phenothiazines, which block dopamine receptors,[28, 29] or catecholamine synthesis inhibitors, which reduce receptor availability of the neurotransmitters,[21] produce an increase in circulating prolactin levels. Conversely, drugs that increase the intrasynaptic levels of dopamine (i.e., L-dopa)[30] or directly stimulate dopamine receptors (i.e., apomorphine[31, 32]) reduce the release of prolactin (for further details, see References 2 and 23).

Serotonin has also been implicated in the regulation of prolactin secretion from the pituitary gland. Either intraventricular[9] or systemic administration of serotonin[34] stimulates the release of prolactin in the rat (one need not postulate a reflex mechanism to explain the latter finding, because the median eminence is outside the blood-brain barrier and is therefore subject to a direct action of agents that penetrate the brain with varying degrees of facility). The administration of serotonin precursors, namely, 5-HTP or L-tryptophan, also stimulates the release of this hormone within a time interval compatible with the conversion of these amino acids to serotonin.[35-39] Pretreatment of 5-HTP-injected animals with a specific serotonin reuptake blocker, namely, fluoxetine, potentiates the actions of the former compound,[18, 40] presumably by prolonging the duration of elevated intrasynaptic serotonin levels. Furthermore, direct stimulation of serotonin receptors with quipazine also stimulates prolactin release.[41] Indeed, the hypothalamus appears to be among the brain sites most reactive to the serotonin agonistic properties of this drug.[42] (Apparently in this instance, a release of hormone induced by stimulation of serotonin receptors predominates over an inhibition of hormone release that results from dopamine receptor activation; that is, quipazine also possesses dopamine agonistic properties.[43]) Interference with the reactivity of serotonin or its agonists at serotonin receptor sites with the peripheral serotonin antagonists cyproheptadine and methysergide produces a blockade of the prolactin-stimulating effects of such receptor activation.[41] The latter results suggest that serotonin released at the median eminence, or at another pituitary site, as opposed to a more distal central location, such as the hypothalamus, may be regulating those effects. Such a qualified hypothesis is based on the observation that cyproheptadine and methysergide may not act as central serotonin antagonists, because they neither induce a compensatory rise of 5-hydroxyindole synthesis associated with receptor blockade[44] nor accelerate the neuronal firing rate of raphe neurons.[45] Support for this hypothesis is derived from the ability of peripheral serotonin administration, presumably impermeant to the central nervous system, to stimulate prolactin release.[34] Some reports based on depletion of brain serotonin levels have also tended to support a facilitatory role of this indole in prolactin release. For example, reduction of brain serotonin with either *p*-CPA or *p*-chloroamphetamine (*p*-CA) subsequently reduces prolactin levels in adult male[46] or in estrogen-primed ovariectomized rats.[35,36] 5,7-Dihydroxytryptamine injection, with or without DMI pretreatment, also reduces plasma prolactin levels.[46]

More recent data presented in this symposium have failed to support earlier observations made with the use of serotonin neurotoxins. Neither 5,7-DHT nor *p*-CA impairs prolactin release in male rats;[18,20] these drugs also do not influence the stress-induced[20] or proestrous surge of this hormone.[18] However, 5,7-DHT does inhibit the afternoon rise of prolactin seen during estrus;[18] similar results are achieved with methysergide.[41] These findings suggest that a different mechanism mediates the estrous and proestrous release of prolactin. 5,7-Dihydroxytryptamine is also ineffective in blocking the suckling-induced rise of prolactin.[18] The latter result, while in conflict with the observations that *p*-CPA[47] and methysergide[48] do interfere successfully with this hormonal stimulus, may actually be due to a recovery of function, because a sufficient time interval for such a regenerative process was allowed to transpire between drug administration and the onset of suckling. In support of this possibility is Clemens'[18] observation that supersensitivity to 5-HTP (i.e., an enhanced prolactin release) occurs in the chemically indolectomized rat.

Growth Hormone

Considerable evidence exists to suggest a regulatory role for both the catecholamines and serotonin in the regulation of growth hormone secretion. Whether these biogenic amines act on the release of a growth hormone-releasing or -inhibiting (somatostatin) factor or by some other mechanism remains to be elucidated.

Pharmacologic manipulations that increase the synaptic availability of catecholamines, that is, by L-dopa administration[49] or by intrahypothalamic noradrenaline injection,[50] or that otherwise act to stimulate specific catecholamine receptors, for example, α-adrenergic receptors with clonidine [51] or dopaminergic receptors with apomorphine [52] (although there is evidence to indicate that this response is not universal [53, 54]) act to promote the release of growth hormone (GH). Use of agents that act to block α- or β-adrenergic receptors have also revealed that the α-receptor appears to be facilitatory and the β-receptor inhibitory to GH release.[55] However, the rat, which appears to have an atypical hypothalamic-pituitary growth hormone axis (unless, of course, one opines that it is the nonrodent species that have deviated from the norm) responds to intraventricular administration of dopamine with a reduction of basal hormone levels.[56] This complex subject is comprehensively covered in a recent review.[57]

Early and more recent studies of the role of serotonin in the regulation of growth hormone in primates, including man, indicate that enhanced synaptic availability of this amine, achieved by 5-HTP administration, results in an increase of circulating GH levels.[53, 58, 59] Studies that attempted to increase serotonin stores with L-tryptophan have not yielded the same obvious results.[53, 60] However, increasing synaptic indole levels by electrically stimulating raphe neurons does enhance the release of GH after an initial inhibition.[57] Collu,[61] attempting to correlate GH rhythmicity with changes in brain serotonin content (not turnover), could find no such relationship, although a slight lowering of the light-associated GH rise was noted after *p*-CPA treatment. More recently, decreasing receptor availability of serotonin with cyproheptadine or methysergide (recall that one must interpret results with caution when attributing a central mode of action to these peripheral serotonin antagonists [44, 45]) has also been shown to impair physiologic and provocative stimuli for GH release.[62–66]

Studies that utilized serotonin neurotoxins have not supported the hypothesis that serotonin normally exerts a facilitatory role in the release of GH. Collu[19, 61] has shown that treatment of male prepubertal rats with 5,6-DHT at 3 weeks of age does not influence either pituitary stores or circulating GH levels. The retardation of growth in such drug-treated animals can be attributed to a drug-induced reduction of food intake. A similar lack of effect on GH occurs after serotonin depletion with *p*-CA.[18] A recent review article [67] has referred to unpublished findings that indicate that, at least in the young rat, a nonspecific depletion of brain serotonin with 5,7-DHT may lead to an elevation of plasma GH levels.

Adrenocorticotrophic Hormone

The regulation of corticotropin-releasing factor (CRF) and subsequent adrenocorticotrophic hormone (ACTH) and adrenal steroid secretion involves interactions among numerous control factors, both excitatory and inhibitory, which include stress, circadian rhythms, and feedback mechanisms that are integrated at the level of the hypothalamus.

Catecholamines appear to exert an inhibitory influence on the secretion of ACTH via norepinephrine acting on a central α-adrenergic receptor. Much of the investigations that elucidated these mechanisms have been performed by Ganong and coworkers. A discussion of these results is beyond the scope of this article; rather, the reader is referred to articles by Ganong.[68, 69]

Data derived from the study of the actions of serotonin on the hypothalamic-pituitary-adrenal axis have not led to a general agreement concerning the role of the amine on this system. Earlier observations indicated that peripheral serotonin or 5-HTP administration stimulated the release of adrenal corticosteroids or depleted adrenal ascorbic acid levels.[72-80] These results could be attributed to the stress involved in serotonin administration or to a direct effect of this indole on the adrenal gland.[73] Pioneering studies that attempted to correlate central serotonergic function with ACTH release indicated that serotonin normally inhibits the release of this tropin, because amine depletion with reserpine resulted in an increase of basal circulating corticosteroid levels in the rat.[74] Prolonged *p*-CPA treatment in the same species, while without effect on basal steroid levels,[75, 76] enhanced the subsequent response to stress.[75, 77] Depletion of serotonin with *p*-CPA also enhances adrenal growth subsequent to unilateral adrenalectomy.[77] Furthermore, intraventricular or intracerebral injection of rather large doses of serotonin did suppress ether-induced elevations (although not resting levels) of corticosterone [78] and prevented adrenal hypertrophy and hypersecretion after unilateral adrenalectomy.[79] (The significance of a serotonin involvement in this phenomenon must be reexamined, because there is evidence that these events are mediated by spinal neuronal, and not traditionally thought of central neurohumoral mechanisms.[80]) Indeed, the Hungarian group of investigators have also shown that serotonin blocks the hypothalamic release of ACTH *in vitro*,[81] that complete and frontal deafferentation of the hypothalamus, which results in a fall of hypothalamic serotonin, results in an elevation of basal corticosterone levels but impairs the hormonal response to stress,[82] and that lesions of the raphe nuclei increase resting levels of corticosterone and, as concluded by these investigators, increase the subsequent response to stress.[83] (A closer look at the latter finding suggests another conclusion: raphe lesions do not alter the stress-induced hypersecretion of corticosterone. While steroid levels are higher after either electrical shock or ether stressing in the raphe-lesioned animal, the magnitude of change from resting hormonal levels are not different between surgically denervated and intact animals.) Fuxe *et al.*[84] have also noted an elevation of corticosterone after lesions of the raphe nuclei induced with 5,6-DHT. However, this effect was noted only 1 day, but not 1 week, after drug administration, making interpretation of these results hazardous. A similar inhibition of surgery-induced secretion of adrenal steroids has been obtained with large intraventricular doses of 5-hydroxytryptamine in dogs.[85] Earlier studies that attempted to examine the effects of a tryptophan-induced elevation of brain serotonin were conflicting: a "lower" dose (204 mg/kg) enhanced the response, whereas a higher dose (408 mg/kg) suppressed the response.[76] Very recently, the same group [77] has shown that lower and somewhat more proximate to a "physiologic" dose of this amino acid suppresses the subsequent corticosterone rise in response to stress. Intraventricular injection of several indoles has also been reported to suppress the pituitary-adrenal axis.[86]

More recent studies, which employed more newly introduced pharmacologic agents, have confirmed earlier observations that 5-HTP induces a dose-related increase in circulating corticosteroids,[54, 87, 88] an effect that can be enhanced by preventing serotonin reuptake with fluoxetine [76,88] and that can be blocked by drugs that prevent conversion of 5-HTP to serotonin (RO-4–4602) or that block serotonin receptors, namely, cyproheptadine and methergoline.[8] (Indeed, these drugs have been shown to suppress the ACTH response to several provocative and physiologic stimuli in man [65, 89-91] and have recently been introduced for the therapy of Cushing's disease; [92] whether such effects are exerted via central serotonin receptor antagonism is open to debate.) Furthermore, removal of the pituitary gland abolishes the corticosteroid-stimulating effect of 5-HTP.[87] Other supportive evidence for a stimulatory role of serotonin in ACTH release are the observations (in apparent contradiction of Vermes and Telegdy's [75] findings) that this amine (again, in large doses) implanted in various areas of the hypothalamus, midbrain, or forebrain also increases adrenal steroid release in the guinea pig or cat.[93, 94] Naumenko's group [95] has also presented some evidence for a stimulatory role of serotonin at a lower hypothalamic or pituitary site, because 5-HTP can still increase the release of corticosteroids after total interruption of neural afferents to the medial basal hypothalamus. If such is the case, one might expect, as already demonstrated,[65, 89, 91] that peripheral serotonin antagonists may alter the hypothalamic-pituitary-adrenal axis without necessarily acting on serotonin receptors within the central nervous system. *In vitro* studies that employed rat hypothalami have also supported a CRH-stimulating role for serotonin.[96]

Studies of serotonergic effects on steroid rhythmicity have shown that drugs, including *p*-CA, that alter the synaptic availability of serotonin reduce the diurnal elevation of corticosteroids in cats, rats, and man [76, 97, 98] and elevate nadir levels in the rat.[98] Similar effects can be achieved by lesioning the raphe nuclei or the fornix [68] (although there is evidence that hormonal changes are due to a lack of prior adaptation and not to the lesion-induced depletion of serotonin).[99]

Other than the latter study (which utilized the cat [97] in an attempt to examine the effects of *p*-CA at a time when brain serotonin levels were probably reduced after an initial discharge of this amine), we are aware of only a single study that employed a serotonin neurotoxin to study adrenal rhythmicity. The results of this study,[100] which, unfortunately, are based on doses of 5,6-DHT without DMI pretreatment and at levels that may be neurotoxic to catecholamine-containing neurons, indicate that indolectomy of rats at 3 days of age did not alter the subsequent development of circadian rhythmicity when tested at 30 days of age. Development of steroid periodicity proceeded normally, even after intentional chemical sympathectomy with 6-hydroxydopamine.

Thyroid-Stimulating Hormone

The role of biogenic amines in the regulation of the hypothalamic-pituitary-thyroid axis has not been studied intensively and remains a fertile field for future investigation. While reduction of synaptic catecholamine availability appears to have no effects on baseline thyroid-stimulating hormone (TSH) levels, an impairment of the ability to secrete this hormone is noted during subsequent exposure to a cold environment.[101, 102] Furthermore, clonidine, an α-adrenergic agonist, corrects this impaired responsivity,[102] suggesting that norepinephrine mediates this effect.

Early investigations that linked serotonin with thyroid function indicated that systemic administration of this amine interfered with thyroid hormone synthesis and release,[103] an effect that diminished with time. However, Melander *et al.*[104–106] have presented convincing data that suggest a facilitatory role of serotonin, released from mast cells present in the thyroid [106–108] by the action of TSH. Increasing serotonin levels by precursor loading with 5-HTP appears to have no effect on baseline levels of TSH in intact rats,[36, 101] although it does stimulate release of this hormone in the estrogen-primed ovariectomized rat [36] and in rats or mice that had previously been subjected to reduction of TSH levels with reserpine.[36] However, by use of more specific means of increasing neuronal serotonin levels, namely, tryptophan loading, Mueller *et al.*[39] were able to show that TSH levels were subsequently reduced by this manipulation. The latter finding is compatible with a serotonin inhibition of TRH release from hypothalamic tissue [109] and with results obtained after intraventricular or intrahypothalamic administration of massive doses of serotonin.[110] A similar inhibitory effect with 5-HTP is not seen unless the animal is exposed to cold.[101]

Conversely, reduction of brain serotonin levels with *p*-CPA does not alter the cold-induced increase of TSH release but does reduce circulating basal levels of this tropin.[101] (Unfortunately, low doses of the drug were used, and serotonin levels were not measured.) *p*-Chloroamphetamine, utilized at a time prior to the onset of neurotoxicity (16 hr postinjection), also reduced resting levels of TSH in estrogen-primed ovariectomized rats.[36] We are not aware of any studies that employed 5-hydroxytryptamine neurotoxins as a tool to study thyroid function. Indeed, the use of these agents need not be limited to the study of central mechanisms that regulate the hypothalamic-pituitary-thyroid axis, because the thyroid serotonin pool appears to share at least some of the uptake characteristics of serotonin-containing neurons.[111]

Other Hormones

Little effort has thus far been expended in determining the interaction of brain monoamines with pituitary secretion of melanocyte-stimulating hormone, although MSH, in a manner similar to prolactin, appears to be under an inhibitory dopaminergic tone [112-114] and, as has been recently shown in the frog, can be released after serotonin or 5-HTP administration.[115] Furthermore, the role of biogenic amines in the regulation of posterior pituitary secretion is yet to be examined intensively. Certainly, the utilization of serotonin neurotoxins along with other specific pharmacologic agents could prove valuable in future studies of these regulatory mechanisms.

Over the past several years, there has been a resurgence of interest in the metabolic effects of serotonin. For example, serotonin neurons have been shown to originate in the myenteric plexus,[116] and a role of this transmitter in the regulation of gut motility and digestive hormone secretion may be postulated. Furthermore, both insulin- and glucagon-secreting cells of the pancreas have been shown to contain serotonin or to have the ability to synthesize this amine; [117, 118] the possible role of these serotonin reservoirs in mediating the apparent effects of serotonin or, for that matter, other stimuli on insulin (see References 119 and 120 for review) and glucagon [121-123] secretion are unknown. Should serotonin neurotoxins prove to exert a cytotoxic effect on these tissue stores (serotonin-containing neurons of the gut seem to be a logical candidate for such a study), we can expect to gain further insight into the physiologic role of these peripheral indoleamines.

Conclusions

The introduction of serotonin neurotoxins represents a potent new tool in elucidating the role of serotonin in neuroendocrine regulation. Thus far, utilization of chemical indolectomy has not yet yielded the dramatic insights into hypothalamic-pituitary function that one might have expected at the onset. Results obtained with these agents have not always corroborated findings accumulated by depleting brain serotonin by other means. Few normally occurring endocrine regulatory mechanisms appear to be altered, except some rhythmic phenomena, such as the estrous rise of prolactin and corticosteroid rhythms, the physiologic functions of which remain unclear. Perhaps these initial results with this newly introduced pharmacologic approach have already provided the answers we do not necessarily want to hear, that is, that a lack of serotonin, or the neuronal population influenced by these agents, is not very critical in the homeostatic mechanisms thus far studied. Perhaps neuronal degeneration, being less than total, allows these surviving neuronal substrates to maintain and subsequently recover functional capabilities within a relatively short period of time. More likely, we have not yet addressed ourselves to many of the questions that chemical indolectomy is quite willing and able to answer.

References

1. Frohman, L. A. & M. E. Stachura. 1975. Metabolism **24:** 211–234.
2. MacLeod, R. M. 1976. *In* Mammary Cancer and Neuroendocrine Therapy. B. A. Stoll, Ed. : 138–159. Butterworths. London, England.
3. McCann, S. M. & S. R. Ojeda. 1976. *In* Reviews of Neuroscience. S. Ehrenpreis & I. J. Kopin, Eds. Vol. 2: 92–109. The Raven Press. New York, N.Y.
4. Reichlin, S., R. Saperstein, I. M. D. Jackson, A. E. Boyd & Y. Patel. 1976. Annu. Rev. Physiol. **38:** 389–424.
5. Versteeg, D. H. G. & R. J. Wurtman. 1976. *In* Molecular and Functional Neurobiology. W. H. Gispen. Ed. : 201–234. Elsevier. Amsterdam, The Netherlands.
6. Wuttke, W. 1976. Rev. Physiol. Biochem. Pharmacol. **76:** 62–101.
7. O'Steen, W. K. 1965. Endocrinology **77:** 937–939.
8. Domanski, E., F. Przekop, B. S. Kubiszewski & E. Wolinska. 1976. Neuroendocrinology **17:** 265–273.
9. Kamberi, I. A., R. S. Mical & J. C. Porter. 1971. Endocrinology **88:** 1288–1293.
10. Kamberi, I. A., R. S. Mical & J. C. Porter. 1970. Endocrinology **87:** 1–12.
11. Schneider, H. P. G. & S. M. McCann. 1970. Endocrinology **86:** 1127–1133.
12. Kordon, C., F. Javoy, G. Vassent & J. Glowinski. 1968. Eur. J. Pharmacol. **4:** 169–174.
13. Carrer, H. F. & S. Taleisnik. 1970. J. Endocrinol. **48:** 527–539.
14. Fraschini, F., R. Collu & E. E. Muller. 1973. *In* Hormones and Brain Function. K. Lissak, Ed. : 225–230. Plenum Publishing.
15. Wurtman, R. J., J. Axelrod & E. W. Chu. 1963. Science **141:** 277–278.
16. McIsaac, W. M., R. G. Taborsky & G. Farrell. 1964. Science **145:** 63–64.
17. Martin, J. E., J. N. Engel & D. C. Klein. 1977. Endocrinology **100:** 675–680.
18. Clemens, J. A. This monograph.
19. Collu, R. This monograph.
20. Wuttke, W., J. L. Hancke, K. G. Hohn & H. G. Baumgarten. This monograph.
21. Ladosky, W. & J. Noronha. 1974. J. Endocrinol. **62:** 677–678.
22. Saller, C. F. & E. M. Stricker. 1976. Science **192:** 385–387.
23. Briesch, S. T., F. P. Zemlan & B. G. Hoebel. 1976. Science **192:** 382–384.
24. Donoso, A. O., W. Bishop, C. P. Fawcett, L. Krulich & S. M. McCann. 1971. Endocrinology **89:** 774–784.
25. Fenske, M., F. Ellendorff & W. Wuttke. 1975. Exp. Brain Res. **22:** 495–507.

26. Hery, M., E. La Plante & C. Kordon. 1976. Endocrinology **99:** 496–503.
27. Fajer, A. B., D. Hoffman & E. Shillito. 1970. J. Reprod. Fertil. **22:** 379–389.
28. Ben-David, M., A. Danon & F. G. Sulman. 1970. Neuroendocrinology **6:** 336–342.
29. Lu, K. H., Y. Amenomori, C. L. Chen & J. Meites. 1970. Endocrinology **87:** 667–672.
30. Lu, K. H. & J. Meites. 1972. Endocrinology **91:** 868–872.
31. MacLeod, R. M., E. H. Fontham & J. E. Lehmeyer. 1970. Neuroendocrinology **6:** 283–294.
32. Smalstig, E. B., B. D. Sawyer & J. A. Clemens. 1974. Endocrinology **95:** 123–129.
33. Clemens, J. A. 1976. *In* Hypothalamus and Endocrine Functions. F. Labrie, J. Meites & G. Pelletier, Eds. : 283–301. Plenum Publishing Corporation. New York, N.Y.
34. Lawson, D. M. & R. R. Gala. 1975. Endocrinology **96:** 313–318.
35. Caligaris, L. & S. Taleisnik. 1974. J. Endocrinol. **62:** 25–33.
36. Chen, H. J. & J. Meites. 1975. Endocrinology **96:** 10–14.
37. Kato, Y., Y. Naokai, H. Imura, K. Chihara & S. Ohgo. 1974. J. Clin. Endocrinol. Metab. **38:** 695–697.
38. MacIndoe, J. H. & R. W. Turkington. 1973. J. Clin. Invest. **52:** 1972–1978.
39. Mueller, G. P., C. P. Twohy, H. T. Chen, J. P. Advis & J. Meites. 1976. Life Sci. **18:** 715–724.
40. Krulich, L. 1975. Life Sci. **17:** 1141–1144.
41. Clemens, J. A., B. D. Sawyer & B. Cerimele. 1977. Endocrinology **100:** 692–698.
42. Jacoby, J. H, R. A. Howd, M. S. Levin & R. J. Wurtman. 1976. Neuropharmacology **15:** 529–534.
43. Grabowska, M., L. Antkiewicz & J. Michaluk. 1974. J. Pharm. Pharmacol. **26:** 74–76.
44. Jacoby, J. H., J. J. Poulakos & G. F. Bryce. 1978. Neuropharmacology. In press.
45. Haigler, H. & G. Aghajanian. 1974. J. Neural Transmission **35:** 257–273.
46. Gil-Ad, I., F. Zambotti, M. O. Carruba & E. E. Muller 1976. Proc. Soc. Exp. Biol. Med. **151:** 512–518.
47. Kordon, C., C. A. Blake, J. Terkel & C. H. Sawyer. 1973–4. Neuroendocrinology **13:** 213–222.
48. Gallo, R. V., J. Rabii & C. P. Moberg. 1975. Endocrinology **97:** 1069–1105.
49. Boyd, A. E., H. E. Lebovitz & J. B. Pfeiffer. 1970. N. Engl. J. Med. **283:** 1425–1429.
50. Toivola, P. T. K. & C. C. Gale. 1972. Endocrinology **90:** 895–902.
51. Lal, S., G. Tolis, J. B. Martin, G. M. Brown & H. Guyda. 1975. J. Clin. Endocrinol. Metab. **41:** 827–832.
52. Lal, S., C. de la Vega, T. L. Sourkes & H. G. Freisen. 1973. J. Clin. Endocrinol. Metab. **37:** 719–724.
53. Jacoby, J. H., M. Greenstein, J. F. Sassin & E. D. Weitzman. 1974. Neuroendocrinology **14:** 95–102.
54. Chambers, J. W. & G. M. Brown. 1976. Endocrinology **98:** 420–428.
55. Blackard, W. G. & S. A. Heidingsfelder. 1968. J. Clin. Invest. **47:** 1407–1414.
56. Collu, R., F. Fraschini & L. Martini. 1972. Endocrinology **90:** 1231–1237.
57. Martin, J. B. 1976. *In* Frontiers in Neuroendocrinology. L. Martin & W. F. Ganong, Eds. : 129–168. The Raven Press. New York, N.Y.
58. Imura, H., Y. Nakai & T. Hoshimi. 1973. J. Clin. Endocrinol. Metab. **36:** 204–206.
59. Smythe, G. A. & L. Lazarus. 1973. Nature (London) **244:** 230–231.

60. MULLER, E. E., F. BRAMBILLA, F. CAVAGNINI, M. PERACCHI & A. PANERAI. 1974. J. Clin. Endocrinol. Metab. **39:** 1–5.
61. COLLU, R. 1975. *In* Hypothalamic Hormones. M. Motta, P. G. Crosignani & L. Martini, Eds. : 205–216. The Raven Press. New York, N.Y.
62. NAKAI, Y., H. IMURA, H. SAKURAI, H. KURAHACHI & Y. YOSHIMI. 1974. J. Clin. Endocrinol. Metab. **38:** 446–449.
63. NAKAI, Y. & H. IMURA. 1975. Endocrinol. Japon. **22:** 357–360.
64. DELITALA, G., A. MASALA, S. ALAGNA & L. DE VILLA. 1975. Biomedicine **23:** 406–409.
65. CHIHARA, K., Y. KATO, K. MAEDA, S. MATSUKARA & H. IMURA. 1976. J. Clin. Invest. **57:** 1393–1402.
66. ARNOLD, M. A. & J. D. FERNSTROM. 1976. Soc. Neurosci. 6th Annu. Meet., Abst. 944: 664.
67. BAUMGARTEN, H. G. & A. BJÖRKLUND. 1976. Annu. Rev. Pharmacol. Toxicol. **16:** 101–111.
68. GANONG, W. F. 1974. The Neurosciences Third Study Program. F. O. Schmitt & F. G. Worden, Eds. : 549–564. The MIT Press. Cambridge, Mass.
69. GANONG, W. F. 1977. Ann. N.Y. Acad. Sci. **297:** 370–372.
70. MIYAKAWI, H., M. UI & B. KOBAYASHI. 1961. Endocrinol. Japon. **8:** 148–152.
71. MOUSSATCHE, H. & H. ALVARES-PEREIRO. 1957. Acta Physiol. Lat. Amer. **7:** 71–75.
72. NAUMENKO, E. V. 1968. Brain Res. **11:** 1–10.
73. VERDESCA, A. S., C. D. WESTERMANN, R. S. CRAMPTON, W. C. BLACK, R. I. NEDELJKOVIC & J. G. HILTON. 1961. Amer. J. Physiol. **201:** 1065–1067.
74. WESTERMANN, E. O. 1962. J. Pharmacol. Exp. Ther. **138:** 208–217.
75. VERMES, I. & G. TELEGDY. 1973. Acta Physiol. Acad. Sci. Hung. **43:** 105–114.
76. VERNIKOS-DANELLIS, J. & P. A. BERGER. 1973. *In* Serotonin and Behavior. J. Barchas & E. Usdin, Eds. : 173–178. Academic Press, Inc. New York, N.Y.
77. VERNIKOS-DANELLIS, J., K. J. KELLAR, D. KENT, C. GONZALES, P. A. BERGER & J. D. BARCHAS. 1977. Ann. N.Y. Acad. Sci. **297:** 518–526.
78. VERMES, I. & G. TELEGDY. 1972. Acta. Physiol. Acad. Sci. Hung. **42:** 49–59.
79. VERMES, I., D. MOLNAR & G. TELEGDY. 1973. Acta Physiol. Acad. Sci. Hung. **43:** 27–32.
80. ENGELAND, W. C., J. SHINSAKO & M. F. DALLMAN. 1975. Amer. J. Physiol. **229:** 1461–1464.
81. VERMES, I., G. TELEGDY & K. LISSAK. 1972. Acta Physiol. Acad. Sci. Hung. **41:** 95–98.
82. VERMES, I., D. MOLNAR & G. TELEGDY. 1973. Acta Physiol. Acad. Sci. Hung. **43:** 239–245.
83. VERMES, I., G. TELEGDY & K. LISSAK. 1974. Acta Physiol. Acad. Sci. Hung. **45:** 217–224.
84. FUXE, K., HOKFELT, G. JONSSON, S. LEVINE & P. LIDBRINK. 1973. *In* Brain-Pituitary-Adrenal Interrelationships. A. Brodish & E. S. Redgate, Eds. : 239–269. S. Karger. Basel, Switzerland.
85. VAN LOON, G. R., U. SCAPAGNINI, R. COHEN & W. F. GANONG. 1971. Neuroendocrinology **8:** 257–272.
86. PIVA, F., O. SCHIAFFINI, M. MOTTA & L. MARTINI. 1973. *In* Hormones and Brain Function. K. Lissak, Ed. : 231–236. Plenum Publishing Corporation. New York, N.Y.
87. FULLER, R. W., H. D. SNODDY & B. B. MOLLOY. 1976. Life Sci. **19:** 336–346.
88. MEYER, J. S., N. S. BUCKHOLTZ & W. O. BOGGAN. 1976. Soc. Neurosci. 6th Annu. Meet., Abst. 930: 651.
89. PLONK, J. W., C. H. BIVENS & J. M. FELDMAN. 1974. J. Clin. Endocrinol. Metab. **38:** 836–840.
90. CAVAGNINI, F., A. E. PANERAI, F. VALENTINI, P. BULGHERONI, M. PERRACHI & M. PINTO. 1975. J. Clin. Endocrinol. Metab. **41:** 143–148.
91. PLONK, J. & J. M. FELDMAN. 1976. J. Clin. Endocrinol. Metab. **42:** 291–295.

92. Krieger, D. T., L. Amorosa & F. Linick. 1975. N. Engl. J. Med. **293:** 893–896.
93. Naumenko, E. V. 1969. Neuroendocrinology **5:** 81–88.
94. Krieger, H. P. & D. T. Krieger. 1970. Amer. J. Physiol. **218:** 1632–1641.
95. Popova, N. K., L. N. Maslova & E. V. Naumenko. 1972. Brain Res. **47:** 61–67.
96. Jones, M. T., E. Hillhouse & J. J. Burden. 1976. *In* Frontiers in Neuroendocrinology. L. Martini & W. F. Ganong, Eds. Vol. **4:** 195–226. The Raven Press. New York, N.Y.
97. Krieger, D. T. & F. Rizzo. 1969. Amer. J. Physiol. **217:** 1703–1707.
98. Scapagnini, U., G. P. Moberg, G. R. Van Loon, J. De Groot & W. F. Ganong. 1971. Neuroendocrinology **7:** 90–96.
99. Balestrery, F. G. & G. P. Moberg. 1976. Brain Res. **118:** 503–508.
100. Krieger, D. T. 1975. Neuroendocrinology **17:** 62–74.
101. Annunziato, L., G. Di Renzo, G. Lombardi, F. Scopacasa, G. Schettini, P. Preziosi & U. Scapagnini. 1977. Endocrinology **100:** 738–744.
102. Tuomisto, J., T. Ranta, P. Mannisto, A. Saarinen & J. Leppaluoto. 1975. Eur. J. Pharmacol. **30:** 221–229.
103. Szanto, L., A. L. Reviczky & T. Grynaeus. 1966. Acta Physiol. Acad. Sci. Hung. **29:** 183–194.
104. Melander, A. 1969. Acta Endocrinol. **62:** 565–576.
105. Melander, A., C. Owman & F. Sundler. 1971. Endocrinology **89:** 528–533.
106. Melander, A. & F. Sundler. 1972. Endocrinology **90:** 802–807.
107. Clayton, J. & C. M. Szego. 1967. Endocrinology **80:** 689–698.
108. Paasonen, M. K. 1958. Experientia **14:** 95–96.
109. Grimm, Y. & S. Reichlin. 1973. Endocrinology **93:** 626–631.
110. Mess, B. & L. Peter. 1975. Endocrinol. Exp. **9:** 105–113.
111. Gershon, M. D. & E. A. Nunez. 1976. Mol. Cell. Endocrinol. **5:** 169–180.
112. Kastin, A. J. & A. V. Schally. 1966. Endocrinology **79:** 1018–1021.
113. Scott, G. J. & W. A. Stillings. 1972. Endocrinology **90:** 545–551.
114. Tilders, F. J. H., A. H. Mulder & P. G. Smelik. 1975. Neuroendocrinology **18:** 125–130.
115. Thornton, V. F. 1974. J. Endocrinol. **63:** 55P–56P.
116. Gershon, M. D., R. G. Robinson & L. L. Ross. 1976. J. Pharmacol. Exp. Ther. **198:** 548–561.
117. Cegrell, L. 1968. Acta Physiol. Scand. Suppl. **314:** 1–111.
118. Lundquist, I. 1971. Acta Physiol. Scand. Suppl. **372:** 3–47.
119. Woods, S. C. & D. Porte. 1974. Physiol. Rev. **54:** 596–619.
120. Smith, P. H. & D. Porte. 1976. Annu. Rev. Pharmacol. Toxicol. **16:** 269–285.
121. Joost, H. G., J. Beckmann, S. Holze, S. Lenzen, W. Poser & A. Hasselblatt. 1976. Diabetologia **12:** 201–206.
122. Marco, J., J. A. Hedo, J. Martinell, C. Calle & M. L. Villaneuva. 1976. J. Clin. Endocrinol. Metab. **42:** 215–221.
123. Jacoby, J. H. & G. F. Bryce. 1977. Endocrine Soc. 59th Annu. Meet., Abst. 31: 72.

EFFECTS OF SEROTONIN NEUROTOXINS ON PITUITARY HORMONE RELEASE

James A. Clemens

The Lilly Research Laboratories
Eli Lilly and Company
Indianapolis, Indiana 46206

The participation of serotonergic neuronal systems in the control of anterior pituitary hormone secretion has been suggested by several authors. Inhibitory effects of serotonin on luteinizing hormone (LH) release have been observed after direct infusion of the amine into the cerebrospinal fluid.[1] In the presence of a monoamine oxidase inhibitor, 5-hydroxytryptophan (5-HTP) administration blocked ovulation.[2] The monoamine oxidase inhibitor and 5-HTP had to be given for 2 consecutive days to be effective. In contrast, the LH surges that are induced in ovariectomized rats by estrogen administration were shown to be blocked by *p*-chlorophenylalanine (*p*-CPA) injection, provided that *p*-CPA was given 24 hr before the surge of LH was to occur.[3] Kordon[4] has proposed that the inhibitory effects of serotonin on LH release result from the action of serotonin on neurons in the medial-basal hypothalamus, while serotonin acts on more rostral structures to facilitate the surges of LH. Thus, the exact nature of serotonergic involvement in LH control is not clear and remains to be determined.

Stimulation of prolactin release was observed after intracerebroventricular injection of serotonin, after administration of 5-HTP[6] and treatment with serotonin agonists.[7] The estrogen-induced surges of serum prolactin can be blocked by *p*-CPA or methysergide,[8] and the elevation of serum prolactin observed after the suckling stimulus in lactating rats can be blocked by *p*-CPA[9] or methysergide.[10] Most of the evidence, therefore, is in favor of the existence of a serotonin pathway stimulatory to prolactin release.

Evidence that growth hormone (GH) in the rat is stimulated via serotonergic pathways has been reported.[11, 12] In the human, 5-HTP has been shown to elevate GH[13] or have no effect.[14] In humans with serotonin-secreting tumors, GH levels are elevated,[15] and, furthermore, GH release can be blocked by treatment with serotonin antagonists.[14, 17]

There have been few studies on the effects of serotonin neurotoxins on pituitary hormone release. In prepubertal male rats, intraventricular administration of 5,6-dihydroxytryptamine (5,6-DHT) appeared to retard sexual maturation,[18] whereas another report indicated that administration of 50 μg of 5,6-DHT caused massive increases in serum gonadotropins.[19] We therefore decided to pursue further the role of serotonin in the control of pituitary hormone release by use of the serotonin neurotoxins as tools.

Materials and Methods

In the following studies, *p*-chloroamphetamine hydrochloride (*p*-CA) and 5,7-dihydroxytryptamine creatinine sulfate (5,7-DHT) were purchased from Regis Chemical Company and were used as the serotonin neurotoxins. In most

0077-8923/78/0305-0399 $01.75/0

studies, *p*-CA was dissolved in saline and injected by the intraperitoneal route. In one study, the drug was injected into the lateral ventricle in a volume of 5 μl over a period of 2 min. To administer 5,7-DHT, the rats were anesthetized with sodium pentobarbital, and 200 μg of 5,7-DHT dissolved in 15 μl of 0.5% ascorbic acid-saline were injected slowly into the lateral ventricle over a period of 5 min. All rats were pretreated with 25 mg/kg of desmethylimipramine 3 hr and again 1 hr before 5,7-DHT injection.

At the end of each study, the rats were decapitated, and the blood was collected for radioimmunoassay of LH, prolactin, or GH. Radioimmunoassay kits provided by the National Institute of Arthritis, Metabolism, and Digestive Diseases (NIAMDD) were used to assay serum GH and prolactin; LH was assayed by use of antiserum provided by Dr. Gordon Niswender, and purified prolactin for iodination was provided by Dr. L. E. Reichert, Jr. Serum levels of hormones were expressed in terms of the reference standards for LH, prolactin, and GH supplied by the NIAMDD. Brain serotonin and 5-hydroxyindoleacetic acid levels were determined by a spectrofluorometric assay.[20] Student's *t* test was used as a measure of statistical significance.

Animals were kept under controlled laboratory conditions and provided food and water ad lib. The animal rooms were kept on a 14/10-hr light/dark cycle, with lights on at 4 AM.

Results and Discussion

Effect of p-*Chloroamphetamine on Luteinizing Hormone, Prolactin, and Growth Hormone Release in Male Rats*

The effects of *p*-CA on LH, prolactin, and GH secretion in rats were determined. In the first experiment, 600 μg of *p*-CA were given by injection into the lateral ventricle. *p*-Chloroamphetamine was dissolved in 0.5% ascorbic acid and injected in a volume of 5 μl over a period of 2 min. Control animals received injections of the ascorbic acid vehicle. Rats were killed by decapitation 4 days after treatment, and the blood was collected and assayed by radioimmunoassay for LH, prolactin, and GH. The brains were removed and frozen on dry ice and assayed for serotonin and 5-hydroxyindoleacetic acid (5-HIAA). Intraventricular administration of *p*-CA lowered significantly brain serotonin and 5-HIAA but had little effect on serum levels of LH, GH, and prolactin (Table 1). Serum prolactin levels were significantly lower in the first experiment, but in the second experiment, as in subsequent experiments, no change was observed. Thus, the neuronal damage that resulted from *p*-CA treatment did not alter basal levels of LH, prolactin, and GH in male rats.

Effect of p-*CA on the Proestrous Surges of Prolactin and Luteinizing Hormone in Female Rats*

Although *p*-CA treatment had no effect on basal levels of LH, prolactin, and GH, serotonin has been implicated in prolactin and LH surges in female rats. Therefore, it was of interest to examine the influence of *p*-CA on the proestrous surge of LH and prolactin. Adult female Sprague-Dawley rats that showed normal estrous cycles were selected for experimentation. All rats re-

TABLE 1

EFFECT OF *p*-CA ON SERUM LEVELS OF LH, PROLACTIN, AND GH IN MALE RATS

Treatment	Serum LH Levels (ng/ml)	Serum Prolactin Levels (ng/ml)	Serum GH Levels (ng/ml)	Serotonin Concentration (μg/g)	5-HIAA Concentration (μg/g)
Experiment 1					
Control, ascorbic acid vehicle (5)*	27.1±7.5	23.1±4.9	23.7±9.1	0.66±0.03	0.44±0.01
p-CA, μg (5)	18.4±1.8	11.2±1.7	18.3±7.6	0.43±0.02	0.29±0.01
		($p < 0.05$)		($p < 0.001$)	($p < 0.001$)
Experiment 2					
Control, ascorbic acid vehicle (5)	14.3±1.3	14.3±1.9	17.6±8.1	—	—
p-CA, 600 μg (5)	15.9±0.3	13.2±2.6	14.4±7.6	—	—

* Numbers of rats are shown in parentheses.

ceived an intraperitoneal injection of 5 mg/kg of *p*-CA on the day of estrus and were killed by decapitation on the following proestrous day (3–4 days later), during which time surges normally occur. Serum was collected and assayed for prolactin and LH by radioimmunoassay. The surges of LH and prolactin were not altered by *p*-CA administration (TABLE 2). In view of the compelling evidence that serotonin plays a role in the control of pituitary hormone release, it appears that either not enough serotonin neurons were destroyed by *p*-CA to cause a deficit in serotonergic neurotransmission or that the neurons destroyed by *p*-CA are not involved in pituitary control.

Effect of 5,7-Dihydroxytryptamine on Prolactin and Luteinizing Hormone Release

Administration of 5,7-dihydroxytryptamine (5,7-DHT) in rats pretreated with desmethylimipramine (DMI) has been shown to selectively destroy serotonin neurons. Quite a large reduction in brain serotonin concentration can be achieved by use of this method.[21] We decided to employ this approach to examine further the serotonergic control of LH and prolactin release.

In our animal colony, a significant rise in serum prolactin occurs in female rats on the afternoon of estrus. This surge is usually of a lower magnitude than is the proestrous surge. The exact meaning of this estrous diurnal rise is not clear. Perhaps it is a diurnal rhythm that is only present on the days of proestrus and estrus or perhaps it is stress related. The animals may be more susceptible to stress on the afternoon of estrus than at other times. The only stress involved would be our presence in the animal room and handling the rat prior to decapitation. An attempt is always made to keep this time to a minimum. TABLE 3 shows a comparison between morning and afternoon levels on the day of estrus. A highly significant rise in serum prolactin was observed at 6 PM compared to 8 AM. A small rise was seen in serum LH levels; however, the LH rise is variable and not seen repeatedly. The estrous afternoon rise in serum prolactin was obliterated by 5,7-DHT treatment (TABLE 4). The rats were decapitated at 6 PM on the afternoon of estrus between 21 and 24 days after intraventricular injection of 200 μg of 5,7-DHT. Brain norepinephrine levels were unchanged, but brain serotonin levels were markedly reduced. No effect was seen on serum levels of LH. Inhibition of the estrous afternoon rise of serum prolactin was one of the few effects observed after administration of 5,7-DHT. If 5,7-DHT were exerting its effects through destruction of serotonin neurons, we should see the same inhibition after treatment with a serotonin antagonist or synthesis inhibitor. We chose to treat rats on the day of estrus

TABLE 2

EFFECT OF *p*-CA ON THE PROESTROUS SURGES OF LH AND PROLACTIN

Treatment	Number of Rats	Serum LH Levels (ng/ml)	Serum Prolactin Levels (ng/ml)
Control, saline	10	380.5±80.2	281.0±67.5
p-CA, 5 mg/kg	10	362.1±74.6	257.1±54.3

TABLE 3

MORNING AND AFTERNOON VALUES FOR PROLACTIN AND LH ON THE DAY OF VAGINAL ESTRUS

Time *	Serum Prolactin Levels (ng/ml)	Serum LH Levels (ng/ml)
8 AM (10)†	25.8±5.1	10.5±1.0
3 PM (10)	247.4±34.9	19.9±4.1
	($p < 0.001$)	

* The animal room is on a 14-hr light/10-hr dark cycle, with lights on at 4 AM and off at 6 PM.

† Numbers of rats are shown in parentheses.

with methysergide to ascertain whether the drug was capable of exerting an action similar to that of 5,7-DHT, namely, inhibition of the estrous rise of serum prolactin. Adult female rats that showed normal 4-day estrous cycles were chosen for experimentation on the day of estrus. The control group received an intraperitonal injection of saline at 2 PM; the treated group received an injection of 15 mg/kg of methysergide in saline at 2 PM. The rats were decapitated starting at 4 PM, and the serum was collected and assayed for prolactin. Animals treated with methysergide exhibited a highly significant reduction in serum prolactin levels (TABLE 5). This finding supports the hypothesis that 5,7-DHT reduces serum prolactin levels on the day of estrus through destruction of serotonin neurons.

Effect of 5,7-DHT on the Proestrous Surges of Luteinizing Hormone and Prolactin

Kordon[4] had emphasized the importance of serotonin in induction of LH surges observed after estrogen treatment. It is believed that the surges of LH and prolactin seen after daily estrogen treatment in ovariectomized animals are identical in nature to the surges of LH and prolactin observed on the afternoon of proestrus. In this experiment, we wanted to determine whether 5,7-DHT administration would alter these surges, even though we saw no effect after *p*-CA treatment. Adult cycling female rats were pretreated with DMI (25 mg/kg) 3 hr and again 1 hr before 5,7-DHT (200 μg) administration. Vaginal smears were recorded after treatment, and all animals were found to show normal estrous cycles. Rats were decapitated at 4 PM on proestrus about 28 days after 5,7-DHT treatment, and the blood was collected for radioimmunoassay of prolactin and LH. As shown in TABLE 6, there was no inhibition of the prolactin surge; in fact, the LH surge appeared to be enhanced, although the difference was not statistically significant. Although serotonin has been reported to be implicated in these surges, destruction of serotonin neurons by 5,7-DHT treatment did not alter these surges, as was the case with *p*-CA treatment. Despite a 66% reduction in brain serotonin, enough serotonin neurons may have survived to maintain relatively normal endocrine function. However, the large amount of LH released suggests that 5,7-DHT treatment removed a

TABLE 4

EFFECTS OF 5,7-DHT ON SERUM LEVELS OF PROLACTIN AND LH ON THE AFTERNOON OF ESTRUS

Treatment	Serum Prolactin Levels (ng/ml)	Serum LH Levels (ng/ml)	Brain Norepinephrine Levels ($\mu g/g$)	Brain Serotonin Levels ($\mu g/g$)
Control, Ascorbic acid * (6) †	134.9 ± 46.0	15.1 ± 3.4	0.53 ± 0.03	0.72 ± 0.02
5,7-DHT (7)	28.4 ± 7.6 ($p < 0.05$)	12.2 ± 2.9	0.51 ± 0.02	0.27 ± 0.01 ($p < 0.001$)

* Ascorbic acid was the vehicle for 5,7-DHT.
† Numbers of rats are shown in parentheses.

TABLE 6

EFFECT OF 5,7-DHT ON THE PROESTROUS SURGES OF LH AND PROLACTIN

Treatment	Serum LH Levels (ng/ml)	Serum Prolactin Levels (ng/ml)	Brain Serotonin Levels ($\mu g/g$)	Brain Norepinephrine Levels ($\mu g/g$)
Control ascorbic acid	282.4 ± 77.7	222.3 ± 11.9	0.77 ± 0.02	0.52 ± 0.03
5,7-DHT (200 μg of free base)	410.0 ± 63.0	192.1 ± 16.4	0.26 ± 0.01 ($p < 0.001$)	0.58 ± 0.02

TABLE 5

EFFECT OF METHYSERGIDE ON SERUM PROLACTIN LEVELS ON THE AFTERNOON OF ESTRUS

Treatment	Number of Rats	Serum Prolactin Levels (ng/ml)
Saline	10	143.2±32.9
Methysergide (15 mg/kg)	10	35.2± 6.6 ($p < 0.01$)

serotonergic inhibitory influence on LH release. Another alternative is that serotonin may play no role in the proestrous release of LH and prolactin. The rise of serum prolactin levels on the afternoon of estrus certainly appears to depend on a set of serotonin neurons different from those involved in the proestrous prolactin surge, because the estrous rise is obliterated by 5,7-DHT injection.

Effect of 5,7-DHT on the Suckling-Induced Prolactin Rise

Adult female Sprague-Dawley rats received intraventricular injections of 5,7-DHT and were mated with males 2 weeks later. Controls received intraventricular injections of the vehicle. All rats in both groups became pregnant; however, many of the 5,7-DHT-treated rats experienced difficulty during delivery of their pups. Some litters were dead, and some females did not deliver any pups. Rats with normal litters were selected for the experiment. Litter size was reduced to six per female, and the pups were separated from the mother for a period of 4 hr, after which time they were returned to their mothers. After 15 min of nursing, the mothers were decapitated, and the serum was assayed for prolactin. It is evident from TABLE 7 that reduction of brain serotonin did not impair prolactin release in response to the suckling stimulus. Earlier studies [9, 10] demonstrated that *p*-CPA or methysergide was able to block the suckling-induced rise in prolactin levels. This finding is an example of a situation where a synthesis inhibitor or a receptor blocker exerted an effect that

TABLE 7

EFFECTS OF 5,7-DHT ON THE SUCKLING-INDUCED RISE IN PROLACTIN

Treatment	Serum Prolactin Levels (ng/ml)	Brain Serotonin Levels (μg/g)
Control, ascorbic acid (3)*	164.0±26.3	0.75±0.02
5,7-DHT (5)	150.4±19.1	0.39±0.04 ($p < 0.01$)

* Numbers of rats are shown in parentheses.

could not be obtained with a serotonin neurotoxin. Apparently, *p*-CPA or methysergide is more effective in blocking serotonergic function than are the serotonin neurotoxins; however, there is not complete agreement on the specificity of *p*-CPA or on the ability of methysergide to block central serotonin receptors.

Effects of 5-Hydroxytryptophan and Fluoxetine on Prolactin Release in 5,7-DHT-Treated Rats

Earlier, we demonstrated that 5-HTP (30 mg/kg) was a potent releaser of prolactin in rats pretreated with the serotonin uptake inhibitor fluoxetine.[7] Fluoxetine and 5-HTP demonstrated a marked inability to release prolactin when given separately. We decided to examine the effects of a combination of fluoxetine and 5-HTP on prolactin release in diestrous female rats. Adult female Sprague-Dawley rats received intraventricular injections of 5,7-DHT or ascorbic acid vehicle. Daily vaginal smears were recorded, and after 2 weeks had passed, animals were selected for treatment on the first day of diestrus. The 5,7-DHT-treated and control groups were further subdivided into two groups. One subgroup of control and 5,7-DHT-treated rats received an intraperitoneal injection of saline, followed 1 hr later by another injection of saline. The other subgroup received 10 mg/kg of fluoxetine intraperitoneally, followed 1 hr later by 30 mg/kg of intraperitoneal 5-HTP. Rats were killed by decapitation 45 min after the last injection. Hypothalamic concentrations of serotonin and norepinephrine and the concentrations of serotonin and norepinephrine in the remainder of the brain were measured. The results are shown in TABLE 8. Rats treated with 5,7-DHT displayed a significantly greater elevation of serum prolactin after treatment with fluoxetine and 5-HTP than did the control group. These rats appeared to be hypersensitive to the serotonergic stimulus. It was also evident that the reduction of serotonin observed in the hypothalamus after 5,7-DHT treatment is much less than that seen in the remainder of the brain. Perhaps the hypothalamus contains neurons that are more resistant to the action of 5,7-DHT than are neurons found in other brain areas. In the 5,7-DHT-lesioned rats, 5-HTP increased serotonin to a much greater extent in the hypothalamus than in the rest of the brain. This observation indicates that there were more serotonin neurons in the hypothalamus capable of synthesizing serotonin. In each group, there was no change in total brain norepinephrine; however, there was a small, barely significant decrease in the hypothalamic norepinephrine concentration after fluoxetine and 5-HTP combination treatment in the 5,7-DHT-pretreated group.

Effect of 5-Hydroxytryptophan Administration on Prolactin Release in 5,7-DHT- and p-*CA-Treated Rats*

Recent studies have shown that animals treated with 5,7-DHT may develop a serotonin denervation hypersensitivity similar to that produced for dopamine neurons after 6-hydroxydopamine administration.[22, 23] Trulson *et al.*[22] reported that 5,7-DHT-treated rats became supersensensitive to the lysergic acid diethylamide-induced behavioral syndrome. Stewart *et al.*[23] reported an increase in sensitivity to 5-HTP-induced myoclonus after 5,7-DHT administra-

TABLE 8

EFFECTS OF 5-HTP AND FLUOXETINE ON SERUM PROLACTIN LEVELS AND BRAIN AND HYPOTHALAMIC SEROTONIN AND NOREPINEPHRINE CONCENTRATIONS

Group and Treatment			Serum Prolactin Levels (ng/ml)	Hypothalamic Serotonin Levels (μg/g)	Hypothalamic Norepinephrine Levels (μg/g)	Brain Serotonin Levels (μg/g)	Brain Norepinephrine Levels (μg/g)
	Time 0	1 hr					
Control	Saline	Saline	25.5±4.3 *	0.69±0.03 *	1.33±0.1 *	0.60±0.02 *	0.40±0.02 *
	Fluoxetine (10 mg/kg)	5-HTP (30 mg/kg)	138.3±21.5 †	2.64±0.1 †	1.17±0.1 *	2.46±0.2 †	0.38±0.01 *
5,7-DHT	Saline	Saline	33.7±5.2 *	0.45±0.03 ‡	1.82±0.3 †	0.23±0.02 ‡	0.38±0.02 *
	Fluoxetine (10 mg/kg)	5-HTP (30 mg/kg)	228.3±15.9 ‡	2.60±0.1 †	1.58±0.05 ‡	1.88±0.1 §	0.39±0.02 *

*†‡§ Values with different footnote symbols are statistically significantly different from each other ($p < 0.01$).

TABLE 9

EFFECT OF 5-HTP ON PROLACTIN RELEASE IN RATS PRETREATED WITH 5,7-DHT

Group and Treatment	Number of Rats	Serum Prolactin Levels (ng/ml)
Control, ascorbic acid plus saline	6	44.2±5.9 *
Control, ascorbic acid plus 30 mg/kg of 5-HTP	6	98.05±13.1 †
5,7-DHT-treated plus saline	6	44.0±4.7 *
5,7-DHT-treated plus 30 mg/kg of 5-HTP	6	261.0±45.5 ‡

*†‡ Values with different footnote symbols are statistically significantly different from each other ($p < 0.01$).

tion in rats. These studies and our above results on a hypersensitivity to 5-HTP in fluoxetine-pretreated 5,7-DHT-lesioned rats prompted us to examine the effect of 5-HTP alone in male rats pretreated with 5,7-DHT and *p*-CA. Adult male Sprague-Dawley rats were divided into three groups. The first group received a lateral ventricular injection of ascorbic acid vehicle, the second group underwent intraventricular injection of saline, and the third group was injected intraperitoneally with 10 mg/kg of *p*-CA. Rats were treated with 30 mg/kg of 5-HTP or saline on the 14th day after 5,7-DHT or *p*-CA injection. All rats were killed by decapitation 45 min after injection of 5-HTP. Serum was collected and assayed for prolactin. TABLE 9 shows the results of 5-HTP administration to 5,7-DHT-lesioned rats. In the controls, 5-HTP injection caused only a modest rise of serum prolactin. In the rats treated with 5,7-DHT, significantly more prolactin was released than in the controls treated with 5-HTP.

The results of 5-HTP administration to rats treated earlier with *p*-CA are shown in TABLE 10. In the rats treated with *p*-CA, there was a rise in serum prolactin after 5-HTP injection that was statistically significantly greater than the rise seen in control rats after 5-HTP. Thus, rats treated with 5,7-DHT or *p*-CA appear to develop a serotonin denervation hypersensitivity, according to which they demonstrate an exaggerated release of prolactin after administration of 5-HTP. The hypersensitivity appeared to be greater after treatment with 5,7-DHT than *p*-CA. Furthermore, the increased sensitivity to 5-HTP after

TABLE 10

EFFECT OF 5-HTP ON PROLACTIN RELEASE IN RATS PRETREATED WITH *p*-CA

Group and Treatment	Number of Rats	Serum Prolactin Levels (ng/ml)
Control plus saline	9	37.1±3.8 *
Control plus 30 mg/kg of 5-HTP	10	79.5±10.1 †
p-CA-treated plus saline	8	28.8±3.0 *
p-CA-treated plus 30 mg/kg of 5-HTP	9	148.34±21.2 ‡

*†‡ Values with different footnote symbols are statistically significantly different from each other ($p < 0.01$).

selective destruction of serotonin neurons provides evidence that the dose of 5-HTP used in this study to increase prolactin release is acting via serotonergic neurons.

Conclusions

The results of this study indicate that *p*-CA and 5,7-DHT have little influence on basal serum levels of LH, prolactin, and GH. There are aspects of prolactin release that *p*-CPA and methysergide can block that the serotonin neurotoxins, however, cannot block. This difference is presumably due to incomplete destruction of serotonin neurons and higher resistance of hypothalamic serotonin neurons to the serotonin neurotoxins. The elevation of serum prolactin levels observed on the afternoon of estrus is totally abolished by treatment with 5,7-DHT. This finding suggests that the serotonin pathway involved in the estrous afternoon rise is destroyed, perhaps completely, by this serotonin neurotoxin. In addition, after treatment with serotonin neurotoxins, especially 5,7-DHT, a serotonin denervation hypersensitivity is produced, and animals become hypersensitive to the prolactin-releasing effects of 5-HTP. The serotonin denervation hypersensitivity produced by 5,7-DHT should prove useful in studying the influence of serotonin on several other neuroendocrine parameters.

Acknowledgments

Materials for radioimmunoassays were provided by the National Institute of Arthritis, Metabolism, and Digestive Diseases, Rat Pituitary Hormone Distribution Program. We thank Dr. Gordon Niswender for luteinizing hormone antiserum, Dr. L. E. Reichert, Jr. for purified ovine luteinizing hormone for iodination, and Dr. Ray Fuller for providing the assay for serotonin and norepinephrine.

References

1. Schneider, H. P. G. & S. M. McCann. 1970. Endocrinology **86:** 1127–1133.
2. Labhsetwar, A. P. 1972. J. Endocrinol. **52:** 269–275.
3. Hery, M., E. Laplante & C. Kordon. 1976. Endocrinology **99:** 496–503.
4. Kordon, C. 1976. *In* Hypothalamus and Endocrine Functions. F. Labrie, J. Meites & G. Pelletier, Eds. : 51–61. Plenum Publishing Corporation. New York, N.Y.
5. Kamberi, I., R. S. Mical & J. C. Porter. 1971. Endocrinology **88:** 1288–1293.
6. Mueller, G. P., C. P. Twohy, H. T. Chen, J. P. Advis & J. Meites. 1976. Life Sci. **18:** 715–724.
7. Clemens, J. A. 1975. *In* Hypothalamus and Endocrine Functions. F. Labrie, J. Meites & G. Pelletier, Eds. : 283–302. Plenum Publishing Corporation. New York, N.Y.
8. Caligaris, L. & S. Taleisnik. 1974. J. Endocrinol. **62:** 25–33.
9. Kordon, C., C. Blake, J. Terkel & C. Sawyer. 1973/4. Neuroendocrinology **13:** 213–217.
10. Gallo, R. V., J. Rabbi & G. Moberg. 1975. Endocrinology **97:** 1096–1105.

11. Collu, R., F. Fraschini, P. Visconti & L. Martini. 1972. Endocrinology **90:** 1231–1237.
12. Smythe, G. A. & L. Lazarus. 1973. Nature (London) **244:** 230–231.
13. Imura, H., Y. Nakai & T. Yoshimi. 1973. J. Clin. Endocrinol. Metab. **36:** 204–206.
14. Handwerger, S., J. Plonk, H. Lebovitz, C. H. Bivens & J. Feldman. 1975. Hormone Metab. Res. **7:** 214–216.
15. Feldman, J., J. Plonk, C. Bivens, H. Lebovitz & S. Hanwerger. 1975. Amer. J. Med. Sci. **269:** 333–347.
16. Nakai, Y., H. Imura, H. Sakurai, H. Kurahachi & T. Yoshimi. 1974. J. Clin. Endocrinol. Metab. **38:** 446–449.
17. Bivens, C., H. Lebovitz & J. Feldman. 1973. N. Engl. J. Med. **289:** 236–239.
18. Collu, R., J.-C. Jéquier, J. Letarte, G. Leboeuf & J. R. Ducharme. 1974. Neuroendocrinology **14:** 139–150.
19. Ladosky, W. & J. Noronha. 1974. J. Endocrinol. **62:** 677–678.
20. Maickel, R. P., R. Cox, Jr., J. Saillant & F. Miller. 1968. Int. J. Neuropharmacol. **7:** 275–281.
21. Gerson, S. & R. Baldessarini. 1975. Brain Res. **85:** 140–145.
22. Trulson, M., C. Ross & B. Jacobs. 1976. Psychopharmacol. Commun. **2:** 149–164.
23. Stewart, R., J. Growdon, D. Cancian & R. Baldessarini. 1976. Neuropharmacology **15:** 449–455.

Discussion

Dr. R. M. Stewart: I thought it was particularly instructive that 5-HTP induced a rise of prolactin in both *p*-CA- and 5,7-DHT-pretreated animals but that associated movements were absent in the *p*-CA-treated group. This would eliminate the stress reaction as the origin of prolactin release.

Dr. J. Meyer (*Rockefeller University, New York, N.Y.*): Have you tried to block the 5-HTP effects with carbidopa or some other peripheral decarboxylase inhibitor to ensure that the effects are occurring centrally?

Dr. Clemens: No.

Dr. W. Wuttke: Do you think that hypothalamic 5-HT neurons or terminals are more resistant to the action of the neurotoxins? What is the evidence for serotonergic perikarya in the hypothalamus? Why do you sometimes see a rise of prolactin with 5-HTP, whereas other times you do not?

Dr. Clemens: The evidence for cell bodies or neurons in the basal hypothalamus is weak; perhaps at this time one should only say that there are hypersensitive serotonin terminals. While most evidence indicates that such innervation is from ascending structures, some recent reports demonstrate the presence of cell bodies in the basal hypothalamus. We are unclear about the inconsistent effect of 5-HTP on prolactin release; perhaps this effect is due to a change of sensitivity in the animals.

ENDOCRINE EFFECTS OF BRAIN SEROTONIN DEPLETION BY 5,6-DIHYDROXYTRYPTAMINE IN PREPUBERTAL MALE RATS *

Robert Collu

Pediatric Research Center
Sainte-Justine Hospital
and University of Montreal
Montreal, Quebec H3T 1C5, Canada

Brain serotonin [5-hydroxytryptamine (5-HT)] has been demonstrated to play a role in the regulation of several bodily functions, such as sleep,[1] temperature,[2] and mental activity.[3] In addition, more recent data have clearly indicated that the secretion of several anterior pituitary hormones is modulated by this biogenic amine.[4] In particular, it has been reported that brain 5-HT activates the secretion of ACTH,[5-7] prolactin,[8-10] and growth hormone [11-14] (GH) and inhibits the secretion of gonadotropins.[15, 16]

The present work was undertaken to evaluate the endocrine effects that result from depletion of brain 5-HT content in prepubertal male rats by treatment with 5,6-dihydroxytryptamine [17] (5,6-DHT).

MATERIALS AND METHODS

Experiment I

Male Sprague-Dawley rats were injected intraventricularly at 21 and 23 days of age, under ether anesthesia, with either 50 μg of 5,6-DHT creatinine sulfate (Regis Chemical Company; calculated as the free base) or with the diluent (20 μl of 0.9% sodium chloride solution that contained 0.1 mg of ascorbic acid per milliliter). The rats injected with the diluent (controls) were either fed ad lib. or pair fed. Pair feeding was performed by daily measurement of the food intake of 5,6-DHT-treated rats and providing pair-fed animals with the same amount of rat pellets consumed by experimental animals during the preceding 24 hr. All animals had free access to water and were housed in a temperature- (22° C) and light-controlled room (14 hr of light and 10 hr of darkness). Groups of rats were sacrificed in the morning by decapitation 1, 2, and 4 weeks after the first injection.

Body weight, tail length, and the weights of brain, anterior pituitary, adrenals, testes, prostate, and seminal vesicles were recorded. The brains were stored at −20° C until determination of the levels of 5-HT noradrenaline (NA), and dopamine (DA). These determinations were performed on the same sample, after extraction in acidified butanol,[18] by the method of Maickel *et al.*,[19] as modified by Thompson *et al.*[20] for 5-HT, and by the method of Ansell and Beeson [18] for NA and DA. Fluorescence values were measured with an Aminco-Bowman spectrophotofluorometer. Trunk blood was collected

* Supported by Grant MA 4691 from the Medical Research Council of Canada.

0077–8923/78/0305–0411 $01.75/0

in heparinized tubes. The plasma was then separated and stored at −20° C until determination of growth hormone (GH) levels with the radioimmunoassay method of Schalch and Reichlin [21] by use of NIAMDD-Rat-GH-RP-1 as the standard preparation. Anterior pituitaries were homogenized in 0.5 ml of 0.01 N NaOH. The homogenate was then centrifuged, and the supernatant was diluted with 1% bovine serum albumin in pH 7.6 phosphosaline buffer and stored at −20° C until determination of GH levels by radioimmunoassay. All GH determinations were performed on a single batch.

Experiment II

Male Sprague-Dawley rats were treated, fed, and housed as in Experiment I. One, 2, and 4 weeks after the first injection, groups of rats were sacrificed by decapitation. Body weight, tail length, and organ weights were recorded as in Experiment I. Brain levels of 5-HT, NA, and DA were measured as in Experiment I. Trunk blood was collected in heparinized tubes. The plasma was then separated and stored at −20° C until determination of follicle-stimulating hormone (FSH) levels with the radioimmunoassay method of Parlow *et al.*,[22] by use of NIAMDD-Rat-FSH-RP-1 as the standard preparation. Anterior pituitaries were homogenized in 5 ml of saline, and the supernatant was kept at −20° C until determination of FSH levels by radioimmunoassay. All FSH determinations were performed on a single batch. The testes of the animals sacrificed at 4 weeks were fixed in formalin, sectioned at 4 μm, and stained with hemalum-floxine-saffran for histologic examination.

Statistical Analysis

Statistical evaluation of the numerical data was performed by analysis of variance and individual between-groups nonorthogonal comparisons (Student's *t* test) with a level of significance set at 0.01. Covariance analyses of testicular weights (adjusted for body weights) and total pituitary GH contents (adjusted for body weights) were also performed.

Results

Experiment I

Intraventricular injection of 5,6-DHT had no obvious effect on the behavior of the animals, although food intake was significantly decreased. In effect, 5,6-DHT-treated rats ate a mean amount of 9.4 g/rat/day, as compared to the mean consumption of 16.2 g/rat/day of the control group fed ad lib. Pair-fed rats were given a mean amount of 9.4 g/rat/day. A few animals of the 5,6-DHT- and of the pair-fed control groups died during the first 2 weeks postinjection. As shown in Figure 1, brain 5-HT levels were significantly depleted by the 5,6-DHT treatment and were 40.5 and 50.2% of those of the control group fed ad lib., respectively, 1 and 2 weeks after the first intraventricular injection. Four weeks after treatment, brain 5-HT stores were replenished. Brain levels of NA were not modified by 5,6-DHT treatment,

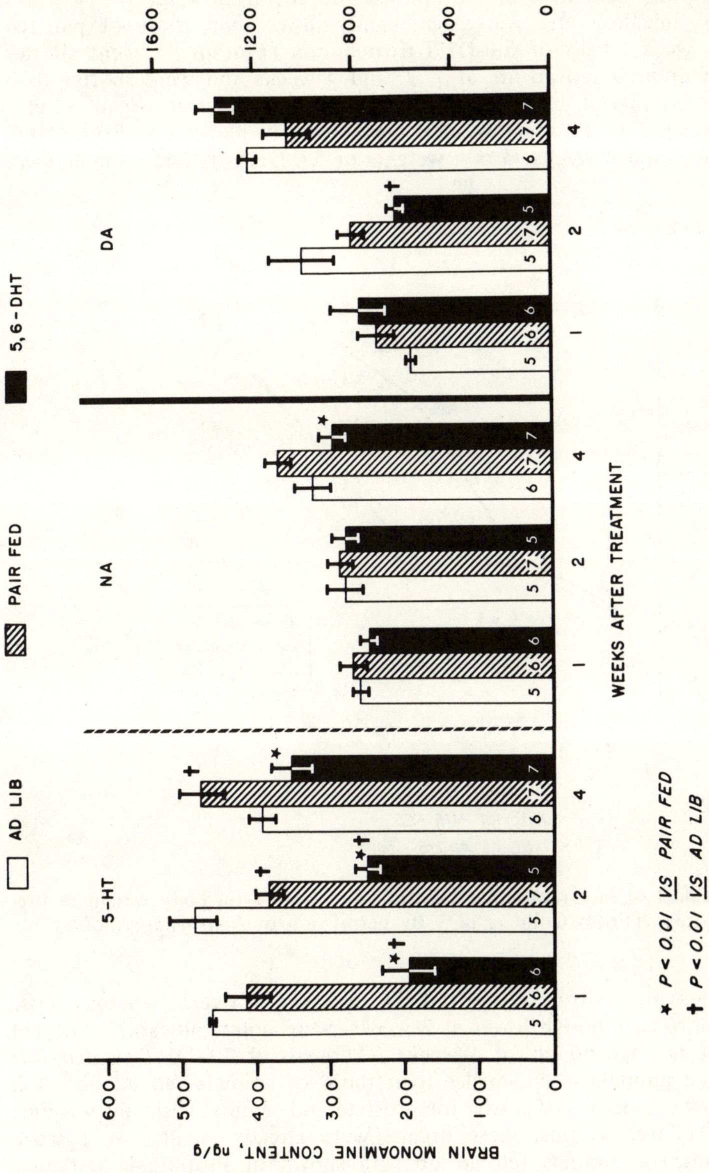

FIGURE 1. Effect of intraventricular injection of 5,6-DHT on brain levels of 5-HT, NA, and DA in prepubertal male rats. (From Collu *et al.*[23] By permission of *Neuroendocrinology*.)

but a significant decrease in DA levels was observed at 2 weeks. The pair-fed group showed a significant decrease of 5-HT levels at 2 weeks and a significant increase at 4 weeks. Body weights of 5,6-DHT-treated rats (FIGURE 2) were significantly lower than those of the animals fed ad lib. at 1, 2, and 4 weeks after the first injection and were significantly lower than those of pair-fed animals at 1 week. Tails of 5,6-DHT-treated rats (FIGURE 3) were shorter than those of animals fed ad lib. at 1, 2, and 4 weeks and were shorter than those of pair-fed rats at 1 and 2 weeks. Brain and endocrine organ weights are given in TABLE 1. Pituitary weights were lower in pair-fed and 5,6-DHT-treated rats at 2 and 4 weeks. Brain weights of 5,6-DHT-treated animals were

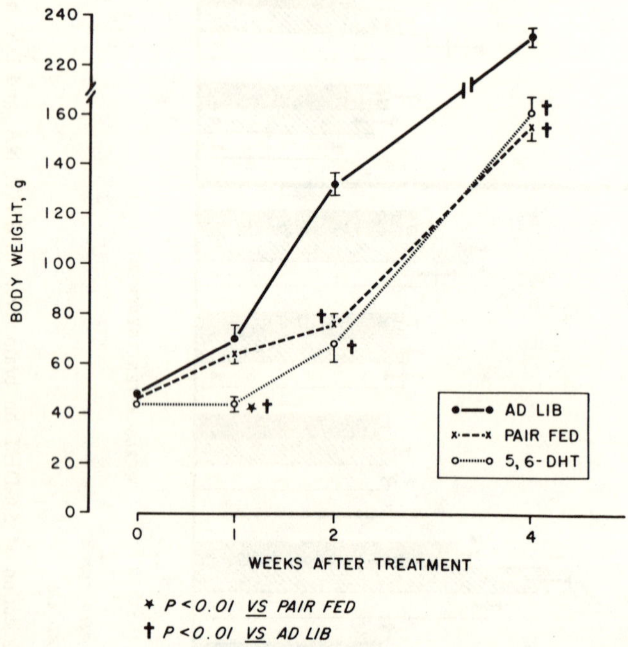

FIGURE 2. Effect of intraventricular injection of 5,6-DHT on body weight in prepubertal male rats. (From Collu *et al.*[23] By permission of *Neuroendocrinology*.)

lower than those of control rats fed ad lib. at 2 and 4 weeks, whereas those of pair-fed animals, although lower at 2 weeks, were not significantly different from those of rats fed ad lib. at 4 weeks. Adrenals of 5,6-DHT-treated rats and of pair-fed animals were smaller than those of animals fed ad lib. at 2 and 4 weeks; the same was also true for prostate and seminal vesicular weights, but in 5,6-DHT-treated rats, these organs were already smaller at 1 week than were those of animals fed ad lib. As shown in FIGURE 4, testicular weights of 5,6-DHT-treated rats were lower than those of rats fed ad lib. at 1, 2, and 4 weeks and were lower than those of pair-fed rats at 2 and 4 weeks. Covariance analysis of testicular weights (corrected for body weights) showed that at 4 weeks, a statistically significant difference ($p < 0.01$) existed

TABLE 1

ENDOCRINE ORGAN WEIGHTS OF MALE RATS INJECTED INTRAVENTRICULARLY AT 21 AND 23 DAYS OF AGE WITH 5,6-DIHYDROXYTRYPTAMINE (5,6-DHT) OR VEHICLE AND SACRIFICED 1, 2, OR 4 WEEKS LATER *

Groups †	Pituitary (mg)	Brain (mg)	Adrenals (mg)	Prostate (mg)	Seminal Vesicles (mg)
One week					
Vehicle-fed ad lib. (5)	2.09±0.07 ‡	1106±13	18.3±1.8	45.8±5.6	18.2±2.5
Vehicle, pair fed (6)	1.57±0.08	1055±27	16.0±1.3	27.5±4.8	10.8±2.0
5,6-DHT (6)	1.35±0.15	995±18	13.3±1.0	12.6±1.9 §	7.6±0.7 §
Two weeks					
Vehicle-fed ad lib. (5)	3.82±0.08	1245±34	26.9±2.7	87.4±3.9	40.3±2.5
Vehicle, pair fed (7)	1.71±0.14 §	1081±17 §	18.1±0.8 §	36.7±6.3 §	19.7±2.3 §
5,6-DHT (5)	1.31±0.14 §	1020±31 §	15.1±1.7 §	14.6±3.5 §	12.0±2.8 §
Four weeks					
Vehicle-fed ad lib. (6)	6.81±0.38	1347±9	35.4±1.5	224.4±19.0	293.7±28.1
Vehicle, pair fed (7)	4.84±0.21 §	1243±21	28.3±1.8 §	119.1±10.9 §	144.0±15.9 §
5,6-DHT (7)	4.54±0.35 §	1182±63 §	27.6±1.2 §	119.7±18.7 §	125.2±24.6 §

* From Collu *et al.*[23]

† Numbers of rats are shown in parentheses.

‡ Values are means ± SE.

§ $p < 0.01$ versus vehicle-fed ad lib.

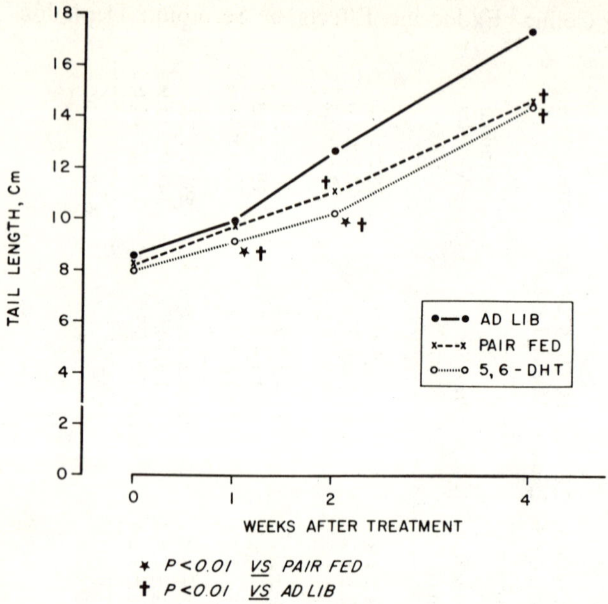

FIGURE 3. Effect of intraventricular injection of 5,6-DHT on tail length in prepubertal male rats. (From Collu *et al.*[23] By permission of *Neuroendocrinology*.)

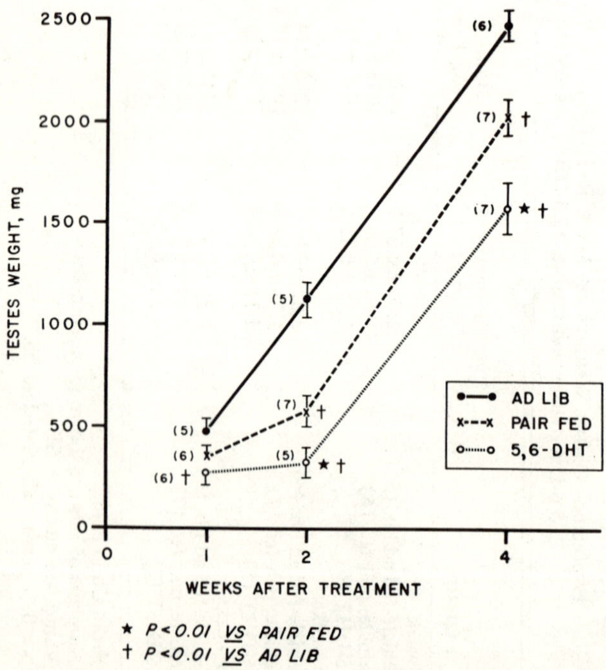

FIGURE 4. Effect of intraventricular injection of 5,6-DHT on testicular weight in prepubertal male rats. (From Collu *et al.*[23] By permission of *Neuroendocrinology*.)

only between 5,6-DHT-treated rats and rats fed ad lib. (ad lib., 1864.8 mg; pair-fed, 2295.0 mg; 5,6-DHT, 1793.5 mg). Plasma GH levels were extremely variable from one animal to the other; the data, entirely inconclusive, are therefore omitted for simplicity. Pituitary GH levels are shown in FIGURE 5. Total content was decreased at 2 weeks in both 5,6-DHT- and pair-fed groups and only in the pair-fed group at 4 weeks. Covariance analysis of total GH contents (corrected for body weights) revealed a statistically significant difference ($p < 0.01$) between rats fed ad lib. and 5,6-DHT-treated and pair-fed animals at 2 weeks (ad lib., 133.78 μg; pair-fed, 68.18 μg; 5,6-DHT, 63.89 μg).

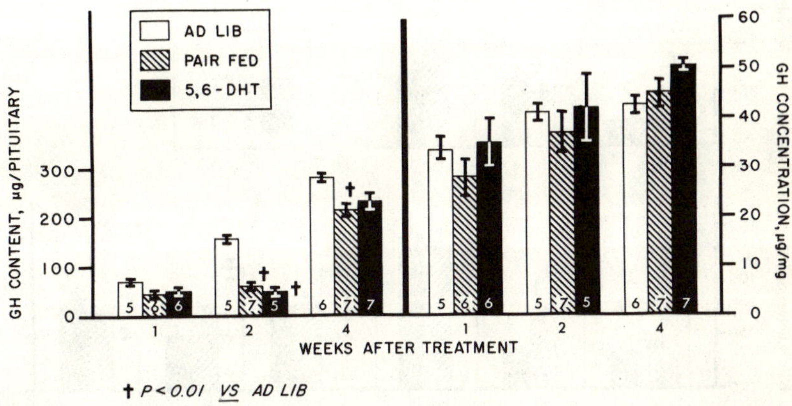

FIGURE 5. Effect of intraventricular injection of 5,6-DHT on pituitary GH levels in prepubertal male rats. (From Collu *et al.*[23] By permission of *Neuroendocrinology*.)

Experiment II

Brain monoamine levels, body and organ weights, and tail lengths of the animals in this experiment were comparable at all times with those of the animals of Experiment I. These data are omitted for simplicity.

Follicle-stimulating hormone levels are shown in FIGURE 6. Plasma hormone levels were undetectable at 1 week in both 5,6-DHT- and pair-fed groups and were significantly lower at 2 weeks in pair-fed rats. Total pituitary hormone content was reduced at 1 and 2 weeks in 5,6-DHT-treated and pair-fed rats, whereas FSH concentration was significantly reduced only in the 5,6-DHT-treated group at 2 weeks and rose to significantly higher levels in the same group at 4 weeks.

Histologic examination of the testes showed that rats fed ad lib. had reached full spermatogenesis 4 weeks after the first injection (51 days of age). Full spermatogenesis was also observed in testes of pair-fed animals, although the tubules were somewhat smaller and the spermatozoa less abundant. Testes of animals treated with 5,6-DHT had much smaller tubules, and spermatogenesis, although reaching the spermatid stage in some tubules, generally did not develop beyond the pachytene spermatocyte stage.

Discussion

Rats injected intraventricularly with 5,6-DHT showed a highly significant depletion of brain 5-HT stores after 1 and 2 weeks; 4 weeks later, the stores were replenished. These results are in accord with those of Baumgarten *et al.*[17] and represent a selective chemical degeneration of indoleamine-containing nerve terminals.[24] A significant decrease of DA levels was also observed at 2 weeks, which was probably not specific, because pair-fed animals also exhibited

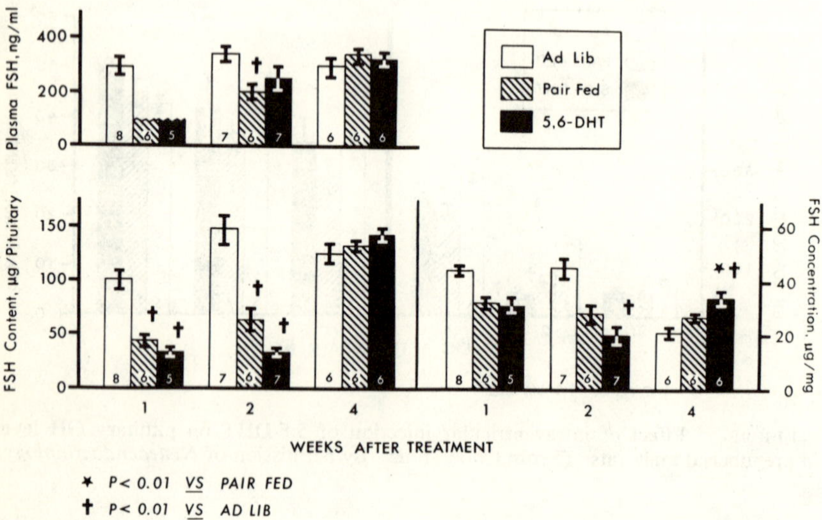

FIGURE 6. Effect of intraventricular injection of 5,6-DHT on plasma and pituitary FSH levels in prepubertal male rats. (From Collu *et al.*[23] By permission of *Neuroendocrinology*.)

a small decrease at the same time, and no statistically significant difference was found between the DA levels of 5,6-DHT- and pair-fed groups.

The depletion of brain 5-HT levels had a strong anorexigenic effect, and 5,6-DHT-treated animals ate on the average 41.8% less food than did rats fed ad lib. This effect has also been reported by Victor *et al.*[25] and suggests that, like catecholamines,[26] 5-HT is involved in the control of ingestive behavior. In contrast with these data, it has recently been reported that hyperphagia and increased growth followed the intraventricular injection of 5,7-DHT in association with desmethylimipramine (DMI), an inhibitor of noradrenaline (NA) uptake, in adult male rats.[27] According to the authors, this phenomenon is observed only when selective depletion of brain 5-HT levels is obtained. However, daily food intake decreased by 24–90% during the first 2 days after administration of 5,7-DHT plus DMI and remained low for several days thereafter. As a consequence, treated animals weighed up to 26% less than did the controls during the first week. Only after at least 50 days did treated rats gain significantly more weight than that shown by controls. Hyperphagia and obesity may therefore be the latter part of a biphasic modification of ingestive behavior induced by brain 5-HT depletion. Decreased protein intake by 5,6-DHT-treated animals is probably responsible for the lower weights of pituitary, adrenals, prostate, and seminal vesicles, because no significant difference was found between 5,6-DHT-treated and pair-fed animals at any posttreatment period. Although decreased protein intake was also probably responsible for lower body weights and smaller tails in 5,6-DHT-treated rats, the existence of statistically significant differences between this group of animals and pair-fed rats, at certain time periods, indicates that an additional, more specific mechanism was also responsible for the impaired growth of 5,6-DHT-treated rats, at least during the first 2 posttreatment weeks. It is interesting to note that this period coincides with the time of brain 5-HT depletion. Recently, it has been demonstrated that this amine is involved in the modulation of GH secretion.[11-14] Pituitary GH contents were significantly reduced at 2 weeks in 5,6-DHT-treated and pair-fed animals. It is therefore difficult to attribute this effect to anything more specific that the well-known pituitary-depleting effect of starvation.[28, 29] However, the differences in growth parameters observed during the first 2 weeks between 5,6-DHT-treated and pair-fed rats could be due to alterations of GH secretion rates induced by the depletion of brain 5-HT stores. In effect, although no statistically significant difference in plasma GH values was found in the present experiment among the three animal groups, published and unpublished data demonstrate clearly that severe depletion of brain 5-HT stores induces low circulating GH levels.[15, 30]

Brain weights of 5,6-DHT-treated *and* of pair-fed rats were significantly reduced at 2 weeks, suggesting that malnutrition was responsible for this effect. However, at 4 weeks, the brains of pair-fed animals were not significantly different from those of rats fed ad lib., indicating, as has already been reported,[31] that malnutrition, when induced after 21 days of age, produces a more pronounced decrease in body weight than in brain weight. On the contrary, at 4 weeks, the brains of 5,6-DHT-treated rats were still significantly smaller than those of animals fed ad lib. This observation is in agreement with the results of Baker and Quay [32] and of Hole,[33] which indicate that 5-HT may be of importance in brain development.

An interesting finding was the very significant decrease in testicular weight caused by 5,6-DHT treatment. This decrease was not due solely to malnutrition, because, at 4 weeks, covariance analysis of testicular weight (corrected for body weight) revealed a statistically significant difference only between 5,6-DHT-treated rats and rats fed ad lib. In addition, at 4 weeks (51 days of age), the testes of control rats fed ad lib. and of pair-fed rats had reached full spermatogenesis, while those of 5,6-DHT-treated animals were still immature. These results suggest that depletion of brain 5-HT stores in the prepubertal period might have prevented sexual maturation through inhibition of the secretion of pituitary gonadotropins, particularly FSH. Although 5-HT has generally been considered to exert an inhibitory influence on FSH secretion,[15, 16] a few published reports suggest that 5-HT may also play an excitatory role.[34, 35] No definite evidence could be found in the present experiment to indicate that the decrease in circulating and pituitary FSH levels was due to anything more specific than malnutrition, because this phenomenon was observed in both 5,6-DHT-treated and in pair-fed rats. However, it is interesting to note that at 2 weeks, only 5,6-DHT-treated rats had significantly reduced pituitary FSH concentrations, which rose to significantly higher values at 4 weeks. This effect could be interpreted as an indication that an additional factor, namely, brain 5-HT depletion, inhibited FSH synthesis and that a rebound phenomenon occurred at the time of 5-HT store repletion.

In conclusion, depletion of brain 5-HT inhibited body and brain growth and sexual maturation partly through the ensuing anorexia and malnutrition, but possibly also through a specific decrease in synthesis and/or secretion of GH and FSH.

Acknowledgments

We thank the National Pituitary Agency of the National Institute of Arthritis, Metabolism, and Digestive Diseases and Dr. A. F. Parlow for the generous gift of materials for the radioimmunoassays of rat GH and FSH. The outstanding technical assistance of Mrs. H. Guillet and the secretarial help of Miss F. Gingras are also gratefully acknowledged.

References

1. Jouvet, M. 1969. Biogenic amines and the states of sleep. Science **163:** 32–41.
2. Beckman, A. L. & J. S. Eisenman. 1970. Microelectrophoresis of biogenic amines on hypothalamic thermosensitive cells. Science **170:** 334–336.
3. Wooley, D. W. & T. Van Der Hoeven. 1965. Serotonin deficiency in infancy as a cause of a mental defect in experimental phenyl ketonuria. Int. J. Neuropsychiat. **1:** 529–544.
4. Collu, R. 1978. Role of central cholinergic and aminergic neurotransmitters in the control of anterior pituitary hormone secretion. *In* Clinical Neuroendocrinology. L. Martini & G. M. Besser, Eds. Academic Press, Inc. New York, N.Y. In press.
5. Krieger, H. P. & D. T. Krieger. 1970. Chemical stimulation of the brain: effect on adrenal corticoid release. Amer. J. Physiol. **218:** 1632–1641.
6. Scapagnini, U., G. P. Moberg, G. R. Van Loon, J. De Groot & W. F. Ganong. 1971. Relation of brain 5-hydroxytryptamine content to the diurnal variation in plasma corticosterone in the rat. Neuroendocrinology **7:** 90–96.
7. Imura, H., Y. Nakai & T. Yoshimi. 1973. Effect of 5-hydroxytryptophan (5-HTP) on growth hormone and ACTH release in man. J. Clin. Endocrinol. Metab. **36:** 204–206.
8. Kamberi, I. A., R. S. Mical & J. C. Porter. 1971. Effects of melatonin and serotonin in the release of FSH and prolactin. Endocrinology **88:** 1288–1293.
9. Lu, K. H. & J. Meites. 1973. Effects of serotonin precursors and melatonin on serum prolactin release in rats. Endocrinology **93:** 152–155.
10. MacIndoe, J. H. & P. W. Turkington. 1973. Stimulation of human prolactin secretion by intravenous infusion of L-tryptophan. J. Clin. Invest. **52:** 1972–1978.
11. Collu, R., F. Fraschini, P. Visconti & L. Martini. 1972. Adrenergic and serotoninergic control of growth hormone secretion in adult male rats. Endocrinology **90:** 1231–1237.
12. Smythe, C. A., J. F. Brandstater & L. Lazarus. 1975. Serotoninergic control of rat growth hormone secretion. Neuroendocrinology **17:** 245–257.
13. Bivens, C. H., H. E. Lebovitz & J. M. Feldman. 1973. Inhibition of hypogly-

cemia-induced growth hormone secretion by the serotonin antagonists cyproheptadine and methysergide. N. Engl. J. Med. **289:** 236–239.

14. COLLU, R., J. C. JEQUIER, J. LETARTE, G. LEBOEUF & J. R. DUCHARME. 1975. L-dopa and pituitary hormone secretion: influence of cyproheptadine. Hormone Metab. Res. **7:** 96–97.
15. COLLU, R. 1975. Role of indolamines in the regulation of growth hormone and gonadotropin secretion. *In* Hypothalamic Hormones. M. Motta, P. G. Crosignani & L. Martini, Eds. : 205–216. Academic Press, Inc. New York, N.Y.
16. KAMBERI, I. 1973. The role of brain monoamines and pineal indoles in the secretion of gonadotrophins and gonadotrophin-releasing factors. Progr. Brain Res. **39:** 261–276.
17. BAUMGARTEN, H. G., A. BJÖRKLUND, L. LACHENMAYER, A. NOBIN & U. STENEVI. 1971. Long-lasting selective depletion of brain serotonin by 5,6-dihydroxytryptamine. Acta Physiol. Scand. Suppl. **373:** 1–15.
18. ANSELL, G. B. & M. F. BEESON. 1968. A rapid and sensitive procedure for the combined assay of noradrenaline, dopamine, and serotonin in a single brain sample. An. Biochem. **23:** 196–206.
19. MAICKEL, R. P., R. H. COX, JR., J. SAILLANT & F. P. MILLER. 1968. A method for the determination of serotonin and norepinephrine in discrete areas of rat brain. Int. J. Neuropharmacol. **7:** 275–281.
20. THOMPSON, J. H., C. A. SPEZIA & H. ANGULO. 1970. Fluorometric detection of serotonin using o-phthaldialdehyde: an improvement. Experientia **26:** 327–329.
21. SCHALCH, D. S. & S. REICHLIN. 1966. Plasma growth hormone concentration in the rat determined by radioimmunoassay: influence of sex, pregnancy, lactation, anesthesia, hypophysectomy and extrasellar pituitary transplants. Endocrinology **79:** 275–280.
22. PARLOW, A. F., T. A. DAANE & A. V. SCHALLY. 1969. Quantitative, differential measurement of rat serum FSH, LH and hypothalamic FSH-RH and LH-RH with specific radioimmunoassays. Progr. 51st Meet. Endocrine Soc. **88:** 83 (Abst.).
23. COLLU, R., J. C. JEQUIER, J. LETARTE, G. LEBOEUF & J. R. DUCHARME. 1974. Endocrine effects of brain serotonin depletion by 5,6-dihydroxytryptamine in prepubertal male rats. Neuroendocrinology **14:** 139–150.
24. BAUMGARTEN, H. G., L. LACHENMAYER & H. G. SCHLOSSBERGER. 1972. Evidence for a degeneration of indoleamine-containing nerve terminals in rat brain, induced by 5,6-dihydroxytryptamine. Z. Zellforsch. **125:** 553–569.
25. VICTOR, S. J., H. G. BAUMGARTEN & W. LOVENBERG. 1974. Depletion of tryptophan hydroxylase by 5,6-dihydroxytryptamine in rat brain. Time course and regional differences. J. Neurochem. **22:** 541–546.
26. GROSSMAN, S. P. 1960. Direct adrenergic and cholinergic stimulation of hypothalamic mechanisms. Science **132:** 301–302.
27. SALLER, C. F. & E. M. STRICKER. 1976. Hyperphagia and increased growth in rats after intraventricular injection of 5,7-dihydroxytryptamine. Science **192:** 385–387.
28. FRIEDMAN, R. C. & S. REICHLIN. 1965. Growth hormone content of the pituitary gland of starved rats. Endocrinology **76:** 787–788.
29. DICKERMAN, E., A. NEGRO-VILAR & J. MEITES. 1969. Effects of starvation on plasma GH activity, pituitary GH and GH-RF levels in the rat. Endocrinology **84:** 814–819.
30. DU RUISSEAU, P., Y. TACHE & R. COLLU. 1978. Role of putative neurotransmitters in PRL, GH and LH response to acute stress in male rats. Neuroendocrinology. In press.
31. DOBBING, J. 1968. Vulnerable periods in developing brain. *In* Applied Neurochemistry. A. N. Davison & J. Dobbing, Eds. : 287–316. Blackwell Scientific Publications. Oxford, England.

32. Baker, P. C. & W. B. Quay. 1969. 5-Hydroxytryptamine metabolism in early embryogenesis, and the development of brain and retinal tissues. A review. Brain Res. **12:** 273–295.
33. Hole, K. 1972. Reduced 5-hydroxyindole synthesis reduces postnatal brain growth in rats. Eur. J. Pharmacol. **18:** 361–366.
34. Fajer, A. B., D. Hoffman & E. Shillito. 1970. Inhibitory effect of p-chlorophenylalanine on the sexual maturation of female rats. J. Reprod. Fertil. **22:** 379–380.
35. Brown, P. S. & L. Fawke. 1972. Effects of reserpine, p-chlorophenylalanine, α-methyltyrosine, thymoxamine or methallibure on pituitary FSH in male rats. J. Reprod. Fertil. **28:** 167–175.

Discussion

Dr. J. Meyer (*Rockefeller University, New York, N.Y.*): Many investigators have shown that rat plasma growth hormone levels are reduced by stress. Could the values that you have reported be due to some nonspecific stress response? Do animals that show low basal growth hormone levels respond to classic challenges with growth hormone surges?

Dr. Collu: Our conditions were well controlled; we had two groups of controls treated exactly the same way as 5,6-DHT-injected animals. We found absolutely no difference among the three groups. The major cause for the decrease of pituitary growth hormone levels was apparently malnutrition.

The rat responds to hypoglycemia in a manner unlike humans; that is, growth hormone release is inhibited in this animal. The only stimuli that effect rat growth hormone release are pentobarbital, morphine, and, as has recently been shown, *met*-enkephalin and β-endorphin.

Dr. H. Tamir: Might not the rhythmic pattern of 5-HT-mediated growth hormone release be altered by manipulating this brain amine? Wouldn't this manipulation obviate the need for taking repeated measurements during the day and night?

Dr. Collu: Absolutely. The plasma growth hormone changes in our experiments are confusing. Dr. Joseph Martin has shown that rat growth hormone is secreted in spontaneous bursts that occur approximately every 3.3 hr. *p*-Chlorophenylalanine abolishes these bursts.

Dr. G. Smith (*Cornell University, Ithaca, N.Y.*): Have you attempted to factor out malnutrition versus a more specific effect of 5-HT by maintaining a normal intake by rats after such treatments?

Dr. Collu: We are going to try that.

EFFECT OF INTRAVENTRICULAR INJECTION OF 5,7-DIHYDROXYTRYPTAMINE ON SERUM GONADOTROPINS AND PROLACTIN *

W. Wuttke, J. L. Hancke, K. G. Höhn, and H. G. Baumgarten

Max Planck Institute for Biophysical Chemistry
D 34 Göttingen-Nikolausberg, Federal Republic of Germany

A rich innervation of limbic and hypothalamic areas with monoaminergic terminals has been demonstrated by means of the histochemical fluorescence technique.[1-3] There is little doubt that mesencephalic noradrenergic fibers that enter the hypothalamus play a key role in stimulating the preovulatory luteinizing hormone (LH) surge.[4-6] The function of the tuberoinfundibular dopaminergic neurons is less clear. Because this system is the primary source of intrahypothalamic dopamine (DA) content,[7] the extrahypothalamic DA systems do not play an important role, if any, in controlling anterior pituitary hormone secretion. The spectrum of experimental findings ranges from inhibitory to stimulatory effects of DA on gonadotropin section.[5, 9-12] Dopamine, however, has a precise role in that it inhibits prolactin secretion at the pituitary.[13, 14] The literature on the function of serotonin is even more confusing. Intraventricular injection of 5-hydroxytryptamine (5-HT) was shown to inhibit gonadotropin release [10] and to stimulate prolactin release in male rats.[15] Inhibition of 5-HT synthesis with *p*-chlorophenylalanine (*p*-CPA) gave inconclusive results, because in some authors hands, such treatment inhibited prolactin or LH release,[16] and in others it did not.[18] Systemic injection of 5-hydroxytryptophan, the immediate precursor of 5-HT, was able to block the preovulatory LH discharge in intact rats, indicating an inhibitory function on LH release.[19] The LH surges induced in castrated rats by estrogen treatment, which are believed to have the same underlying neural mechanisms as those responsible for the preovulatory LH peaks, could be blocked by inhibition of 5-HT synthesis, suggesting a stimulatory function of the neurotransmitter on LH release.[17] Until recently, relatively consistent data were available on the role of 5-HT in prolactin release. There was good evidence that this monoamine stimulates pituitary prolactin release.[15, 20-22] However, Gallo *et al.*,[18] who used different approaches, could not confirm these results; we also demonstrated recently that the stress-induced release of prolactin may be mediated via the noradrenergic system.[6]

These highly confusing results may indicate that 5-HT has dual functions, depending on the animal used for experimentation. It was therefore interesting to investigate the endocrine effects of reducing brain 5-HT content with the serotonin neurotoxin 5,7-dihydroxytryptamine [23, 24] (5,7-DHT).

Materials and Methods

Adult male and female Sprague-Dawley rats that weighed 200–250 g were used in this study. The animals were housed under controlled conditions (light

* Supported by the Deutsche Forschungsgemeinschaft.

0077-8923/78/0305-0423 $01.75/0

on from 5 AM to 7 PM, 25 ± 1° C) and fed ad lib. Four experiments were performed on 107 male, 80 ovariectomized female, and 16 intact female rats. All animals were pretreated with desmethylimipramine (DMI) 30 min prior to intraventricular injection of 5,7-DHT.

Thirty-two male rats each received 100 μg of 5,7-DHT; another 14 were each injected intraventricularly with 150 μg of 5,7-DHT. Forty-six of the remaining male rats served as controls; they were also pretreated with DMI and injected intraventricularly with 20 μl of "artificial" cerebrospinal fluid, the carrier solution for 5,7-DHT. The brains of these animals were quickly removed; the forebrain and brain stem were separated and stored under liquid nitrogen. The dopamine, norepinephrine (NE), and 5-HT contents of each specimen were determined fluorometrically.[25, 26] The rats treated with 100 μg of 5,7-DHT and the appropriate controls were divided into subgroups of eight animals each and decapitated 5, 12, 26, or 55 days after treatment. The other rats treated with 150 μg of 5,7-DHT and their controls were bled by heart puncture under ether anesthesia on Day 5, 12, 26, or 42. Fifteen control rats were rapidly decapitated and served as unstressed controls.

Forty chronically ovariectomized rats were injected intraventricularly with 50 μg of 5,7-DHT after DMI pretreatment; another 40 served as cerebrospinal fluid-treated controls. Subgroups of these animals were implanted with a jugular vein catheter 5, 12, 26, or 55 days after treatment. After recovery from the light ether anesthesia, that is, 3–4 hr later, the animals were heparinized, and consecutive blood samples (100 μl) were drawn every 10 min over a period of 4 hr.

Seven ovariectomized rats treated identically with 5,7-DHT were decapitated in the animal rooms; another seven were ether anesthetized and then decapitated. The blood was quickly removed from the trunk and the brain was frozen. After freezing, the brains were mounted on a cryostat, and 1.0 mm thick sections that included the anterior part of the mediobasal hypothalamus were cut. The left mediobasal hypothalamus (part of the median eminence, the arcuate nucleus, and the ventromedial nucleus) was punched with a sharp hypodermic needle (inner diameter, 1 mm) and placed in 100 μl of 0.1 N $HClO_4$. The right counterpart was also punched and placed in 100 μl of 0.1 N HCl. After homogenization, 10 μl were pipetted for later protein determination, according to the procedure of Lowry *et al.*[27] The rest of the homogenate was centrifuged at 10,000*g*, and the supernatant was frozen. Serotonin was measured by the radioenzymatic hydroxyindole *O*-methyltransferase method,[28] and DA and NE were determined by the radioenzymatic catechol *O*-methyltransferase method.[29, 30] Determinations were made in pairs with internal standards.

Sixteen intact female rats with regular 4- or 5-day estrous cycles were also injected intraventricularly with 50μg of 5,7-DHT after DMI pretreatment, and the regularity of the estrous cycle was checked by reading vaginal smears daily. If vaginal cytologic findings indicated a proestrous day, blood samples were collected by heart puncture from these and from control animals.

The serum or blood content of LH, follicle-stimulating hormone (FSH), and prolactin were measured by specific radioimmunoassays.[31–33] Statistical significance was estimated with Student's *t* test and, where necessary, by an analysis of variance, followed by Duncan's multiple-range test.

Results

Effects of 5,7-Dihydroxytryptamine on Hormones and Monoamines in Male Rats

FIGURE 1 shows serum prolactin, LH, and FSH levels in "mildly stressed" animals. This mild stress implies removal of the animals from the animal

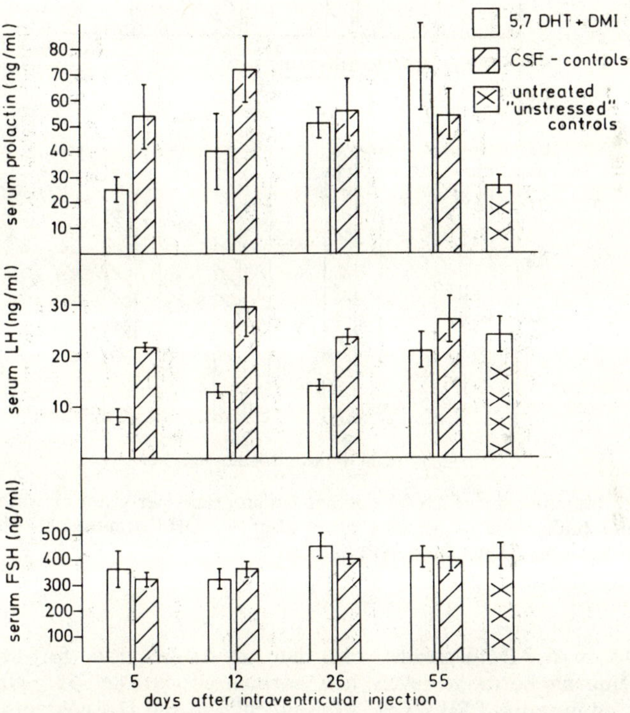

FIGURE 1. Serum prolactin, LH, and FSH levels at various times after 5,7-DHT treatment in "mildly stressed" male rats. Note the low LH levels seen until Day 26 and the slight impairment of prolactin levels due to the stress response at Day 5. Standard error of the mean (SEM) is shown above each bar. Data from Wuttke *et al.*[37]

room into a neighboring room where they were decapitated. The time between entering the room and decapitation ranged from 2 to 3 min. As a response to this stress, serum prolactin levels were elevated in all control animals and also in 5,7-DHT-treated rats (100 μg/rat) from Day 12 after treatment onward when compared with the levels in unstressed control rats. By Day 5, the serum levels were significantly lower ($p < 0.01$) than those in cerebrospinal fluid-treated controls. Serum LH levels were significantly lower

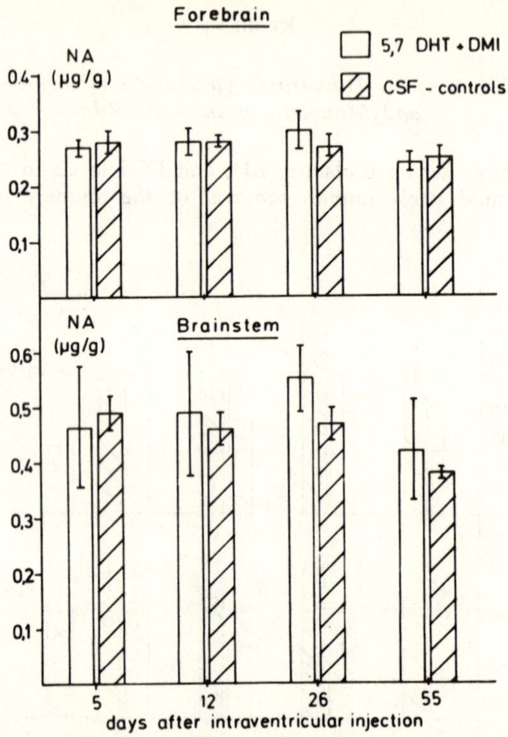

FIGURE 2. Norepinephrine (NA) content (micrograms per gram of wet weight) of forebrain and brain stem at various times after 5,7-DHT treatment. The SEM is shown above each bar. Data from Wuttke *et al.*[37]

($p < 0.001$) in 5,7-DHT-treated rats than in control rats for at least 26 days. Luteinizing hormone levels had normalized by Day 55. No changes were observed in serum FSH levels. Norepinephrine-and DA contents (FIGURES 2 and 3, respectively) of the forebrain and brainstem in 5,7-DHT-injected animals were not different from those in control animals. However, the 5-HT content in the drug-treated animals (FIGURE 4) was significantly reduced at all times tested, even at Day 55 after 5,7-DHT treatment. The 5-HT content in the drug-treated animals increased slightly with time in the brain stem, but the forebrain concentration remained low at all times. It should be mentioned here that hormone levels had completely normalized 55 days after drug treatment.

In the second series of experiments that involved injection of 150 μg of 5,7-DHT, rats were stressed purposely by means of ether anesthesia. Serum prolactin levels were elevated in all animals, regardless of whether they were treated with the drug or were pretreated with DMI (FIGURE 5). These high serum prolactin levels are significantly different ($p < 0.001$) from the levels in unstressed rats. Serum LH levels in these animals were significantly suppressed by 5,7-DHT injection for at least 26 days ($p < 0.001$) when compared with those in vehicle-injected and untreated control animals. As in the first experiment, serum FSH levels remained unaffected by 5,7-DHT treatment.

Effects of 5,7-Dihydroxytryptamine on Hormones and Monoamines in Ovariectomized Rats

In the chronically ovariectomized, jugular vein-cannulated rats, only blood LH levels were measured. Five days after 5,7-DHT treatment (379 ± 63 ng/ml), the mean LH levels were significantly lower than in control animals (536 ± 79 ng/ml; $p < 0.01$). Graphic presentation of the data revealed that LH levels in control animals fluctuated regularly, with peaks occurring approximately every hour (FIGURE 6). The mean interpeak time was 62.7 ± 4.9 min in these control animals. Regularly occurring LH peaks (of the same magnitude as those in control animals) were also observed 5 days after 5,7-DHT treatment; however, they occurred significantly less frequently (FIGURE 7). The interpeak time in these animals was 126.3 ± 7.2 min, or more than twice as long as in controls ($p < 0.005$). By Day 12 or 28 after 5,7-DHT treatment, the time between LH peaks was still longer than in controls. A return to the oscillatory pattern seen in control rats was found 55 days after 5,7-DHT treatment (FIGURE 8) (interpeak time, 56.9 ± 4.8 min).

TABLE 1 shows hormonal levels in the 5,7-DHT- and cerebrospinal fluid-treated, ovariectomized rats, which were decapitated 12 days after treatment. Some of these animals were ether anesthesia stressed prior to decapitation. The others were killed immediately in the animal rooms. Stress caused by ether anesthesia did not alter serum LH levels in either group. However, there was a significant difference in LH levels, between drug-treated and cerebrospinal fluid-injected animals (338.6 ± 14.3 and 462.5 ± 26.5 ng/ml, respectively; $p < 0.01$).

Serum prolactin levels in 5,7-DHT-treated and control rats 12 days after

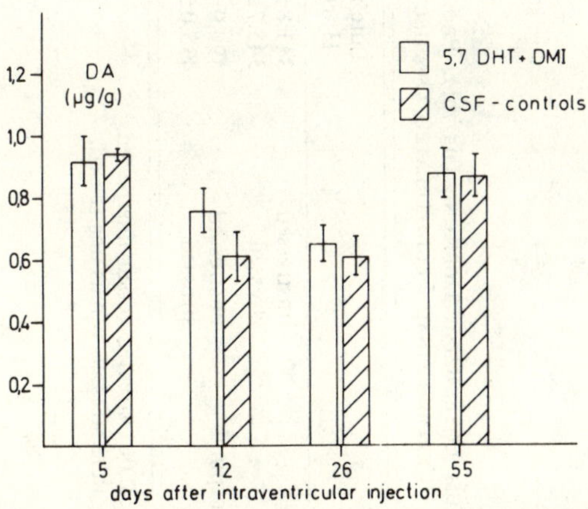

FIGURE 3. Dopamine content (micrograms per gram of wet weight) in the forebrain of 5,7-DHT-treated rats. The SEM is shown above each bar. Data from Wuttke *et al.*[37]

TABLE 1

SERUM LUTEINIZING HORMONE (ng/ml), PROLACTIN (ng/ml), AND MONOAMINE (ng/mg OF PROTEIN) LEVELS IN 5,7-DHT- OR CEREBROSPINAL FLUID-TREATED OVARIECTOMIZED RATS IN THE ANTERIOR PART OF THE MEDIOBASAL HYPOTHALAMUS

		Luteinizing Hormone	Prolactin	5-Hydroxytryptamine	Norepinephrine	Dopamine
5,7-DHT	unstressed	341.8±12.8	34.1±3.8	10.4±1.4	75.9±9.2	28.2±9.9
	stressed	345.7±30.1	114.9±23.8 *	12.2±1.2	46.7±11.6 †	26.5±4.5
Cerebrospinal fluid	unstressed	490.0±30.8	31.5±2.7	17.2±1.3	79.7±10.8	27.2±5.1
	stressed	435.0±42.7	127.0±17.2 *	17.5±3.4	37.5±6.6 †	21.0±1.7

* $p < 0.001$ stressed versus unstressed.
† $p < 0.01$ stressed versus unstressed.

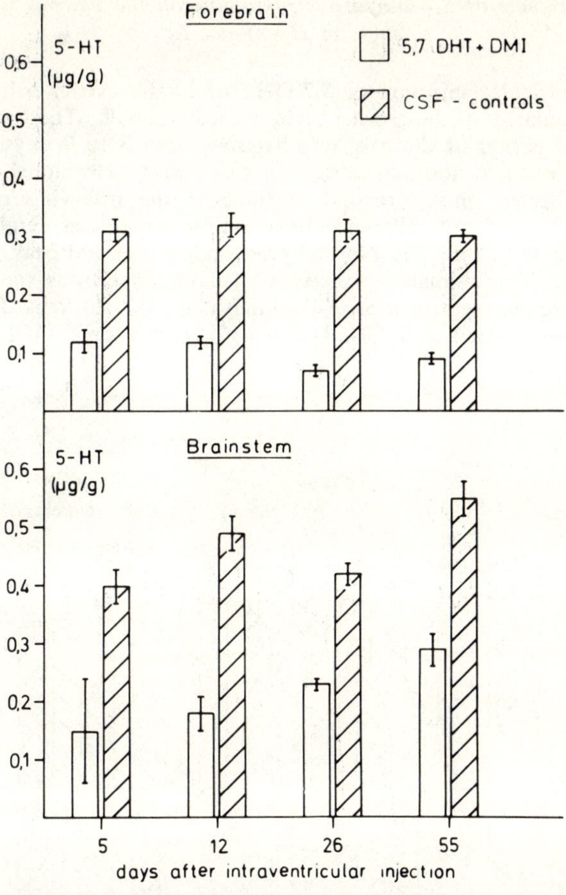

FIGURE 4. Serotonin (5-HT) content (micrograms per gram of wet weight) at various times after 5,7-DHT treatment. Note the low forebrain levels at Day 55. The SEM is shown above each bar. Data from Wuttke *et al.*[37]

drug injection were not different from each other when the stressed or unstressed groups were compared. Ether anesthesia stressing resulted in a significant rise of prolactin levels in both the drug-treated and sham-treated groups.

The hypothalamic 5-HT content 12 days after 5,7-DHT administration was lower than in controls but was not different from that in animals subjected to ether anesthesia stressing. Norepinephrine levels in 5,7-DHT- and cerebrospinal fluid-injected animals did not differ significantly. However, both groups responded to ether anesthesia stressing with a reduced hypothalamic NE content. Neither 5,7-DHT nor ether anesthesia stressing produced significant changes in hypothalamic DA content.

Effects of 5,7-Dihydroxytryptamine on the Estrous Cycle and on Hormonal Levels

Intraventricular injection of 5,7-DHT or of the carrier solution interfered with the regularity of the estrous cycle in most animals. These animals showed a prolonged period of diestrus, which lasted from 3 to 7 days. However, all animals returned to normal cycles after this period. Hormonal levels in blood samples collected on the second or third of the following proestrous days revealed no significant difference between the drug- and cerebrospinal fluid-treated animals (TABLE 2). Normal preovulatory LH and peak prolactin levels were found. The animals were decapitated on an estrous morning, and the oviducts were checked for ova. All animals had ovulated, regardless of whether they had been treated with 5,7-DHT or with cerebrospinal fluid.

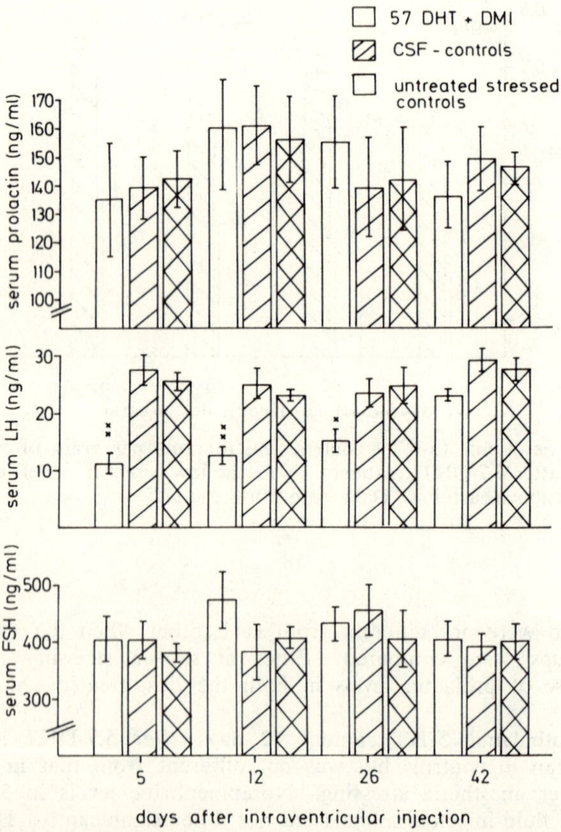

FIGURE 5. Serum prolactin, LH, and FSH levels at various times after 5,7-DHT treatment in ether-stressed male rats. Note the suppressed LH levels and normal prolactin levels. The SEM is shown above each bar.

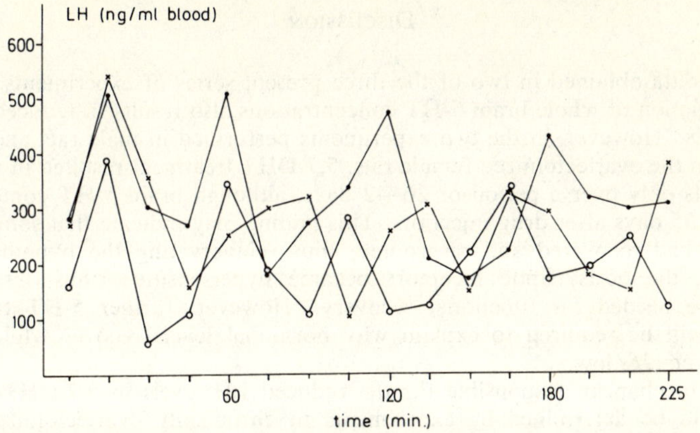

FIGURE 6. Circhoral LH pattern in three ovariectomized, cerebrospinal fluid-injected control rats 6 days after injection. Note the occurrence of LH peaks approximately every 60 min.

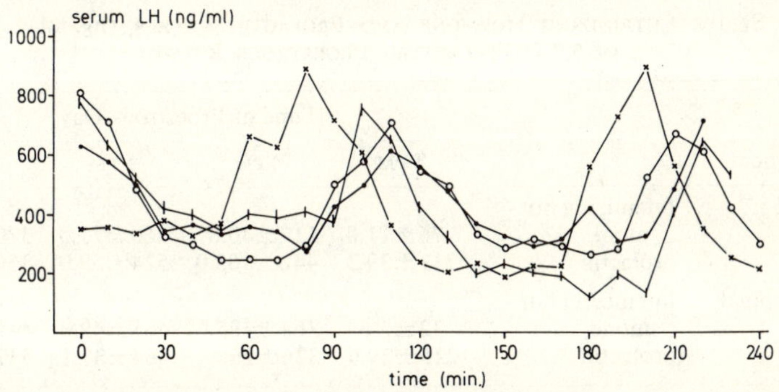

FIGURE 7. Oscillatory LH pattern in four ovariectomized rats 5 days after 5,7-DHT treatment. Note the LH peaks, which occur approximately every 2 hr.

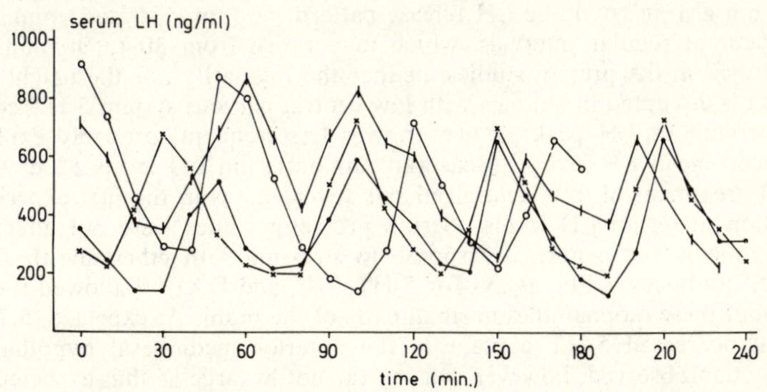

FIGURE 8. Restored circhoral LH pattern in five ovariectomized rats 55 days after 5,7-DHT injection. Note the occurrence of LH peaks approximately every 60 min.

Discussion

The data obtained in two of the three present series of experiments suggest that reduction of whole brain 5-HT concentrations also results in reduced serum LH levels. However, in the two experiments performed in male rats and in the studies in the ovariectomized female rats, 5,7-DHT treatment resulted in reduced LH levels only over a period of 26–42 days, although brain 5-HT content was still low 55 days after drug injection. This finding may indicate that some 5-HT neurons had recovered and resprouted, thus reinnervating the hypothalamus. Possibly, the postsynaptic receptors became hypersensitive; thus, less 5-HT would be needed for functional recovery. However, further 5-HT turnover studies will be required to explain why hormonal levels recover while 5-HT content remains low.

The mechanism responsible for the reduced LH levels in 5,7-DHT-treated rats could be determined by experiments in chronically ovariectomized rats. It is known that removal of the gonads results not only in increased gonado-

Table 2

Serum Luteinizing Hormone and Prolactin Levels (ng/ml) in 5,7-DHT-Treated Proestrous Rats

Treatment		Time of Proestrous Day 3 PM	4 PM	5 PM	6 PM
5,7-DHT	luteinizing hormone	76±11.0	117±40.0	460±71.0	370±52.0
	prolactin	211±39.3	442±103.0	524±59.0	356±31.0
Cerebrospinal fluid	luteinizing hormone	32±5.6	364±185.0	465±49.0	441±92.0
	prolactin	225±29.6	326±26.4	464±37.1	337±14.3

tropin release due to the lack of negative feedback control exercised by gonadal steroids but also in rhythmic LH release pattern.[34, 35] In castrated animals, LH peaks occur at regular intervals, which may range from 30 to 90 min.[36] As demonstrated in the present studies, neither the regularity nor the height of the LH peaks is disrupted in animals with low central nervous system 5-HT content; the occurrence of LH peaks, however, was less frequent, probably explaining the reduced mean LH levels. Measurements of serum LH levels 12 days after 5,7-DHT treatment of ovariectomized rats revealed (as in the first experiment) a reduction in serum LH levels. Serum prolactin values were not affected by drug treatment, and a normal response to stressing with ether anesthesia was observed. Radioenzymatic assays for 5-HT, NE, and DA[28-30] allowed the measurement of these monoamines in small parts of the brain. As expected, 5,7-DHT treatment decreased 5-HT content in the anterior mediobasal hypothalamus. The reduction observed, however, was by far not as large as that expected from previous studies in which 5-HT was measured fluorometrically in whole brains or larger portions of the brain of 5,7-DHT-treated rats.[23, 24, 37] As reported by others, the specificity of 5,7-DHT in reducing 5-HT content was exhibited after protection of the noradrenergic system by DMI.[23, 24, 37]

It was interesting to observe that NE content in 5,7-DHT- and also in cerebrospinal fluid-treated control rats was reduced after a 2-min exposure to ether anesthesia. It should be emphasized again that serum prolactin levels in these stressed animals were increased when compared to the unstressed control values. This observation provides further support to our previous hypothesis that stress-induced prolactin release is governed primarily by the noradrenergic system. Our NE values in the hypothalamus of stressed animals coincide with those reported by Palkovits *et al.*[29] However, we observed higher NE levels in unstressed rats, which indicates that extreme care must be exercised in handling the rats or removing them from their environment.[6] Otherwise, prolactin levels increase dramatically, apparently in association with a massive release of NE, resulting in decreased hypothalamic NE content.

The temporal irregularities of the estrous cycles in 5,7-DHT-treated rats were probably not mediated by the drug, because several control animals also had prolonged diestrous periods. After 3–7 days, normal cycles were recorded in both the drug-treated and control groups. Proestrous and estrous hormonal levels were also normal, although the 5,7-DHT-injected animals still had significantly lower hypothalamic 5-HT concentrations at the termination of the experiment. These results indicate that the serotonergic system is not involved in the mechanisms that control preovulatory LH release. This view is in disagreement with those of several previous reports.[10, 15, 19] It was reported that either administration of 5-HT into the third ventricle[10, 15] or pharmacologic intervention to increase brain 5-HT content blocks LH release.[19] Recently, however, evidence has appeared that 5-HT may be involved differently in LH release-controlling mechanisms.[17]

There is ample evidence that 5-HT has a stimulatory function on prolactin release when central nervous system 5-HT content is increased either by direct injection into the ventricular system[15] or by drug administration.[20, 22] However, decreased 5-HT levels did not interfere with normal pituitary prolactin release in the present study. Pituitary prolactin release is highly susceptive to any kind of stress.[6, 37, 38] Even removal of the animals from their usual environment (mild stress) results in increased serum prolactin levels.[6] This stress-induced prolactin release was not prevented by 5,7-DHT treatment. Only 5 days after treatment could a slight impairment of serum prolactin levels between mildly stressed 5,7-DHT- and cerebrospinal fluid-treated rats be observed. The stronger stress of ether anesthesia was capable of increasing serum prolactin levels, as in controls, 5 days after treatment, although these animals received a higher dose of 5,7-DHT.

The preovulatory prolactin discharge in the proestrous female rats was also not affected by reduced 5-HT levels, which may also indicate that 5-HT does not play an important role. Recently, Gallo *et al.*[18] demonstrated the failure of *p*-CPA to block suckling-induced prolactin release; these results could also indicate that 5-HT is not involved in the control of pituitary prolactin release.

The evidence of an inhibitory effect of 5-HT on FSH release was derived from experiments in which 5-HT was injected intraventricularly.[15] On the basis of the present experiments, these earlier results cannot be confirmed. No effect of selective depletion of 5-HT on serum FSH could be observed.

The present experiments allow the following conclusions to be made.

The drug 5,7-DHT reduces 5-HT brain content selectively after protection of the noradrenergic system by pretreatment with DMI.

The endocrine effects of reduction of 5-HT content are such that one can conclude that 5-HT is stimulatory to hypothalamic luteinizing hormone-releasing factor release. Serotonin appears to have little effect on prolactin and no effect on FSH release.

The time course of the endocrine changes observed does not parallel that of reduced 5-HT content; that is, serum hormonal levels are normal at a time when brain 5-HT is still greatly reduced.

Reduced NE content of the anterior hypothalamus in response to an acute stress suggests that the noradrenergic system mediates the stress message to the hypothalamus.

If the high selectivity of the serotonin neurotoxin 5,7-DHT can be confirmed for all putative neurotransmitter systems, the drug can provide a useful tool to study the biologic effects of reduced 5-HT levels in the brain.

References

1. Falck, B., N.-A. Hillarp, G. Thieme & A. Torp. 1962. Fluorescence of catecholamines and related compounds condensed with formaldehyde. J. Histochem. Cytochem. **10:** 348–354.
2. Fuxe, K. & U. Ungerstedt. 1968. Histochemical studies on the distribution of catecholamines and 5-hydroxy-tryptamine after intraventricular injections. Histochemie **13:** 16–28.
3. Ungerstedt, U. 1971. Stereotaxic mapping of the monoamine pathways in the rat brain. Acta Physiol. Scand. Suppl. **367:** 1–59.
4. Rubinstein, L. & C. H. Sawyer. 1970. Role of catecholamines in stimulating the release of pituitary ovulating hormone(s) in rats. Endocrinology **8:** 988–995.
5. Ojeda, S. R. & S. M. McCann. 1973. Evidence for participation of a catecholaminergic mechanism in the postcastration rise in plasma gonadotropins. Neuroendocrinology **12:** 295–315.
6. Fenske, M. & W. Wuttke. 1975. Effects of intraventricular 6-hydroxydopamine injections on serum prolactin release. Brain Res. **104:** 63–70.
7. Weiner, R. I., J. E. Stergue, R. A. Gorski & C. H. Sawyer. 1972. Changes in catecholamine content of the rat hypothalamus following deafferentation. Endocrinology **90:** 867–873.
8. Brownstein, M. J., M. Palkovits, M. L. Tappaz, J. M. Saavedra & J. S. Kizer. 1976. Effect of surgical isolation of the hypothalamus on its neurotransmitter content. Brain Res. **117:** 287–295.
9. Schneider, H. P. G. & S. M. McCann. 1970. Release of LH-releasing-factor (LRF) into the peripheral circulation of hypophysectomized rats by dopamine and its blockade by estradiol. Endocrinology **87:** 249–253.
10. Kamberi, I. A., R. S. Mical & J. C. Porter. 1970. Effect of anterior pituitary perfusion and intraventricular injection of catecholamines and indolamines on LH release. Endocrinology **87:** 1–12.
11. Fuxe, K. & T. Hökfelt. 1969. Catecholamines in the hypothalamus and the pituitary gland. *In* Frontiers in Neuroendocrinology. L. Martini & W. F. Ganong, Eds. : 237–290. Oxford University Press. New York, N.Y.
12. Fuxe, K., A. Löfström, P. Eneroth, J.-A. Gustafsson, T. Hökfelt, P. Scett, W. Wuttke, H. Fraser & S. Jeffcoate. 1975. Interactions between hypothalamic catecholamine nerve terminals and LRF containing neurons. Further evidence for an inhibitory dopaminergic and facilitatory noradrenergic influence. Excerpta Med. Int. Congr. Ser. **374:** 165–177.

13. McLeod, R. M. 1969. Influence of norepinephrine and catecholamine-depleting agents on the synthesis and release of prolactin and growth hormone. Endocrinology **85:** 916–923.
14. Shaar, C. J. & J. A. Clemens. 1974. The role of catecholamines in the release of anterior pituitary prolactin in vitro. Endocrinology **95:** 1202–1212.
15. Kamberi, I. A., R. S. Mical & J. C. Porter. 1971. Effects of melatonin and serotonin on the release of FSH and prolactin. Endocrinology **88:** 1012–1020.
16. Kordon, C. & J. Glowinski. 1972. Role of hypothalamic monoaminergic neurones in the gonadotrophin release-regulating mechanisms. Neuropharmacology **11:** 153–162.
17. Héry, M., E. Laplante & C. Kordon. 1976. Participation of serotonin in the phasic release of LH. I. Evidence from pharmacological experiments. Endocrinology **99:** 496–503.
18. Gallo, R. V., J. Rabii & G. P. Moberg. 1975. Effect of methysergide, a blocker of serotonin receptors, on plasma prolactin levels in lactating and ovariectomized rats. Endocrinology **97:** 1069–1105.
19. Kordon, C., F. Javoy, G. Vassent & J. Glowinski. 1968. Blockade of superovulation in the immature rat by increased brain serotonin. Eur. J. Pharmacol. **4:** 169–174.
20. Kordon, C., C. A. Blake, J. Terkel & C. H. Sawyer. 1973–74. Participation of serotonin-containing neurons in the suckling-induced rise in plasma prolactin levels in lactating rats. Neuroendocrinology **13:** 213–222.
21. Mena, F., A. Enjalbert, L. Carbonell, M. Priam & C. Kordon. 1976. Effect of suckling on plasma prolactin and hypothalamic monoamine levels in the rat. Endocrinology **99:** 445–451.
22. Lu, K.-H. & J. Meites. 1973. Effects of serotonin precursors and melatonin on serum prolactin release in rats. Endocrinology **93:** 152–155.
23. Gerson, S. & R. J. Baldessarini. 1975. Selective destruction of serotonin terminals in rat forebrain by high doses of 5,7-dihydroxytryptamine. Brain Res. **85:** 140–145.
24. Björklund, A., H. G. Baumgarten & A. Rensch. 1975. 5,7-Dihydroxytryptamine: improvement of its selectivity for serotonin neurons in the CNS by pretreatment with desimipramine. J. Neurochem. **24:** 833–835.
25. Bertler, A. 1961. Effect of reserpine on the storage of catecholamines in brain and other tissues. Acta Physiol. Scand. **51:** 75–83.
26. Häggendahl, J. 1963. An improved method for fluorometric determination of small amounts of adrenaline and noradrenaline in plasma and tissue. Acta Physiol. Scand. **59:** 242–254.
27. Lowry, O. H., M. J. Rosebrough, A. L. Farr & R. J. Randall. 1951. Protein measurement with the Folin phenol reagent. J. Biol. Chem. **193:** 265–275.
28. Coyle, J. T. & D. Henry. 1973. Catecholamines in fetal and newborn rat brain. J. Neurochem. **21:** 61–67.
29. Palkovits, M., M. Brownstein, J. M. Saavedra & J. Axelrod. 1974. Norepinephrine and dopamine content of hypothalamic nuclei of the rat. Brain Res. **77:** 137–149.
30. Saavedra, J. M., M. Palkovits, M. Brownstein & J. Axelrod. 1974. Serotonin distribution in the nuclei of the rat hypothalamus and preoptic region. Brain Res. **77:** 157–165.
31. Monroe, S. E., R. W. Rebar, V. L. Gay & A. R. Midgley, Jr. 1969. Radioimmunoassay for determination of luteinizing hormone during the estrous cycle of the rat. Endocrinology **89:** 1448–1453.
32. Niswender, G. D., C. L. Chen, A. R. Midgley, Jr., J. Meites & S. Ellis. 1969. Radioimmunoassay for rat prolactin. Proc. Soc. Exp. Biol. Med. **130:** 793–797.
33. Daane, T. A. & A. F. Parlow. 1971. Serum FSH and LH in constant light-induced persistent estrus: short-term and long-term studies. Endocrinology **88:** 964–968.

34. GAY, V. L. & N. A. SHETH. 1972. Evidence for a periodic release of LH in castrated male and female rats. Endocrinology **90:** 158–162.
35. BECK, W., S. ENGELBART, M. GELATO & W. WUTTKE. 1977. Antigonadotrophic effect of prolactin in adult castrated and in immature female rats. Acta Endocrinol. **84:** 62–71.
36. BLAKE, C. H. 1974. Temporal patterns in plasma luteinizing hormone concentration in ovariectomized rats during different times of the day. Endocrinology **95:** 813–817.
37. WUTTKE, W., A. BJÖRKLUND, H. G. BAUMGARTEN, L. LACHENMAYER & M. FENSKE. 1977. De- and regeneration of brain serotonin neurons following 5,7-dihydroxytryptamine treatment: Effects on serum LH, FSH and prolactin levels in male rats. Brain Res. **134:** 317–331.

DISCUSSION

DR. R. COLLU: We have also found that severe depletion of brain 5-HT with *p*-chlorophenylalanine plus 5,6-DHT does not modify the increment of plasma prolactin in rats immobilized for 30 min. On the contrary, treatment with 6-hydroxydopamine plus α-methyl-*p*-tyrosine reduces this increment by 50%.

DR. WUTTKE: In our paper that showed that 6-hydroxydopamine prevents prolactin release, we noted that hypothalamic dopamine neurons were remarkably resistant to the drugs; thus, we concluded that norepinephrine causes prolactin release.

DR. J. A. CLEMENS: The change in luteinizing hormone (LH) rhythm that you observe after ovariectomy plus 5,7-DHT injection is of interest. Blake has reported that these circoral surges of LH release in ovariectomized animals were normal after complete hypothalamic deafferentation.

Do you think that 5-HT cell bodies within the basal hypothalamus may be participating in this regulation?

DR. WUTTKE: We have no evidence that 5-HT perikarya are located within the hypothalamus. I'm skeptical about these complete cuts. Not many animals survive this procedure, and I don't know how fast the hypothalamus is reinnervated by extrahypothalamic neurons.

DR. L. V. O'BRIEN (*Hershey Medical Center, New York, N.Y.*): Kennedy and coworkers and I have shown in organ cultures of rat and mink hypothalami that 5-HT was produced by ependymal and tanycyte cells; these organ cultures did not show any evidence of neuronal bodies or terminals after 8 days, although 5-HT was still produced. Thus, we have evidence that 5-HT is produced in the hypothalamus, which is not involved with nerve terminals.

CHANGES IN SEIZURE SUSCEPTIBILITY AFTER INTRACEREBRAL TREATMENT WITH 5,7-DIHYDROXYTRYPTAMINE: ROLE OF SEROTONERGIC NEURONS *

Ronald A. Browning, William E. Hoffmann, and Ronald L. Simonton

School of Medicine and Department of Physiology
Southern Illinois University
Carbondale, Illinois 62901

The possible importance of serotonin [5-hydroxytryptamine (5-HT)] as a modulator of seizure susceptibility was suggested more than 20 years ago when Chen *et al.*[1] observed seizure-facilitating effects of reserpine in mice subjected to electroshock- or pentylenetetrazol (PTZ)- induced convulsions. Since then, many investigators (for reviews, see References 2–5) employing several species have provided evidence that the seizure-facilitating effects of reserpine are due to the depletion of both catecholamines and 5-HT.

Most attempts to elucidate the putative role of 5-HT in the convulsive process have involved pharmacologic manipulations of this amine. In these studies, seizure susceptibility has usually been examined after drug treatments designed either to increase or decrease the availability of 5-HT at central 5-HT receptors. Several laboratories have shown that treatments believed to increase the synaptic concentration of 5-HT reduce the susceptibility to experimentally induced seizures. For example, the systemic administration of 5-hydroxytryptophan (5-HTP), the precursor of 5-HT, in conjunction with a monoamine oxidase inhibitor[6–8] or a peripheral decarboxylase inhibitor,[9] has been shown to either reduce seizure susceptibility or antagonize the reserpine-induced facilitation of seizures.[10, 11]

Other investigators have focused their attention on the relationship between depletion of brain 5-HT and the sensitivity of animals to convulsive stimuli. In these studies, inhibition of 5-HT synthesis by *p*-chlorophenylalanine (*p*-CPA), an inhibitor of tryptophan hydroxylase, has been reported to exacerbate seizures induced by electrical,[12–14] chemical,[9, 15] audiogenic,[16, 17] or photic stimulation.[18]

Although the pharmacologic studies generally favor a seizure inhibitory function for 5-HT, interpretation of those findings is confounded by the nonspecificity of the treatments employed. For example, *p*-CPA in doses that profoundly reduce the concentration of 5-HT usually produces some depletion of catecholamines,[8, 19] while injection of 5-HTP can lead to the accumulation of 5-HT in both catecholamine-containing and serotonin-containing neurons.[20]

Further confusion has arisen regarding the seizure-antagonizing role for 5-HT, because several conflicting observations have also been reported. Perfusion of the rat hippocampus with 5-HT was found to increase seizure activity in this structure.[21] Other workers have reported that the administration of high doses of 5-HTP to guinea pigs[22] or moderate doses to rats already depleted of

* Supported by a grant from the Epilepsy Foundation of America.

0077–8923/78/0305–0437 $01.75/0

brain 5-HT[23] can produce myoclonic seizures in these species. In addition, seizures have been observed in rats that received tryptophan in conjunction with a monoamine oxidase inhibitor,[24] a treatment believed to increase the availability of 5-HT at the receptors.[25] Finally, *p*-CPA has been reported to prevent rather than enhance audiogenic seizures in one strain of mice.[26] Thus, the current status of 5-HT as a modulator of seizure susceptibility remains somewhat equivocal.

The studies described below were designed to evaluate seizure susceptibility in animals whose serotonergic neurons had been removed by chemical or surgical lesioning. It was felt that this approach would provide a more selective means of producing a transmitter deficit at central 5-HT receptors. If 5-HT does, in fact, exert anticonvulsant effects, such animals would be expected to exhibit an enhanced sensitivity to convulsive stimuli. The extensive studies by Baumgarten *et al.*[27-29] (and described elsewhere in this monograph), which demonstrated the neurotoxic effects of the dihydroxylated tryptamines on monoamine neurons, provided the necessary chemical tools to perform such a study. 5,7-Dihydroxytryptamine (5,7-DHT) was chosen over 5,6-dihydroxytryptamine because of its lower toxicity and the fact that it could be administered in doses that produce a more complete depletion of brain 5-HT.[29] The chief limitation in the use of 5,7-DHT, toxicity to noradrenergic neurons, was overcome by pretreatment of the animals with the norepinephrine (NE) uptake blocker protriptyline.

The accessibility of the dorsal and median raphe nuclei to surgical destruction provided another means for removing the serotonergic neurons that innervate the forebrain and thus for assessing the functional role of such neurons in seizure susceptibility.

Although some investigators have suggested an important role for 5-HT in the anticonvulsant action of phenytoin,[30] phenobarbital,[31] and the carbonic anhydrase inhibitors,[13] their studies are difficult to interpret because of the nonspecificity of the treatments employed to alter brain 5-HT and because of the possible direct interaction between 5-HT-depleting drugs and the anticonvulsants. It was felt that the destruction of serotonergic neurons by chemical or surgical means would provide greater insight into the mechanism(s) of action of these drugs because of improved specificity and the ability to use 5-HT-deficient animals that are drug free, except for the anticonvulsant.

Effect of 5,7-DHT or a Raphe Lesion on Seizure Susceptibility

Materials and Methods

Male Sprague-Dawley rats that weighed 260–312 g were injected with protriptyline hydrochloride (20 mg/kg, ip), except in preliminary studies, where the effects of protriptyline were being examined. Two hours after injection of protriptyline, animals were anesthetized with ether, placed in a stereotaxic head holder, and dosed intraventricularly by use of the procedure of Noble *et al.*[32] Rats received either 30 μl of vehicle (saline that contained 0.1 mg/ml of ascorbic acid) or vehicle that contained 200 μg of 5,7-DHT creatinine sulfate (calculated as the free base). All animals also received 15 mg/kg of pentobarbital sodium (ip) immediately after the intraventricular injection to prevent the convulsions that usually accompany the intraventricular administration of 5,7-DHT.

Brain monoamine analyses were performed on animals killed by decapitation at times described in the Figure legends or Tables. After removal, the brains were immediately dissected into forebrain and brain stem regions by a cut between the mammillary bodies of the hypothalamus and the inferior colliculus. Forebrains were further divided into right and left halves by a midsaggital cut. The individual regions were immediately frozen on dry ice and stored at −20° C until monoamine determinations were performed. All analyses were performed within 1 week after the animals were killed. Half of the forebrain, the whole brain stem (midbrain and pons-medulla), and the spinal cord (in some studies) from each animal were analyzed for 5-HT by use of a modification of the procedure of Bogdanski *et al.*[33] The other half of the forebrain was analyzed for NE and dopamine (DA) according to the method of Anton and Sayre.[34]

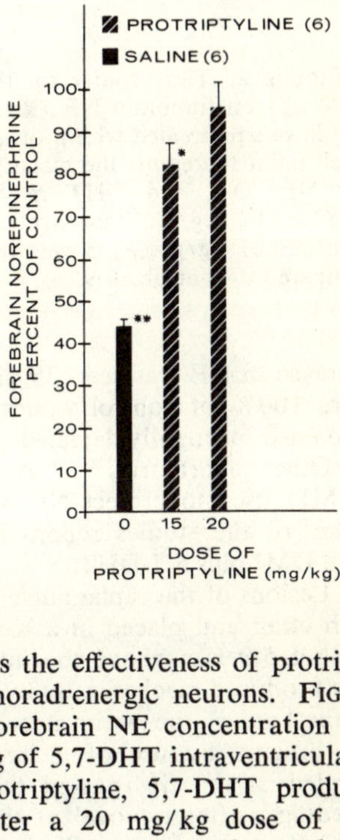

FIGURE 1. Effects of intraventricular injection of 5,7-DHT (200 μg) on rat forebrain NE concentration. Brains were analyzed 11 days after injection. Rats received either protriptyline hydrochloride (15 or 20 mg/kg) or saline 2 hr prior to injection of 5,7-DHT. Control value for NE was 0.307±0.01 μg/g. *, $p < 0.05$ compared to control (protriptyline and vehicle); **, $p < 0.001$ compared to control (saline and vehicle).

Preliminary studies were conducted to assess the effectiveness of protriptyline in preventing the effects of 5,7-DHT on noradrenergic neurons. FIGURE 1 shows that in the absence of protriptyline, forebrain NE concentration was reduced by 60% in animals that received 200 μg of 5,7-DHT intraventricularly. However, after injection of 15 mg/kg of protriptyline, 5,7-DHT produced only 18–20% reductions in forebrain NE; after a 20 mg/kg dose of protriptyline, the effects of 5,7-DHT on forebrain NE were prevented completely.

To further evaluate the effectiveness of protriptyline plus 5,7-DHT as a means for selective lesioning of serotonergic neurons, we examined the changes in brain monoamines 4–45 days after treatment with protriptyline (20 mg/kg) and 5,7-DHT (200 μg). Forebrain 5-HT was maximally reduced to 12% of control values by 10 days and showed no significant recovery after 45 days (FIGURE 2). Brain stem 5-HT was less severely depleted and showed significant recovery 25–45 days after treatment, a finding that has also been reported by other investigators.[29] A slight (14%) reduction in NE was observed at 4 days and in DA (17%) at 10 days after treatment. Moreover, a small (11%)

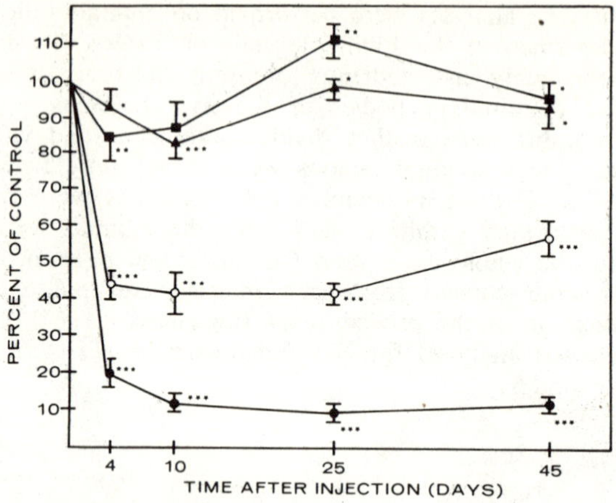

FIGURE 2. Time course for the effect of an intraventricular injection of 5,7-DHT (200 μg) on forebrain NE (■), DA (▲), and 5-HT (●) and on brain stem 5-HT (○) in rats pretreated with protriptyline (20 mg/kg) 2 hr prior to injection of 5,7 DHT. Each point represents the mean±SEM for four to six determinations. Control values for NE, DA, and 5-HT in the forebrain were 0.333±0.02, 0.880±0.05, and 0.683±0.02 μg/g, respectively. The control value for brain stem 5-HT was 1.061±0.01 μg/g. *, p was not significant compared to control; **, $p < 0.05$ compared to control; ***, $p < 0.001$ compared to control.

increase in NE was seen 25 days after treatment. However, both NE and DA were 100% of control values 45 days after treatment, while forebrain 5-HT remained maximally depleted.

Other laboratories [35, 36] reported similar findings with desmethylimipramine (DMI) used to protect NE neurons from the neurotoxic effects of 5,7-DHT. Many of the studies reported elsewhere in this monograph were performed with DMI plus 5,7-DHT.

Lesions of the raphe nuclei were produced in rats (270–300 g) anesthetized with ether and placed in a Kopf stereotaxic instrument with the upper incisor bar set 5.0 mm above the interaural plane. Destruction of the raphe nuclei was produced mechanically by lowering a stainless steel rod (1.2 mm diameter) to the base of the brain at coordinates 6.0, 6.5, 7.0, and 7.5 mm posterior to the bregma by use of a lateral coordinate of zero. The rod, which was beveled at the tip, passed through the midsaggital sinus, causing extensive bleeding. After removal of the rod, bleeding was stopped by the application of Gelfoam® sponges. Control animals were sham operated by removing the skull and breaking the midsaggital sinus to produce a comparable amount of bleeding without lowering the stainless steel rod. Histologic examination of the lesion was made on 40 μm thick frozen sections stained with thionine.[37] Damage consisted of a lesion tract that extended through the midline of the mesencephalon at the level of the inferior colliculus. The lesions extended 2.5–3.0 mm in the rostrocaudal direction and was about 1.0 mm wide, although severe damage was restricted to a zone about 0.5 mm wide. Both the dorsal and median raphe nuclei were ablated by the lesion. The lesion was also observed to produce damage to the medial aspects of the medial longitudinal fasciculus, the brachium conjunctivum, and the very medial aspects of the medial lemniscus. The posterior portion of the nucleus interpeduncularis was also damaged by the rostral extension of the lesion in several animals. All lesions also interrupted the commissure of the inferior colliculus, enlarged the cerebral aqueduct, and produced extensive damage to the central gray matter. A typical section through the lesion at the level of the raphe nuclei is shown in FIGURE 3.

Pentylenetetrazol seizures were induced by an intraperitoneal injection of 30 mg/kg of the drug. After injection, each animal was placed in a clear plastic cage and observed by an individual who had no knowledge of the prior treatment. The time from injection to the onset of the myoclonic jerk and the generalized seizure was measured by use of an electric interval timer or a stopwatch. Seizure severity was also scored by the observer who used a

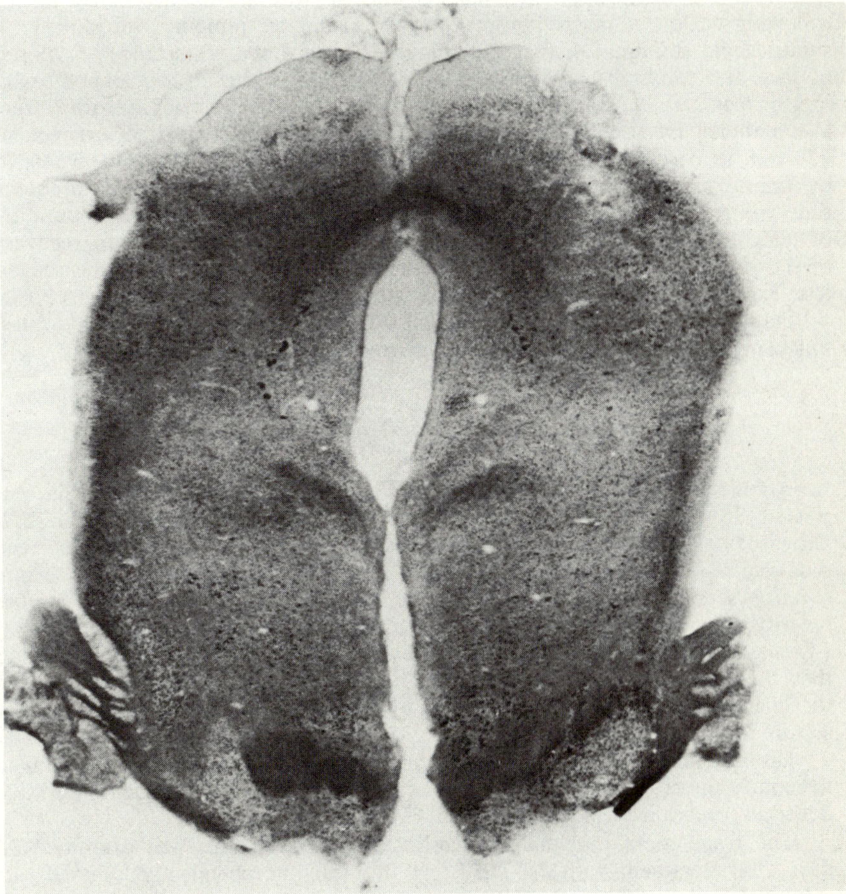

FIGURE 3. Photomicrograph of a 40 μm thick section of rat midbrain stained with thionine shows a lesion that involves the dorsal and median raphe nuclei.

modification of the scale employed by De La Torre *et al.*[9] The procedure used to assign seizure severity scores was as follows: 1, no seizure (no response); 2, preclonic jerk only; 3, minimal clonic seizure without loss of posture; 4, clonic seizure with loss of posture; 5, clonic-tonic seizure that recovered within 1 min; 6, clonic-tonic seizure that recovered within 2 min; 7, clonic-tonic seizure that recovered within 3 min; 8, clonic-tonic seizure that recovered within 5 min; 9, multiple clonic-tonic seizures with brief interictal periods; 10, death.

All electroshock seizures were induced by corneal stimulation with 60 cycle alternating current. The electroshock seizure threshold, that is, the current necessary to elicit at least 5 sec of facial clonus, was calculated from the responses of each rat to a current at or near threshold according to the method of Woodbury and Davenport.[38] Each rat received a minimum of three shocks with a 3-day interval between tests to establish threshold. In addition, the entire group was subjected to a current estimated as the mean threshold for the group, and seizure severity was scored according to the following scale: 1, no response; 2, furor (running, jumping, squealing); 3, subthreshold stunning; 4, facial clonus that lasted 5 sec (threshold); 5, clonus of face and forelimbs without loss of posture; 6, clonus of whole body with loss of posture; 7, tonic-clonic seizure. Maximal electroshock seizures were also induced by use of corneal electrodes with a supramaximal current of 150 mA as described by Toman *et al.*[39] In this test, animals were evaluated by timing of three components of the maximal (tonic-clonic) seizure: the time from the stimulus to full extension of the hindlimbs (latency), duration of tonic extension, and the time from the end of tonic extension to recovery of the righting reflex (recovery). Each rat was subjected to only one seizure test, except the animals used for electroshock threshold determinations.

Statistical comparisons were made by use of a two-tailed Student's *t* test unless otherwise stated in the Tables or Figure legends.

Results and Discussion

An increase in the severity of PTZ-induced seizures was evident in rats treated 35 days earlier with protriptyline and 5,7-DHT (FIGURE 4). 5,7-Dihydroxytryptamine-treated rats also displayed an earlier seizure onset than did vehicle-injected controls. Although the time of onset and the seizure severity scores in the PTZ test varied from group to group, the seizure severity of 5,7-DHT-treated rats was always greater than that in controls (TABLE 1). This variability in response to PTZ administered by routes other than the intravenous one has also been noted by Adler.[40] The effectiveness of the chemical lesioning was confirmed by brain amine analysis 1 week after seizure testing (TABLE 2).

On the other hand, destruction of the dorsal and median raphe nuclei failed to modify the response of rats to PTZ (FIGURE 4), though producing an 80% deficit in forebrain 5-HT (TABLE 3).

The response to maximal electroshock-induced seizures was examined 35 days after treatment with 5,7-DHT. In this test, the chemically lesioned rats exhibited a decrease in the latency and an increase in the duration of the tonic extensor phase of the seizure (FIGURE 5). Changes of this nature are believed to be associated with an increased seizure severity.[41] However, raphe-

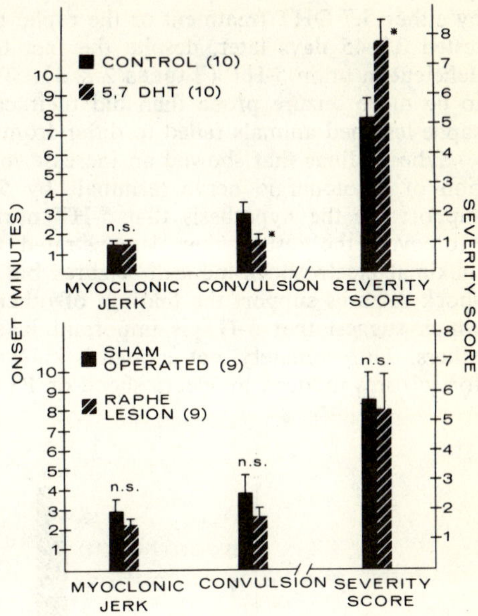

FIGURE 4. Response of rats to PTZ (30 mg/kg, ip) 35 days after treatment with 5,7-DHT (200 μg) or a raphe nuclei lesion. Animals that received 5,7-DHT were pretreated with protriptyline (20 mg/kg) 2 hr earlier. Control rats received protriptyline plus vehicle. The procedures employed for producing the raphe lesion and the sham operation are described in MATERIALS AND METHODS. Vertical lines indicate the SEM. The number of animals studied is indicated in parentheses. ∗, $p < 0.05$ compared to control.

lesioned animals, unlike 5,7-DHT-treated rats, did not differ from sham-operated animals in their response to maximal electroshock (FIGURE 5). The effectiveness of 5,7-DHT and the raphe lesion was confirmed by brain monoamine analysis performed 1 week after seizure testing (TABLES 2 & 3). Histologic verification of lesion placement was also performed on the brain stems of raphe-lesioned rats.

The threshold for minimal electroshock seizures was found to be unaltered

TABLE 1

RESPONSE OF RATS TO PTZ AFTER TREATMENT WITH 5,7-DHT *

Experiment	Seizure Severity Scores: Vehicle	Seizure Severity Scores: 5,7-DHT	Increase (%)	p Value †
1	2.2±0.3 (9)	5.6±0.8 (9)	154	< 0.001
2 ‡	5.2±0.7 (10)	7.8±0.6 (10)	50	< 0.05
3	3.5±0.4 (8)	5.4±0.6 (8)	54	< 0.01
4	2.8±0.7 (7)	4.7±0.8 (7)	68	< 0.05

* All animals received PTZ (30 mg/kg) 35 days after treatment with protriptyline (20 mg/kg) plus 5,7-DHT (200 μg), and their seizure severity was scored according to the procedure described in MATERIALS AND METHODS. Control animals received protriptyline (20 mg/kg) plus vehicle (saline that contained 0.1 mg/ml of ascorbic acid). Details of treatment are described more fully in MATERIALS AND METHODS. Seizure severity scores are reported as the mean ± SEM. The number of animals studied is indicated in parentheses.

† p Values indicate statistical difference between vehicle- and 5,7-DHT-treated rats.

‡ Data from this group of animals are also displayed in FIGURE 4.

by either 5,7-DHT treatment or the raphe nuclei lesion (FIGURE 6) in animals tested 35–45 days later, despite the fact that these animals were chronically deficient in brain 5-HT (TABLES 2 & 3). Thus, 5,7-DHT-treated rats appeared to be more seizure prone than did their controls in two of three tests, while raphe-lesioned animals failed to differ from their controls in any of these tests.

The findings that showed an increase in seizure susceptibility after destruction of serotonergic nerve terminals by 5,7-DHT would appear to provide support for the hypothesis that 5-HT normally exerts anticonvulsant effects. Moreover, the results that demonstrated that 5,7-DHT facilitated PTZ and maximal electroshock-induced seizures but had no effect on minimal electroshock seizures support the findings of others with *p*-CPA [8, 14, 15] and 5-HTP,[42] which suggest that 5-HT is important in some types of seizures and not in others. Unfortunately, not enough is known about the mechanisms responsible for seizures induced by electroshock or PTZ to speculate on the reason(s) for

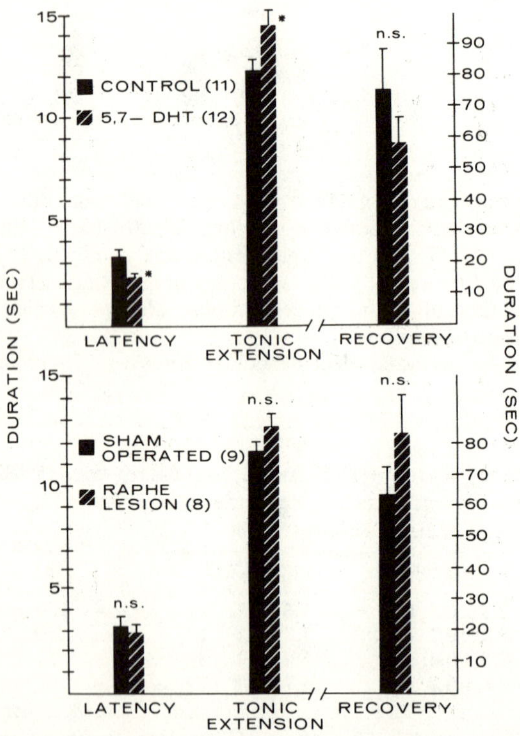

FIGURE 5. Response of rats to maximal electroshock 35 days after treatment with 5,7-DHT (200 μg) or a raphe nuclei lesion. Animals that received 5,7-DHT were pretreated with protriptyline (20 mg/kg) 2 hr earlier. Control rats received protriptyline plus vehicle. The procedures employed for producing the raphe lesion and the sham operation are described in MATERIALS AND METHODS. Maximal electroshock seizures were induced by stimulation with 150 mA through corneal electrodes. Vertical lines indicate the SEM. The number of animals studied is shown in parentheses. *, $p < 0.05$ compared to control.

TABLE 2

MONOAMINE ANALYSES IN 5,7-DHT-TREATED RATS *

Group	Number of Rats	Forebrain 5-HT (μg/g)	Forebrain NE (μg/g)	Brain Stem 5-HT (μg/g)	Spinal Cord 5-HT (μg/g)
Vehicle-PTZ	9	0.657±0.03	0.378±0.01		
5,7-DHT-PTZ	9	0.079±0.003 †	0.369±0.08		
Change (%)		−88			
Vehicle-PTZ	7	0.666±0.01			0.600±0.04
5,7-DHT-PTZ	7	0.110±0.01 †			0.113±0.01 †
Change (%)		−84			−82
Vehicle-MES	11	0.648±0.02	0.379±0.01	1.144±0.01	
5,7-DHT-MES	12	0.079±0.003 †	0.385±0.01	0.736±0.04 †	
Change (%)		−88		−36	
Vehicle-EST	10	0.619±0.01	0.397±0.02	1.035±0.03	
5,7-DHT-EST	8	0.095±0.002 †	0.363±0.03	0.678±0.03 †	
Change (%)		−85		−35	

* Monoamine analyses were performed 45–50 days after treatment with 5,7-DHT (1 week after seizure test). All values are reported as mean ± SEM. Treatment consisted of protriptyline (20 mg/kg, ip) 2 hr prior to intraventricular 5,7-DHT (200 μg). The responses of these groups to PTZ, maximal electroshock (MES), and electroshock seizure threshold (EST) testing are shown in FIGURES 4, 5, and 6, respectively.

† $p < 0.001$ compared to vehicle-injected control.

the differential effects of 5,7-DHT. However, it is of interest that some anticonvulsant drugs exhibit differential effects in their ability to antagonize electrically and chemically induced seizures, and these differences are sometimes predictive of their clinical usefulness for a particular type of epilepsy.[43]

The discrepancy between the effects of 5,7-DHT and a raphe lesion on experimentally induced convulsions was somewhat surprising in view of the fact that both procedures produced a comparable reduction in forebrain 5-HT. However, there are known anatomic differences between the effects of the two procedures on central nervous system 5-HT that could conceivably account for the disparate findings. For example, the mechanical lesion damages primarily the ascending serotonergic fibers that arise from nuclei B-7, B-8, and the medial parts of B-9 (as designated by Dahlström and Fuxe [44]), while the chemical lesion destroys both the ascending and descending systems [29] (TABLE 3). One could argue that the depletion of spinal cord 5-HT after 5,7-DHT treatment is responsible for the enhanced seizure susceptibility and therefore should not be seen in animals after lesioning restricted to ascending serotonergic neurons. This argument, however, is somewhat weakened by the observation that stimulation of the ascending serotonergic system protects animals against PTZ-induced convulsions [45] and by the finding that bulbospinal serotonergic neurons exert excitatory influences on the motoneuron.[46]

The 5,7-DHT treatment would also presumably destroy the terminals of 5-HT neurons that arise from nuclei B-3 and B-5, which are believed to

TABLE 3
MONOAMINE ANALYSES IN RAPHE-LESIONED RATS *

Group	Number of Rats	Forebrain 5-HT (μg/g)	Forebrain NE (μg/g)	Forebrain DA (μg/g)	Brain Stem 5-HT (μg/g)	Spinal Cord 5-HT (μg/g)
Sham-PTZ	9	0.828±0.04	0.452±0.02	1.000±0.07		
Raphe-PTZ	8	0.174±0.01 †	0.442±0.04	1.130±0.08		
Change (%)		−79				
Sham-MES	9	0.707±0.02	0.406±0.01	0.999±0.03		
Raphe-MES	8	0.137±0.01 †	0.398±0.01	0.890±0.03 ‡		
Change (%)		−81		−10		
Sham-EST	9	0.794±0.03	0.392±0.02	0.877±0.05		
Raphe-EST	9	0.169±0.03 †	0.369±0.01	0.795±0.04		
Change (%)		−79				
Sham §	6	0.649±0.03			1.327±0.03	0.535±0.01
Raphe	6	0.114±0.01			0.841±0.03 †	0.556±0.02
Change (%)		−82			−37	

* Monoamine analyses were performed 37–45 days after a mechanical lesion was placed in the midbrain (see MATERIALS AND METHODS) or 1 week after seizure testing. All values are reported as mean±SEM. The responses of these groups to PTZ, maximal electroshock (MES), and electroshock seizure threshold (EST) testing are shown in FIGURES 4, 5, and 6, respectively.

† $p < 0.001$ compared to sham-operated controls.

‡ $p < 0.05$ compared to sham-operated controls.

§ A group of raphe-lesioned and sham-operated animals that were used only for monoamine analyses.

innervate the brain stem and cerebellum, whereas the raphe lesion employed would not be expected to damage these neurons. While differences in the depletion of brain stem 5-HT after 5,7-DHT treatment and the mechanical lesion may account for their different effects on seizure susceptibility, this hypothesis was not supported by monamine analysis of this region. Both procedures produced 35–40% reductions in brain stem 5-HT (TABLES 2 & 3). However, it is likely that the 5-HT depletion in this area, while similar after the two procedures, is the result of damage to different groups of neurons and

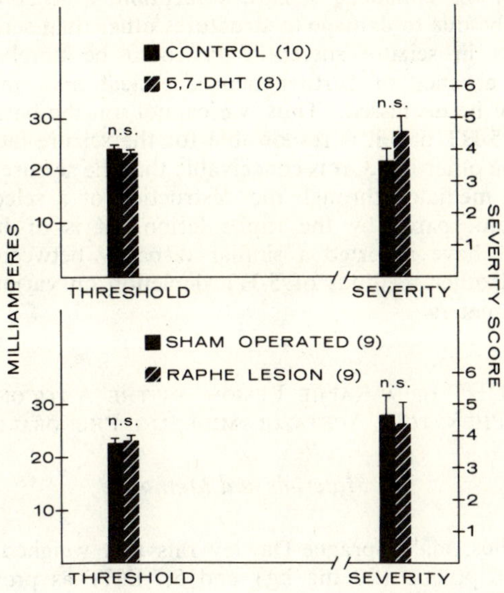

FIGURE 6. Electroshock seizure threshold in rats 35–45 days after treatment with 5,7-DHT (200 μg) or a raphe nuclei lesion. Animals that received 5,7-DHT were pretreated with protriptyline (20 mg/kg) 2hr earlier. Control rats received protiptyline plus vehicle. The procedures employed for producing the raphe lesion and the sham operation are described in MATERIALS AND METHODS. Seizure thresholds were determined by the method of Woodbury and Davenport.[38] The actual procedure is described more fully under the section mentioned above. Seizure severity scores were obtained by subjecting the entire group to the mean threshold current for that group. Vertical lines indicate the SEM. The number of animals studied is shown in parentheses.

could still account for the different effects on seizure activity. For example, the deficit in brain stem 5-HT observed after 5,7-DHT injection probably results from damage to functionally important nerve terminals in this area, while the deficit after the mechanical lesion is more likely due to the destruction of the cell bodies of 5-HT neurons that innervate the forebrain and the concomitant reduction in cell body 5-HT content. It is also possible that a specific group of ascending serotonergic fibers damaged by the chemical lesion, but spared by the mechanical lesion, are involved in altering seizure susceptibility. It is known, for example, that the lateral aspects of the B-9 nuclei are not destroyed by the mechanical lesion.

Another possibility is that any increase in seizure susceptibility that would have resulted from 5-HT depletion after a raphe lesion was masked by damage to other midbrain structure(s) that normally facilitate seizures. This possibility seems very likely in view of the fact that the acetylcholine content of the forebrain can be reduced by fairly discrete lesions of the dorsal and median raphe nuclei[47] and the fact that acetylcholine appears to be a facilitator of convulsions.[5] Thus, damage to nonserotonergic neurons may have opposed the effects of a 5-HT deficit.

Alternatively, the enhanced seizure susceptibility observed after 5,7-DHT treatment could be due to damage to structures other than serotonergic neurons, and the changes in seizure susceptibility could be unrelated to the 5-HT deficit. In the absence of further neurochemical and anatomic data, this possibility cannot be dismissed. Thus, we cannot, on the basis of these studies, conclude that a 5-HT deficit is responsible for the seizure-facilitating effects of 5,7-DHT. On the other hand, it is conceivable that the seizure-facilitating effects of 5,7-DHT are mediated through the destruction of a select group of 5-HT neurons that were spared by the raphe lesion. It is of interest that other investigators[48, 49] have reported a similar disparity between the effects of a raphe lesion and other methods of 5-HT depletion on various behavioral and physiologic parameters.

Effect of 5,7-DHT or a Raphe Lesion on the Anticonvulsant Action of Phenytoin, Acetazolamide, and Phenobarbital

Materials and Methods

In these studies, male Sprague-Dawley rats that weighed 250–300 g were treated with protriptyline (20 mg/kg) and 5,7-DHT as previously described, except that the dose of 5,7-DHT was reduced to 150 μg because of a 20–25% mortality rate observed in rats that received 200 μg of 5,7-DHT. Thirty days later, the rats were subjected to maximal electroshock 2 hr after pretreatment with phenytoin, acetazolamide, or phenobarbital. The ability of the anticonvulsants to abolish hindleg extension in the maximal electroshock test was assessed by use of three or four doses of anticonvulsant with at least a 1-week interval between doses. This test is widely used to evaluate the anticonvulsant activity of drugs.[39, 43] Separate groups of rats were used for each anticonvulsant, and all studies were repeated in at least one separate group of rats where the order of dosing was reversed to determine whether a given dose altered the response to a subsequent dose 1 week later. With all three drugs, reversing the order of dosage did not influence the dose-response curve obtained. Each animal also received one electroshock without anticonvulsant, and rats that did not display full hindlimb extension were classified as nonextenders and eliminated from the study. Values for the ED_{50} were calculated by use of the method of Litchfield and Wilcoxon.[50] Phenytoin and phenobarbital were dissolved in 0.9% NaCl and administered intraperitoneally. Acetazolamide was dissolved in the same vehicle and administered intravenously via the femoral vein.

Brain monoamine analyses were performed as described on pages 438–442.

1–2 weeks after the last seizure test. Lesioning of the raphe nuclei was also accomplished by the procedure described on those pages.

Results and Discussion

In rats tested 30–50 days after treatment with 5,7-DHT, the anticonvulsant activity of phenytoin and acetazolamide was found to be reduced (FIGURE 7). With phenytoin, the ED_{50} was increased from 10.5 mg/kg (95% confidence limits, 8.5–12.9) in vehicle-treated rats to 20 mg/kg (13.8–29.0) in 5,7-DHT-treated rats ($p < 0.05$). The greatest difference between the two groups was observed with 12 mg/kg of phenytoin, which protected 83% of the control rats that received maximal electroshock from extension but which failed to protect any rat in the 5,7-DHT-treated group (FIGURE 7). Similar

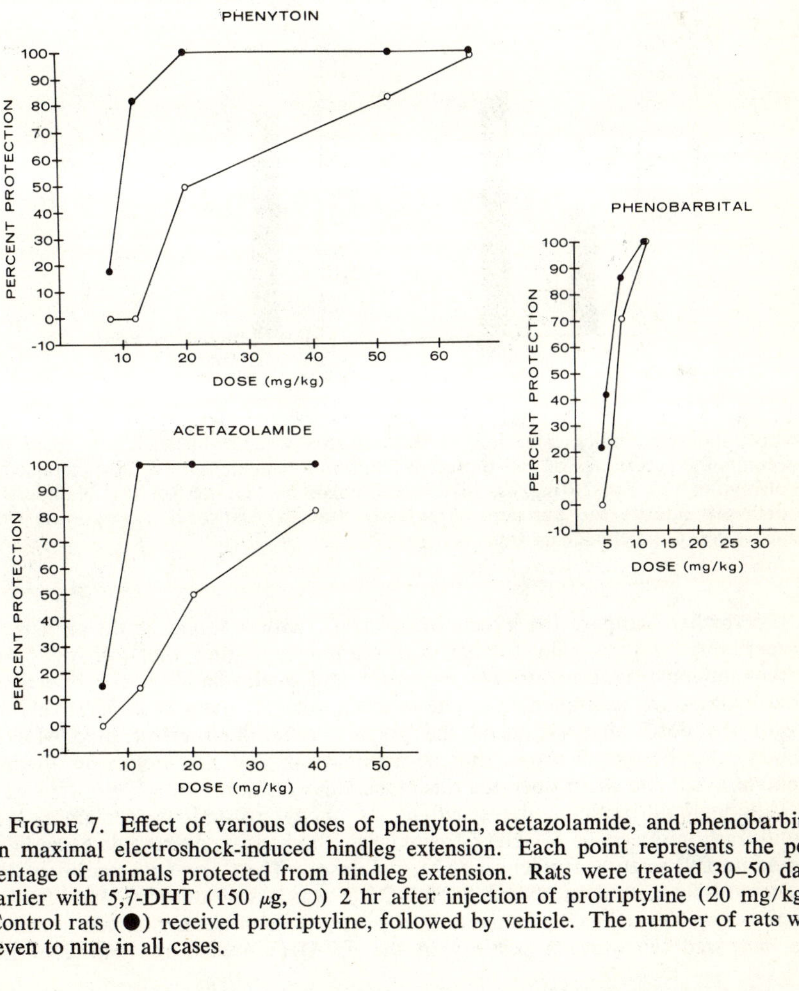

FIGURE 7. Effect of various doses of phenytoin, acetazolamide, and phenobarbital on maximal electroshock-induced hindleg extension. Each point represents the percentage of animals protected from hindleg extension. Rats were treated 30–50 days earlier with 5,7-DHT (150 μg, ○) 2 hr after injection of protriptyline (20 mg/kg). Control rats (●) received protriptyline, followed by vehicle. The number of rats was seven to nine in all cases.

findings were obtained with acetazolamide, where a 12 mg/kg dose protected 100% of the vehicle-injected rats from hindleg extension, yet only 14% of the 5,7-DHT-treated rats. The ED_{50} for acetazolamide was 8 mg/kg (6.6–9.8) in control rats and 23 mg/kg (14.8–35.7) in 5,7-DHT-treated rats ($p < 0.05$). The response to phenobarbital, however, was not significantly altered in 5,7-DHT-treated rats (FIGURE 7). The phenobarbital ED_{50} for abolishing hindleg extension was found to be 6 mg/kg (4.7–7.2) and 7.2 mg/kg (6.2–8.4) for control and 5,7-DHT-treated rats, respectively. Monoamine analyses performed 2 weeks after the last seizure test revealed a significant deficit in the forebrain 5-HT content of 5,7-DHT-treated rats but no deficit in NE (TABLE 4).

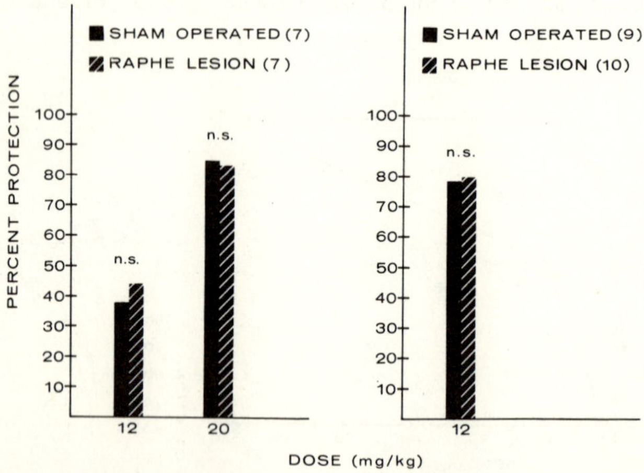

FIGURE 8. Effect of phenytoin (*left*) and acetazolamide (*right*) on maximal electroshock-induced hindleg extension 30–35 days after a raphe nuclei lesion. Each bar represents the percentage of rats protected from hind leg extension 2 hr after injection of phenytoin (12 and 20 mg/kg, ip) or acetazolamide (12 mg/kg, iv). Numbers in parentheses indicate the numbers of animals studied. Statistical comparisons were made by use of the chi-square test.

To further compare the effects of 5,7-DHT with a raphe lesion on seizure susceptibility and to gain further information regarding the role of 5-HT in the anticonvulsant action of phenytoin and acetazolamide, the effects of these drugs were examined in raphe-lesioned rats 30 days postoperatively by use of the doses that produced the greatest differential effect in 5,7-DHT-treated rats. FIGURE 8 shows that both phenytoin and acetazolamide affected raphe-lesioned and sham-operated rats in a similar fashion.

Although the anticonvulsant activity of phenytoin and acetazolamide was significantly reduced in 5,7-DHT-treated rats, both drugs were still capable of providing full seizure protection at higher doses. Thus, the antagonism of the anticonvulsant action produced by 5,7-DHT was surmountable. This finding suggests that the reduced effectiveness of phenytoin and acetazolamide is due to the increased seizure susceptibility in the 5,7-DHT-treated rat and does not

TABLE 4

MONOAMINE ANALYSES IN 5,7-DHT-TREATED AND RAPHE-LESIONED RATS AFTER ANTICONVULSANT TESTING *

Group	Number of Rats	Forebrain 5-HT (μg/g)	Forebrain NE (μg/g)	Forebrain DA (μg/g)	Brain Stem 5-HT (μg/g)
Vehicle-phenytoin	7	0.776±0.02	0.427±0.02		1.227±0.07
5,7-DHT-phenytoin	8	0.232±0.04 †	0.385±0.02		0.992±0.02 ‡
Change (%)		−71			−20
Vehicle AZM	7	0.812±0.02	0.452±0.01		
5,7-DHT-AZM	7	0.174±0.02 †	0.410±0.02		
Change (%)		−79			
Vehicle Pb	7	0.804±0.04	0.445±0.03		1.089±0.03
5,7-DHT-Pb	7	0.240±0.02 †	0.437±0.04		0.921±0.03 ‡
Change (%)		−71			
Sham-phenytoin	7	0.678±0.02	0.434±0.01	1.039±0.06	
Raphe-phenytoin	7	0.143±0.01 †	0.418±0.009	1.104±0.05	
Change (%)		−79			
Sham-AZM	7	0.631±0.02	0.392±0.02	1.180±0.05	
Raphe-AZM	7	0.139±0.03 †	0.370±0.03	1.044±0.05	
Change (%)		−78			

* Monoamine analysis in rats used for anticonvulsant testing were performed 1–2 weeks after the last seizure test or 50–70 days after treatment with 5,7-DHT (150 μg) or a raphe nuclei lesion. Details for the treatment with 5,7-DHT and the method of producing a raphe lesion are described in MATERIALS AND METHODS. Control rats in the 5,7-DHT experiment received protriptyline (20 mg/kg) 2 hr prior to the intraventricular injection of vehicle. All values are reported as the mean±SEM. The responses of these groups to phenytoin, acetazolamide (AZM), and phenobarbital (Pb) are shown in FIGURES 7 AND 8.

† $p < 0.001$ compared to vehicle-injected or sham-operated controls.

‡ $p < 0.05$ compared to vehicle-injected or sham-operated controls.

imply a key role for 5-HT in the mechanism of action of these drugs. On the other hand, if the reduced effectiveness of phenytoin and acetazolamide is due merely to a nonspecific antagonism, that is, an increased central nervous system excitability in the 5,7-DHT-treated rat, one would expect to see a similar decrease in the action of phenobarbital. Therefore, we cannot exclude the possibility that the altered anticonvulsant activity in 5,7-DHT-treated rats reflects some involvement of 5-HT in the action of these drugs. Other investigators have, for example, implicated 5-HT in the anticonvulsant action of phenytoin,[30] phenobarbital,[31] and the carbonic anhydrase inhibitors.[13] However, comparison of these studies with one another or with the present one is greatly confounded by the use of different species, different types of experimental seizures, and different anticonvulsants and would not be worthwhile.

Moreover, any discussion of the specific role of 5-HT in the action of phenytoin or acetazolamide seems premature, because comparable data were not obtained in raphe-lesioned rats. Obviously, any further interpretation of these findings is subject to the same kinds of speculation described on page 445 regarding anatomic differences in the effect of two procedures on central 5-HT neurons, and any firm conclusions regarding the role of 5-HT in the anticonvulsant-antagonizing effects of 5,7-DHT must be withheld until we can account for the failure to obtain similar results in raphe-lesioned rats.

Summary and Conclusions

Protriptyline-pretreated rats injected intraventricularly with 5,7-DHT (200 μg) exhibited an increased sensitivity to seizures induced by PTZ or maximal electroshock 35 days after treatment. Analyses of brain 5-HT in the 5,7-DHT-treated rats revealed marked deficits in the forebrain (−84%) and spinal cord (−82%) with a moderate deficit (−42%) in the brain stem. The forebrain concentrations of NE and DA were unaltered in the 5,7-DHT-treated rats at the time of seizure testing. The ED_{50} values of two anticonvulsants, phenytoin and acetazolamide, for preventing hindleg extension in the maximal electroshock test were significantly increased ($p < 0.05$) in the 5,7-DHT-treated rats, providing further evidence of an increased propensity for seizures in the chemically lesioned animals. The anticonvulsant action of phenobarbital, however, was unaltered by 5,7-DHT treatment. The ability of both phenytoin and acetazolamide to provide essentially complete protection at higher doses demonstrates their effectiveness in 5-HT-deficient animals and argues against a key role for 5-HT in the mechanism of action of these drugs, as some authors have proposed.[13, 31, 32] On the other hand, failure to observe an increase in the ED_{50} of phenobarbital in 5,7-DHT-treated rats leaves open the possibility of differences in the mechanism by which these drugs abolish the electroshock-induced hindleg extension and suggests that 5-HT may influence the action of these drugs to different extents.

The demonstration of a seizure-facilitating effect of 5,7-DHT by the present study supports the hypothesis that 5-HT normally exerts an inhibitory influence on the production or manifestation of experimentally induced generalized seizures and is thus in agreement with the majority of the literature on this subject. Moreover, due to the greater specificity of protriptyline and 5,7-DHT, the present study appears to strengthen the concept that 5-HT normally exerts anticonvulsant effects.

However, any firm conclusions regarding the role of 5-HT in the effects of 5,7-DHT must account for the finding that a lesion of the raphe nuclei, which produced an 80% deficit in forebrain 5-HT, failed to alter the response of animals to either chemically or electrically induced seizures. This observation is compatible with several interpretations that may be invoked to explain the disparity in results. First, whereas the chemical lesion reduced the spinal cord 5-HT content by 82%, the mechanical lesion was without effect. Conceivably, removal of the descending serotonergic neurons could account for the increased seizure susceptibility in the chemically lesioned rats. Second, it is possible that both the ascending and descending 5-HT fibers must be damaged before changes in seizure susceptibility are observed. These possibilities cannot be dismissed, but they do not harmonize well with what is known about the effects of the descending system on the excitability of the spinal cord. Third, it seems likely that, although producing a similar reduction in the content of brain stem 5-HT, the two procedures damaged different groups of brain stem 5-HT neurons (*vide supra*). Thus, there may be significant regional differences in the depletion of 5-HT produced by the two procedures that could explain the discrepancy. Fourth, the mechanical lesion undoubtedly destroyed many nonserotonergic neurons. It is conceivable that one or more of these neurons may have opposed the effects of serotonergic neurons on seizure susceptibility and that its removal therefore compensated for any changes that would have been observed. It has been shown, for example, that a raphe lesion can lower the forebrain concentration of acetylcholine,[47] a neurotransmitter that is believed to facilitate seizures.[5] Finally, 5,7-DHT may have seizure-facilitating effects unrelated to its action on serotonergic neurons.

While we cannot reconcile the confusion concerning the role of 5-HT in seizure susceptibility, we have demonstrated that 80% of forebrain 5-HT can be removed without altering the response to convulsive stimuli. Moreover, this study emphasizes the fact (reported by others[48,49] and by Lorens in this monograph) that different methods of 5-HT depletion do not always produce comparable changes in function. Recent research has also shown that individual serotonergic pathways may be responsible for a specific behavior or physiologic function. For example, Jacobs *et al.*[51] provided evidence that loss of the 5-HT neurons that innervate the hippocampus is responsible for the increase in locomotor behavior observed after a raphe lesion. By analogy, it seems likely that a specific group of 5-HT neurons is responsible for antagonizing seizures. Presumably, the neurons of this group were not among those damaged by lesioning of the dorsal and median raphe nuclei.

Acknowledgments

The authors thank Dr. C. A. Stone of Merck Sharp and Dohme for generously supplying the protriptyline hydrochloride. Thanks are also due to Drs. D. R. Wade and L. A. Baez for helpful suggestions made during the preparation of this manuscript.

References

1. Chen, G., C. R. Ensor & B. Bohner. 1954. A facilitation action of reserpine on the central nervous system. Proc. Soc. Exp. Biol. Med. **86:** 507–510.

2. Maynert, E. W. 1969. The role of biochemical and neurohumoral factors in the laboratory evaluation of antiepileptic drugs. Epilepsia **10:** 145–162.
3. Lovell, R. A. 1971. Some neurochemical aspects of convulsions. *In* Handbook of Neurochemistry. A. Lajtha, Ed. Vol. **6:** 63–100. Plenum Publishing Corporation. New York, N.Y.
4. Boggan, W. O. 1973. Serotonin and convulsions. *In* Serotonin and Behavior. J. Barachas & E. Usdin, Eds. : 167–172. Academic Press, Inc. New York, N.Y.
5. Maynert, E. W., T. J. Marczynski & R. A. Browning. 1975. The role of the neurotransmitters in the epilepsies. *In* Advances in Neurology. W. J. Friedlander, Ed. Vol. **13:** 79–147. The Raven Press. New York, N.Y.
6. Lessin, A. W. & M. W. Parkes. 1959. The effects of reserpine and other agents upon leptazol convulsions in mice. Brit. J. Pharmacol. **14:** 108–111.
7. Chen, G., C. R. Ensor & B. Bohner. 1968. Studies of drug effects on electrically induced extensor seizures and clinical implications. Arch. Int. Pharmacodyn. **172:** 183–218.
8. Rudzik, A. D. & G. A. Johnson. 1970. Effects of amphetamine and amphetamine analogs on convulsive thresholds. *In* International Symposium on Amphetamines and Related Compounds. E. Costa & S. Garattini, Eds. : 715–728. The Raven Press. New York, N.Y.
9. De La Torre, J. C., H. M. Kawanaga & S. Mullan. 1970. Seizure susceptibility after manipulation of brain serotonin. Arch. Int. Pharmacodyn. **188:** 298–304.
10. Azzaro, A. J., G. R. Wenger, E. R. Craig & R. E. Stitzel. 1972. Reserpine-induced alterations in brain amines and their relationship to changes in the incidence of minimal electroshock seizures in mice. J. Pharmacol. Exp. Ther. **108:** 558–568.
11. Boggan, W. O. & L. S. Seiden. 1973. 5-Hydroxytryptophan reversal of reserpine enhancement of audiogenic seizure susceptibility in mice. Physiol. Behav. **10:** 9–12.
12. Chen, G., C. R. Ensor & B. Bohner. 1968. Drug effects on the disposition of active biogenic amines in the CNS. Life Sci. **7:** 1063–1074.
13. Gray, W. D. & C. E. Rauh. 1971. The relation between monoamines in brain and the anticonvulsant action of inhibitors of carbonic anhydrase. J. Pharmacol. Exp. Ther. **177:** 206–218.
14. Kilian, M. & H. H. Frey. 1973. Central monoamines and convulsive thresholds in mice and rats. Neuropharmacology **12:** 681–692.
15. De La Torre, J. C. & S. Mullan. 1970. A possible role for 5-hydroxytryptamine in drug-induced seizures. J. Pharm. Pharmacol. **22:** 858–859.
16. Lehmann, A. 1968. Modification de l'intensité de la arise audiogène par des substances active sur le métabolisme des amines biogènes du cerveau de souris. Compt. Rend. Soc. Biol. **162:** 24–27.
17. Jobe, P. C., A. L. Picchioni & L. Chin. 1973. Role of brain 5-hydroxytryptamine in audiogenic seizure in the rat. Life Sci. **13:** 1–13.
18. Wada, J. A., E. Balzumo, B. S. Meldrum & R. Naquet. 1972. Behavioral and electrographic effects of L-5-hydroxytryptophan and DL-parachlorophenylalanine on epileptic Senegalese baboon (Papio papio). Electroencephalogr. Clin. Neurophysiol. **33:** 520–526.
19. Koe, B. K. & A. Weissman. 1966. p-Chlorophenylalanine: a specific depletor of brain serotonin. J. Pharmacol. Exp. Ther. **154:** 499–516.
20. Chase, T. N. & D. L. Murphy. 1973. Serotonin and central nervous system function. Annu. Rev. Pharmacol. **13:** 181–197.
21. Nasello, A. G. & E. S. Marichich. 1973. Effects of some cholinergic, adrenergic and serotoninergic compounds, glutamic acid and GABA on hippocampal seizures. Pharmacology **9**(4)**:** 233–239.
22. Klawans, H. L., C. Goetz & W. J. Weiner. 1973. 5-Hydroxytryptophan-induced myoclonus in guinea pigs and the possible role of serotonin in infantile myoclonus. Neurology **23:** 1234–1240.

23. STEWART, R. M., J. H. GROWDON, D. CANCIAN & R. J. BALDESSARINI. 1976. 5-Hydroxytryptophan-induced myoclonus: increased sensitivity to serotonin after intracranial 5,7-dihydroxytryptamine in the adult rat. Neuropharmacology **15:** 449–455.
24. HERDBERG, L. J. & K. B. J. FRANKLIN. 1976. The 'stimulant' action of tryptophan-monoamine oxidase inhibitor combinations: suppression of self-stimulation. Neuropharmacology **15:** 349–351.
25. GREEN, A. R. & D. G. GRAHAME-SMITH. 1976. Effects of drugs on the process regulating the functional activity of brain 5-hydroxytryptamine. Nature (London) **260:** 487–491.
26. ALEXANDER, G. J., L. M. KOPELOFF & R. B. ALEXANDER. 1971. Anticonvulsant effect of p-chlorophenylalanine in audiosensitive mice. Life Sci. **10:** 877–882.
27. BAUMGARTEN, H. G., L. LACHENMAYER & H. G. SCHLOSSBERGER. 1972. Evidence for a degeneration of indolamine-containing nerve terminals in rat brain induced by 5,6-dihydroxytryptamine. Z. Zellforsch. **115:** 553–569.
28. BAUMGARTEN, H G. & L. LACHENMAYER. 1972. 5,7-Dihydroxytryptamine: improvement in chemical lesioning of indoleamine neurons in the mammalian brain. Z. Zellforsch. **135:** 399–414.
29. BAUMGARTEN, H. G., A. BJÖRKLUND, L. LACHENMAYER & A. NOBIN. 1973. Evaluation of the effects of 5,7-dihydroxytryptamine on serotonin and catecholamine neurons in the rat CNS. Acta Physiol. Scand. Suppl. **391:** 1–19.
30. JENNER, P., D. CHADWICK, E. H. REYNOLDS & C. D. MARSDEN. 1975. Altered 5-HT metabolism with clonazepam, diazepam and diphenylhydantoin. J. Pharm. Pharmacol. **27:** 707–710.
31. MEYER, H. & H. H. FREY. 1973. Dependence of anticonvulsant drug action on central monoamines. Neuropharmacology **12:** 939–947.
32. NOBLE, E. P., R. J. WURTMAN & J. AXELROD. 1967. A simple and rapid method for injecting H^3-norepinephrine into the lateral ventricle of the rat brain. Life Sci. **6:** 281–291.
33. BOGDANSKI, D. F., A. PLETSCHER, B. B. BRODIE & S. UDENFRIEND. 1956. Identification and assay of serotonin in brain. J. Pharmacol. Exp. Ther. **117:** 82–88.
34. ANTON, A. H. & D. F. SAYRE. 1964. The distribution of dopamine and dopa in various animals and a method for their determination in diverse biological material. J. Pharmacol. Exp. Ther. **45:** 326–336.
35. GERSON, S. & R. J. BALDESSARINI. 1975. Selective destruction of serotonin terminals in rat forebrain by high doses of 5,7-dihydroxytryptamine. Brain Res. **85:** 140–145.
36. BJÖRKLUND, A., H. G. BAUMGARTEN & A. RENSCH. 1975. 5,7-Dihydroxytryptamine: improvement of its selectivity for serotonin neurons in the CNS by pretreatment with desipramine. J. Neurochem. **24:** 833–835.
37. DONOVICK, P. J. 1974. A metachromatic stain for neural tissue. Stain Technol. **49:** 49–51.
38. WOODBURY, L. A. & V. D. DAVENPORT. 1952. Design and use of a new electroshock seizure apparatus and analysis of factors altering seizure threshold and pattern. Arch. Int. Pharmacodyn. **92:** 97–107.
39. TOMAN, J. E. P., E. A. SWINYARD & L. S. GOODMAN. 1946. Properties of maximal seizures and their alteration by anticonvulsant drugs and other agents. J. Neurophysiol. **9:** 231–240.
40. ADLER, M. W. 1969. Laboratory evaluation of antiepileptic drugs. The chronic use of lesions. Epilepsia **10:** 263–280.
41. LAFFAN, R. J., E. A. SWINYARD & L. S. GOODMAN. 1957. Stimulus intensity, maximal electroshock seizures and potency of anticonvulsants in rats. Arch. Int. Pharmacodyn. **111:** 60–69.
42. JOBE, P. C., R. E. STULL & P. F. GEIGER. 1974. The relative significance of norepinephrine, dopamine and 5-hydroxytryptamine in electroshock seizure in the rat. Neuropharmacology **13:** 961–968.

43. SWINYARD, E. A. 1969. Laboratory evaluation of antiepileptic drugs: review of laboratory methods. Epilepsia **10:** 107–119.
44. DAHLSTRÖM, A. & K. FUXE. 1964. Evidence for the existence of monoamine containing neurons in the central nervous system: demonstration of monoamines in the cell bodies of the brain stem neurons. Acta Physiol. Scand. **62**(Suppl. 232): 1–55.
45. KOVACS, D. A. & J. G. ZOLL. 1974. Seizure inhibition by median raphe nucleus stimulation in rat. Brain Res. **70:** 165–169.
46. ANDERSON, E. G. 1972. Bulbospinal serotonin-containing neurons and motor control. Fed. Proc. **31:** 107–112.
47. PEPEU, G., L. GARAU & M. L. MULAS. 1974. Does 5-hydroxytryptamine influence cholinergic mechanisms in the central nervous system. *In* Advances in Biochemical Psychopharmacology. E. Costa, G. L. Gessa & M. Sandler, Eds. Vol. **10:** 247–252. The Raven Press. New York, N.Y.
48. SHEARD, M. H. 1973. Brain serotonin depletion by p-chlorophenylalanine or lesions of raphe neurons in rats. Physiol. Behav. **10:** 809–811.
49. HARVEY, J. A., A. J. SCHLOSBERG & L. M. YUNGER. 1974. Effect of p-chlorophenylalanine and brain lesions on pain sensitivity and morphine analgesia in the rat. *In* Advances in Biochemical Psychopharmacology. E. Costa, G. L. Gessa & M. Sandler, Eds. Vol. **10:** 233–245. The Raven Press. New York, N.Y.
50. LITCHFIELD, J. T. & F. WILCOXON. 1949. A simplified method of evaluating dose effect experiments. J. Pharmacol. Exp. Ther. **96:** 99–113.
51. JACOBS, B. L., W. D. WISE & K. M. TAYLOR. 1974. Differential and neurochemical effects following lesions of the dorsal or median raphe nuclei in rats. Brain Res. **79:** 353–361.

DISCUSSION

DR. D. V. COSCINA: It may be important to consider the model of convulsion you use.

DR. BROWNING: I agree with you. For example, in audiogenic seizures or the threshold to electric shock, we see no effect with 5,7-DHT. Others have reported the same thing with *p*-CPA.

CARDIOVASCULAR EFFECTS OF NEUROTOXIC INDOLETHYLAMINES *

Manfred Göthert and Niels Klupp

Institute of Pharmacology
University of Hamburg
2000 Hamburg 20, Federal Republic of Germany

Intracerebroventricular injection of 5,6-dihydroxytryptamine (5,6-DHT),[1-4] 5,7-dihydroxytryptamine (5,7-DHT),[5,6] or related tryptamine derivatives[7] has been shown to cause toxic damage to 5-hydroxytryptamine (5-HT) -containing neurons and, to a lesser degree, also to catecholamine-containing neurons. These compounds can be used as pharmacologic tools for the evaluation of the physiologic role of central 5-HT-secreting neurons. Thus, intracerebroventricular or intracisternal administration of 5,6-DHT has been used to investigate the role of serotonergic neurons in the central regulation of blood pressure.[8-12]

However, the cytotoxic effects of these compounds are not restricted to the central monoamine neurons. After systemic administration, certain indolethylamines produce toxic effects on myocardium[13] and on sympathetic nerve terminals of the heart.[14-16] In addition, it has been shown in rats that intravenous injection of neurotoxic tryptamine derivatives induces an increase in blood pressure;[17,18] evidence has been presented that the pressor effect of 5,6-DHT is due to an activation of 5-HT receptors in the vascular bed.[17] On the basis of these results, it appeared to be of interest to perform additional experiments in isolated vascular and heart preparations to compare more precisely the affinity of the neurotoxic indolethylamines and of 5-HT to the 5-HT receptors.

Recently, it was demonstrated in rabbit hearts that the positive chronotropic effect of 5-HT is due to norepinephrine release from the sympathetic nerves, which, in turn, is caused by an activation of 5-HT receptors on the nerve terminals.[19,20] It is therefore possible to investigate in isolated rabbit hearts whether the toxic indolethylamines also interfere with neuronal receptors.

During the last few years, we have performed investigations on the cardiovascular effects of indolethylamines, and this paper will review our findings. Most of these results have not yet been published.

Effects on Cardiovascular 5-HT Receptors

Intravenous injection of 5,6-DHT induced a dose-dependent increase in blood pressure of pithed rats.[17] Neither pretreatment with reserpine nor administration of cocaine decreased the pressor effect of 5,6-DHT; therefore, an indirect sympathomimetic effect does not contribute to the pressor effect of the compound, although the norepinephrine content of the heart was reduced by 68% within 3 hr after intraperitoneal injection of 5,6-DHT (30 mg/kg).[17] The pressor effect of 5,6-DHT was not altered by 2-[β-(4-hydroxyphenyl)ethyl-

* Supported by the Deutsche Forschungsgemeinschaft.

0077–8923/78/0305–0457 $01.75/0

aminomethyl]tetralone (HEAT),[17] which has been shown to be a competitive antagonist of α-adrenoceptors.[21, 22] However, phentolamine, which blocks not only α-adrenoceptors but also 5-HT receptors,[23, 24] inhibited the effect of 5,6-DHT.[17] The selective, competitive 5-HT antagonist methysergide blocked the pressor effect of 5,6-DHT in a competitive fashion. These results indicate that 5,6-DHT stimulates the 5-HT receptors of the cardiovascular system. However, the 5,6-DHT dose that produces 50% of the maximal response (170 μg/kg) was 14 times higher than that of 5-HT, suggesting a lower affinity of 5,6-DHT for the 5-HT receptors.[17]

The pressor effect of 5,7-DHT was considerably less pronounced than that of 5,6-DHT; neither methysergide nor HEAT altered the increase in blood pressure induced by this amine.[18] Therefore, 5,6-DHT and 5,7-DHT seem to differ in their mechanism of action underlying the pressor effect. It is tempting to speculate that the vasoconstrictor effect of 5,7-DHT may be so weak that it does not contribute substantially to the pressor effect. In this case, the rise in blood pressure may be induced by a marked increase in cardiac output due to positive chronotropic and inotropic effects, which are caused by an activation of 5-HT receptors in the heart. Such receptors are blocked neither by HEAT nor by methysergide (see below). Contrary to that observation, the pressor effects of 5,6-DHT and 5-HT appear to be caused mainly by vasoconstriction, elicited by stimulation of vascular 5-HT receptors. However, only *in vitro* experiments in isolated vascular and heart preparations can provide more detailed information regarding the relative potency and mechanism of action of the various tryptamines.

Since it appeared possible that the marked pressor effect of the indolethylamines, particularly of 5-HT and 5,6-DHT, contributed to the acute toxicity of the compounds (e.g., by inducing acute heart failure), we investigated whether pretreatment of mice with methysergide decreased their acute toxicity. The LD_{50}s of 5-HT, 5,6-DHT, and 5,7-DHT in male albino mice (body weight, 17–23 g) amounted to 135, 105, and 200 mg/kg, respectively (determined as described by Behrens and Kärber;[25] injection of the compounds into a tail vein; duration of injection, 1 min). All mice that did not die within 15 min (severe convulsions before death) survived for more than 6 days. Intravenous injection of methysergide (0.2 and 2.0 mg/kg) 30 min before administration of 5-HT increased the LD_{50} of this amine to values of 212 and 238 mg/kg, respectively. However, the LD_{50} of 5,6-DHT was only slightly increased (to 125 mg/kg) by previous intravenous injection of methysergide (2 mg/kg). It may be concluded that vasoconstriction or other effects of 5,6-DHT that are mediated by 5-HT receptors susceptible to blockade with methysergide do not contribute substantially to the acute toxicity. Death seemed to be caused by respiratory failure, which was also observed in rats anesthetized with urethane after intravenous injection of 5,6-DHT.[17]

The LD_{50} of 5,6-DHT after subcutaneous injection has been shown by Creveling *et al.*[13] to be about 50% higher than the value that we determined after intravenous injection. However, most of the animals died on Day 4 or 5 after subcutaneous injection, suggesting different mechanisms of toxicity after subcutaneous (see below) and intravenous administration routes, respectively. It is interesting to note that no animal died within 5 days after subcutaneous injection of 5,7-DHT at doses up to 400 mg/kg.[13]

The vasoconstrictor effect of the indolethylamines was studied in more detail in isolated strips of the rabbit aorta (prepared according to Furchgott and Bhadrakom[26]). The methods used have been described in a previous paper.[27] Cocaine (30 μM) was present in the bathing solution throughout the experiments to inhibit the amine uptake mechanism located in the membrane of adrenergic neurons. Thus, the possibility of a tyramine-like indirect sympathomimetic effect was excluded. Preliminary experiments revealed that cocaine did not attenuate the effects of 5-HT, 5,6-DHT, and 5,7-DHT in this preparation. This finding is in contrast to the results obtained in guinea pig and rabbit hearts (see below).

Both 5,6-DHT and 5,7-DHT caused a concentration-dependent increase in tension (FIGURE 1). The slopes of the concentration-response curves and the maximal responses induced by the amines were not significantly different from

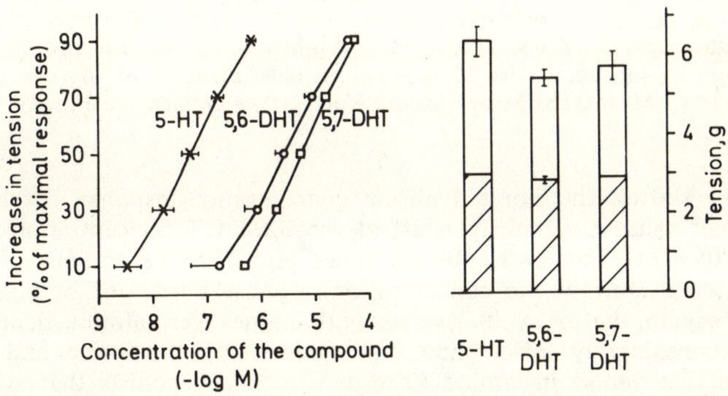

FIGURE 1. Cumulative concentration-response curves of tryptamine derivatives in isolated strips of rabbit aorta. The bathing solution was Tyrode (37°C) that contained cocaine (30 μM). In each strip, only one concentration-response curve was determined. Means±SEM are given. *Left* (graphs): Increase in tension expressed as percentage of the maximal response. The drug concentrations that caused 10–90% of the maximum effect were determined graphically for each individual strip (according to Van Rossum[28]). 5-HT (×): 5-Hydroxytryptamine; $n=6$. 5,6-DHT (○): 5,6-Dihydroxytryptamine; $n=4$. 5,7-DHT (□): 5,7-Dihydroxytryptamine; $n=6$. *Right* (histograms): Shaded areas of columns represent basal tension before addition of the amines; shaded plus open areas indicate maximal tension evoked by the amine; open areas denote maximal increase in tension evoked by the amine (which corresponds to 100% of the left scale).

those of 5-HT. The negative logarithms of the concentrations of 5,6-DHT, 5,7-DHT, and 5-HT that caused 50% of the maximum effect (pD_2 values, determined as described by Van Rossum[28]) amounted to 5.57, 5.27, and 7.33, respectively. Methysergide induced a concentration-dependent parallel shift of the 5,6-DHT concentration-response curve, indicating that methysergide is a competitive antagonist of 5,6-DHT (FIGURE 2). The inhibitory potency of methysergide can be expressed in terms of its pA_2 value (negative logarithm of the antagonist concentration that causes a shift of the concentration-response curve of the agonist by a factor of two, determined as previously described[27]). The pA_2 value of methysergide versus 5,6-DHT was almost identical to that versus 5-HT (TABLE 1). In contrast, a low concentration of HEAT, which was roughly equivalent to the highest methysergide concentration used and which

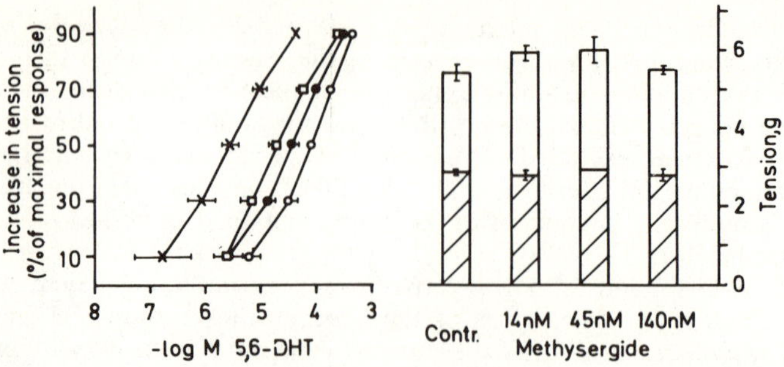

FIGURE 2. Effect of methysergide on the cumulative concentration-response curve of 5,6-dihydroxytryptamine (5,6-DHT) in isolated strips of rabbit aorta. In each strip, only one concentration-response curve was determined. Controls (Contr., ×; $n=4$). Methysergide: 14 nM (○; $n=4$); 45 nM (●; $n=4$); 140 nM (□; $n=4$); administration of the antagonist 20 min before determination of the concentration-response curves. Means±SEM are given. For further details, see FIGURE 1 legend.

considerably shifted the norepinephrine concentration-response curve to the right, did not cause a significant shift of the 5,6-DHT concentration-response curve (FIGURE 3). However, 30–300 times higher concentrations of HEAT also produced a shift of the concentration-response curves of both 5,6-DHT and 5-HT. Again, the pA_2 values versus both amines were almost identical, but they were considerably lower than that against norepinephrine and slightly higher than that versus histamine (TABLE 1). Since agonists that act at the same receptor site produce the same pA_2 value of an antagonist, we conclude that the vasoconstrictor effect of 5,6-DHT is due to activation of the 5-HT receptors in aortic smooth muscle. Neither α-adrenoceptors nor histamine receptors are involved.

FIGURE 3 suggests that the pattern of effects of low concentrations of methysergide and HEAT on the increase in tension produced by administration of a single concentration of an amine should also be suitable to differentiate

TABLE 1

ANTAGONISM OF METHYSERGIDE AND 2-[β-(4-HYDROXYPHENYL)ETHYLAMINOMETHYL]-TETRALONE (HEAT) TOWARD THE CONTRACTILE RESPONSE TO VASOCONSTRICTOR AMINES IN ISOLATED STRIPS OF RABBIT AORTA

Agonist	pA_2 Values of Methysergide	pA_2 Values of HEAT
5-Hydroxytryptamine	8.63±0.22 *	5.69±0.09
5,6-Dihydroxytryptamine	8.45±0.09	5.65±0.13
Norepinephrine	not determined	9.05±0.01 †
Histamine	not determined	5.02±0.13 †

* All values given are the means±SEM of 16–24 aortic strip determinations for each series of experiments.

† Significantly different from the pA_2 value of HEAT versus 5-hydroxytryptamine ($p < 0.02$ or less in nonpaired Student's t test).

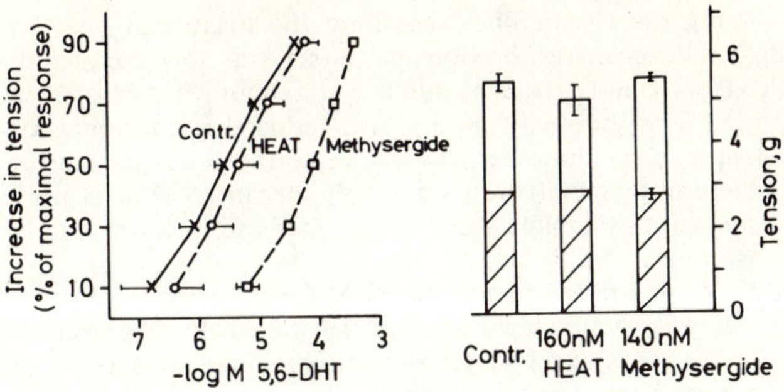

FIGURE 3. Comparison of the effects of 2-[β-(4-hydroxyphenyl)ethylaminomethyl] tetralone (HEAT, 160 nM; $n=4$) and of methysergide (140 nM; $n=4$) on the cumulative concentration-response curve of 5,6-dihydroxytryptamine (5,6-DHT) in isolated strips of rabbit aorta. In each strip, only one concentration-response curve was determined. Contr.: controls (solid line, $n=4$). Administration of the antagonist 20 min before determination of the concentration-response curves. Means±SEM are given. For other details, see FIGURE 1 legend.

whether the amine interferes with 5-HT receptors or α-adrenoceptors. Since the increase in tension induced by 57 μM 5,7-DHT was inhibited by 140 μM methysergide by 94% ($p < 0.001$) but was not affected by 160 μM HEAT, the conclusion may be drawn that 5,7-DHT probably also stimulates the 5-HT receptors. However, the affinities of 5,7-DHT and 5,6-DHT for the receptor are 115 and 58 times lower, respectively, than that of 5-HT (FIGURE 1).

Spontaneously beating right guinea pig atria were used to investigate the cardiac effects of the neurotoxic indolethylamines. Usually, the Tyrode solution (37° C) did not contain cocaine, because this drug almost completely abolished the effects of the tryptamine derivatives (see below). A detailed description of the methods employed has appeared in a previous paper.[27] All indolethylamines studied (5-HT, 5,6-DHT, 5,7-DHT) caused a concentration-dependent increase in heart rate. However, the pD_2 values, which were determined by means of cumulative concentration-response curves, were low (TABLE 2), and the decreasing order of potency was 5-HT > 5,7-DHT > 5,6-DHT. There was no

TABLE 2

pD_2 VALUES AND MAXIMUM INCREASE IN HEART RATE INDUCED BY TRYPTAMINE DERIVATIVES IN ISOLATED STRIPS OF GUINEA PIG RIGHT ATRIA (MEANS±SEM)

Agonist	pD_2 Values	Maximum Increase in Heart Rate (beats/min)
5-Hydroxytryptamine ($n=4$)	5.51±0.08	28±2
5,6-Dihydroxytryptamine ($n=4$)	4.72±0.08 *	23±6
5,7-Dihydroxytryptamine ($n=4$)	5.05±0.08 *	26±2

* Significantly different from the pD_2 value of 5-hydroxytryptamine ($p < 0.01$ or less in nonpaired Student's t test).

difference among the tryptamines regarding the maximum heart rate induced by the drugs. The positive chronotropic effect was low compared to that of, for example, isoprenaline, which caused a maximum increase in heart rate of 117 beats/min. The indolethylamines also induced a concentration-dependent positive inotropic effect; however, the effects of the drugs on tension developed by the atria were not evaluated quantitatively, because the atria were not driven electrically, and inotropic effects due to the increase in heart rate could not be estimated.

To investigate the mechanism responsible for the positive chronotropic effect of the tryptamine derivatives, we studied the influence of various drugs on the increase in heart rate induced by the indolethylamines. Reserpine pretreatment, which almost completely abolished the positive chronotropic effect of tyramine, left the increase in heart rate induced by the tryptamine derivatives isoprenaline and histamine unaffected (TABLE 3). Therefore, the positive chronotropic effects of 5-HT, 5,6-DHT, and 5,7-DHT are not due to an indirect sympathomimetic effect. This conclusion is supported by our finding that propranolol, at a concentration that reduced the effect of isoprenaline by more than 60%, also did not alter the effect of the tryptamine derivatives (TABLE 4). It is interesting to note that methysergide and cyproheptadine also failed to decrease significantly the stimulant effect of the indolethylamines; however, the latter drug diminished slightly the tachycardia induced by histamine (TABLE 4). Obviously, the increase in heart rate evoked by 5-HT, 5,6-DHT, and 5,7-DHT is not mediated by the type of 5-HT receptors that can be blocked by those antagonists. In contrast to the findings obtained with methysergide and cyproheptadine, cocaine almost completely inhibited the positive chronotropic effects of 5-HT, 5,6-DHT, and 5,7-DHT in atria from both reserpine-pretreated (TABLE 3) and untreated animals (TABLE 4). Inhibition of the effect of 5-HT in guinea pig atria by cocaine and morphine was also reported by Trendelenburg.[29] This inhibitory influence cannot be attributed to the unspecific membrane effect of cocaine, because the effects of isoprenaline and histamine were only slightly decreased in atria of untreated guinea pigs and were unaffected in atria of reserpine-pretreated animals. Therefore, the possibility may be considered that cocaine may block 5-HT receptors in cardiac tissue and that the positive chronotropic effect of the neurotoxic indolethylamines is probably due to an activation of 5-HT receptors in pacemaker fibers. However, the affinities of 5,6-DHT and 5,7-DHT for these receptors are low (TABLE 2). In this connection, it is interesting to note that in this preparation, 5,7-DHT has a slightly higher affinity than does 5,6-DHT ($p < 0.05$), whereas the opposite is true in isolated strips of rabbit aorta (FIGURE 1).

ACUTE EFFECTS ON SYMPATHETIC NERVES OF THE RABBIT HEART

The results reported so far provide evidence that 5,6-DHT and 5,7-DHT are capable of activating 5-HT receptors in both vascular and cardiac muscle. However, it would be more interesting to know whether the neurotoxic indolethylamines also activate 5-HT receptors in the membrane of nerve cells.

The positive chronotropic effect of 5-HT in rabbit hearts was shown to be due to an indirect sympathomimetic effect;[29, 30] the release of norepinephrine (NE) appears to be caused by an activation of presynaptic 5-HT receptors on sympathetic nerve terminals.[19, 20] Therefore, we investigated in isolated rabbit

TABLE 3

INCREASE IN HEART RATE (BEATS/MIN) INDUCED BY TRYPTAMINE DERIVATIVES AND SOME OTHER AMINES IN ISOLATED STRIPS OF GUINEA PIG RIGHT ATRIA (MEANS±SEM)

	5-Hydroxy-tryptamine (5.7 μM)	5,6-Dihydroxy-tryptamine (52 μM)	5,7-Dihydroxy-tryptamine (17 μM)	Tyramine (73 μM)	Isoprenaline (4.7 nM)	Histamine (9.0 μM)
Controls ($n=11$)	27±7	26±7	21±5	71±6	96±5	104±6
Reserpine pretreatment ($n=5$) *	27±4	32±6	22±4	3±3 †	118±13	105±10
Reserpine pretreatment plus cocaine ‡ ($n=5$)	1±1 §	12±5 §	1±1 §	4±2	108±7	98±9

* Influence of reserpine pretreatment (4 mg/kg 24 hr before the the experiments) on the positive chronotropic effect of the amines.

† Significantly different from controls ($p < 0.001$) in nonpaired Student's *t* test).

‡ Influence of cocaine (30 μM) on the positive chronotropic effect in atrial strips of reserpine-pretreated guinea pigs.

§ Significantly different from the corresponding effect in reserpine-pretreated animals before injection of cocaine ($p < 0.05$ or less in paired Student's *t* test).

TABLE 4

INFLUENCE OF VARIOUS ANTAGONISTS ON THE POSITIVE CHRONOTROPIC EFECTS OF TRYPTAMINE DERIVATIVES AND SOME OTHER AMINES IN ISOLATED STRIPS OF GUINEA PIG RIGHT ATRIA *

	5-Hydroxy-tryptamine (5.7 μM)	5,6-Dihydroxy-tryptamine (52 μM)	5,7-Dihydroxy-tryptamine (17 μM)	Isoprenaline (4.7 nM)	Histamine (9.0 μM)
Controls ($n=6$)	61 ± 7	97 ± 10	89 ± 15	97 ± 3	100 ± 4
Methysergide (140 nM; $n=6$)	41 ± 8	71 ± 15	57 ± 10	118 ± 14	111 ± 11
Cyproheptadine (174 nM; $n=6$)	86 ± 16	73 ± 10	81 ± 9	83 ± 10	82 ± 5 †
Propranolol (193 nM; $n=11$)	85 ± 5	89 ± 12	98 ± 7	32 ± 2 ‡	96 ± 2
Cocaine (30 μM; $n=5$)	11 ± 5 ‡	5 ± 5 ‡	6 ± 6 ‡	74 ± 7 †	79 ± 6 †

* Each amine was administered twice: the first time before, the second time 20 min after injection of the antagonist or vehicle [0.9% NaCl (controls)] into the organ bath. The positive chronotropic effects induced by the second administration of the amines are expressed as the percentages of the effects induced by the first injection.

† Significantly different from controls ($p < 0.02$ in nonpaired Student's t test).

‡ Significantly different from controls ($p < 0.005$ in nonpaired Student's t test).

hearts perfused with Tyrode solution (33° C) according to the Langendorff technique whether the neurotoxic indolethylamines also induce a positive chronotropic effect by a mechanism similar to that of 5-HT. The details of all methods utilized have been published previously.[31, 32] The experiments were performed with 5,7-DHT. Heart rate, tension developed by the hearts, and NE output from the heart were measured. The term "NE output" is used to denote the amount of NE that escapes uptake into the sympathetic nerves and that diffuses into the perfusion fluid; this fraction can be determined experimentally. The term "release" describes the efflux of NE from the interior of the nerve across the cell membrane into the extracellular space.

Continuous perfusion of the heart with 5,7-DHT evoked a transient increase in both heart rate and NE output (FIGURE 4 & TABLE 5). The maximum of both effects occurred within the first 2 min of drug perfusion. The increase in NE output was clearly related to the concentration of 5,7-DHT, whereas the maximum positive chronotropic effect was reached with the lowest concentration

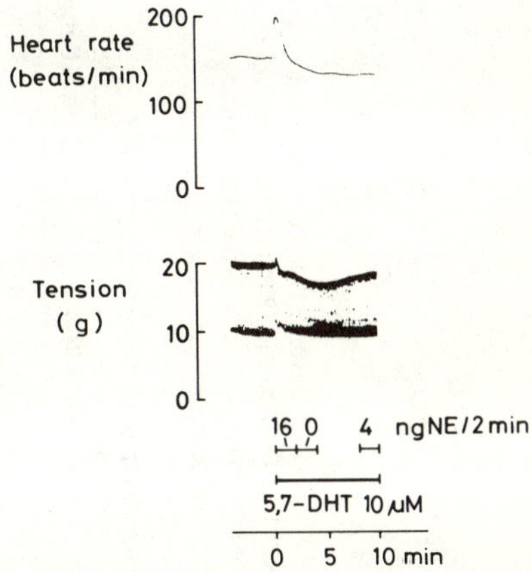

FIGURE 4. Effects of continuous infusion of 5,7-dihydroxytryptamine (5,7-DHT) into the aortic cannula of an isolated rabbit heart perfused with Tyrode solution. The concentration of 5,7-DHT indicated is the final concentration in the perfusion fluid. Abscissa: time after onset of 5,7-DHT infusion. NE: norepinephrine output from the heart.

tested (TABLE 5). The NE output during the second 2-min period after the onset of perfusion with 5,7-DHT was more than 75% less than that measured in the first 2 min and was not significantly different from that determined in the perfusate sampled 8–10 min after the onset of 5,7-DHT administration (FIGURE 4 & TABLE 5). In the latter time period, NE output in the presence of 32 μM 5,7-DHT was still increased. Ten minutes after the onset of perfusion with 5,7-DHT, heart rate did not differ significantly from that in controls.

5,7-Dihydroxytryptamine produced only a very small, transient increase in tension, which was followed by a negative inotropic effect (FIGURE 5). It is evident that the initial increase in tension is almost negligible when the effect of 5,7-DHT is compared to that of electrical stimulation of the postganglionic sympathetic nerves (FIGURE 5; prepared as described by Huković and Muscholl [33]): the positive chronotropic effect of 5,7-DHT was more pronounced than that of electrical nerve stimulation, whereas the increase in tension induced by electrical stimulation of the sympathetic nerves was five times higher than that evoked by 5,7-DHT. Also in this preparation, we decided not to perform

TABLE 5

EFFECTS OF 5,7-DIHYDROXYTRYPTAMINE (5,7-DHT, ADMINISTERED CONTINUOUSLY FOR 20 MIN) ON NOREPINEPHRINE (NE) OUTPUT, HEART RATE, AND POSITIVE CHRONOTROPIC EFFECT OF EXOGENOUS NE IN ISOLATED RABBIT HEARTS PERFUSED WITH TYRODE SOLUTION (MEANS±SEM)

	Norepinephrine Output (ng/2 min)			Maximum Increase in Heart Rate (beats/min) Evoked by	
	0–2 min *	2–4 min *	8–10 min *	5,7-DHT †	NE (59 nM) ‡
Controls ($n=10$)	3.7±1.3	—	1.9±0.5	—	17±2
5,7-DHT (10 μM; $n=6$)	21.1±5.8 §	4.6±4.6	3.4±1.2	65±4	19±7
5,7-DHT (32 μM; $n=5$)	52.2±22.2 §	10.0±2.3	8.4±2.8 §	63±4	6±4 §

* Time after onset of 5,7-DHT administration.

† The maximum occurs within the first minute of 5,7-DHT administration.

‡ 5,7-Dihydroxytryptamine was present in the Tyrode solution 10 min before and during perfusion of the hearts with exogenous NE (duration of perfusion with NE, 10 min).

§ Significantly different from controls ($p < 0.02$ or less in nonpaired Student's t test).

a detailed quantitative evaluation of the inotropic effects of 5,7-DHT, because the hearts were not driven electrically.

The site and mechanism of action responsible for the initial effects of 5,7-DHT were studied in hearts that were stimulated with bolus injections of 5,7-DHT. Pretreatment of rabbits with reserpine abolished both the positive chronotropic effect and the NE output induced by the amine (FIGURE 6). The β-adrenoreceptor antagonist pindolol also blocked the positive chronotropic effect of 5,7-DHT but left NE output unaffected (TABLE 6). Therefore, we conclude that the observed increase in heart rate is the result of 5,7-DHT-induced NE release from sympathetic nerves. Perfusion of the heart with Ca^{2+}-free Tyrode solution drastically decreased the NE output in response to 5,7-DHT (TABLE 6). Contrary to that finding, desipramine at a concentration that was

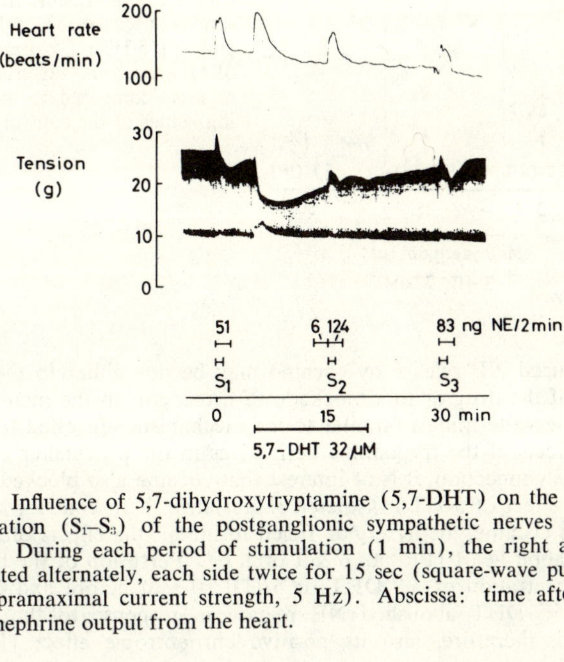

FIGURE 5. Influence of 5,7-dihydroxytryptamine (5,7-DHT) on the effects of electrical stimulation (S_1–S_3) of the postganglionic sympathetic nerves of an isolated rabbit heart. During each period of stimulation (1 min), the right and left nerves were stimulated alternately, each side twice for 15 sec (square-wave pulses of 3 msec duration; supramaximal current strength, 5 Hz). Abscissa: time after onset of S_1. NE: norepinephrine output from the heart.

shown to inhibit the uptake of NE into sympathetic nerves[34] increased the NE output evoked by the indolethylamine (TABLE 6). These results indicate that the indirect sympathomimetic effect of 5,7-DHT is not tyramine-like, because the effect of this drug is not decreased by omission of Ca^{2+} ions[35] but is dependent on the unimpaired function of the amine uptake mechanism in the membrane of the sympathetic nerves.[36]

Cocaine considerably decreased the NE output elicited by 5,7-DHT (TABLE 6), although this drug also inhibits NE uptake into sympathetic nerves.[36] The inhibition of NE uptake appears to be responsible for the fact that the positive chronotropic effect of 5,7-DHT is only partially blocked despite the strong inhibition of NE output (TABLE 6): the small amount of NE that is still released can only slowly escape from the synaptic cleft. The inhibition of

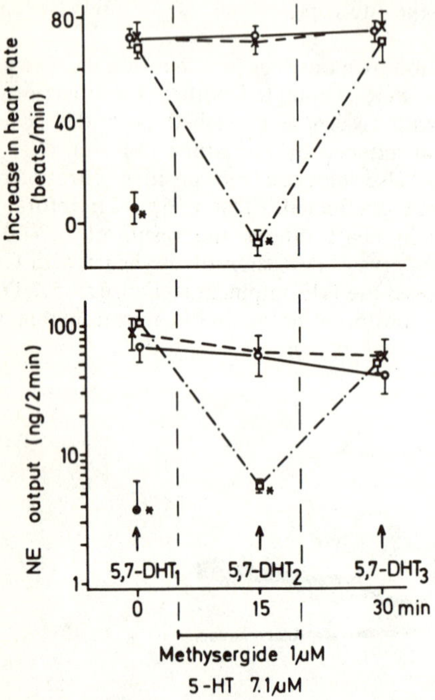

FIGURE 6. Effects of various drugs on the norepinephrine (NE) output from isolated rabbit hearts and on the increase in heart rate induced by bolus injections of 5,7-dihydroxytryptamine ($5,7\text{-DHT}_1$–$5,7\text{-DHT}_3$ 0.5 μmol) into the aortic cannula. Controls (○; $n=6$). Reserpine pretreatment (●, 5 mg/kg twice, ip, 48 and 24 hr before the experiment; $n=4$). Methysergide (×; $n=5$). 5-Hydroxytryptamine [(5-HT) □; $n=4$). Means ± SEM are given. Asterisks: $p < 0.05$ or less (compared to the corresponding values of the control experiments).

5,7-DHT-induced NE release by cocaine may be due either to the local anesthetic effect of the drug or to a blockade of a receptor in the membrane of the sympathetic nerve terminals (similar to the mechanism suggested for the inhibition of the effects of the tryptamine derivatives in the pacemaker of guinea pig atria). In this connection, it is of interest that cocaine also blocked the positive chronotropic effect of 5-HT in isolated rabbit atria.[29] In contrast to the results obtained with cocaine, methysergide failed to inhibit the effects of 5,7-DHT on the isolated rabbit heart (FIGURE 6 & TABLE 6). Perfusion of the hearts with a constant concentration of 5,7-DHT or 5-HT 10 min before and during bolus injection of 5,7-DHT abolished NE release in response to the injection of 5,7-DHT and, therefore, also its positive chronotropic effect (FIGURE 6 & TABLE 6).

In conclusion, the pattern of the effects that result from interactions among various drugs and 5,7-DHT is identical to that previously obtained with 5-HT.[20] It was suggested that 5-HT causes NE release from the cardiac sympathetic nerves by a brief depolarization of the neuronal membrane (due to an activation of 5-HT receptors) that is rapidly terminated by "autoinhibition."[20] A similar mechanism has been reported for nicotinic drugs that stimulate nicotinic receptors on cardiac sympathetic nerves.[37, 38] Since our results with 5,7-DHT (regarding interactions with other drugs and autoinhibition) are very similar to those obtained with 5-HT, and because cross-tachyphylaxis occurred between 5,7-DHT and 5-HT, we suggest that both tryptamine derivatives probably act at the same receptor sites on the cardiac sympathetic nerves.

In addition to these findings, 5,7-DHT has been shown to decrease the elimination of exogenous NE from the perfusion fluid (TABLE 7), indicating that the amine uptake mechanism in the sympathetic nerves is inhibited by the drug (extraneuronal uptake is negligible in the rabbit heart[39]). The positive chronotropic effect of the NE infused was considerably decreased by 32 μM 5,7-DHT (TABLE 5).

In isolated rabbit hearts with an intact postganglionic sympathetic nerve supply, we studied the influence of 32 μM 5,7-DHT on the effects of electrical stimulation of the nerves (FIGURES 5 & 7). Norepinephrine output in response to electrical nerve stimulation was more than doubled in the presence of 5,7-DHT and was still increased 10 min after termination of 5,7-DHT administration. In contrast, during perfusion with 5,7-DHT, the positive chronotropic effect of endogenous NE tended to be lower than that in control hearts and was significantly reduced after withdrawal of the amine.

The increase in NE output evoked by electrical nerve stimulation in the presence of 5,7-DHT probably reflects the inhibition of neuronal reuptake of NE, leading to an increased diffusion of the transmitter into the perfusion fluid. This effect does not seem to be due to irreversible damage to the uptake mechanism, because the effect was clearly reduced 10 min after withdrawal of the drug. In this connection, it is of interest to discuss our finding that NE output was still increased 8–10 min after the onset of continuous administration of 5,7-DHT at a concentration of 32 μM (TABLE 5). The late release of NE

TABLE 6

EFFECTS OF VARIOUS DRUGS AND OF OMISSION OF Ca^{2+} IONS, RESPECTIVELY, ON NOREPINEPHRINE (NE) OUTPUT FROM ISOLATED RABBIT HEARTS AND ON THE POSITIVE CHRONOTROPIC EFFECT EVOKED BY BOLUS INJECTION OF 0.5 μMOL 5,7-DIHYDROXYTRYPTAMINE (5,7-DHT)*

	Norepinephrine Output		Increase in Heart Rate	
	S_2	S_3	S_2	S_3
Controls ($n=6$)	83±11	54±4	101±3	103±4
Methysergide (1 μM; $n=5$)	76±10	55±4	98±6	101±4
Pindolol (2 μM; $n=4$)	107±22	61±5	3±4 †	29±6 †
Cocaine (15 μM; $n=4$)	16±5 †	44±11	75±5 †	109±4
Desipramine (150 nM; $n=4$)	127±14 †	108±12 †	94±4	100±8
Ca^{2+}-free Tyrode ($n=4$)	20±10 †	25±19	—	—
5,7-DHT (7.1 μM; $n=4$)	7±3 †	46±13	−1±1 †	106±5
5-Hydroxytryptamine (7.1 μM; $n=4$)	8±4 †	51±4	−11±7 †	102±8

* Three successive injections of 5,7-DHT (S_1–S_3) at intervals of 15 min. The drugs were present in the perfusion fluid (and Ca^{2+} ions were omitted, respectively) from 10 min before until 5 min after S_2. The NE output and positive chronotropic effect evoked by S_2 and S_3 are given. The values are expressed as the percentage of those evoked by S_1 (means±SEM). The NE output and the increase in heart rate evoked by S_1 (all experiments shown in the Table) amounted to 66±8 ng/2 min and 69±2 beats/min, respectively.

† Significantly different from controls ($p < 0.05$ or less in nonpaired Student's *t* test).

TABLE 7
INFLUENCE OF 5,7-DIHYDROXYTRYPTAMINE (5,7-DHT) ON THE REMOVAL OF EXOGENOUS NOREPINEPHRINE BY ISOLATED RABBIT HEARTS

	Removal of Norepinephrine (percent of amount infused *)
Controls ($n=10$)	41.0±3.5
5,7-DHT (10 μM; $n=6$)	31.8±7.3
5,7-DHT (32 μM; $n=6$)	9.7±2.4 †

* Means±SEM. Norepinephrine was infused into the aortic cannula for 10 min to give a final concentration of 59 nM. 5,7-Dihydroxytryptamine was present in the perfusion fluid 10 min before and during norepinephrine infusion.
† Significantly different from controls ($p < 0.001$) in nonpaired Student's t test.

may be caused by displacement of NE from its storage sites by 5,7-DHT after uptake of the indoleamine into the sympathetic nerve terminals. This mechanism is entirely different from that responsible for the initial release of NE and plays a role only at very high 5,7-DHT concentrations. The decrease in the positive chronotropic effect of both exogenous and endogenous NE by 5,7-DHT may be caused by an impairment of membrane function of pacemaker fibers that was not reversible within 10 min (FIGURES 5 & 7). In most preparations, premature beats occurred during and after perfusion with 5,7-DHT (FIGURE 5).

Similar to our results with 5,7-DHT in isolated rabbit hearts, the four isomeric dihydroxytryptamines have been shown by Creveling *et al.*[16] to inhibit uptake of [^{3}H]NE into mouse heart *in vivo* and to induce release of [^{3}H]NE from prelabeled cardiac storage sites. Inhibition of monoamine oxidase (MAO) with iproniazid increased the releasing potency of 4,7-dihydroxytryptamine,

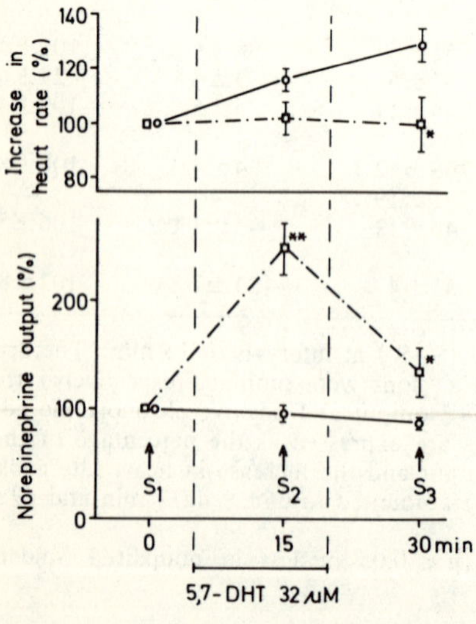

FIGURE 7. Influence of 5,7-dihydroxytryptamine (5,7-DHT) on the norepinephrine output and on the increase in heart rate evoked by electrical stimulation (S_1–S_3) of the postganglionic sympathetic nerves of isolated rabbit hearts. The norepinephrine output and the positive chronotropic effect are expressed as percentages of the values induced by the first stimulation. Controls (○; $n=9$). 5,7-DHT (□; $n=5$). Means±SEM are given. *, $p < 0.05$; **, $p < 0.001$ (compared to the corresponding values of the control experiments). For further information, see FIGURE 5 legend.

5,7-DHT, and 6,7-dihydroxytryptamine, whereas cocaine prevented the releasing effect of 5,7-DHT. Obviously, the compounds have an intraneuronal site of action and displace [^{3}H]NE from the storage sites by a mechanism similar to that of tyramine.

Subacute and Long-Term Effects on Heart and Cardiac Sympathetic Nerves

After intravenous or intraperitoneal injection of mice with 5,7-DHT, long-lasting reductions in NE content of the heart[15] and in uptake of [^{3}H]NE into the heart[16] were found. These effects are due to degeneration of the sympathetic nerve terminals, as verified by fluorescence and electron microscopic observations.[15, 16] The cytotoxic effects of 5,7-DHT were blocked by cocaine and thus were dependent on active uptake into the neuron.[16] In animals in which MAO had been inhibited, 5,7-DHT did not cause any long-term effects, suggesting a dependency of the neurodegenerative effect on MAO-catalyzed oxidative metabolism of 5,7-DHT, perhaps to a cytotoxic metabolite.[16] In rats pretreated with dibenamine, 5,6-DHT caused degenerative changes in the ultrastructure of the cardiac sympathetic nerves,[14] but such effects were not evident in mice.[16] It has been suggested[16] that these differences may be due to the inhibition of vasoconstriction by dibenamine; thus, higher amounts of 5,6-DHT may have reached the uptake sites.

Four to 5 days after subcutaneous injection of 5,6-DHT (60–400 mg/kg), 25 of 60 mice died.[13] The toxicity of this compound was related to a time- and dose-dependent induction of subepicardial lesions. It has been suggested that 5,6-DHT may induce this toxic effect by irreversible binding to proteins or by an interaction with cardiac Ca^{2+} kinetics, leading to intracellular Ca^{2+} overload in myocardial fibers.[13] In contrast to these results, subcutaneous injection of 5,7-DHT (doses up to 400 mg/kg) did not produce cardiomyopathy.[13]

Cardiovascular Effects of Intracerebroventricular or Intracisternal Administration of Indolethylamines

The effects of intracerebroventricular or intracisternal injection of 5,6-DHT on the blood pressure of normotensive and hypertensive animals have been studied by various groups of investigators in an attempt to evaluate the role of central serotonergic pathways in the control of blood pressure. Intracisternal injection of rabbits with 5,6-DHT prevented the development of neurogenic, but not renal, hypertension.[8, 9] When given to rabbits with sustained neurogenic hypertension, 5,6-DHT caused a fall in arterial pressure.[8, 9] In normotensive rats, intracisternal injection of 5,6-DHT did not affect arterial blood pressure.[10] Whereas Myers *et al.*[10] found that 5,6-DHT failed to influence either the development or maintenance of Doca® (deoxycorticosterone acetate) -saline hypertension in the rat, Finch[11] reported that 5,6-DHT induced a rapid and long-lasting fall in blood pressure of Doca-saline hypertensive rats. Some details of the methods used by these investigators differed: Myers *et al.*[10] injected the drug intracisternally in light ether anesthesia and measured systolic blood pressure by means of a tail cuff method; Finch[11] determined the mean arterial pressure by means of an implanted aortic cannula after intracerebroventricular injection of 5,6-DHT in conscious rats (implanted with a permanent brain ventricular cannula). One possible explanation for the discrepancy between the results of these investigations may be that depletion of central 5-HT evokes a more pronounced decrease in mean arterial pressure than in systolic blood pressure. In spontaneously hypertensive rats, intracerebroventricular injection of 5,6-DHT into 6-week-old animals retarded the development of hypertension for at least 6 weeks; however, intracerebroventricular injection of the drug into 14–15-week-old animals with established hypertension produced a drop in blood pressure that lasted only for less than 24 hr.[12]

The failure of 5,6-DHT to cause a long-lasting decrease in blood pressure of rats with established hypertension may be due to the fact that the degree and selectivity of degeneration of 5-HT neurons after intracerebroventricular administration of 50 μg of 5,6-DHT are not sufficient. The selectivity of destruction of central indoleamine-containing neurons can be improved by desipramine pretreatment prior to intracerebroventricular injection of 5,7-DHT.[40] Therefore, we decided to perform additional experiments in spontaneously hypertensive male rats (Wistar/Okamoto strain; age, 14 weeks) that were injected with desipramine (25 mg/kg, ip) 30 min before intracerebroventricular injection of 5,7-DHT or vehicle [20 μl of 0.9% saline that contained 0.1 mg/ml of ascorbic acid (controls)]. Injections into the lateral ventricle were performed under methohexital anesthesia (40 mg/kg, ip).

The systolic blood pressure was measured in the conscious unrestrained animals by means of a tail-cuff sphygmomanometer connected to a pressure transducer. A microphone crystal was placed over the tail distal to the cuff to record arterial pulsations that induce piezoelectricity in the crystal. The pressure in the cuff and the arterial pulsations were displayed on a two-channel recorder. When the pressure in the inflated cuff was lowered, the appearance of arterial pulsations indicated systolic blood pressure. Pulse rate was determined from high-speed tracings.

The initial levels of blood pressure in control rats and in animals injected with 5,7-DHT were not significantly different from each other (FIGURE 8). No change in systolic blood pressure occurred in control animals within the first 6 days after intracerebroventricular injection of the vehicle, whereas the values measured after injection of 5,7-DHT were slightly, but significantly, decreased compared to the initial level. The fall in blood pressure was accompanied by an increase in heart rate. Also, when the changes in blood pressure that were determined after administration of the vehicle and of 5,7-DHT (60 or 120 μg), respectively, were compared to each other, it was evident that the compound produced a dose-dependent fall in systolic blood pressure during the first 6 days and that the differences between the animals injected with 5,7-DHT and the control group were significant until the 10th day after injection (FIGURE 9). On Days 8 and 10, the pressure in control rats was higher than the initial level (FIGURES 8 & 9); this increase was not observed in 5,7-DHT-treated animals. In these rats, the loss of body weight that occurred in all animals persisted until the 10th day. Interestingly, Saller and Stricker,[41] who used a similar regimen for long-lasting 5-HT depletion in adult male Sprague-Dawley rats, obtained results opposite to ours regarding food intake and body weight gain. However, these changes were not apparent until about 20 days after combination treatment with 5,7-DHT and desipramine.

The decrease in blood pressure of rats and rabbits with established hypertension that was measured after intracerebroventricular or intracisternal injec-

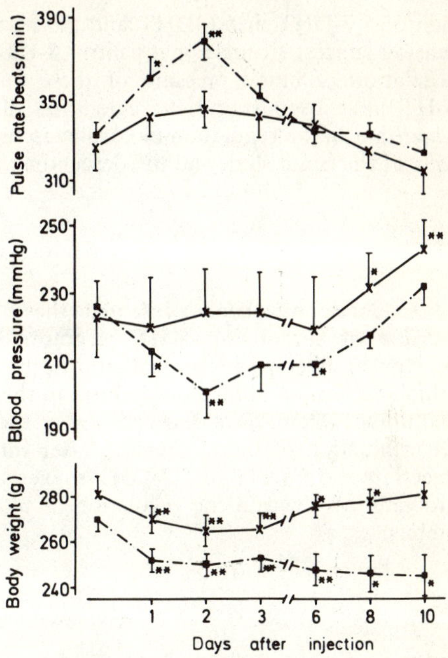

FIGURE 8. Effects of intracerebroventricular injection of 5,7-dihydroxytryptamine (120 μg) on systolic blood pressure, pulse rate, and body weight of conscious spontaneously hypertensive rats. Controls (20 μl vehicle, $\times$; $n=8$). 5,7-Dihydroxytryptamine (■; $n=8$). Means±SEM are given. $*$, $p < 0.05$; $**$, $p < 0.005$ (significantly different from the initial levels in paired Student's t test).

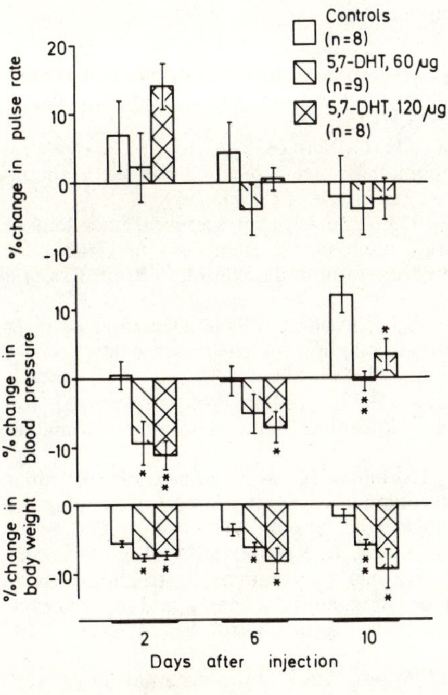

FIGURE 9. Change in systolic blood pressure, pulse rate, and body weight after intracerebroventricular injection of 5,7-dihydroxytryptamine (5,7-DHT) in spontaneously hypertensive rats. The changes are expressed as percentages of the initial values. Means±SEM are given. $*$, $p < 0.05$; $**$, $p < 0.005$ (significantly different from changes in the control animals; nonpaired Student's t test).

tion of 5,7-DHT or 5,6-DHT and the retardation of the development of hypertension suggest strongly that central 5-HT-secreting neurons are involved in the regulation of blood pressure of these animals. Nerve terminals that contain 5-HT have been found in regions of the central nervous system considered important in the regulation of cardiovascular function, namely, the medulla-pons area of the brain stem and the descending bulbospinal tracts in the spinal cord.[42]

Conclusions

Evidence has been presented that the neurotoxic indolethylamines are capable of stimulating 5-HT receptors in vascular smooth muscle, cardiac pacemaker fibers, and sympathetic nerve terminals of the rabbit heart. They inhibit the amine uptake mechanism in the membrane of the sympathetic nerves and induce a long-term degeneration of the cardiac sympathetic nerve terminals. Determination of blood pressure after intracerebroventricular or intracisternal injection of 5,6-DHT or 5,7-DHT suggests that central serotonergic neurons are probably involved in the regulation of blood pressure of hypertensive rats and rabbits.

Acknowledgment

The technical assistance of Mrs. G. Thielecke is gratefully acknowledged.

References

1. Baumgarten, H. G., A. Björklund, L. Lachenmayer, A. Nobin & U. Stenevi. 1971. Long lasting selective depletion of brain serotonin by 5,6-dihydroxytryptamine. Acta Physiol. Scand. Suppl. **373:** 1–15.
2. Baumgarten, H. G., L. Lachenmayer & H. G. Schlossberger. 1972. Evidence for a degeneration of indoleamine containing nerve terminals in rat brain induced by 5,6-dihydroxytryptamine. Z. Zellforsch. **125:** 553–569.
3. Daly, J. W., K. Fuxe & G. Jonsson. 1973. Effects of intracerebral injections of 5,6-dihydroxytryptamine on central monoamine neurons: evidence for selective degeneration of central 5-hydroxytryptamine neurons. Brain Res. **49:** 476–482.
4. Victor, S. J., H. G. Baumgarten & W. Lovenberg. 1974. Depletion of tryptophan hydroxylase by 5,6-dihydroxytryptamine in rat brain—time course and regional differences. J. Neurochem. **22:** 541–546.
5. Baumgarten, H. G. & L. Lachenmayer. 1972. 5,7-Dihydroxytryptamine: improvement in chemical lesioning of indoleamine neurons in the mammalian brain. Z. Zellforsch. **135:** 399–414.
6. Baumgarten, H. G., A. Björklund, L. Lachenmayer & A. Nobin. 1973. Evaluation of the effects of 5,7-dihydroxytryptamine on serotonin and catecholamine neurons in the rat CNS. Acta Physiol. Scand. Suppl. **391:** 1–19.
7. Baumgarten, H. G., A. Björklund, A. Nobin, E. Rosengren & H. G. Schlossberger. 1975. Neurotoxicity of hydroxylated tryptamines: structure-activity relationships. 1. Long-term effects on monoamine content and fluorescence morphology of central monoamine neurons. Acta Physiol. Scand. Suppl. **429:** 5–27.
8. Chalmers, J. P., J. L. Reid & L. M. H. Wing. 1973. Intra-cisternal 6-hydroxydopamine (6-OH-DA) and 5,6-dihydroxytryptamine (5,6-DHT) in experi-

mental hypertension. *In* Frontiers in Catecholamine Research. E. Usdin & S. H. Snyder, Eds. : 875–877. Pergamon Press, Oxford, England.

9. Wing, L. M. H. & J. P. Chalmers. 1974. Participation of central serotonergic neurones in the control of the circulation of the unanaesthetized rabbit. Circulation Res. **35:** 504–513.
10. Myers, M. G., J. L. Reid & P. J. Lewis. 1974. The effect of central serotonin depletion on DOCA/saline hypertension in the rat. Cardiovasc. Res. **8:** 806–810.
11. Finch, L. 1975. The cardiovascular effects of intraventricular 5,6-dihydroxytryptamine in conscious hypertensive rats. Clin. Exp. Pharmacol. Physiol. **2:** 503–508.
12. Buckingham, R. E., T. C. Hamilton & D. Robson. 1976. Effect of intracerebroventricular 5,6-dihydroxytryptamine on blood pressure of spontaneously hypertensive rats. Eur. J. Pharmacol. **36:** 431–437.
13. Creveling, C. R., J. Lundstrom & B. Highman. 1976. Cardiovascular toxicity of dihydroxytryptamines. Res. Commun. Chem. Pathol. Pharmacol. **14:** 1–12.
14. Baumgarten, H. G., M. Göthert, A. F. Holstein & H. G. Schlossberger. 1972. Chemical sympathectomy induced by 5,6-dihydroxytryptamine. Z. Zellforsch. **128:** 115–134.
15. Baumgarten, H. G., H. P. Groth, M. Göthert & A. A. Manian. 1974. The effect of 5,7-dihydroxytryptamine on peripheral adrenergic nerves in the mouse. Arch. Pharmakol. Exp. Pathol. **282:** 245–254.
16. Creveling, C. R., J. Lundstrom, E. T. McNeal, L. Tice & J. W. Daly. 1975. Dihydroxytryptamines: effects on noradrenergic function in mouse heart in vivo. Mol. Pharmacol. **11:** 211–222.
17. Baumgarten, H. G., M. Göthert, H. G. Schlossberger & P. Tuchinda. 1972. Mechanism of pressor effect of 5,6-dihydroxytryptamine in pithed rats. Arch. Pharmakol. Exp. Pathol. **274**: 375–384.
18. Göthert, M., P. Tuchinda & H. G. Baumgarten. 1973. Pressor effect of indolethylamines. Eur. J. Pharmacol. **21:** 242–245.
19. Fozard, J. & G. Mwaluko. 1975. The stimulant response of 5-hydroxytryptamine on cardiac sympathetic nerves: resistance to blockade by classical tryptamine receptor antagonists. *In* Abstracts of the Sixth International Congress of Pharmacology, Helsinki, Finland. : 299. Sanomaprint. Helsinki, Finland.
20. Fozard, J. & G. Mwaluko. 1976. Mechanism of the indirect sympathomimetic effect of 5-hydroxytryptamine on the isolated heart of the rabbit. Brit. J. Pharmacol. **57:** 115–125.
21. Benthe, H. F., M. Göthert & P. Tuchinda. 1972. Noradrenalin-antagonistische Wirkung verschiedener Phenylaethylamin- und Phenoxyaethylamin-Derivate. Arzneim.-Forsch. **22:** 1468–1474.
22. Benthe, H. F., M. Göthert & J. Krug. 1975. Beeinflussung der Wirkungsstärke von α-Rezeptorenblockern durch Reserpin. Arzneim.-Forsch. **25:** 402–407.
23. Page, I. H. & J. W. McCubbin. 1953. Modification of vascular response to serotonin by drugs. Amer. J. Physiol. **174:** 436–444.
24. Krishnamurty, V. S. R. 1971. Receptors for sympathomimetic amines and 5-hydroxytryptamine in the rat aorta. Arch. Int. Pharmacodyn. **189:** 90–99.
25. Behrens, B. & G. Kärber. 1935. Wie sind Reihenversuche für biologische Auswertungen am zweckmäßigsten anzuordnen? Arch. Pharmakol. Exp. Pathol. **177:** 379–388.
26. Furchgott, R. F. & S. Bhadrakom. 1953. Reactions of strips of rabbit aorta to epinephrine, isopropylarterenol, sodium nitrate and other drugs. J. Pharmacol. Exp. Ther. **108:** 129–143.
27. Göthert, M., F. K. Thies & N. Veth. 1976. Effects of droperidol on cardiovascular adrenoceptors. Arch. Int. Pharmacodyn. **224:** 199–214.
28. Van Rossum, J. M. 1963. Cumulative dose-response curves. II. Technique for the making of dose-response curves in isolated organs and the evaluation of drug parameters. Arch. Int. Pharmacodyn. **143:** 299–330.

29. TRENDELENBURG, U. 1960. The action of histamine and 5-hydroxytryptamine on isolated mammalian atria. J. Pharmacol. Exp. Ther. **130:** 450–460.
30. JACOB, J. & M. POITE-BEVIERRE. 1960. Actions de la sérotonine et de la benzyl-1-demethyl 2,5-sérotonine sur le coeur isolé de lapin. Arch. Int. Pharmacodyn. **127:** 11–26.
31. GÖTHERT, M. 1974. Effects of halothane on the sympathetic nerve terminals of the rabbit heart. Differences in membrane actions of halothane and tetracaine. Arch. Pharmakol. Exp. Pathol. **286:** 125–143.
32. GÖTHERT, M. & G. THIELECKE. 1976. Inhibition by ethanol of noradrenaline output from peripheral sympathetic nerves: possible interaction of ethanol with neuronal receptors. Eur. J. Pharmacol. **37:** 321–328.
33. HUKOVIĆ, S. & E. MUSCHOLL. 1962. Die Noradrenalin-Abgabe aus dem isolierten Kaninchenherzen bei sympathischer Nervenreizung und ihre pharmakologische Beeinflussung. Arch. Pharmakol. Exp. Pathol. **244:** 81–96.
34. BARTH, N., M. MANNS & E. MUSCHOLL. 1975. Arrhythmias and inhibition of noradrenaline uptake caused by tricyclic antidepressants and chlorpromazine on the isolated perfused rabbit heart. Arch. Pharmakol. Exp. Pathol. **288:** 215–231.
35. LINDMAR, R., K. LÖFFELHOLZ & E. MUSCHOLL. 1967. Unterschiede zwischen Tyramin und Dimethylphenylpiperazin in der Ca^{++}-Abhängigkeit und im zeitlichen Verlauf der Noradrenalinfreisetzung am isolierten Kaninchenherzen. Experientia **23:** 933–934.
36. TRENDELENBURG, U. 1972. Classification of sympathomimetics. *In* Handbook of Experimental Pharmacology. H. Blaschko & E. Muscholl, Eds. Vol. 33: 336–362. Springer Verlag. Berlin, Heidelberg, New York.
37. HAEUSLER, G., H. THOENEN, W. HAEFELY & A. HUERLIMAN. 1968. Electrical events in cardiac adrenergic nerves and noradrenaline release from the heart induced by acetylcholine and KCl. Arch. Pharmakol. Exp. Pathol. **261:** 389–411.
38. LÖFFELHOLZ, K. 1970. Autoinhibition of nicotinic release of noradrenaline from postganglionic sympathetic nerves. Arch. Pharmakol. Exp. Pathol. **267:** 49–63.
39. LINDMAR, R. & K. LÖFFELHOLZ. 1974. Neuronal and extraneuronal uptake and efflux of catecholamines in the isolated rabbit heart. Arch. Pharmakol. Exp. Pathol. **284:** 63–92.
40. BJÖRKLUND, A., H. G. BAUMGARTEN & A. RENSCH. 1975. 5,7-Dihydroxytryptamine: improvement of its selectivity for serotonin neurons in the CNS by treatment with desipramine. J. Neurochem. **24:** 833–835.
41. SALLER, C. F. & E. M. STRICKER. 1976. Hyperphagia and increased growth in rats after intraventricular injection of 5,7-dihydroxytryptamine. Science **192:** 385–387.
42. DAHLSTRÖM, A. & K. FUXE. 1965. Evidence for the existence of monoamine containing neurons in the central nervous system. I. Demonstration of monoamines in the cell bodies of brain stem neurones. Acta Physiol. Scand. **62** (Suppl. 232): 1–55.

DISCUSSION

MR. J. F. REINHARD, JR. (*Massachusetts Institute of Technology, Cambridge, Mass.*): Might the positive chronotropic effect of the dihydroxytryptamines be due to their actions as reuptake blockers? Could we enhance this effect with Lilly 110,140?

DR. GÖTHERT: 5,7-Dihydroxytryptamine is actually an inhibitor of norepinephrine uptake at 32 μM; hence, there may be an indirect sympathomimetic effect, as already described by Creveling and coworkers in mice. However, this effect does not become evident until several minutes after the initiation of infusion.

GENERAL DISCUSSION

Jacob H. Jacoby, *Moderator*

Department of Pharmacology
College of Medicine and Dentistry of New Jersey
New Jersey Medical School
Newark, New Jersey 07103

DR. R. D. MYERS: Dr. Hoebel, your animals showed an increased food intake with no change in body weight. Since *p*-CPA may cause a transient lowering of body temperature, this effect suggests that the animals are ingesting more calories to offset this hypothermia.

DR. B. G. HOEBEL: That is an interesting possibility. We also noticed hyperactivity, especially in 5,7-DHT-treated rats, whether given the drug in the posterior median forebrain bundle or intraventricularly.

DR. M. O. CARRUBA: I think that *p*-chlorophenylalanine does not produce hypothermia in rabbits, fowl, and rat.

Since fenfluramine and *p*-chloroamphetamine can induce anorexia, it appears that stimulation of the serotonergic system is involved in this behavior. We should be careful about interpreting the acute anorexia following 5,6-DHT injection to the rat, fowl, and rabbit, because this effect may also be due to an initial stimulatory phenomenon.

DR. C. T. GIAMBALVO: Dr. Pujol, you mentioned that there is no change in tyrosine hydroxylase in the substantia nigra. Were these injections made in the median raphe nucleus or the locus ceruleus?

DR. J. F. PUJOL: The injections were intracisternal. There was a decrease of tyrosine hydroxylase activity (20%) in the substantia nigra during the first few days after a 50-mg injection of 5,6-DHT, which is probably nonspecific. This decrease was followed by an increase in activity, which may be due to subsequent regulation by the locus ceruleus. Another possibility is the onset of a compensatory mechanism in the remaining cells. This problem could be further investigated by injecting 6-hydroxydopamine into the locus ceruleus. The striatum changed in a manner similar to the substantia nigra.

DR. GIAMBALVO: Would you care to speculate on whether such changes are due to chronic enzyme induction or a change in the configuration of the enzyme?

DR. PUJOL: There could be two forms of enzyme, an active and an inactive form; activation of the enzyme could increase the number of active sites. It could be due to increased enzyme synthesis or to a decrease in degradation of the enzyme. We do not have any direct evidence, but we believe that this effect is achieved by a control of synthesis.

DR. JACOBY: Dr. Clemens, might the failure of 5-HT depletion to prevent the induced prolactin rise in females treated with 5,7-DHT prior to pregnancy be due to the long time allotted between injection and study, a sufficient time interval to allow for a recovery of function?

DR. J. A. CLEMENS: Yes. We allowed about a 1-week recovery period after 5,7-DHT treatment, about 1 week for mating, then about 3 weeks for pregnancy.

Dr. D. V. Coscina: Dr. Hoebel, since you used *p*-CPA methyl ester hydrochloride, did you employ an equimolar dose of a hydrochloride or a similar pH as a control? The doses of 3–4 mg of *p*-CPA were high. Could your effects be due to nonspecific destruction of nearby ventricular structures? Have you done any histologic studies?

Dr. Hoebel: We've injected acidic vehicle and have neutralized both *p*-CPA and the vehicle and have observed the same phenomenon.

Dr. B. Natelson (*New Jersey Medical School, Newark, N.J.*): Dr. Wuttke, I noticed that the periodicity of the luteinizing hormone rhythm returned after about 3 weeks and that blood pressure recovered after about 10 days. Why is this reversal so rapid?

Dr. W. Wuttke: At about 10–15 days after 5,7-DHT injection, there is some resprouting; furthermore, the previously denervated 5-HT receptors become hypersensitive.

PART VI. BEHAVIORAL EFFECTS

BEHAVIORAL EFFECTS OF SEROTONIN NEUROTOXINS: AN OVERVIEW *

Rita B. Messing

Department of Psychobiology
University of California
Irvine, California 92714

Douglas J. Pettibone, Nathan Kaufman, and Loy D. Lytle

Department of Nutrition and Food Science
Massachusetts Institute of Technology
Cambridge, Massachusetts 02139

It has been only 20 years since Brodie and Shore [1] first suggested that brain serotonin neurons might play an important role in mediating the behavioral states of sleeping and waking. In the relatively brief period of time that has since elapsed, many investigations, which utilized a variety of techniques, much too numerous to even list, have confirmed the importance of brain serotonergic neurons for many behavioral states other than sleep and arousal. The initial report concerning the possible selective neurotoxic action of 5,6-dihydroxytryptamine (5,6-DHT) on brain serotonergic nerves by Baumgarten *et al.*[2] appeared only six years ago; since that time, several other drugs [5,7-dihydroxytryptamine (5,7-DHT),[3] *p*-chloroamphetamine (*p*-CA),[4] fenfluramine [5]] with neurotoxic effects on these brain neurons have also been identified. It now seems evident that the mechanisms by which these various compounds produce lesions of brain serotonergic nerve tracts differ from one another (see, for example, the papers by Creveling and Rotman, Schlossberger, Klemm and Baumgarten, and Horn in this monograph); moreover, these drugs do not appear to exert the specific and selective neurotoxic effects on brain serotonin-containing neurons that they had been thought to possess previously. Despite these shortcomings, the use of serotonergic neurotoxic drugs has contributed substantially to our understanding of the behavioral importance of brain neurons that release serotonin as their neurotransmitter.

This paper will briefly review the evidence regarding the nature of the various behavioral changes that result after administration of neurotoxic drugs and will compare the behavioral effects of these drugs to those seen in animals subjected to other surgical, pharmacologic, or dietary manipulations of brain serotonergic neurotransmission.

MOTOR BEHAVIOR

Locomotor Activity

As mentioned previously, several years ago, Brodie and Shore [1] postulated that two distinct, antagonistic neurochemical systems in the brain, one mediated

* Supported in part by United States Public Health Service Grants DA-01685 and MH-25075 and by an Alfred P. Sloan Foundation Fellowship in Neurosciences to L.D.L.

0077–8923/78/0305–0480 $01.75/0

by nerves that release norepinephrine as their neurotransmitter and the othe: mediated by nerves that release serotonin, might integrate autonomic functions for the control of various behavioral states. In this schema, an *ergotrophic* system, modulated by noradrenergic cells, prepared the body for action; predominance of this system was thought to be associated with increased behavioral arousal and stimulation, increased sympathetic activity and skeletal muscle tone, and an enhanced responsivity to external stimulation. The *trophotrophic* system, when activated by brain serotonergic nerves, was associated with behavioral sedation and sleep, increased parasympathetic output in the periphery, decreased locomotor activity, and reduced responsiveness to external stimulation. Although this hypothesis is a gross oversimplification in light of more recent data, the results of several experiments suggest that manipulations that increase brain serotonergic neurotransmission usually produce behavioral sedation and depression. For example, direct injection of the neurotransmitter into feline cerebral ventricles results in lethargy and drowsiness;[6] furthermore, the systemic administration of the immediate amino acid precursor in serotonin biosynthesis, 5-hydroxytryptophan,[7, 8] or electrical stimulation of serotonin perikarya in the raphe nuclei[9] produces behavioral depression and generalized decreases in the locomotor activities of rats. Conversely, inhibition of serotonin biosynthesis after injections of the tryptophan hydroxylase inhibitor *p*-chlorophenylalanine results in hyperirritability[10] and increased psychomotor stimulation.[11, 12] These behavioral abnormalities return to normal if the animals are treated with 5-hydroxytryptophan. Finally, animals with electrolytic lesions of the raphe nuclei,[9, 13] especially in the median raphe nucleus of the mesencephalon,[14–18] are also made hyperactive for a brief period of time.

It should be noted that not all data are consistent with the hypothesis that there is an inverse correlation between brain serotonergic neurotransmission and behavioral arousal; although some of these discrepancies can be accounted for on the basis of differences in experimental technique (see, for example References 11, 19, and 20), others cannot. For example, injections of *l*-tryptophan, the essential amino acid precursor for serotonin biosynthesis,[21] do not result in noticeable changes in locomotor activity unless the injections occur in animals pretreated with monoamine oxidase inhibitors.[22, 23] Somewhat surprisingly, this drug treatment combination makes animals *hyper*active, rather than *hypo*active or somnolent.[22, 23] It should be noted that the tryptophan and monoamine oxidase inhibitor treatment combination also results in the formation of tryptamine in brain tissue;[24] however, the role of brain tryptamine in mediating locomotor activity remains to be determined.

The effects of serotonin neurotoxic drugs on locomotor activity are complex and appear to depend largely on the particular neurotoxic drug administered, the testing paradigm used to assess activity, and the temporal interval that elapses between drug administration and testing. Moreover, the complex behavioral effects of the various neurotoxic drugs may be determined, in large part, by their relative effects on different neuroanatomic and/or neurochemical systems in the brain. For example, most,[25–31] but not all,[36, 37] studies have demonstrated that animals injected with *p*-CA become hyperactive shortly after the injection and remain so for as long as 3 days posttreatment. A portion of the hyperactivity induced by *p*-CA may result from an action of the drug on brain catecholaminergic neurons (*p*-CA, similar to other amphetamine-like drugs, increases catecholamine release and turnover and blocks the uptake of these compounds into nerve terminals[26, 32, 33]). However, the *p*-CA-induced changes in catecholaminergic neurotransmission do not account for all of the stimulant effects of the drug: whereas most of the measurable effects of the drug on catecholaminergic neurotransmission are absent by 24 hr after injection, the locomotor stimulation associated with *p*-CA treatment persists for at least 3 days;[31] the stimulant effects of *p*-CA can be blocked in animals pretreated with fluoxetine,[31] a drug that inhibits serotonin, but not catecholamine, uptake into brain neurons;[34] and animals injected with *p*-CA exhibit reduced psychomotor stimulation after subsequent treatment with the drug (a finding compatible with the notion that the initial *p*-CA treatment destroys brain serotonergic nerves, thus attenuating the behavioral effects of the subsequent drug injection).[35]

In contrast to the hyperactivity generally seen after *p*-CA treatment, initial studies indicated that animals injected with the other known neurotoxic halogenated amphetamine derivative, fenfluramine,[5] showed no increases, and even slight depressions, in locomotor activity.[38-43] However, other studies indicate that the method of assessing changes in locomotor activity, the dosage schedule used, or the species tested may alter the classification of fenfluramine as a depressant or nonstimulant drug.[44-48] For example, humans[45, 46] or animals[44, 49, 50] given large doses of fenfluramine display marked psychomotor stimulation; even at lower doses, fenfluramine appears to increase some types of locomotor activity via an action of the drug on brain serotonergic neurons.[48] The possible neurotoxic effects of fenfluramine on serotonin-containing neurons have only recently been described,[5] and interpretation of the behavioral effects of the drug in light of this finding have not yet been forthcoming.

Animals injected intraventricularly with one of the dihydroxytryptamine serotonergic neurotoxins (5,6-DHT or 5,7-DHT) show behavioral depression and sedation within the first 2 hr posttreatment.[51, 52] This initial period is followed for the next few days by bouts of hyperactive activity, explosive running (see Maickel *et al.*, this monograph), and disorganized locomotion and other abnormal behavior; thereafter, it is difficult to distinguish dihydroxytryptamine-treated animals from normal ones. Neonatal rats injected with 5,7-DHT (with or without monoamine oxidase inhibitor pretreatment to antagonize the toxic effects of 5,7-DHT on brain norepinephrine-containing neurons[53]) also demonstrate a period of hyperactivity that disappears within 1 month posttreatment (Breese and Mueller, this monograph).

The relative abilities of the various serotonergic neurotoxic drugs to alter locomotor activity may depend, at least in part, on differences in their toxic effects on specific populations of brain serotonin-containing neurons, but little evidence presently exists that would support or refute this hypothesis. For example, some investigators report that animals are hypoactive after intrabrain stem injections of 5,7-DHT into the medial and/or lateral serotonin fiber tracts;[54] others report no change in activity after administration of smaller doses of this drug into the same brain regions or into the dorsal raphe nucleus.[55] Although differences in drug dosage alone may account for these discrepancies, this alternative cannot explain the fact that intrabrain stem injections of 5,7-DHT reduce forebrain concentrations of serotonin in a manner comparable to that which occurs after mesencephalic raphe nuclei electrolytic lesions, which are usually associated with hyperactivity.[55]

Motor Movement Impairments

Although the locomotor activity patterns of animals injected with the various serotonergic neurotoxins appear relatively normal shortly after treatment, these animals continue to show a characteristic syndrome of motor response abnormalities when challenged with drugs known to alter brain serotonergic neurotransmission. Rats pretreated intraventricularly with 5,6-DHT or 5,7-DHT and then injected with drugs that increase brain serotonin biosynthesis (e.g., 5-hydroxytryptophan) display a syndrome of motor movement abnormalities characterized by tremor, rigidity, and muscle twitches (myoclonus);[56-61] a slightly different, but comparable, motor impairment syndrome has also been described in animals injected with tryptophan and a monoamine oxidase inhibitor[22, 23] or with various doses of *p*-CA.[62] This particular neurologic syndrome may prove to be a promising model for studying the functional changes that occur in brain serotonergic nerves injured by neurotoxic drugs (for a detailed discussion of the various features of the syndrome, see the papers by Trulson and Jacobs, Stewart *et al.*, and Growdon in this monograph); moreover, animal models of this kind ultimately may aid in an understanding of what roles brain serotonergic nerves may play in motor behavior.

Sleeping and Waking Behavior

Sleep patterns in man and other mammals alternate among several states, with the two most characteristic being rapid eye movement (REM) and slow-wave sleep. During REM sleep (behaviorally the deepest part of sleep), brain electroencephalograms show desynchronized electrical activity, eye movements are rapid, and neck muscle tonus is absent. Slow-wave sleep episodes are characterized electrophysiologically by a synchronized, high-voltage pattern with some neck muscle electromyographic activity. A variety of evidence suggests that brain serotonin-containing neurons may mediate the initiation of slow-wave sleep. Injections of tryptophan or 5-hydroxytryptophan increase total and slow-wave sleep time in rats[63, 64] and cats;[65, 66] moreover, cats are made lethargic and show signs of slow-wave sleep after injections of serotonin into the ventricles[6] or into the area postrema.[67] Conversely, raphe nuclei lesions produce behavioral and electrophysiologic signs of insomnia in cats, with the magnitude and duration of sleeplessness directly related to the extent of the lesion damage.[68-71] Not surprisingly, electrical stimulation of cells in the raphe nuclei produces behavioral depression and lethargy in rats.[9] Finally, *p*-chlorophenylalanine-induced reductions in brain serotonin are associated with insomnia, with the extent of decreased slow-wave sleep time clearly related to the degree of neurotransmitter depletion.[72-74] Brain serotonin levels and normal sleep patterns are restored in *p*-chlorophenylalanine-pretreated animals after the administration of 5-hydroxytryptophan.[65]

Although relatively few studies have examined changes in sleep patterns after injections of the serotonergic neurotoxic drugs, the results of these experiments are consistent with the notion that brain serotonin neurons play a functional role in slow-wave sleep. Intraventricular injections of 5,6-DHT[75] or systemic injections of *p*-CA[76] in cats, but not intraventricular injections of 5,7-DHT in rats,[77] cause long-term reductions in slow-wave and REM sleep time. Interestingly, 5,6-DHT injections directly into the dorsal raphe or superior centralis nuclei of cats result in transient decreases in slow-wave sleep and significant increases in REM sleep time for the first 24 hr posttreatment; thereafter, sleep patterns return to normal.[75]

It should be noted that dose-related reductions in REM sleep time, increased restlessness, and increased slow-wave sleep time have been observed in obese human patients chronically administered various doses (40–120 mg/day) of fenfluramine; after termination of drug treatment, most of the subjects experience lethargy and show increased REM sleep time.[78] Clearly, it is still not possible to determine whether these drug-induced alterations in sleep patterns were caused by fenfluramine's acute effects on brain serotonergic neurotransmission or whether they might result from a neurotoxic action of the drug on these cells.

Feeding and Drinking Behavior

Intraventricular[79] or intrahypothalamic[80, 81] injections of serotonin generally produce no observable change in feeding or drinking behavior in rodents; however, relatively larger doses of the neurotransmitter injected into various diencephalic sites in monkeys,[82] or into the ventromedial or anterior hypothalamic area of sheep,[83] cause small but delayed increases in food consumption. Other serotonergic drug manipulations also produce inconsistent changes in feeding behavior. For example, injections of 5-hydroxytryptophan increase brain serotonin synthesis and suppress food intake in rats;[84] somewhat paradoxically, injections of *p*-chlorophenylalanine decrease brain serotonin biosynthesis but also cause anorexia and hypodipsia.[85, 86] The meaning of the latter findings is difficult to interpret, inasmuch as animals injected with *p*-chlorophenylalanine appear to be sick immediately after the injection.[86] Moreover, the food intake patterns of *p*-chlorophenylalanine-treated animals return to normal, even though brain serotonin concentrations are still reduced as a result of drug treatment.[85] Although injection of 5-hydroxytryptophan restores serotonin concentrations to normal in animals pretreated with *p*-chlorophenylalanine, it does not prevent the drug-induced reduction in food intake. Finally, different results are obtained when *p*-chlorophenylalanine is injected into the lateral ventricles of rats: long-lasting hyperphagia and weight gain parallel the drug-induced reduction in brain serotonin synthesis.[87] Since systemic injections of cyproheptadine, a drug thought to block serotonin receptors, also increase appetite and weight gain in humans,[88] cats,[89] and rats,[90, 91] there is some support for the notion that a brain serotonergic mechanism may normally serve to suppress eating behavior.

Animals injected with fenfluramine or *p*-CA are anorexic immediately after the injection,[92-97] corresponding to the time interval when these drugs appear to increase serotonergic neurotransmission acutely by increasing the release and blocking the reuptake of the neurotransmitter.[98-101] The short-term anorexic effects of fenfluramine and *p*-CA probably result from the latter neurochemical effects of the drugs and are not due to their neurotoxic actions, which appear to occur somewhat later (see the paper by Sanders-Bush and Steranka in this monograph). Both animals and humans[102, 103] show rapid tolerance to the anorexic effects of these drugs if they are administered chronically. Lesions of the midbrain raphe nuclei[97] or intraventricular injections of 5,6-DHT[134] or

5,7-DHT[101] antagonize the anorexia usually observed after single injections of fenfluramine. It therefore seems possible that the rapid development of tolerance to the anorexic effects of fenfluramine when it is administered chronically may result, at least in part, from a progressive neurotoxic action of the drug on brain serotonergic neurons. Certainly this possibility should be tested, inasmuch as fenfluramine and *p*-CA have been used for the regulation of body weight in humans.[102, 103]

Although both of the neurotoxic halogenated amphetamine derivatives seem to produce short-term reductions in food intake, the dihydroxytryptamine neurotoxins appear to have mixed, or no, effects on eating and drinking behaviors. In almost all studies reported,[105-109] intraventricular or intracerebral injections of the dihydroxytryptamines appear to produce a transient period of anorexia, hypodipsia, and weight loss immediately after treatment; however, animals rapidly regain normal patterns of food and water intake (although, in some cases,[110, 111] preferences for different flavored solutions may be permanently altered) and body weight gain. It has recently been suggested[107] that the failure to observe consistent changes in eating and drinking behaviors after treatment with the dihydroxytryptamines may be due to nonspecific neurotoxic effects of these compounds on brain norepinephrine-containing neurons. Thus, it was reported that animals pretreated with desmethylimipramine and then injected intraventricularly with 5,7-DHT became hyperphagic and obese, whereas animals injected with 5,7-DHT alone showed no change or slight reductions in eating, drinking, and body weight gain.[107] The latter observation may have important ramifications for studies undertaken in an attempt to assess behavioral changes in animals injected with serotonergic neurotoxic drugs; however, the hyperphagia and obesity described in animals with the combined desmethylimipramine-5,7-DHT treatment have not been replicated by others (see, for example, the papers by Myers and by Breese and Mueller in this monograph).

Sexual Behavior

Meyerson *et al.*[112-116] have provided evidence to suggest that brain serotonin-containing neurons may play a tonic inhibitory role in mediating the copulatory behavior of female animals. Injections of reserpine or tetrabenazine, drugs that deplete brain monoaminergic neurotransmitters by interfering with their storage in nerve terminals, increase the incidence of the lordosis response in estrogen-primed female rats. In contrast, injections of a monoamine oxidase inhibitor (pargyline or nialamide), or injections of 5-hydroxytryptophan, decrease lordosis response frequency. Injections of *p*-chlorophenylalanine also increase the number of receptive sexual responses, as do injections of the putative serotonin receptor antagonists cinanserin or methysergide directly into the medial forebrain bundle, posterior hypothalamus, or medial preoptic-anterior hypothalamic area.[117, 118] Taken together, these results suggest that increases in brain serotonergic neurotransmission inhibit the occurrence of the lordosis response in female animals.

These same increases in brain serotonergic nerve activity also appear to inhibit male sexual behavior. Injections of *p*-chlorophenylalanine increase the compulsive sexual activity of normal or hormonally treated, castrated male rats,[119, 120] cats,[121] or rabbits.[119] Male rats treated with this drug also show increased homosexual activity, manifested by repeated mountings of other males and rapid copulatory movements.[119] The drug-induced sexual behavior enhancement can be reversed by injecting the animals with 5-hydroxytryptophan or can be potentiated in animals injected with a monoamine oxidase inhibitor.[122] The enhanced sexual behavior of male rats is dependent on the availability of testosterone, inasmuch as castration blocks the drug-induced excitation and injections of testosterone potentiate it.[123] Interestingly, sexually sluggish male rats show increased sexual performance after an injection of *p*-chlorophenylalanine,[124] but heterosexually active male rats do not.[125]

Relatively few studies have yet examined the extent to which serotonergic neurotoxic drugs might alter sexual behavior; however, the results of the available studies are compatible with the hypothesis that brain serotonin nerves inhibit the expression of sexual responses. Male or female animals injected systemically with *p*-CA or intraventricularly with 5,7-DHT appear to be more sexually receptive than untreated animals.[126, 127] Moreover, the increased mounting behavior seen in rats of both sexes treated with intraventricular injections of 5,6-DHT or 5,7-DHT can be reversed with injections of 5-hydroxytryptophan [128] (see also Da Prada *et al.*, this monograph).

Aggressive Behavior

Pharmacologic experiments conducted during the past few years have provided evidence that brain serotonergic neurons may have an important function in predatory aggression (i.e., aggressive behavior by one animal of a species directed against an animal of another species [129]) and, possibly, in irritable aggression (i.e., intra- or interspecies aggression elicited by an aversive stimulus [129]) as well. For example, rats pretreated with *p*-chlorophenylalanine [129, 130] show increased incidences of mouse killing, whereas animals injected with 5-hydroxytryptophan show fewer episodes of the behavior.[131] Similar increases in feline aggressive behavior after *p*-chlorophenylalanine treatment have also been observed.[121] Some,[132] but not all,[129] studies have also found that *p*-chlorophenylalanine-treated rats appear to show more numerous episodes of shock-induced fighting.

Animals injected with serotonergic neurotoxic drugs show increased aggressiveness, similar to that observed after *p*-chlorophenylalanine treatment. Thus, animals pretreated with *p*-CA [133, 134] exhibit increased irritable aggression for as long as 4 weeks postinjection. Along similar lines, rats injected intraventricularly with 5,6-DHT or 5,7-DHT progress through a transient period of increased intraspecies aggressiveness immediately after treatment; increased predatory aggression toward mice has been observed in the same animals for as long as 3 weeks postinjection.[105, 135]

Behavioral Responses to Novel or Noxious Stimuli

Sensitization and Habituation to Novel Stimuli

The repetitive presentation of a novel stimulus to an animal maintained in a well-controlled environment eventually causes habituation to the stimulus. Initial presentation of a novel stimulus elicits increased exploratory activity and other signs of behavioral alertness; subsequent presentations of the stimulus lead to a progressive attenuation of the arousal pattern, until it becomes very weak and disappears.[136] The process of behavioral habituation does not appear to be caused by a simple fatigue of sensory receptor or motor effector mechanisms, inasmuch as animals that exhibit habituation to a *familiar* stimulus display all of the signs of behavioral alerting when they are presented with a *novel* stimulus of equal intensity.[137] Some investigators hypothesize that two neural processes occur when stimuli are repeatedly presented: one serves to decrease subsequent responsivity (habituation), whereas the other serves to increase subsequent responsivity (sensitization).[138]

Experiments in rats pretreated with *p*-chlorophenylalanine suggest that brain serotonergic neurons may participate in the process of sensitization: these animals show large increases in the amplitude of the startle response (sensitization) after the presentation of a novel stimulus.[138-142] Along similar lines, animals with large lesions of the median and dorsal raphe nuclei also show increased startle response amplitudes when presented with novel acoustic stimuli or with electroshock.[139] Furthermore, animals injected with drugs (*d*-lysergic acid diethylamide, *N,N*-dimethyltryptamine) that that inhibit the unit firing rates of midbrain raphe nuclei also display larger startle responses to novel stimuli.[143]

Animals injected with *p*-CA show differential changes in the amplitude of the startle response, depending on the time interval between drug treatment and testing.[144] The rate of startle response sensitization is inhibited immediately after the injection of *p*-CA, at a time when the drug increases the release of serotonin from presynaptic nerve terminals. Shortly thereafter, *p*-CA-treated animals show enhanced startle response sensitization to presentations of acoustic stimuli, which persist for at least 4 weeks posttreatment. The latter period of enhanced startle response sensitization correlates with the time when brain serotonergic nerves are destroyed by the drug.[144]

Responses to Noxious Stimuli

Surgical, pharmacologic, or dietary manipulations that modify brain serotonin levels are usually associated with changes in the behavioral response to noxious stimuli.[145, 146] For example, surgical (lesions in the dorsomedial midbrain tegmentum, medial forebrain bundle, septum, or nucleus accumbens[147-153]), pharmacologic (injections of *p*-chlorophenylalanine[154-158] or valine[156]), or dietary (feeding animals a tryptophan-deficient, corn-based diet[156, 159]) manipulations that decrease brain serotonergic neurotransmission in animals produce an increased sensitivity and/or reactivity (hyperalgesia) to noxious stimuli. The surgical, pharmacologic, or diet-induced hyperalgesia and reductions in brain serotonin can be restored by treating these animals with tryptophan or 5-hydroxytryptophan.

Electrophysiologic (electrical stimulation of the nucleus raphe dorsalis[160] or the nucleus raphe magnus[161]) or pharmacologic (injections of fluoxetine[156, 157] or quipazine[162]) manipulations that increase brain serotonergic neurotransmission produce analgesia. The stimulation or fluoxetine-induced analgesia can be blocked in animals pretreated with *p*-chlorophenylalanine,[157, 163, 164] whereas the analgesia associated with quipazine is antagonized by a putative serotonin receptor-blocking drug.[162]

It should be noted that not all data are compatible with the notion that brain serotonergic nerves normally inhibit nociceptive responses (see, for example, References 145, 146, 165, and 166 and the review paper by Lorens, this monograph); the reasons for these various discrepancies are not easily reconciliable. Thus far, animals injected with any of the serotonin neurotoxic drugs fail to show much change in nociception. For example, intraventricular injections of 5,6-DHT[167, 168] or injections of 5,7-DHT directly into the midbrain[169] (or raphe nuclei electrolytic lesions[54, 55, 149]) do not alter the behavioral response to presentations of noxious stimuli. However, each of these manipulations do antagonize the analgesic effectiveness of some antinociceptive drugs.

Responses to Antinociceptive Drugs

As might be anticipated from the previous discussion of the role of brain serotonergic nerves in nociception, surgical,[161, 170–172] pharmacologic,[173–178] or dietary[146] manipulations that tend to decrease brain serotonergic neurotransmission also antagonize the analgesic potency of morphine; in contrast, electrophysiologically[160] or pharmacologically[157, 179–183] induced increases in serotonergic neurotransmission potentiate morphine analgesia. Again, not all experimental results easily fit this simple relationship between brain serotonergic nerve activity and antinociceptive drug action (see, for example, References 156 and 157 and Lorens, this monograph); however, the bulk of experiments that employed serotonin neurotoxic drugs have provided data compatible with this hypothesis. For example, *p*-CA or fenfluramine injections have a biphasic influence on the antinociceptive properties of morphine: immediately after injection, when these drugs increase the release of serotonin from brain presynaptic nerve terminals, morphine analgesia is enhanced;[174, 185] somewhat later, after *p*-CA or fenfluramine has destroyed serotonin-containing neurons, morphine analgesia is antagonized.[175, 179, 184] Similarly, animals pretreated with 5,6-DHT also show an attenuated behavioral response to the analgesic effects of morphine.[168, 185]

Temperature Regulation

Although brain neurons that contain serotonin clearly participate in the control of body temperature, their exact role remains enigmatic, due to data conflicts. Elevations in ambient temperature that increase body temperature stimulate the rate of serotonin turnover[186, 187] and increase the firing rates of cells in the raphe nuclei[188] in rats. However, the increased body temperatures of animals placed in hot environments do not appear to be directly related to the heightened activity in brain serotonin-containing nerves: animals pretreated with *d*-lysergic acid diethylamide and placed in a hot environment show no increase in brain serotonin turnover or neuronal electrical activity, even though their body temperatures are elevated.[189] However, pyrogen-induced fevers in rabbits are enhanced after pretreatment with *p*-chlorophenylalanine.[190]

Although these data suggest indirectly that brain serotonin neurons may participate in a physiologic system that stimulates heat loss, other data suggest an opposite role for these nerves. Microinjection of serotonin into the hypothalamus of monkeys kept at room temperature[191] or in hot environments[192] raises body temperature, as does electrical stimulation of neurons in the raphe nuclei.[193]

The thermogenic effects of serotonin neurotoxic drugs also vary and depend on a variety of factors, including possible species differences and the time interval between drug injection and testing. Immediately after an intraventricular injection of 5,6-DHT[105, 106] or 5,7-DHT,[205] rats become transiently hypothermic; the degree of 5,7-DHT-induced hypothermia becomes attenuated with repeated 75-μg intraventricular injections of the drug at 10-day intervals.[205] Taken together, these data provide tentative evidence that the dihydroxytryptamines might decrease body temperature immediately after injection by releasing serotonin from presynaptic nerves.

Other data argue against this interpretation: the 5,7-DHT-induced hypothermia is neither enhanced nor antagonized by pretreating rats with various drugs (*p*-chlorophenylalanine, tryptophan, methysergide, cyproheptadine) thought to alter brain serotonergic neurotransmission.[205] Moreover, Myers *et al.*[194–196] find that 5,6-DHT injections into the anterior hypothalamus result in a transient *hyper-*, rather than *hypo*, -thermia; repeated hypothalamic 5,6-DHT injections also produce less hyperthermia than does the initial treatment (see Myers, this monograph). Rats or monkeys that receive intrahypothalamic 5,6-DHT injections show impaired abilities to maintain body heat after exposure to either a hot or cold environment for at least 48 hr postinjection, again suggesting that brain serotonergic nerve terminals in this brain region may be important in the regulation of body heat.

Conclusions

In general, many of the behavioral changes seen in animals injected with serotonin neurotoxic drugs are similar to those observed after other surgical, pharmacologic, or dietary manipulations of brain serotonergic neurotransmission. However, some of the behavioral alterations associated with these neurotoxins are not easily understood in light of other data, even though similar experimental protocols have often been employed. For example, even a cursory review of the literature reveals that the behavioral changes observed after chemical lesions of brain serotonin-containing neurons often differ from those associated with electrolytic lesions of these cells. Moreover, the behavior of animals injected with one of the serotonin neurotoxic drugs is frequently different from that of animals treated with other drugs in this group.

To fully understand why these discrepancies occur, future studies will have to determine to what extent the neuroanatomic and neurochemical changes that follow treatment with neurotoxic drugs are similar to, or different from, those induced by other manipulations of brain serotonergic neurotransmission. For example, it no longer seems clear that electrolytic lesions of the median or dorsal raphe nuclei most likely damage *ascending* serotonergic nerves but leave serotonin-containing neurons that send axons and terminals into the spinal cord relatively intact. In contrast, intraventricular injections of the dihydroxytryptamine neurotoxins produce relatively greater damage to serotonin nerves that *descend* in the spinal cord rather than to those that *ascend* into forebrain regions.[2, 168] We know relatively little about the neurotoxic effects of the halogenated amphetamines (*p*-CA or fenfluramine), although some evidence indicates that these drugs may destroy a set of serotonergic nerves different from those that originate in the median or dorsal raphe nuclei.[5, 6] Thus, it is not terribly surprising to find that the behavioral effects of the various neurotoxic drugs are sometimes different from one another and often different from the changes in behavior observed after electrolytic brain lesions.

Few studies have yet determined the extent to which behavioral changes that result from neurotoxic drug treatment are related to specific effects on brain serotonin-containing neurons or to their effects on other cells. Unfortunately, we do not yet know about all of the neurochemical changes induced by the neurotoxins; however, many studies relate the behavioral changes seen after injections of these drugs to their putative *specific* serotonergic actions. Although it is known, for example, that 5,7-DHT has mixed effects on both brain serotonin- and norepinephrine-containing cells,[197] few studies have yet determined which of the behavioral changes associated with this particular drug treatment are related to its *serotonergic* or *noradrenergic* action. It will be impossible to fully understand the behavioral changes observed after administration of the neurotoxins until a more complete profile is forthcoming that describes how each of these drugs resembles, or differs from, one another in terms of their relative neurochemical potencies.

Many of the behavioral changes induced by these drugs are present for only 2–3 days postinjection, even though their neurotoxic effects are permanent and often irreversible. Since many of these drugs produce a variety of toxic side effects that may interfere with behavioral analyses, it is important to make certain that many of the short-term changes in behavior are not merely a consequence of these side effects. Moreover, many of the neurotoxic drugs (especially *p*-CA and fenfluramine) also alter catecholaminergic neurotransmission; therefore, many of the short-term behavioral effects of these drugs may not be related to their neurotoxic effects on brain serotonergic nerves. Too often, many studies fail to control for these other possible actions of the drug.

The lack of permanent change in many of the behavioral effects of the neurotoxins is surprising in view of the permanency of their destructive action. It has been suggested that behavioral function recovers as a result of the regeneration of injured brain serotonergic nerves, but no direct evidence in support of this attractive hypothesis has been forthcoming. Alternative hypotheses (such as the possibility that the initial behavioral alterations observed shortly after serotonin neurotoxic drug administration may not be related to drug-induced changes in serotonergic neurotransmission) are rarely entertained and even less frequently tested.

Finally, it should be noted that the halogenated amphetamine neurotoxins have been administered to human patients for the treatment of obesity [102] and for other neurologic or psychiatric disorders.[198] Some patients under medication have been hospitalized for treatment of fenfluramine-induced *schizophreniform* or *paranoid* symptoms; [199] others [200-204] report that humans treated with fenfluramine show signs of postdrug withdrawal depression, irritability, sleeplessness, fearfulness, or loss of sexual drive. The evidence that shows that fenfluramine or *p*-CA produces neurotoxic effects on serotonin-containing neurons in animals is relatively new, and additional studies should be conducted to confirm and extend these results in animals. Although not yet established, these drugs may also produce comparable neurotoxic effects in humans. Since many of the symptoms observed in humans treated in the past with these drugs are strikingly similar to the behavioral changes seen in animals, safety requires that their use in humans be terminated immediately.

REFERENCES

1. BRODIE, B. B. & P. A. SHORE. 1957. Ann. N.Y. Acad. Sci. **66:** 631–642.
2. BAUMGARTEN, H. G., A. BJÖRKLUND, L. LACHENMAYER, A. NOBIN & U. STENEVI. 1971. Acta Physiol. Scand. Suppl. **373:** 1–15.
3. BAUMGARTEN, H. G., A. BJÖRKLUND, L. LACHENMAYER & A. NOBIN. 1973. Acta Physiol. Scand. Suppl. **391:** 1–19.
4. HARVEY, J. A., S. E. MCMASTER & L. M. YUNGER. 1975. Science **187:** 841–843.
5. HARVEY, J. A. & S. E. MCMASTER. 1975. Psychopharmacol. Commun. **1:** 217–228.
6. FELDBERG, W. & S. L. SHERWOOD. 1953. J. Physiol. (London) **120:** 12P.
7. APRISON, M. H. & C. B. FERSTER. 1961. J. Pharmacol. Exp. Ther. **131:** 100–107.
8. JOYCE, D. & N. MROSOVKY. 1964. Psychopharmacologia **5:** 417–423.
9. KOSTOWSKI, W., E. GIACOLONE, S. GARATTINI & L. VALZELLI. 1968. Eur. J. Pharmacol. **4:** 371–376.
10. DEMENT, W. C. 1969. *In* Sleep, Physiology and Pathology, A Symposium. A. Kaler, Ed. : 245–265. J. B. Lippincott Company. Philadelphia, Pa.
11. FIBIGER, H. C. & B. A. CAMPBELL. 1971. Neuropharmacology **10:** 25–32.
12. MABRY, P. D. & B. A. CAMPBELL. 1973. Brain Res. **49:** 381–391.
13. LORENS, S. A., J. P. SORENSEN & L. M. YUNGER. 1971. J. Comp. Physiol. Psychol. **77:** 48–52.
14. JACOBS, B. L., W. D. WISE & K. M. TAYLOR. 1974. Brain Res. **79:** 353–361.
15. SREBRO, B. & S. A. LORENS. 1975. Brain Res. **89:** 303–325.
16. GEYER, M. A., A. PUERTO, D. B. MENKES, D. S. SEGAL & A. J. MANDELL. 1976. Brain Res. **106:** 257–270.
17. JACOBS, B. L. & A. COHEN. 1976. J. Comp. Physiol. Psychol. **90:** 102–108.
18. JACOBS, B. L., C. TRIMBACH, E. E. EUBANKS & M. TRULSON. 1975. Brain Res. **94:** 253–261.
19. VOLICER, L. 1969. Int. J. Neuropharmacol. **8:** 361–364.
20. TENEN, S. S. 1967. Psychopharmacologia **10:** 214–219.
21. WURTMAN, R. J. & J. D. FERNSTROM. 1976. Biochem. Pharmacol. **25:** 1691–1696.
22. GRAHAME-SMITH, D. G. 1971. J. Neurochem. **18:** 1053–1066.
23. GRAHAME-SMITH, D. G. 1971. Brit. J. Pharmacol. **43:** 856–864.
24. SAAVEDRA, J. M. & J. AXELROD. 1973. J. Pharmacol. Exp. Ther. **185:** 523–529.
25. STRADA, S. J., E. SANDERS-BUSH & F. SULSER. 1970. Biochem. Pharmacol. **19:** 2621–2629.
26. COSTA, E. L., K. M. NAIMZADA & A. REVUELTA. 1971. Brit. J. Pharmacol. **43:** 570–579.
27. LASSEN, J. B. 1974. Psychopharmacologia **34:** 243–254.
28. SCHEEL-KRUGER, J. 1972. Eur. J. Pharmacol. **18:** 63–73.
29. FREY, J. & M. MAGNUSSEN. 1968. Biochem. Pharmacol. **17:** 1299–1307.
30. TESSEL, R. E., J. H. WOODS, R. E. COUNSELL & M. LU. 1975. J. Pharmacol. Exp. Ther. **192:** 310–318.
31. MESSING, R. B., L. PHEBUS, L. A. FISHER & L. D. LYTLE. 1976. Neuropharmacology **15:** 157–163.
32. GORLITZ, B. D. & H. H. FREY. 1972. Eur. J. Pharmacol. **20:** 171–180.
33. SANDERS-BUSH, E., J. BUSHING & F. SULSER. 1975. J. Pharmacol. Exp. Ther. **192:** 33–41.
34. FULLER, R. W., K. W. PERRY & B. B. MOLLOY. 1975. Eur. J. Pharmacol. **33:** 119–124.
35. EL-YOUSEF, M. K., L. R. STERANKA & E. SANDERS-BUSH. 1977. Psychopharmacology. In press.
36. SCHAEFER, G. J., R. J. BARRETT, E. SANDERS-BUSH & C. V. VORHEES. 1974. Pharmacol. Biochem. Behav. **2:** 783–789.

37. VORHEES, C. V., G. J. SCHAEFER & R. J. BARRETT. 1975. Pharmacol. Biochem. Behav. **3:** 279–284.
38. BOISSIER, J. R., P. SIMON, J. FICHELLE & F. HERVOUET. 1965. Therapie **20:** 297–209.
39. DUHAULT, J. 1972. Paper presented at the International Symposium on Fenfluramine, Montreal, Canada.
40. FINK, M., D. M. SHAPIRO & M. I. TURAN. 1971. Psychopharmacologia **22:** 269–383.
41. ELLIOT, B. W. 1970. Curr. Ther. Res. **12:** 502–515.
42. HILL, R. C. & P. TURNER. 1967. J. Pharm. Pharmacol. **19:** 337–338.
43. LEDOUAREC, J. C. & C. NEVEU. 1970. *In* Amphetamines and Related Compounds. E. Costa & S. Garattini, Eds. : 75–105. The Raven Press. New York, N.Y.
44. TAYLOR, M., A. J. COUDIE & A. WILLIAMS. 1973. Psychopharmacologia **31:** 63–76.
45. CAMPBELL, D. B. & B. W. R. MOORE. 1969. Lancet **2:** 1307.
46. FLEISCHER, M. R. & D. B. CAMPBELL. 1969. Lancet **2:** 1306.
47. EVERITT, B. J. & N. B. HACKET. 1972. Brit. J. Pharmacol. **44:** 342–343.
48. CLINESCHMIDT, B. V., J. C. MCGUFFIN, A. B. PFLUEGER & J. A. TOTARO. 1975. Neuropharmacology **14:** 301–311.
49. YELNOSKY, T. & R. B. LAWLOR. 1970. Arch. Int. Pharmacodyn. **184:** 374–388.
50. SOUTHGATE, P. J., S. R. MAYER, E. BOXALL & A. B. WILSON. 1971. J. Pharm. Pharmacol. **23:** 600–605.
51. MASSOTTI, M., A. S. DECAROLIS & G. LONGO. 1974. Pharmacol. Biochem. Behav. **2:** 769–775.
52. BAUMGARTEN, H. G. & H. G. SCHLOSSBERGER. 1973. *In* Serotonin and Behavior. J. Barchas & E. Usdin, Eds. : 209–224. Academic Press, Inc. New York, N.Y.
53. BREESE, G. R. & B. R. COOPER. 1975. Brain Res. **98:** 517–527.
54. HOLE, K., K. FUXE & G. JONSSON. 1976. Brain Res. **107:** 385–399.
55. LORENS, S. A., H. C. GLUDBERG, K. HOLE, C. KOHLER & B. SREBRO. 1976. Brain Res. **108:** 97–113.
56. TRULSON, M. E., E. E. EUBANKS & B. L. JACOBS. 1976. J. Pharmacol. Exp. Ther. **198:** 23–32.
57. TRULSON, M. E., C. A. ROSS & B. L. JACOBS. 1976. Psychopharmacol. Commun. **2:** 149–164.
58. TRULSON, M. E., A. D. STARK & B. L. JACOBS. 1977. J. Pharm. Pharmacol. **29:** 382–383.
59. JACOBS, B. L. 1976. Life Sci. **19:** 777–786.
60. STEWART, R. M., J. H. GROWDON, D. CANCIAN & R. J. BALDESSARINI. 1976. Neurology **26:** 690–692.
61. STEWART, R. M., J. H. GROWDON, D. CANCIAN & R. J. BALDESSARINI. 1976. Neuropharmacology **15:** 449–455.
62. GROWDON, J. H. 1977. Neurology **11:** 1074–1077.
63. HARTMANN, E. 1967. Psychonom. Sci. **8:** 295–296.
64. HARTMANN, E. 1968. Psychophysiology **4:** 390.
65. JOUVET, M. 1967. Res. Pub. Ass. Nervous Mental Dis. **45:** 86–126.
66. DELORME, F., M. JEANNEROD & M. JOUVET. 1965. C.R. Soc. Biol. **159:** 900.
67. KOELLA, W. P. & J. CZICMAN. 1963. Amer. J. Physiol. **204:** 873–880.
68. JOUVET, M. 1973. *In* Serotonin and Behavior. J. Barchas & E. Usdin, Eds. : 385–400. Academic Press, Inc. New York, N.Y.
69. JOUVET, M. & J. RENAULT. 1966. C.R. Soc. Biol. **160:** 1461–1465.
70. JOUVET, M. 1969. Science **163:** 32–41.
71. JOUVET, M. 1967. Physiol. Rev. **47:** 117–177.
72. DELORME, F., J. L. FROMENT & M. JOUVET. 1966. C.R. Soc. Biol. **160:** 2347–2351.
73. MOURET, J., P. BOBILLER & M. JOUVET. 1968. Eur. J. Pharmacol. **5:** 17–22.

74. Weitzman, E. D., M. Rapport, P. McGregor & J. Jacoby. 1968. Science **160:** 1361–1363.
75. Jouvet, M. & J. Pujol. 1974. Advan. Biochem. Pharmacol. **11:** 199–209.
76. Jouvet, M. 1972. Ergebnisse der Physiolog. **64:** 166–307.
77. Ross, C. A., M. E. Trulson & B. L. Jacobs. 1976. Brain Res. **114:** 517–523.
78. Lewis, S. A., I. Oswald & D. L. F. Dunleavy. 1971. Brit. Med. J. **3:** 67–70.
79. Myers, R. D. & T. L. Yaksh. 1968. Physiol. Behav. **3:** 917–928.
80. Slangen, J. L. & N. E. Miller. 1969. Physiol. Behav. **4:** 543–552.
81. Wagner, J. W. & J. DeGroot. 1963. Amer. J. Physiol. **204:** 483–487.
82. Seoane, J. R. & C. A. Baile. 1973. Physiol. Behav. **10:** 915–923.
83. Beyea, J. S. & C. A. Baile. 1973. Fed. Proc. **32:** 304.
84. Blundell, J. E. & M. B. Leshem. 1974. Eur. J. Pharmacol. **28:** 81–88.
85. Biggio, G., G. P. Mereu, M. P. Piccardi, G. Demontis, P L. Caruso, M. Oliana & L. Vardiu. 1973. Riv. Farmacol. Ter. **4:** 183–188.
86. Panksepp, J. & D. M. Nance. 1974. Physiol. Psychol. **2:** 360–364.
87. Breisch, S. T., F. P. Zemlan & B. G. Hoebel. 1976. Science **192:** 382–385.
88. Shah, N. M. 1968. Curr. Med. Pract. **12:** 961–964.
89. Chakrabarty, A. S., R. V. Pillai, B. K. Anand & B. Singh. 1967. Brain Res. **6:** 561–569.
90. Baxter, M. G., A. A. Miller & F. E. Soroko. 1970. Brit. J. Pharmacol. **39:** 229P–230P.
91. Ghosh, M. N. & S. Parvathy. 1973. Brit. J. Pharmacol. **48:** 328–329.
92. Bizzi, A., A. Bonaccorsi, S. Jespersen, A. Jori & S. Garattini. 1970. *In* Amphetamines and Related Compounds. E. Costa & S. Garattini, Eds. : 577–595. The Raven Press. New York, N.Y.
93. Clineschmidt, B. V. 1973. Eur. J. Pharmacol. **24:** 405–409.
94. Clineschmidt, B. V., J. C. McGuffin &A B. Werner. 1974. Eur. J. Pharmacol. **27:** 313–323.
95. Frey, H.-H. 1970. *In* Amphetamines and Related Compounds. E. Costa & S. Garattini, Eds. : 343–347. The Raven Press. New York, N.Y.
96. Leshem, M. B. & J. E Blundell. 1974. J. Pharm. Pharmacol. **26:** 905–906.
97. Samanin, R., D. Ghezzi, L. Valzelli & S. Garattini. 1972. Eur. J. Pharmacol. **19:** 318–322.
98. Pletscher, A., M. Da Prada & Y. Burkard. 1970. *In* Amphetamines and Related Compounds. E. Costa & S. Garattini, Eds. : 331–341. The Raven Press. New York, N.Y.
99. Pletscher, A., M. Da Prada, Y. Burkard, G. Bartholini, F. Steiner, H. Bruderer & F. Bigler. 1966. J. Pharmacol. Exp. Ther. **154:** 64–72.
100. Wong, D., J. Horng & R. W. Fuller. 1973. Biochem. Pharmacol. **22:** 311–322.
101. Sanders-Bush, E., D. A. Gallager & F. Sulser. 1974. Advan. Biochem. Pharmacol. **10:** 185–194.
102. Shoulson, I. & T. N. Chase. 1975. Clin. Pharmacol. Ther. **17:** 616–621.
103. Woodward, E. 1970. *In* Amphetamines and Related Compounds. E. Costa & S. Garattini, Eds. : 685–691. The Raven Press. New York, N.Y.
104. Fuxe, K., B. Hamberger, L.-O. Farnebo & S.-O. Ögren. 1975. Postgrad. Med. J. **51:** 35–45.
105. Breese, G. R., B. R. Cooper, L. D. Grant & R. D. Smith. 1975. Neuropharmacology **13:** 177–187.
106. Ho, A. K S., C. S. Tsai, T. C. A. Chen, H. Begleiter & B. Kissin. 1974. Psychopharmacologia **40:** 101–107.
107. Saller, C. F. & E. M. Stricker. 1976. Science **192:** 385–387.
108. Lytle, L. D., J. H. Jacoby, M. F. Nelson & H. G. Baumgarten. 1974. Life Sci. **15:** 1203–1217.
109. Myers, R. D. 1975. Brain Res. **94:** 491–506.
110. Myers, R. D. & C. L. Melchior. 1975. Res. Commun. Chem. Pathol. Pharmacol. **10:** 363–378.

111. Melchior, C. L. & R. D. Myers. 1976. Pharmacol. Biochem. Behav. **5:** 63–72.
112. Meyerson, B. J. 1964. Arch. Int. Pharmacodyn. **150:** 4–33.
113. Meyerson, B. J. 1964. Acta Physiol. Scand. Suppl. **241:** 1–32.
114. Meyerson, B. J. 1966. Acta Pharmacol. Toxicol. **24:** 363–376.
115. Meyerson, B. J. 1966. Acta Physiol. Scand. **67:** 411–422.
116. Meyerson, B. J. 1970. Life Sci. **9:** 661–671.
117. Crowley, W. R., I. L. Ward & D. L. Margules. 1975. J. Comp. Physiol. Psychol. **88:** 62–68.
118. Grabowska, M. 1974. *In* Sexual Behavior: Pharmacology and Biochemistry. M. Sandler & G. L. Gessa, Eds. : 59–62. The Raven Press. New York, N.Y.
119. Tagliamonte, A., P. Tagliamonte, G. L. Gessa & B. B. Brodie. 1969. Science **166:** 1433–1435.
120. Sheard, M. 1969. Brain Res. **15:** 524–528.
121. Ferguson, J., S. Henrikson, H. Cohen, G. Mitchell, J. Barchas & W. C. Dement. 1970. Science **168:** 499–501.
122. Tagliamonte, A., P. Tagliamonte & G. L. Gessa. 1971. Nature (London) **230:** 244–245.
123. Gessa, G. L., A. Tagliamonte, P. Tagliamonte & B. B. Brodie. Nature (London) **227:** 616–617.
124. Sjoerdsma, A., W. Lovenberg, K. Engelman, W. T. Carpenter, R. J. Wyatt & G .L. Gessa. 1970. Ann. Intern. Med. **73:** 607–629.
125. Whalen, R. E. & W. G. Luttge. 1970. Science **169:** 1000–1001.
126. Trulson, M. E. & F. P. Zemlan. 1975. Paper presented at Eastern Reproductive Behavior Conference, Nags Head, N.C.
127. Everitt, B. J., K. Fuxe & G. Jonsson. 1975. J. Pharmacol. **6:** 25–32.
128. Da Prada, M., M. Carruba, R. A. O'Brien, A. Saner & A. Pletscher. 1972. Eur. J. Pharmacol. **19:** 288–290.
129. Eichelman, B. S. & N. B. Thoa. 1973. Biol. Psychiat. **6:** 143–164.
130. DiChiara, G., R. Camba & P. F. Spano. 1971. Nature (London) **233:** 272–273.
131. Kulkarni, A. S. 1968. Pharmacologist **10:** 180.
132. Conner, R. L., S. Levine, G. A. Wertheim & J. F. Cummer. 1970. Physiol. Behav. **5:** 1221–1224.
133. Gianutsos, G. & H. Lal. 1975. Res. Commun. Pathol. Pharmacol. **10:** 379–382.
134. Sheard, M. H. & M. Davis. 1976. Eur. J. Pharmacol. **40:** 295–302.
135. Breese, G R. & B. R. Cooper. 1975. Brain Res. **98:** 517–527.
136. Sharpless, S. & H. Jasper. 1956. Brain **79:** 655–680.
137. Groves, P. M. & R. F. Thompson. 1970. Psychol. Rev. **77:** 419–450.
138. Davis, M. & M. H. Sheard. 1974. Physiol. Behav. **12:** 425–431.
139. Carlton, P. L. & C. Advokat. 1973. Physiol. Biochem. Behav. **1:** 657–663.
140. Rosecrans, J. A. 1970. Physiol. Behav. **5:** 453–458.
141. Swonger, A. K. & R. H. Rech. 1972. J. Comp. Physiol. Psychol. **85:** 509–522.
142. Brody, J. F. 1970. Psychopharmacologia **17:** 14–33.
143. Davis, M. & H. D. Bear. 1972. Psychopharmacologia **27:** 29–44.
144. Davis, M. & M. H. Sheard. 1976. Eur. J. Pharmacol. **35:** 261–273.
145. Messing, R. B. & L. D. Lytle. 1977. J. Pain **4:** 1–21.
146. Lytle, L. D., L. Phebus, L A. Fisher & R. B. Messing. 1977. *In* Factors Affecting the Action of Narcotics. M. Adler, L. Manara & R. Samanin, Eds. The Raven Press. New York, N.Y. In press.
147. Harvey, J. A. & C. E. Lints. 1971. J. Comp. Physiol. Psychol. **74:** 28–36.
148. Harvey, J. A., A. J. Schlosberg & L. M. Yunger. 1975. Fed. Proc. **34:** 1796–1801.
149. Harvey, J. A., A. J. Schlosberg & L. M. Yunger. 1974. Advan. Biochem. Pharmacol. **10:** 232–246.
150. Lints, C. E. & J. A. Harvey. 1969. J. Comp. Physiol. Psychol. **67:** 23–31.
151. Lints, C. E. & J. A. Harvey. 1969. Physiol. Behav. **4:** 29–31.

152. Lorens, S. A., J. P. Sorensen & J. A. Harvey. 1970. J. Comp. Physiol. Psychol. **75:** 284–290.
153. Yunger, L. M. & J. A. Harvey. 1973. J. Comp. Physiol. Psychol. **83:** 173–183.
154. Tenen, S. S. 1967. Psychopharmacologia **10:** 214–219.
155. Fibiger, H. C., P. H. Mertz & B. A. Campbell. 1972. Physiol. Behav. **8:** 259–263.
156. Messing, R. B., L. Fisher, L. Phebus & L. D. Lytle. 1976. Life Sci. **18:** 707–714.
157. Messing, R. B., L. Phebus, L. Fisher & L. D. Lytle. 1975. Psychopharmacol. Commun. **1:** 511–521.
158. Sicuteri, F., B. Anselmi & P. L. Del Bianco. 1973. Psychopharmacologia **29:** 347–356.
159. Lytle, L. D., R. B. Messing, L. Fisher & L. Phebus. 1975. Science **190:** 692–694.
160. Samanin, R. & L. Valzelli. 1971. Eur. J. Pharmacol. **16:** 298–302.
161. Proudfit, H. K. & E. G. Anderson. 1975. Brain Res. **98:** 612–618.
162. Samanin, R., S. Bernasconi & A. Quattrone. 1976. Psychopharmacologia **46:** 219–222.
163. Akil, H. & J. C. Liebeskind. 1975. Brain Res. **94:** 279–296.
164. Akil, H. & D. J. Mayer. 1972. Brain Res. **44:** 692–697.
165. Harvey, J. A. & L. M. Yunger. 1973. *In* Serotonin and Behavior. J. Barchas & E. Usdin, Eds. : 232–246. Academic Press, Inc. New York, N.Y.
166. Hole, K. & C. A. Marsden. 1975. Pharmacol. Biochem. Behav. **3:** 307–309.
167. Blasig, J., K. Reinhold & A. Herz. 1973. Psychopharmacologia **31:** 111–119.
168. Vogt, M. 1974. Physiol. (London) **236:** 483–498.
169. Hole, K. & S. A. Lorens. 1975. Pharmacol. Biochem. Behav. **3:** 95–102.
170. Samanin, R., D Ghezzi, C. Mauron & L. Valzelli. 1973. Psychopharmacologia **33:** 365–368.
171. Adler, M., W. Kostowski, M. Recchia & R. Samanin. 1975. Eur. J. Pharmacol. **32:** 39–44.
172. Liebeskind, J. C., G. J. Giesler & G. Urca. 1976. *In* Sensory Function of the Skin in Primates, with Special Reference to Man. Y. Zotterman, Ed. : 239–251. Pergamon Press. Oxford, England.
173. Tenen, S. S. 1968. Psychopharmacologia **12:** 278–285.
174. Gorlitz, B.-D. & J. J. Frey. 1972. Eur. J. Pharmacol. **20:** 171–180.
175. Tulanay, F. C., I. Yano & A. E. Takemori. 1976. Eur. J. Pharmacol. **35:** 285–292.
176. Schneider, J. A. 1954. Proc. Soc. Exp. Biol. Med. **87:** 614–615.
177. Takagi, H., T. Takashima & K. Kimura. 1964. Arch. Int. Pharmacodyn. **149:** 484–492.
178. Yaksh, T. L., J. C. Du Chateau & T. A. Rudy. 1976. Brain Res. **104:** 367–372.
179. Takemori, A. E., F. C. Tulanay & I. Yano. 1975. Life Sci. **17:** 21–28.
180. Calcutt, C. R., S. L. Handley, C. G. Sparkes & P. S. J. Spencer. 1971. *In* Agonist and Antagonist Actions of Narcotic Drugs. H. W. Kosterlitz, H. O. J. Collier & S. E. Villarreal, Eds. : 176–181. University Park Press. Baltimore, Md.
181. Sewell, R. D. E. & P. S. J. Spencer. 1974. Brit. J. Pharmacol. **51:** 140P–141P.
182. Ross, S. B. & A. L. Renyi. 1977. Neuropharmacology **16:** 57–63.
183. Sugrue, M. F. & I. McIndewar. 1976. J. Pharm. Pharmacol. **28:** 447–448.
184. Duncan, C. & P. S. J. Spencer. 1973. J. Pharm. Pharmacol. **25:** 124P–125P.
185. Genovese. E., N. Zonta & P. Mantegazza. 1973. Psychopharmacologia **32:** 359–364.
186. Corrodi, H., K Fuxe & T. Hokfelt. 1967. Acta Physiol. Scand. **71:** 224–232.
187. Reid, W. D., L. Volicer, J. Smookler, M. A. Beaven & B. B. Brodie. 1968. Pharmacology **1:** 329–344.

188. WEISS, B. L. & G. AGHAJANIAN. 1971. Brain Res. **26:** 37–48.
189. AGHAJANIAN, G. K. & B. L. WEISS. 1968. Nature (London) **220:** 795–796.
190. GIARMAN, N. J., C. TANAKA, J. MOONEY & E. ATKINS. 1968. Advan. Pharmacol. **6A:** 307–317.
191. MYERS, R. D. 1968. Advan. Pharmacol. **6A:** 318–321.
192. SIMMONDS, M. A. 1970. J. Physiol. (London) **211:** 93–108.
193. SHEARD, M. H. & G. K. AGHAJANIAN. 1968. J. Pharmacol. Exp. Ther. **163:** 425–430.
194. MYERS, R. D. 1975. Brain Res. **94:** 491–506.
195. WALLER, M. B., R. D. MYERS & G. E. MARTIN. 1976. Neuropharmacology **15:** 61–68.
196. WALLER, M. B. & R. D. MYERS. 1974. Res. Commun. Chem. Pathol. Pharmacol. **9:** 421–429.
197. JACOBY, J. H., M. F. NELSON & L. D. LYTLE. 1974. Life Sci. **14:** 909–919.
198. SHOULSON, I. & T. N. CHASE. 1974. N. Engl. J. Med. **291:** 850–851.
199. SHANNON, P. J., D. LEONARD & M. A. KIDSON. 1974. Brit. Med. J. **3:** 576.
200. LEVIN, A. 1973. Brit. Med. J. **2:** 49.
201. HARDING, T. 1972. Brit. J. Psychiat. **121:** 338–339.
202. OSWALD, I., S. A. LEWIS, D. L. F. DUNLEAVY, V. BREZINOVA & M. BRIGGS. 1971. Brit. Med. J. **3:** 70–73.
203. GOLDING, D. 1970. Lancet **1:** 238.
204. STEEL, J. M. & M. BRIGGS. 1972. Brit. Med. J. **3:** 26–27.
205. JACOBY, J. H. & L. D. LYTLE. 1975. Unpublished observations.

BEHAVIORAL EVIDENCE FOR DENERVATION SUPERSENSITIVITY AFTER DESTRUCTION OF CENTRAL SEROTONERGIC NERVE TERMINALS *

Michael E. Trulson and Barry L. Jacobs

Department of Psychology
Princeton University
Princeton, New Jersey 08540

Development of supersensitivity after interruption of neuronal impulse flow is a ubiquitous phenomenon in the nervous system. It was first described and investigated most intensively in the peripheral nervous system;[1-4] however, several recent studies have demonstrated supersensitivity in the central nervous system (CNS),[5-10] especially in dopamine-mediated synapses.

Until recently, supersensitivity in central serotonergic synapses had not been investigated systematically. This was attributable both to the lack of a specific, measurable index of activity in central serotonin-mediated synapses and to the fact that a method for producing specific and extensive destruction of central serotonergic nerve terminals had not been available. This paper reviews our research on this subject during the past two years.[11-13] Other laboratories have also made significant contributions to this field in recent years.[14-16]

A Method for Studying Supersensitivity in the Central Serotonergic System

A complex behavioral syndrome in the rat, which consists of tremor, rigidity, reciprocal forepaw treading, Straub tail, hindlimb abduction, and lateral head weaving, is a reflection of enhanced activity in central serotonin-mediated synapses and thus provides a potentially valuable tool for assessing supersensitivity in the serotonergic system. The conclusion that this syndrome is specifically serotonin mediated is based on the following evidence: administration of the serotonin precursors L-tryptophan (after monamine oxidase inhibition) and L-5-hydroxytryptophan (L-5-HTP) both produce the syndrome [17-20]; these precursor effects are blocked by inhibition of serotonin synthesis via blockade of L-5-HTP decarboxylation or L-tryptophan hydroxylation [18]; monoamine oxidase inhibition (MAOI), which produces a large increase in cerebral serotonin, produces the syndrome, and this effect is blocked selectively by serotonin synthesis inhibition [21]; selective inhibition of serotonin reuptake via chlorimipramine administration potentiates the L-5-HTP-induced syndrome [22]; the serotonin agonist 5-methoxy-*N*,*N*-dimethyltryptamine produces the syndrome [23, 24]; and the serotonin-releasing agents *p*-chloroamphetamine and fenfluramine produce the syndrome, and this effect is blocked by prior serotonin synthesis inhibition.[24] This syndrome, together with the specific destruction of central serotonergic terminals produced by intraventricular injection of 5,7-

* Supported by Grants MH 23433 and MH 13445 from the National Institute of Mental Health.

0077–8923/78/0305–0497 $01.75/0

dihydroxytryptamine (5,7-DHT) in combination with a catecholamine uptake blocking agent,[25, 26] provides the opportunity to study denervation supersensitivity in central serotonin-mediated synapses.

We predicted that if denervation supersensitivity occurred in central serotonergic synapses after 5,7-DHT administration, it would be manifested by an increased behavioral responsivity or sensitivity to exogenously administered precursors (e.g., L-tryptophan after MAOI or L-5-HTP), because they markedly increase the level of brain serotonin[18, 27, 28] but not in response to compounds whose primary mode of action is the release of the endogenous stores of serotonin (e.g., fenfluramine).[29] We also predicted that supersensitivity to 5-methoxy-*N*,*N*-dimethyltryptamine (5-MDMT) and lysergic acid diethylamide (LSD), direct-acting serotonin agonists,[30, 31] would develop after 5,7-DHT administration. For comparison, we examined the dose-response relationships for the elicitation of the syndrome by L-5-HTP, 5-MDMT, and LSD in animals treated chronically with *p*-chlorophenylalanine (*p*-CPA). *p*-Chlorophenylalanine inhibits serotonin synthesis at the tryptophan hydroxylase step[32] and thus results in marked depletion of CNS serotonin, whereas serotonin nerve terminals remain morphologically intact. This finding allowed us to compare the behavioral responses to serotonin precursors and agonists in animals with a degree of serotonin depletion approximately equivalent to that of those that received 5,7-DHT but with intact presynaptic serotonin nerve terminals. Chronic *p*-CPA administration might be expected to induce supersensitivity at postsynaptic serotonin receptors, because a parallel phenomenon has been demonstrated in central dopamine receptors after synthesis inhibition.[33-35] In addition, we examined the dose-response relationships for the elicitation of the syndrome by L-5-HTP in rats treated chronically with a serotonin receptor blocker, methysergide. Chronic receptor blockade might also be expected to induce supersensitivity at postsynaptic serotonin receptors, because a similar effect occurs in central dopamine receptors after chronic receptor blockade.[36, 37]

Supersensitivity After Destruction of Serotonergic Nerve Terminals

Male Sprague-Dawley rats that weighed 250–300 g were administered desmethylimipramine hydrochloride (25 mg/kg, ip, 1 hr prior to surgery), then anesthetized with sodium pentobarbital (40 mg/kg, ip). A 20 μl saline solution (plus 0.02% ascorbic acid) that contained 50 μg of 5,7-DHT creatinine sulfate (dose calculated as free base) was injected into the lateral ventricle via a stereotaxically guided cannula. Control rats were treated in an identical manner, except that only the vehicle was injected into the lateral ventricle. 5,7-Dihydroxytryptamine produced 37 and 82% decreases in serotonin levels in whole brain and spinal cord, respectively, while norepinephrine and dopamine levels were not altered significantly (Table 1).

All behavioral observations were made with the animals in round plastic buckets (20 cm high $\times$ 35 cm in diameter) with metal screen lids and a layer of sawdust covering the floor. The rats were examined for the presence of the following signs: resting tremor (especially of the head and forelimbs), rigidity or hypertonicity (evaluated by grasping the rat around the torso and by extending and flexing the hindlimbs), hindlimb abduction (a splaying out of the hindlimbs), Straub tail, lateral head weaving (slow side to side movements of

TABLE 1

MONOAMINE LEVELS (μg/g) AFTER 5,7-DHT OR *p*-CPA TREATMENT *

		Brain			Spinal Cord		
Group	*N*	5-Hydroxy-tryptamine	Norepinephrine	Dopamine	5-Hydroxy-tryptamine	Norepinephrine	Dopamine
Vehicle	11	0.317±0.009 †	0.377±0.007	0.507±0.006	0.345±0.009	0.222±0.007	0.025±0.003
5,7-DHT	12	0.203±0.024 ‡	0.368±0.009	0.510±0.008	0.063±0.005 ‡	0.218±0.007	0.026±0.002
Percent of control		64	98	101	18	98	104
Saline	5	0.281±0.005	0.307±0.014	0.478±0.015	0.302±0.014	0.179±0.005	0.020±0.001
p-CPA	5	0.039±0.003 ‡	0.232±0.019 §	0.344±0.023 ¶	0.063±0.023 ‡	0.150±0.014	0.016±0.002
Percent of control		14	77	72	21	84	80

* 5,7-Dihydroxytryptamine creatinine sulfate (50 μg of the free base in 20 μl of a saline plus ascorbic acid solution) was injected into the lateral ventricle. Monoamine assays were performed 20 days after the injections. *p*-CPA methyl ester hydrochloride (400 mg/kg, ip, of the salt) was injected every third day. Assays were performed 24 days after the first injection. Serotonin was assayed by use of a modification [44] of the method of Bogdanski *et al.*,[43] and catecholamines were measured by the method of Snyder and Taylor.[45] Monoamine values have not been corrected for recoveries, which were as follows: serotonin, 84.5±2.3%; norepinephrine, 82.2±2.8%; dopamine, 78.7±3.4%. (From Trulson *et al.*[11])

† Each value is the mean ± SEM.

‡ Significance of the difference between means of drug-injected group and control groups: ‡, $p < 0.001$; §, $p < 0.01$; ¶, $p < 0.05$ (two-tailed Student's *t* tests).

the head), and reciprocal forepaw treading (rhythmic dorsoventral movements of the forelimbs).[24, 38-41] If at least four of these six signs were present, the animal was considered to have displayed the syndrome. The ED_{50} values were estimated by probit analysis.[42] The statistical significance between control and experimental groups was evaluated with Student's *t* test.

To determine the time course for the development of supersensitivity, 5,7-DHT-pretreated groups of rats were injected with L-5-HTP (125 mg/kg, which is the highest noneffective dose for eliciting the behavioral syndrome in control animals) either 12, 24, 48, or 96 hr after 5,7-DHT injection. Supersensitivity had developed in all rats tested 96 hr after 5,7-DHT administration. At 12 hr after 5,7-DHT injection, none of the rats showed increased responsivity, while 40 and 80% of the animals showed evidence of supersensitivity at 24 and 48 hr, respectively. Therefore, all of the data described below were obtained in experiments in which the rats were tested a minimum of 7 days after 5,7-DHT injection.

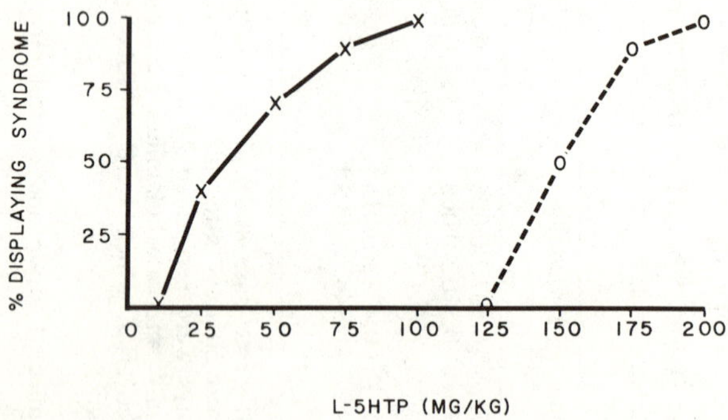

FIGURE 1. Effect of various doses of L-5-HTP in producing the syndrome in either 5,7-DHT- (×) or vehicle- (○) injected rats. Each point represents the percentage of animals that displayed the syndrome at each dose of L-5-HTP. *N*, 10–12 in all cases. (From Trulson *et al.*[11] By permission of *Journal of Pharmacology and Experimental Therapeutics.*)

Of all the compounds tested, L-5-HTP produced the greatest degree of supersensitivity in 5,7-DHT-treated rats (FIGURE 1). The ED_{50} value for 5,7-DHT-treated rats (37 mg/kg) was approximately 20% of the ED_{50} value in control animals (163 mg/kg), and the overall response to L-5-HTP in 5,7-DHT-treated rats was significantly different from that of control rats ($p < 0.01$). Some of the rats with the greatest degree of supersensitivity displayed a full syndrome at a dose of 12.5 mg/kg of L-5-HTP, making the effective dose in these cases approximately 10% of the dose required to induce the syndrome in control animals. There also appeared to be a decrease in the latency to onset of the syndrome after L-5-HTP administration in some 5,7-DHT-treated rats. While control rats typically showed the syndrome within 20–30 min after L-5-HTP administration, some 5,7-DHT-treated rats displayed the syndrome within 10 min postinjection.

5,7-Dihydroxytryptamine-treated rats displayed a marked supersensitivity to 5-MDMT and LSD (FIGURES 2 & 3). The ED_{50} values for 5,7-DHT-treated

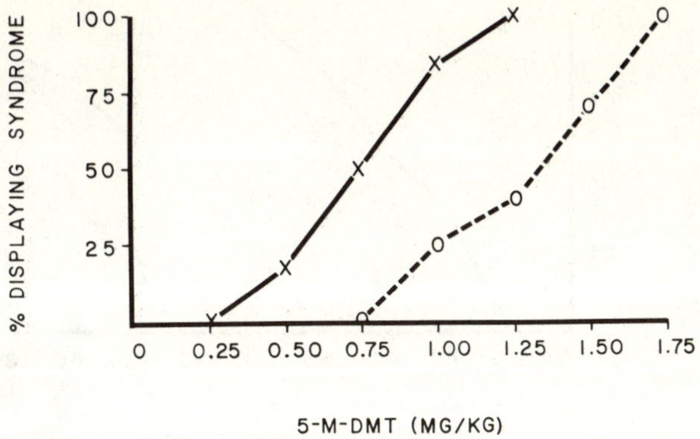

FIGURE 2. Effects of various doses of 5-methoxy-*N,N*-dimethyltryptamine (5-MDMT) in producing the syndrome in either 5,7-DHT- (×) or vehicle- (○) injected rats. Each point represents the percentage of animals that displayed the syndrome at each dose of 5-MDMT. *N*, 10–12 in all cases. (From Trulson *et al.*[11] By permission of *Journal of Pharmacology and Experimental Therapeutics.*)

rats (0.75 mg/kg and 200 μg/kg) were approximately 55 and 40% of the ED_{50} values in control animals (1.35 mg/kg and 520 μg/kg) for 5-MDMT and LSD, respectively, and the overall response to 5-MDMT or LSD in 5,7-DHT-treated rats was significantly different from that of control rats ($p < 0.05$). There appeared to be no change in the latency to onset of the 5-MDMT- or LSD-induced syndrome after 5,7-DHT treatment as compared to control injection, with the latency being 3–5 min in each group. The 5,7-DHT-treated rats also showed supersensitivity to L-tryptophan after pargyline pretreatment (FIGURE 4). The ED_{50} value for 5,7-DHT-treated rats (55 mg/kg) was approximately 40% of the ED_{50} value in control animals (145 mg/kg), and the overall response to L-tryptophan in 5,7-DHT-treated rats was significantly different from that of control rats ($p < 0.025$). The latency to onset of the syndrome did not appear to be decreased in 5,7-DHT-treated rats; the syndrome appeared between 30 and 45 min in both 5,7-DHT-injected and control rats.

The 5,7-DHT-treated rats showed a marked subsensitivity to fenfluramine (FIGURE 5). The ED_{50} value for 5,7-DHT-treated rats (20 mg/kg) was ap-

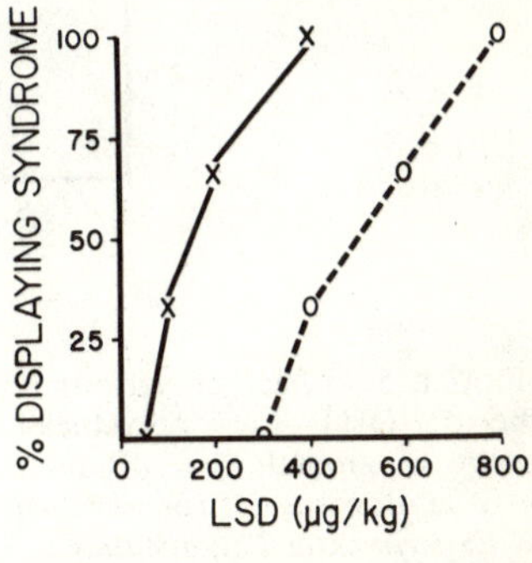

FIGURE 3. Effects of various doses of LSD in producing the syndrome in either 5,7-DHT- (×) or vehicle- (○) injected rats. Each point represents the percentage of animals that displayed the syndrome at each dose of LSD. *N*, six in all cases. (From Trulson *et al.*[12] By permission of *Psychopharmacology Communications.*)

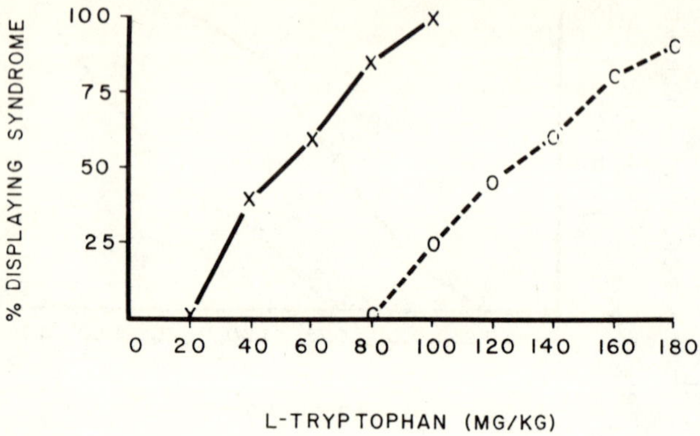

FIGURE 4. Effect of various doses of L-tryptophan (after MAOI) in producing the syndrome in either 5,7-DHT- (×) or vehicle- (○) injected rats. Each point represents the percentage of animals that displayed the syndrome at each dose of L-tryptophan. *N,* 10–12 in all cases. (From Trulson *et al.*[11] By permission of *Journal of Pharmacology and Experimental Therapeutics.*)

proximately 250% of the ED_{50} value in control animals (8 mg/kg), and the overall response to fenfluramine in 5,7-DHT-treated rats was significantly different from that of control rats ($p < 0.01$). Furthermore, only about 50% of the 5,7-DHT-treated rats displayed the syndrome when a dose greater than twice the ED_{100} for control rats was given. The effects of fenfluramine in 5,7-DHT-injected rats were potentiated by pretreatment with pargyline but still did not produce a supersensitive response. The latency to onset of the syndrome was typically 3–5 min after fenfluramine administration in both control and 5,7-DHT-injected rats and was unchanged by prior treatment with pargyline.

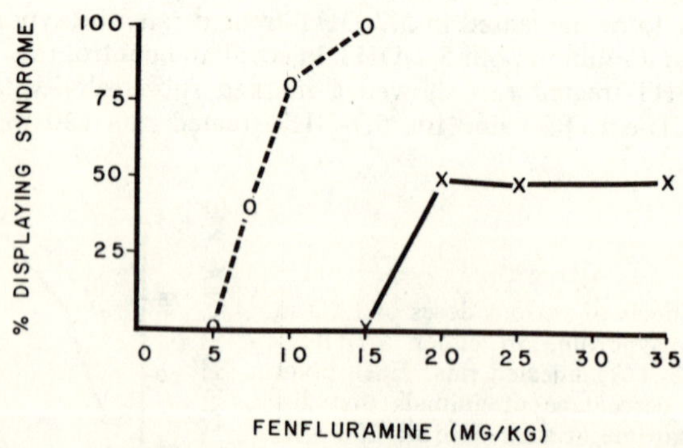

FIGURE 5. Effect of various doses of fenfluramine in producing the syndrome in either 5,7-DHT- (×) or vehicle- (○) injected rats. Each point represents the percentage of animals that displayed the syndrome at each dose of fenfluramine. *N,* 10–12 in all cases. (From Trulson *et al.*[11] By permission of *Journal of Pharmacology and Experimental Therapeutics.*)

Lack of Supersensitivity After Chronic Serotonin Synthesis Inhibition or Receptor Blockade

To determine whether supersensitivity occurs when serotonin is depleted without nerve terminal destruction, L-5-HTP, 5-MDMT, and LSD were administered to groups of rats that had received chronic administration of *p*-CPA methyl ester hydrochloride (400 mg/kg, ip, of the salt, every 3 days for a total of 24 days) or vehicle injections (1 ml of saline, ip, every 3 days). This *p*-CPA treatment produced 86 and 79% decreases in serotonin levels in whole brain and spinal cord, respectively, and slight, but statistically significant, reductions of norepinephrine and dopamine in brain, while catecholamine levels in the spinal cord were not changed significantly (Table 1).

Since *p*-CPA depletes serotonin but leaves nerve terminals morphologically intact, it is possible that a portion of the administered L-5-HTP would be utilized to replenish these empty terminals, thus masking any supersensitivity effect that might exist. To examine this possibility, the *p*-CPA experiment

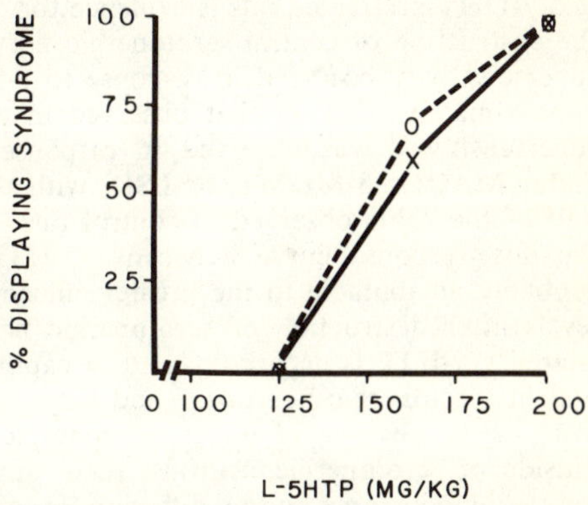

Figure 6. Effect of various doses of L-5-HTP in producing the syndrome in rats chronically administered either *p*-CPA (×) or vehicle (○). Each point represents the percentage of animals that displayed the syndrome at each dose of L-5-HTP. *N*, 10–12 in all cases. (From Trulson *et al.*[11] By permission of *Journal of Pharmacology and Experimental Therapeutics.*)

described above was repeated, and, in addition, each rat received an injection of a serotonin reuptake blocking agent, 3-(*p*-trifluoromethylphenoxy)-*N*-methyl-3-phenylpropylamine (Lilly 110140, 10 mg/kg, ip), 1 hr before L-5-HTP treatment. This treatment has been shown to significantly decrease serotonin reuptake.[46] An additional group of saline-pretreated rats received vehicle injection, rather than the reuptake blocker, 1 hr prior to L-5-HTP.

Chronic *p*-CPA administration did not induce supersensitivity to L-5-HTP (Figure 6), 5-MDMT, or LSD administered 1 day after the final *p*-CPA injection. For L-5-HTP, the ED_{50} value for *p*-CPA-treated rats (162 mg/kg) was approximately 105% of the ED_{50} value in control animals (154 mg/kg). For 5-MDMT and LSD, respectively, the ED_{50} values for *p*-CPA-treated rats (1.20 mg/kg and 480 μg/kg) were 109 and 96% of the ED_{50} values in control animals (1.10 mg/kg and 500 μg/kg). Administration of a specific serotonin reuptake blocker prior to L-5-HTP injection did not alter its behavioral effects in *p*-CPA-treated rats, as compared to control rats, which also received the reuptake blocker. The reuptake blocker, however, did potentiate the behavioral response when compared to that of rats that did not receive it.

Chronic administration of a serotonin receptor blocker, methysergide (1 or 10 mg/kg per day for 14 consecutive days), produced no evidence for supersensitivity to L-5-HTP. When tested 24 hr after the final injection of methysergide, the ED_{50} values for methysergide-treated rats (162 and 169 mg/kg) were 98 and 102% of the ED_{50} value in control animals (165 mg/kg) for 1 and 10 mg/kg doses of methysergide, respectively. When tested 72 hr after the last dose of methysergide, the ED_{50} values for methysergide-treated rats (150 and 161 mg/kg) were 98 and 105% of the ED_{50} value in control animals (153 mg/kg) for 1 and 10 mg/kg doses of methysergide, respectively. The overall response to L-5-HTP in rats chronically administered methysergide was not significantly different from that of control rats in any of the four groups ($p > 0.2$).

Nature of Supersensitivity in the Central Serotonergic System

The shift to the left in the dose-response curve that relates dose of L-5-HTP, 5-MDMT, LSD, or L-tryptophan (after MAOI) to elicitation of the syndrome in 5,7-DHT-pretreated rats is evidence for denervation supersensitivity following the destruction of central serotonergic nerve terminals. The greatest degree of supersensitivity occurred in response to L-5-HTP, with the ED_{50} decreasing to approximately 20% of that observed in control animals. A lesser degree of supersensitivity was observed in response to administration of L-tryptophan (after MAOI), 5-MDMT, or LSD, with the ED_{50} in each case approximately half of the value observed in control rats. The difference in degree of shift in the dose-response curve between L-5-HTP and MAOI plus L-tryptophan is probably attributable to the greater efficacy of L-5-HTP in elevating serotonin levels after destruction of serotonergic nerve terminals. Exogenously administered L-5-HTP is decarboxylated in capillaries and catecholaminergic neurons and in serotonergic neurons,[47] and large increases in cerebral serotonin occur, even in the absence of intact serotonergic terminals.[48] The serotonin formed outside of serotonergic neurons, then, apparently reaches serotonin receptors via the blood supply and by diffusion through extracellular spaces. On the other hand, L-tryptophan is converted to serotonin exclusively in serotonergic neurons, and when the majority of serotonergic terminals are destroyed, little serotonin is formed from exogenously administered L-tryptophan.[49] However, when serotonin catabolism is prevented via MAOI, serotonin accumulates in the remaining intact serotonergic neurons[50] and apparently is of sufficient magnitude to "spill over"[18] onto adjacent denervated serotonin receptors and elicit a supersensitive response. The lower degree of supersensitivity to 5-MDMT and LSD as compared to L-5-HTP may be due to the fact that supersensitivity in the central serotonergic system may be, at least in part, presynaptic in nature (see below). Thus, destruction of the presynaptic terminal would have relatively little effect on the efficacy of 5-MDMT and LSD, because these compounds have a direct postsynaptic action.[30, 31]

In contrast to the supersensitivity to L-tryptophan (after MAOI), L-5-HTP, 5-MDMT, or LSD, a marked subsensitivity to fenfluramine was observed following destruction of serotonergic nerve terminals with 5,7-DHT. The behavioral response to fenfluramine was potentiated by MAOI pretreatment, but still no supersensitivity was found. These findings support the hypothesis that a major effect of fenfluramine is to release stored serotonin[29]: since little stored serotonin remains to be released after destruction of the majority of serotonergic nerve terminals with 5,7-DHT, no syndrome occurs in many instances. Inhibition of MAO potentiates the response to fenfluramine, because inhibition of serotonin catabolism increases the amount of stored serotonin available for release; however, in the absence of increased synthesis rate, such as that which follows MAOI plus L-tryptophan administration, no supersensitivity is apparent. These results that demonstrate subsensitivity to fenfluramine parallel those with the catecholamine releaser amphetamine that occur after lesioning of central catecholamine neurons with the neurotoxin 6-hydroxydopamine.[7, 51, 52] Similarly, the L-5-HTP, 5-MDMT, and LSD data also parallel those obtained in recent studies of the catecholamine system in which behavioral evidence for denervation supersensitivity was reported after the administration of agonists or precursors to animals whose catecholamine nerve terminals had been destroyed by previous 6-hydroxydopamine administration.[5-7, 53]

In contrast to the supersensitivity that occurred after 5,7-DHT administration, no evidence was obtained for the development of supersensitivity, that is, a shift to the left in the dose-response curve, following chronic pretreatment with either *p*-CPA or methysergide. The discrepancy between the behavioral effects observed in 5,7-DHT- and *p*-CPA-treated animals is even more surprising when we consider the fact that the depletion of brain serotonin produced by *p*-CPA was substantially greater than that produced by 5,7-DHT. However, since the results of a previous study from our laboratory indicated that the syndrome is mediated largely by the brain stem and spinal cord,[41] it may be more relevant to compare *p*-CPA- and 5,7-DHT-induced serotonin depletions in the spinal cord. Even here, the depleting effects of *p*-CPA were equal to those of 5,7-DHT. Administration of a specific serotonin reuptake blocker prior to L-5-HTP in animals pretreated chronically with *p*-CPA still produced no evidence for supersensitivity, indicating that the lack of supersensitivity is not attributable to presynaptic uptake and sequestration of L-5-HTP and/or serotonin. One possibility is that even a small functional pool of serotonin present in the intact nerve terminals of *p*-CPA-treated animals may be sufficient to prevent the development of supersensitivity. Similarly, chronic receptor blockade may not completely abolish activity in serotonin-mediated synapses. Another possibility, discussed more fully below, is that the supersensitivity produced by 5,7-DHT is attributable specifically to the degeneration and disappearance of the presynaptic axon terminal, which, of course, does not occur after *p*-CPA or methysergide injection.

Failure to observe supersensitivity after depletion of serotonin by *p*-CPA-induced synthesis inhibition or chronic receptor blockade is not paralleled in the catecholamine system. Several studies have reported that catecholamine depletion by synthesis inhibition[33, 35] or chronic catecholamine receptor blockade[36, 37] potentiates the behavioral response to catecholamine precursors and agonists. Whether these data reflect a fundamental difference between serotonergic and catecholaminergic synapses or merely methodologic and/or pharmacologic differences in the test or treatment drugs in these experiments is a subject for future investigation.

The enhanced efficacy of L-tryptophan (after MAOI), L-5-HTP, 5-MDMT, and LSD in 5,7-DHT pretreated animals may be attributable to presynaptic, rather than postsynaptic, changes. The absence of a presynaptic uptake mechanism for serotonin and/or easier access of serotonin, 5-MDMT, or LSD to the postsynaptic receptor after presynaptic terminal degeneration[54] may account for the supersensitivity observed. At present, there is no readily available method for distinguishing between pre- and postsynaptic mechanisms for supersensitivity in the serotonin system. For example, the time course for the onset of supersensitivity (1–4 days) corresponds to the morphologic degeneration of the presynaptic nerve terminal after treatment with 5,7-DHT.[55-57] The time course for the onset of behavioral supersensitivity is also closely paralleled in central dopaminergic neurons after destruction of presynaptic terminals.[5-7] However, postsynaptic supersensitivity also develops within a few days. The latter phenomenon is exemplified by decentralization supersensitivity, which occurs via an actual increase in the sensitivity of the postsynaptic receptor or via proliferation of postsynaptic receptor sites,[1] and by the supersensitivity that follows axon terminal destruction in central catecholamine neurons (as measured by increased adenyl cyclase activity[58-60]). Thus, time course data alone do not provide sufficient information to distinguish between the pre- and postsynaptic nature of supersensitivity in central serotonergic neurons. Until direct evidence for postsynaptic supersensitivity is provided, the possibility remains that supersensitivity in the central serotonergic system may be purely presynaptic in nature, without an accompanying change in postsynaptic receptor sensitivity.

References

1. Fleming, W. W., J. J. McPhillips & D. P. Westfall. 1973. Postjunctional supersensitivity and subsensitivity of excitable tissues to drugs. Ergebn. Physiol. **68:** 55–119.
2. Sharpless, S. K. 1964. Reorganization of function in the nervous system—use and disuse. Annu. Rev. Physiol. **26:** 357–388.
3. Trendelenburg, U. 1963. Supersensitivity and subsensitivity to sympathomimetic amines. Pharmacol. Rev. **15:** 225–276.
4. Trendelenburg, U. 1966. Mechanisms of supersensitivity and subsensitivity to sympathomimetic amines. Pharmacol. Rev. **18:** 629–640.
5. Schoenfeld, R. & N. Uretsky. 1972. Altered response to apomorphine in 6-hydroxydopamine-treated rats. Eur. J. Pharmacol. **19:** 115–118.
6. Schoenfeld, R. I. & N. J. Uretsky. 1973. Enhancement by 6-hydroxydopamine of the effects of dopa upon the motor activity of rats. J. Pharmacol. Exp. Ther. **186:** 616–624.
7. Ungerstedt, U. 1971. Striatal dopamine release after amphetamine or nerve degeneration revealed by rotational behavior. Acta Physiol. Scand. Suppl. **367:** 49–68.
8. Bird, S. J. & G. K. Aghajanian. 1975. Denervation supersensitivity in the cholinergic septo-hippocampal pathway: a microiontophoretic study. Brain Res. **100:** 355–370.
9. Zis, A. P. & H. C. Fibiger. 1975. Functional evidence for postsynaptic supersensitivtiy of central noradrenergic receptors after denervation. Nature (London) **256:** 659–660.
10. Fibiger, H. C. & D. S. Grewaal. 1974. Neurochemical evidence for denervation supersensitivity: the effect of unilateral substantia nigra lesions on apomorphine-induced increases in neostriatal acetylcholine levels. Life Sci. **15:** 57–63.
11. Trulson, M. E., E. E. Eubanks & B. L. Jacobs. 1976. Behavioral evidence for supersensitivity following destruction of central serotonergic nerve terminals by 5,7-dihydroxytryptamine. J. Pharmacol. Exp. Ther. **198:** 23–32.
12. Trulson, M. E., C. A. Ross & B. L. Jacobs. 1976. Behavioral evidence for the stimulation of CNS serotonin receptors by high doses of LSD. Psychopharmacol. Commun. **2:** 149–164.

13. Trulson, M. E., A. D. Stark & B. L. Jacobs. 1977. Lack of supersensitivity to L-5-hydroxytryptophan following chronic methysergide treatment. J. Pharm. Pharmacol. **29:** 382–384.

14. Longo, V. G., A. Scotti de Carolis, A. Liuzzie & M. Massotti. 1974. A study of the central effects of 5,6-dihydroxytryptamine. *In* Advances in Biochemical Psychopharmacology, Serotonin: New Vistas. E. Costa, G. L. Gessa & M. Sandler, Eds. Vol. **10:** 109–120. The Raven Press. New York, N.Y.

15. Nygren, L. G., K. Fuxe, G. Jonsson & L. Olson. 1974. Functional regeneration of 5-hydroxytryptamine nerve terminals in the rat spinal cord following 5,6-dihydroxytryptamine induced degeneration. Brain Res. **78:** 377–394.

16. Steward, R. M., J. H. Growdon, D. Cancian & R. J. Baldessarini. 1976. 5-Hydroxytryptophan induced myoclonus: increased sensitivity to serotonin after intracranial 5,7-dihydroxytryptamine in the adult rat. Neuropharmacology **15:** 449–455.

17. Butcher, L. L., J. Engle & K. Fuxe. 1972. Behavioral, biochemical and histochemical analyses of the central effects of monamine precursors after peripheral decarboxylase inhibition. Brain Res. **41:** 387–411.

18. Grahame-Smith, D. G. 1971. Studies *in vivo* on the relationships between brain tryptophan, brain 5-HT synthesis and hyperactivity in rats treated with a monomine oxidase inhibitor and L-tryptophan. J. Neurochem. **18:** 1053–1066.

19. Jacobs, B. L., E. E. Eubanks & W. D. Wise. 1974. Effects of indolealkylamine manipulations on locomotor activity in the rat. Neuropharmacology **13:** 575–585.

20. Modigh, K. 1972. Central and peripheral effects of 5-hydroxytryptophan on motor activity in mice. Psychopharmacology **23:** 48–54.

21. Modigh, K. & T. H Svensson. 1972. On the role of central nervous system catecholamines and 5-hydroxytryptamine in the nialamide-induced behavioral syndrome. Brit. J. Pharmacol. **46:** 32–45.

22. Modigh, K. 1973. Effects of chlorimipramine and protriptyline on the hyperactivity induced by 5-hydroxytryptophan after peripheral decarboxylase inhibition in mice. J. Neural Transmission **34:** 101–109.

23. Grahame-Smith, D. G. 1971. Inhibitory effect of chlorpromazine on the syndrome of hyperactivity produced by L-tryptophan or 5-methoxy-N,N-dimethyltryptamine in rats treated with a monoamine oxidase inhibitor. Brit. J. Pharmacol. **43:** 856–864.

24. Trulson, M. E. & B. L. Jacobs. 1976. Behavioral evidence for the rapid release of CNS serotonin by p-chloramphetamine and fenfluramine. Eur. J. Pharmacol. **36:** 149–154.

25. Gerson, S. & R. J. Baldessarini. 1975. Selective destruction of serotonin terminals in rat forebrain by high doses of 5,7-dihydroxytryptamine. Brain Res. **85:** 140–145.

26. Björklund, A., H. G. Baumgarten & A. Rensch. 1975. 5,7-Dihydroxytryptamine: improvement of its selectivity for serotonin neurons in the CNS by pretreatment with desipramine. J. Neurochem. **25:** 833–835.

27. Hess, S. M. & W. Doepfner. 1961. Behavioral effects and brain amine content in rats. Arch. Int. Pharmacodyn. **134:** 89–99.

28. Moir, A. T. B. & D. Eccleston. 1968. The effects of precursor loading in the cerebral metabolism of 5-hydroxyindoles. J. Neurochem. **15:** 1093–1108.

29. Costa, E., A. Groppetti & A. Revuelta. 1971. Action of fenfluramine on monoamine stores of rat tissues. Brit. J. Pharmacol. **41:** 57–64.

30. Fuxe, K., B. Holmstedt & G. Jonsson. 1972. Effects of 5-methoxy-N,N-dimethyltryptamine on central monoamine neurons. Eur. J. Pharmacol. **19:** 25–34.

31. Andén, N. E., H. Corrodi, K. Fuxe & T. Hokfelt. 1968. Evidence for a central 5-hydroxytryptamine receptor stimulation by lysergic acid diethylamide. Brit. J. Pharmacol. **34:** 1–7.

32. Koe, B. K. & A. Weissman. 1966. p-Chlorophenylalanine: a specific depletor of brain 5-HT. J. Pharmacol. Exp. Ther. **154:** 499–516.
33. Dominic, J. A. & K. E. Moore. 1969. Supersensitivity to the central stimulant action of adrenergic drugs following discontinuation of a chronic diet of α-methyltyrosine. Psychopharmacology **15:** 96–101.
34. Moore, K. E. & J. E. Thornburg. 1975. Drug-induced dopaminergic supersensitivity. *In* Advances in Neurology. D. B. Caline, T. N. Chase & A. Barbeau, Eds. Vol. 9: 93–104. The Raven Press. New York, N.Y.
35. Thornburg, J. E. & K. E. Moore. 1973. The relative importance of dopaminergic and noradrenergic neuronal systems for the stimulation of locomotor activity induced by amphetamine and other drugs. Neuropharmacology **12:** 853–866.
36. Klawans, H. L. & R. Rubovits. 1972. An experimental model of tardive dyskinesia. J. Neural Transmission **33:** 235–246.
37. Tarsy, D. & R. J. Baldessarini. 1974. Behavioral supersensitivity to apomorphine following chronic treatment with drugs which interfere with the synaptic function of catecholamines. Neuropharmacology **13:** 927–940.
38. Jacobs, B. L. 1974. Effect of two dopamine receptor blockers on a serotonin-mediated behavioral syndrome in rats. Eur. J. Pharmacol. **27:** 363–366.
39. Jacobs, B. L. 1974. Evidence for the functional interaction of two central neurotransmitters. Psychopharmacology **39:** 81–86.
40. Jacobs, B. L. 1976. An animal behavior model for studying central serotonergic synapses. Life Sci. **19:** 777–786.
41. Jacobs, B. L. & H. Klemfuss. 1975. Brainstem and spinal cord mediation of a serotonergic behavioral syndrome. Brain Res. **100:** 450–457.
42. Bliss, C. I. 1952. The Statistics of Bioassays. Academic Press, Inc. New York, N.Y.
43. Bogdanski, D. F., A. Pletscher, B. B. Brodie & S. Udenfriend. 1956. Identification and assay of serotonin in brain. J. Pharmacol. Exp. Ther. **117:** 82–88.
44. Jacobs, B. L., W. D. Wise & K. M. Taylor. 1974. Differential behavioral and neurochemical effects following lesions of the dorsal or median raphe nuclei in rat. Brain Res. **79:** 353–361.
45. Snyder, S. H. & K. M. Taylor. 1972. Assay of biogenic amines and their deaminating enzymes in animal tissues. *In* Research Methods in Neurochemistry. N. Marks & R. Rodnight, Eds. Vol. 1: 287–315. Plenum Publishing Corporation. New York, N.Y.
46. Fuller, R. W., K. W. Perry & B. B. Molloy. 1974. Effect of an uptake inhibitor on serotonin metabolism in rat brain: studies with 3-(p-trifluoromethylphenoxy)-N-methyl-3-phenylpropylamine (Lilly 110140). Life Sci. **15:** 1161–1171.
47. Lichtensteiger, W., U. Mutzner & H. Langemann. 1967. Uptake of 5-hydroxytryptamine and 5-hydroxytryptophan by neurons of the central nervous system normally containing catecholamines. J. Neurochem. **14:** 489–497.
48. Korf, J., K. Venema & F. Postema. 1974. Decarboxylation of exogenous L-5-hydroxytryptophan after destruction of the cerebral raphe system. J. Neurochem. **23:** 249–252.
49. Kuhar, M. J., G. K. Aghajanian & R. H. Roth. 1972. Tryptophan hydroxylase activity and synaptosomal uptake of serotonin in discrete brain regions after midbrain raphe lesions: correlations with serotonin levels and histochemical fluorescence. Brain Res. **44:** 165–176.
50. Aghajanian, G. K. & I. M. Asher. 1971. Histochemical fluorescence of raphe neurons: selective enhancement by tryptophan. Science **172:** 1159–1162.
51. Breese, G. R., B. R. Cooper & R. D. Smith. 1973. Biochemical and behavioral alterations following 6-hydroxydopamine administration into brain. *In* Frontiers in Catecholamine Research. E. Usdin & S. H. Snyder, Eds. : 701–706. Pergamon Press. New York, N.Y.
52. Creese, I. & S. D. Iversen. 1973. Blockage of amphetamine induced motor

stimulation and stereotypy in the adult rat following neonatal treatment with 6-hydroxydopamine. Brain Res. **55:** 369–382.

53. UNGERSTEDT, U. 1971. Postsynaptic supersensitivity after 6-hydroxydopamine induced degeneration of the nigrostriatal dopamine system. Acta Physiol. Scand. Suppl. **367:** 69–93.
54. WAGNER, K. & U. TRENDELENBERG. 1971. Development of degeneration contraction and supersensitivity in the cat's nictitating membrane after 6-hydroxydopamine. Arch. Pharmakol. **270:** 215–236.
55. BAUMGARTEN, H. G. & L. LACHENMAYER. 1972. 5,7-Dihydroxytryptamine: improvement in chemical lesioning of indoleamine neurons in the mammalian brain. Z. Zellforsch. **135:** 399–414.
56. NOBIN, A., H. G. BAUMGARTEN, A. BJÖRKLUND, L. LACHENMAYER & U. STENEVI. 1973. Axonal degeneration and regeneration of the bulbo-spinal indoleamine neurons after 5,6-dihydroxytryptamine treatment. Brain Res. **56:** 1–24.
57. GERSON, S., R. J. BALDESSARINI & S. C. WHEELER. 1974. Biochemical effects of dihydroxylated tryptamine on central indoleamine neurons. Neuropharmacology **13:** 987–1004.
58. PALMER, G. C. 1972. Increased cyclic AMP response to norepinephrine in the rat brain following 6-hydroxydopamine. Neuropharmacology **11:** 145–149.
59. HUANG, M., A. K. S. HO & J. W. DALY. 1973. Accumulation of adenosine cyclic 3′,5′-monophosphate in rat cerebral cortical slices. Mol. Pharmacol. **9:** 711–717.
60. KALISKER, A., C. O. RUTLEDGE & J. P. PERKINS. 1973. Effect of nerve degeneration by 6-hydroxydopamine on catecholamine-stimulated adenosine 3′,5′-monophosphate formation in rat cerebral cortex. Mol. Pharmacol. **9:** 619–629.

CHANGES IN MOTOR BEHAVIOR FOLLOWING THE ADMINISTRATION OF SEROTONIN NEUROTOXINS *

John H. Growdon †

Laboratory of Neuroendocrine Regulation
Department of Nutrition and Food Science
Massachusetts Institute of Technology
Cambridge, Massachusetts 02139

Numerous anatomic,[1, 2] physiologic,[3, 4] and metabolic [5] studies have greatly expanded our knowledge of serotonin [5-hydroxytryptamine (5-HT)] neurons in the central nervous system. These neurons affect a wide variety of behaviors in animals, including sleep,[6] feeding,[7, 8] temperature control,[9, 10] and pain sensitivity [11, 12]; in the human, these neurons influence sleep,[13] mood,[14] hypothalamic peptides,[15] and myoclonus.[16, 17] Serotonin neurons in the rat apparently participate in motor activity as well, because aggressive behavior occurred after *p*-chlorophenylalanine depleted brain 5-HT content; [18] increased locomotor activity occurred after the administration of *p*-chloroamphetamine (*p*-CA) [19–22]; hind limb reflexes were enhanced by drugs that stimulated 5-HT receptors [23–25]; and hyperactivity, opisthotonus, ataxia, and clonic fits occurred after administration of the 5-HT precursor tryptophan, following pretreatment with tranylcypromine, a monoamine oxidase inhibitor.[26] The last effect has attracted new interest, because it may define a specific serotonergic behavior [27]; if confirmed, this symptom complex may have the potential research applications for studying central 5-HT neurons that amphetamine [28] and apomorphine [29] stereotyped and rotational behaviors [30] have had for dopaminergic mechanisms. Newer techniques to stimulate 5-HT receptors pharmacologically include the administration of 5-hydroxytryptophan (5-HTP) intraperitoneally to rats pretreated with intracranial injections of 5,7-dihydroxytryptamine (5,7-DHT) [31, 32] and injections of *p*-CA into normal rats.[33, 34] Both procedures elicited altered motor behaviors that reproduced and further defined features of the serotonergic syndrome. This paper will review the features of 5-HTP-induced myoclonus and fully describe the *p*-CA-induced syndrome.

Myoclonic Syndrome Induced by 5-HTP in Rats Previously Treated with Desmethylimipramine and 5,7-DHT

The dihydroxylated tryptamines, including 5,6-dihydroxytryptamine (5,6-DHT) and 5,7-DHT, destroy cells that contain both catecholamines and serotonin,[35–38] but pretreatment with desmethylimipramine (DMI) apparently limits their toxicity to serotonergic neurons.[39, 40] Thus, Stewart *et al.*[31, 32] gave DMI (25 mg/kg, ip) to rats 4 and, again, 1 hr prior to cisternal injection of 5,7-DHT

* Supported by the Massachusetts General Hospital Parkinson Fund and by Grant NGR–22–009–627 from the National Aeronautics & Space Administration and Grant AM–14228 from the United States Public Health Service.

† John R. Whittier Fellow of the Committee to Combat Huntington's Disease.

0077–8923/78/0305–0510 $01.75/0

(200 μg, calculated as the free base) to lesion selectively serotonin-containing neurons. 5,7-Dihydroxytryptamine was administered in 25 μl of artificial cerebrospinal fluid with added ascorbic acid (0.5% w/v) and gassed with nitrogen to retard oxidation; control animals received vehicle alone. Many rats developed generalized seizures immediately after 5,7-DHT injection and required pentobarbital (20 mg/kg) to control these seizures. After recovery, rats were temporarily irritable and aggressive (biting and rearing), but after several days, they were difficult to distinguish from controls. Subsequent injections of 5-HTP, the biochemical precursor of 5-HT, produced a characteristic pattern of behavior that consisted of immobile postures (FIGURE 1), ataxic gait, body and head tremors, and myoclonus (FIGURE 2), whereas single injections of 5-HTP into normal rats never produced these effects. This response was dose related; a dose of 65 mg/kg was optimal for eliciting the full syndrome, while dosages ≧100 mg/kg caused seizures and death. The syndrome began 5 min after 5-HTP injection and lasted 60–120 min before subsiding. Autonomic signs were often prominent and included salivation, lacrimation, piloerection, diarrhea, and ejaculation. Piloerection and increased muscular activity (tremors, myoclonus) might be expected if the rats were hypothermic and shivering; these responses were paradoxic and inappropriate, however, because the body temperature rose dramatically immediately after 5-HTP injection (FIGURE 3).

Pharmacologic testing indicated that the myoclonic syndrome was dependent on serotonergic mechanisms. 5-Hydroxytryptophan administration (50–500 mg/kg) did not affect normal rats (although it will produce myoclonus and seizures in normal guinea pigs [41]), whereas much smaller amounts did elicit the full syndrome in rats pretreated with 5,7-DHT. This effect apparently depends on the conversion of 5-HTP to 5-HT; pretreatment with high doses of RO4–4602 to prevent this decarboxylation step also blocked the syndrome (TABLE 1). On the other hand, drugs that act as 5-HT agonists (quipazine, 5-methoxy-*N*,*N*-dimethyltryptamine [42]) produced the syndrome in rats pretreated with 5,7-DHT, but catecholamine agonists (levodopa, apomorphine) were ineffective. Prior treatment with drugs that are thought to antagonize 5-HT receptors (lysergic acid diethylamide, 2-bromolysergic acid diethylamide, methysergide) also blocked the syndrome, whereas drugs that interfere with dopaminergic, noradrenergic, and cholinergic neurotransmission did not.

Based on these observations, we postulated that the myoclonic syndrome was related specifically to central indoleamine neurons and probably reflected synaptic supersensitivity to 5-HT that was induced by prior 5,7-DHT administration. 5-Hydroxytryptophan also produced the main features of the syndrome in rats pretreated with intracranial doses of 5,6-DHT, and we expected similar results after pretreatment with another 5-HT neurotoxin, *p*-CA, as well. This expectation was not realized, although a single dose of *p*-CA *did* cause a similar syndrome immediately after its systemic injection into normal rats.

Postural Abnormalities, Tremor, and Myoclonus Induced by *p*-CA in the Rat

A single systemic injection of *p*-CA lowers brain levels of indoleamines within 8 hr, and these levels remain depressed for months.[43–46] Even though the extent of 5-HT depletion was comparable to that measured after administration of DMI plus 5,7-DHT (TABLE 2), a 5-HTP challenge dose did not produce

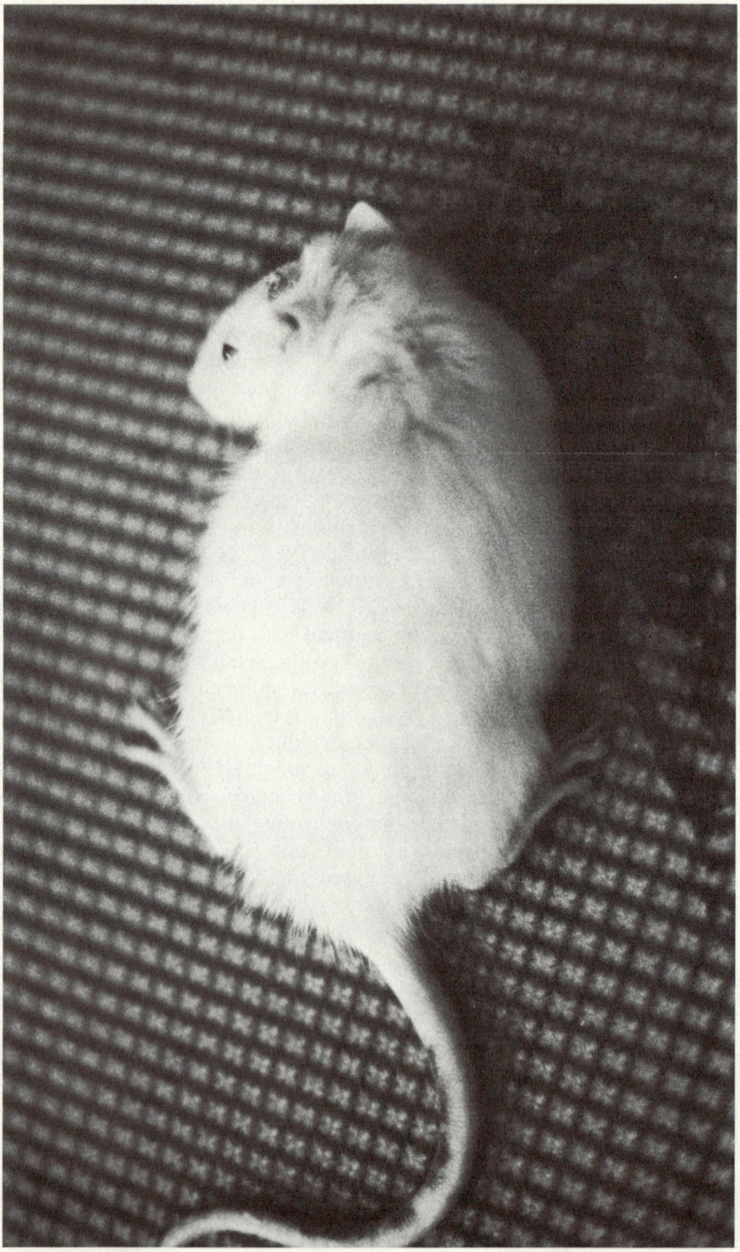

FIGURE 1. Photograph of a rat 30 min after an injection of 5-HTP (65 mg/kg, ip); DMI and intracranial 5,7-DHT had been given 3 weeks earlier. Note the immobile rigid posture with abducted hind limbs and stiffly curled tail.

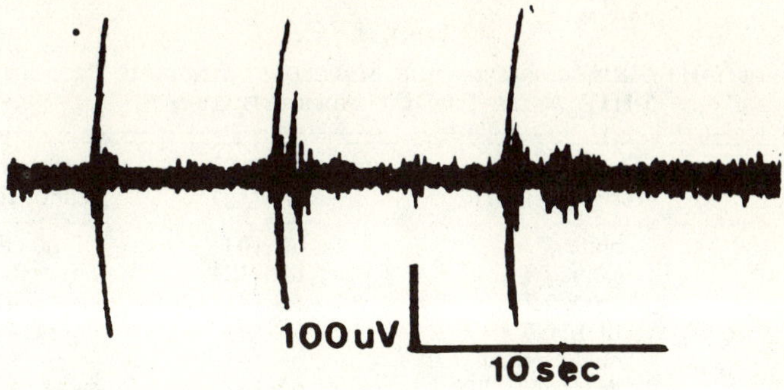

FIGURE 2. Representative record of myoclonus induced by 5-HTP in a rat recorded from the posterior cervical region with intramuscular electrodes. The animal had received 200 μg of 5,7-DHT 7 days previously and 65 mg/kg of 5-HTP 20 min earlier. Note the brief bursts of high-amplitude muscle activity. During peak activity (10–30 min after injection of 65 mg/kg of 5-HTP), from eight to 13 contractions per minute were recorded. (From Stewart *et al.*[31] By permission of *Neuropharmacology*.)

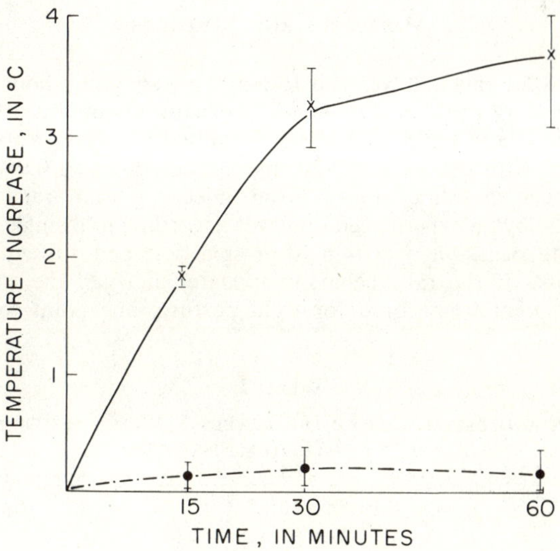

FIGURE 3. The effect of 5-HTP on rectal temperature in rats pretreated with DMI and 5,7-DHT. Each point represents the mean, and the vertical bars indicate the SEM of five rats per group. In rats pretreated with DMI and 5,7-DHT, there was a marked temperature increase after 5-HTP injection (65 mg-kg, × and —), whereas temperature did not rise after administration of (● and —•—).

TABLE 1

EFFECT OF 5-HT ANTAGONISTS ON THE MYOCLONIC SYNDROME PRODUCED BY 5-HTP AFTER 5,7-DHT ADMINISTRATION *

5,7-DHT (μg/rat)	Pretreatment (mg/kg)	5-HTP (mg/kg)	Syndrome (per-cent of control)
None	none	65 (4)	no effect
200	saline	65 (12)	100
200	LSD, 0.2	65 (4)	47 †
200	BOL, 0.5	65 (4)	42 †
200	methysergide, 1.0	65 (4)	61 †
200	RO4–4602, 200	65 (4)	17 †

* Rats were injected with 5,7-DHT intracranially 1 month prior to the 5-HTP challenge dose. Pretreatment drugs, lysergic acid diethylamide (LSD), 2-bromolysergic acid diethylamide (BOL), methysergide, and RO4–4602, were given intraperitoneally 30–60 min before 5-HTP. The numbers of rats tested with each agent are noted in parentheses. (Based on data from Stewart *et al.*[32])

† Differs significantly ($p < 0.01$ or less by Student's *t* test) from rats pretreated with saline.

any motor abnormality. Rigid postures, tremor, myoclonus, and autonomic signs did occur, however, immediately after a single injection of *p*-CA (10 mg/kg, ip) in normal rats.

MATERIALS AND METHODS

Sprague-Dawley rats that weighed 150–200 g were group housed, maintained on a 12-hr light-dark cycle in a room with constant temperature (22° C), given free access to food and water, and tested in individual cages. Rats were injected intraperitoneally with *d,l-p*-CA (2–20 mg/kg) dissolved in 0.9% saline, while control rats received saline alone. After injection, each rat's behavior was scored for 30 sec by an experienced observer according to a single-blind protocol on five separate occasions: 5 min after injection and subsequently at four 15-min intervals. If the rat's behavior appeared normal, the score was zero; otherwise, one point was scored for rigid posture, one point for tremor, one

TABLE 2

BRAIN AND SPINAL CORD 5-HT LEVELS 3 WEEKS AFTER *p*-CA AND 5,7-DHT ADMINISTRATION

	Brain 5-HT * (μg/g)	Spinal Cord 5-HT (μg/g)
Control	0.44±0.02	0.61±0.02
p-CA	0.21±0.03 †	0.39±0.02 †
5,7-DHT	0.20±0.02 †	0.19±0.03 †

* Mean±SEM of five rats per group.

† $p < 0.001$, compared with control group (Student's *t* test).

point for myoclonus, and one point for autonomic signs (salivation, piloerection, ejaculation). Thus, the maximum score for each grading period was four; a rat's total score was the sum of the five interval scores (maximal score of 20) and reflected the extent of the *p*-CA-induced behavior. Different groups of rats were injected with *p*-CA (10 mg/kg) 15–30 min after pretreatment with either fluoxetine (10 mg/kg), methysergide (1 mg/kg), methiothepin (5–20 mg/kg), methergoline (1.0 mg/kg), cyproheptadine (0.4 mg/kg), phenoxybenzamine (10 mg/kg), propranolol (0.2 mg/kg), haloperidol (1 mg/kg), pimozide (10 mg/kg), DMI (25 mg/kg), atropine (1–2 mg/kg), or methylscopolamine (1 mg/kg) or with an equivalent volume of the appropriate vehicle. Other groups were pretreated with *p*-chlorophenylalanine (300 mg/kg) 48 hr before *p*-CA or with α-methyl-*p*-tyrosine (70 mg/kg) 3 hr and, again, 1 hr before *p*-CA injection (see TABLE 4). Behavior was rated according to the scoring system described above. Brain and spinal cord levels of 5-HT and 5-hydroxyindoleacetic acid (5-HIAA) were measured[47] during the syndrome (30 min after *p*-CA injection) and also 1 day later.

RESULTS

Rats injected with 10 mg/kg of *p*-CA developed a reproducible syndrome of altered behavior. Five minutes after injection, they paced about the cage; vibrissae twitched rapidly, and respiratory rate increased. Within 10–15 min, the pacing ceased, and the rats assumed a crouched immobile posture, with the abdomen flat on the cage floor (FIGURE 4). The hind limbs were abducted and extended rigidly, while the forelimbs made small, rapid, repetitive prancing movements. Once in this position, most rats pivoted 45–90 degrees around their hind quarters, and some rats circled a full 360 degrees. In addition, most rats tended to inch backward and would fall from a table top unless checked. Between 15 and 30 min after injection, they developed a fine, rapid head tremor that often spread to involve the entire body, and their tails stiffened and curled. Muscle tone increased, and myoclonic jerks developed in the limbs and trunk, especially when the animals attempted to walk but also when they were immobile. By 30 min, autonomic nervous system signs were prominent and consisted of piloerection, lacrimation, salivation, loose stools, and ejaculation. The entire syndrome lasted 60–90 min and occurred in every rat (23/23) injected with 10 mg/kg of *p*-CA.

The intensity of the syndrome was proportional to the dose of *p*-CA injected. In separate experiments, rats received individual doses of *p*-CA that ranged from 2 to 20 mg/kg. The resultant behavioral syndrome was graded, and the total scores were plotted against the *p*-CA dose (FIGURE 5). The manifestations of the syndrome increased linearly with increasing doses up to 10 mg/kg, but 20 mg/kg did not appreciably enhance the syndrome.

Whole brain levels of 5-HT and its major metabolite 5-HIAA measured 30 min after *p*-CA injection, when the syndrome was at its peak, were (mean ± SEM) 0.50 ± 0.05 and 0.45 ± 0.3 μg/g of tissue, respectively; neither differed significantly from control values (TABLE 3). One day after *p*-CA injection, however, the 5-HT level had fallen to 0.18 ± 0.3 μg/g ($p < 0.001$) and the 5-HIAA level to 0.30 ± 0.03 μg/g ($p < 0.01$).

p-Chloroamphetamine-induced behavioral changes were all significantly antagonized by drugs known to alter serotonergic neurotransmission (TABLE 4).

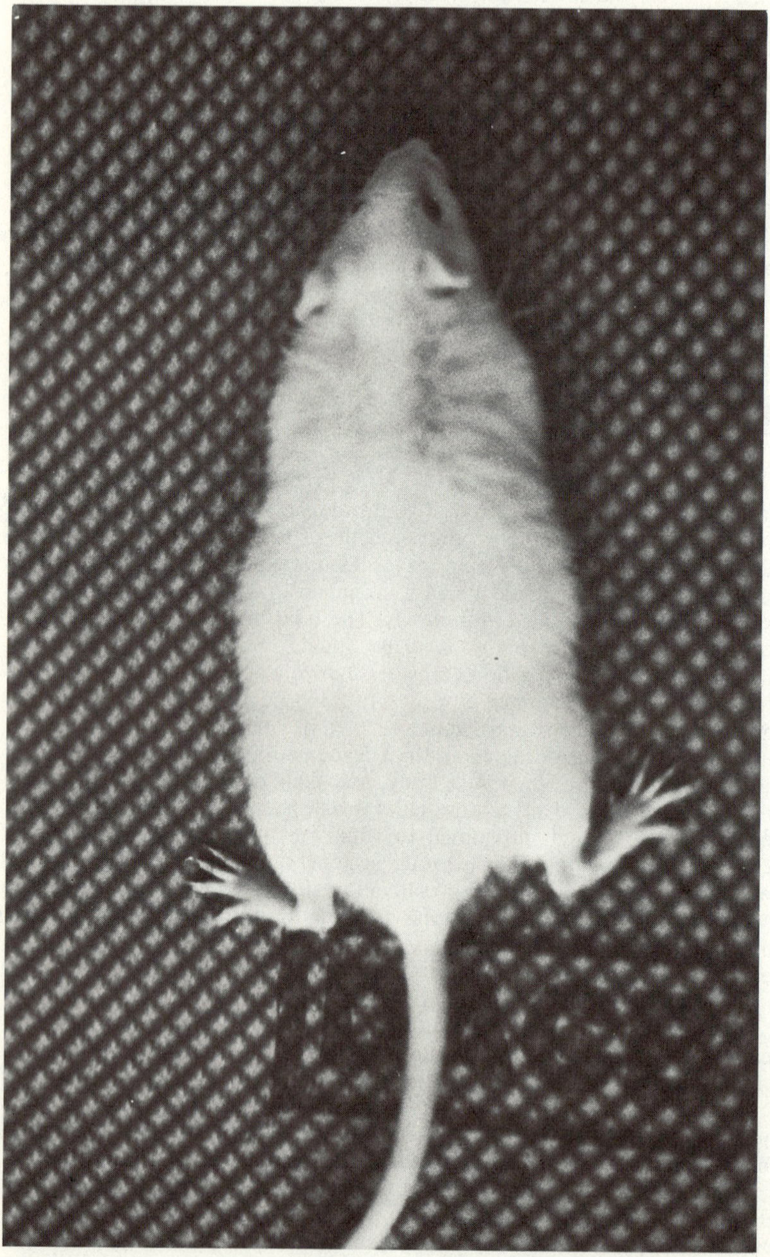

FIGURE 4. Photograph of a rat 30 min after a single injection of *p*-CA (10 mg/kg, ip). The rat's abdomen rested on the mat, and the hind limbs were abducted and extended. Piloerection and a stiff tail were also evident.

TABLE 3

BRAIN LEVELS OF 5-HT AND 5-HIAA MEASURED 30 MIN AND 1 DAY AFTER AN INTRAPERITONEAL INJECTION OF *p*-CA (10 MG/KG)*

	5 HT † (μg/g)	5-HIAA † (μg/g)
Control	0.48±0.04	0.44±0.02
30 min after injection	0.50±0.05 ‡	0.45±0.03 ‡
1 day after injection	0.18±0.03 §	0.30±0.03 ¶

* From Growdon.[34]
† Mean ±SEM of five rats per group.
‡ No significant difference compared with control groups.
§ $p < 0.001$ compared with the control group (Student's *t* test).
¶ $p < 0.01$ compared with the control group (Student's *t* test).

TABLE 4

EFFECT OF ACUTE DRUG TREATMENT ON THE *p*-CA-INDUCED POSTURAL SYNDROME IN THE RAT *

Drug	Dose (mg/kg)	Number of Rats	Score (mean±SEM)
p-CA alone	10	23	11.30±0.5
p-CA plus			
pretreatment with			
Fluoxetine	10	5	2.1±0.4 †
p-Chlorophenylalanine	300	3	1.3±0.6 †
Methergoline	1.0	4	3.4±0.6 †
Methiothepin	5–20	7	0
Methysergide	1.0	3	10.5±0.9
Cyproheptadine	0.4	3	9.9±0.3
Desmethylimipramine	25	3	8.3±1.3
α-Methyl-*p*-tyrosine	70	3	14.1±1.7
Haloperidol	1.0	3	12.2±0.6
Pimozide	10	3	13.5±0.9
Phenoxybenzamine	10	3	10.0±1.5
Propranolol	0.2	3	9.8±0.4
Atropine	1.0	5	9.1±0.6
Methylscopolamine	1.0	3	9.8±1.3

* Groups of rats were given *p*-CA (10 mg/kg, ip) after pretreatment with one of the drugs listed above. The resultant behavior was scored for the presence of tremor, myoclonus, rigid postures, and autonomic signs (see text for details) (From Growdon.[34])
† $p < 0.01$, one-way analysis of variance and the Scheffé test.

For example, fluoxetine, a drug that selectively blocks *p*-CA and 5-HT uptake into presynaptic nerve terminals, significantly blocked all aspects of the *p*-CA-induced syndrome. Pretreatment with the tryptophan hydroxylase inhibitor *p*-chlorophenylalanine or with the putative serotonergic receptor-blocking drugs methiothepin and methergoline also antagonized the *p*-CA-induced rigidity, tremor, myoclonus, and autonomic activity. In contrast to the other drugs, methiothepin sedated the rats; this effect may partly account for the fact that it blocked the *p*-CA syndrome so completely. Interestingly, among other drugs postulated to block 5-HT receptors, cyproheptadine and methysergide had no observable effects.

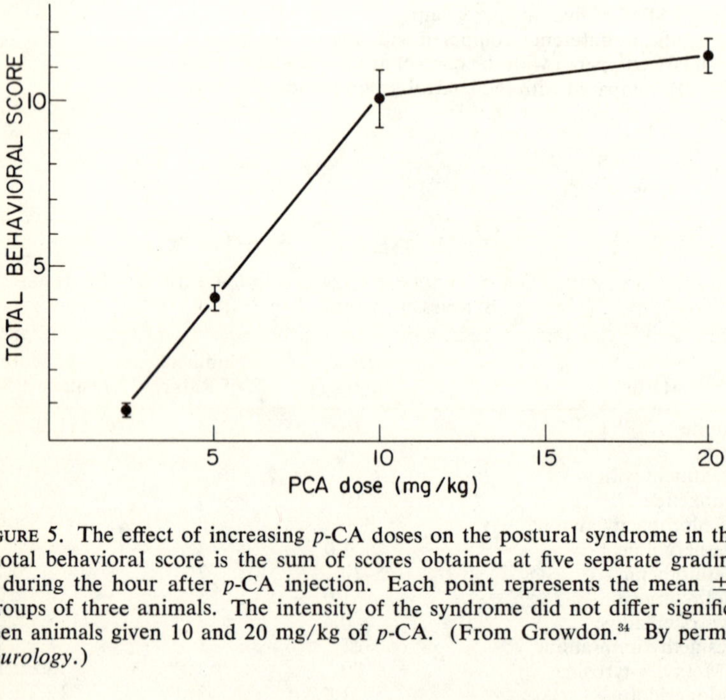

FIGURE 5. The effect of increasing *p*-CA doses on the postural syndrome in the rat. The total behavioral score is the sum of scores obtained at five separate grading periods during the hour after *p*-CA injection. Each point represents the mean ±SEM for groups of three animals. The intensity of the syndrome did not differ significantly between animals given 10 and 20 mg/kg of *p*-CA. (From Growdon.[34] By permission of *Neurology*.)

In contrast to drugs that alter 5-HT metabolism, the cholinergic receptor-blocking drugs atropine and methylscopolamine did not affect the abnormal posture, tremor, or myoclonus. The syndrome was also unaffected by drugs that alter the metabolism of other neurotransmitters (the catecholamine uptake inhibitor DMI, the tyrosine hydroxylase inhibitor α-methyl-*p*-tyrosine, the adrenergic receptor-blocking drugs phenoxybenzamine and propranolol, or the dopamine receptor-blocking drugs haloperidol or pimozide). Pretreatment with α-methyl-*p*-tyrosine, haloperidol, or pimozide intensified the tremor and myoclonus, and the total scores of rats that received these drugs plus *p*-CA were somewhat higher than those of animals that received *p*-CA alone, but these differences were not statistically significant.

Discussion

The immediate *p*-CA syndrome was distinct from the increased locomotor activity reported several days after *p*-CA injection.[19-22] Hyperactivity did occur briefly at the beginning of the immediate postural syndrome but ceased within 5 min as the rats developed the immobility, tremor, and myoclonus that are not part of delayed hypermotility. The *p*-CA postural syndrome also differed from the stereotyped licking, biting, rearing, and repetitive head movements induced by amphetamine; conversely, rigid immobile postures, tremor, and myoclonus induced by *p*-CA were not reported after amphetamine administration.[28] Furthermore, pretreatment with drugs that inhibit amphetamine-induced stereotypic behavior (α-methyl-*p*-tyrosine, haloperidol[48]) did not prevent the immediate *p*-CA syndrome, although α-methyl-*p*-tyrosine apparently did block the hyperactivity syndrome induced by tranylcypromine and tryptophan.[49]

The rigid postures, tremor, myoclonus, and autonomic signs that occurred immediately after a single injection of *p*-CA were most similar to behaviors previously noted in rats challenged with 5-HTP after pretreatment with 5,7-DHT.[31, 32] Both responses began 5 min after the appropriate drug challenge, were dose dependent, and lasted 60–120 min. Extended hind limbs, repetitive forepaw prancing, and circling were more characteristic of the *p*-CA-induced syndrome, while tremor and myoclonus were more prominent after 5-HTP injection. Autonomic signs were present equally in both syndromes, but only 5-HTP caused hyperthermia. Both behaviors involve changes in serotonergic neurotransmission: in the 5-HTP myoclonic syndrome, 5-HT, or a 5-HT agonist, apparently stimulates supersensitive 5-HT receptors; *p*-CA-induced behavior seems to result from increased 5-HT release from intact presynaptic terminals onto normal 5-HT receptors. Thus, pretreatment with *p*-chlorophenylalanine to reduce the number of 5-HT molecules stored in the presynaptic terminal antagonized the syndrome, whereas pretreatment with α-methyl-*p*-tyrosine did not. *p*-Chloroamphetamine itself does not act as a postsynaptic agonist to produce the postural syndrome but may function as an indirect receptor stimulant by releasing 5-HT stored in presynaptic terminals. Fluoxetine, a drug that specifically blocks 5-HT and *p*-CA uptake into presynaptic terminals,[50] also blocked the postural syndrome, whereas DMI, which was given to block uptake into catecholaminergic neurons, did not suppress it. The effects of drugs that antagonize 5-HT receptors in the periphery are less predictable in the central nervous system: methiothepin and methergoline significantly blocked the *p*-CA syndrome, but methysergide and cyproheptadine did not. Drugs believed to block central postsynaptic dopamine, norepinephrine, and acetylcholine receptor sites did not diminish the *p*-CA syndrome. These observations indicate the similarities between the *p*-CA and 5-HTP syndromes and emphasize their common association with serotonergic mechanisms. The syndromes described by Grahame-Smith,[26] Trulson and Jacobs,[33] and Jacobs[51] may be similar to the 5-HTP- and *p*-CA-induced behaviors; all seem to involve heightened 5-HT stimulation, although the neuronal structures that subserve these behaviors remain obscure. It is clear, however, that elevated brain 5-HT content does not by itself cause altered behavior.[26]

Brain levels of 5-HT and 5-HIAA were significantly reduced 1 day after injection, because *p*-CA releases stored 5-HT,[52] inactivates tryptophan hydroxylase,[53, 54] and destroys serotonergic cell bodies,[55] especially in the mesencephalic B9 group.[56] In contrast, brain 5-HT content did not differ significantly from control values 30 min after *p*-CA injection, when the syndrome was most prominent; 5-HIAA levels, which might be expected to rise as a result of increased 5-HT release, were also unchanged. Thus, there was no correlation between static brain indoleamine content and altered motor behavior. Other *p*-CA effects, including inhibition of monoamine oxidase[57] and reuptake blockade,[58] may account for this finding; they certainly could potentiate the effects of increased 5-HT release. *p*-Chloroamphetamine may release stored catecholamines as well, but such an action is not critical, because drugs that block the effects of dopamine and norepinephrine failed to block the *p*-CA-induced postural syndrome.

References

1. Ungerstedt, U. 1971. Stereotoxic mapping of the monoamine pathway in the rat brain. Acta Physiol. Scand. Suppl. **367:** 1–48.
2. Descarries, L., A. Beaudet & K. C. Watkins. 1975. Serotonin nerve terminals in adult rat neocortex. Brain Res. **100:** 563–588.
3. Aghajanian, G. K., H. J. Haigler & F. E. Bloom. 1972. Lysergic acid diethylamide and serotonin: direct actions on serotonin-containing neurons in rat brain. Life Sci. **11** (1): 615–622.
4. Aghajanian, G. K. & H. J. Haigler. 1974. L-Tryptophan as a selective histochemical marker for serotonergic neurons in single-cell recording studies. Brain Res. **81:** 364–372.
5. Wurtman, R. J. & J. D. Fernstrom. 1976. Control of brain neurotransmitter synthesis by precursor availability and nutritional state. Biochem. Pharmacol. **25:** 1691–1696.
6. Jouvet, M. 1969. Biogenic amines and the states of sleep. Science **163:** 32–41.
7. Breisch, S. T., F. P. Zemlan & B. G. Hoebel. 1976. Hyperphagia and obesity following serotonin depletion by intraventricular p-chlorophenylalanine. Science **192:** 382–384.
8. Saller, C. F. & E. M. Strickler. 1976. Hyperphagia and increased growth in rats after intraventricular injection of 5,7-dihydroxytryptamine. Science **192:** 385–387.
9. Horita, A. & J. H. Gogerty. 1958. The pyretogenic effect of 5-HTP and its comparison with that of LSD. J. Pharmacol. Exp. Ther. **122:** 195–200.
10. Myers, R. D. & T. L. Yaksh. 1969. Control of body temperature in the unanesthetized monkey by cholinergic and aminergic systems in the hypothalamus. J. Physiol. (London) **202:** 483–500.
11. Harvey, J. A. & C. E. Lints. 1965. Lesions in the medial forebrain bundle: delayed effects on sensitivity to electric shock. Science **148:** 250–252.
12. Lytle, L. D., R. B. Messing, L. Fisher & L. Phebus. 1975. Effects of long-term corn consumption on brain serotonin and the response to electric shock. Science **190:** 692–694.
13. Wyatt, R. J., D. J. Kupfer, A. Sjoerdsma, K. Engelman, D. H. Fram & F. Snyder. 1970. Effects of 1-tryptophan (a natural sedative) on human sleep. Lancet **2:** 842–845.
14. Coppen, A., P. C. Whybrow, M. B. R. Noguera, R. Maggs & A. J. Prange. 1972. The comparative anti-depressant value of l-tryptophan and imipramine with and without attempted potentiation by liothyronine. Arch. Gen. Psychiat. **26:** 234–241.
15. Macindoe, J. H. & R. W. Turkington. 1973. Stimulation of human prolactin secretion by intravenous infusion of l-tryptophan. J. Clin. Invest. **52:** 1972–1978.
16. Growdon, J. H., R. R. Young & B. T. Shahani. 1976. L-5-hydroxytryptophan in treatment of several different syndromes in which myoclonus is prominent. Neurology **26:** 1135–1140.

17. VanWoert, M. H., D. Rosenbaum, J. Howieson & M. B. Bowers. 1977. Long-term therapy of myoclonus and other neurologic disorders with L-5-hydroxy-tryptophan and carbidopa. N. Engl. J. Med. **296:** 70–75.

18. Sheard, M. H. 1969. The effect of p-chlorophenylalanine on behavior in rats: relation to brain serotonin and 5-hydroxyindoleacetic acid. Brain Res. **15:** 524–528.

19. Strada, S. J., E. Sanders-Bush & F. Sulser. 1970. p-Chloroamphetamine. Temporal relationship between psychomotor stimulation and metabolism of brain norepinephrine. Biochem. Pharmacol. **19:** 2621–2629.

20. Costa, E., K. M. Naimzada & A. Revuelta. 1971. Effect of phenmetrazine, aminorex and (±) p-chloroamphetamine on the motor activity and turnover rate of brain catecholamines. Brit. J. Pharmacol. **43:** 570–579.

21. Lassen, J. B. 1974. The effect of p-chloroamphetamine on motility in rats after inhibition of monoamine synthesis, storage, uptake, and receptor interaction. Psychopharmacologia **34:** 243–254.

22. Messing, R. B., L. Phebus, L. A. Fisher & L. D. Lytle. 1976. Effects of p-chloroamphetamine on locomotor activity and brain 5-hydroxyindoles. Neuropharmacology **15:** 157–163.

23. Meek, J., K. Fuxe & N. E. Andén. 1970. Effects of antidepressant drugs of the imipramine type on central 5-hydroxytryptamine neurotransmission. Eur. J. Pharmacol. **9:** 325–332.

24. Daly, J., K. Fuxe & G. Jonsson. 1974. 5,7-Dihydroxytryptamine as a tool for the morphological and functional analysis of central 5-hydroxytryptamine neurons. Res. Commun. Chem. Pathol. Pharmacol. **7:** 175–187.

25. Maj, J., W. Palider & L. Baran. 1976. The effects of serotonergic and anti-serotonergic drugs on the flexor reflex of spinal rat: a proposed model to evaluate the action on the central serotonin receptor. J. Neural Transmission **38:** 131–147.

26. Grahame-Smith, D. G. 1971. Studies *in vivo* on the relationship between brain tryptophan, brain 5-HT synthesis and hyperactivity in rats treated with a mono-amine oxidase inhibitor and L-tryptophan. J. Neurochem. **18:** 1053–1066.

27. Jacobs, B. L. 1976. An animal behavior model for studying central serotonergic synapses. Life Sci. **19:** 777–786.

28. Randrup, A., I. Munkvad & P. Udsen. 1963. Adrenergic mechanisms and amphetamine induced abnormal behavior. Acta Pharmacol. Toxicol. **20:** 145–157.

29. Ernst, A. M. 1966. Mode of action of apomorphine and dextroamphetamine on gnawing compulsion in rats. Psychopharmacologia **10:** 316–323.

30. Ungerstedt, U. 1971. Striatal dopamine release after amphetamine or nerve degeneration revealed by rotational behaviour. Acta Physiol. Scand. Suppl. **367:** 49–68.

31. Stewart, R. M., J. H. Growdon, D. Cancian & R. J. Baldessarini. 1976. 5-Hydroxytryptophan-induced myoclonus: increased sensitivity to serotonin after intracranial 5,7-dihydroxytryptamine in the adult rat. Neuropharmacology **15**: 449–455.

32. Stewart, R. M., J. H. Growdon, D. Cancian & R. J. Baldessarini. 1976. Myoclonus after 5-hydroxytryptophan in rats with lesions of indoleamine neurons in the central nervous system. Neurology **26:** 690–692.

33. Trulson, M. E. & B. L. Jacobs. 1976. Behavioral evidence for the rapid release of CNS serotonin by PCA and fenfluramine. Eur. J. Pharmacol. **36:** 149–154.

34. Growdon, J .H. 1977. Postural changes, tremor, and myoclonus in the rat immediately following injections of p-chloroamphetamine. Neurology **27:** 1074–1077.

35. Baumgarten, H. G., A. Björklund, L. Lachenmayer, A. Nobin & U. Stenevi. 1971. Long-lasting selective depletion of brain serotonin by 5,6-dihydroxy-tryptamine. Acta Physiol. Scand. Suppl. **373:** 1–15.

36. Baumgarten, H. G. & L. Lachenmayer. 1972. 5,7-Dihydroxytryptamine: improvement in chemical lesioning of indoleamine neurons in the mammalian brain. Z. Zellforsch. **135:** 399–414.
37. Baumgarten, H. G., L. Lachenmayer & H. G. Schlossberger. 1972. Evidence for a degeneration of indoleamine containing nerve terminals in rat brain induced by 5,6-dihydroxytryptamine. Z. Zellforsch. **125:** 553–599.
38. Baumgarten, H. G., A. Björklund, L. Lachenmayer & A. Nobin. 1973. Evaluation of the effects of 5,7-dihydroxytryptamine on serotonin and catecholmine neurons in the rat CNS. Acta Physiol. Scand. Suppl. **931:** 1–19.
39. Gerson, S. & R. J. Baldessarini. 1975. Selective destruction of serotonin terminals in rat forebrain by high doses of 5,7-dihydroxytryptamine. Brain Res. **85:** 140–145.
40. Björklund, A., H. G. Baumgarten & A. Rensch. 1975. 5,7-Dihydroxytryptamine: improvement of its selectivity for serotonin neurons in the CNS by pretreatment with desipramine. J. Neurochem. **24:** 833–835.
41. Klawans, H. L., Jr., C. Goetz & W. J. Weiner. 1973. 5-Hydroxytryptophan-induced myoclonus in guinea pigs and the possible role of serotonin in infantile myoclonus. Neurology **23:** 1234–1240.
42. Trulson, M. E., E. E. Eubanks & B. L. Jacobs. 1976. Behavioral evidence for supersensitivity following destruction of central serotonergic nerve terminals by 5,7-dihydroxytryptamine. J. Pharmacol. Exp. Ther. **198:** 23–32.
43. Pletscher, A., G. Bartholini, H. Bruderer, W. P. Burkard & K. F. Gey. 1964. Chlorinated arylalkylamines affecting the cerebral metabolism of 5-hydroxytryptamine. J. Pharmacol. Exp. Ther. **145:** 344–350.
44. Fuller, R. W., C. W. Hines & J. Mills. 1965. Lowering of brain serotonin level by chloroamphetamines. Biochem. Pharmacol. **14:** 483–488.
45. Fuller, R. W. & H. D. Snoddy. 1974. Long-term effects of p-chloroamphetamine on brain 5-hydroxyindole metabolism in rats. Neuropharmacology **13:** 85–90.
46. Sanders-Bush, E., J. A. Bushing & F. Sulser. 1975. Long-term effects of p-chloroamphetamine and related drugs on central serotonergic mechanisms. J. Pharmacol. Exp. Ther. **192:** 33–41.
47. Curzon, G. & A. R. Green. 1970. Rapid method for the determination of serotonin and 5-hydroxyindoleacetic acid in small regions of rat brain. Brit. J. Pharmacol. **39:** 653–655.
48. Randrup, A. & I. Munkvad. 1966. Role of catecholamines in the amphetamine excitory response. Nature (London) **211:** 540.
49. Green, A. R. & D. G. Grahame-Smith. 1974. The role of brain dopamine in the hyperactivity syndrome produced by increased 5-hydroxytryptamine synthesis in rats. Neuropharmacology **13:** 949–959.
50. Fuller, R. W., K. W. Perry & B. B. Molloy. 1975. Effect of 3-(p-trifluoromethylphenoxy)-N-methyl-3-phenylpropylamine on the depletion of brain serotonin by 4-chloroampetamine. J. Pharmacol. Exp. Ther. **193:** 796–803.
51. Jacobs, B. L. 1974. Evidence for the functional interaction of two central transmitters. Psychopharmacology **39:** 81–86.
52. Sanders-Bush, E., D. A. Gallagher & F. Sulser. 1974. On the mechanism of brain 5-hydroxytryptamine depletion by p-chloroamphetamine and related drugs and the specificity of their actions. Advan. Biochem. Psychopharmacol. **10:** 185–194.
53. Sanders-Bush, E., J. A. Bushing & F. Sulser. 1972. Long-term effects of p-chloroamphetamine on tryptophan hydroxylase activity and on the levels of 5-hydroxytryptamine and 5-hydroxyindoleacetic acid in brain. Eur. J. Pharmacol. **20:** 385–388.
54. Sanders-Bush, E., J. A. Bushing & F. Sulser. 1972. p-Chloroamphetamine-inhibition of cerebral tryptophan hydroxylase. Biochem. Pharmacol. **21:** 1501–1510.

55. Gál, E. M., P. A. Christiansen & L. M. Yunger. 1975. Effect of p-chloroamphetamine on cerebral tryptophan-5-hydroxylase in vivo: a reexamination. Neuropharmacology **14:** 31–39.
56. Harvey, J. A., S. E. McMaster & L. M. Yunger. 1975. p-Chloroamphetamine: selective neurotoxic action in brain. Science **187:** 841–843.
57. Fuller, R. W. 1966. Serotonin oxidation by rat brain monoamine oxidase: inhibition by 4-chloroamphetamine. Life Sci. **5:** 2247–2252.
58. Wong, D. T., J. Horng & R. W. Fuller. 1973. Kinetics of serotonin accumulation into synaptosomes of rat brain: effects of amphetamines and chloroamphetamines. Biochem. Pharmacol. **22:** 311–322.

EFFECTS OF SEROTONIN NEUROTOXINS ON ROTATIONAL BEHAVIOR IN THE RAT *

C. T. Giambalvo, S. R. Snodgrass, and N. J. Uretsky

Department of Neurology and Neuropathology
Harvard Medical School
and Department of Neuroscience
The Children's Hospital Medical Center
Boston, Massachusetts 02115

The model behavior of rotation after unilateral lesions of the nigrostriatal pathway in rats was developed by Andén *et al.*[1] and Ungerstedt[2,3] and has been useful for studying the effects of drugs on motor functions. In unilaterally lesioned rats, amphetamine is thought to induce ipsilateral turning by releasing dopamine (DA) from the intact nigrostriatal nerve endings onto DA receptors,[1] while apomorphine is thought to induce contralateral turning by preferential stimulation of supersensitive DA receptors in the denervated striatum.[3] The effects of amphetamine and apomorphine are blocked by haloperidol and other DA receptor-blocking agents. Recent studies have shown that amphetamine can induce rotation after induction of unilateral lesions in sites outside the nigrostriatal pathway.[4] For example, Costall and Naylor[5] have recently shown that unilateral electrolytic lesions of the median raphe nucleus (MRN) cause both spontaneous rotation and rotation in response to amphetamine and apomorphine. The relationship between these brain lesions and the DA neurons is not clear. It may be explained by a release of tonic inhibition of the serotonergic MRN over the nigrostriatal pathway, because 5-hydroxytryptamine (5-HT) inhibits firing in the substantia nigra (SN) and other forebrain areas that receive a raphe input.[6–9] To clarify the relationship among MRN lesions, drug-induced rotation, and the DA neurons, we have studied the effects of unilateral injections of two serotonin neurotoxins, *p*-chloroamphetamine (*p*-CA) and 5,7-dihydroxytryptamine (5,7-DHT) on rotational behavior and DA turnover in the striatum.

Materials and Methods

Adult rats were pretreated with desmethylimipramine (25 mg/kg) 1 hr before the injection to protect norepinephrine (NE) neurons from degeneration.[10] 5,7-Dihydroxytryptamine (28 μg) or *p*-CA (82 μg), dissolved in 2 μl of vehicle, was injected in a course of 2 min, while the rats were under chloral hydrate anesthesia. Sham controls received 2 μl of the vehicle. The coordinates for the unilateral MRN lesions were[11]: A, +0.3 mm; V, −2.6 mm; L, 0.3 mm; for the SN, they were: A, +1.95 mm; V, −2.1 mm; L, 1.8 mm; for the medial forebrain bundle (MFB), they were: A, 3.75 mm; V, −3.0 mm; L, 1.3 mm. The injected rats were tested for rotation with amphetamine (1.5 mg/kg) or

* Supported by Project Grant NS 12368 from the National Institute of Neurological and Communication Disorders and Stroke.

0077–8923/78/0305–0524 $01.75/0

apomorphine (0.5 mg/kg) on the third day. They were sacrificed on the eighth day after an injection of the decarboxylase inhibitor RO4–4602 (800 mg/kg) to measure the turnover of DA and 5-HT.[12] The dihydroxyphenylalanine (DOPA) and 5-hydroxytryptophan (5-HTP) formed were separated on Dowex® 50 columns and then assayed fluorometrically by the method of Kehr *et al.*[13] The brain stem and SN were fixed with buffered formaldehyde for histologic examination.

Tissue indoles were extracted with acid-butanol and then partitioned with heptane/hydrochloric acid that contained 1% cysteine. The 5-HT in the aqueous phase was assayed fluorometrically.[14] The 5-hydroxyindoleacetic acid (5-HIAA) -containing organic phase was reextracted with a 30 mM NaH_2PO_4-$NaHCO_3$ solution (pH 7.6), and 5-HIAA was assayed fluorometrically.[14]

For the uptake studies, brain tissues were homogenized in 0.32 M sucrose and incubated with [^{3}H]-5-HT, [^{3}H]NE, or [^{3}H]DA at a final concentration of 0.1 μM in a Hepes-Tris medium (pH 7.4) for 5 min at 37° C. Blanks were run at 0° C. Incubations were terminated by Millipore® filtration, followed by washing with 5 ml of saline. The radioactivity in the filters was counted in a Packard liquid scintillation counter.

All biochemical results were expressed as percentage change of the right, or operated, side compared to the left, or unoperated, side of the same animal.

TABLE 1

EFFECT OF *p*-CA OR 5,7-DHT INJECTED INTO THE MRN (MIDLINE) ON 5-HT AND 5-HIAA LEVELS AND ON 5-HT AND NE UPTAKES 7 DAYS AFTER THE OPERATION

	Percentage Decrease Compared to Vehicle Controls *	
	p-CA	5,7-DHT
5-HT in cortex	27.6±0.5 (5)†	63.7±2.8 (3)
5-HIAA in cortex	47.4±1.3 (8)	64.7±0.7 (3)
5-HT in striatum	35.4±2.0 (5)	65.5±2.5 (4)
5-HIAA in striatum	45.0±5.6 (6)	64.2±2.8 (4)
5-HT uptake in cortex	43.3±1.6 (6)	74.5±4.2 (4)
NE uptake in cortex	8.0±5.7 (6)	0.0 (3)

* The control levels of 5-HT were 0.47±0.01 and 0.38±0.01 μg/g in the cortex and striatum, respectively. The contro levels of 5-HIAA were 0.59±0.05 and 0.47±0.02 μg/g in the cortex and striatum, respectively.

† Results are the means ± SEM for the number of determinations shown in parentheses.

BIOCHEMICAL AND BEHAVIORAL EFFECTS OF UNILATERAL MEDIAN RAPHE NUCLEUS INJECTIONS

p-Chloroamphetamine and 5,7-DHT produced lesions in the MRN that were marked by necrosis or gliosis, whereas vehicle injections produced no observable histologic changes. The two drugs also caused decreases of 5-HT and 5-HIAA levels in the cortex and striatum and a decreased uptake of [^{3}H]-5-HT into the cortex, without affecting the uptake of NE in this tissue (TABLE 1). This finding agrees with Gerson and Baldessarini's observation that pretreatment with desmethylimipramine will protect NE neurons from the effect of 5,7-DHT.[10] *p*-Chloroamphetamine and 5,7-DHT also produced a decrease in 5-HT turnover in both the cortex and striatum.

Both *p*-CA and 5,7-DHT-lesioned rats showed an initial spontaneous contralateral rotation (away from the side of the injection), which declined by the third day. Animals tested after this period exhibited a contralateral rotation in response to amphetamine and apomorphine that could be blocked by haloperidol treatment. The observation that amphetamine and apomorphine both caused rotation in the same direction suggests that the rotational behavior is not due to the development of DA receptor supersensitivity, as postulated by Ungerstedt after unilateral 6-hydroxydopamine (6-OHDA) lesions.[3] Also, the time course of our rotational behavior differs from that of receptor supersensitivity: the 5,7-DHT plus *p*-CA-injected rats demonstrated spontaneous rotation immediately after the operation, which declined over a few days, at which time the animals were still responsive to amphetamine and apomorphine. Behavior due to receptor supersensitivity takes longer to develop and should be increasing rather than decreasing at 7 days after the lesion.[16, 17] Furthermore, since we destroyed 5-HT rather than DA neurons, supersensitivity, if present at all, would most likely be due to 5-HT rather than to DA.

In the *p*-CA- or 5,7-DHT-injected rats, there is a significant correlation between the size of the lesion in the MRN and the extent of the drug-induced rotation: the greater the size of the lesion, the more vigorous the rotation. However, the rotation is not related to nonspecific mechanical damage to the MRN, because sham-operated animals that received a vehicle injection did not rotate. There was a significant correlation between lesion size and decrease in 5-HT turnover in the cortex. Thus, the decrease in cortical 5-HT turnover appears to be a biochemical index of the extent of the lesion.

The observation that these serotonin lesions produced contralateral rotation after injection of amphetamine or apomorphine that could be blocked by haloperidol suggests that the rotation might be due to an increase in activity of DA neurons on the side of the lesion (Ungerstedt's model). This hypothesis is supported by our observation that such rotational behavior (after *p*-CA or 5,7-DHT injections) was accompanied by an increase in DA turnover in the ipsilateral striatum. Moreover, there was a significant correlation between the increase in striatal DA turnover and the rate of rotation: the greater the increase in DA turnover, the more vigorous the rotation after apomorphine administration. Also, there was a significant correlation between the extent of the lesion (as measured by the decrease in cortical 5-HT turnover) and the increase in striatal DA turnover (FIGURE 1). These correlations among rotational behavior, the decrease in cortical 5-HT turnover, and the increase in striatal DA turnover suggest that a unilateral lesion of this inhibitory raphe projection would result in an increased activity of the dopaminergic nigrostriatal pathway on the side ipsilateral to the lesion. This increase in neuronal activity is reflected biochemically by an increase in DA turnover and behaviorally by contralateral rotation.

Effects of Unilateral Injections of 5,7-Dihydroxytryptamine into the Substantia Nigra

Similar to the effects observed after the unilateral injection of 5,7-DHT into the MRN, unilateral injections of 5,7-DHT (28 μg) into the SN also caused contralateral rotation in response to both apomorphine and amphetamine, a decrease in cortical 5-HT turnover, and an increase in striatal DA turnover. The correlation between the decrease in 5-HT turnover in the cortex and the increase in striatal DA turnover was qualitatively the same as that resulting from unilateral MRN lesions (FIGURE 1). This finding suggests that the rotation and the striatal DA turnover changes seen after MRN lesions might be mediated by a projection from the MRN to the SN.

The rotation observed after 5,7-DHT injection into the SN is probably due to destruction of 5-HT terminals in the SN and not to damage to the DA neurons for the following reasons: 1. We observed an *increase* in DA turnover in the striatum that presumably reflects an increase in neuronal firing rate or functional state of the neuron; degeneration of the DA neuron should produce

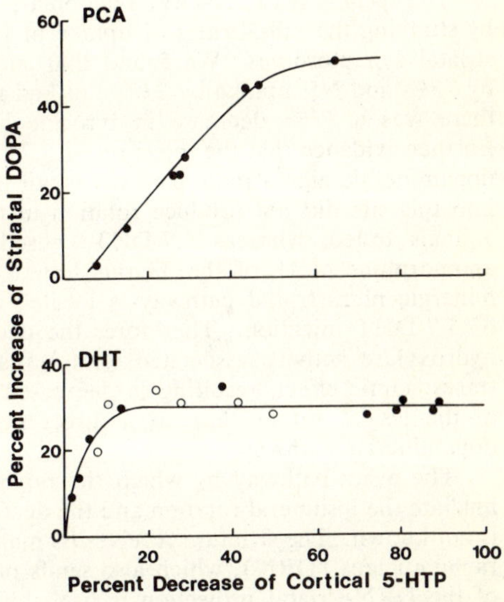

FIGURE 1. Correlation between the decrease in cortical 5-HT turnover and the increase in striatal DA turnover in rats injected unilaterally with *p*-CA or 5,7-DHT into the MRN (●) or the SN (○). Results are expressed as percentage change of levels of 5-HTP or Dopa in the operated side compared to the unoperated side after injection of RO4–4602; $p < 0.001$ for *p*-CA- and $p < 0.05$ for 5,7-DHT-treated animals.

the opposite effect. 2. The contralateral rotation seen after amphetamine injection indicates a hyperfunction of the nigrostriatal pathway on the side of the lesion (Ungerstedt's model) rather than a degeneration of the DA neurons in the SN. 3. We found that injection of 6-OHDA into this area did not produce rotation in response to amphetamine. This lack of response is not an artifact, because the coordinates we used for the 5,7-DHT injection are +0.45 mm more posterior than are Understedt's coordinates.[2] When we injected 6-OHDA into the SN, according to Ungerstedt's coordinates, we did observe the ipsilateral rotation with amphetamine and contralateral rotation with apomorphine reported by Ungerstedt and others. Thus, we conclude that the rotation and increase in DA turnover observed after injection of DHT into the SN are not due to damage to DA neurons.

Effects of Unilateral Injections of 5,7-Dihydroxytryptamine into the Medial Forebrain Bundle

Unilateral injection of 5,7-DHT into the medial forebrain bundle (MFB), rostral to the SN but caudal to the striatum, caused *ipsilateral* rotation in response to amphetamine and apomorphine, a decrease in cortical 5-HT turnover, and a *decrease* in striatal DA turnover in the operated side. There were significant correlations between the rate of rotation and the decrease in cortical 5-HT turnover and between the rate of rotation and the decrease in striatal DA turnover. Moreover, there was a significant correlation between the decrease in cortical 5-HT turnover and the decrease in striatal DA turnover (Figure 2). The changes in rotation and in DA turnover are opposite to those observed after MRN and SN lesions, therefore suggesting that the effects of the MRN lesions are mediated via a direct projection to the SN and not through a projection to the striatum.

The specificity of 5,7-DHT's action after injection into the MFB was assessed by studying the initial rates of uptake of [^{3}H]-5-HT, [^{3}H]DA, and [^{3}H]NE into striatal synaptosomes. We found that such injections decreased 5-HT uptake by 75% and NE uptake by 23% but had no effect on DA uptake, even though there was a 77% decrease in tyrosine hydroxylase activity in the striatum. Further evidence that the injection of 5,7-DHT into the MFB did not injure the dopaminergic nigrostriatal pathway is our observation that injection of 6-OHDA into this site did not produce rotation in response to apomorphine in the nine animals tested, whereas 5,7-DHT caused ipsilateral rotation in response to apomorphine in 18 of the 33 rats tested. This finding implies that the dopaminergic nigrostriatal pathway is located a small distance away from the site of 5,7-DHT injection. Therefore, the decrease in DA turnover and tyrosine hydroxylase activity associated with ipsilateral rotation is probably due to a transsynaptic effect, resulting in decreased activity of DA neurons on the side of the lesion, rather than to a direct neurotoxic effect of 5,7-DHT on the dopaminergic pathway.

The exact pathway by which the raphe nucleus projects to the striatum to mediate the ipsilateral rotation and the decrease in striatal DA turnover observed is not known. The striatum receives its major serotonergic input from the dorsal raphe nucleus (DRN), which also sends projections to the SN.[18] Degeneration of the DRN-striatal projection is probably responsible for the behavioral and biochemical changes we observed after unilateral injection of 5,7-DHT into the MFB. Costall and Naylor reported that electrolytic lesions of the DRN did not produce any spontaneous or drug-induced rotation.[5] We have confirmed these

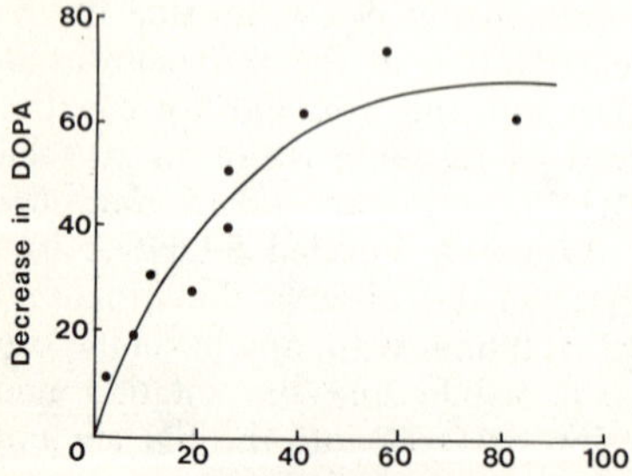

Figure 2. Correlation between the decrease in cortical 5-HTP and the decrease in striatal Dopa after injection of RO4–4602. Results are expressed as percentage changes of the operated sides compared to the unoperated sides; $p < 0.001$.

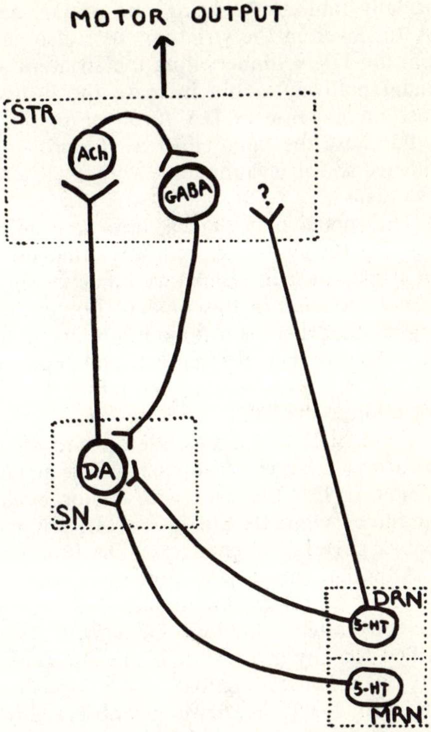

FIGURE 3. Diagrammatic model of serotonergic raphe inputs to the nigrostriatal pathway.

results after unilateral injection of 5,7-DHT or *p*-CA into the DRN. These results could be explained by destruction of the DRN projections to both the striatum and the SN, which normally exert opposite effects on DA neuronal activity.

To study possible interactions between the ascending 5-HT neurons in the MFB and cholinergic neurons known to exist in the striatum,[19] and possibly also in the SN,[19] we have investigated the effect of atropine pretreatment (10 mg/kg) on the rotational behavior of rats lesioned unilaterally at the MFB with DHT. We found that atropine facilitated the amphetamine-induced rotation observed after MFB lesions: there was a decreased latency to maximal rotation after administration of amphetamine (from 23 to 7 min), though the maximal amount of turning and the direction of rotation were unchanged. This finding is compatible with the observation that high doses of atropine, like MFB lesions, reduce the utilization of newly synthesized DA.[19]

MODEL

We would like to propose the following model for the role of serotonin neurons on the nigrostriatal pathway, based on our observations and those reported by others (FIGURE 3): In addition to the dopaminergic-cholinergic-γ-aminobutyric acidergic feedback loop proposed by various investigators, we suggest that there is a monosynaptic projection from the MRN to the SN that normally inhibits the firing rate of DA neurons in the SN and thus decreases DA turnover in the striatum. We also suggest that there are two projections from the DRN: innervating the striatum and the SN, respectively. The DRN-striatal pathway would increase the firing rate of DA neurons and therefore cause an increase in DA turnover in the striatum. The DRN-SN projection would cause the same effect as the projection from the MRN to the SN. Our evidence so far is compatible with such a model, but further studies are needed to validate it.

The model only depicts how serotonin neurons might regulate the nigrostriatal pathway. Other pathways that utilize other neurotransmitters probably can also exert their regulatory influence on the nigrostriatal system, though they are not included in this model. For example, the work of Donaldson *et al.*[20] suggests that there is a tonic inhibitory pathway from the locus ceruleus, which also interacts with the nigrostriatal bundle. These workers have presented no data on DA turnover nor have they proven that the relevant pathway is the projection to the SN.

The model also draws attention to the motor output from the neostriatum, which arises largely from the globus pallidus.[21, 22] In attempting to correlate changes in DA turnover with motor behavior, we are, of course, presuming that motor efferents are intact. If a lesion destroys motor efferents, the behavioral correlate of changes in DA function cannot occur.

Although chemical neurotoxins produce some nonspecific damage, they can selectively lesion specific monoamine neurons when *carefully* utilized. Even so, specificity after induction of chemical lesions cannot be assumed and must be verified for any given system. In our experiments, we show biochemical specificity by demonstrating that 5,7-DHT has no effect on the uptake of monoamines other than 5-HT. We have also observed that 5,7-DHT and 6-OHDA produced different effects on rotational behavior when injected into the SN or MFB, implying selective actions of these two neurotoxins. Similarly, Costall *et al.*[23] have observed that electrolytic lesions of the SN produce a different direction of rotation after apomorphine- than 6-OHDA-induced lesions at the same site, implying a more specific action of 6-OHDA. This selective action of neurotoxins makes it possible to analyze the behavioral consequences of damage to specific types of neurons in a given anatomic region.

The recent introduction of kainic acid, an agent that appears to destroy cell bodies but not axons and nerve terminals after local injection into the brain,[24, 25] further increases the ability of pharmacologists to separately analyze the role of various neurotransmitter systems within a given brain region. If the effects of 5,7-DHT injections into the MFB are mediated via interneurons in the striatum that alter nigrostriatal DA neuron firing through the striatonigral pathway, kainic acid injections into the striatum, which do not destroy DA terminals, should abolish the effect of MFB injections of 5,7-DHT.

Summary

Injections of *p*-CA or 5,7-DHT into the MRN caused decreases in 5-HT and 5-HIAA levels in the cortex and striatum. They also caused a decrease in 5-HT turnover in the cortex and an increase in striatal DA turnover. The unilaterally injected rats showed contralateral rotation after amphetamine or apomorphine administration. There was a significant correlation among the rate of rotation, the decrease in cortical 5-HT turnover, and the increase in striatal DA turnover. Injection of 5,7-DHT into the SN produced the same biochemical and behavioral changes as did MRN lesions, suggesting that the changes induced by MRN lesions might be due to a direct projection from the raphe to the substantia nigra. Injection of 5,7-DHT into the MFB caused ipsilateral rotation after injection of amphetamine or apomorphine and a decrease in DA turnover in the striatum. There was a significant correlation among the rate of rotation, the decrease in 5-HT turnover in the cortex, and the decrease in striatal DA turnover in the MFB-lesioned rats. These effects might be due to a projection from the DRN to cell bodies in the striatum.

References

1. Andén, N. E., A. Dahlström, K. Fuxe & K. Larsson. 1966. Acta Pharmacol. **24:** 263–274.
2. Ungerstedt, U. 1971. Acta Physiol. Scand. **32:** 49–68.
3. Ungerstedt, U. 1971. Acta Physiol. Scand. **32:** 69–93.
4. Glick, S. P., R. P. Jerussi & L. N. Fleisher. 1976. Life Sci. **18:** 889–893.
5. Costall, B. & R. J. Naylor. 1974. Eur. J. Pharmacol. **29:** 203–222.
6. Dray, A., T. J. Gonye, N. R. Oakley & T. M. Tanner. 1976. Brain Res. **113:** 44–57.
7. Haigler, J. J. & G. K. Aghajanian. 1974. J. Pharmacol. Exp. Ther. **188:** 688–699.
8. Miller, J. J., T. L. Richardson, H. C. Fibiger & H. McLennan. 1975. Brain Res. **97:** 133–138.
9. Wang, R. Y. & G. K. Aghajanian. 1977. Brain Res. **120:** 85–102.
10. Gerson, S. & R. J. Baldessarini. 1965. Brain Res. **85:** 140–145.
11. Konig, J. F. R. & R. A. Klippel. 1963. A Stereotaxic Atlas of the Forebrain and Lower Parts of the Brain Stem. The Williams & Wilkins Co. Baltimore, Md.
12. Carlsson, A. & M. Lindqvist. 1970. J. Pharm. Pharmacol. **22:** 725–727.
13. Kehr, W., A. Carlsson & M. Lindqvist. 1972. Naunyn-Schmiedeberg's Arch. Pharmakol. **274:** 273–280.
14. Atack, C. & M. Lindqvist. 1973. Naunyn-Schmiedeberg's Arch. Pharmakol. **279:** 267–284.
15. Giambalvo, C. T. & S. R. Snodgrass. 1977. Brain Res. In press. cation.
16. Trendelenburg, U. 1966. Pharmacol. Rev. **18:** 629–640.
17. Fleming, W. W. 1976. Rev. Neurosci. **2:** 43–90.
18. Bobillier, P., S. Seguin, F. Petitjean, D. Salvert, M. Touret & M. Jouvet. 1976. Brain Res. **113:** 449–486.
19. Javoy, F., A. Yves & J. Glowinski. 1975. J. Pharm. Pharmacol. **27:** 677–681.
20. Donaldson, I. MacG., A. Dolphin, P. Jenner, C. D. Marsden & C. Pycock. 1976. Brain **99:** 427–446.
21. Nauta, W. J. H. & W. R. Mehler. 1966. Brain Res. **1:** 3–42.
22. Carpenter, M. B. 1976. *In* The Basal Ganglia. M. D. Yahr, Ed. : 1–35. The Raven Press. New York, N.Y.
23. Costall, B., C. D. Marsden, R. J. Naylor & C. J. Pycock. 1976. Brain Res. **118:** 87–113.
24. Coyle, J. T. & R. Schwartz. 1976. Nature (London) **263:** 244–246.
25. McGeer, E. G. & P. L. McGeer. 1976. Nature (London) **263:** 517–518.

SOME BEHAVIORAL EFFECTS OF SEROTONIN DEPLETION DEPEND ON METHOD: A COMPARISON OF 5,7-DIHYDROXYTRYPTAMINE, *p*-CHLOROPHENYLALANINE, *p*-CHLOROAMPHETAMINE, AND ELECTROLYTIC RAPHE LESIONS *

Stanley A. Lorens

Department of Pharmacology
Loyola University
Stritch School of Medicine
Maywood, Illinois 60153

Thirteen years ago, Dahlström and Fuxe,[1] using the Falck-Hillarp method, distinguished nine groups (B1–9) of serotonin- [5-hydroxytryptamine (5-HT)] containing perikarya in the rat, which, for the most part, were localized within the brain stem raphe.[2] Concurrent lesion studies by Heller, Harvey, and Moore[3,4] suggested that both ascending 5-HT and noradrenergic fibers traveled within the lateral hypothalamic portion of the medial forebrain bundle (MFB) to innervate a variety of forebrain structures. Subsequent lesion[5–7] and stimulation[8,9] studies indicated that the mesencephalic dorsal (B7) and median (B8) raphe nuclei were the primary origin of these ascending 5-HT projections.

As this evidence accumulated, we began focusing our[10,11] attention on the midbrain dorsal and median raphe nuclei. We felt, like other workers,[5,6] that electrocoagulation of the B7 and B8 cell groups offered a technique by which to deplete forebrain 5-HT without affecting peripheral 5-HT levels, a problem encountered with the use of such drugs as *p*-chlorophenylalanine (*p*-CPA). Electrolytic destruction of the midbrain raphe, thus, seemed an excellent method by which to investigate the behavioral effects of forebrain 5-HT depletion and thereby gain insight into the functional role of central 5-HT systems.

Anatomy

The literature through 1973 gave one the impression that the B7–9 cell groups (and probably B5 and B6 also) were the common source of an overlapping, diffuse 5-HT input to forebrain regions. The medullary B1–3 groups were viewed as the primary origin of spinalopetal 5-HT fibers. A major difficulty, however, confronted investigators interested in the organization of central 5-HT pathways: fluorescing 5-HT neurons rapidly decompose during irradiation. This problem was overcome in part by tryptophan loading[12] and by the use of 5,7-dihydroxytryptamine[13] (5,7-DHT). The recent development of an immunohistochemical technique for the visualization of tryptophan hydroxylase,[14,15] in combination with other methods, however, hopefully will result in a major advance in our knowledge of the organization of central 5-HT pathways.

* Supported in part by grants from the National Institute on Drug Abuse and the Norwegian Research Council.

0077–8923/78/0305–0532 $01.75/0

Preliminary observations from behavioral studies [16] convinced us that we should examine the possibility that the 5-HT cell groups selectively innervated different central regions. This work was predicated on the principle that destruction of the cell bodies of origin would produce terminal degeneration and an associated loss of transmitter substance in the innervated regions. On the basis of these [17, 18] and other [19, 20] electrolytic lesion studies, it became clear that the 5-HT projections from the B5–9 groups were not uniform and that they differentially innervated distinct brain regions (FIGURE 1 & TABLE 1).

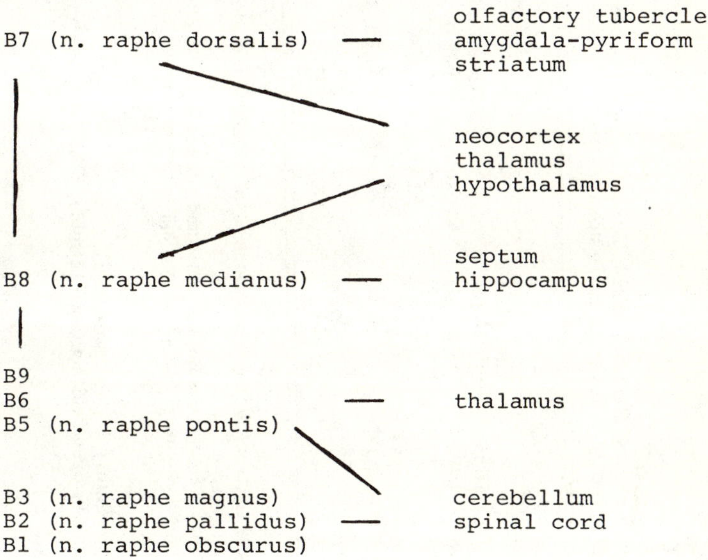

FIGURE 1. Schematic representation of projections from 5-HT-containing cell groups (B1–9). The B7 and B8 groups appear to have overlapping projections to the neocortex and diencephalon, although these fibers may reach different nuclei or cortical laminae. The 5-HT inputs to such structures as the amygdala-striatum and septum-hippocampus seem to originate exclusively in the B7 and B8 groups, respectively. Cerebellar and spinal 5-HT fibers arise principally in the B1–3 and B5 groups; B5, B6, and, perhaps, B9 provide an additional source of thalamic 5-HT fibers. The targets for B4 axons are at present unknown. It is likely, however, that all 5-HT-containing cell groups send collaterals and/or direct projections to other brain stem structures, including different raphe nuclei. Finally, it appears that the raphe is the origin of 5-HT fibers to the ventricular ependyma and vasculature of the meninges and brain stem.

The B7 group [midbrain dorsal raphe (DR) nucleus] appears to be the major origin [16] of ascending 5-HT fibers and contains the highest [21] (or second highest [22]) 5-HT concentration and tryptophan hydroxylase [23–25] activity in the brain stem. Lesions (FIGURE 2) restricted to B7 lead to 5-HT, 5-hydroxyindoleacetic acid (5-HIAA), and/or tryptophan hydroxylase reductions in the cerebral cortex, striatum, amygdala-pyriform region, olfactory tubercle, thalamus, and hypothalamus but *not* in the septum or hippocampus [17–20, 164] (see also

TABLE 1

BEHAVIORAL CHANGES AND PERCENT REDUCTIONS IN REGIONAL 5-HT CONCENTRATIONS 4–5 WEEKS AFTER DISCRETE ELECTROLYTIC BRAIN STEM LESIONS *

	Region Assayed ‡							Behavior ‖	
Lesion Site †	Spinal Cord	Hypothalamus	Thalamus	Hippocampus	Amygdala-Pyriform Cortex	Striatum	Telencephalon	OF	CAR
dr (5)	6 §	33 ¶	50 ¶	15	69 ¶	52 ¶	58 ¶	0	0
mr (7)	5 §	37 ¶	24 ¶	35 ¶	25	28	31	↑	0
vtg (5)	—	29	32 ¶	44 ¶	19	11	20 ¶	↑	↑
dtg (4)	—	15	21	18	6	14	15	0	↑
pr (5)	21	25	35 ¶	21	24	8	21	0	↑

* Adapted from Lorens *et al.*[18]

† dr, Dorsal raphe nucleus; mr, median raphe nucleus; vtg, ventral tegmental nucleus of Gudden; dtg, dorsal tegmental nucleus of Gudden; pr, pontine raphe (B5 and B6 groups).

‡ Control ($n=8$) 5-HT values (mean ± SD; ng/g, wet weight); results not corrected for recovery, ~ 60%: spinal cord, 222±57; hypothalamus, 394±70; dorsal thalamus, 304±43; hippocampus, 170±43; amygdala-pyriform cortex, 315±84; striatum, 293±61; remaining telencephalon, 256±39.

§ Percent change from Lorens and Guldberg.[17]

¶ Significant difference from control ($p < 0.05$, Mann-Whitney U test, two-tailed).

‖ Behavior: 0, no change; ↑, enhanced or facilitated; OF, open-field activity (12 min); CAR, two-way (shuttle) conditioned avoidance acquisition.

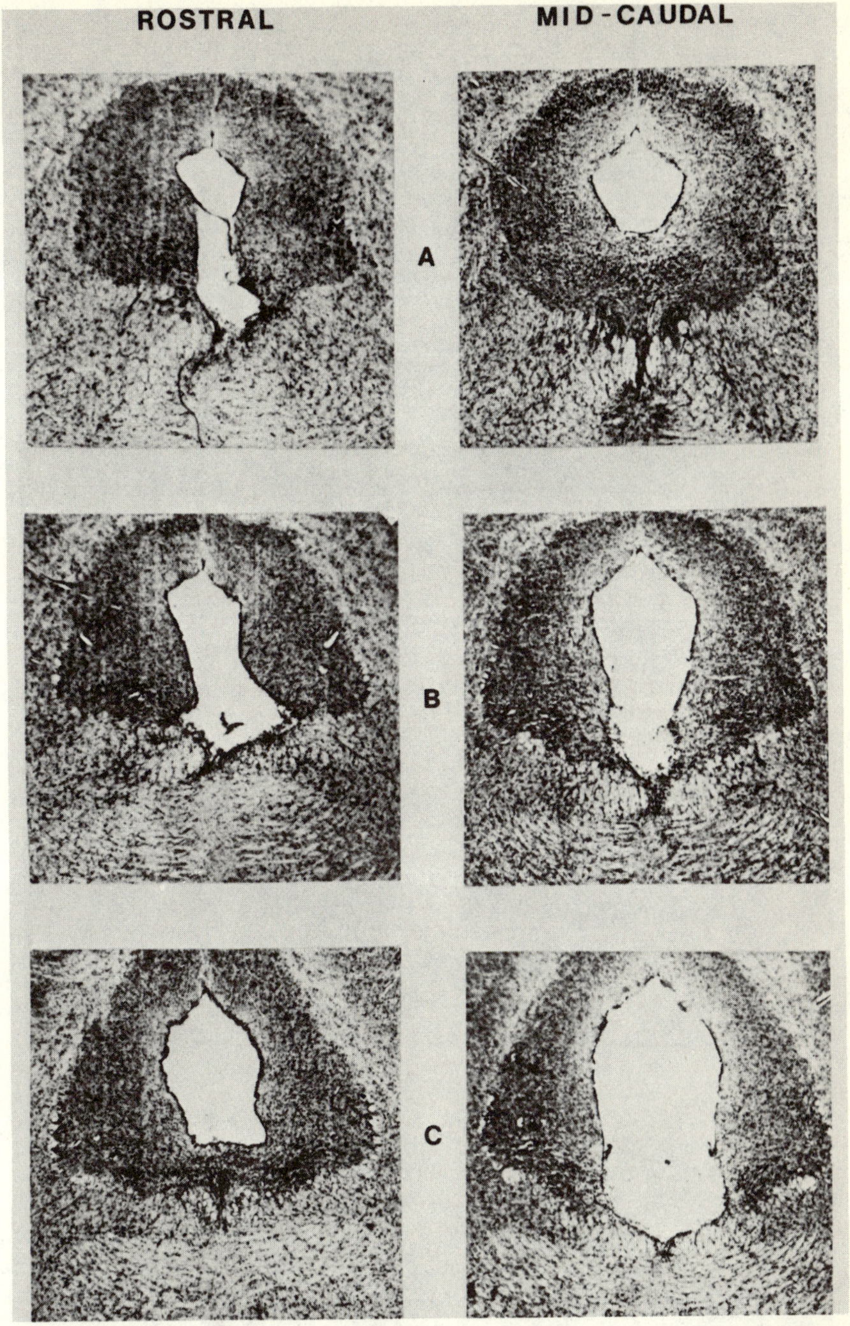

FIGURE 2. Photomicrographs, each separated by about 0.6 mm, of representative electrolytic lesions in the dorsal raphe nucleus. The "rostral" lesion reaches the level (A) of the trochlear nucleus. Both lesions produced 51–52% and 22–31% reductions in cortical and diencephalic 5-HT, respectively. The "rostral" lesion, however, produced a significantly greater (66%) reduction in striatal 5-HT content than did the "midcaudal" lesion (37%). Neither lesion affected hippocampal or spinal 5-HT concentration. Adapted from Lorens and Guldberg,[17] Table V. ×10.

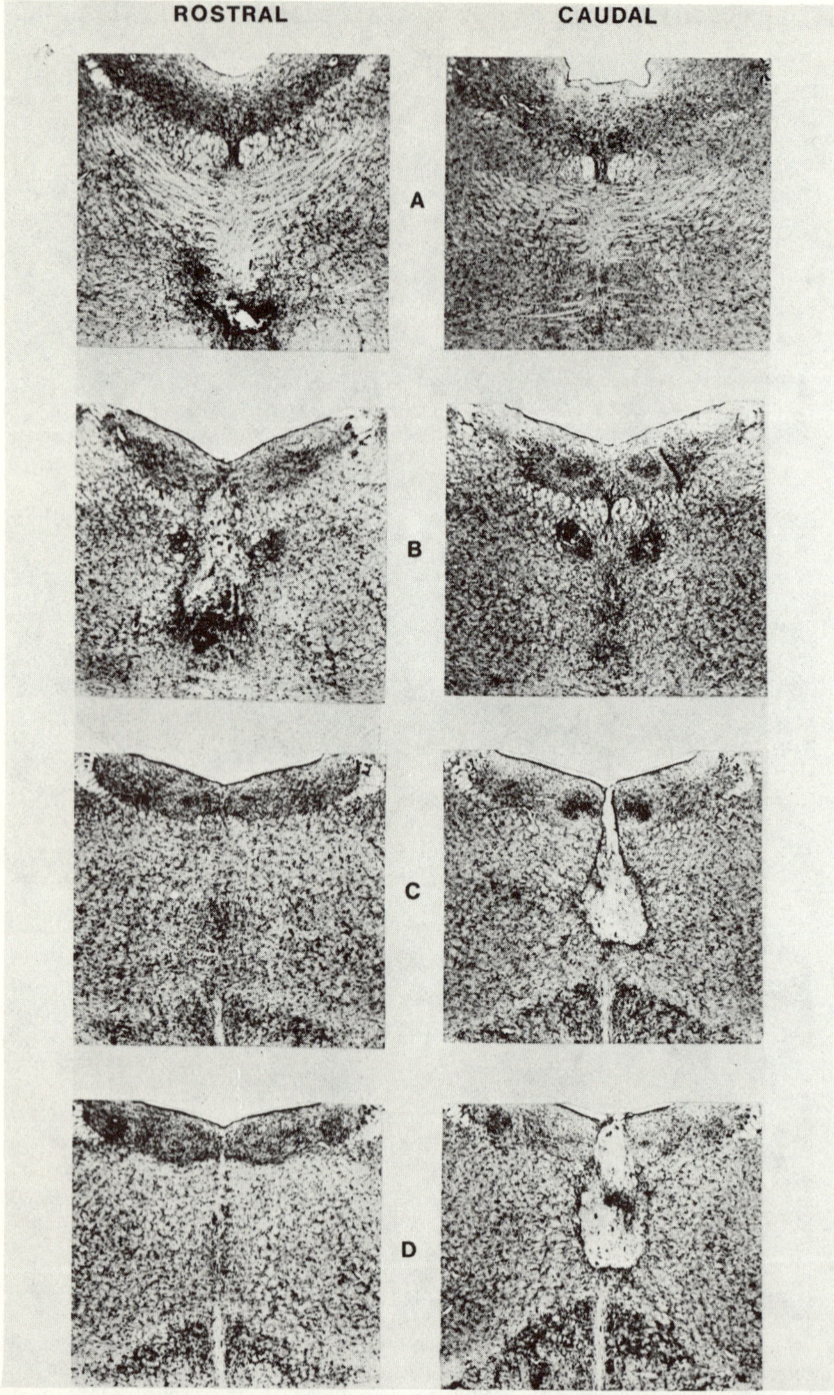

FIGURE 3. Representative electrolytic lesions in the mesencephalic median raphe nucleus. Each successive photomicrograph is separated by approximately 0.6 mm. The "caudal" lesion, which can be seen to extend into the pontine raphe (level D), reduced (27%) 5-HT only in the hippocampus. The "rostral" lesion not only pro-

TABLE 1). These conclusions have been supported by recent elegant studies that combined electrophysiologic and anatomic methods.[26, 27] Results from autoradiographic,[28, 29] retrograde degeneration,[30] and horseradish peroxidase[31] (HRP) studies, however, suggest that the DR does not send fibers to the cerebellum. Chan-Palay,[32] on the other hand, has found labeling in all raphe nuclei, except in the linear nuclei and nucleus raphe pallidus, after HRP injections into rat and monkey cerebellar nuclei and cortex.

The B8 group [mesencephalic median raphe (MR) nucleus] also contains high levels of 5-HT and tryptophan hydroxylase[21-25] and appears to be the primary origin of septal and hippocampal 5-HT,[17-20, 164] a conclusion supported by autoradiographic[28, 33] and HRP[34, 35] tracing studies. Lesion (FIGURE 3) studies also suggest that the B8 group sends 5-HT fibers to the cortex and diencephalon but *not* to the striatum, amygdala-pyriform region, or olfactory tubercle[17-20] (see also TABLE 1). Recently, Dray *et al.*[36] provided evidence for a B8 projection to the substantia nigra, an observation supported by autoradiographic[28] but not by HRP[37] work. The MR also appears to supply fibers to the DR[38, 39] but sends few, if any, projections to the cerebellum, at least in the cat.[28-31] It should be noted, however, that Kuhar *et al.*[7] found that large mesencephalic raphe lesions, which involved both the MR and the DR, reduced cerebellar 5-HT levels in the rat. It is highly possible, nevertheless, that these lesions extended caudally into the B5 and B6 groups. Conversely, Chan-Palay[32] has provided HRP evidence for an MR projection to the cerebellum.

The projections from the B9 group are not known but, at present, do not seem to reach forebrain structures[20, 40] other than, possibly, the thalamus.[41] Massari and Sanders-Bush[41] found that electrolytic B9 lesions resulted in a 35% reduction in thalamic synaptosomal 5-HT uptake. Geyer *et al.*[20] and Meek and Bertilsson,[40] on the other hand, did not find any significant change in dorsal thalamic and habenular tryptophan hydroxylase activity, respectively, after B9 lesions. In the few rats with B9 destruction that we have studied (unpublished data), no change in regional forebrain 5-HT concentration was observed. It may be that the B9 group innervates principally brain stem structures, including the MR.[38]

5-Hydroxytryptamine fibers from the B5 and B6 groups probably reach the thalamus but not other forebrain areas.[18] Lesions that damage primarily the B5 and B6 cell groups result in a significant 5-HT reduction in the dorsal thalamus but not in other forebrain areas[18] (TABLE 1). Autoradiographic,[28] retrograde degeneration,[30] and HRP[31, 32] studies suggest that fibers from the B5 region (pontine raphe) innervate the cerebellum, other brain stem regions, and the spinal cord.

The B1–3 cell groups appear to be the principal source of spinal 5-HT fibers[1, 30] and also the origin of cerebellar afferents.[30-32] The projections from B4 are largely unknown. It is quite likely, moreover, that these groups give rise to intrabrain stem 5-HT fibers.

It is now clear that in addition to their projections to other neural structures, the raphe nuclei innervate the ventricular ependyma[42-45] and the vasculature of the leptomeninges[45] and the brain stem.[46] Some of these 5-HT fibers arise in the midbrain raphe[44] and ascend in the MFB.[43] Recent work,[152, 165-167]

---

duced a significantly greater reduction (76%) in hippocampal 5-HT but also lowered cortical (23%) and diencephalic (17%) 5-HT content. Neither lesion affected striatal or spinal 5-HT concentration. Adapted from Lorens and Guldberg,[17] Table IV. ×8.5.

furthermore, has provided evidence suggesting the presence of 5-HT-containing perikarya in the paraventricular hypothalamus.

The projections of the B1–9 cell groups (summarized in FIGURE 1) thus appear to reach different central regions. Some of these projections, such as those to the diencephalon, may be overlapping. Recent work [152] performed with the Saavedra *et al.*[47] microassay and Palkovits' [48] "punch" technique, however, indicates that 5-HT fibers from the DR and MR innervate different hypothalamic nuclei and areas, as has been suggested by the autoradiographic work of Bobillier *et al.*[28] Thus, while both the MR and DR send fibers to the arcuate nucleus and anterolateral hypothalamic area, only MR lesions resulted in significant reductions in the 5-HT concentration of the medial preoptic area, anterior hypothalamic area, and suprachiasmatic nucleus. No change was seen in the ventromedial and dorsomedial hypothalamic nuclei. Recently, Palkovits *et al.*[163] reported that lesions in the DR led to significant 5-HT reductions in several hypothalamic nuclei. However, the same lesions resulted in significant reductions in hippocampal tryptophan hydroxylase activity, suggesting that a large number of 5-HT fibers from the MR were also damaged.[20, 164] Our [152] MR but *not* DR lesions reduced hippocampal 5-HT levels. Future studies employing a combination of lesion, microassay, and immunohistochemical techniques will undoubtedly lead to a revision and extension of our present understanding of the different 5-HT projections.

It should be noted, furthermore, that the raphe nuclei contain perikarya that utilize transmitter substances other than 5-HT. Acetylcholinesterase has been localized in the raphe,[49–52] implying the presence of cholinergic perikarya and/or fibers. If cholinergic neurons are located in the mesencephalic raphe, evidence for their projection to forebrain areas has yet to be obtained. For example, we [18] did not find any changes in regional forebrain choline acetyltransferase activity after lesions in either the DR, MR, or the tegmental nuclei of Gudden. If cholinergic raphe perikarya do exist, their projections may be limited to brain stem structures, such as the interpeduncular nucleus.[49] Several workers have suggested that some raphe neurons may employ tryptamine or 5-methoxytryptamine (5-MT) as a transmitter agent.[53, 54] Small amounts of 5-MT (16% of the concentration of 5-HT) have been detected in the hypothalamus.[55] The presence of "B-type" [53] indolealkylamine perikarya in the midbrain raphe, however, has been debated.[56] It is also possible that substance P[57] and other peptides may serve as transmitter agents for some raphe cells.

That some raphe perikarya may utilize transmitter substances other than 5-HT has considerable bearing on morphologic studies. Other than the inherent problems associated with the HRP, autoradiographic, and Fink-Heimer tracing techniques (such as the degree and locus of "label" uptake and survival time), these methods are not "transmitter" specific. Results from studies that use such "nonspecific" methods should thus be interpreted cautiously with respect to 5-HT and other monoaminergic pathways.

The afferent connections of the raphe have recently become the subject of vigorous reinvestigation. Previous anterograde degeneration studies have suggested inputs to the MR and DR from the agranular frontal cortex,[58] the lateral hypothalamus,[59] and the lateral habenula.[60] A septohippocampal raphe projection has been reported [61, 62] but not confirmed.[63] The results of these studies have been supported by recent HRP studies,[38, 64, 65] one [38] of which, interestingly, did not find evidence for midbrain raphe afferents from the basal ganglia and amygdala. It should be recalled, nevertheless, that it was on the basis of their presumed reciprocal "limbic" connections that Nauta[62] included the MR and DR as part of his "limbic midbrain area." The raphe nuclei, in addition, receive fibers from the locus ceruleus and other brain stem catecholamine (CA) -containing cell groups.[66, 67] Early degeneration work,[68] moreover, suggested that the pontine and medullary raphe receive afferents from the spinal cord and cerebellum. It should be noted, furthermore, that Walberg[69] found a large number of reticuloreticular fibers crossing through the raphe.

The organization of central 5-HT systems and their afferent supply, not surprisingly, is highly complex. On the other hand, the distinct 5-HT cell groups appear to innervate different brain regions. These findings have important implications for lesion and pharmacologic studies that seek to elucidate the role of 5-HT in behavior. More specifically, they raise such questions as: Do the drug treatments or lesions affect one or more of the 5-HT cell groups and their projections? Are the treatments specific to 5-HT neurons or do they involve other transmitter and/or physiologic systems? And are both peripheral and central 5-HT processes affected? In evaluating studies aimed at determining the functional role of 5-HT, these questions should be kept in mind.

Behavioral Studies

Serotonin has been implicated in many psychologic and physiologic processes, including sleep,[6] sexual behavior,[70] and temperature regulation.[153] In this review, however, we shall restrict our comments to studies concerning the role of 5-HT in pain sensitivity, activity level, and avoidance behavior.

Pain Sensitivity

There is an apparent consensus that 5-HT is directly or indirectly involved in determining pain threshold and in regulating the perceptual and/or behavioral response to painful stimuli. In general, this viewpoint derives from three lines of investigation: decreased 5-HT levels after certain brain lesions,[71, 72] *p*-CPA administration,[73-75] and tryptophan-poor diets[76] lead to increased pain sensitivity that can be reversed by treatment with 5-HTP;[73, 74, 77] evidence that suggests that 5-HT is involved in brain stimulation[78] and morphine-induced analgesia; and the effectiveness of 5-HT antagonists, such as methysergide and cyproheptadine (also a potent antihistamine agent), in the prophylactic treatment of migraine.[79]

Electrolytic lateral hypothalamic (LH) -MFB, septal, dorsomedial tegmental, and nucleus accumbens lesions and *p*-CPA (300 mg/kg) administration reduce telencephalic 5-HT content to varying degrees and lower the jump threshold.[71-74] 5-Hydroxytryptophan, furthermore, has been shown to reverse the LH-MFB lesion- and *p*-CPA-induced hyperalgesia.[73, 74, 77] In contrast, mesencephalic raphe lesions and intramidbrain 5,7-DHT (10 μg) injections, which reduce forebrain 5-HT by 70–75%, do not affect pain sensitivity.[80, 81] Lesions in Gudden's tegmental nuclei, which produce reductions in regional forebrain 5-HT, except in the striatum, also failed to alter jump thresholds.[82] In addition, intraventricular administration of 5,6-dihydroxytryptamine (5,6-DHT; 75 μg), which lowered brain 5-HT by 31%, as well as raphe lesions, were reported to be without effect on the vocalization threshold to electrical stimulation of the tail.[83] Thus, a reduction in the brain concentration of 5-HT is not a sufficient condition for the development of hyperalgesia.

It should be noted that lesions that have been reported to induce hyperalgesia do not exclusively affect 5-HT neurons. Thus, MFB lesions transect ascending CA, cholinergic,[49, 84] and histaminergic[85] pathways, not to mention the damage sustained by interstitial pathway neurons and descending MFB fibers. The septal area and nuclei accumbens both are targets for mesolimbic dopamine (DA) projections.[86] The septum also is the major source of the cholinergic input to the hippocampus.[49] Both septal and dorsomedial tegmentum lesions, furthermore, produce significant reductions in brain norepinephrine (NE) content.[71] In contrast, lesions restricted to the DR or MR and intrabrain stem 5,7-DHT injections have not been found to affect striatal or frontal cortical DA content and/or regional forebrain choline acetyltransferase.[17, 87-89] On the other hand, mesencephalic raphe lesions can reduce brain tryptamine content[54] and can enhance tyrosine hydroxylase activity in the NE-rich locus ceruleus (A6), although not in the DA A9 and A10 groups.[90] By their very nature, furthermore, electrolytic lesions do not specifically damage any one transmitter system. Thus, MR and DR lesions destroy both non-5-HT and 5-HT perikarya, as well as fibers of passage.

With few exceptions, *p*-CPA consistently has been found to induce hyperalgesia. Its effects, however, are not restricted to the central nervous system.[91] It has been estimated that well over 90% of the 5-HT in the body is localized in the intestinal wall and blood platelets. 5-Hydroxytryptamine clearly is a potent "autacoid," able to stimulate gastrointestinal motility, to produce vasoconstriction or dilatation, and to affect cutaneous sensory processes.[92] In addition to its effects on tryptophan hydroxylase, *p*-CPA inhibits phenylalanine hydroxylase,[91, 93] increases tryptamine levels, and decreases CA synthesis and concentration.[94, 95] Thus, *p*-CPA affects both peripheral and central 5-HT and tryptamine synthesis and also has an effect on phenylalanine and tyrosine hydroxylases.

If reductions in 5-HT were associated with an increased sensitivity to painful stimuli, one might expect that elevations in 5-HT levels would produce hypoalgesia. Neither 5-HTP (37.5–100 mg/kg)[74] nor tryptophan (100–200 mg/kg),[96] however, appears to affect pain sensitivity. The 5-HT-depleting agent *p*-chloroamphetamine (*p*-CA; 4 or 8 mg/kg), both acutely and 24 hr postinjection, has been reported to increase the pain threshold, as measured by the paw compression test.[97] Harvey,[154] furthermore, did not find a change in the flinch-jump threshold 30 days after *p*-CA (20 mg/kg) administration. It is interesting, in this regard, that *p*-CA does not appear to affect peripheral (intestinal) 5-HT levels or hepatic tryptophan hydroxylase[147] and has only a short-term reversible effect on spinal 5-HT.[131, 155] Recently, on the other hand, the selective 5-HT uptake inhibitor fluoxetine (Lilly 110140) was reported to elevate jump threshold and to reverse the hyperalgesia associated with a low-tryptophan diet.[76] These data are of particular interest in that the use of 5-HTP to normalize the jump threshold of LH-MFB-lesioned and *p*-CPA-treated rats has been criticized on the grounds that 5-HTP can be taken up and metabolized by DA neurons and that it results in an abnormal distribution of 5-HT.[98-101] The reputed 5-HT receptor agonist quipazine also has been reported to produce hypoalgesia.[102]

It appears that 5-HT plays a role in both morphine- and brain stimulation-induced analgesia. Electrolytic midbrain raphe lesions,[104-106] as well as *p*-CPA[107] and intraventricular 5,6-DHT (50 μg)[108] administration, have been reported to antagonize the analgesic property of morphine but not of other analgesic agents.[109] Conversely, several reports[83, 88, 110-113] have failed to confirm these findings. Lorens and Yunger,[111] in fact, were unable to replicate the observations of Samanin *et al.*,[104] even though they employed identical methods. Only the time of testing postoperatively differed, 1[104] versus 3[111] weeks. Assuming that this difference accounts for the discrepancy in results, it would appear that the effects of electrolytic median raphe lesions on morphine analgesia are short in duration. Another possibility is that the lesions of Samanin *et al.*[104] and others[105, 106] extended caudally into the nucleus raphe magnus, because lesions in that nucleus have been reported to block morphine-induced analgesia.[103, 168, 169] The 5-HT uptake inhibitors fluoxetine and zimelidine (H102/09),[114] furthermore, have been found to potentiate morphine-induced analgesia, as measured by the hot plate technique.[115, 116] On the other hand, intraventricular injection of 6-hydroxydopamine (6-OHDA; 200–250 μg) has been reported to block morphine-induced analgesia,[83, 117] an effect that is reversible by pretreatment with L-Dopa (50 mg/kg). Samanin and Bernasconi,[118] in contrast, reported that 6-OHDA potentiated morphine-induced analgesia. Price and Fibiger[119] may have provided a solution to this dilemma by demonstrating that intracerebral injection of 6-OHDA into the substantia nigra blocks morphine analgesia, whereas injection into the dorsal NE bundle (after desipramine pretreatment) potentiates it. These data suggest that both 5-HT and the catecholamines are involved in mediating morphine analgesia, a conclusion also reached by Akil and Liebeskind[78] with respect to brain stimulation-induced analgesia.

Attention has focused on the medullary nucleus raphe magnus, which forms part of the B3 5-HT cell group.[1, 132] Lesions in the nucleus raphe magnus have been reported to produce hyperalgesia[103] and attenuate morphine analgesia.[103, 168] Unfortunately, spinal monoamine levels were not measured in these studies. We recently were able to replicate these findings.[169] However, contrary to our expectations, the effective electrolytic lesions reduced spinal cord NE (20%) but had no effect on spinal cord 5-HT content. The lesions damaged the dorsal three-fourths of the raphe magnus nucleus and extended ±0.5 mm from the midline. Thus, the B3 group, whose cells extend laterally over the dorsal surface of the pyramidal tracts, sustained only minor damage. Subsequent work employing intramedullary injections of 5,7-DHT and 6-OHDA indicated that only treatments that lead to reduced spinal cord NE levels were accompanied by hyperalgesia and attenuation of the antinociceptive action of morphine.

Some of the evidence reviewed above suggests that 5-HT is involved in regulating the response to painful stimuli. However, the mechanism by which 5-HT exerts its influence on pain sensitivity is far from clear. As with other behaviors, the catecholamine systems also appear to play a significant role. Interpretation, furthermore, is made difficult by the many conflicting reports. This disagreement may well be related to the diverse measures of pain sensitivity used, species differences, and the employment of "nonspecific" drug and lesioning methods. At present, however, little correlation can be found between forebrain or spinal cord 5-HT levels and pain sensitivity. The bulk of the evidence suggests, moreover, that peripheral 5-HT stores play a major role in mediating the response to pain. Finally, in view of its relevance to the discussion below, it should be noted that mesencephalic raphe lesions and intramidbrain 5,7-DHT administration have not been found to affect pain sensitivity [80, 81, 88]; that *p*-CPA, with few exceptions, has been reported to enhance pain sensitivity [73, 81]; and that although the effects of *p*-CA on the response to painful stimuli require additional investigation, this drug has been reported to elevate the pain threshold, as measured by the paw compression test.[97]

Activity Level

There is considerable evidence to indicate that the catecholamines are directly involved in regulating locomotor activity. 5-Hydroxytryptamine, however, appears to play a "homeostatic" role vis-à-vis the CA systems. Accordingly, a reduction in 5-HT level should induce hyperactivity and potentiate the locomotor effects of amphetamine.[120]

Large midbrain raphe lesions have been reported to increase home cage, running wheel, and open-field activity.[5, 10, 16] This effect of raphe lesions seems to be due to damage to the MR, because increased activity is observed after MR but *not* DR lesions.[16, 19, 121 123] However, lesions in the ventral (VTG) and dorsal (DTG) tegmental nucleus of Gudden [18, 82, 124] (TABLE 1) also can produce enhanced open-field activity. Since MR (and VTG) lesions, but not DR lesions, reduce hippocampal 5-HT,[17, 19] it has been suggested that the locomotor effects of MR lesions are mediated via an influence on the hippocampus.[123]

Beginning 24 hr after injection, *p*-CPA (300–400 mg/kg) also has been observed to elevate activity level in animals residing in stabilimeter cages.[123, 125, 126] Increased activity in a novel environment (open field [125] or circular runway [127]) also has been reported following *p*-CPA administration. This effect, however, does not appear until about 30 min after the animal has been placed in the apparatus and seems to reflect a change in "habituation" rate. Thus, while the controls were virtually inactive 1.5 hr after the beginning of the session, the *p*-CPA animals maintained a low but relatively constant level of activity.[125] Tenen [73] and Brody [128] did not observe differences between their *p*-CPA-treated and control rats during 15–30 min open-field tests, unless the animals were stimulated by sudden changes in lighting condition or noise level. Recently, we [129] were unable to demonstrate differences between saline- and *p*-CPA- (300 mg/kg) treated rats tested in the open field for 50 min beginning 24, 48, or 72 hr postinjection. In contrast, midbrain MR plus DR-lesion rats, tested at the same time and in the same apparatus, showed enhanced locomotion, which was evident almost immediately after the animals were placed in the center of the field. All of the rats exhibited "habituation" (a within-session reduction in locomotion), and no group difference in defecation was found. However, five MR plus DR raphe-lesioned rats, which also had received *p*-CPA, were more active than the saline-treated raphe-lesioned animals during the first 5 min and last 25 min of the session. This finding suggests that *p*-CPA can potentiate the locomotor effects of raphe lesions. On the other hand, it is clear that there are major differences between raphe-lesion and *p*-CPA-treated animals in their locomotor response to a novel environment.

p-Chloroamphetamine has been reported to enhance or decrease activity, depending on the testing conditions, time after injection, and dose. Vorhees *et al.*[130] tested separate groups of rats in an open field 3 min/day for 5 consecutive days beginning 1, 10, 20, or 30 days after three daily injections of *p*-CA

(5 mg/kg/day) or distilled water. The locomotor activity of the *p*-CA-treated groups was lower than control at all postinjection times. Defecation scores, on the other hand, were higher for the rats that received *p*-CA. This schedule of *p*-CA administration reduced whole brain 5-HT by 70 and 41% 1 and 38 days postinjection, respectively. Köhler *et al.*[155] also have found a reduction in open-field activity after *p*-CA (20 mg/kg) treatment. Messing *et al.*,[131] in contrast, found an increase in home cage (stabilimeter) activity for up to 3 days after *p*-CA (5.0 mg/kg) administration. This dose of *p*-CA lowered whole brain 5-HT by 28 and 14% 2 and 14 days postinjection, respectively. A lower dose of *p*-CA (2.5 mg/kg) failed to significantly increase home cage activity despite a 22% reduction in whole brain 5-HT 2 days after injection. On the other hand, pretreatment with the selective 5-HT uptake blocker fluoxetine (10 mg/kg) antagonized the effect of *p*-CA (5.0 mg/kg) on home cage activity. In a recent experiment (unpublished), we tested *p*-CA- (10 mg/kg) treated rats (previously used in a self-stimulation experiment) on the 18th postinjection day in a photocell cage ($38.5 \times 15.5 \times 45.0$ cm high) illuminated by a 15-W bulb. Activity counts were recorded each 3 min for 21 min. The *p*-CA-injected animals ($n = 4$) were significantly more active than the saline controls ($n = 4$) during the initial 9 min of the session (total session mean ± SEM activity scores: control, 1417 ± 183; *p*-CA, 2045 ± 90; $U = 0$, $p < 0.28$; two-tailed). Habituation was observed in all rats, and no difference in rearing activity was found. This increase in activity after *p*-CA treatment might be a function of transient changes in background noise level.

We recently reported[87] that intramidbrain 5,7-DHT (10 μg) administration failed to affect open-field activity 16–21 days postinjection despite 62–83% reductions in regional forebrain 5-HT. Regional forebrain CA concentrations were not affected. In contrast, electrolytic MR plus DR-lesion rats, tested at the same time, evidenced hyperactivity. Hole *et al.*[88] observed hypoactivity in rats injected with 5,7-DHT (after protryptyline pretreatment) into the ventromedial tegmentum 3–10 days earlier. On the other hand, we[89] found no change in open-field activity 4, 9, and 29 days after bilateral 5,7-DHT (2 μg) injections into the ventromedial midbrain tegmentum. Although receptor supersensitivity[156] and/or regenerative sprouting[157] may account in part for these findings, it is clear that the long-term reduction in central 5-HT after 5,7-DHT treatment does not produce the behavioral changes seen after *p*-CA and *p*-CPA treatment or electrolytic midbrain raphe destruction.

In summary, changes in activity level after reduction in 5-HT levels depend on the behavioral test (novel versus familiar environment) and method of depletion employed:

Electrolytic mesencephalic raphe lesions enhance locomotion in a variety of situations, including the open field, running wheel, and home cage. Although these behavioral changes may be due to an influence on hippocampal mechanisms,[123] damage to adjacent tegmental nuclei or fibers of passage cannot be excluded.[16, 17, 82, 124]

p-Chlorophenylalanine enhances home cage activity, especially during dark (nocturnal) hours. In contrast, *p*-CPA does not greatly affect activity level in a novel environment,[129] depending on the level of illumination[125, 127, 129] and the duration of the test session.

p-Chloroamphetamine has been found to enhance activity in both the home cage[131] and in a novel environment. In other studies[130, 155] *p*-CA was observed to decrease activity in an open field.

Either no change[87, 89] or a reduction[88] in open-field activity has been observed in rats after intramidbrain 5,7-DHT administration. These results are in contrast to the effects of electrolytic midbrain raphe lesions and systematic administration of *p*-CA or *p*-CPA.

Avoidance Behavior

In 1971, we[10] found, and later confirmed,[15, 101, 119] that large DR plus MR lesions facilitated the acquisition of a signaled two-way (shuttle) conditioned avoidance response (CAR). Selective MR or DR lesions, however, did not affect two-way CAR acquisition.[16, 18, 111] These data suggested that if the raphe lesion-induced change in avoidance behavior was related to a reduction in forebrain 5-HT, the projections from both the DR and the MR had to be interrupted.

Lesions confined to Gudden's tegmental nuclei, on the other hand, also were found to facilitate two-way CAR acquisition.[82] These nuclei do not appear to contain 5-HT perikarya.[132] Nevertheless, the ventral tegmental nucleus in the rat appears to be perforated by ascending 5-HT fibers (see FIGURE 2 in Reference 133), probably arising in B8 and in the pontine raphe (B5 and B6). Thus, lesions in the ventral tegmental nuclei produce decreases in regional forebrain 5-HT similar to those seen after lesions in the MR (TABLE 1). In addition, Gudden's tegmental nuclei give rise to fibers that pass through and immediately lateral to the midbrain raphe nuclei.[134-136] Recent work has demonstrated, furthermore, that whereas lesions restricted to the dorsal tegmental nucleus lead to facilitated two-way CAR acquisition, they fail to affect regional forebrain 5-HT (TABLE 1). A similar enhancement of two-way CAR acquisition was seen after electrolytic destruction of the pontine raphe (B5 and B6), although the only significant reduction in regional forebrain 5-HT content observed was in the dorsal thalamus (TABLE 1). These findings led us to the conclusion that the effects of electrolytic raphe lesions on avoidance behavior are not due to an interruption of ascending 5-HT pathways and are due to damage to adjacent tegmental nuclei and/or their projections. This viewpoint is supported by studies that demonstrated that two-way CAR acquisition is not affected after 5-HT depletion produced by the following treatments: *p*-CPA (300 mg/kg, 72 hr prior to testing),[129] 5,7-DHT (10 μg injected into the midbrain, 2 weeks previously; unpublished data), and *p*-CA (10 mg/kg, 7 and 8 days before testing[158]). Interestingly, if rats are given escape training prior to conditioning, *p*-CA can greatly suppress two-way CAR acquisition.[127] It should be noted, however, that Breese and Cooper[138] reported that intracisternal 5,7-DHT (200 μg) administration after pargyline (50 mg/kg) pretreatment led to facilitated two-way CAR acquisition 21–30 days postoperatively. Intracisternal 5,7-DHT alone or after desipramine (30 mg/kg) pretreatment, on the other hand, did not significantly alter the acquisition of the two-way CAR. Previously, Breese *et al.*[139] had reported that intracisternal 5,6-DHT administration [two 75-μg injections, separated by 7 days; the first injection followed pargyline (50 mg/kg) pretreatment] enhanced two-way CAR acquisition 30 days after the second injection. Unfortunately, the long-term effects of pargyline alone were not examined.

Both large and selective MR lesions retard the acquisition and forced extinction of an *unsignaled* one-way CAR.[16, 80, 87] Lesions restricted to the DR do not produce similar effects. Likewise, intramidbrain 5,7-DHT (10 μg, 21–23 days prior to testing),[87] *p*-CPA (300 mg/kg, 72 hr before testing),[129] and *p*-CA (10 mg/kg, 18 days prior to testing; unpublished data) all failed to affect one-way CAR acquisition. These data suggest, as those cited above, that the effects of electrolytic midbrain raphe lesions on avoidance learning are not related to their effects on 5-HT pathways.

MR lesions[140] and *p*-CA (5.0 mg/kg/day for 3 consecutive days) administration[130] have been reported to facilitate the acquisition of a discriminated Y-maze CAR. In this situation, a light was on continuously in the "safe" arm. Enhanced acquisition of a *signaled* one-way CAR (such as pole climbing[141] and jumping onto a platform[73, 128]) also has been observed after *p*-CPA administration. All of these studies employed a conditional stimulus (CS), whereas in our studies on one-way CAR acquisition,[16, 80, 87, 129] no CS was used. It is known that electrolytic destruction of the midbrain raphe,[142] *p*-CPA administration,[143] and *p*-CA treatment[144] can produce sensitization to sensory stimuli. *p*-Chlorophenylalanine, but not raphe lesions or *p*-CA, also increases pain sensitivity. Increased sensitivity to a CS and/or shock could be expected to facilitate one-way CAR learning.[145]

Interestingly, electrolytic raphe lesions appear to disrupt an animal's ability to use spatial cues. Thus, rats with large DR plus MR lesions show deficits in unsignaled, but not signaled, spatial reversal learning.[146] Such a deficit would benefit a rat in a shuttlebox avoidance conditioning procedure in which the animal is required to return to the place where it previously received shock and in which it is to make a hurdle jump response during the presentation of a CS, regardless of its position in the apparatus. Such a deficit, however, would prove detrimental in a one-way "place" learning avoidance paradigm. It is clear, therefore, that the parameters of the avoidance conditioning procedure and apparatus, in addition to the method of 5-HT depletion, are critical variables for any study that purports to evaluate the role of 5-HT in avoidance conditioning.

Conclusions

Attempts to elucidate the functional role of central 5-HT systems have been complicated by the nonequivalence and nonselectivity of the pharmacologic and lesion methods used to deplete 5-HT, or otherwise alter its metabolism, and by the variations in the behavioral testing procedures and apparatus employed.

It is clear that electrolytic or thermal (radiofrequency) lesions of the raphe nuclei or their fiber systems (for example, those ascending in the lateral hypothalamic portion of the medial forebrain bundle) are not selective for 5-HT. Thus, fibers of passage from adjacent tegmental nuclei, reticulorecticular projections,[69] and non-5-HT perikarya within the raphe are damaged. Differences in behavioral and 5-HT effects can be observed, depending on the raphe nucleus or 5-HT cell group destroyed. The distinct 5-HT cell groups innervate different brain and spinal regions, suggesting that they subserve different functions. Thus, 5-HT itself does not appear to play a unitary role in behavioral and physiologic processes.

The tryptophan hydroxylase (TrH) inhibitor *p*-CPA depletes both peripheral and central 5-HT. It also inhibits both tyrosine and phenylalanine hydroxylase, resulting in transient reductions in catecholamine synthesis and in an elevation in phenylalanine and tryptamine concentrations. *p*-Chloroamphetamine, in contrast, does not appear to affect intestinal 5-HT or hepatic TrH activity [147] but produces long-term reductions in brain 5-HT concentration and TrH activity.[148, 159] These changes in central 5-HT content and TrH activity, however, do not appear to be uniform throughout the brain [22, 40, 149] and are of short duration in the spinal cord [131] and brain stem.[155] Although *p*-CA can produce cellular

TABLE 2

IMPORTANCE OF METHODOLOGY IN DETERMINING THE BEHAVIORAL EFFECTS OF 5-HT DEPLETION

Behavioral Measure	Midbrain Raphe Lesions *	5,7-DHT	*p*-CPA	*p*-CA
Pain Sensitivity	0 [80, 81] †	0 [80, 88]	↑ [73, 81]	0 [154] or ↓ [97]
Locomotor Activity				
Home cage (familiar)	↑ [5, 19, 121, 123]		↑ [123, 125, 126]	↑ [131]
Open field (novel)	↑ [16, 18, 87, 122]	0 [87–89]	0 [73, 128, 129] or (↑) [125, 127] ‡	↓ [130, 155]
Avoidance Acquisition				
One-way	↓ [16, 80, 87]	0 [80, 88]	0 [129] or (↑) [73, 128, 141] §	0 [162]
Shuttle	↑ [16, 111, 129] ¶	0 [138]	0 [129]	0 [158] ‖
Discriminated	↑ [140]			↑ [130]

* Numerical superscripts indicate references; 0, no effect; ↑, increased or facilitated, ↓, decreased or deficit. Most of the behavioral effects noted have been seen after damage to the median raphe nucleus (B8) alone; lesions restricted to the dorsal raphe nucleus (B7) produce few behavioral effects.

† Raphe magnus (B3) lesions have been reported to produce hyperalgesia,[103] an effect associated with spinal cord NE reductions.[169]

‡ Increases in open-field activity were observed beginning 30 min after the start of the test session;[125, 127] rapid changes in illumination [73, 128] also induce bursts of open-field activity in *p*-CPA-treated rats.

§ Depends on the use of a conditional stimulus (CS).

¶ Requires combined median plus dorsal raphe nucleus lesions and has been ascribed to damage of Gudden's tegmental nuclei and the pontine raphe (B5 and B6).[18, 82]

‖ If animals receive escape pretraining, a deficit is observed.[137]

changes suggestive of retrograde degeneration, these effects do not seem to be restricted to 5-HT cell bodies (they are observed, for example, in the non-5-HT-containing perikarya of the nucleus reticularis tegmenti pontis [160]). Recent data suggest, furthermore, that 5-HT terminals do not undergo degeneration after injection of 4-chloro-*N*-methylamphetamine.[150] Moreover, *p*-CA acutely affects catecholamine uptake and turnover [94, 148, 151] and can elevate activity levels 6–8 hr after injection.[151]

Of the neurotoxic tryptamines, 5,7-DHT appears to provide an effective means for depleting central 5-HT. Compared with 5,6-DHT, 5,7-DHT is much less toxic and has a greater affinity for 5-HT neurons and a slower rate of oxidation. 5,7-Dihydroxytryptamine does not appear to affect "B-type" perikarya[13] or raphe neurons,[87, 89] except at very high doses (200 μg, intraventricularly[161]). Desipramine (10–30 mg/kg) pretreatment, furthermore, blocks the effects of 5,7-DHT on NE systems, and if 5,7-DHT is injected into the brain stem caudal to DA cell bodies, it does not affect striatal or frontal cortical DA concentrations.[87, 89] Intrabrain stem injections, however, are accompanied by "nonspecific" damage due to cannula insertion and the injection process. The extent of this damage can be greatly reduced, nevertheless, by very slow injection rates and by excluding ascorbic acid (as an antioxidant) from the solution.

As seen in TABLE 2, the different methods of 5-HT depletion do not necessarily produce similar effects on the same behavioral measure. It should be noted, furthermore, that the same method of 5-HT depletion can affect activity level and conditioned avoidance acquisition in different ways, depending on the parameters of the behavioral tests. Thus, the behavioral effects of 5-HT depletion depend on both the method of depletion and the behavioral test employed.

Acknowledgments

The author expresses his gratitude to Drs. Gerald Gebhart and Walter Randall for critically reviewing the manuscript and making several helpful suggestions and to Mrs. Laurie Larsen and Vicky Moothart for their patience and typing efforts. The author acknowledges the contributions of his several collaborators, in particular, Professor Hans C. Guldberg of the University of Trondheim, Norway, for his encouragement and continued support.

References

1. Dahlström, A. & K. Fuxe. 1964. Evidence for the existence of monoamine-containing neurons in the central nervous system. Acta Physiol. Scand. **62**(Suppl. 232): 1–55.
2. Taber, E., A. Brodal & F. Walberg. 1960. The raphe nuclei of the brainstem in the cat. I. Normal topography and cytoarchitecture and general discussion. J. Comp. Neurol. **114:** 161–188.
3. Heller, A. & R. Y. Moore. 1968. Control of brain serotonin and norepinephrine by specific neural systems. Advan. Pharmacol. **6A:** 191–206.
4. Moore, R. Y. 1970. Brain lesions and amine metabolism. Int. Rev. Neurobiol. **13:** 67–92.
5. Kostowski, W., E. Giacolone, S. Garattini & L. Valzelli. 1968. Studies on behavioral and biochemical changes in rats after lesions of midbrain raphe. Eur. J. Pharmacol. **4:** 371–376.
6. Jouvet, M. 1972. The role of monoamines and acetylcholine-containing neurons in the regulation of the sleep-waking cycle. Ergeb. Physiol. Biol. Chem. Exp. Pharmakol. **64:** 166–307.
7. Kuhar, M. J., G. K. Aghajanian & R. H. Roth. 1972. Tryptophan hydroxylase activity and synaptosomal uptake of serotonin in discrete brain regions after midbrain raphe lesions: correlation with serotonin levels and histochemical fluorescence. Brain Res. **44:** 165–176.
8. Sheard, M. H. & G. K. Aghajanian. 1968. Stimulation of the midbrain raphe: effect on serotonin metabolism. J. Pharmacol. Exp. Ther. **163:** 425–430.

9. HOLMAN, R. B. & M. VOGT. 1972. Release of 5-hydroxytryptamine from caudate nucleus and septum. J. Physiol. (London) **23:** 243–254.
10. LORENS, S. A., J. P. SORENSEN & L. M. YUNGER. 1971. Behavioral and neurochemical effects of lesions in the raphe system of the rat. J. Comp. Physiol. Psychol. **77:** 48–52.
11. LORENS, S. A. 1971. Effect of lesions in the raphe system on self-stimulation in the rat. Physiol. Behav. **7:** 815–818.
12. AGHAJANIAN, G. K. & I. M. ASHER. 1971. Histochemical fluorescence of raphe neurons: selective enhancement by tryptophan. Science **172:** 1159–1161.
13. BJÖRKLUND, A., A. NOBIN & U. STENEVI. 1973. The use of neurotoxic dihydroxytryptamines as tools for morphological studies and localized lesioning of central indoleamine neurons. Z. Zellforsch. **145:** 479–501.
14. JOH, T. H., T. SHIKIMI, V. M. PICKEL & D. J. REIS. 1975. Brain tryptophan hydroxylase: purification of, production of antibodies to, and cellular and ultrastructural localization in serotonergic neurons of rat midbrain. Proc. Nat. Acad. Sci. USA **72:** 3575–3579.
15. PICKEL, V. M., T. H. JOH & D. J. REIS. 1976. Monoamine-synthesizing enzymes in central dopaminergic, noradrenergic and serotonergic neurons. Immunocytochemical localization by light and electron microscopy. J. Histochem. Cytochem. **24:** 792–806.
16. SREBRO, B. & S. A. LORENS. 1975. Behavioral effects of selective midbrain raphe lesions in the rat. Brain Res. **89:** 303–325.
17. LORENS, S. A. & H. C. GULDBERG. 1974. Regional 5-hydroxytryptamine following selective midbrain raphe lesions in the rat. Brain Res. **78:** 45–56.
18. LORENS, S. A., O. J. BROCH, H. C. GULDBERG, C. KÖHLER & B. SREBRO. 1978. Regional 5-hydroxytryptamine and behavior following selective brainstem lesions. Submitted for publication.
19. JACOBS, B. L., W. D. WISE & K. M. TAYLOR. 1974. Differential behavioral and neurochemical effects following lesions of the dorsal or median raphe nuclei in rats. Brain Res. **79:** 353–362.
20. GEYER, M. A., A. PUERTO, W. J. DAWSEY, S. KNAPP, W. P. BULLARD & A. J. MANDELL. 1976. Histological and enzymatic studies on the mesolimbic and mesostriatal serotonergic pathways. Brain Res. **106:** 241–256.
21. PALKOVITS, M., M. BROWNSTEIN & J. M. SAAVEDRA. 1974. Serotonin content of the brainstem nuclei in the rat. Brain Res. **80:** 237–249.
22. BERTILSSON, L., S. H. KOSLOW & E. COSTA. 1975. 5-Hydroxytryptamine depletion in mesencephalic nuclei of rat brain following a single injection of *p*-chloroamphetamine. Brain Res. **91:** 348–350.
23. KAN, J.-P., N. BUDA & J.-F. PUJOL. 1975. Tryptophan-5-hydroxylase activity in the raphe system of the rat brainstem. Brain Res. **93:** 353–357.
24. BROWNSTEIN, M. J., M. PALKOVITS, J. M. SAAVEDRA & J. S. KIZER. 1975. Tryptophan hydroxylase in the rat brain. Brain Res. **97:** 163–166.
25. MEEK, J. L. & L. N. NECKERS. 1975. Measurement of tryptophan hydroxylase in single brain nuclei by high pressure liquid chromatography. Brain Res. **91:** 336–340.
26. MILLER, J. J., T. L. RICHARDSON, H. C. FIBIGER & H. MCLENNAN. 1975. Anatomical and electrophysiological identification of a projection from the mesencephalic raphe to the caudateputamen in the rat. Brain Res. **97:** 133–138.
27. WANG, R. Y. & G. K. AGHAJANIAN. 1977. Inhibition of neurons in the amygdala by dorsal raphe stimulation: mediation through a direct serotonergic pathway. Brain Res. **120:** 85–102.
28. BOBILLIER, P., S. SEGUIN, F. PETITJEAN, D. SALVERT, M. TOURET & M. JOUVET. 1976. The raphe nuclei of the cat brainstem: a topographical atlas of their efferent projections as revealed by autoradiography. Brain Res. **113:** 449–486.
29. CONRAD, L. C., C. M. LEONARD & D. W. PFAFF. 1974. Connections of the median and dorsal raphe nuclei in the rat: an autoradiographic and degeneration study. J. Comp. Neurol. **156:** 179–206.

30. Brodal, A., E. Taber & F. Walberg. 1960. The raphe nuclei of the brainstem in the cat. II. Efferent connections. J. Comp. Neurol. **114:** 239–260.
31. Shinnar, S., R. J. Maciewicz & R. J. Shofer. 1973. A raphe projection to cat cerebellar cortex. Brain Res. **97:** 139–143.
32. Chan-Palay, V. 1977. Cerebellar Dentate Nucleus. Organization, Cytology and Transmitters. Springer-Verlag. New York, N.Y.
33. Moore, R. Y. & A. E. Halaris. 1975. Hippocampal innervation by serotonin neurons of the midbrain raphe in the rat. J. Comp. Neurol. **164:** 171–184.
34. Segal, M. & S. C. Landis. 1974. Afferents to the hippocampus of the rat studied with the method of retrograde transport of horseradish peroxidase. Brain Res. **82:** 263–268.
35. Segal, M. & S. C. Landis. 1974. Afferents to the septal area of the rat studied with the method of retrograde axonal transport of horseradish peroxidase. Brain Res. **82:** 263–268.
36. Dray, A., T. J. Gonye, N. R. Oakley & T. Tanner. 1976. Evidence for the existence of a raphe projection to the substantia nigra in rat. Brain Res. **113:** 45–57.
37. Bunney, B. S. & G. K. Aghajanian. 1976. The precise localization of nigral afferents in the rat as determined by a retrograde tracing technique. Brain Res. **117:** 423–425.
38. Aghajanian, G. K. & R. Y. Wang. 1977. Habenular and other midbrain raphe afferents demonstrated by a modified retrograde tracing technique. Brain Res. **122:** 229–242.
39. Mosko, S. S., D. Haubrich & B. L. Jacobs. 1977. Serotonergic afferents to the dorsal raphe nucleus: evidence from HRP and synaptosomal uptake studies. Brain Res. **119:** 269–290.
40. Meek, J. L & L. Bertilsson. 1975. Comparison of the effects of lesion in the 'B9' cell body group and p-chloroamphetamine on tryptophan hydroxylase and 5-hydroxytryptamine in rat brain nuclei. Brain. Res. **100:** 140–144.
41. Massari, V. J. & E. Sanders-Bush. 1975. Synaptosomal uptake and levels of serotonin in rat brain areas after p-chloroamphetamine or B-9 lesions. Eur. J. Pharmacol. **33:** 419–422.
42. Lorez, H. P. & J. G. Richards. 1973. Distribution of indolealkylamine nerve terminals in the ventricles of the rat brain. Z. Zellforsch. **144:** 511–522.
43. Lorez, H. P., L. Pieri & J. G. Richards. 1975. Disappearance of supra-ependymal 5-HT axons in the rat forebrain after electrolytic and 5,6-DHT-induced lesions of the medial forebrain bundle. Brain Res. **100:** 1–12.
44. Aghajanian, G. K. & D. W. Gallagher. 1975. Raphe origin of serotonergic nerves terminating in the cerebral ventricles. Brain Res. **88:** 221–231.
45. Chan-Palay, V. 1976. Serotonin axons in the supra- and subependymal plexuses and in the leptomeninges; their roles in local alterations of cerebrospinal fluid and vasomotor activity. Brain Res. **102:** 103–130.
46. Scheibel, M. E., U. Tomiyasu & A. B. Scheibel. 1975. Do raphe nuclei of the reticular formation have a neurosecretory or vascular sensor function? Exp. Neurol. **47:** 316–329.
47. Saavedra, J. M., M. Brownstein & J. Axelrod. 1974. A specific and sensitive enzymatic-isotopic microassay for serotonin in tissues. J. Pharmacol. Exp. Ther. **186:** 505–515.
48. Palkovits, M. 1973. Isolated removal of hypothalamic or other brain nuclei of the rat. Brain Res. **59:** 449–450.
49. Lewis, P. R. & C. C. D. Shute. 1967. The cholinergic limbic system: projections to hippocampal formation, medial cortex, nuclei of the ascending cholinergic reticular system, and the subfornical organ and supra-optic crest. Brain **90:** 521–542.
50. Ramon-Moliner, E. 1972. Acetylthiocholinesterase distribution in the brainstem of the cat. Ergeb. Anat. Entwicklungsgesch. **46:** 1–53.
51. Palkovits, M. & D. M. Jacobowitz. 1974. Topographic atlas of catechol-

amine and acetylcholinesterase-containing neurons in the rat brain. II. Hindbrain (mesencephalon, rhombencephalon). J. Comp. Neurol. **157:** 29–42.
52. Deikel, S. M. & L. L. Butcher. 1976. Neuronal somata in the B7 5-hydroxytryptamine (5-HT) cell body group also contain acetylcholinesterase (AChE, EC3.1.7)? Neurosci. Abst. **2**(1): 579.
53. Björklund, A., B. Falck & U. Stenevi. 1971. Classification of monoamine neurones in the rat mesencephalon: distribution of a new monoamine neurone system. Brain Res. **32:** 1–17.
54. Knott, P., C. A. Marsden & G. Curzon. 1974. Comparative studies of brain 5-hydroxytryptamine and tryptamine. Advan. Biochem. Psychopharmacol. **11:** 109–114.
55. Green, A. R., S. H. Koslow & E. Costa. 1973. Identification and quantitation of a new indolealkylamine in rat hypothalamus. Brain Res. **51:** 371–374.
56. Jonsson, G., P. Einarsson, K. Fuxe & H. Hallman. 1975. Microspectrofluorimetric analysis of the formaldehyde induced fluorescence in midbrain raphe neurons. Med. Biol. **53:** 25–39.
57. Nilsson, G., T. Hokfelt & B. Pernow. 1974. Distribution of substance P-like immunoreactivity in the rat central nervous system as revealed by immunohistochemistry. Med. Biol. **52:** 424–427.
58. Leonard, C. M. 1969. The prefrontal cortex of the rat. I. Cortical projection of the mediodorsal nucleus. II. Efferent connexions. Brain Res. **12:** 321–343.
59. Guillery, R. W. 1957. Degeneration in the hypothalamic connexions of the albino rat. J. Anat. **91:** 91–115.
60. Akagi, K. & E. W. Powell. 1968. Differential projections of habenular nuclei. J. Comp. Neurol. **132:** 263–274.
61. Nauta, W. J. H. 1956. An experimental study of the fornix system in the rat. J. Comp. Neurol. **104:** 247–272.
62. Nauta, W. J. H. 1958. Hippocampal projections and related neural pathways to the midbrain in the cat. Brain **81:** 319–340.
63. Meibach, R. C. & A. Siegal. 1977. Efferent connexions of the septal area in the rat: an analysis utilizing retrograde and anterograde transport methods. Brain Res. **119:** 1–20.
64. Llamas, A., F. Reinoso-Suarez & E. Martinez-Moreno. 1975. Projections to the gyrus proreus from the brainstem tegmentum (locus coeruleus, raphe nuclei) in the cat, demonstrated by retrograde transport of horseradish peroxidase. Brain Res. **89:** 331–336.
65. Pasquier, D. A., C. Anderson, W. B. Forbes & P. J. Morgane. 1976. Horseradish peroxidase tracing of the lateral habenular-midbrain raphe nuclei connexions in the rat. Brain Res. Bull. **1:** 443–451.
66. Lindvall, O. & A. Björklund. 1974. The organization of the ascending catecholamine neuron systems in the rat brain as revealed by the glyoxylic acid fluorescence method. Acta Physiol. Scand. Suppl. **412:** 1–48.
67. Saavedra, J. M., H. Grobecker & J. Zivin. 1976. Catecholamines in the raphe nuclei of the rat. Brain Res. **114:** 339–345.
68. Brodal, A., F. Walberg & E. Taber. 1960. The raphe nuclei of the brainstem in the cat. III. Afferent connexions. J. Comp. Neurol. **114:** 261–282.
69. Walberg, F. 1974. Crossed reticulo-reticular projections in the medulla, pons and mesencephalon. An autoradiographic study in the cat. Z. Anat. Entwicklungsgesch. **143:** 127–134.
70. Gessa, G. L. & A. Tagliamonte. 1974. Role of brain monoamines in male sexual behavior. Life Sci. **14:** 425–436.
71. Lints, C. E. & J. A. Harvey. 1969. Altered sensitivity to foot shock and decreased brain content of serotonin following brain lesions in the rat. J. Comp. Physiol. Psychol. **67:** 23–31.
72. Lorens, S. A., J. P. Sorensen, Jr. & J. A. Harvey. 1970. Lesions in the nuclei accumbens septi of the rat: behavioral and neurochemical effects. J. Comp. Physiol. Psychol. **75:** 284–290.

73. TENEN, S. S. 1967. The effects of p-chlorophenylalanine, a serotonin depletor, on avoidance acquisition, pain sensitivity and related behavior in the rat. Psychopharmacologia **10:** 204–219.

74. HARVEY, J. A. & C. E. LINTS. 1971. Lesions in the medial forebrain bundle: relationship between pain sensitivity and telencephalic content of serotonin. J. Comp. Physiol. Psychol. **74:** 28–36.

75. SICUTERI, F., B. ANSELMI & P. L. DEL BIANCO. 1973. 5-Hydroxytryptamine supersensitivity as a new theory of headache and central pain: a clinical, pharmacological approach with p-chlorophenylalanine. Psychopharmacologia **29:** 347–356.

76. MESSING, R. B., L. A. FISHER, L. PHEBUS & L. D. LYTLE. 1976. Interaction of diet and drug in the regulation of brain 5-hydroxyindoles and the response to painful electric shock. Life Sci. **18:** 707–714.

77. LINTS, C. E. & J. A. HARVEY. 1969. Drug-induced reversal of brain damage in the rat. Physiol. Behav. **4:** 29–31.

78. AKIL, H. & J. C. LIEBESKIND. 1975. Monoaminergic mechanisms of stimulation-produced analgesia. Brain Res. **94:** 279–296.

79. FOZARD, J. R. 1975. The animal pharmacology of drugs used in the treatment of migraine. J. Pharm. Pharmacol. **27:** 297–312.

80. HOLE, K. & S. A. LORENS. 1975. Response to electric shock in the rat: effects of selective midbrain raphe lesions. Pharmacol. Biochem. Behav. **3:** 95–102.

81. HARVEY, J. A., A. J. SCHLOSBERG & L. M. YUNGER. 1974. Effects of p-chlorophenylalanine and brain lesions on pain sensitivity and morphine analgesia in the rat. Advan. Biochem. Psychopharmacol. **10:** 233–245.

82. LORENS, S. A., C. KÖHLER & H. C. GULDBERG. 1975. Lesions in Gudden's tegmental nuclei produce behavioral and 5-HT effects similar to those after raphe lesions. Pharmacol. Biochem. Behav. **3:** 653–659.

83. BLÄSIG, J., K. REINHOLD & A. HERZ. 1973. Effect of 6-hydroxydopamine, 5-hydroxytryptamine, 5,6-dihydroxytryptamine and raphe lesions on the anti-nociceptive actions of morphine in rats. Psychopharmacologia **31:** 111–119.

84. ROMMELSPACHER, H. & M. J. KUHAR. 1975. Effects of dopaminergic drugs and acute medial forebrain bundle lesions on striatal acetylcholine levels. Life Sci. **16:** 65–70.

85. SCHWARTZ, J.-C. 1975. Histamine as a transmitter in brain. Life Sci. **17:** 503–518.

86. LINDVALL, O. 1975. Mesencephalic dopaminergic afferents to the lateral septal nucleus of the rat. Brain Res. **87:** 89–95.

87. LORENS, S. A., H. C. GULDBERG, K. HOLE, C. KÖHLER & B. SREBRO. 1976. Activity, avoidance learning and regional 5-hydroxytryptamine following intra-brainstem 5,7-dihydroxytryptamine and electrolytic midbrain raphe lesions in the rat. Brain Res. **108:** 97–113.

88. HOLE, K., K. FUXE & G. JONSSON. 1976. Behavioral effects of 5,7-dihydroxytryptamine lesions of ascending 5-hydroxytryptamine pathways. Brain Res. **107:** 385–399.

89. TYE, N. C. & S. A. LORENS. 1978. Medial frontal and lateral hypothalamic self-stimulation in the rat: differential effect of chlordiazepoxide, amphetamine, and morphine following 5,7-dihydroxytryptamine induced serotonin depletion. Submitted for publication.

90. LEWIS, B. D., B. RENAUD, M. BUDA & J.-F. PUJOL. 1976. Time-course variations in tyrosine hydroxylase activity in the rat locus coeruleus after electrolytic destruction of the nuclei raphe dorsalis or raphe centralis. Brain Res. **108:** 339–349.

91. KOE, B. K. & A. WEISSMAN. 1966. P-chlorophenylalanine: a specific depletor of brain serotonin. J. Pharmacol. Exp. Ther. **154:** 499–516.

92. DOUGLAS, W. W. 1966. 5-Hydroxytryptamine and antagonists; polypeptides-angiotensin and kinins. *In* The Pharmacological Basis of Therapeutics. L. S.

Goodman & A. Gilman, Eds. 3rd edit. Chapter **30:** 644–664. Macmillan. New York, N.Y.

93. Jequier, E., W. Lovenberg & A. Sjoerdsma. 1967. Tryptamine hydroxylase inhibition: the mechanism by which p-chlorophenylalanine depletes rat brain serotonin. Mol. Pharmacol. **3:** 274–278.
94. Miller, F. P., R. H. Cox, W. R. Snodgrass & R. P. Maickel. 1970. Comparative effects of p-chlorophenylalanine, p-chloroamphetamine, and p-chloro-N-methylamphetamine on rat brain norepinephrine, serotonin and 5-hydroxyindole-3-acetic acid. Biochem. Pharmacol. **19:** 435–442.
95. Tagliamonte, A., P. Tagliamonte, G. U. Corsini, G. P. Mereu & G. L. Gessa. 1973. Decreased conversion of tyrosine to catecholamines in the brain of rats treated with p-chlorophenylalanine. J. Pharm. Pharmacol. **25:** 101–103.
96. Hole, K. & C. A. Marsden. 1975. Unchanged sensitivity to electric shock in L-tryptophan treated rats. Pharmacol. Biochem. Behav. **3:** 307–309.
97. Görlitz, B.-D. & H.-H. Frey. 1972. Central monoamines and antinociceptive drug action. Eur. J. Pharmacol. **20:** 171–180.
98. Moir, A. T. B. & D. Eccleston. 1968. The effects of precursor loading in the cerebral metabolism of 5-hydroxyindoles. J. Neurochem. **15:** 1093–1108.
99. Fuxe, K., L. L. Butcher & J. Engel. 1971. DL-5-hydroxytryptophan-induced changes in central monoamine neurons after peripheral decarboxylase inhibition. J. Pharm. Pharmacol. **23:** 420–424.
100. Coscina, D. V., J. J. Warsh, D. P. Godse & H. S. Stancer. 1974. Nonspecificity of 5-hydroxytryptamine in repleting brain serotonin in rats with MFB lesions. Res. Commun. Chem. Pathol. Pharmacol. **7:** 617–620.
101. Yunger, L. M. & J. A. Harvey. 1976. Behavioral effects of L-5-hydroxytryptophan after destruction of ascending serotonergic pathways in the rat: the role of catecholaminergic neurons. J. Pharmacol. Exp. Ther. **196:** 307–315.
102. Samanin, R., S. Bernasconi & A. Quattrone. 1976. Antinociceptive action of quipazine: relation to central serotonergic receptor stimulation. Psychopharmacologia **46:** 219–222.
103. Proudfit, H. K. & E. G. Anderson. 1975. Morphine analgesia: blockade by raphe magnus lesions. Brain Res. **98:** 612–618.
104. Samanin, R., W. Gumulka & L. Valzelli. 1970. Reduced effect of morphine in midbrain raphe lesioned rats. Eur. J. Pharmacol. **10:** 339–343.
105. Adler, M., W. Kostowski, M. Recchia & R. Samanin. 1975. Anatomical specificity as the critical determinant of the interrelationship between raphe lesions and morphine analgesia. Eur. J. Pharmacol. **32:** 39–44.
106. Garau, L., M. L. Mulas & G. Pepeu. 1975. The influence of raphe lesions on the effect of morphine on nociception and cortical ACH output. Neuropharmacology **14:** 259–263.
107. Tenen, S. S. 1968. Antagonism of the analgesic effect of morphine and other drugs by p-chlorophenylalanine, a serotonin depletor. Psychopharmacologia **12:** 278–285.
108. Genovese, E., N. Zonta & P. Mantegazza. 1973. Decreased antinociceptive activity of morphine in rats pretreated intraventricularly with 5,6-dihydroxytryptamine, a long lasting selective depletor of brain serotonin. Psychopharmacologia **32:** 359–364.
109. Samanin, R., D. Ghezzi, C. Mouron & L. Valzelli. 1973. Effect of midbrain raphe lesion on the antinociceptive action of morphine and other analgesics in rats. Psychopharmacologia **33:** 365–368.
110. Harvey, J. A. & L. M. Yunger. 1973. Relationship between telencephalic content of serotonin and pain sensitivity. *In* Serotonin and Behavior. J. Barchas & E. Usdin, Eds. : 179–189. Academic Press, Inc. New York, N.Y.
111. Lorens, S. A. & L. M. Yunger. 1974. Morphine analgesia, two-way avoidance, and consummatory behavior following lesions in the midbrain raphe nuclei of the rat. Pharmacol. Biochem. Behav. **2:** 215–221.

112. Reinhold, K., J. Bläsig & A. Herz. 1973. Changes in brain concentration by biogenic amines and the antinociceptive effect of morphine in rats. Naunyn-Schmiedeberg's Arch. Pharmakol. **278:** 69–80.
113. Buxbaum, D. N., G. G. Yarbrough & M. E. Carter. 1973. Biogenic amines and narcotic effects. I. Modification of morphine-induced analgesia and motor activity after alteration of cerebral amine levels. J. Pharmacol. Exp. Ther. **185:** 317–327.
114. Ross, S. B. & A. L. Renyi. 1977. Inhibition of the neuronal uptake of 5-hydroxytryptamine and noradrenaline in rat brain by (Z)- and (E)-3-(4-bromophenyl)-N,N-dimethyl-3-(3-pyridyl)allylanines and their secondary analogues. Neuropharmacology **16:** 57–63.
115. Sugrue, M. F. & I. McIndewar. 1976. Effect of blockade of 5-hydroxytryptamine re-uptake on drug-induced antinociception in the rat. J. Pharm. Pharmacol. **28:** 447–448.
116. Lorens, S. A. & G. F. Gebhart. 1977. Serotonin reuptake inhibition potentiates morphine but not meperidine analgesia. Submitted for publication.
117. Ayhan, I. H. 1972. Effect of 6-hydroxydopamine on morphine analgesia. Psychopharmacologia **25:** 183–188.
118. Samanin, R. & S. Bernasconi. 1972. Effect of intraventricularly injected 6-OH-dopamine or midbrain raphe lesions on morphine analgesia in rats. Psychopharmacologia **25:** 175–182.
119. Price, M. T. C. & H. C. Fibiger. 1975. Ascending catecholamine systems and morphine analgesia. Brain Res. **99:** 189–193.
120. Jacobs, B. L., W. D. Wise & K. M. Taylor. 1975. Is there a catecholamine-serotonin interaction in the control of locomotor activity? Neuropharmacology **14:** 501–506.
121. Geyer, M. A., A. Puerto, D. B. Menkes, D. S. Segal & A. J. Mandell. 1976. Behavioral studies following lesions of the mesolimbic and mesostriatal serotonergic pathways. Brain Res. **106:** 257–270.
122. Jacobs, B. L. & A. Cohen. 1976. Differential behavioral effects of lesions of the median or dorsal raphe nuclei in rats: open field and pain-elicited aggression. J. Comp. Physiol. Psychol. **90:** 102–108.
123. Jacobs, B. L., C. Trimbach, E. E. Eubanks & M. Trulson. 1975. Hippocampal mediation of raphe lesion- and PCPA-induced hyperactivity in the rat. Brain Res. **94:** 253–261.
124. Galey, D. & M. Le Moal. 1976. Locomotor activity after various radiofrequency lesions of the limbic midbrain area in the rat. Evidence for a particular role of the ventral mesencephalic tegmentum. Life. Sci. **19:** 677–684.
125. Fibiger, H. C. & B. A. Campbell. 1971. The effect of *para*-chlorophenylalanine on spontaneous locomotor activity in the rat. Neuropharmacology **10:** 25–32.
126. Mabry, P. D. & B. A. Campbell. 1974. Ontogeny of serotonergic inhibition of behavioral arousal in the rat. J. Comp. Physiol. Psychol. **86:** 193–201.
127. Pirch, J. H. 1969. Stimulation of locomotor activity by *p*-chlorophenylalanine and a low dose of reserpine. Arch. Int. Pharmacodyn. **181:** 434–440.
128. Brody, J. F., Jr. 1970. Behavioral effects of serotonin depletion and of *p*-chlorophenylalanine (a serotonin depletor) in rats. Psychopharmacologia **17:** 14–33.
129. Köhler, C. & S. A. Lorens. 1978. Open field activity and avoidance behavior following serotonin depletion: a comparison of the effects of *para*-chlorophenylalanine and electrolytic midbrain raphe lesions. Pharmacol. Biochem. Behav. In press.
130. Vorhees, C. V., G. J. Schaefer & R. J. Barrett. 1975. *p*-Chloroamphetamine: behavioral effects of reduced cerebral serotonin in rats. Pharmacol. Biochem. Behav. **3:** 279–284.
131. Messing, R. B., L. Phebus, L. A. Fisher & L. D. Lytle. 1976. Effects of *p*-chloroamphetamine on locomotor activity and brain 5-hydroxyindoles. Neuropharmacology **15:** 157–163.

132. FUXE, K. 1965. Evidence for the existence of monoamine neurons in the central nervous system. IV. Distribution of monoamine nerve terminals in the central nervous system. Acta Physiol. Scand. **64**(Suppl. 247): 37–85.
133. FUXE, K. & G. JONSSON. 1974. Further mapping of central 5-hydroxytryptamine neurons: studies with the neurotoxic dihydroxytryptamines. Advan. Biochem. Psychopharmacol. **10:** 1–12.
134. MOREST, D. K. 1961. Connexions of the dorsal tegmental nucleus in rat and rabbit. J. Anat. **95:** 229–246.
135. BRIGGS, T. L. & W. W. KAELBER. 1971. Efferent fiber connections of the dorsal and deep tegmental nuclei of Gudden. An experimental study in the cat. Brain Res. **29:** 17–29.
136. PETROVICKY, P. 1973. Note on the connections of Gudden's tegmental nuclei. I. Efferent ascending connections in the mammillary peduncle. Acta Anat. **86:** 165–190.
137. ÖGREN, S.-O., S. B. ROSS & L. BAUMANN. 1975. 5-Hydroxytryptamine and learning: long term effect of *p*-chloroamphetamine on acquisition. Med. Biol. **53:** 165–168.
138. BREESE, G. R. & B. R. COOPER. 1975. Behavioral and biochemical interactions of 5,7-dihydroxytryptamine with various drugs when administered intracisternally to adult and developing rats. Brain Res. **98:** 517–527.
139. BREESE, G. R., B. R. COOPER, L. D. GRANT & R. D. SMITH. 1974. Biochemical and behavioral alterations for 5,6-dihydroxytryptamine administration into brain. Neuropharmacology **13:** 177–187.
140. STERANKA, L. R. & R. J. BARRETT. 1974. Facilitation of avoidance acquisition by lesion of the median raphe nucleus: evidence for serotonin as a mediator of shock-induced suppression. Behav. Biol. **11:** 205–213.
141. SCHLESINGER, K., R. A. SCHREIBER & G. T. PRYOR. 1968. Effects of *p*-chlorophenylalanine on conditioned avoidance learning. Psychonom. Sci. **11:** 225–226.
142. DAVIS, M. & M. H. SHEARD. 1974. Habituation and sensitization of the rat startle response: effects of raphe lesions. Physiol. Behav. **12:** 425–431.
143. CONNER, R. L., J. M. STOCK, J. D. BARCHAS & S. LEVINE. 1970. Parachlorophenylalanine and habituation to repetitive auditory startle stimuli in rats. Physiol. Behav. **5:** 1215–1219.
144. DAVIS, M. & M. H. SHEARD. 1976. *p*-Chloroamphetamine (PCA): acute and chronic effects of hibituation and sensitization of the acoustic startle response in rats. Eur. J. Pharmacol. **35:** 261–273.
145. MOYER, K. E. & J. H. KORN. 1966. Effect of UCS intensity on the acquisition and extinction of a one-way avoidance response. Psychonom. Sci. **4:** 121–122.
146. SREBRO, B., F. JELLESTAD & S. A. LORENS. 1975. Activity, avoidance behavior, and spatial reversal learning after midbrain raphe lesions. Exp. Brain Res. **23**(Suppl.): 193.
147. SANDERS-BUSH, E., J. A. BUSHING & F. SULSER. 1972. *p*-Chloroamphetamine—inhibition of cerebral tryptophan hydroxylase. Biochem. Pharmacol. **21:** 1501–1510.
148. SANDERS-BUSH, E., J. A. BUSHING & F. SULSER. 1975. Long-term effects of *p*-chloroamphetamine and related drugs on central serotonergic mechanisms. J. Pharmacol. Exp. Ther. **192:** 33–41.
149. NECKERS, L. M., L. BERTILSSON, S. H. KOSLOW & J. L. MEEK. 1975. Reduction of tryptophan hydroxylase activity and 5-hydroxytryptamine concentration in certain rat brain nuclei after *p*-chloroamphetamine. J. Pharmacol. Exp. Ther. **196:** 333–338.
150. LOREZ, H. P., A. SANER, J. G. RICHARDS & M. DA PRADA. 1976. Accumulation of 5-HT in non-terminal axons after p-chloro-N-methylamphetamine without degeneration of identified 5-HT nerve termnials. Eur. J. Pharmacol. **38:** 79–88.

151. Strada, S. J., E. Sanders-Bush & F. Sulser. 1970. *p*-Chloroamphetamine: temporal relationship between psychomotor stimulation and metabolism of brain norepinephrine. Biochem. Pharmacol. **19:** 2621–2629.
152. van de Kar, L. D. & S. A. Lorens. 1978. Differential serotonergic innervation of individual hypothalamic nuclei and other forebrain regions by the dorsal and median midbrain raphe nuclei. Submitted for publication.
153. Myers, R. D. This monograph.
154. Harvey, J. A. 1977. Personal communication.
155. Köhler, C., S. B. Ross, B. Srebro & S.-O. Ögren. This monograph.
156. Trulson, M. E. & B. L. Jacobs. This monograph.
157. Nobin, A. & A. Björklund. This monograph.
158. Ögren, S.-O. & C. Köhler. 1977. Personal communication.
159. Sanders-Bush, E. & L. R. Steranka. This monograph.
160. Harvey, J. A. This monograph.
161. Nobin, A. 1977. Personal communication.
162. Lorens, S. A. 1977. Unpublished results.
163. Palkovits, M., J. M. Saavedra, D. M. Jacobowitz, J. S. Kizer, L. Zaborszky & M. J. Brownstein. 1977. Serotonergic innervation of the forebrain: effect of lesions on serotonin and tryptophan hydroxylase levels. Brain Res. **130:** 121–134.
164. Kellar, K. J., P. A. Brown, J. Madrid, M. Bernstein, J. Vernikos-Dannelis & W. R. Mehler. 1977. Origins of serotonin innervation of forebrain structures. Exp. Neurol. **56:** 52–62.
165. Brownstein, M. J., M. Palkovits, M. L. Tappaz, J. M. Saavedra & J. S. Kizer. 1976. Effect of surgical isolation of the hypothalamus on its neurotransmitter content. Brain Res. **117:** 287–295.
166. Beaudet, A. & L. Descarries. 1975. A serotonin-containing cell group in rat hypothalamus. Neurosci. Abst. **1:** 678.
167. Chan-Palay, V. 1977. Indoleamine neurons and their processes in the normal rat brain and in chronic diet-induced thiamine deficiency demonstrated by uptake of ^{3}H-serotonin. J. Comp. Neurol. **176:** 467–494.
168. Yaksh, T. L., R. L. Plant & T. A. Rudy. 1977. Studies on the antagonism by raphe lesions of the antinociceptive action of systemic morphine. Eur. J. Pharmacol. **41:** 399–408.
169. Lorens, S. A. & G. F. Gebhart. 1978. Pain sensitivity, morphine analgesia and the nucleus raphe magnus. Submitted for publication.

HYPOTHALAMIC ACTIONS OF 5-HYDROXYTRYPTAMINE NEUROTOXINS: FEEDING, DRINKING, AND BODY TEMPERATURE *

R. D. Myers

Departments of Psychological and Biological Sciences
Purdue University
Lafayette, Indiana 47907

Although serotonin [5-hydroxytryptamine (5-HT)] -containing neurons have been implicated in numerous autonomic and behavioral functions since the mid-1950s (c.f. Reference 8), it was not until almost 10 years later that firm experimental support was provided for its central role in energy metabolism, temperature control, and food intake. In 1964, it was first proposed that 5-HT within the anterior hypothalamus mediates heat production [12] by way of a pathway that originates in this part of the diencephalon.[24] During the same era, the idea that 5-HT was also involved in ingestive behavior arose because 5-hydroxytryptophan (5-HTP) administered peripherally altered both food and water intakes of the rat.[18] Whereas the role of 5-HT in the thermoregulatory mechanism within the hypothalamus has been documented repeatedly over the years, the specificity of brain-stem 5-HT underlying ingestive behavior is not yet clear. Nevertheless, several observations favor the view that 5-HT is involved in the central nervous system processes required for the ingestion of food and fluids.

As early as 1967, Veale, in this laboratory, showed that *p*-chlorophenylalanine (*p*-CPA), which inhibits tryptophan hydroxylase and thereby depletes brain 5-HT, causes a marked reduction in food intake of the rat and a concomitant loss in its body weight.[40, 41] Wapnir *et al.*[44] not only confirmed this finding in the rat but also demonstrated the dose-dependent nature of *p*-CPA's anorexic effect when this depletor of cerebral 5-HT is administered in the animal's diet over a 2-week period. Similar reports of anorexia of varying intensity and duration have been published for the rat and other species under a variety of experimental conditions.[6, 13, 19, 31, 32] Even the human patient given *p*-CPA relates the symptom of a notable loss in appetite.[39]

As a direct corollary of these investigations, fenfluramine, generally considered to act on serotonergic neurons in the central nervous system, is a powerful anorexic agent, possibly functioning via an affinity for postsynaptic 5-HT receptors.[1, 10, 38] Evidence related to this point is that in the rat, 5-HT microinjected in the lateral or other part of the hypothalamus produces a transient and slight anorexia.[15, 37] On the other hand, Sharpe and Myers [36] found that 5-HT microinjected in areas scattered widely throughout the hypothalamus of the rhesus monkey often evokes feeding behavior or drinking. However, this primate's eating response is usually delayed by as much as 15–30 min. This finding suggests a kind of secondary or indirect agonistic action of

* Supported in part by Office of Naval Research Contract N–00014–75–C–0203 and by Grant BMS 75–18441 from the National Science Foundation.

5-HT, or, perhaps, one of its metabolites, because this indoleamine is degraded rapidly on contact with brain tissue after intracerebral application.[30]

With respect to thermoregulation, four types of evidence support the concept that 5-HT release is of functional significance in the heat production pathway: serotonergic nerve endings terminate in the thermosensitive region within the anterior hypothalamus of several species;[14, 17] a rise in body temperature occurs in nearly every species when 5-HT is applied directly to the thermosensitive zone within the anterior hypothalamus under careful experimental control;[24] the resting release of 5-HT into perfusates collected from the serotonin-rich anterior hypothalamus is augmented significantly when the trunk of an animal is exposed to cold air but not warm air;[26] and a lesion to the anterior hypothalamic, preoptic area, which is produced by a microinjection of a 5-HT neurotoxin, severely impairs thermoregulation in both the rat and monkey.[25, 43] A detailed discussion of the latter finding comprises a major portion of the present paper.

Induction of 5,6-Dihydroxytryptamine Lesions

With the advent of new neurochemical tools for producing degenerative changes in the presynaptic elements of indoleamine-containing neurons, the role of 5-HT in the control of body temperature has been investigated from an entirely different standpoint. A powerful and long-term reduction in cerebral 5-HT can be achieved by introducing such compounds as 5,6-dihydroxytryptamine (5,6-DHT) or 5,7-dihydroxytryptamine (5,7-DHT) into the brain via either the cerebral ventricular system or, more directly, by tissue injection.[2] This reduction is caused by the subsequent degeneration of axons and terminals that contain the indoleamine, as revealed by histochemical fluorescence microscopy.[3]

In our laboratory, we have undertaken a series of investigations in which the serotonergic neurotoxin is infused directly into hypothalamic structures that, historically, are identified as principal areas that mediate the general functions of energy metabolism, temperature control, and ingestive behavior. The rationale for using the microinjection procedure rather than the intraventricular route of administration rests solely on the fundamental necessity for ascertaining the precise morphologic region of the serotonergic mechanism involved. That is, only by a reasonably selective destruction of the monoamine system within a given area is one able to delineate the anatomic site of the specific functional system (see chapter 2 in Reference 24).

In these studies, rats and monkeys were prepared under surgical anesthesia with permanently indwelling guide cannulae, which were implanted well above the animal's hypothalamus to minimize cytologic damage to this structure. On postoperative recovery, microinjections of the control vehicle, which consists of an osmotically balanced artificial cerebrospinal fluid, plus 5-HT or 5,6-DHT were administered according to procedures described previously.[23] In the rat, no more than 0.5 μl of a test solution was infused into the hypothalamus; in the monkey, the microinjection volume ranged between 1.0 and 1.5 μl due to the substantially larger dimension of this primate's hypothalamus. The vehicle solution, which contained either 5,6-DHT or 5-HT, was prepared in pyrogen-free cerebrospinal fluid or saline to which 0.1 mg/ml of ascorbic acid was added to retard autoxidation of the respective amine.

For each animal, food and water intakes were monitored on a daily basis, whereas body temperature was recorded continuously for as long as 12–24 hr during and after a period associated with a specific thermal challenge to the animal. In both species, a thermistor bead was inserted into the colon to measure internal "core" temperature; however, cerebral temperature was recorded in several monkeys by an indwelling thermistor bead that was positioned against the sagittal sinus during the surgical implantation of the guide cannulae.

Agonistic Action of 5,6-DHT on Serotonergic Thermosensitive Regions

As mentioned earlier, 5-HT exerts a selective hyperthermic action on neurons in the rostral hypothalamus when the indoleamine is injected directly into this thermoreactive region.[24] It should be emphasized that the hyperthermic response is clearly dose dependent, because a transient fall in body temperature always precedes the typical hyperthermia if too large a dose is administered. This temporary inhibition of heat production is probably due to the nonspecific nature of an amine blockade effected by swamping of receptor sites or some other membrane effect.[24]

Figure 1 illustrates the nearly identical increase in the body temperature of a monkey produced by a direct injection, on different days, of 5-HT and 5,6-DHT into the same site within the anterior hypothalamus. Because the time course and magnitude of the rise parallel each other so closely, and because the febrile response is antagonized identically by a systemic injection of the antipyretic acetaminophen, one may conclude that 5,6-DHT exerts an agonistic action on the serotonergic receptor process that underlies the heat production mechanism in the rostral hypothalamus.[42, 43]

In the rat, the same type of agonistic action of 5,6-DHT on the hypothalamus is observed.[25] Moreover, the results of successive injections of 5,6-DHT into the anterior preoptic region provide evidence of a gradual dysfunction of the serotonergic system as the progressive effects of the neurotoxin advance over time. In a representative experiment illustrated in Figure 2, 5,6-DHT was microinjected bilaterally in a volume of 0.5 μl at the sites indicated in the histologic inset. The neurotoxin was given in this way at weekly intervals and in successive doses of 1.25, 2.5, and 5.0 μg.

From the declining intensity of the three individual hyperthermic responses, it is clear that a gradual degeneration of the neuronal elements that trigger heat production occurs, even with a dose more than three times greater than that given during the third week. It is essential to note, however, that the functional lack of reactivity to 5,6-DHT, both in terms of magnitude and latency, would be localized to the circumscribed region reached by the injected solution of neurotoxin. Nevertheless, these observations demonstrate that within a specific site in the hypothalamus, there is no evidence of supersensitivity after lesioning of the 5-HT system by 5,6-DHT.

Additional evidence against the notion of denervation supersensitivity was provided by a series of nine experiments, presented in Table 1, in which 5-HT was reinjected within the anterior hypothalamus at the same serotonin-sensitive site at which 5,6-DHT had been injected 12–16 days earlier. The difference between the mean rise of 0.93° C before and 1.03° C after the 5,6-DHT lesion was only 0.1° C, not statistically different ($t = 0.84$; df $= 8$).

Thermoregulatory Incapacitation After Hypothalamic 5,6-DHT Injection

When exposed to cold air of 8° C blown into its test cage or, alternatively, warm air heated to 35° C, the unanesthetized rat thermoregulates surprisingly well when either thermal challenge is delivered for a duration of 1 hr. As shown in Figure 3, the average decline in the rat's body temperature in response to cold air was about 0.2° C, whereas the rise in response to warm air challenge was less than 0.5° C. Following a bilateral microinjection of 1.25 μg of 5,6-DHT into the anterior preoptic area, regulatory deficits of nearly equal magnitude in

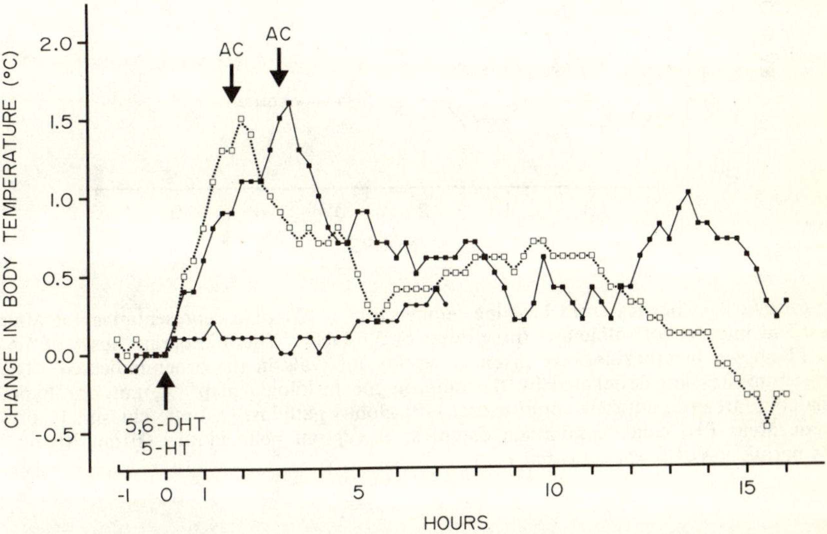

Figure 1. Deviations from the baseline body temperature of a monkey after intrahypothalamic injection at the first arrow of 3.5 μg of 5-HT (■) or 6.0 μg of 5,6-DHT (□). The amine-induced fevers were attenuated by 150 mg of acetaminophen given orally at the second and third arrows (AC). An anatomic control injection of 3.5 μg of 5-HT at a 5-HT-insensitive site is also shown (●). (From Waller & Myers.[42] By permission of *Research Communications in Chemical Pathology and Pharmacology*.)

response to both peripheral cold and warm exposures were evident 24–48 hr after injection. This finding is illustrated in Figure 3 and implies that 5,6-DHT produces an initial nonspecific cytologic destruction of the thermoregulatory mechanism. When the dose of 5,6-DHT injected into precisely the same site was doubled and the rat was again exposed to cold or warm air for 1 hr, thermoregulatory impairment was even more pronounced (Figure 3). However, 15–18 days after the second hypothalamic microinjection of 5,6-DHT, the rats had discovered their capacity to thermoregulate in the warm at virtually the same degree as that prior to injection. Conversely, in response to the same cold air challenge, the fall in body temperature was three times that of the rat's normal temperature change during the exposure to cold. These results indicate

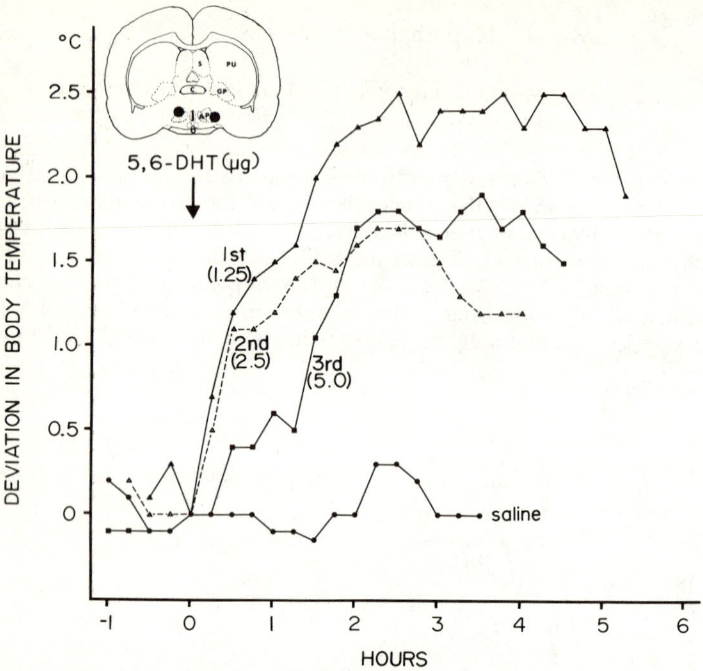

FIGURE 2. Changes from baseline temperature (° C) of a representative rat after a 0.5 μl injection of saline and three doses of 5,6-DHT. The microgram doses of 5,6-DHT shown in brackets were given at weekly intervals in the order indicated. The injection sites are designated by the dots on the histologic map. A, anterior hypothalamic area; C, anterior commissure; GP, globus pallidus; O, optic chiasm; P, preoptic area; PU, caudate putamen complex; S, septum pellucidum. (From Myers.[25] By permission of *Brain Research*.)

TABLE 1

MAXIMUM RISE IN INDIVIDUAL RAT'S TEMPERATURE INDUCED BY 2.5 μG OF 5-HT INJECTED UNILATERALLY IN THE ANTERIOR HYPOTHALAMUS BEFORE AND AFTER 5,6-DHT INJECTIONS AT THE SAME SITE *

Animal	Side †	Pre-5,6-DHT	Post-5,6-DHT
A	L	1.0	0.9
	R	1.2	1.9
B	L	0.9	0.5
	R	1.1	1.2
C	L	0.8	0.9
	R	0.8	1.4
D	L	0.9	0.7
	R	1.0	1.0
E	L	0.7	0.8
		$\overline{X}$ 0.93±0.05	$\overline{X}$ 1.03±0.14

* From Myers.[25]
† L, left hypothalamus; R, right hypothalamus.

clearly that the impairment produced by the localized 5,6-DHT lesion is functionally selective. The lesion produced by 5,6-DHT interferes with the mechanism for heat production, for which 5-HT release within the anterior hypothalamic, preoptic area is thought to be required.

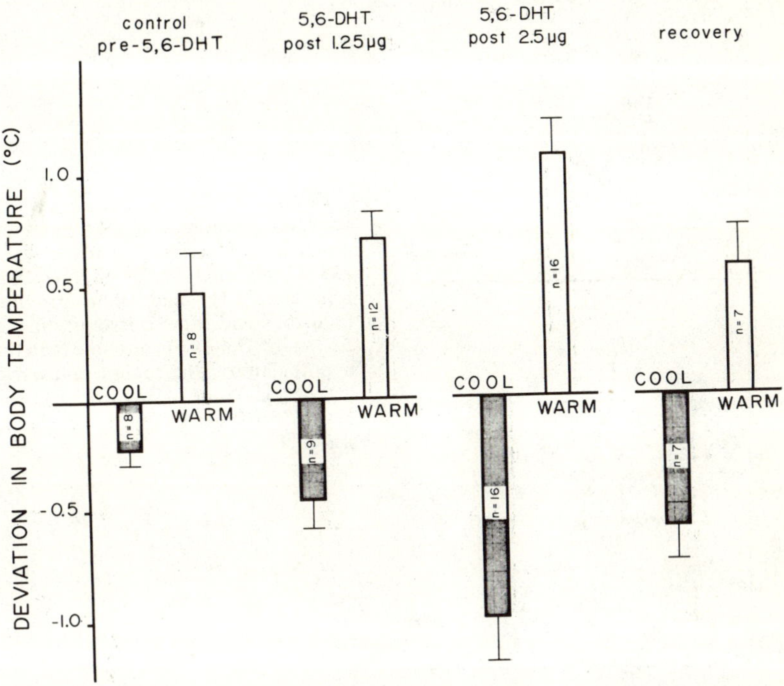

FIGURE 3. Average deviation in body temperature (° C±SEM) in response to a 1-hr exposure to cold (8° C) or warm (35° C) air. The cooling and warming procedures were separated by an interval of from 1 to 24 hr and given 24–48 hr after microinjection of 5,6-DHT into the anterior hypothalamus at the doses indicated. The control responses to cooling and warming were obtained after the vehicle and 5-HT microinjections. The recovery interval was 15–18 days after the last microinjection of 5,6-DHT. N, number of experiments. (From Myers.[25] By permission of *Brain Research*.)

Responses of the Rat

The differential effects of a 5,6-DHT microinjection on the thermoregulatory response of a representative rat are presented in FIGURE 4. In these experiments, the rat had received a microinjection of 2.5 μg of 5,6-DHT into the anterior hypothalamus 24 hr before its environmental temperature was varied. Whereas the animal's temperature rose by nearly 1.0° C during the 1-hr exposure to a temperature of 35° C, its colonic temperature declined by more than 3.0° C during an interval at which the air temperature was lowered to 8° C.

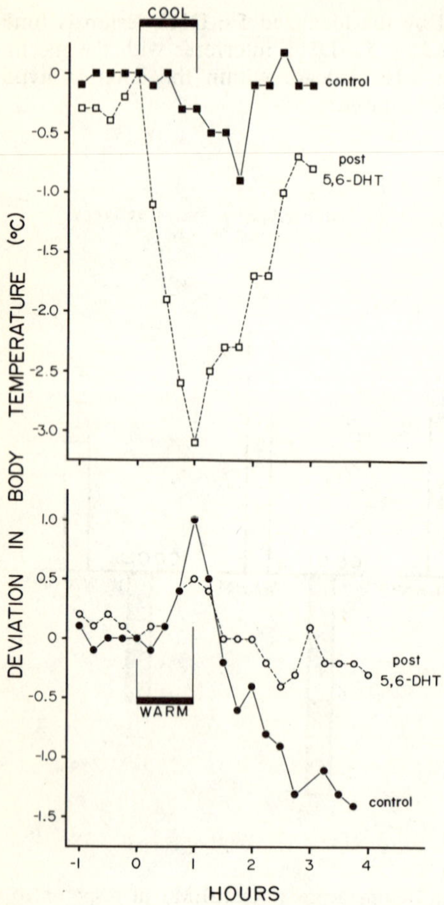

FIGURE 4. Deviation from baseline body temperature (° C) of a rat exposed to cold (8° C, *top*) and warm (35° C, *bottom*) air for an interval of 1 hr after 5-HT and saline injections (control) and after microinjection of 2.5 μg of 5,6-DHT into the anterior hypothalamus. The cooling and warming periods were 4 hr apart. (From Myers.[25] By permission of *Brain Research*.)

Responses of the Monkey

In the rhesus monkey, the effect of this 5-HT neurotoxin on the animal's capacity to thermoregulate depends on two major factors: the dose of 5,6-DHT injected and the anatomic region of the hypothalamus to which 5,6-DHT was applied. A major similarity between the rodent and primate species was the relatively miniscule dose of neurotoxin required to alter the animal's response to a thermal challenge.

Although repeated microinjections of up to 6.0 μg of 5,6-DHT at sites in the dorsal and caudal areas of the hypothalamus usually had little effect, a single low-dose injection of the amine in the rostral hypothalamus caused severe and selective impairment of the animal's ability to maintain its body temperature. An example of this result is presented in FIGURE 5. When 3.0 μg of 5,6-DHT were injected simultaneously at sites in coronal planes AP 14.0 and AP 17.0, as indicated in the histologic insets, the subsequent exposure of the monkey to an ambient temperature of 4° C on the following day caused a precipitous decline in its colonic temperature. This inability to maintain body heat persisted throughout the 3-hr exposure to cold air, which surrounded the monkey's trunk. In contrast, exposure of the primate on the following day to an air temperature of 40° C for 3 hr induced a rise in body temperature of only 0.5° C.

Fatal Hypothermia

Terminal hypothermia was inadvertently and unexpectedly produced in two monkeys when 5,6-DHT was microinjected into the hypothalamus while each

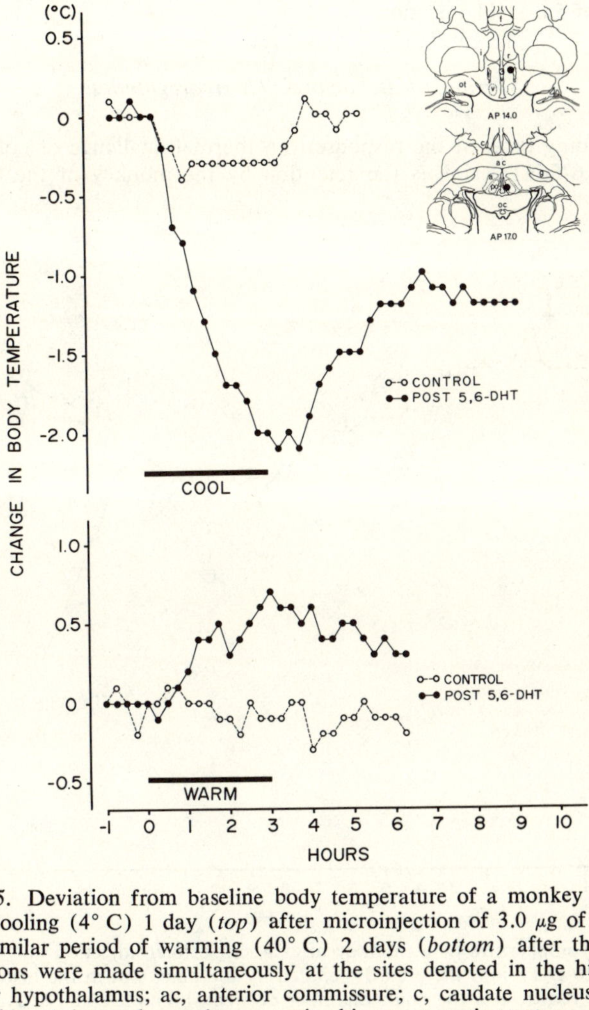

FIGURE 5. Deviation from baseline body temperature of a monkey during a 3-hr period of cooling (4° C) 1 day (*top*) after microinjection of 3.0 μg of 5,6-DHT and during a similar period of warming (40° C) 2 days (*bottom*) after these injections. The injections were made simultaneously at the sites denoted in the histologic inset. a, Anterior hypothalamus; ac, anterior commissure; c, caudate nucleus; f, fornix; g, globus pallidus; i, internal capsule; oc, optic chiasm; ot, optic tract; po, preoptic area; r, reticular nucleus; v, ventromedial nucleus. (From Waller *et al.*[43] By permission of *Neuropharmacology*.)

animal was maintained at its usual laboratory temperature, without any thermal challenge. In one case, a dose of only 2.7 μg of the neurotoxin, infused at two sites adjacent to each other in the dorsal and ventral portions of the anterior hypothalamic, preoptic area at AP 17.0, was sufficient to cause death. As shown in FIGURE 6, the short-lived hyperthermia typically observed after a punctate injection of 5,6-DHT was followed by a slow decline in the animal's body temperature over time. In an attempt to ameliorate this hypothermia, warm air from a heater was blown into the monkey's chair chamber for 3 hr, approximately 9–12 hr after the microinjection. During this interval, the animal's temperature tended to stabilize; however, on discontinuation of the external source of warmth, the monkey's temperature resumed a steep decline. The animal died when its colonic temperature dropped below 33° C, occurring 14–18 hr after 5,6-DHT injection.

Sparing of Behavioral Thermoregulation

An intriguing facet of the response to a thermal challenge of cold or warm air after a 5,6-DHT lesion is the retention by the monkey of the capacity to

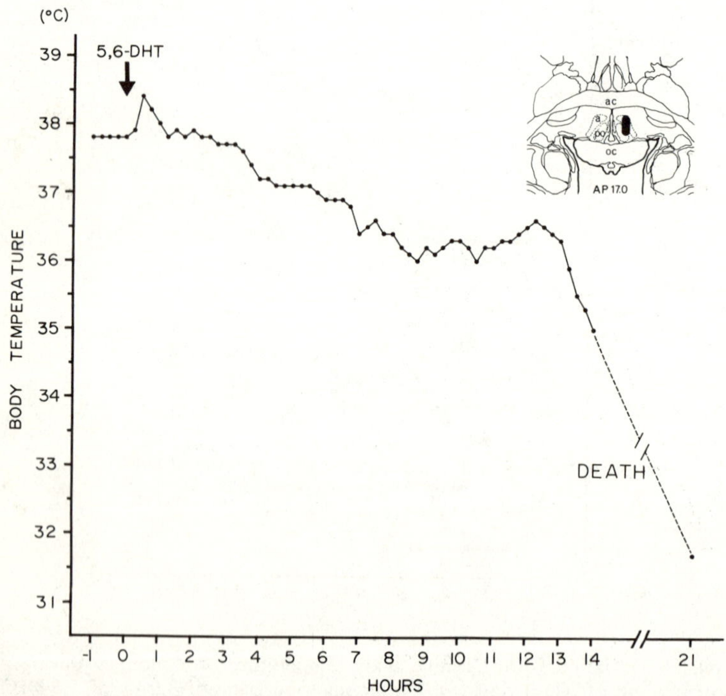

FIGURE 6. Temperature of a monkey after a single microinjection at 0 hr (arrow) of 2.7 μg of 5,6-DHT into the area designated by the black dot in the histologic inset. Warm air was blown into the chair chamber between hours 9 and 12. Anatomic abbreviations are the same as in FIGURE 5. (From Waller *et al.*[43] By permission of *Neuropharmacology*.)

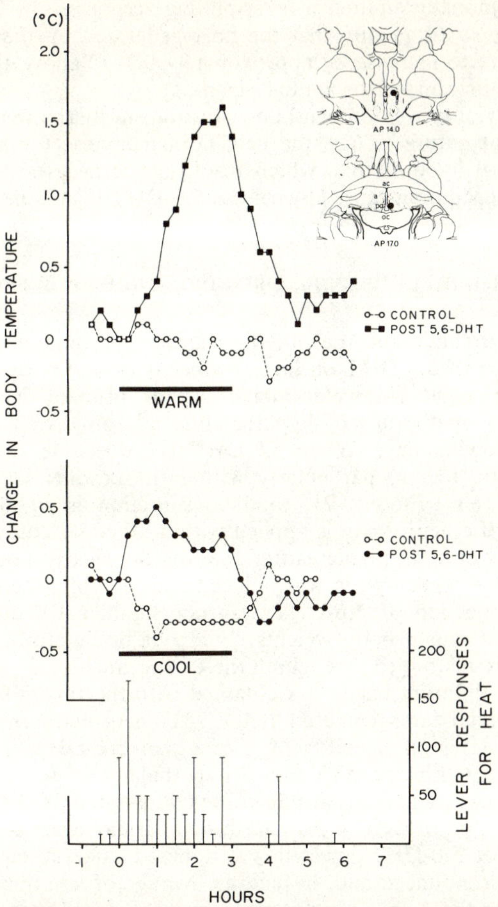

FIGURE 7. Deviation from baseline body temperature of a monkey recorded during a 3-hr period of warming (*top*) 2 days after injection of 6.0 μg of 5,6-DHT and during a similar period of cooling (*bottom*) 1 day after the injection. 5,6-DHT was injected at the sites shown on the inset. Lever responses emitted to turn on a heat lamp on the monkey's trunk for 4 sec are indicated for the cooling experiment. Anatomic abbreviations (except for l, lateral hypothalamus) are the same as in FIGURE 5. (From Waller *et al.*[43] By permission of *Neuropharmacology*.)

thermoregulate behaviorally. When we injected 6.0 μg of 5,6-DHT again 1 month later at the same site in the same monkey described in the experiments illustrated in FIGURE 5, an even more severe thermoregulatory deficit was produced. FIGURE 7 (*top*) shows that a rise in body temperature of more than 1.5° C occurred during the 3-hr interval when the animal's trunk was exposed to warm air. As depicted in FIGURE 7 (*bottom*), however, the monkey pulled a lever with its paw to obtain warmth from a heat lamp during a corresponding interval when cold air was blown into the chair chamber. The lamp, located about 30 cm from the monkey's trunk, was automatically switched on for 4 sec every time the monkey emitted a lever-pulling response. In fact, the monkey worked its lever so frequently that the heat generated by the lamp caused its body temperature to increase by approximately 0.5° C above the baseline, even though cold air surrounded the animal's trunk.

Such a behavioral overcompensation illustrates a thermoregulatory deficit of a rather different nature. Thus, the behavioral response for heat entirely prevented the typical hypothermia, which would occur necessarily because of the physiologic incapacitation caused by central 5,6-DHT injection.

Changes in Feeding, Drinking, and Body Weight

In both the monkey and the rat, the transient yet potent effect of a hypothalamic injection of 5,6-DHT on ingestive behavior and body weight is unmistakably evident. Food and water intakes of the primate decline in most instances, not only on the day of hypothalamic microinjection of 5,6-DHT but also on the following day. When 5,6-DHT is infused in the anterior hypothalamic, preoptic region, particularly within the coronal level of stereotaxic plane AP 17.0 (e.g., Figure 7), food consumption is significantly reduced. However, when the neurotoxin is applied within other stereotaxic planes rostral or caudal to this coronal plane, eating patterns are nearly normal. Moreover, even though water consumption was reduced considerably, the amount of water taken in after injection of 5,6-DHT was not significantly different from the baseline intake. Primate body weights could not be determined daily because the monkeys were restrained throughout these experiments.

A somewhat different result was obtained with the rat, probably because of the more widespread damage caused by the 5-HT neurotoxin in the smaller-sized hypothalamus. After an injection of 1.25 μg of 5,6-DHT in one of several anatomic regions within the rat's rostral hypothalamus, the average amount of food consumed by all animals declined. As illustrated in Figure 8, the hypophagia persisted until usually the fourth or fifth day after the microinjection. When the dose of 5,6-DHT injected was doubled, the subsequent hypophagia was even more pronounced and, in fact, an average of less than 10 g was eaten by the rats during the 3 days after neurotoxin injection. Thereafter, food intake increased slowly but did not begin to approach the normal level ingested until much later. The anatomic region of 5,6-DHT sensitivity is homologous to one of the principal areas within which injections of norepinephrine evoke spontaneous feeding.[24]

The intake of water was likewise affected in an almost identical manner to that of food intake. Figure 9 presents the magnitude of the hypodypsia that was produced by both the 1.25 and 2.5 μg doses of 5,6-DHT injected into the rostral hypothalamus. The volume of water consumed declined from an average of approximately 40 ml/day/animal to about half of that amount after injection of the higher dose of the neurotoxin.

Overall reductions in ingestion of both food and water were reflected by a concomitant decline in the average body weight of the 5,6-DHT-treated rats. After administration of the lower dose of 5,6-DHT, body weight declined by over 30 g but following the higher dose was almost 50 g lower.

With respect to food intake and change in body weight, the effects of central neurochemical lesioning by an injection of 5,6-DHT and 5,7-DHT [or 6-hydroxydopamine (6-OHDA)] can be differentiated on the basis of the genetic constitution of the animal. In fact, the magnitude of a given shift in body

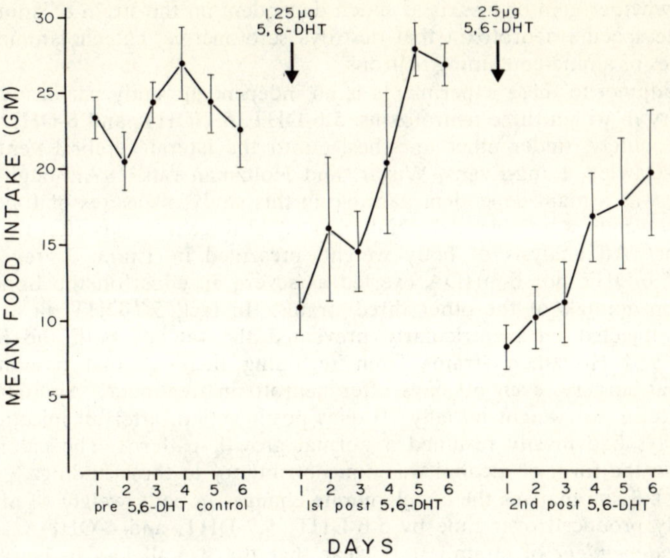

FIGURE 8. Mean daily food intake in grams (±SEM) recorded for eight rats for 6 days after control and 5-HT microinjections and for 6 days after bilateral microinjections of 1.25 and then 2.5 μg of 5,6-DHT into sites located in the rostral hypothalamus. (From Myers.[25] By permission of *Brain Research.*)

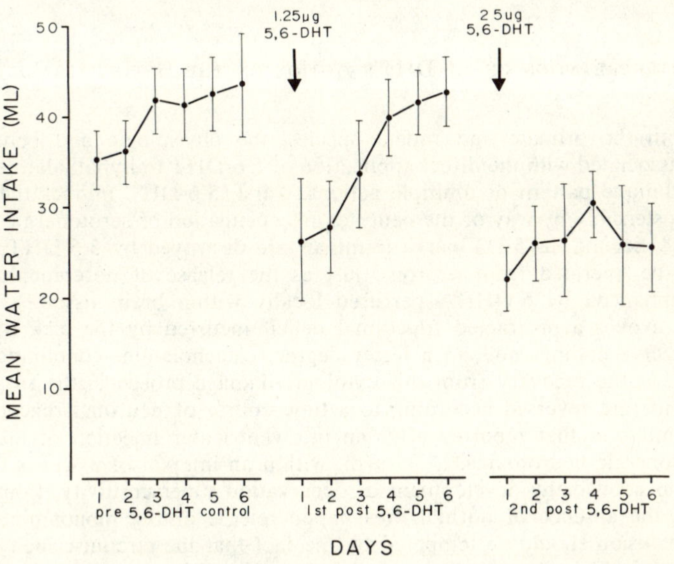

FIGURE 9. Mean daily water intake in grams (±SEM) recorded for eight rats for 6 days after control and 5-HT microinjections and for 6 days after bilateral microinjections of 1.25 and then 2.5 μg of 5,6-DHT into sites located in the rostral hypothalamus. (From Myers.[25] By permission of *Brain Research.*)

weight, whether gain or loss, is as much dependent on the strain of animal used as on the specific neurotoxin that destroys serotonergic, catecholaminergic, or both types of amine-containing neurons.

An adjunct to these experiments is an independent study performed in our laboratory in which three neurotoxins, 5,6-DHT, 5,7-DHT, and 6-OHDA, were injected acutely, under ether anesthesia, into the lateral cerebral ventricle of Sprague-Dawley, Long-Evans, Wistar, and Holtzman rats.[22] Although alcohol drinking was a main dependent variable in this study, measures of food intake and body weight were obtained simultaneously at three intervals after neurotoxin treatment. An analysis of body weight, presented in TABLE 2, reveals that neither 5,6-DHT nor 6-OHDA exerted as severe an effect on the Long-Evans rats as on animals of the other three strains. In fact, 5,7-DHT, in a dose of 100 μg, injected intraventricularly prevented the rats of both the Sprague-Dawley and Holtzman strains from regaining their original baseline body weights at surgery, even 60 days after neurotoxin treatment. Animals of the Wistar strain lost weight initially 10 days postinjection, after all injections, but by 60 days had mostly resumed a normal growth pattern. The caloric level derived in the form of alcohol was noncontributory to these findings.[22]

TABLE 3 summarizes the conglomerate changes in body weight of all rats in this study produced over time by 5,6-DHT, 5,7-DHT, and 6-OHDA as determined *independent* of strain. It is clear that the overall loss in body weight produced by the injections was not vastly different between any treatment conditions, with one exception. Animals of the strains given 5,7-DHT intraventricularly showed the greatest loss in body weight at 10 and 20 days, and only after 60 days had they regained the weight lost in the postinjection period. This result documents the clear-cut depressant effect on caloric intake and, perhaps, on metabolism that is produced by this serotonin neurotoxin.

INTERPRETATION OF 5,6-DHT'S ACTION ON THE HYPOTHALAMUS

In both the primate and rodent species, the physiologic and behavioral changes associated with the direct application of 5,6-DHT to hypothalamic tissue follow a unique pattern of multiple actions. First, 5,6-DHT mimics the effect of 5-HT ostensibly by way of the neurotoxin's occupation of serotonergic receptor sites.[42] Second, as 5-HT nerve terminals are destroyed by 5,6-DHT, 5-HT seems to be liberated from neurons, just as the release of norepinephrine is sharply enhanced by 6-OHDA perfused locally within brain tissue.[21] Third, 5,6-DHT evokes a protracted functional deficit incurred by the lack of serotonergic nerve endings and, to a lesser degree, catecholamine-containing axon terminals; as the recovery from this cytologic damage proceeds, the respective impairments are reversed according to a time course of neuronal regeneration that is similar to that reported after an intraventricular injection of either of two serotonergic neurotoxins.[2, 4] Fourth, within an interval of 3 weeks or less, there is no noteworthy development of denervation supersensitivity, thought to evolve in the absence of normal presynaptic release of the monoamine. The latter conclusion should be tempered by the fact that the circumscribed site of injection of 5-HT in the rat's diencephalon is precisely the same as that at which both 5-HT and 5,6-DHT were applied in tests to elevate the animal's body temperature.

TABLE 2

AVERAGE LOSS OR GAIN IN BODY WEIGHT OF FOUR STRAINS OF RATS AT THREE INTERVALS AFTER A SINGLE ACUTE INJECTION OF 5,6-DHT, 5,7-DHT, 6-OHDA, OR CEREBROSPINAL FLUID SOLUTION IN THE CEREBRAL VENTRICLE *

Strain	Substance Injected	Body Weight (g) Baseline (day of injection)	10 Days Post	Loss or Gain	20 Days Post	Loss or Gain	60 Days Post	Loss or Gain
Sprague-Dawley	5,6-DHT	545.6	525.8	−19.8	539.8	−5.8	563.4	+17.8
	5,7-DHT	553.0	514.2	−38.8	521.0	−32.0	545.4	−7.6
	6-OHDA	557.0	533.2	−23.8	539.8	−17.2	574.3	+17.3
	CSF †	534.6	520.4	−14.2	516.6	−18.0	529.4	−5.2
Long-Evans	5,6-DHT	541.4	532.4	−9.0	548.6	+7.2	589.0	+47.6
	6-OHDA	497.2	515.0	+17.8	526.6	+29.4	546.8	+49.6
	CSF	492.0	508.2	+16.2	510.6	+18.6	527.6	+35.6
Wistar	5,6-DHT	571.7	526.5	−45.2	563.3	−8.4	608.0	+36.3
	5,7-DHT	615.8	587.4	−28.4	602.8	−13.0	635.5	+19.7
	6-OHDA	596.2	560.8	−35.4	570.6	−25.6	613.8	+17.6
	CSF	540.7	525.7	−15.0	530.5	−10.2	576.5	+35.8
Holtzman	5,6-DHT	509.0	465.3	−43.7	485.5	−23.5	530.0	+21.0
	5,7-DHT	479.2	432.7	−46.5	443.5	−35.7	477.8	−1.4
	6-OHDA	495.3	456.0	−39.3	472.8	−22.5	516.8	+21.5
	CSF	504.5	502.5	−2.0	498.0	−6.5	507.0	+2.5

* Based on data from Melchior and Myers.[22]

† Cerebrospinal fluid.

Deficits in Feeding and Water Balance

Because 5,6-DHT microinjected in the anterior and other areas of the rostral hypothalamus interferes with the typical feeding pattern in both the monkey and the rat, the possibility that some norepinephrine-containing fibers are simultaneously damaged by 5,6-DHT could explain this transient anorexia. A 5-HT agonistic action of the neurotoxin is equally plausible, because 5-HT injected in the hypothalamus of the rat actually produces a mild but short-lived anorexia in some cases.[37] Thus, the brief decline in food intake observed in the present experiments could be related in part to the more immediate action of 5,6-DHT in mimicking serotonin's attachment to postsynaptic receptors.[5] Clearly, however, the intensity of the anorexia is not nearly as great as that seen when 6-OHDA is microinjected in homologous areas of the hypothalamus in a correspondingly small volume and range of doses.[20, 28] Since 5,6-DHT

TABLE 3

COMPOSITE AVERAGE LOSS OR GAIN IN BODY WEIGHT OF FOUR STRAINS OF RATS INJECTED INTRAVENTRICULARLY WITH 50 μG OF 5,6-DHT, 200 μG OF 6-OHDA, OR ARTIFICIAL CEREBROSPINAL FLUID AND OF THREE STRAINS (EXCEPT LONG-EVANS) GIVEN 100 μG OF 5,7-DHT SIMILARLY *

		Body Weight (g)					
Substance Injected	Baseline (day of injection)	10 Days Post	Loss or Gain	20 Days Post	Loss or Gain	60 Days Post	Loss or Gain
5,6-DHT	541.0	512.5	−29.4	534.3	−7.6	572.6	+30.7
5,7-DHT	549.3	511.4	−37.9	522.4	−26.9	552.9	+3.6
6-OHDA	536.4	516.3	−20.1	527.5	−8.9	562.9	+26.5
Cerebrospinal fluid	517.9	514.2	−3.7	513.9	−4.0	533.9	+16.0

* Based on data from Melchior and Myers.[22]

injected directly into the hypothalamus reduces the content of dopamine,[3, 34] the findings of hypophagia and hypodypsia in the present experiments are likewise compatible with the view that a portion of the dopaminergic pathway that controls feeding and drinking may also have been affected.

Because water intake declined only in the rat after a 5,6-DHT microinjection, but not significantly in the monkey, the nature of the animal's feeding response may be the determinant of this result. That is, prandial drinking during feeding characteristically accompanies the rat's ingestion of water, at least under laboratory conditions in which dry food pellets are provided. Since the rats reduced their food intake, logically, they should likewise drink less water. On the other hand, water intake in the primate is not so closely dependent on food consumption. Therefore, the 5,6-DHT lesion may exert an effect on water balance secondary to that on food intake.

A conflicting view of the role of 5-HT in food and water intakes has arisen recently in several quarters.[9, 16] According to this idea, a deficiency in 5-HT content in the forebrain or, perhaps, midbrain structures causes a regulatory disturbance that is reflected by enhanced intake of food and water, not by reduced intake, as we report here. For example, if *p*-CPA is injected intraventricularly in the rat, overeating reportedly occurs for a few days, and a slight and transitory parallel increase in body weight accompanies this overeating.[7] In another paper, an intraventricular injection of 5,7-DHT in Sprague-Dawley rats purportedly produces hyperphagia and a sharp rise in body weight to as much as 900 g or more;[33] however, this phenomenon is not consistent and happens, for some unknown reason, in only some animals. As revealed by Saller and Stricker in a footnote to their paper, ". . . many rats given 200 μg of 5,7-DHT after DMI treatment would not maintain elevated rates of body weight gain despite the absence of significant depletions of brain norepinephrine. . . ."[33] Further, Diaz *et al.* report that repeated intraventricular doses of 10 μg of 5,6-DHT cause no change in body weight and only a very slight increase in food intake per day.[11]

Therefore, the concept that 5-HT-containing neurons are involved in the mechanisms that subserve satiety rather than hunger would require a very specific and detailed morphologic identification of a local site of action and/or regions of such 5-HT involvement. This task has not yet been accomplished. In our experiments with Sprague-Dawley rats, 5,6-DHT has been found to cause a pronounced reduction in hypothalamic serotonin levels without prior pretreatment with a central monoamine oxidase inhibitor or desmethylimipramine.[22] Yet neither a concomitant hyperphagia nor an increase in body weight occurs. On the contrary, we find that animals of this and other strains given intraventricular injections of 5,6-DHT or 5,7-DHT are usually differentiated in terms of their subsequent weight gain on the basis of their growth pattern *prior* to the central administration of a given neurotoxin. This finding is demonstrated in FIGURE 10, which portrays the separation in growth curves of Long-Evans animals. The weight gain in each instance is directly related to the rat's earlier growth history.

Thermoregulatory Deficits

Of paramount importance is the precise anatomic region at which 5,6-DHT exerts its neurotoxic effect. When the neurotoxin is injected at sites in other parts of the hypothalamus in either the rat or monkey, deficits in other functions are not noted, although it is possible that they may indeed occur. Such morphologic concordance of these 5,6-DHT-sensitive sites corresponds with the outcome of other experiments dealing with 5-HT and thermoregulation. The application of the indoleamine directly to the rostral hypothalamus evokes heat production and a subsequent rise in temperature.[29] Similarly, when a monkey is exposed to cold air, so that heat production is required, the release of 5-HT from an anterior hypothalamic site of perfusion is markedly enhanced. Since serotonergic nerve terminals are in dense numbers in this self-same region,[14] the totality of these results supports the idea that intact 5-HT nerve terminals form a vital link in the neuronal network of cells that mediate heat conservation and heat production.[24] Without this network, the animal cannot withstand a severe thermal challenge and, as was witnessed in two monkeys, may not survive even at a euthermic ambient temperature.

The fact that the monkey thermoregulates behaviorally quite adequately in the cold after a 5,6-DHT lesion leads to the conclusion that another area of the hypothalamus is responsible for the sensorimotor component of this response. In this connection, an electrolytic lesion of the anterior hypothalamus produces a regulatory impairment to both cold or heat exposure, yet a rat will depress a lever to obtain heat when cold; however, a lesion of the lateral hypothalamus inhibits lever responding but leaves intact the capacity of the animal to thermoregulate physiologically.[35]

In the monkey and rat, the results of our physiologic experiments with warming and cooling offer support for the hypothesis that there is initial damage to both catecholaminergic and serotonergic nerve fibers, which is followed by

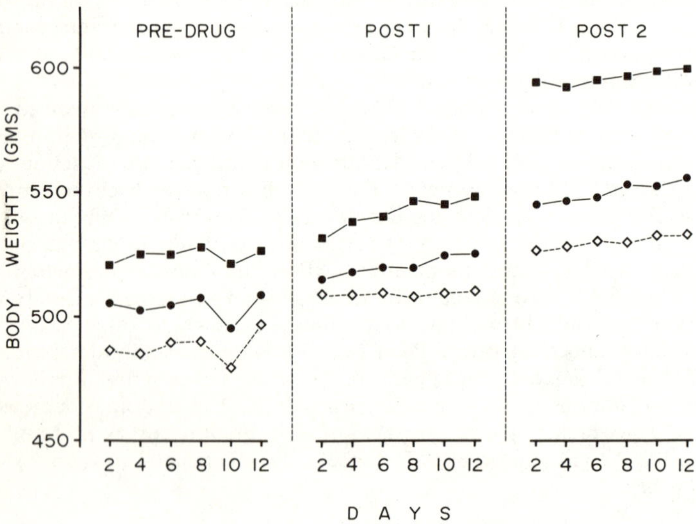

FIGURE 10. Mean body weights for successive 2-day periods during each 12-day ethanol preference test obtained for the 5,6-DHT (■), 5-ion control (◇), and 6-OHDA (●) injected animals of the Long-Evans strain. Ten days elapsed between "pre-drug" and "post 1," whereas 60 days elapsed between "post 1" and "post 2." (From Melchior & Myers.[22] By permission of *Pharmacology, Biochemistry, and Behavior.*)

more selective damage to 5-HT-containing neurons. In the immediate aftermath of a cold or warm challenge, both the heat production and the heat loss mechanisms of these two species are noticeably impaired in the tests performed shortly after the neurotoxin is injected. However, the recovery of the capacity of the animal to regulate its heat-dissipating capacity almost normally suggests that the norepinephrine-release mechanism, thought to underlie the heat loss process,[27] recovers almost totally within a period of 2 weeks or less. In support of Baumgarten *et al.*'s contention,[3] it seems, therefore, that the protracted action of 5,6-DHT is directed more specifically toward serotonergic elements of the heat production pathway in the hypothalamus and that nonspecific damage to the essential catecholamine system in this part of the diencephalon is minimal.

Conclusions

Overall, the destruction of 5-HT nerve terminals and the consequent loss of serotonergic activity in the principal region of the hypothalamus responsible for control of feeding and drinking processes do exert a short-term but, nonetheless, deleterious effect. From this functional impairment, the animal recovers relatively rapidly in most cases. The recovery ensues presumably because noradrenergic neurons are largely spared and because ancillary 5-HT neurons may assume the functionality of serotonergic elements rendered dysfunctional by the neurotoxin. Thus, the patterns of feeding and weight gain resume their normal course.

In a similar fashion, the thermosensitive region of the anterior hypothalamic preoptic area requires an intact 5-HT-noradrenergic neuronal network, a conclusion based on a vast amount of research in the field of thermoregulation.[24] Damage to 5-HT nerve terminals by virtue of locally injected 5,6-DHT disrupts primarily the heat production/heat conservation component of the thermoregulatory system. Thus, 5,6-DHT and 5,7-DHT, or indole analogs that possess similar pharmacologic properties, now open the door to the neuroscientist for obtaining a greater understanding of the role of serotonergic neurons in the control of these vital processes of ingestive behavior and temperature regulation.

Acknowledgments

The author is indebted to his colleagues, Drs. G. E. Martin, C. L. Melchior, and M. B. Waller, for their contributions.

References

1. Barchas, J. & E. Usdin. (Eds.) 1973. Serotonin and Behavior. Academic Press. New York and London.
2. Baumgarten, H. G., L. Lachenmayer, A. Björklund, A. Nobin & E. Rosengren. 1973. Long-term recovery of serotonin concentrations in the rat CNS following 5,6-dihydroxytryptamine. Life Sci. **12:** 357–364.
3. Baumgarten, H. G., A. Björklund, A. Nobin, E. Rosengren & H. G. Schlossberger. 1975. Neurotoxicity of hydroxylated tryptamines: structure-activity relationships. 1. Long-term effects on monoamine content and fluorescence morphology of central monoamine neurons. Acta Physiol. Scand. Suppl. **429:** 7–27.
4. Björklund, A., A. Nobin & U. Stenevi. 1973. Regeneration of central serotonin neurons after axonal degeneration induced by 5,6-dihydroxytryptamine. Brain Res. **50:** 214–220.
5. Blundell, J. E. 1977. Is there a role for serotonin (5-hydroxytryptamine) in feeding? Int. J. Obesity **1:** 15–42.
6. Borbély, A. A., J. P. Huston & P. G. Waser. 1973. Physiological and behavioral effects of parachlorophenylalanine in the rat. Psychopharmacologia **31:** 131–142.
7. Breisch, S. T., F. P. Zemlan & B. G. Hoebel. 1976. Hyperphagia and obesity following serotonin depletion by intraventricular *p*-chlorophenylalanine. Science **192:** 382–385.
8. Brodie, B. B. & P. A. Shore. 1957. A concept for a role of serotonin and norepinephrine as chemical mediators in the brain. Ann. N.Y. Acad. Sci. **66:** 631–642.

9. Coscina, D. V., L. D. Grant, S. Balagura & S. P. Grossman. 1972. Hyperdipsia after serotonin-depleting midbrain lesions. Nature [New Biol.] **235:** 63–64.
10. Clineschmidt, B. V., J. C. McGuffin & A. B. Werner. 1974. Role of monoamines in the anorexigenic actions of fenfluramine, amphetamine and *p*-chloromethamphetamine. Eur. J. Pharmacol. **27:** 313–323.
11. Diaz, J., G. Ellison & D. Masuoka. 1974. Opposed behavioral syndromes in rats with partial and more complete central serotonergic lesions made with 5,6-dihydroxytryptamine. Psychopharmacologia **37:** 67–79.
12. Feldberg, W. & R. D. Myers. 1963. A new concept of temperature regulation by amines in the hypothalamus. Nature (London) **200:** 1325.
13. Funderburk, W. H., J. C. Hazelwood, R. T. Ruckart & J. W. Ward. 1971. Is 5-hydroxytryptamine involved in the mechanism of action of fenfluramine? J. Pharm. Pharmacol. **23:** 468–470.
14. Fuxe, K., T. Hökfelt & U. Ungerstedt. 1968. Localization of indolealkylamines in CNS. *In* Advances in Pharmacology. S. Garattini & P. A. Shore, Eds. Vol. **6:** 235–251. Academic Press. New York and London.
15. Goldman, H. W., D. Lehr & E. Friedman. 1971. Antagonistic effects of alpha and beta-adrenergically coded hypothalamic neurones on consummatory behaviour in the rat. Nature (London) **231:** 453–455.
16. Hoebel, B. G. 1976. Satiety: hypothalamic stimulation, anorectic drugs, and neurochemical substrates. *In* Hunger: Basic Mechanisms and Clinical Implications. D. Novin, W. Wyrwicka & G. Bray, Eds. : 33–50. The Raven Press. New York, N.Y.
17. Jouvet, M. 1972. The role of monoamines and acetylcholine-containing neurons in the regulation of the sleep-waking cycle. Ergeb. Physiol. **64:** 166–307.
18. Joyce, D. & N. Mrosovsky. 1964. Eating, drinking and activity in rats following 5-hydroxytryptophan (5-HTP) administration. Psychopharmacologia **5:** 417–423.
19. McFarlain, R. A. & J. M. Bloom. 1972. The effects of para-chlorophenylalanine on brain serotonin, food intake, and U-maze behavior. Psychopharmacologia **27:** 85–92.
20. Martin, G. E. & R. D. Myers. 1974. Action of intrahypothalamic 6-hydroxydopamine on motivated responding for food and water in the rat. Pharmacol. Biochem. Behav. **2:** 393–399.
21. Martin, G. E., R. D. Myers & D. C. Newberg. 1976. Catecholamine release by intracerebral perfusion of 6-hydroxydopamine and desipramine. Eur. J. Pharmacol. **36:** 299–311.
22. Melchior, C. L. & R. D. Myers. 1976. Genetic differences in ethanol drinking of the rat following injection of 6-OHDA, 5,6-DHT or 5,7-DHT into the cerebral ventricles. Pharmacol. Biochem. Behav. **5:** 63–72.
23. Myers, R. D. 1971. Methods for chemical stimulation of the brain. *In* Methods in Psychobiology. R. D. Myers, Ed. Vol. **I:** 247–280. Academic Press, Ltd. London, England.
24. Myers, R. D. 1974. Handbook of Drug and Chemical Stimulation of the Brain. Van Nostrand Reinhold. New York, N.Y.
25. Myers, R. D. 1975. Impairment of thermoregulation, food and water intakes in the rat after hypothalamic injections of 5,6-dihydroxytryptamine. Brain Res. **94:** 491–506.
26. Myers, R. D. & D. B. Beleslin. 1971. Changes in serotonin release in hypothalamus during cooling or warming of the monkey. Amer. J. Physiol. **220:** 1746–1754.
27. Myers, R. D. & C. Chinn. 1973. Evoked release of hypothalamic norepinephrine during thermoregulation in the cat. Amer. J. Physiol. **224:** 230–236.
28. Myers, R. D. & G. E. Martin. 1973. 6-OHDA lesions of the hypothalamus: interaction of aphagia, food palatability, set-point for weight regulation, and recovery of feeding. Pharmacol. Biochem. Behav. **1:** 329–345.

29. MYERS, R. D. & T. L. YAKSH. 1969. Control of body temperature in the unanaesthetized monkey by cholinergic and aminergic systems in the hypothalamus. J. Physiol. (London) **202:** 483–500.

30. MYERS, R. D., M. TYTELL, A. KAWA & T. RUDY. 1971. Micro-injection of ^{3}H-acetylcholine, ^{14}C-serotonin and ^{3}H-norepinephrine into the hypothalamus of the rat: diffusion into tissue and ventricles. Physiol. Behav. **7:** 743–751.

31. PANKSEPP, J. & D. M. NANCE. 1974. Effects of para-chlorophenylalanine on food intake in rats. Physiol. Psychol. **2:** 360–364.

32. PÉREZ-CRUET, J., A. TAGLIAMONTE, P. TAGLIAMONTE & G. L. GESSA. 1972. Changes in brain serotonin metabolism associated with fasting and satiation in rats. Life Sci. **11:** 31–39.

33. SALLER, C. F. & E. M. STRICKER. 1976. Hyperphagia and increased growth in rats after intraventricular injection of 5,7-dihydroxytryptamine. Science **192:** 385–387.

34. SANER, A., L. PIERI, J. MORAN, M. DA PRADA & A. PLETSCHER. 1974. Decrease of dopamine and 5-hydroxytryptamine after intracerebral application of 5,6-dihydroxytryptamine. Brain Res. **76:** 109–117.

35. SATINOFF, E. & S. Y. Y. SHAN. 1971. Loss of behavioral thermoregulation after lateral hypothalamic lesions in rats. J. Comp. Physiol. Psychol. **77:** 302–312.

36. SHARPE, L. G. & R. D. MYERS. 1969. Feeding and drinking following stimulation of the diencephalon of the monkey with amines and other substances. Exp. Brain Res. **8:** 295–310.

37. SINGER, G., I. SANGHVI & S. GERSHON. 1972. Exploration of certain behavioral patterns induced by psychoactive agents in the rat. Commun. Behav. Biol. **6:** 307–314.

38. SUGRUE, M. F., I. GOODLET & I. MCINDEWAR. 1975. Failure of depletion of rat brain 5-hydroxytryptamine to alter fenfluramine-induced anorexia. J. Pharm. Pharmacol. **27:** 950–953.

39. VAN WOERT, M. H., L. M. AMBANI & R. J. LEVINE. 1972. Clinical effects of para-chlorophenylalanine in Parkinson's disease. Disease Nervous System **33:** 777–780.

40. VEALE, W. L. 1968. Effects of brain monoamine depletion in the rat. Unpublished Masters Thesis. Purdue University. Lafayette, Ind.

41. VEALE, W. L. & R. D. MYERS. 1970. Decrease in ethanol intake in rats following administration of *p*-chlorophenylalanine. Neuropharmacology **9:** 317–326.

42. WALLER, M. B. & R. D. MYERS. 1974. Similar actions of intra-hypothalamic 5,6-DHT and 5-HT in evoking hyperthermia in the monkey. Res. Commun. Chem. Pathol. Pharmacol. **9:** 421–429.

43. WALLER, M. B., R. D. MYERS & G. E. MARTIN. 1976. Thermoregulatory deficits in the monkey produced by 5,6-dihydroxytryptamine injected into the hypothalamus. Neuropharmacology **15:** 61–68.

44. WAPNIR, R. A., R. L. HAWKINS, J. H. STEVENSON & S. P. BESSMAN. 1970. Effects of oral administraiton of *p*-chlorophenylalanine to experimental animals. Biochem. Med. **3:** 397–403.

5,6-DIHYDROXYTRYPTAMINE AS A TOOL FOR STUDYING SLEEP MECHANISMS AND INTERACTIONS BETWEEN MONOAMINERGIC SYSTEMS *

J. F. Pujol, P. Keane, P. Bobillier, B. Renaud, and M. Jouvet

Department of Experimental Medicine
Claude Bernard University
69373 Lyon Cedex 2, France

The role of serotonin [5-hydroxytryptamine (5-HT)] in the mechanisms of sleep in the cat is supported by several observations. Thus, total insomnia follows inhibition of 5-HT biosynthesis with *p*-chlorophenylalanine or destruction of the raphe system (see references in Reference 14). However, in the latter experiment, the lesion may destroy raphe neurons other than those that contain 5-HT. The possible selective destruction of either 5-HT perikarya or terminals with 5,6-dihydroxytryptamine (5,6-DHT) was therefore an interesting method to explore 5-HT mechanisms responsible for sleep in the cat.[10] Since preliminary data had also shown that some interactions could exist between 5-HT neurons and catecholaminergic neurons,[13] 5,6-DHT was also used to study more intimately this possible intermonoaminergic regulation in the rat.

5,6-Dihydroxytryptamine and Sleep in the Cat

These experiments were performed in 29 cats that were continuously recorded with chronically implanted electrodes for 1 week to obtain control data concerning the relative amounts of waking and sleep time. The animals were then subdivided into four groups according to the experimental procedure: group 1, direct stereotaxic injection of the solvent (Ringer plus 0.1 mg/ml of ascorbic acid) into the nuclei raphe dorsalis and centralis superior under nembutal anesthesia in 13 different locations with each location receiving 2 μl (1 μl/min) (Figure 1); group 2, 13 μg of 5,6-DHT (Regis Chemical Co.) dissolved into 26 μl of the solvent and injected as above; group 3, 150 μl of the solvent (Ringer and 0.1 mg/ml of ascorbic acid) injected in each lateral ventricle through a chronic cannula; group 4, 1 mg of 5,6-DHT diluted into 200 μl of the solvent injected in both lateral ventricles.

After 2 weeks of continuous polygraphic recording, the cats were sacrificed. The brain stems of groups 1 and 2 were examined histologically, while other structures were analyzed biochemically for tryptophan, 5-HT, 5-hydroxyindoleacetic acid (5-HIAA), and noradrenaline.[9] Dopamine was also measured only in the caudate nuclei. The results of these experiments can be summarized briefly as follows.

* Supported by Institut National de la Santé et de la Recherche Médicale (INSERM) Grant 71–5–1408 (U52), Centre National de la Recherche Scientifique (CNRS) Grant L.A. 162, and Direction de la Recherche et des Moyens d'Essais (DRME) Grant 75–024.

0077–8923/78/0305–0576 $01.75/0

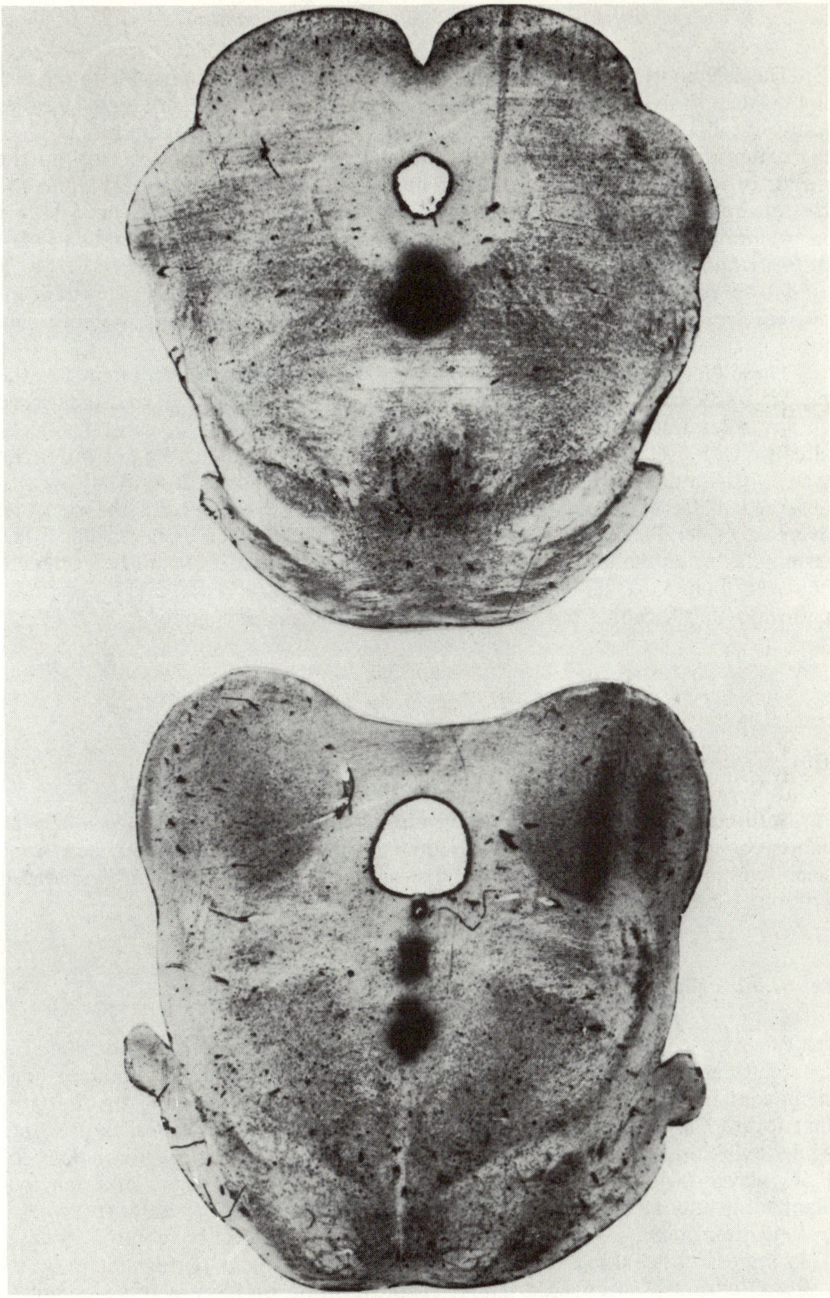

FIGURE 1. Topography of the area perfused with 5,6-DHT. A solution of 26 μl of Evans blue has been injected stereotaxically according to the same method and coordinates as the 5,6-DHT solution, and the animal was sacrificed 1 hr later. The nuclei raphe dorsalis and centralis superior were partially injected.

Effects of 5,6-DHT on Brain Monoamines

The selectivity of 5,6-DHT for 5-HT neurons is quite remarkable in the cat, in that there was no alteration of brain tyrosine or noradrenaline (NA) in either group in all brain structures analyzed and no alteration of dopamine (DA) in the caudate nucleus. However, the alterations of 5-HT depend mostly on the means of introduction of 5,6-DHT. Thus, direct injection of 5,6-DHT into the rostral raphe does not alter the concentration of tryptophan, 5-HT, or 5-HIAA in any brain structure, except the cerebellum, where significant decreases (47%) of 5-HT and 5-HIAA occur.

However, after intraventrivular injection of 1 mg of 5,6-DHT, significant decreases of both 5-HT and 5-HIAA were observed in all nervous system structures, except the spinal cord (see TABLE 1).

These biochemical results are somewhat different from those obtained in the rat:[6, 7] in the cat, we have not observed any decrease of 5-HT in the spinal cord or any alteration in endogenous DA levels. The absence of any significant alteration of 5-HT in the forebrain after direct injection in the rostral raphe system is another remarkable difference: with a similar dose, Daly *et al.*[7] observed a decrease of cerebral 5-HT in the rat. We interpret the selective decrease of cerebellar 5-HT as possible capture of 5,6-DHT by cerebellar 5-HT terminals during the stereotaxic lowering of the cannula. It cannot be excluded, however, that 5,6-DHT may have destroyed specifically the 5-HT cerebellar pathways that ascend from the nuclei raphe centralis and pontis.[2]

Effects of 5,6-HT on Sleep (FIGURES 2 & 3)

Direct Injection into the Rostral Raphe

Both control (Ringer injection, group 1) and experimental cats (group 2) exhibited significant increases of paradoxical sleep (PS) and stage-2 slow-wave sleep within 24 hr. Normal amounts of sleep were subsequently observed (FIGURES 2 & 3).

Intraventricular Injection

Whereas control injections of Ringer (group 3) had no immediate or secondary effect on sleep, administration of 5,6-DHT (group 4) produced both significant immediate and secondary alterations of sleep. During the 2–10 hr that follow the injection, the cat, after a short period of polypnea, is prostrate while exhibiting continuous fast cortical activity. Permanent discharges of PGO waves (which can be recorded also after reserpine or *p*-chlorophenylalanine injections) occurred in only 50% of the cats. During the next 2 weeks, waking time increased by 80% (compared with preinjection control values). This increase of waking time was due to a striking and permanent decrease (50%) of stage-2 sleep ($p < 0.05$). The evolution of PS was more complex, with an initial total disappearance, followed by a return to preinjection levels after 4–5 days, then by a 50% decrease up to the 15th day ($p < 0.01$). There was a significant correlation between the fall in stage-2, or paradoxical sleep and

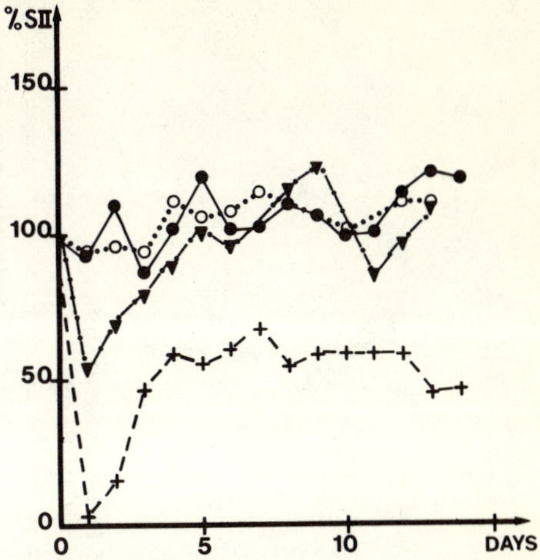

FIGURE 2. Evolution of the amount of stage-2 sleep after injection of 5,6-DHT. Stage 2 values are expressed as percentages of control values for each day of recording. Abscissa, days after injection; ○, injection of Ringer solution into the ventricles; ●, injection of Ringer solution into the raphe system; ▼, injection of 5,6-DHT into the raphe system. +, intraventricular injection of 5,6-DHT.

the decrease of 5-HT and 5-HIAA at the cortical level ($r = 0.77$; $p < 0.001$). The demonstration of such a correlation creates several problems.

Indeed, destruction of 5-HT terminals after intraventricular injection of 5,6-DHT is diffuse, because it occurs mostly in the tele-diencephalon and to a lesser extent in the brain stem. Thus, it is likely that the terminals from several raphe nuclei (the rostral group and also from the raphe magnus and pontis) may have been subjected to the toxic effect of 5,6-DHT. It is also likely that this toxic effect was partial and far less drastic than the secondary degeneration that follows total destruction of raphe nuclei.[14] Thus, it is difficult to draw any conclusions concerning the localization of the 5-HT terminals that are most important in relation to sleep mechanisms. However, the significant correlation between the decrease of stage-2 sleep and the alteration of telencephalic 5-HT suggests that the most rostral parts of the raphe (raphe dorsalis and centralis superior), which send their efferents to the cortex, might be involved in the

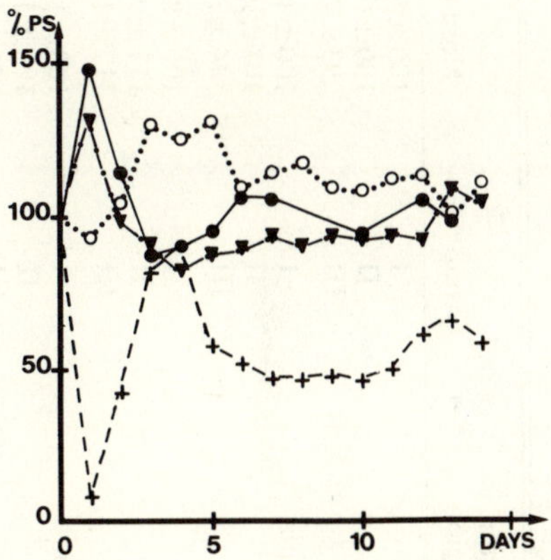

FIGURE 3. Evolution of the amount of paradoxical sleep after administration of 5,6-DHT. Symbols are the same as in FIGURE 2.

TABLE 1

EFFECTS OF 5,6-DHT ON ENDOGENOUS NORADRENALINE, DOPAMINE, 5-HYDROXYTRYPTAMINE, AND 5-HYDROXYINDOLEACETIC ACID IN DIFFERENT PARTS OF THE CAT BRAIN

Brain Region	Group *	Noradrenaline	Dopamine	5-Hydroxytryptamine	5-Hydroxyindoleacetic Acid
Cortex	I	100±5(0.523)		100±6(0.377)	100±4(0.286)
	II	100±6		85±5	54±5 †
	III	93±7		60±11 †	31±8 ‡
Cerebellum	I	100±12(0.360)		100±10(0.168)	100±16(0.111)
	II	102±9		53±6 ‡	18±0 §
	III	94±7		30±6 §	
Caudate	I	100±8(0.426)	100±7(9.369)	100±18(0.597)	100±12(0.579)
	II	117±9	127±12	96±9	74±7
	III	117±10	102±8	49±5 †	49±4 ‡
Thalamus	I	100±15(0.443)		100±12(0.559)	100±11(0.873)
	II	106±5		97±16	73±11
	III	106±4		56±5 †	53±2 §

Mesencephalon	I	100±10(0.999)	100±3(1.202)	100±5(1.336)
	II			
	III	98±7	67±3 †	67±6 ‡
Pons and medulla	I	100±7(0.864)	100±5(0.672)	100±2(0.781)
	II			
	III	98±4	89±6	59±5 ‡
Spinal cord	I	100±13(0.246)	100±14(0.366)	100±7(0.276)
	II	109±6	90±4	82±7
		110±4	74±9	70±8

* Group-I cats were controls and received no injection. In group II, 13 μg of 5,6-DHT were injected in the rostral part of the raphe system. In group III, 1 mg of 5,6-DHT was injected in the lateral ventricles. In groups II and III, the animals were killed 15 days after the injection. The results are expressed as percentages of the mean values obtained in the control group ± SEM. The numbers in parentheses are the absolute values (micrograms per gram) obtained in the control group. No monoamine determinations were performed in the pons and mesencephalon of group II due to the histologic control of the injection. The results of the cats injected intracerebrally or intraventricularly with control solution (Ringer and ascorbic acid) are not significantly different from those of group I for both catecholamines and indoleamines. In all groups, there was no significant alteration of either tyrosine or tryptophan in the same brain structure.

† $p < 0.05$.
‡ $p < 0.01$.
§ $p < 0.001$.

mechanisms that control stage-2 sleep. This hypothesis is in accord with the results of more recent studies that indicate that specific destruction of the nucleus raphe dorsalis is followed mainly by a specific decrease of stage-2 sleep.

Thus, our study has demonstrated that the selective but diffuse destruction of 5-HT terminals leads to a significant increase of waking time. This insomnia is mainly a consequence of decreases of both stage-2 and paradoxical sleep. However, it may also be due to a parallel increase of waking mechanisms. The fact that cortical arousal (low-voltage fast activity) may depend on noradrenergic pathways that ascend from the locus ceruleus [12, 17] has prompted us to study the mechanism by which 5-HT neurons may control these pontine catecholaminergic neurons.

5,6-Dihydroxytryptamine as a Tool for Demonstrating Serotonergic Control of the Locus Ceruleus

Direct evidence for a serotonergic control of the dorsal noradrenergic system was first suggested by experiments performed in the cat: electrolytic destruction of the mesencephalic part of the raphe system, including the nuclei raphe dorsalis and centralis, was found to induce an increase in NA turnover in the cortex and cerebellum, suggesting activation of utilization of NA in the terminals of the dorsal noradrenergic system.[24] Similar lesions, performed in the rat, were found to produce a significant increase of 3-methoxy-4-hydroxyphenylglycol sulfate (MOPEG-SO_4) in the cortex and hippocampus, indicating a greater metabolism of NA in these structures. This increase reached a maximum at about 4 days after lesioning.[15] Recently, highly significant increases of tyrosine hydroxylase (TH) activity were observed specifically in the locus ceruleus (LC) of the rat after lesioning of the nucleus raphe dorsalis, centralis, or pontis. The time course of this alteration exhibited a slow rise of activity, with a maximum at about 4 days and a progressive return to control values between 8 and 15 days after coagulation. Moreover, the relative increases in enzymatic activity were found to be different after destruction of each of these nuclei, suggesting a functional independence of the interaction between each of these nuclei and the LC.[16, 23]

These results suggested strongly that LC activity may be controlled by neurons of the raphe system. However, considering the anatomic heterogeneity of the structures affected by such lesions, it is difficult to discuss their functional effects in terms of serotonergic mechanisms only. To obtain further evidence for a direct relationship between specific inactivation of 5-HT cells and alterations observed in NA metabolism, 5,6-DHT was used as a tool for producing global or local degeneration of 5-HT terminals.

Intracisternal Injection of 5,6-DHT

Intracisternal injections of 5,6-DHT were performed to induce a global inactivation of 5-HT terminals.[25] As shown in Figure 4, a highly significant increase of TH was observed in the LC. This increase became significant at 1 day, reached its maximum between 4 (97%; $p < 0.001$) and 6 days (92%; $p < 0.001$), and returned to control values between 8 and 15 days.

In the frontal cortex, as in the LC, TH activity decreased to a minimum at

6 days, rose again to a second peak of activity at 8 days (56%; $p < 0.05$), and remained elevated at 15 days (40%; $p < 0.05$). No significant increase in TH activity was found in A9 and A10 dopaminergic groups,[25] suggesting that changes in cortical TH activity could occur in terminals from the neurons of the LC.

Recent kinetic studies performed at 4 days in the LC revealed significant increases in maximum velocities determined with tyrosine and 6,7-dimethyl-5,6,7,8-tetrahydropterine ($DMPH_4$). The affinities of TH for substrate and cosubstrate were not altered (see TABLE 2). These results suggest that the

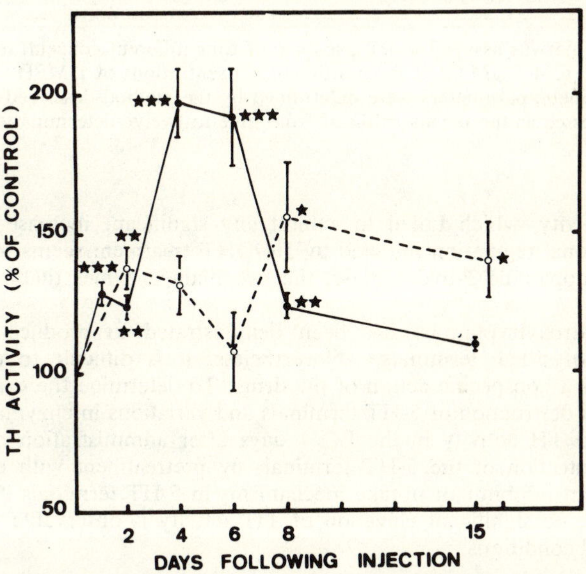

FIGURE 4. Effect of intracisternally injected 5,6-DHT on TH activity in the locus ceruleus (●) and frontal cortex (○). Tyrosine hydroxylase was assayed by a modification of the method of Nagatsu *et al.*,[18] with a tyrosine concentration of 20 μM and a $DMPH_4$ concentration of 1.14 mM. Each point represents the mean ± SEM of from four to eight experimental determinations. The absolute TH values in sham-injected controls were: locus ceruleus, 93.5 pmol of Dopa/hr/structure (±7.5 SEM); frontal cortex, 24.7 pmol of Dopa/hr/mg of protein (±5.9 SEM). ★, $p < 0.05$; ★★, $p < 0.01$; ★★★, $p < 0.001$.

elevated TH activity that occurs in the LC after 5,6-DHT administration resulted from an increased number of enzymatic sites, possibly due to an increase in the synthesis of the enzymatic protein. Thus, the two peaks of activity observed in the frontal cortex could result from different forms of the enzyme[11] transported by separate components of the axoplasmic flow.[9]

A significant alteration of dopamine β-hydroxylase activity can be observed simultaneously to increase TH activity (see FIGURE 5), but there is no observable significant elevation of monoamine oxidase (MAO) activity (see FIGURES 5 & 6). Similar results were obtained recently by our group for Dopa decar-

TABLE 2

KINETIC PROPERTIES OF TYROSINE HYDROXYLASE IN THE LOCUS CERULEUS OF RATS 4 DAYS AFTER INTRACISTERNAL TREATMENT WITH 75 μg OF 5,6-DHT

	Control	5,6-DHT
Tyrosine		
K_m ($M \times 10^{-4}$)	0.78±0.01	0.76±0.01
V_{max} (picomoles of Dopa/hour/structure)	2140±94	3057±171 †
$DMPH_4$		
K_m ($M \times 10^{-4}$)	3.27±0.9	2.6±0.4
V_{max} (picomoles of Dopa/hour/structure)	1635±163	2586±134 †

* The enzyme was assayed in the presence of four different concentrations of tyrosine (from 20 to 400 μM) and three different concentrations of $DMPH_4$ (from 0.1 to 1.0 mM). Kinetic parameters were determined by the method described by Cleland [4] and are expressed as the means ±SD of from nine to twelve determinations.

† $p < 0.002$.

boxylase activity, which failed to exhibit any significant increase in activity. Thus, functional regulation initiated by 5,6-DHT treatment seems to be limited to TH and dopamine β-hydroxylase, the two main enzymes that regulate NA synthesis.

5,6-Dihydroxytryptamine has been demonstrated to produce a selective destruction of 5-HT terminals.[1] Nevertheless, it is difficult to exclude the possibility of a nonspecific action of the drug. To determine the exact relationship between destruction of 5-HT terminals and variations in enzymatic activity, we examined TH activity in the LC 4 days after administration of 5,6-DHT but after protection of the 5-HT terminals by pretreatment with chlorimipramine, a potent inhibitor of uptake mechanisms in 5-HT terminals.[26] As shown in FIGURE 7, no significant elevation of TH activity is observable under these experimental conditions.

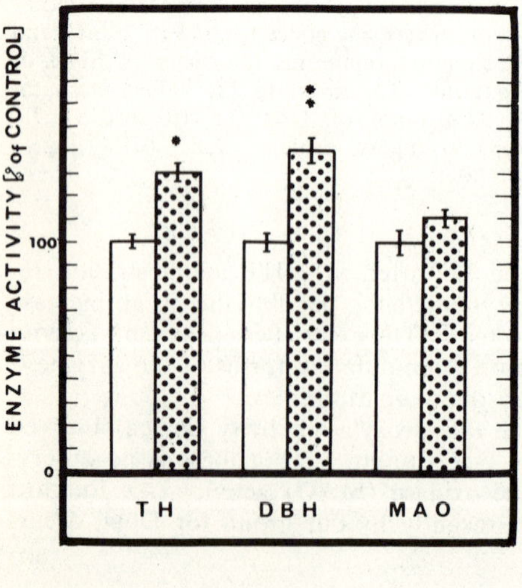

FIGURE 5. Activity of various enzymes in the locus ceruleus 8 days after intracisternal injection of 75 μg of 5,6-DHT. Tyrosine hydroxylase was determined as in FIGURE 4. Dopamine β-hydroxylase activity was determined by a modification of the technique of Nagatsu *et al.*[10] [absolute value of sham-injected animals was 1.6 nmol of *p*-hydroxybenzaldehyde formed/hr/structure ±0.09 (SEM)]. Monoamine oxidase activity was measured according to the technique of Parvez and Parvez [21] with 5-HT used as the substrate [absolute value of sham-injected controls was 75 pmol of deaminated 5-HT/hr/structure ±5.8 (SEM)]. Each point represents the mean ± SEM of from seven to eight determinations. *, $p < 0.05$; **, $p < 0.001$.

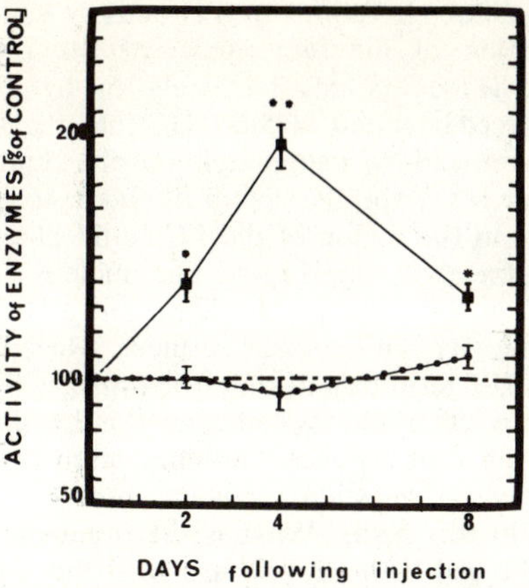

FIGURE 6. Comparison of the time course of the effects of intracisternal injection of 75 μg of 5,6-DHT on TH and MAO (●) activities in the locus ceruleus. The enzymes were analyzed as described in the legends to FIGURES 4 and 5. Each point represents the mean ± SEM of from four to eight determinations. *, p <0.05; **, p <0.001.

Local Injection of 5,6-DHT

Bilateral injection of 5,6-DHT directly into the LC also increased TH activity to an extent similar to that observed after intracisternal administration (see FIGURE 8). This elevation did not appear after chlorimipramine pretreatment.

In summary, destruction of 5-HT-containing nerve terminals by 5,6-DHT was found to induce a gradual increase of TH activity in the LC, which reached a maximum between 4 and 6 days, with a total duration of 10 days. This increase does not result from a change in enzymatic affinity for substrate or cosubstrate but, rather, from an elevation of the number of active sites. Dopamine β-hydroxylase activity is also altered, but Dopa decarboxylase (DDC) and MAO activities are unchanged. Increased TH activity is also observed at the level of terminals in the frontal cortex. Similar results are obtained by intracisternal administration of 5,6-DHT or by its direct bilateral injection into the LC. When 5-HT terminals are protected by chlorimipramine, 5,6-DHT injection does not provide any significant change in enzymatic activity.

The use of 5,6-DHT as a tool for producing selective inactivation of 5-HT terminals has thus provided much evidence for direct control of the locus

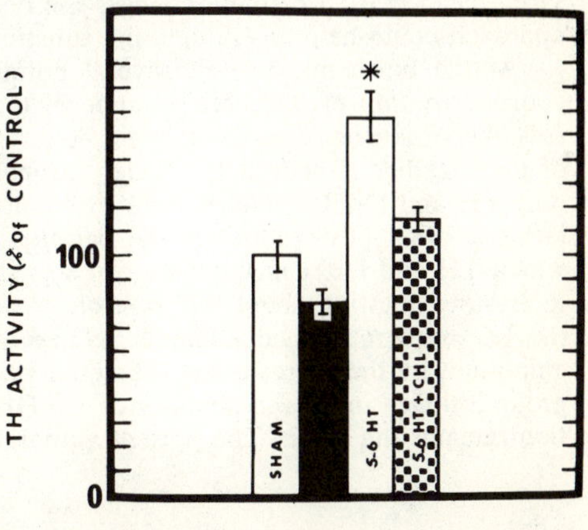

FIGURE 7. Inhibition by chlorimipramine of the increase of TH activity in the locus ceruleus by 5,6-DHT treatment. Chlorimipramine (20 mg/kg) was injected intraperitoneally 1 hr before intracisternal injection of 50 μg of 5,6-DHT. Each point represents the mean ± SEM of eight determinations. *, p <0.02.

ceruleus by serotonergic innervation. Similar alterations of TH activity can be observed in this nucleus after coagulation of different nuclei of the raphe system,[16, 23] by global inactivation of 5-HT-containing terminals, or by local destruction of these nerve endings by direct injection of 5,6-DHT into the LC. The fact that chlorimipramine pretreatment leads to a suppression of the changes induced by 5,6-DHT is a good argument for a specific action of this cytotoxic drug; it seems likely that 5-HT terminals in the region of the LC could play an important role in the regulation of the metabolic activity of this nucleus.

Biochemical studies have demonstrated the presence of significant amounts of 5-HT[20] and tryptophan hydroxylase activity[3] in the locus ceruleus. Recently, anatomic evidence for the presence of 5-HT terminals in this structure has been reported.[8, 22] Direct connections between the raphe system and the LC have also been demonstrated.[2] Thus, the existence of serotonergic innervation of the LC is now demonstrated. However, several questions remain unsolved and could help to orientate future research in this field: What is the organization of these nerve endings with respect to the cell body group located in the raphe

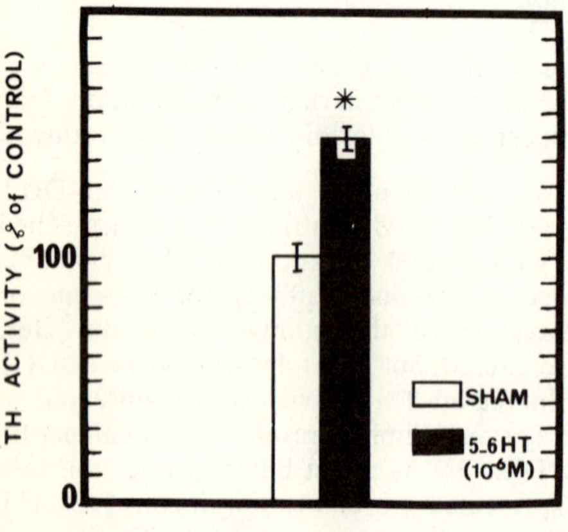

FIGURE 8. Increase in TH activity in the locus ceruleus 4 days after bilateral injection of 5,6-DHT (10^{-12} mol in 1 μl) in this structure. Each point represents the mean ± SEM of eight determinations. *, $p < 0.01$.

system? Does serotonergic control of the LC affect globally or selectively the CA-containing cells of the nucleus? The fact that TH activity, for example, is altered in this structure after lesioning of different raphe nuclei does not necessarily mean that all of these changes are occurring in the same CA cells. This approach could help to evaluate the functional heterogeneity of the LC.

At the biochemical level, several points should be mentioned. The time course variation of TH activity appears to result from a slow, tonic process. It is likely that such regulation could be due to a tonic change in the metabolism of the enzyme. Kinetic data strongly support this hypothesis, and the fact that only TH and DBH activities seem to be altered in this model indicates a similarity to other observations in the peripheral sympathetic nervous system,[27] in which TH and DBH metabolism can also be controlled transsynaptically. One interesting question about this possible transsynaptic control of TH activity in the LC by serotonergic afferents relates to the duration of the effect. The amplitude of the increase of TH activity varies, depending on the cell body group lesioned or on the quantity of 5,6-DHT injected, but its pattern of duration remains the same. The present authors feel that these similar time courses are a function of enzyme turnover rate. With this interpretation, 5-HT deafferentation could induce a change in the metabolism of the enzyme that involves modification of the concentration of the enzymatic protein. The duration of the resultant enzymatic activity alterations could depend principally on the rate of utilization of the enzyme.[5] These mechanisms might induce tonic changes in enzymatic activity in the locus ceruleus, which could also be observed at the level of nerve terminals after a time lag during which the enzyme is transported from the cell bodies.[25] This control could be exercised at the level of different enzymes (at least TH and DBH) that have different turnover times, which would provide a hypothetical basis for a "molecular clock," which, in turn, could control changes in metabolism of the transmitter for different durations.

Considering the patterns of increase of TH activity observed in the LC and in the frontal cortex after intracisternal injection of 5,6-DHT, it seems likely that more than one regulation process occurs under these conditions. In the absence of any direct experimental evidence concerning modification of LC cell firing rate, it is difficult to determine whether changes in cellular electrical activity are a prerequisite for the metabolic alterations or whether 5-HT is able to alter TH activity without changing neuronal firing. Two different kinds of 5-HT terminals have been found in the LC. Some terminals have true synaptic contacts with dendrites of the LC cells, but the majority fail to make true synaptic connections.[8] Thus, it is possible that these different nerve ending types control different aspects of the functional activity of the CA cells localized in the LC. It is also possible that 5-HT secreted by terminals that possess no specific synaptic contacts exercises diffuse neurohumoral control over CA cells, while more specific serotonergic synapses participate in the control of neuronal firing.

Further investigations must be undertaken to demonstrate the relative importance of these two kinds of regulatory processes in sleep mechanisms. It should be emphasized that the tonic regulation of catecholaminergic neurons that we have demonstrated with the use of 5,6-DHT could be an interesting model to explore some long-lasting alterations of the sleep-waking cycle, such as circadian rhythm or the duration of the paradoxical sleep rebound.

References

1. Baumgarten, H. G., A. Björklund, L. Lachenmayer, A. Nobin & U. Stenevi. 1971. Long-lasting selective depletion of brain serotonin by 5,6-dihydroxytryptamine. Acta Physiol. Scand. Suppl. **373:** 1–15.
2. Bobillier, P., S. Seguin, F. Petitjean, D. Salvert, M. Touret & M. Jouvet. 1976. The raphe nuclei of the cat brain stem: a topographical atlas of their efferent projections as revealed by autoradiography. Brain Res. **113:** 449–486.
3. Brownstein, M. J., M. Palkovits, J. M. Saavedra & J. S. Kizer. 1975. Tryptophan hydroxylase in the rat brain. Brain Res. **97:** 163–166.
4. Cleland, W. W. 1967. The statistical analysis of enzyme kinetics data. Advan. Enzymol. **29:** 1–32.
5. Costa, E., D. M. Chuang, A. Guidotti & P. Uzunov. 1975. Cyclic 3′,5′ adenosine monophosphate dependent molecular mechanisms in the transsynaptic induction of tyrosine hydroxylase in rat. *In* Chemical Tools in Catecholamine Research. O. Almgren, A. Carlsson & J. Engel, Eds. Vol. **II:** 283–292. North-Holland Publishing Company/American Elsevier. Amsterdam and New York.

6. Costa, E., H. Lefever, J. Meek, A. Revuelta, F. Spano, S. Strada & J. Daly. 1972. Serotonin and catecholamine concentrations in brain of rats injected intracerebrally with 5,6-dihydroxytryptamine. Brain Res. **44:** 304–308.
7. Daly, J., K. Fuxe & G. Jonsson. 1973. Effects of intracerebral injections of 5,6-dihydroxytryptamine on central monoamine neurons: evidence for selective degeneration of central 5-hydroxytryptamine neurons. Brain Res. **49:** 476–482.
8. Descarries, L. & L. Leger. 1976. Serotonin nerve terminals in adult rat locus coeruleus. *In* Interactions between Putative Neurotransmitters in the Brain. S. Garattini, J. F. Pujol & R. Samanin, Eds. The Raven Press. New York, N.Y. In press.
9. Fibiger, H. C., R. E. Pudritz, P. L. McGeer & E. G. McGeer. 1972. Axonal transport in nigro-striatal and nigro-thalamic neurons: effects of medial forebrain bundle lesions and 6-hydroxydopamine. J. Neurochem. **19:** 1697–1708.
10. Froment, J. L., F. Petitjean, N. Bertrand, C. Cointy & M. Jouvet. 1974. Effets de l'injection intracérébrale de 5–6 hydroxytryptamine sur les monoamines cérébrales et les états de sommeil du chat. Brain Res. **67:** 405–417.
11. Joh, T. H. & D. J. Reis. 1975. Different forms of tyrosine hydroxylase in central dopaminergic and noradrenergic neurons and sympathetic ganglia. Brain Res. **85:** 146–151.
12. Jones, B., P. Bobillier, C. Pin & M. Jouvet. 1973. The effect of lesions of catecholamine-containing neurons upon monoamine content of the brain in electroencephalographic and behavioural waking in the cat. Brain Res. **58:** 157–177.
13. Jouvet, M. 1971. L'alpha-methyl-P-tyrosine supprime l'insomnie consécutive à la lésion du système du Raphé. C.R. Soc. Biol. **165**(Suppl. 11)**:** 2128–2131.
14. Jouvet, M. 1972. The role of monoamines and acetylcholine containing neurons in the regulation of the sleep waking cycle. Ergeb. Physiol. Biol. Chem. Exp. Pharmakol. **64:** 166–307.
15. Kostowski, W., R. Samanin, S. R. Bareggi, S. Garattini & L. Valzelli. 1974. Biochemical aspects of the interaction between midbrain raphe and locus coeruleus in the rat. Brain Res. **82:** 178.
16. Lewis, B. D., B. Renaud, M. Buda & J. F. Pujol. 1976. Time-course variations in tyrosine hydroxylase activity in the rat locus coeruleus after electrolytic destruction of the nuclei raphe dorsalis or raphe centralis. Brain Res. **108:** 339–349.
17. Lidbrink, P. 1974. The effect of lesions of ascending noradrenaline pathways on sleep and waking in the rat. Brain Res. **74:** 19–40.
18. Nagatsu, T., M. Levitt & S. Udenfriend. 1964. A rapid and simple radioassay for tyrosine hydroxylase activity. Anal. Biochem. **9:** 122–126.
19. Nagatsu, T., P. Thomas, R. Rush & S. Udenfriend. 1973. A radioassay for dopamine-β-hydroxylase activity in rat serum. Anal. Biochem. **55:** 615–619.
20. Palkovits, M., M. Brownstein & J. M. Saavedra. 1975. Serotonin content of the brain stem nuclei in the rat. Brain Res. **80:** 237–249.
21. Parvez, H. & S. Parvez. 1973. Microradioisotopic determination of enzyme catechol-O-methyltransferase, phenylethanolamine-N-methyltransferase and monoamine oxidase in a single concentration of tissue homogenate. Clin. Chim. Acta **46:** 85–90.
22. Pickel, V. M., T. H. Joh & D. J. Reis. 1976. Immunocytochemical evidence for a serotonergic innervation of noradrenergic neurons in nucleus locus coeruleus. *In* Interactions between Putative Neurotransmitters in the Brain. S. Garattini, J. F. Pujol & R. Samanin, Eds. The Raven Press. New York, N.Y. In press.
23. Pujol, J. F., P. E. Keane, A. McRae, D. Lewis & B. Renaud. 1976. Biochemical evidence for serotonergic control of the locus coeruleus. *In* Interactions between Putative Neurotransmitters in the Brain. S. Garattini, J. F. Pujol & R. Samanin, Eds. The Raven Press. New York, N.Y. In press.

24. PUJOL, J. F., D. STEIN, C. BLONDAUX, F. PETITJEAN, J. L. FROMENT & M. JOUVET. 1973. Biochemical evidences for interaction phenomena between noradrenergic and serotoninergic system in the cat brain. *In* Frontiers in Catecholamine Research. E. Usdin & S. H. Snyder, Eds. : 771–772. Pergamon Press. New York, N.Y.
25. RENAUD, B., M. BUDA, B. D. LEWIS & J. F. PUJOL. 1975. Effects of 5–6 dihydroxytryptamine on tyrosine-hydroxylase activity in central catecholaminergic neurons of the rat. Biochem. Pharmacol. **24:** 1739–1742.
26. ROSS, S. B., A. L. RENYI & S. O. ÖGREN. 1972. Inhibition of the uptake of noradrenaline and 5-hydroxytryptamine by chlorphentermine and chlorimipramine. Eur. J. Pharmacol. **17:** 107–112.
27. THOENEN, M. 1974. Trans-synaptic enzyme induction. Life Sci. **14:** 223–235.

DISCUSSION

DR. C. T. GIAMBALVO: We found that injection of *p*-CA and DHT in the median raphe nucleus can increase dopamine turnover, as measured by Dopa accumulation after decarboxylase inhibition with RO-4–4602. Thus, our results complement the increase of tyrosine hydroxylase activity that you have observed.

Donaldson and Pycock lesioned the locus ceruleus and found unilateral rotational behavior; one can speculate that there is a serotonergic, noradrenergic, and dopaminergic interaction here.

DR. PUJOL: Our laboratory has recently shown that there is a direct connection between the locus ceruleus and the substantia nigra; thus, it is possible that your effects are mediated by the locus ceruleus.

DIFFERENTIAL EFFECTS OF *p*-CHLOROPHENYLALANINE AND 5,7-DIHYDROXYTRYPTAMINE ON FEEDING IN RATS

B. G. Hoebel, F. P. Zemlan, M. E. Trulson, R. G. MacKenzie, R. P. DuCret, and C. Norelli

Department of Psychology
Princeton University
Princeton, New Jersey 08540

The possibility that serotonin inhibits feeding [1–5] is supported by our finding that *p*-chlorophenylalanine (*p*-CPA) injected intraventricularly induces hyperphagia. Further investigation of the phenomenon has revealed that 5-hydroxytryptophan (5-HTP) reverses *p*-CPA-induced hyperphagia, which conforms with the view that the hyperphagia results from serotonin depletion. However, depletion of serotonin with a neurotoxin, 5,7-dihydroxytryptamine (5,7-DHT), failed to produce hyperphagia. This finding is similar to the results of others who have found differential effects of *p*-CPA and serotonin neurotoxins.[5–8]

p-Chlorophenylalanine methyl ester (4 mg) injected intraventricularly at pH 2 or 6 caused hyperphagia that reached a maximum in 4 days and lasted at least 2 weeks (FIGURE 1). Some of the female rats gained weight in accord with their hyperphagia, but others showed no weight gain despite clear hyperphagia. In male rats, hyperphagia was also observed, but obesity was never seen.

If hyperphagia was the result of serotonin depletion, repletion with the serotonin precursor 5-HTP should reverse the effect. We found that 5-HTP (25 mg/kg, ip) in combination with a peripheral decarboxylase inhibitor (MK-486) did cause food intake to normalize. This observation is summarized in FIGURE 2, which shows composite data from several groups of rats. All rats in FIGURE 2, except the control group, received *p*-CPA, and each of the subgroups indicated by a dotted line received 5-HTP on a different day. As a result of the 5-HTP treatment, hyperphagia was suppressed significantly in each case. The same dose of 5-HTP administered to control rats had a slight, but not statistically significant, anorectic effect.

Neurochemical analysis revealed that injection of *p*-CPA (4 mg, intraventricularly) depleted whole brain 5-HT to 25% of normal, and 5-HTP (25 mg/kg, ip) on Day 7 restored serotonin levels to normal 24 hr after drug administration (FIGURE 3). Norepinephrine and dopamine were not depleted. These behavioral and neurochemical data suggest that serotonin depletion disinhibits feeding, and serotonin repletion corrects this abnormality.

Given that *p*-CPA causes a transient depletion of serotonin, chronic depletion with 5,7-DHT might lead to chronic hyperphagia. This finding has been reported by Saller and Stricker;[9] however, we have not been able to reproduce it with essentially the same procedure. Female rats received 5,7-DHT (200 μg of free base) intraventricularly 1 hr after desmethylimipramine (25 mg/kg) pretreatment. Serotonin in the forebrain was depleted to 17%, but norepinephrine (NE) also decreased to 79%, and dopamine (DA) was reduced to

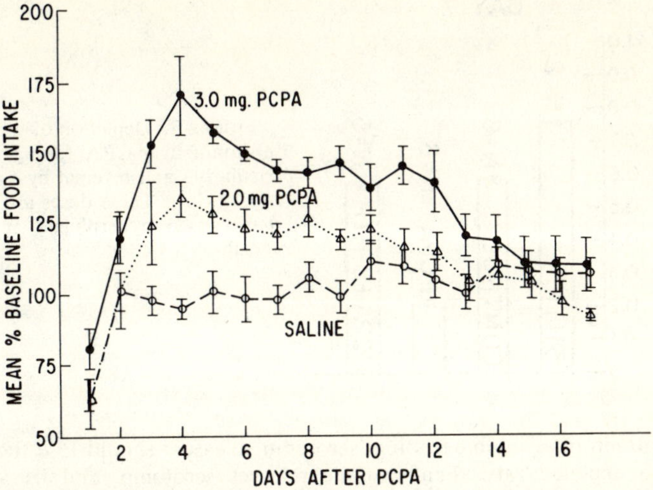

FIGURE 1. Food intake (percentage of mean baseline ± SEM) after intraventricular injection of *p*-CPA methyl ester.

71% (TABLE 1). Local injection of 5,7-DHT (8 μg of free base) in the posterior medial forebrain bundle also did not produce hyperphagia, even though forebrain serotonin was depleted to 21% of normal without NE or DA depletion. This new evidence demonstrates that serotonin depletion does not necessarily disinhibit feeding.

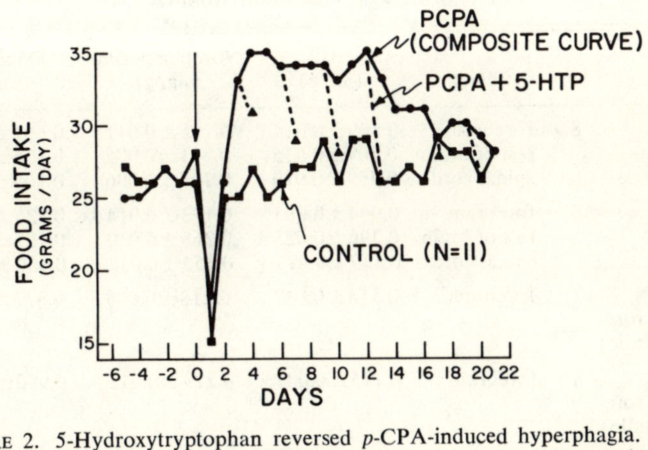

FIGURE 2. 5-Hydroxytryptophan reversed *p*-CPA-induced hyperphagia. On Day 0 (arrow), rats were injected with *p*-CPA or sham injected. *p*-Chlorophenylalanine-pretreated rats received 5-HTP (25 mg/kg, ip) and a peripheral decarboxylase inhibitor either on Day 3, 6, 9, 12, 16, or 19. These groups are combined in the composite curve that shows suppression of hyperphagia in each of the 5-HTP groups (all except Day 3 data are statistically significant, unlike normal rats, in which feeding suppression with this dose of 5-HTP was not statistically significant).

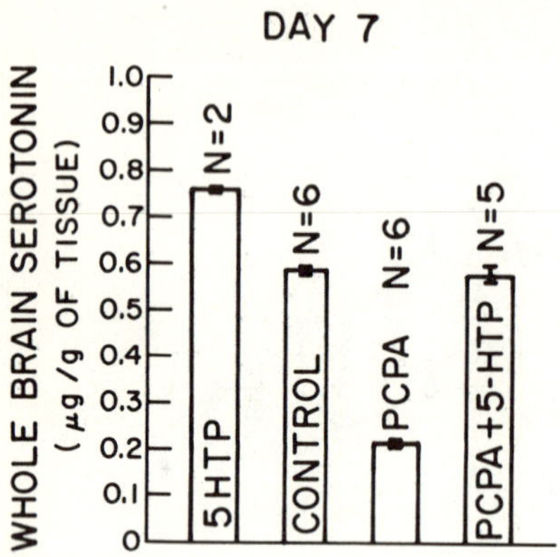

FIGURE 3. Depletion of whole brain serotonin by *p*-CPA (3 mg, intraventricularly) and reversal by 5-HTP (25 mg/kg, ip). These doses are the same that caused hyperphagia and its reversal.

If serotonin does inhibit feeding, serotonin releasers should lose their effectiveness in depleted rats. Fenfluramine releases serotonin, and its anorectic potency has been attributed to this effect.[1-4] If fenfluramine-induced anorexia is mediated by the release of forebrain serotonin, serotonin depletion effected by intraventricular *p*-CPA or 5,7-DHT would be expected to reduce fenfluramine anorexia. This expectation was not confirmed; on the contrary, fenfluramine-induced anorexia was enhanced. For these tests, rats were injected

TABLE 1

MEAN CONCENTRATIONS OF MONOAMINES AFTER ADMINISTRATION OF 200 μG OF 5,7-DHT INTRAVENTRICULARLY AND 8 μG OF 5,7-DHT IN THE MEDIAN FOREBRAIN BUNDLE

Group	*N*	Region	5-HT (μg/g)	Norepinephrine (μg/g)	Dopamine (μg/g)
Vehicle (intraventricular)	8	forebrain	0.509±0.014 *	0.218±0.011	0.601±0.038
		rest of brain	0.709±0.015	0.291±0.005	0.125±0.055
		spinal cord	0.479±0.019	0.145±0.006	0.106±0.021
5,7-DHT (intraventricular)	8	forebrain	0.084±0.005 †	0.173±0.014 ‡	0.425±0.043 §
		rest of brain	0.296±0.025 †	0.268±0.030	0.107±0.036
		spinal cord	0.139±0.013 †	0.155±0.012	0.137±0.062
Vehicle (median forebrain bundle)	7	forebrain	0.518±0.017	0.218±0.006	0.877±0.059
5,7-DHT (median forebrain bundle)	8	forebrain	0.115±0.007 †	0.229±0.012	0.990±0.026

* All values are uncorrected for recovery, which ranged from 89.3±1.0% to 99.0±1.8%.

† Difference from vehicle-injected control: $p < 0.001$.

‡ Difference from vehicle-injected control: $p < 0.01$.

§ Difference from vehicle-injected control: $p < 0.05$.

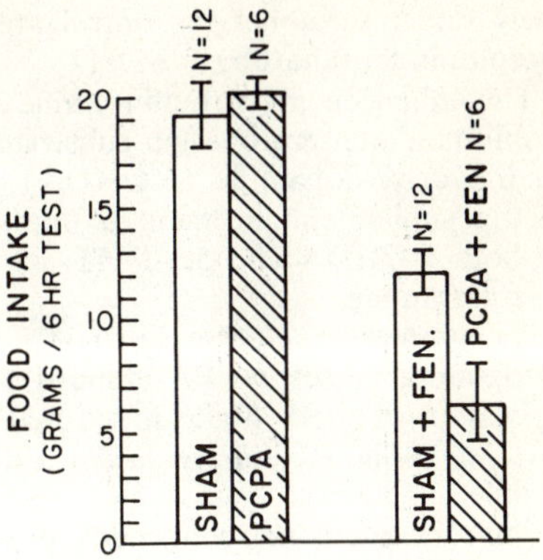

FIGURE 4. Food intake (mean ± SEM) in a restricted 6 hr/day test was the same for *p*-CPA- and sham-injected rats (left bars). Fenfluramine (4 mg/kg) suppressed feeding significantly more in the *p*-CPA group than in the sham-operated control group.

with fenfluramine 30 min before access to powdered food during the last 6 hr of the light period of a 14-hr light/10-hr dark cycle.

On Day 6 after *p*-CPA treatment, enhancement of fenfluramine-induced anorexia occurred (FIGURE 4). After 5,7-DHT treatment, fenfluramine was again enhanced, as demonstrated previously by Hollister *et al.*[10] 5,7-Dihydroxytryptamine did not affect amphetamine-induced anorexia in the same rats. 5-Hydroxytryptophan-induced anorexia in the same feeding schedule was similarly enhanced in 5,7-DHT-injected rats but only when the animals showed signs of motor impairment typical of serotonin excess[8] (FIGURE 5).

The absence of potentiated 5-HTP-induced anorexia after 5,7-DHT administration suggests two important points. First, it implies that the mechanisms for fenfluramine potentiation that operate after serotonin depletion are not effected through a direct action of the compound on "supersensitive" 5-HT receptors, because such "supersensitivity" also should have enhanced 5-HTP-induced anorexia. Second, the data suggest that 5-HTP-induced anorexia is not mediated at central 5-HT synapses, because the effects of 5-HTP on behaviors

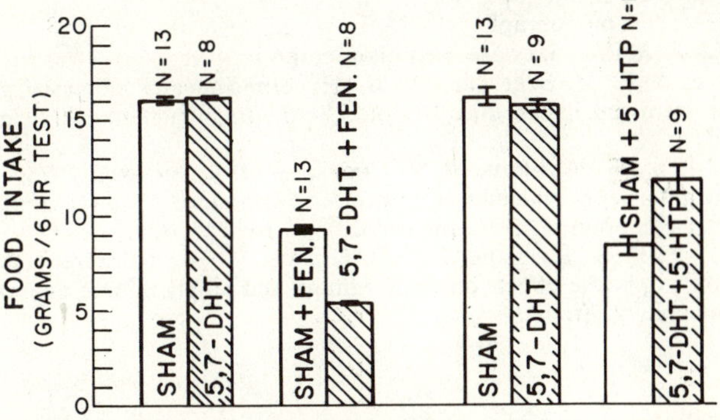

FIGURE 5. As with *p*-CPA (FIGURE 4), fenfluramine-induced anorexia was enhanced after 5,7-DHT treatment (left half of graph). 5-Hydroxytryptophan- (30 mg/kg) induced anorexia was not enhanced over this 6-hr test.

clearly under serotonergic control are invariably potentiated after destruction of serotonin terminals by 5,7-DHT.

The enhancement of fenfluramine anorexia, therefore, implies that 5,7-DHT disinhibits a nonserotonergic substrate for fenfluramine. We have also found that 6-hydroxydopamine (6-OHDA), injected in the ventral tegmentum, produced a similar enhancement of fenfluramine-induced anorexia,[2, 11] suggesting that both 6-OHDA and 5,7-DHT are disinhibiting a nonserotonergic substrate for fenfluramine.

To summarize, *p*-CPA, but not 5,7-DHT, caused hyperphagia. 5-Hydroxytryptophan reverses *p*-CPA-induced hyperphagia without causing comparable suppression of food intake in normals. Neither *p*-CPA nor 5,7-DHT reduces fenfluramine-induced anorexia but, rather, enhances it in rats on a 6-hr feeding schedule. Anorexia caused by 5-HTP administered according to the same feeding schedule is not enhanced, suggesting that fenfluramine and 5-HTP have different mechanisms of action.

We conclude that if there is a serotonin system that suppresses feeding, as supported by our *p*-CPA data, the 5,7-DHT procedure we used does not sufficiently deplete the necessary components of the system to produce hyperphagia. On the other hand, *p*-CPA may induce overeating by a mechanism different from that which affects serotonin systems.

References

1. Hoebel, B. G. 1977. Pharmacology of feeding. *In* Handbook of Psychopharmacology. L. L. Iversen, S. D. Iversen & S. H. Snyder, Eds. Plenum Publishing Corporation. New York, N.Y.
2. Hoebel, B. G. 1977. Pharmacologic control of food intake. Annu. Rev. Pharmacol. Toxicol. **17:** 605–621.
3. Blundell, J. 1977. Is there a role for serotonin (5-HT) in feeding? Int. J. Obesity **1:** 15–42.
4. Garattini, S., A. Bizzi, G. deGaetano, A. Jori & R. Samanin. 1975. Recent advances in the pharmacology of anorectic agents. *In* Recent Advances in Obesity Research: 1. A. Howard, Ed. : 354–367. Newman Publishers. London, England.
5. Coscina, D. V. This monograph.
6. Lorens, S. A. This monograph.
7. Myers, R. D. This monograph.
8. Trulson, M. E. & B. L. Jacobs. This monograph.
9. Saller, C. F. & E. M. Stricker. 1976. Hyperphagia and increased growth in rats after intraventricular injection of 5,7-dihydroxytryptamine. Science **192:** 385–397.
10. Hollister, A. S., G. N. Ervin, B. R. Cooper & G. R. Breese. 1975. The role of monoamine neural systems in the anorexia induced by (+) amphetamine and related compounds. Neuropharmacology **14:** 715–723.
11. Ahlskog, J. E., P. K. Randall & B. G. Hoebel. 1978. Noradrenergic neuron destruction: opposite effects on fenfluramine and amphetamine-anorexia. Psychopharmacology. In press.

EFFECTS OF 5,6-DIHYDROXYTRYPTAMINE AND 5,7-DIHYDROXYTRYPTAMINE ON THE RAT CENTRAL NERVOUS SYSTEM AFTER INTRAVENTRICULAR OR INTRACEREBRAL APPLICATION AND ON BLOOD PLATELETS *IN VITRO*

M. Da Prada, L. Pieri, H. H. Keller, M. Pieri, and E. P. Bonetti

Pharmaceutical Research Department
F. Hoffmann-La Roche & Co., Ltd.
CH 4002 Basel, Switzerland

Among the various dihydroxylated tryptamines synthesized and studied so far,[1] 5,6-dihydroxytryptamine (5,6-DHT) and 5,7-dihydroxytryptamine (5,7-DHT) seem, under well-defined conditions, to be rather selective in inducing specific chemical degeneration of tryptaminergic pathways in the central nervous system.[2–4] Intracerebroventricular injection of 50–75 μg of 5,6-DHT induces a quite selective but moderate reduction (30–50%) of cerebrospinal 5-hydroxytryptamine (5-HT). This reduction is accompanied by enhancement of sexual activity in male rats.[5]

No selectivity could be detected by injecting 5,6-DHT stereotaxically into the medial forebrain bundle (MFB). In fact, by this mode of application, we have found a considerable decrease of both 5-HT and dopamine (DA), whereas the noradrenaline (NA) content of the rat telediencephalon was not markedly affected. Therefore, we have proposed these unilaterally lesioned animals as a model suitable for elucidating the controversial role of 5-HT in the turning behavior induced by DA agonists and releasers.[6]

The subsequently synthesized serotonin neurotoxin 5,7-DHT has been found to be less toxic than 5,6-DHT but even less specific. In fact, it has been reported that 5,7-DHT given intracerebroventricularly in rats has pronounced depleting effects not only on 5-HT but also on NA fibers.[7, 8]

A rather satisfactory selectivity has been achieved by pretreatment of the animals with monoamine oxidase inhibitors (MAOI)[9] or with catecholamine (CA) uptake blockers, such as desmethylimipramine (DMI) or protriptyline.[3, 4, 10]

In light of the aforementioned points, it is clear that the important issue of the selectivity of these two neurotoxins is still controversial. Therefore, we considered it worthwhile to reinvestigate the biochemical effects of both compounds by applying them intracerebroventricularly or stereotaxically in the MFB and by measuring not only the monoamine levels but also levels of the corresponding metabolites. In fact, changes in metabolite levels, a suitable and sensitive index of alterations in monoamine turnover, are a much better indication of even minimal functional changes of aminergic systems.

Furthermore, with the goal of a better understanding of the possible relationship between 5-HT and sexual behavior,[11] we correlated the amine and metabolite levels in the brain, found after injection of 5,6-DHT or 5,7-DHT

0077–8923/78/0305–0595 $01.75/0

intracerebroventricularly, with sexual behavior in male and female rats. For this purpose, we also compared mounting behavior induced by intracerebroventricular administration of 5,6-DHT and 5,7-DHT with that observed after injection of *p*-chlorophenylalanine (*p*-CPA)[12]; all of these treatments reduce brain 5-HT levels.

Finally, *in vitro* experiments were performed with rat blood platelets incubated in plasma with 5,6-DHT or 5,7-DHT to investigate the release of endogenous 5-HT induced by these drugs. These platelets were also examined by electron microscopy to assess the subcellular localization that occurs after the uptake of the two dihydroxytryptamines and to determine whether 5-HT release is accompanied by ultrastructural damage.

Materials and Methods

The experiments were performed on SPF rats of both sexes and various ages (Füllinsdorf Albino stock). The creatinine sulfate salts of 5,6-DHT (50 μg of base) and 5,7-DHT (75 μg of base, Regis), which were dissolved in 10 μl of 0.9% NaCl that contained 0.1% ascorbic acid, were injected by means of a Hamilton syringe into the right lateral cerebral ventricle of rats that weighed 180–200 g through a chronically implanted cannula. The injections were performed under light ether anesthesia or sodium pentobarbitone (30 mg/kg, ip) to reduce seizures.

In some experiments, the intracerebroventricular injection of 5,7-DHT was preceded for various periods of time by an intraperitoneal injection of either DMI or maprotiline (MAP), two rather specific NA uptake blockers, Lu 10–171 [1-(3-dimethylamino)propyl-1-(*p*-fluorophenyl-5-phthalancarbonitrile) hydrobromide], a specific 5-HT blocker, or harmaline, a short-acting MAOI.

Mounting behavior was quantified during a 2-hr period between 7 and 10 PM after grouping the formerly individually housed rats into rectangular (27 × 40 × 15 cm) Macrolone® cages (three rats per cage) in a room with artificial diffuse dim light. This grouping and counting of the mountings were conducted 24 hr after intracerebroventricular injection of 5,7-DHT (at variance with earlier experiments, which were performed 9 hr after injection of 5,6-DHT by the same route of administration[5]).

p-Chlorophenylalanine ethyl ester hydrochloride (*p*-CPA, 100 mg/kg of the free amino acid) was injected intraperitoneally once daily for 3 consecutive days in isolated male and female rats, and the behavioral observations were made 24 hr after the last injection.

Stereotaxic injections of 5,6-DHT and 5,7-DHT into the MFB were performed as described earlier[6] in halothane-anesthetized rats that weighed 140–160 g. Here, too, 5,7-DHT (at different doses and concentrations) was injected either alone or after pretreatment with DMI or MAP. Some of these unilaterally lesioned rats were used in studies of turning behavior (automatically recorded by a rotameter)[13] in which lisuride (Lysenyl®, Spofa), a nonhallucinogenic semisynthetic ergot derivative, was compared with *d*-lysergic acid diethylamide (*d*-LSD) and apomorphine for its efficacy in inducing contralateral turning, that is, turning away from the lesioned side.

Monoamines in the telediencephalon or spinal cord and the major amine metabolites in the whole brain (minus the cerebellum) were determined fluorimetrically.[14–17] In some cases, 5-hydroxyindoleacetic acid (5-HIAA), homovanillic acid (HVA), and 3-methoxy-4-hydroxyphenylethyleneglycol sulfate (MOPEG-SO_4) were measured 2 hr after injection of methiothepin, a powerful blocker of cerebral 5-HT, DA, and NA receptors.[18, 19] In unilaterally MFB-lesioned rats, amines and metabolites were assayed separately in the left and right hemispheres.

In some experiments, isolated rat blood platelets were incubated in plasma at 37° C in the presence of 5,6-DHT, 5,7-DHT, or metaraminol (10^{-3} M) for various periods of time. After washing the platelets with a modified Tyrode solution,[20] endogenous 5-HT was measured as previously described.[21]

Electron microscopic studies were performed on ultrathin sections of platelets double fixed with glutaraldehyde and osmium tetroxide and contrasted with uranyl acetate and lead citrate.[22] Statistical significance for differences between experimental groups was calculated with Student's *t* test.

Results

Monoamines and Metabolites in the Rat Central Nervous System After Treatment with 5,6-DHT and 5,7-DHT

Intracerebroventricular Application of 5,6-DHT

Intracerebroventricular injection of 5,6-DHT in rats caused a relatively selective and long-lasting decrease of cerebral and spinal cord 5-HT content.[23, 24] By 12 hr after injection of 50 μg of 5,6-DHT (Table 1), 5-HT had been reduced by about 50%, without any appreciable modifications of DA and NA levels. In such animals, intraperitoneal injection of 5-hydroxytryptophan (100 mg/kg of the free amino acid) restored 5-HT to normal levels, leaving both DA and NA unaffected.

Intracerebroventricular Application of 5,7-DHT

Improved selectivity of intracerebroventricularly injected 5,7-DHT for 5-HT neurons can be attained by pretreating rats with NA uptake blockers[3, 4] or MAOI.[9]

Uptake Blockers. Table 2 shows that MAP (50 mg/kg, ip), due to its rather specific blockade of NA uptake, had practically the same effect as DMI (30 mg/kg, ip) in protecting NA fibers and did not enhance 5-HT depletion. In agreement with previous findings,[1, 2, 25] 5,7-DHT given alone (Table 2) markedly reduced both 5-HT and NA levels in the brain and spinal cord.

Study of the major monoamine metabolites, determined at various periods after intracerebroventricular treatment with DMI plus 5,7-DHT, showed that 5-HIAA decreased progressively, reaching about 50% of control values after 10 days (Table 3). MOPEG-SO_4 was elevated 2 and 10 days after neurotoxin injection. At 10 days, HVA was found to be reduced, but there was no concomitant decrease of DA (Tables 2 & 3). The increase of both HVA and 5-HIAA observed in control animals after intraperitoneal injection of methiothepin was markedly (5-HIAA) or slightly (HVA) attenuated in rats that had received 5,7-DHT 48 hr earlier (Table 4).

Lu 10–171 prevented a major uptake of 5,7-DHT into 5-HT neurons, as

TABLE 1

EFFECT OF 5,6-DHT ALONE AND IN COMBINATION WITH 5-HYDROXYTRYPTOPHAN OR 6-HYDROXYDOPAMINE ON BRAIN AMINES AND MOUNTING BEHAVIOR OF MALE RATS *

Experimental Group	*n*	Brain Amines (% of control ± SEM)			*n*	Total Mounts per Rat	Animals Mounting (%)	
		5-HT	DA	NA			Once	10 Times
Control	6	100±2	100±2	100±4	60	0	0	0
5,6-DHT	12	49±4	93±1	92±3	90	19	65	37
6-Hydroxydopamine plus 5,6-DHT	6	54±6	45±5	17±1	40	18	75	40
5,6-DHT plus 5-hydroxytryptophan	10	104±9	106±3	83±3	30	0	0	0

* Male rats (200–220 g) were kept isolated for 48 hr. 5,6-Dihydroxytryptamine (50 μg) and 6-hydroxydopamine (three daily doses of 140,140, and 200 μg of base plus an additional injection of 50 μg of 5,6-DHT after the last 6-hydroxydopamine dose) were administered intracerebroventricularly. Nine hours later, the animals were grouped, and their mounting activity was observed for 3 hr. 5-Hydroxytryptophan (100 mg/kg of the free amino acid) was injected intraperitoneally before beginning the observation of mounting. At the end of the observation period, the animals were sacrificed to measure cerebral amines. Brain amine absolute values (μg/g) are: 5-HT, 0.44±0.01; DA, 0.94±0.02; NA, 0.49±0.02.

TABLE 2

CONTENT OF MONOAMINES IN THE RAT CENTRAL NERVOUS SYSTEM 12 DAYS AFTER 5,7-DHT GIVEN ALONE OR TO ANIMALS PRETREATED WITH UPTAKE INHIBITORS *

Drugs	Telediencephalon (% of control ± SEM)			Spinal Cord (% of control ± SEM)	
	NA	DA	5-HT	NA	5-HT
Control, untreated	100.0±2.7	100.0±2.0	100.0±6.3	100.0±3.1	100.0±6.3
5,7-DHT alone, 75 μg	68.0±4.2 †	100.1±6.8	31.6±2.0 †	34.0±9.7 †	34.2±4.8 †
Desmethylimipramine, 30 mg/kg, ip, plus 5,7-DHT	104.0±2.5	96.6±4.3	31.0±1.7	102.8±7.8	22.1±2.4 †
Maprotiline, 50 mg/kg, ip, plus 5,7-DHT	93.0±5.7	102.0±4.7	38.8±6.1 †	76.2±14.2	29.7±8.8 †

* 5,7-Dihydroxytryptamine (75 μg of base) was given intracerebroventricularly both alone and in rats pretreated with desmethylimipramine or maprotiline 1 hr prior to 5,7-DHT. All values are from five to six determinations.

† Significant difference from control: $p < 0.001$.

TABLE 3

CHANGES IN CEREBRAL CONTENT OF MONOAMINE METABOLITES AT VARIOUS TIMES AFTER 5,7-DHT (INTRACEREBROVENTRICULAR) IN RATS PRETREATED WITH DESMETHYLIMIPRAMINE *

Experimental Group	Sex	5-HIAA	HVA	$MOPEG\text{-}SO_4$
Untreated control	M	100.0±2.8(11)	100.0±6.2(11)	100.0±3.2(7)
5,7-DHT, 12 hr	M	70.3±3.8(5)†	113.3±8.7(5)	107.5±7.8(5)
5,7-DHT, 2 days	M	61.1±5.8(5)†	82.5±5.4(5)	133.7±5.9(5)‡
5,7-DHT, 10 days	F	49.7±5.4(10)†	71.8±4.1(10)‡	124.7±3.7(5)‡

* All values are percentages of control values and are means ± SEM. In parentheses are given the numbers of single determinations.

† Significant difference from control: $p < 0.001$.

‡ Significant difference from control: $p < 0.01$.

determined from 5-HT and 5-HIAA levels but, as expected, did not alter the neurotoxic effect on DA and NA neurons, as indicated by a concomitant lowering of amine and metabolite levels (TABLE 5).

Monoamine Oxidase Inhibitors. Relatively long-acting MAOIs, although restricting the neurotoxic action of intracerebroventricularly applied 5,7-DHT on 5-HT pathways,[9] might interfere with the behavioral and biochemical parameters under investigation. Harmaline was therefore chosen on the basis of its short-lasting MAO inhibition.[26] A comparison of the levels of CAs and their metabolites with those of 5-HT and 5-HIAA, given in TABLE 5, indicates that harmaline markedly reduced the neurotoxicity of 5,7-DHT for CA neurons.

TABLE 4

EFFECT OF 5,7-DHT INJECTED INTRACEREBROVENTRICULARLY ON INCREASES OF 5-HIAA AND HVA IN THE RAT TELEDIENCEPHALON INDUCED BY METHIOTHEPIN *

Experimental Group	Metabolite Levels (% of control ± SEM)	
	5-HIAA	HVA
Control (untreated)	100.0±4.1(18)	100.0±1.9(18)
DMI plus 5,7-DHT	43.4±4.4(15)	71.6±2.8(15)
Methiothepin	202.5±10.3(11)	508.3±16.8(11)
DMI plus 5,7-DHT plus methiothepin	79.7±13.3(11)†	428.7±26.3(11)‡

* Methiothepin (20 mg/kg, ip) was injected in male and female rats either alone or after pretreatment with DMI (30 mg/kg, ip, 1 hr) plus 5,7-DHT (75 μg, intracerebroventricularly) 10 days earlier. Rats were killed 2 hr after methiothepin injection. In parentheses are shown the numbers of single determinations.

† Significant difference from methiothepin alone: $p < 0.001$.

‡ Significant difference from methiothepin alone: $p < 0.02$.

TABLE 5

CHANGES IN CEREBRAL MONOAMINES AND THEIR METABOLITES INDUCED WITHIN 48 HR BY 5,7-DHT INJECTED IN RATS PRETREATED WITH HARMALINE OR LU 10–171 *

Experimental Group	Sex	5-HT	DA	NA	5-HIAA	HVA	$MOPEG\text{-}SO_4$
Control (untreated) (8)	M	100.0±1.2 †	100.0±2.4	100.0±3.9	100.0±6.1	100.0±12.2	100.0±7.0
Harmaline plus 5,7-DHT (4,5)	M	74.1±5.1 ‡	109.2±0.9 §	98.3±5.4	64.5±0.9 ‡	83.8±8.9	122.4±7.1
Control (untreated) (8)	F	100.0±1.7	100.0±2.4	100.0±3.9	100.0±2.0	100.0±4.7	100.0±3.7
Lu 10–171 plus 5,7-DHT (4)	F	86.5±4.6 ‡	98.7±4.1	82.1±6.6	82.1±4.5 ‡	67.5±12.3 §	62.1±5.2 ‡

* Rats received one injection of harmaline (10 mg/kg, ip) 1 hr, or two injections of Lu 10-171 (50 mg/kg, ip), 2 and 1 hr prior to 5,7-DHT (75 μg, intracerebroventricularly).

† All results are percentages of control values and are means ± SEM. The numbers of single determinations are given in parentheses.

‡ Significant difference from control: $p < 0.01$.

§ Significant difference from control: $p < 0.05$.

Mounting Behavior After Treatment with 5,6-DHT and 5,7-DHT

Rats injected intracerebroventricularly with 5,6-DHT exhibited hyperexcitability and fighting often associated with so-called bizarre postures.[5, 27] However, as previously reported,[5] mounting activity was the main feature of this behavior. The data given in TABLE 1 indicate that the number of mounts per animal observed after intracerebroventricular treatment with 5,6-DHT or in combination with 6-hydroxydopamine did not differ appreciably, although in the latter case NA and DA were reduced to 17 and 45% of the control values, respectively. On the other hand, the return of 5-HT to normal levels after treatment with 5-hydroxytryptophan completely abolished mounting behavior.

With 5,7-DHT, mountings were found to occur more frequently if the animals were grouped 24 hr, rather than 9–10 hr, after intracerebroventricular injection, probably because of the lesser amount of aggressive and bizarre behavior observed at this time (TABLE 6).

Unexpectedly, after 5,7-DHT treatment, grouped female rats also exhibited a sexual behavior virtually identical to the mounting performance of male rats (TABLE 6). This effect was also induced by 5,6-DHT.

When harmaline was administered prior to intracerebroventricularly injected 5,7-DHT, the number of mountings per animal was found to increase during the observation period, reaching values of more than 20 after harmaline treatment. This increased rate of mounting was accompanied by a diminished irritability, as compared to that seen after pretreatment with DMI. When cerebral 5-HT levels were restored in 5,7-DHT-lesioned rats by injection of its precursor 5-hydroxytryptophan, mounting behavior was abolished, as already reported for 5,6-DHT. When the damage to 5-HT neurons was prevented by treating the animals with the potent 5-HT inhibitor Lu 10–171[28] prior to 5,7-DHT injection, both fighting and mounting behaviors were completely absent (TABLE 6).

TABLE 6

EFFECT OF 5,7-DHT ON MOUNTING BEHAVIOR OF MALE AND FEMALE RATS PRETREATED WITH DESMETHYLIMIPRAMINE (DMI), HARMALINE (HAR), OR LU 10–171 *

Experimental Group	Sex	n	Total Mounts per Rat	Animals Mounting (%)	
				Once	10 Times
DMI plus 5,7-DHT	F	32	6.5	53	19
DMI plus 5,7-DHT	M	35	11.1	23	17
HAR plus 5,7-DHT	M	15	22.1	40	13
Lu 10–171 plus 5,7-DHT	F	6	0.33	17	0

* After implantation with a permanent cannula into the right lateral cerebral ventricle, rats of both sexes (180–200 g) were kept isolated for 5–7 days. The rats then received one injection of harmaline (10 mg/kg, ip) 1 hr, or two injections of Lu 10–171 (50 mg/kg, ip), 2 and 1 hr prior to 5,7-DHT (75 μg, intracerebroventricularly). Twenty-four hours after injection of 5,7-DHT, the animals were grouped together, and their mounting was quantified for 2 hr. Only mountings with pelvic thrusts were counted.

TABLE 7

MOUNTING BEHAVIOR OF MALE, FEMALE, CASTRATED (C), AND OVARIECTOMIZED (O) RATS AFTER *p*-CPA ALONE OR IN COMBINATION WITH 5-HYDROXYTRYPTOPHAN OR HALOPERIDOL (HAL)*

Experimental Group	Sex	Weight (g)	*n*	Total Mounts per Rat	Animals Mounting Once (%)	Animals Mounting 10 Times (%)
p-CPA	F	50	33	13.4	82	55
p-CPA	F	250	24	13.2	50	13
p-CPA	M	50	33	9.3	82	36
p-CPA	M	250	24	4.7	38	21
p-CPA	C	300	6	16.8	50	33
p-CPA	O	150	12	11.5	58	17
p-CPA plus HAL	M	50	6	3.7	33	17
p-CPA plus 5-hydroxytryptophan	M	50	6	0.0	0	0

* *p*-Chlorophenylalanine (100 mg/kg, ip, as the free amino acid) was injected daily for 3 consecutive days in isolated rats, subsequently grouped 24 hr after the last injection. Some of these rats were pretreated with haloperidol (0.05 mg/kg, ip) immediately before grouping. Other animals received benserazide (50 mg/kg, ip), followed 30 min later by 5-hydroxytryptophan (100 mg/kg, ip, as the free amino acid), and finally they were grouped together. Some rats were castrated or ovariectomized 30 days before quantifying sexual behavior. Mounting behavior was observed for 2 hr.

Mounting Behavior After Treatment with p-*Chlorophenylalanine*

In addition to studying adult male rats, we also investigated the effects of *p*-CPA in castrated adult males, in prepubertal males and females, and in normal and ovariectomized adult females (TABLE 7). In all of these different experimental situations, *p*-CPA induced mounting activity. Castration or ovariectomy did not affect mounting behavior. The DA receptor antagonist haloperidol, in noncataleptigenic doses, markedly reduced mounting activity; 5-hydroxytryptophan completely abolished such behavior.

Intracerebral Stereotaxic Application of 5,6-DHT and 5,7-DHT

As already reported,[29] unilateral stereotaxic injection of 2–10 μg of 5,6-DHT in the MFB produced an impairment of not only 5-HT but also DA fibers in the ipsilateral hemitelediencephalon, but NA pathways were unaffected. This unilateral destruction of DA neurons became functionally evident by the turning behavior induced after injection of DA agonists (see below). Uptake studies and fluorescence microscopic investigations confirmed the fact that 5,6-DHT had some neurotoxic effect on the DA system, even after intracerebroventricular application. By this route, some destruction of NA pathways, apparently restricted to the spinal cord, has also been observed.[8, 30] Unilateral injection of 5,7-DHT at different concentrations and doses, after pretreatment with either DMI or MAP, produced a decrement of 5-HT, ranging between 40 and 75% of control values (TABLE 8). Pretreatment with MAP (30 mg/kg, ip, two

TABLE 8

CONTENT OF 5-HT, DA, AND NA IN THE LEFT (L) AND RIGHT (R) TELEDIENCEPHALON 10 DAYS AFTER INJECTION OF 5,7-DHT IN THE RIGHT MFB OF RATS PRETREATED INTRAPERITONEALLY WITH DESMETHYLIMIPRAMNIE (DMI, ONE INJECTION) OR MAPROTILINE (MAP, TWO INJECTIONS)

Experimental Group	5-HT		DA		NA	
	L	R	L	R	L	R
Control (untreated) (6)	100.0±1.1 *	100.0±1.0	100.0±2.3	100.0±2.3	100.0±2.0	100.0±5.6
DMI+5,7-DHT (4 μg/4 μl) (7)	76.5±8.7 †	23.9±3.7 ‡	111.4±1.7	86.0±6.3 †	107.8±2.4	95.1±3.7
Control (untreated) (12)	100.0±5.1	100.0±2.8	100.0±2.2	100.0±2.4	100.0±1.3	100.0±2.9
MAP+5,7-DHT (4 μg/4 μl) (6)	74.1±4.1 †	41.2±5.7 ‡	104.5±3.6	66.1±5.4 ‡	101.6±6.4	82.3±9.2 §
MAP+5,7-DHT (6 μg/4 μl) (4)	76.0±5.8 §	25.8±2.1 ‡	95.5±11.6	115.3±3.1	106.3±4.6	87.1±7.3

* All results are percentages of control values and are means ± SEM. The numbers of single determinations are shown in parentheses.

† Significant difference from control: $p < 0.01$.

‡ Significant difference from control: $p < 0.001$.

§ Significant difference from control: $p < 0.05$.

applications) did not improve selectivity, as did preinjection of DMI (30 mg/kg, ip, one dosage; TABLE 8). Only occasionally did the injection induce a selective and well-localized lesion (TABLE 8). In some cases, injection of 5,7-DHT in the right side also affected the 5-HT fibers of the left MFB. In addition, in some instances, DA and, to a lesser extent, NA neurons in the injected hemisphere were also partially affected, as judged by decreases of DA and NA observed 10 days after 5,7-DHT application. Furthermore, as shown in TABLE 9, reductions of 5-HIAA and HVA were observed both 2 and 10 days after lesioning, while MOPEG-SO_4 values were unchanged. This reduction was more marked at 10 days; at this time, some diminution of the two metabolites was found also in the contralateral injection side. The response to methiothepin is further evidence that this neurotoxin damages predominantly 5-HT neurons and, to a lesser extent, DA neurons. The results in TABLE 10 show that the methiothepin-induced accumulation of 5-HIAA was prevented completely in the right (injected) side and, to a certain extent, in the left side if the animals were pretreated with 5,7-DHT 2 or 10 days earlier.

Turning Behavior After Treatment with 5,6-DHT and 5,7-DHT

It has been previously reported that unilateral injection of 5,6-DHT in the MFB induces selective depletion of 5-HT and DA in the injected hemiteledien-cephalon.[29] In such lesioned animals, turning contralateral to the lesion was induced by DA agonists (apomorphine, piribedil, ergometrine, ergocornine, *d*-LSD), whereas DA releasers (*d*-methamphetamine, methylphenidate) elicited ipsilateral circling. These rotational responses were quite similar to those observed in rats lesioned with 6-OHDA in the MFB when the animals were challenged with the above-mentioned drugs.

Using 5,6-DHT lesioned rats, we have compared the contralateral rotation induced by apomorphine and *d*-LSD with that elicited by lisuride, the DA agonist effect of which has been reported recently.[31-33] It can be seen (FIGURE 1) that lisuride (50 μg/kg, ip) induced a long-lasting (about 4 hr) contralateral rotation, whereas animals injected with apomorphine (1 mg/kg, ip) or *d*-LSD (200 μg/kg, ip) ceased rotating after approximately 1 hr. The drug that induced the maximal rotation within 1 hr was *d*-LSD (about 800 turns), followed by apomorphine (about 600 turns) and lisuride (about 200 turns). When lisuride (50 μg/kg, ip) was injected on 4 consecutive days and rotation recorded on the fourth day, no evidence of tolerance was detected. On the contrary, a greater sensitivity seemed to develop, since rotation reached about 500 turns within 1 hr (FIGURE 2) and thereafter returned to the same number of turns as in rats injected only once. On the other hand, the duration of turning was somewhat longer in rats treated with lisuride for the first time (FIGURE 2). Moreover, after four daily injections of lisuride, application of *d*-LSD on the fifth day did not reveal the presence of either cross tolerance or supersensitivity. Contralateral turning of a similar intensity, after injection of lisuride, has also been recently observed in rats lesioned in the MFB with 6-hydroxydopamine. Such turning behavior could be completely antagonized by administration of haloperidol, a known DA receptor blocker.[33]

The lack of major impairment of DA after injection of 5,7-DHT in the MFB, as determined biochemically, is corroborated by the fact that in such animals, lisuride was unable to induce turning. On the other hand, very weak

TABLE 9

CONTENT OF 5-HIAA, HVA, AND MOPEG-SO_4 IN THE LEFT (L) AND RIGHT (R) TELEDIENCEPHALON OF MALE RATS INJECTED WITH 5,7-DHT IN THE RIGHT MFB AFTER PRETREATMENT WITH DESMETHYLIMIPRAMINE (DMI)*

	5-HIAA		HVA		MOPEG-SO_4	
	L	R	L	R	L	R
Expt 1						
Control (untreated) (6)	100.0±5.1 †	100.0±3.9	100.0±10.2	100.0±8.7	100.0±2.5	100.0±2.4
5,7-DHT, 2 days (4)	100.0±8.6	52.3±3.7 ‡	121.7±17.1	57.7±15.7 §	110.6±4.5	109.7±8.6
Expt 2						
Control (untreated) (6)	100.0±6.4	100.0±10.3	100.0±7.3	100.0±17.6	100.0±2.5	100.0±4.0
5,7-DHT, 10 days (4)	70.4±11.6 §	15.6±4.2 ‡	63.8±2.0 ‖	33.4±10.0 §	97.9±2.1	97.3±5.6

* DMI (30 mg/kg, ip) was given 1 hr prior to 5,7-DHT (6 μg/4 μl), and the rats were killed 2 or 10 days later.

† All results are percentages of control and are means ± SEM. The numbers of single determinations are given in parentheses. Average absolute control tissue levels for both hemispheres (ng/g): 5-HIAA, 262±19; HVA, 59±1; MOPEG-SO_4, 126±4.

‡ Significant difference from control: $p < 0.001$.

§ Significant difference from control: $p < 0.05$.

‖ Significant difference from control: $p < 0.01$.

TABLE 10

EFFECT OF 5,7-DHT INJECTED IN THE RIGHT MFB ON THE METHIOTHEPIN-INDUCED INCREASE OF MONOAMINE METABOLITES IN THE LEFT (L) AND RIGHT (H) TELEDIENCEPHALON OF MALE RATS *

Treatment			5-HIAA		HVA		MOPEG-SO_4	
Pretreatment	Acute	*n*	L	R	L	R	L	R
None	none	12	100.0±3.9 †	100.0±5.3	100.0±6.0	100.0±9.3	100.0±1.7	100.0±2.2
None	methiothepin	6	209.8±10.4	233.9±16.6	619.9±36.4	648.4±35.5	172.7±13.9	185.1±17.1
DMI plus 5,7-DHT, 2 days	methiothepin	4	164.6±23.2	72.3±10.1 ‡	526.3±37.8	194.5±53.3 ‡	172.8±14.0	179.0±13.3
DMI plus 5,7-DHT, 10 days	methiothepin	3,4	166.2±13.7 §	44.7±6.4 ‡	535.0±26.0	247.7±67.9 ‡	187.6±20.7	187.7±14.1

* Methiothepin (20 mg/kg, ip) was injected either alone or after pretreatment with DMI (30 mg/kg, ip, 1 hr) plus 5,7-DHT (6 μg/4 μl) 2 or 10 days earlier.

† All results are percentages of control values and are means ± SEM. The numbers of single determinations are given in parentheses.

‡ Significant difference from methiothepin alone: $p < 0.001$.

§ Significant difference from methiothepin alone: $p < 0.05$.

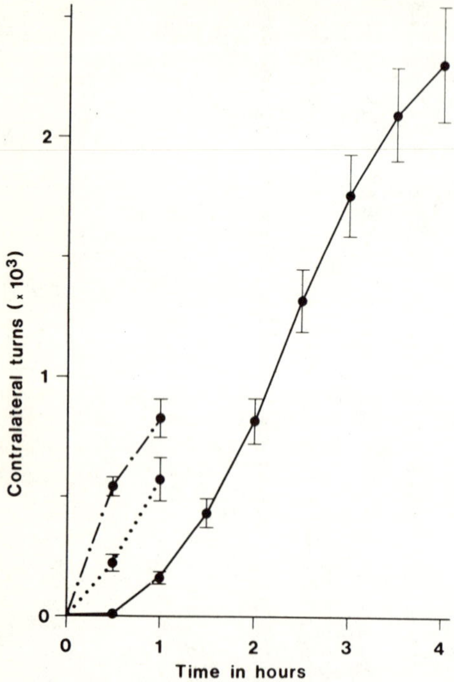

FIGURE 1. Turning, contralateral to the lesioned side, induced in rats by *d*-lysergic acid diethylamide bitartrate (200 μg/kg, —•—), apomorphine hydrochloride (1 mg/kg, •••), and lisuride maleate (50 μg/kg, —). Rats were injected unilaterally with 5,6-DHT (10 μg of base/10 μl) 10 days before testing. By means of a rotameter, turns were quantified cumulatively as long as the rotations lasted. Values are means ± SEM ($n=9$).

ipsilateral turning behavior (about 50 turns in 1 hr) was sporadically observed after *d*-methamphetamine treatment (5 mg/kg, ip). In these animals, a slight decrease of DA on the lesioned side could, in fact, be detected (data not presented).

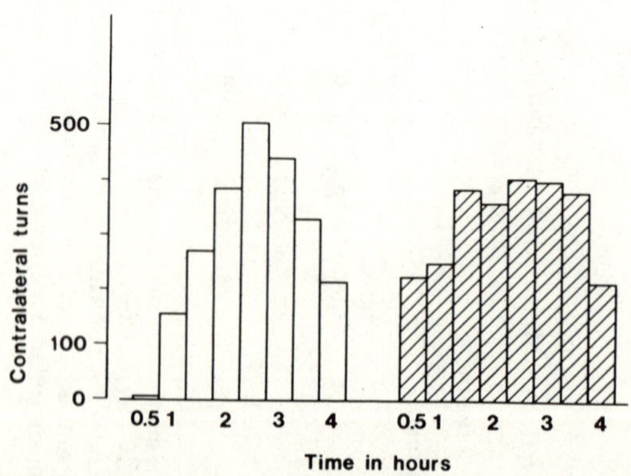

FIGURE 2. Lisuride maleate, given intraperitoneally, was tested for inducing contralateral turning after one injection (50 μg/kg, open bars) and after four consecutive daily injections (50 μg, hatched bars) in the same rats ($n=9$). Note the immediate onset of rotation after the fourth application.

Blood Platelets

Biochemical Changes

In isolated rat platelets in plasma, 5,6-DHT and 5,7-DHT caused a diminution of 5-HT content, which became more pronounced with increasing incubation times. After the platelets were incubated for 1 hr with 5,6-DHT or 5,7-DHT (10^{-3} M), 5-HT levels fell to 45 ± 3% and 50 ± 2%, respectively, of control values, an effect comparable to that induced by metaraminol (36 ± 2%) (FIGURE 3).

Morphologic Changes

Electron micrographs of rat platelets incubated with 5,6-DHT or 5,7-DHT showed significant enlargement and an increased number of osmiophilic cores of 5-HT organelles (FIGURE 4), despite the considerable reduction of endogenous 5-HT levels (FIGURE 3). In contrast, treatment with 5,6-DHT and 5,7-DHT did not induce any significant ultrastructural changes in platelets of rats with storage pool deficiency, that is, fawn-hooded rats (FIGURE 5).

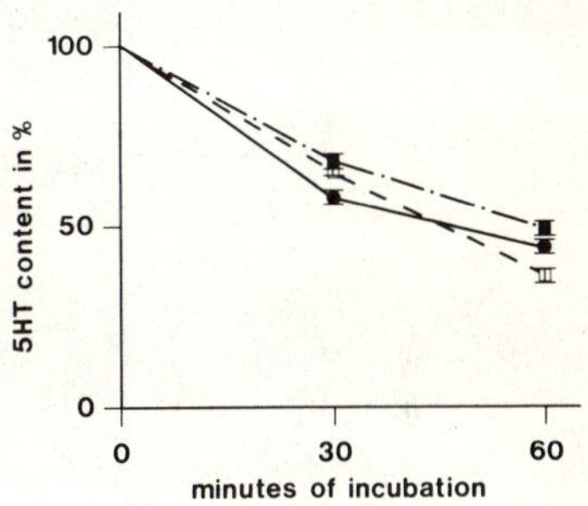

FIGURE 3. Content of 5-hydroxytryptamine (5-HT) in isolated rat blood platelets incubated in plasma with 5,6-DHT (●), 5,7-DHT (■), or metaraminol (□) at a concentration of 10^{-3} M. All results are percentages of control values (5-HT content in platelets incubated in plasma alone) and are means ± SEM of from three to six daily determinations. Absolute 5-HT control value: 11.25±1.64 nmol/mg of protein.

Metaraminol, which lacks intrinsic osmiophilic properties, decreased platelet 5-HT values in both types of platelets, without increasing the osmiophilia of 5-HT organelles (FIGURES 3 & 4). Under all of these experimental conditions, no ultrastructural damage to platelets was detected.

DISCUSSION

Biochemical Findings

An important tool for the study of the functional role of biogenic amines in the central nervous system became available with the introduction of a rather specific neurotoxic agent for CA pathways, 6-hydroxydopamine.[34] Following this approach, two hydroxylated 5-HT analogs, 5,6-DHT and 5,7-DHT, were investigated as possible candidates for specific neurotoxic effects on 5-HT pathways.[1, 2, 4, 9] The degree of selectivity of fiber destruction produced by these

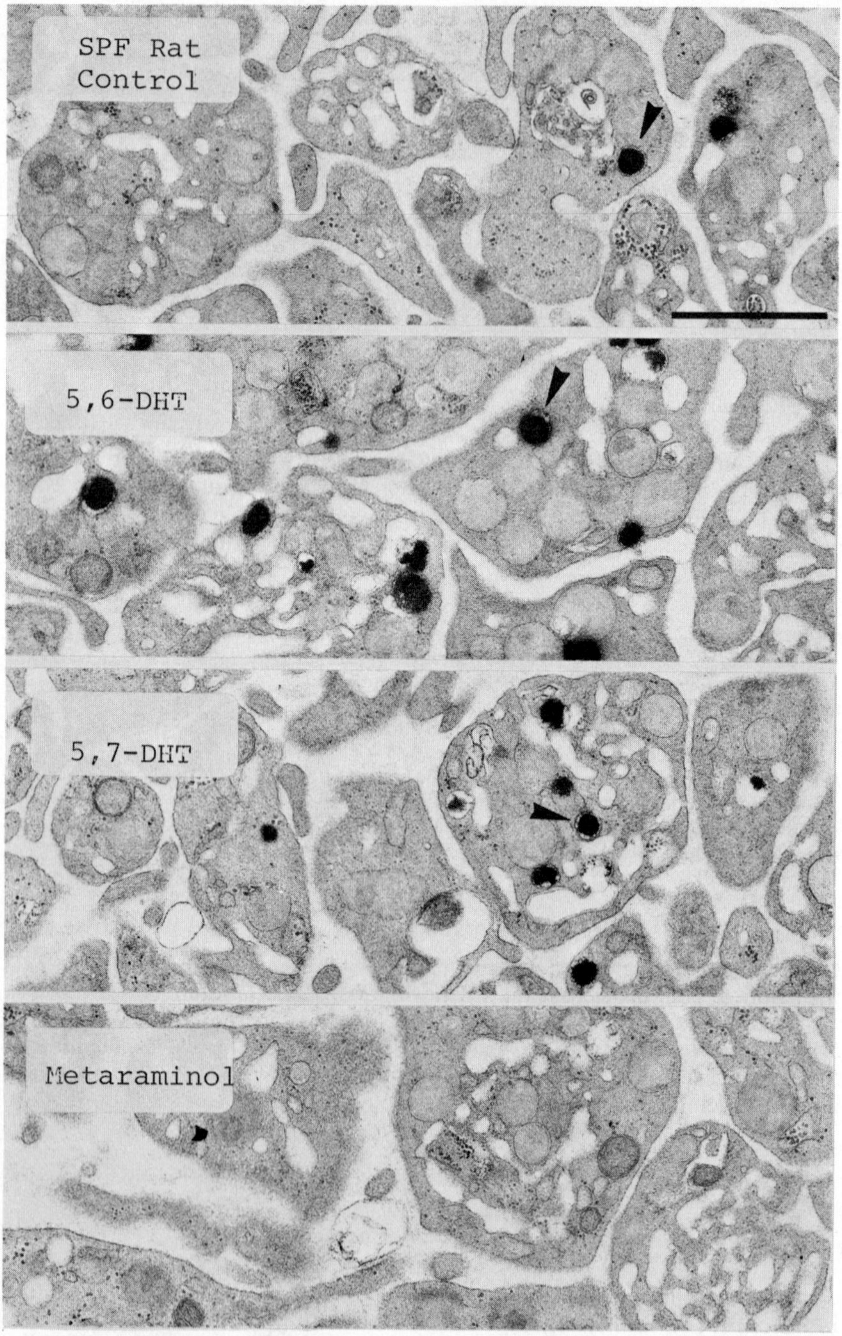

FIGURE 4. Ultrastructural aspects of platelets from SPF albino rats incubated in plasma with 5,6-DHT, 5,7-DHT, or metaraminol (10^{-3} M). Note the increases in size and number of amine storage organelles (arrows) after adding 5,6-DHT, a finding that is not as pronounced after 5,7-DHT; in platelets incubated with metaraminol, such organelles are absent. Bar = 1 μm.

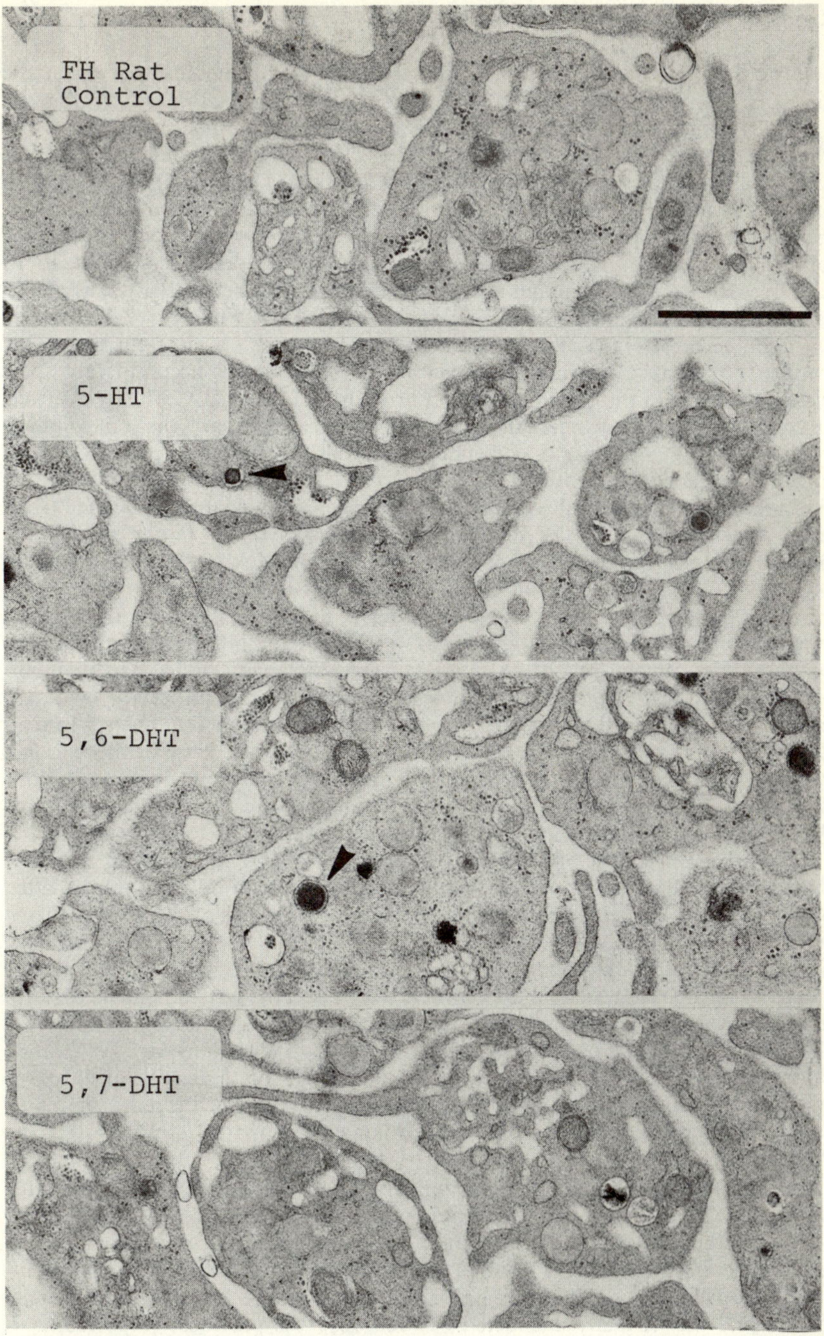

FIGURE 5. Ultrastructural aspects of platelets from fawn-hooded (FH) rats, which have storage pool deficiency, incubated in plasma with 5-HT, 5,6-DHT, or 5,7-DHT (10^{-3} M). Note the virtual absence of amine storage organelles under all experimental conditions. Bar = 1 μm.

two neurotoxins has so far been evaluated mainly by measuring monoamine levels.

By this means, we have observed, in agreement with other authors,[23, 29] that the biochemical damage that occurs after 5,6-DHT is injected intracerebroventricularly seems to be restricted to 5-HT neurons; DA and NA levels remain practically unaltered.[5] We believe, however, that measurement of amine levels per se does not represent a sufficiently sensitive index of specificity in the case of minimal damage. In fact, uptake studies and histofluorescence investigations clearly indicate a concomitant cytotoxic effect on DA neurons after 5,6-DHT treatment in the absence of a significant decrease of DA values.[35, 36] Stereotaxic injection of 5,6-DHT in the MFB confirms this point in a clear-cut way. Our results show that injection of 5,6-DHT in the MFB causes a considerable persistent decrease in the telediencephalon of both 5-HT and DA, while NA levels remain unchanged. Under these experimental conditions, the concomitant reduction of uptake capacity correlates quite well with the marked fall of 5-HT and DA levels due to neuronal degeneration.[29] It seems, therefore, that despite the presence of unspecific mechanical damage, unavoidably associated with local injection (about 10–15%),[29] stereotaxic application in the vicinity of monoaminergic axons is a more sensitive method for assessing the actual selectivity of neurotoxic agents than are injections into ventricular spaces.

With 5,7-DHT, too, stereotaxic injection in the MFB allowed a more accurate evaluation of its noxious action. Even when taking into account the degree of unspecific damage related to many variables and very difficult to maintain within a constant range,[37–39] it follows that this neurotoxin exerts cytotoxic effects on not only 5-HT neurons but also slight effects on DA fibers (TABLES 8 & 9), as observed with 5,6-DHT. Moreover, probably because of the intrinsic variables of the stereotaxic approach,[37, 38] the damage does not seem to be dose dependent, because 5,7-DHT at 4 μg, but not at 6 μg, induced a significant decrease of DA (TABLE 8).

From measurements of 5-HIAA, HVA, and MOPEG-SO_4, it became evident that the extension of damage to monoaminergic neurons is reflected much better by the contents of their main metabolites than by the levels of the amines themselves. We found that on Day 2, and even more so at Day 10, after injection of 5,7-DHT, both 5-HIAA (15% of control values) and HVA (30% of control values) had markedly decreased (TABLE 9). The NA system remained functionally unaffected, as indicated by unaltered levels of MOPEG-SO_4 and NA (TABLES 8 & 9). The occasional minor extension of alterations in 5-HIAA and HVA levels to the side contralateral to the injection, while indicative of diffusion from the injection area, supports the assumption that the decrease of HVA cannot be ascribed merely to mechanical damage but that it is due, in fact, to the neurotoxicity of the compound.

The functional capacity of monoaminergic neurons after chemical lesioning was tested by studying their feedback activation in response to blockade of 5-HT, DA, and NA receptors by administration of methiothepin. The markedly reduced acceleration of 5-HT turnover clearly demonstrates the major impairment of 5-HT neurons; the increase of DA turnover was less pronounced, and that of NA was virtually unaffected (TABLE 10). As previously reported,[1, 3, 4] 5,7-DHT applied intracerebroventricularly destroyed NA neurons, in addition to 5-HT ones, if the animals were not pretreated with compounds that reduce NA uptake or with MAOIs. In the presence of DMI or MAP pretreatment, 5-HT was the only amine affected (TABLE 2); however, when we considered the amine metabolites as indices of functional lesioning, we were able to detect concomitant changes in the DA and NA systems: HVA had decreased to 72% of control values, and MOPEG-SO_4 had slightly increased, even 10 days after 5,7-DHT injection (TABLE 3). The increase of MOPEG-SO_4 is difficult to interpret. It could be due to either a neuronal compensatory activity as a consequence of slight damage or to disinhibition subsequent to suppression of possible inhibition of the NA system by 5-HT.[56] It has been reported that long-acting MAOI prevents the toxic effect of 5,7-DHT on NA fibers. For practical reasons, that is, to reduce the interval between 5,7-DHT application and the time at which a reliable assessment of amine levels can be made, the use of a short-acting MAOI was preferred. Harmaline protected NA and DA fibers from destruction by 5,7-DHT, as indicated by normal NA and DA levels (TABLE 5) and by reductions of 75 and 65%, respectively, of 5-HT and 5-HIAA levels. At present, no reasonable explanation for the protective effect of the MAOI is available.[9]

Chlorimipramine and Lilly 110,140 were unable to protect 5-HT fibers from the neurotoxic effect of 5,7-DHT.[9] In contrast, Lu 10–171, a potent 5-HT uptake inhibitor, was active in this respect. The almost complete protection provided by Lu 10–171 of 5-HT neurons against the neurocytotoxic action of 5,7-DHT lends further evidence for uptake of 5,7-DHT into 5-HT-containing fibers.[1, 4, 9] However, the fact that Lu 10–171 did not prevent decreases of MOPEG-SO_4 and HVA indicates that this compound does not prevent uptake of 5,7-DHT into CA fibers, in contrast to the action of DMI and MAP, and, hence, seems to be a rather selective blocker of 5-HT uptake.

Mounting Behavior

p-Chlorophenylalanine was the first drug reported to induce active male-to-male mounting when injected repeatedly in rats. Since *p*-CPA decreases 5-HT levels, by inhibition of tryptophan hydroxylase, the resultant hypersexuality was interpreted as the consequence of removal of an inhibitory action of 5-HT.[40] Furthermore, it was reported that MAOI potentiated the mounting induced by *p*-CPA, suggesting a stimulatory role of the CA.[11, 12] This hypothesis is supported by the fact that repeated injections of L-Dopa alone or in combination with the peripheral decarboxylase inhibitor benserazide caused marked sexual stimulation in rats. By this treatment, both 5-HT depletion and dopaminergic stimulation were achieved, and, therefore, it was proposed that this animal model could explain the hypersexuality occasionally seen in parkinsonian patients during therapy with L-Dopa.[41]

The reported increased mounting after treatment with apomorphine, a typical DA agonist, further supports the notion of a possible stimulatory effect of DA on sexual activity.[42, 43] The suppression of mounting activity by haloperidol or 5-hydroxytryptophan in *p*-CPA-pretreated rats reinforces the concept that mounting behavior is under dual control, namely, dopaminergic stimulation and serotonergic inhibition.

The increased mounting rate observed in rats injected with 5,6-DHT or 5,7-DHT intracerebroventricularly provides additional evidence for the inhibitory role of 5-HT. In both cases, depletion of 5-HT was associated with mounting, whereas restitution of 5-HT to normal levels by 5-hydroxytryptophan treatment abolished mounting.[5, 41]

Some insight into the possible participation of NA in mounting behavior is provided by experiments in which massive depletion of NA and comparatively minor reduction of DA were induced by 6-hydroxydopamine in combination with 5,6-DHT. In these animals, mounting behavior was not quantitatively different from that observed after treatment with 5,6-DHT alone. These findings, while not excluding the possibility of DA participation, since about 50% of the DA was still present, seem to rule out a role for NA, because the extremely low levels of NA that remained (17% of control values) could hardly exert a normal stimulatory activity (TABLE 1). We also observed mounting after treatment with 5,6-DHT in guinea pigs and rabbits. After injection of 5,6-DHT or 5,7-DHT and *p*-CPA, mounting was also observed in female rats.

The more marked depletion of 5-HT that was obtained by intracerebroventricularly applied 5,7-DHT did not elicit a higher frequency of mounting than that seen after 5,6-DHT treatment, indicating that further reduction of 5-HT is not necessarily reflected by enhanced mounting activity.[41]

Lu 10–171, a potent 5-HT uptake blocker, prevented both 5-HT depletion and the occurrence of mounting behavior. This observation is additional evidence that 5-HT exerts inhibitory control over sexual behavior. The fact that female animals treated with *p*-CPA or with 5,6-DHT and 5,7-DHT exhibit mounting behavior that is very similar to that observed with male animals (TABLES 6 & 7) suggests a common cerebral mechanism.[44, 45] The questions arise as to whether and how this central mechanism is modulated by hormones.

After treatment with *p*-CPA, we have observed the occurrence of mounting in adult and prepubertal male rats (weight 50 g) and in castrated adult male rats (TABLE 7). This finding suggests that testosterone, contrary to previous findings,[11, 40] is apparently not necessary for the expression of mounting behavior. Moreover, the presence of mounting also in prepubertal and ovariectomized adult female rats excludes the role of estrogens as possible triggers of mounting behavior.[46–48]

Finally, it should be noted that mounting behavior is not a general feature in all animal species after treatment with *p*-CPA and 5,6-DHT; these compounds are unable to elicit mounting responses in mice and monkeys.[54]

In conclusion, we must emphasize that this pharmacologically induced mounting response is probably totally unrelated to mating. Moreover, it cannot represent an acquired behavior, because it also occurs in rats isolated 2 weeks after birth. It is possible that primordial genetically transmitted information is present somewhere in the neuronal network, perhaps at the level of the anterior hypothalamus or in some limbic structures.[48, 49]

Turning Behavior

We have proposed the use of rats unilaterally lesioned by 5,6-DHT in the MFB as an alternative model for studying turning behavior and the possible interactions of 5-HT in modulating this response. In rats lesioned with 6-hydroxydopamine or 5,6-DHT in one MFB, rotational behavior was induced by DA receptor agonists and DA releasers that was qualitatively and quantitatively similar. Comparing these two animal models, we were able to demonstrate that *d*-LSD, previously considered to be a pure 5-HT agonist, also stimulates DA receptors.[6, 55] Lisuride, an ergot derivative, which, similar to *d*-LSD, stimulates both 5-HT and DA receptors,[31, 32] was also capable of inducing turning responses away from the lesioned side and of much longer duration than that achieved with *d*-LSD. The ability of lisuride to retard DA turnover, as measured by HVA levels, and to induce turning at a very low dose (50 μg/kg, ip) could be the consequence of either stronger DA receptor stimulation or of a longer half-life in the brain, as compared with that of *d*-LSD. At variance with *d*-LSD, for which there was no change in its effect on circling behavior if administered on 5 consecutive days, daily application of lisuride resulted in a marked enhancement of rotational movements but only during the first hour after injection. In this connection, it is interesting to note that lisuride acts only after a latency of 30 min, when injected for the first time, whereas rotation started almost immediately on repeated applications.

Interestingly, not only was the onset of turning more rapid after repeated applications of lisuride, but also other behavioral signs in intact animals, such as postural tremor and sniffing, which normally appear after a lag time of about 10–15 min following the first injection, occurred almost immediately in response to repeated injections of lisuride. These data suggest that a kind of supersensitivity is induced by lisuride, which is one of the most potent DA receptor agonists available.

In contrast to findings reported by various authors,[50, 51] we were unable, in rats lesioned in the MFB with 5,6-DHT, to demonstrate a possible inhibitory role of 5-HT on turning behavior, since pretreatment with *p*-CPA or 5-hydroxytryptophan did not per se induce circling responses or modify the rotation induced by *d*-methamphetamine, apomorphine, and *d*-LSD, although these pretreatments were quite effective in decreasing or increasing, respectively, rat brain 5-HT levels. In addition, after the unilateral lesion induced by 5,7-DHT, which resulted in a relevant depletion of 5-HT, lisuride, apomorphine, and *d*-methamphetamine were practically ineffective in inducing turning behavior. It appears, therefore, that degeneration of 5-HT neurons in the pathways that course through the MFB does not play a role in eliciting or modulating the turning response, contrary to other reports.[50, 51, 57] Finally, our data that demonstrate a relative lack of specificity for the two neurotoxins indicate that one must interpret with caution results obtained in behavioral studies with animals lesioned by use of 5,6-DHT or 5,7-DHT, because the conclusion that the effects observed are purely serotonergic in nature might be unwarranted.

Blood Platelets

It has been shown previously that 5,6-DHT reduces the 5-HT content of rabbit platelets incubated *in vitro*. Moreover, isolation by density gradient centrifugation of a pure fraction of 5-HT granules from guinea pig platelets incubated with 5,6-DHT revealed that despite extremely low 5-HT levels, the storage granules were strongly electron dense.[52]

These data indicate that 5,6-DHT, after crossing the cytoplasmic membrane, accumulates in the subcellular organelles in which 5-HT is specifically stored.[22, 52] The present findings on isolated rat platelets confirm the ability of 5,6-DHT and 5,7-DHT to displace endogenous 5-HT and to accumulate in the amine storage organelles. In contrast to the hydroxylated 5-HT analogs, metaraminol-induced 5-HT depletion is not accompanied by the appearance of 5-HT electron-dense granules.

Platelets of fawn-hooded rats, which have a hereditary storage pool deficiency, possess very few 5-HT organelles in their platelets, as revealed by electron microscopy. In accordance with the postulated abnormality of 5-HT and nucleotide storage,[53] such platelets display lack of accumulation of 5-HT, 5,6-DHT, or 5,7-DHT (FIGURE 5).

CONCLUSIONS

Cerebral Monoamines and Their Main Metabolites

In the rat, intracerebroventricular injection of both 5,6-DHT and 5,7-DHT (the latter under protection of specific NA uptake blockers) induced a marked decrease of cerebral 5-HT levels, without significantly changing CA values. In contrast, when injected stereotaxically into the MFB, 5,6-DHT markedly and 5,7-DHT slightly decreased cerebral levels of DA, in addition to those of 5-HT. After intracerebroventricular or local injection in the MFB, 5,7-DHT reduced not only 5-HIAA but also HVA levels. These reductions are much more striking in the case of MFB lesions. Further evidence that not only 5-HT but also DA neurons are affected by 5,7-DHT is that in rats lesioned with 5,7-DHT (intracerebroventricularly or stereotaxically injected in the MFB), the enhancement of 5-HT and DA turnover induced by methiothepin was markedly diminished. MOPEG-SO_4 was moderately increased, rather than reduced, in rats injected intracerebroventricularly with 5,7-DHT.

In conclusion, changes in amine metabolites seem to better reflect the extension of damage to monoaminergic neurons than do alterations of the concentrations of the amines themselves.

Mounting Behavior

Enhancement of mounting activity, similar to that seen after systemic injection of *p*-CPA, was induced by intracerebroventricularly applied 5,6-DHT and 5,7-DHT in intact adult male and female rats and in castrated, ovariectomized, or prepubertal rats. Brain 5-HT levels were reduced in all of these experimental situations. Injection of 5-hydroxytryptophan or haloperidol suppressed this enhanced sexual activity.

It is concluded that mounting is a behavioral repertoire that is innate in both male and female animals and is modulated by 5-HT and DA neuronal activity and independently of sex hormones.

Turning Behavior

Lisuride, *d*-LSD, and apomorphine induce contralateral turning behavior in rats injected with 5,6-DHT in one MFB, which can be blocked by neuroleptics. This finding underlines the fact that the three compounds are DA agonists, lisuride being the most potent and long-acting one. Regarding 5-HT, our data are at variance with those of other reports and, rather, exclude a modulatory role of this amine in turning behavior.

Blood Platelets

Electron microscopic studies indicate that 5,6-DHT and 5,7-DHT accumulate *in vitro* in the 5-HT organelles of rat blood platelets. This accumulation is accompanied by release of endogenous 5-HT. Metaraminol also enters platelet 5-HT organelles and liberates endogenous 5-HT but is undetectable by electron microscopy because it lacks osmiophilic properties. Platelets from animals with storage pool deficiency (fawn-hooded rats) exhibit reduced intrinsic storage capacity, because they accumulate *in vitro* much less 5-HT, 5,6-DHT, and 5,7-DHT.

Acknowledgments

It is our pleasure to thank Dr. J. G. Richards, who performed and discussed the electron microscopy studies, and Dr. W. Lotz, who kindly supplied the castrated animals. Thanks are also due to Dr. W. Haefely for critically reading the manuscript. Excellent technical assistance was provided by H. Ehrsam, U. Humbel, I. Kopyto, M. Mangold, B. Ott, and M. Rieffel. We also thank Miss E. Nägelin for her untiring help in typing the manuscript.

References

1. Björklund, A., A. S. Horn, H. G. Baumgarten, A. Nobin & H. G. Schlossberger. 1975. Neurotoxicity of hydroxylated tryptamines: structure-activity relationships. Acta Physiol. Scand. Suppl. **429:** 7–58.
2. Baumgarten, H. G. & A. Björklund. 1976. Neurotoxic indoleamines and monoamine neurons. Annu. Rev. Pharmacol. **16:** 101–111.
3. Björklund, A., H. G. Baumgarten & A. Rensch. 1975. 5,7-Dihydroxytryptamine: improvement of its selectivity for serotonin neurons in the CNS by pretreatment with desipramine. J. Neurochem. **24:** 833–835.
4. Gerson, S. & R. J. Baldessarini. 1975. Selective destruction of serotonin terminals in rat forebrain by high doses of 5,7-dihydroxytryptamine. Brain Res. **85:** 140–145.
5. Da Prada, M., M. Carruba, R. A. O'Brien, A. Saner & A. Pletscher. 1972. The effect of 5,6-dihydroxytryptamine on sexual behavior of male rats. Eur. J. Pharmacol. **19:** 288–290.
6. Pieri, M., L. Pieri, A. Saner, M. Da Prada & W. Haefely. 1975. A comparison of drug-induced rotation in rats lesioned in the medial forebrain bundle with 5,6-dihydroxytryptamine or 6-hydroxydopamine. Arch. Int. Pharmacodyn. **217:** 118–130.
7. Björklund, A., H. G. Baumgarten, L. Lachenmayer & E. Rosengren. 1975. Recovery of brain noradrenaline after 5,7-dihydroxytryptamine induced axonal lesions in the rat. Cell Tissue Res. **161:** 145–155.
8. Gerson, S., R. J. Baldessarini & S. C. Wheeler. 1974. Biochemical effects of dihydroxylated tryptamines on central indolamine neurons. Brain Res. **13:** 987–1004.
9. Breese, G. R. & B. R. Cooper. 1975. Behavioral and biochemical interactions of 5,7-dihydroxytryptamine with various drugs when administered intracisternally to adult and developing rats. Brain Res. **98:** 517–527.
10. Hole, K., K. Fuxe & G. Jonsson. 1976. Behavioral effects of 5,7-dihydroxytryptamine lesions of ascending 5-hydroxytryptamine pathways. Brain Res. **107:** 385–399.

11. GESSA, G. L. & A. TAGLIAMONTE. 1974. Role of brain monoamines in male sexual behavior. Life Sci. **14:** 425–436.
12. TAGLIAMONTE, A., P. TAGLIAMONTE, G. L. GESSA & B. B. BRODIE. 1969. Compulsive sexual activity induced by p-chlorophenylalanine in normal and pinealectomized male rats. Science **166:** 1433–1435.
13. UNGERSTEDT, U., A. AVEMO, E. AVEMO, T. LJUNGBERG & C. RANJE. 1973. Animal models of Parkinsonism. Advan. Neurol. **3:** 257–271.
14. WALDMEIER, P., P. DE HERDT & L. MAÎTRE. 1974. Simultaneously automated estimation of noradrenaline and dopamine in brain tissue. Clin. Chem. **20:** 81–83.
15. SNYDER, S., J. AXELROD & M. ZWEIG. 1965. A sensitive and specific fluorescence assay for tissue serotonin. Biochem. Pharmacol. **14:** 831–835.
16. MEEK, J. L. & N. H. NEFF. 1972. Fluorometric estimation of 4-hydroxy-3-methoxyphenylethyleneglycol sulphate in brain. Brit. J. Pharmacol. **45:** 435–441.
17. KELLER, H. H., R. SCHAFFNER & W. HAEFELY. 1976. Interaction of benzodiazepines with neuroleptics at central dopamine neurons. Arch. Pharmacol. **294:** 1–7.
18. MONACHON, M. A., W. P. BURKARD, M. JALFRE & W. HAEFELY. 1972. Blockade of central 5-hydroxytryptamine receptors by methiothepin. Arch. Pharmacol. **274:** 192–197.
19. KELLER, H. H., G. BARTHOLINI & A. PLETSCHER. 1973. Increase of 3-methoxy-4-hydroxyphenyl ethylene glycol in rat brain by neuroleptic drugs. Eur. J. Pharmacol. **23:** 183–186.
20. DA PRADA, M., G. BARTHOLINI & A. PLETSCHER. 1965. Effect of monoamine liberators on the metabolism of 5-hydroxytryptamine in blood platelets. Biochem. Pharmacol. **14:** 1721–1726.
21. BOGDANSKI, D. F., A. PLETSCHER, B. B. BRODIE & S. UDENFRIEND. 1956. Identification and assay of serotonin in brain. J. Pharmacol. Exp. Ther. **117:** 82–88.
22. TRANZER, J. P., M. DA PRADA & A. PLETSCHER. 1966. Ultrastructural localization of 5-hydroxytryptamine in blood platelets. Nature (London) **212:** 1574–1575.
23. BAUMGARTEN, H. G., D. D. EVETTS, R. B. HOLMAN, L. L. IVERSEN, M. VOGT & G. WILSON. 1972. Effects of 5,6-dihydroxytryptamine on monoaminergic neurons in the central nervous system of the rat. J. Neurochem. **19:** 1587–1597.
24. BAUMGARTEN, H. G., L. LACHENMAYER, A. BJÖRKLUND, A. NOBIN & E. ROSENGREN. 1973. Long-term recovery of serotonin concentrations in the rat CNS following 5,6-dihydroxytryptamine. Life Sci. **12:** 357–364.
25. JACOBY, J. H., L. D. LYTLE & M. F. NELSON. 1974. Long-term effects of 5,7-dihydroxytryptamine on brain monoamines. Life Sci. **14:** 909–919.
26. MANARA, L., C. CERLETTI & T. MANNINI. 1976. Prevention by calcium administration of reserpine action on rat brain noradrenaline stores: a reappraisal. Res. Commun. Chem. Pathol. Pharmacol. **14:** 471–486.
27. BAUMGARTEN, H. G. & H. G. SCHLOSSBERGER. 1973. Effects of 5,6-dihydroxytryptamine on brain monoamine neurons in the rat. *In* Serotonin and Behavior. J. Barches & E. Usdin, Eds. : 209–229. Academic Press, Inc. New York, N.Y.
28. CHRISTENSEN, A. V., B. F. JALLAND, V. PEDERSEN, P. DANNESKIØLD-SAMSØE & O. SVENDSEN. 1977. Pharmacology of a new phthalane (Lu 10–171), with specific 5-HT uptake inhibiting properties. Eur. J. Pharmacol. **41:** 153–162.
29. SANER, A., L. PIERI, J. MORAN, M. DA PRADA & A. PLETSCHER. 1974. Decrease of dopamine and 5-hydroxytryptamine after intracerebral application of 5,6-dihydroxytryptamine. Brain Res. **76:** 109–117.
30. BJÖRKLUND, A., H. G. BAUMGARTEN & A. NOBIN. 1974. Chemical lesioning of central monoamine axons by means of 5,6-dihydroxytryptamine and 5,7-dihydroxytryptamine. Advan. Biochem. Psychopharmacol. **10:** 13–33.
31. KEHR, W. 1977. Effect of lisuride and other ergot derivatives on monoaminergic mechanisms in rat brain. Eur. J. Pharmacol. **41:** 261–273.

32. KELLER, H. H., W. P. BURKARD & M. DA PRADA. 1977. Lisuride and d-LSD: effects on the monoaminergic system in the rat brain. Experientia **33:** 806.
33. PIERI, M., R. SCHAFFNER & W. HAEFELY. 1977. Central dopamine-receptor stimulation by LSD and lisuride. Experientia **33:** 808.
34. THOENEN, H. & J. P. TRANZER. 1968. Chemical sympathectomy by selective destruction of adrenergic nerve endings with 6-hydroxydopamine. Arch. Pharmacol. **261:** 271–288.
35. BAUMGARTEN, H. G., A. BJÖRKLUND, A. S. HORN & H. G. SCHLOSSBERGER. 1974. Studies on the neurotoxic properties of hydroxylated tryptamines. *In* Dynamics of Degeneration and Growth in Neurons. K. Fuxe, L. Olson & Y. Zotterman, Eds. : 153–167. Pergamon Press. New York, N.Y.
36. HEIKKILA, R. E. & G. COHEN. 1973. The inhibition of H^3-biogenic amine uptake by 5,6-dihydroxytryptamine: a comparison with the effects of 6-hydroxydopamine. Eur. J. Pharmacol. **21:** 66–69.
37. WILLIS, G. L., G. SINGER & B. K. EVANS. 1976. Intracranial injections of 6-OHDA. Comparison of catecholamine-depleting effects of different volumes and concentrations. Pharmacol. Biochem. Behav. **5:** 207–213.
38. EVANS, B. K., S. ARMSTRONG, G. SINGER, R. D. COOK & G. BURNSTOCK. 1976. Intracranial injection of drugs: comparison of diffusion of 6-OHDA and guanethidine. Pharmacol. Biochem. Behav. **3:** 205–217.
39. BLOOM, F. E. 1975. Monoaminergic neurotoxins: are they selective? J. Neural Transmission **37:** 183–187.
40. GESSA, G. L. & A. TAGLIAMONTE. 1972. Role of brain monoamines in controlling sexual behavior in male animals. *In* Psychopharmacology, Sexual Disorders and Drug Abuse. T. A. Ban *et al.,* Eds. : 451–462. North-Holland Publishing Company. Amsterdam, The Netherlands.
41. DA PRADA, M., M. CARRUBA, A. SANER, R. A. O'BRIEN & A. PLETSCHER. 1973. The action of L-dopa on sexual behavior of male rats. Brain Res. **55:** 383–389.
42. TAGLIAMONTE, A., W. FRATTA, M. DEL FIACCO & G. L. GESSA. 1973. Evidence that brain dopamine stimulates copulatory behavior in male rats. Riv. Farmacol. Ter. **IV:** 177–181.
43. MALMNÄS, C. O. 1973. Monoaminergic influence on testosterone activated copulatory behavior in the castrated male rat. Acta Physiol. Scand. Suppl. **395:** 1–128.
44. HOYLAND, V. J., E. E. SHILLITO & M. VOGT. 1970. The effect of parachlorophenylalanine on the behavior of cats. Brit. J. Pharmacol. **40:** 659–667.
45. VAN DE POLL, N. E., H. VAN DIS & B. BERMOND. 1977. The induction of mounting behavior in female rats by p-chlorophenylalanine. Eur. J. Pharmacol. **41:** 225–229.
46. CHRISTENSEN, L. W. & L. G. CLEMENS. 1975. Blockade of testosterone-induced mounting behavior in the male rat with intracranial application of the aromatization inhibitor, androst-1,4,6,-triene-3,17-dione. Endocrinology **97:** 1545–1551.
47. WHALEN, R. E. 1974. Estrogen-progesterone induction of mating in female rats. Hormones Behav. **5:** 157–162.
48. CHRISTENSEN, L. W. & L. G. CLEMENS. 1974. Intrahypothalamic implants of testosterone or estradiol and resumption of masculine sexual behavior in long-term castrated male rats. Endocrinology **95:** 984–990.
49. SCHREINER, L. & A. KLING. 1956. Rhinencephalon and behavior. Amer. J. Physiol. **184:** 486–490.
50. BALDESSARINI, R. J., T. T. AMATRUDA, F. F. GRIFFITH & S. GERSON. 1975. Differential effects of serotonin on turning and stereotypy induced by apomorphine. Brain Res. **93:** 158–163.
51. COSTALL, B., R. J. NAYLOR, C. D. MARSDEN & C. J. PYCOCK. 1976. Circling behavior produced by asymmetric medial raphe nuclei lesions in rats. J. Pharm. Pharmacol. **28:** 248–249.

52. Da Prada, M., R. A. O'Brien, J. P. Tranzer & A. Pletscher. 1973. The effect of 5,6-dihydroxytryptamine on uptake, storage and metabolism of 5-hydroxytryptamine by blood platelets. J. Pharmacol. Exp. Ther. **186:** 213–219.
53. Tschopp, T. B. & M. B. Zucker. 1972. Hereditary defect in platelet function in rats. Blood **40:** 217–226.
54. Massotti, M., A. Scotti de Carolis & G. Longo. 1974. Effects of three dihydroxylated derivatives of tryptamine on the behavior and on the brain amine content in mice. Pharmacol. Biochem. Behav. **2:** 769–775.
55. Pieri, L., M. Pieri & W. Haefely. 1974. LSD as an agonist of dopamine receptors in the striatum. Nature (London) **252:** 586–588.
56. Renaud, B., M. Buda, B. D. Lewis & J. F. Pujol. 1975. Effects of 5,6-dihydoxytryptamine on tyrosine-hydroxylase activity in central catecholaminergic neurons of the rat. Biochem. Pharmacol. **24:** 1739–1742.
57. Milson, J. A. & C. J. Pycock. 1976. Effects of drugs acting on cerebral 5-hydroxytryptamine mechanisms on dopamine-dependent turning behavior in mice. Brit. J. Pharmacol. **56:** 77–85.

INFLUENCE OF *p*-CHLOROAMPHETAMINE AND *p*-CHLOROPHENYLALANINE ON FEMALE MATING BEHAVIOR

Frank P. Zemlan

The Rockefeller University
New York, New York 10021

Female sexual receptivity in rodents, and most mammals, depends on gonadal hormones.[23] Castration eliminates sexual behavior, which may then be reinstated with exogenous hormone administration. In the present paper, data are presented that suggest that the serotonin [5-hydroxytryptamine (5-HT)] neurotoxin *p*-chloroamphetamine (*p*-CA) may replace, in part, exogenous hormone administration.

p-Chloroamphetamine has a biphasic effect on neural activity. Initially, *p*-CA releases 5-HT from vesicles[15, 22] and inhibits its reuptake.[1, 22] Both effects tend to increase activity at the 5-HT receptor. This initial increase in serotonergic activity produces a rapid termination of hormone-induced mating behavior (TABLE 1). As can be seen, the termination of ongoing mating behavior is complete by 15 min postinjection. This finding suggests that the release of monoamines, particularly 5-HT, suppresses female mating behavior. Similarly, Meyerson[13] reports that 5-hydroxytryptophan precursor treatment reduces female mating in the rat. Fenfluramine, which, like *p*-CA, releases 5-HT,[3, 17] also suppresses female mating,[7] as does stimulation of 5-HT receptors by high doses of LSD.[6, 7] While no drug, including *p*-CA, is completely specific to serotonergic neurons,[3, 8, 15] the various drugs referred to above have in common the ability to alter 5-HT transmission and together suggest that the stimulation of 5-HT receptors suppresses ongoing mating behavior.

While the short-term effect of *p*-CA is to release monoamines, particularly 5-HT, and suppress female mating behavior, the long-term effect of *p*-CA is to enhance female mating. Serotonin depletion after *p*-CA administration is greatest during the first 2 weeks, and brain levels of 5-HT, 5-hydroxyindoleacetic acid, and tryptophan hydroxylase activity are only partially recovered after 4 months.[15] In the present study, 10 days after *p*-CA treatment, forebrain 5-HT levels were depleted by 66% after injection of 10 mg/kg of *p*-CA and by 80% following a 20 mg/kg dose (TABLE 2). Small, nonsignificant decreases in forebrain norepinepherine (NE) and dopamine (DA) occurred after the lower *p*-CA dose; however, following the higher dose, NE levels were significantly reduced, but little effect on DA levels was observed. Three and 5 days after both 10 and 20 mg/kg of *p*-CA, female mating behavior was facilitated (TABLE 2). This facilitation of mating did not appear to be mediated by adrenal steroids; when the release of adrenal steroids was blocked by dexamethasone (Day 7, TABLE 2), facilitation was still observed.

Whereas a relatively clear-cut facilitation occurred after injection of 10 mg/kg of *p*-CA, the effect of 20 mg/kg was less pronounced. The larger *p*-CA dose significantly decreased NE levels and also appears to inhibit tyrosine hydroxylase activity.[11] This finding may explain why the higher *p*-CA dose produced a smaller facilitation than did the lower dose. While it appears that

0077–8923/78/0305–0621 $01.75/0

TABLE 1

SHORT-TERM EFFECT OF *p*-CA (10 mg/kg) ON FEMALE MATING BEHAVIOR INDUCED BY EB (2 μg, THREE TIMES A WEEK) PLUS PROGESTERONE (1.0 mg)

	Lordosis Quotient *		
	Predrug	15 Min	45 Min
p-CPA (6) †	80.0	0.0 ‡	0.0 ‡
Saline (6)	84.5	82.0	82.0

* Lordosis quotient=number of lordoses × 100/number of mounts.
† Number of animals is shown in parentheses.
‡ Significant difference from saline-treated control; $p < 0.01$.

5-HT suppresses female mating, NE may facilitate it.[7] In this way, the higher *p*-CA dose may have affected two systems that have opposite effects on female mating, thereby diminishing the magnitude of the 5-HT-mediated facilitation (TABLE 2).

Consistent with the ability of the 5-HT neurotoxin *p*-CA to produce a long-term facilitation of mating behavior are data obtained with the tryptophan hydroxylase inhibitor *p*-chlorophenylalanine (*p*-CPA). While ovariectomized, adrenalectomized female rats primed with 10 μg of estradiol benzate (EB, applied intramuscularly) and injected with the carrier solution responded to only 16% of the males' mounts ($N = 8$), similarly treated animals that received

TABLE 2

LONG-TERM NEUROTOXIC EFFECT OF *p*-CA ON FEMALE MATING BEHAVIOR AND FOREBRAIN MONOAMINE LEVELS *

	Day 3	Day 5	Day 7	Day 10 †		
p-CA, 10 mg/kg (6) ‡	38.0 §	66.0 ‖	45.0 **	5-HT	62±4 **	(5)
				NE	349±32	(5)
				DA	344±46	(5)
p-CA, 20 mg/kg (6)	34.0	50.0 §	41.0	5-HT	33±6 **	(4)
				NE	323±22 §	(6)
				DA	365±31	(6)
Saline (6)	0.0	4.0	10.0	5-HT	180±7	(4)
				NE	369±16	(5)
				DA	387±54	(4)

* Female rats were ovariectomized and primed with a subthreshold hormone regime [2 μg of EB, three times per week, plus 250 μg of progesterone (im) 4 hr prior to testing]. On Day 7, females were pretreated with dexamethasone (500 μg/kg, ip) 5 hr before testing to block secretion of adrenal progesterone. Monoamine assays were performed on Day 10 as previously described.[24]
† Monoamine assay data are expressed as group means (ng/g of tissue) ± SE.
‡ Number of animals in each group is shown in parentheses.
§ Significant difference from saline-treated control: $p < 0.05$.
‖ Significant difference from saline-treated control: $p < 0.001$.
** Significant difference from saline-treated control: $p < 0.01$.

p-CPA (316 mg/kg, ip, 3 days prior to testing) responded to 64% of the males' mounts ($N = 8$, statistically significant difference; $p < 0.01$).[25] Others have also noted the ability of *p*-CPA or H 22/54 to deplete 5-HT levels and facilitate female mating behavior.[7, 10]

The above data suggest that a 5-HT system suppresses female mating behavior in the rat. We have examined the location of the 5-HT neurons responsible for this effect. In this regard, it is important to note that naturally occurring estrous behavior results from the synergistic action of the two ovarian hormones estrogen and progesterone. Autoradiographic data suggest that [^{3}H]estradiol is preferentially concentrated in the preoptic-anterior hypothalamic area.[14] Lesions of the anterior hypothalamus or adjacent ventromedial hypothalamus decrease or abolish female mating behavior.[5, 16] These autoradiographic and lesion data taken together suggest that the anteromedial portion of the hypothalamus is important for the ability of estrogen to facilitate female mating behavior. We have demonstrated previously that blockade of 5-HT receptors by methysergide

TABLE 3

EFFECT OF INTRACRANIAL HORMONE ADMINISTRATION ON FEMALE MATING BEHAVIOR *

	N	1. Cholesterol	2. Metycaine	3. Progesterone
Central gray	10	12±4.8 †	9±6.1	17±9.5
Ventral mesencephalon	10	6±2.2	20±5.7	39±7.4

* Ovariectomized, adrenalectomized females were administered a subthreshold amount of hormone (10 μg of EB, 48 hr prior to testing); the experimental substance was then applied intracranially (8–11 μg).

† All values are the maximum postdrug response rate minus the predrug control value (mean ± SE, ventral mesencephalon: 1 vs 2, $p < 0.05$; 1 vs 3, $p < 0.01$; 2 vs 3, $p < 0.05$, central gray: all p's > 0.05).

or cinanserin facilitates female mating in estrogen-primed rats.[19] Histochemical and autoradiographic data suggest that 5-HT terminals exist in the anteromedial portion of the hypothalamus and that cell bodies that give rise to these terminals are predominantly located in the mesencephalon.[2, 9] The mesencephalon appears to be important for the ability of progesterone to influence neural activity. Uptake studies suggest that the ventral mesencephalon shows the highest [^{3}H]progesterone uptake of all brain regions in both the rat[20, 24] and guinea pig.[18] This observation suggests that the 5-HT system that suppresses female mating behavior may have cell bodies in a region that preferentially accumulates progesterone and has terminals in an area important for the mediation of lordosis by estrogen. To examine this possibility, we applied progesterone, Metycaine® (anesthetic control), or cholesterol (steroid control) to either the ventral mesencephalon (near areas, B8 and B9 of Dahlstrom and Fuxe[4]) and dorsal mesencephalon (central gray, B7 area). Female rats were tested before intracranial hormone administration and then 30, 90, and 180 min later. As can be seen in TABLE 3, progesterone and, to a smaller extent, metycaine, when applied to the ventral mesencephalon facilitated female mating behavior. Cholesterol administered to the same area was ineffective, as were all of the intracranially applied substances in the central gray-dorsal raphe area. The histologic features of the effective anatomic placements are illustrated in FIGURE 1. While progesterone was effective in facilitating lordosis when applied to the ventral mesencephalon, metycaine was also effective. If progesterone were acting to disinhibit mating by blocking an ascending 5-HT system that suppresses female mating, metycaine, acting to block neural transmission in the same neurons, would also be expected to facilitate mating.

While the progesterone facilitation of female mating may be due to effects on an ascending 5-HT system, other neural systems also exist in the ventral mesencephalon. For example, progesterone may have affected the mesolimbic DA system or the ventral NE bundle.[4] In this regard, it is interesting to note that neurochemical lesions of the ventral NE bundle block the ability of exogenous progesterone to elevate gonadotropin release.[12]

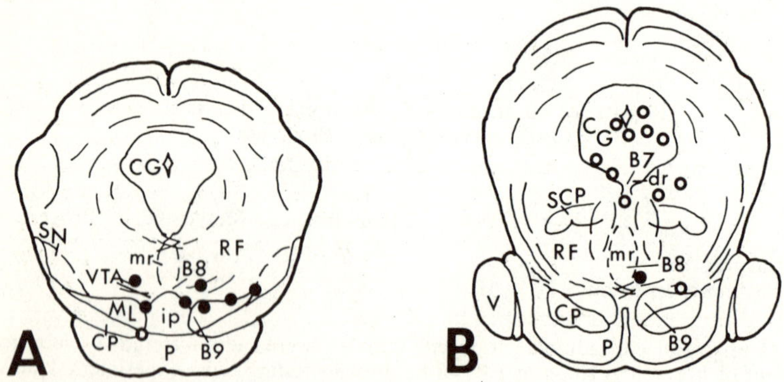

FIGURE 1. Anatomic locations of effective (●) and noneffective (○) progesterone implants in the dorsal (A) and ventral (B) mesencephalon. For further details, see TABLE 3 and the text.

In summary, the present data suggest that 5-HT neurotoxins may be employed to study the neurochemical control of midbrain function. More specifically, female mating behavior, which is dependent on the ovarian hormones estrogen and progesterone, appears to be mediated by a serotonergic system. This system appears to have terminals located in areas shown to preferentially concentrate [^{3}H]estradiol and may have cell bodies in the ventral mesencephalon, an area that preferentially accumulates [^{3}H]progesterone.

REFERENCES

1. CARLSSON, A. 1970. Structural specificity for inhibition of ^{14}C-5-hydroxytryptamine uptake by cerebral slices. J. Pharm. Pharmacol. **22:** 729–731.
2. CONRAD, L. C., C. M. LEONARD & D. W. PFAFF. 1974. Connections of the median and dorsal raphe nuclei in the rat: an autoradiographic and degeneration study. J. Comp. Neurol. **156:** 179–206.
3. COSTA, E., A. GROPPETTI & A. REVUELTA. 1971. Action of fenfluramine on monoamine stores of rat tissue. Brit. J. Pharmacol. **41:** 57–64.

4. Dahlstrom, A. & K. Fuxe. 1964. Evidence for the existence of monoamine containing neurons in the central nervous system: I. Demonstration of monoamines in the cell bodies of brain stem neurons. Acta Physiol. Scand. Suppl. **232:** 1–55.
5. Dorner, G., F. Docke & G. Hinz. 1969. Homo- and hypersexuality in rats with hypothalamic lesions. Neuroendocrinology **4:** 20–24.
6. Eliasson, M., A. Michanek & B. J. Meyerson. 1972. A differential inhibitory action of LSD and amphetamine on copulatory behavior in the female rat. Acta Pharmacol. Suppl. **1:** 22.
7. Everitt, B. J., K. Fuxe, T. Hokfelt & G. Jonsson. 1975. Role of monoamines in the control by hormones of sexual receptivity in the female rat. J. Comp. Physiol. Psychol. **89:** 556–572.
8. Fuller, R. W., C. W. Hines & J. Mills. 1965. Lowering of brain serotonin level by chloroamphetamines. Biochem. Pharmacol. **14:** 483–488.
9. Fuxe, K. 1965. Evidence for the existence of monoamine neurons in the central nervous system. IV. Distribution of monoamine nerve terminals in the central nervous system. Acta Physiol. Scand. Suppl. **247:** 39–85.
10. Gradwell, P. B., B. J. Everitt & J. Herbert. 1975. 5-Hydroxytryptamine in the central nervous system and sexual receptivity of female rhesus monkey. Brain Res. **88:** 281–293.
11. Koe, B. K. & R. F. Corkey. 1976. Inhibition of rat brain tryptophan hydroxylation with *p*-chloroamphetamine. Biochem. Pharmacol. **25:** 31–35.
12. Martinovic, J. V. & S. M. McCann. 1977. Effect of lesions in the ventral noradrenergic tract produced by microinjection of 6-hydroxydopamine on gonadotropin release in the rat. Endocrinology **100:** 1206–1213.
13. Meyerson, B. J. 1964. Central nervous monoamines and hormone induced estrus behavior in the spayed rat. Acta Physiol. Scand. Suppl. **241:** 1–32.
14. Pfaff, D. W. 1968. Uptake of estradiol-17B-H^3 in the female rat brain. An autoradiographic study. Endocrinology **82:** 1149–1155.
15. Sanders-Bush, E., D. A. Gallager & F. Sulser. 1974. On the mechanism of brain 5-hydroxytryptamine depletion by *p*-chloroamphetamine and related drugs and the specificity of their action. Advan. Biochem. Psychopharmacol. **10:** 185–194.
16. Singer, J. J. 1968. Hypothalamic control of male and female sexual behavior in the female rat. J. Comp. Physiol. Psychol. **66:** 738–742.
17. Trulson, M. E. & B. L. Jacobs. 1976. Behavioral evidence for the rapid release of CNS serotonin by PCA and fenfluramine. Eur. J. Pharmacol. **36:** 149–154.
18. Wade, G. N., C. F. Harding & H. H. Feder. 1973. Neural uptake of (1,2-^{3}H) progesterone in ovariectomized rats, guinea pigs and hamsters: correlation with species differences in behavioral responsiveness. Brain Res. **61:** 357–367.
19. Ward, I. L., W. R. Crowley, F. P. Zemlan & D. L. Margules. 1975. Monoaminergic mediation of female sexual behavior. J. Comp. Physiol. Psychol. **88:** 53–60.
20. Whalen, R. E. & B. B. Gorzalka. 1974. Estrogen-progesterone interactions in uterus and brain of intact and adrenalectomized immature and adult rats. Endocrinology **94:** 214–223.
21. Whalen, R. E. & W. G. Luttge. 1971. Role of the adrenal in the preferential accumulation of progestin by mesencephalic structures. Steroids **18:** 141–146.
22. Wong, D. T., J. Horn & R. W. Fuller. 1973. Kinetics of serotonin accumulation into synaptosomes of rat brain—effects of amphetamine and chloramphetamines. Biochem. Pharmacol. **22:** 311–322.
23. Young, W. C. 1961. The hormones and mating behavior. *In* Sex and Internal Secretions. W. C. Young, Ed. Vol. **II:** 1173–1239. The Williams & Wilkins Co. Baltimore, Md.

24. Zemlan, F. P., M. E. Trulson, R. Howell & B. G. Hoebel. 1977. Influence of *p*-chloroamphetamine on female sexual reflexes and brain monoamine levels. Brain Res. **123:** 347–356.
25. Zemlan, F. P., I. L. Ward, W. R. Crowley & D. L. Margules. 1973. Activation of lordotic responding in female rats by suppression of serotonergic activity. Science **179:** 1010–1011.

EFFECTS OF CENTRAL 5,7-DIHYDROXYTRYPTAMINE ON THE MEDIAL HYPOTHALAMIC SYNDROME IN RATS *

Donald V. Coscina †

Section of Neurochemistry
Clarke Institute of Psychiatry
and Departments of Psychology and Psychiatry
University of Toronto
Toronto M5T 1R8, Ontario, Canada

Bilateral damage to the basal aspects of the medial hypothalamic (MH) region is well known for its ability to elicit protracted overeating and subsequent obesity in rats.[1] At the same time, such brain injury predisposes these animals toward dietary "finickiness," that is, overresponsiveness to positive or negative aspects of abrupt sensory changes in the taste or texture of foods or fluids.[2-6] The many behavioral and physiologic similarities between this animal model of obesity and the overfed human condition [7-9] reinforce notions of aberrant functional relationships common to both disorders. Unfortunately, despite extensive research efforts to specify the central nervous system substrates responsible for components of this syndrome, the exact identity of these constituent neural elements remains largely unknown.

One strategy presently in use to help solve this problem is to define the brain neurochemical systems that contribute to this MH syndrome. To this end, evidence has been obtained to suggest that damage to the ventral ascending noradrenergic bundle [10] produces sustained overeating and body weight (BW) gain.[11-18] However, an exclusive role for such norepinephrine (NE) -containing neurons in the MH syndrome seems unlikely.[14-20] In this respect, additional evidence is available to implicate abnormal serotonin [5-hydroxytryptamine (5-HT)] neurotransmission in overeating. More specifically, hyperphagia induced by MH lesions in rats [13-15, 21] or systemic injection of gold thioglucose in mice [16, 17] is associated with reliable depletion of endogenous forebrain 5-HT. For lesioned rats, the magnitude of this depletion correlates inversely with BW gain.[15] In addition, injection of the 5-HT metabolic inhibitor *p*-chlorophenylalanine (*p*-CPA) into the ventricular axis of rats can induce transient hyperphagia and BW gain in a time course compatible with the drug's maximal depletory effects on brain 5-HT.[22] More recent data [23] that demonstrate blockade of such *p*-CPA-induced overeating by systemic injection of 5-HT's precursor 5-hydroxytryptophan support a serotonergic basis for this behavioral change. Also of interest is recent work [23] that shows that this hyperphagia induced by central *p*-CPA seems nonadditive to that produced by MH lesions. This evidence implies that the neuronal systems affected by both treatments are part of the same central network.

Perhaps related to these latter findings are recent observations from this laboratory [21] that prior destruction of the dorsal and median raphe nuclei of

* Supported by funds from the Clarke Institute of Psychiatry.

† Address reprint requests to Section of Biopsychology, Clarke Institute of Psychiatry, 250 College Street, Toronto M5T 1R8, Ontario, Canada.

0077-8923/78/0305-0627 $01.75/0

rats, which alone produced 75% chronic reduction in levels of forebrain 5-HT and its major metabolite 5-hydroxyindoleacetic acid (5-HIAA), substantially blocked the hyperphagia and obesity that followed MH damage. Of particular interest was the apparent specificity of raphe lesions in lowering the level or "set-point"[24] at which BW was regulated. When given a high-fat diet (33% Crisco® Vegetable Shortening plus 67% powdered Purina chow) instead of standard Purina pellet chow, raphe plus MH-lesioned rats displayed the same exaggerated overeating and BW gain as did other rats already made obese by MH lesions. However, even on this highly acceptable diet, the raphe plus MH-lesioned rats did not overeat sufficiently to compensate for their diminished starting BW relative to that of MH-lesioned controls. Other tests in which water supplies were systematically replaced with different sucrose and quinine solutions revealed that raphe plus MH-lesioned rats were apparently as finicky as were obese MH-lesioned rats. These data support earlier work[4] that suggested the presence of two separate but overlapping neural systems in the MH axis that mediate BW regulation and finickiness, respectively. Since raphe lesions alone could account for none of these effects, our data implied that the integrity of ascending 5-HT neurons is important for modulating other brain systems in the selective control of BW regulation.

Histologic and biochemical indices from this previous work[21] indicated that the midbrain damage induced was largely confined, as intended, to 5-HT-containing neurons. However, we could not rule out completely the possibility that such brain insult compromised additional, non-5-HT neural systems. Therefore, the main purpose of the experiments reported here was to determine if chronic depletion of brain 5-HT by a method other than direct raphe lesioning would produce the selective blockade of MH overeating and BW gain seen earlier. To accomplish such depletion, we employed intracisternal injection of the 5-HT neurotoxin 5,7-dihydroxytryptamine (5,7-DHT), in conjunction with systemic desmethylimipramine hydrochloride (desipramine hydrochloride or DMI) pretreatment. The latter agent was used to diminish the nonspecific uptake into, and subsequent neurotoxic action on, NE-containing neurons.[25, 26]

A recent report in the literature states that intraventricular injection of 200 μg of 5,7-DHT after DMI pretreatment in juvenile male rats can produce overeating, apparently to support increased linear growth.[27] A footnote in this report (number 18) claimed that this same drug treatment can also induce overeating and BW gain in adult female rats. Therefore, a second purpose of the present experiment was to determine if we could reliably induce such overeating and weight gain in mature female rats.

Materials and Methods

Subjects

Thirty-six mature (mean BW, 280 g) female albino rats were purchased from Canadian Breeding Farm and Laboratories in Montreal, Quebec. For the duration of the study, all animals were housed in single stainless steel cages with ⅜ in. mesh fronts and bottoms. Cages were isolated in a separate room illuminated from 8 AM to 8 PM daily and maintained at 24 ± 2° C. Except as noted below, rats were given free access throughout these experiments to adequate amounts of fresh Teklad Lab Chow pellets on their cage floors and to fresh tap water in 100-ml calibrated bottles fitted with sipper tubes attached to cage fronts.

Procedure

5,7-Dihydroxytryptamine Injection

Rats were divided into two groups of equal mean BW and designated for intracisternal injection[28] of 5,7-DHT ($n = 20$) or its vehicle ($n = 16$). All rats first received 25 mg/kg of DMI HCl (Norpramin®, Lakeside Laboratories) intraperitoneally 35–55 (mean, 43) min prior to either central injection. After sufficient ether inhalation to permit 30 sec of head immobilization in the ear-bars of a Kopf stereotaxic frame, animals received 200 μg (free base) ‡ of 5,7-DHT creatinine sulfate (Regis) in 20 μl of 1% ascorbic acid-deionized water or 20 μl of this vehicle alone. To minimize the convulsions that can occur after central 5,7-DHT treatment,[29] all rats received 15 mg/kg of sodium pentobarbital (Nembutal®) subcutaneously immediately after intracisternal injections.

Animals were returned to their home cages for 24-hr measurements of food intake, water intake, and BW change for the first week, then once weekly, for a total of 51 days. All intake and BW records were obtained during the midlight cycle (1–3 PM). During this postinjection period, eight 5,7-DHT-injected and three control rats died, leaving 12 and 13 rats, respectively, in each group.

Medial Hypothalamic Lesions

Fifty-two days after intracisternal injections, surviving rats in both groups were further divided to form the final four subgroups: 5,7-DHT plus MH lesion ($n = 7$), vehicle plus MH lesion ($n = 8$), 5,7-DHT plus sham lesion ($n = 5$), and vehicle plus sham lesion ($n = 5$).

Ten minutes after intraperitoneal pretreatment with atropine methyl nitrate (1.5 mg/rat), all animals were anesthetized with Nembutal (40 mg/kg, ip) and placed in the Kopf stereotaxic instrument. Following scalp incision, heads were positioned flat between bregma and lambda skull sutures, and a 3-mm^2 bone flap was removed in the midline along the appropriate rostrocaudal axis. Radiofrequency (Radionics) MH lesions were produced by generating 55° C for 1 min per hemisphere at the following coordinates: 6 mm rostral to the interaural line, 0.5 mm lateral to the midsaggital sinus, and 8.5 mm below the dura mater. After the electrode was withdrawn from the second hemisphere, skull insults were covered with a nonantigenic matrix (Gelfoam®), and incisions were closed with 9-mm stainless steel wound clips. Sham surgery consisted of

‡ This dose of 5,7-DHT after DMI pretreatment was chosen for two reasons. First, pilot work confirmed previous reports[25, 26] of selective brain 5-HT depletion with this regimen. In our hands, 60–70% decrements were obtained 2 weeks after intracisternal administration. This magnitude of depletion compared favorably with that seen in this laboratory after midbrain raphe lesions.[21] Second, the same dose administered intraventricularly has been reported to be effective in eliciting reliable overeating and BW gain 40–80 days later in juvenile male rats or 1–2 weeks later in adult female rats.[27]

all of the above manipulations, except placement of the electrode into the brain.

Animals were again returned to their home cages for assessment of intake and BW changes daily for the first week, then once weekly, for a total of 35 days. In addition to these measures, each rat was tested twice (Days 8 and 36) for 15 min of horizontal locomotory activity in an automated open field. The results of the latter findings will be published elsewhere.[30] During the 5-week postlesioning interval, one 5,7-DHT plus MH-lesioned rat died, reducing the number in this group to six.

Acceptance Tests

Acceptance of drinking water adulterated with varying amounts of sucrose and then quinine sulfate was assessed by use of the design reported in our recent study of rats with raphe plus MH lesions.[21] This procedure involved obtaining initial baseline measures of 24-hr water intake for 3 consecutive days. Then, increasing concentrations of sucrose solutions (1, 2, 4, 8, 16, and 32 g in 100 ml of water) were presented for 24 hr each instead of tap water. Each test was separated by 48 hr of tap water replacement. Following this sequence, decreasing concentrations of quinine sulfate solutions (10, 5, 2.5, and 1.25 mg in 100 ml of water) replaced tap water for 24 hr, again with 48 hr of tap water available between tests. Food pellets were available at all times throughout these tests.

After completing this phase of acceptance testing, rats were weighed and monitored daily for intake of food pellets and tap water for an additional 1 week (baseline) period. Then, all food pellets were removed from home cages, and glass cups were attached by springs to the interior cage fronts. For the next 25 days, these cups were kept amply stocked with a high-fat food source (33% Crisco® Vegetable Shortening plus 67% powdered Teklad Lab Chow) to permit ad lib. intake. Measurements of BW and of 24-hr food and water intakes were made on Days 1, 4, 8, 11, 15, 18, 22, and 25 of this dietary replacement. At the completion of this test, food cups were removed from cages, and standard food pellets were again made freely available to rats for 1 month.

Nasoanal Length

At the end of all behavioral testing but several days prior to sacrificing, all rats were lightly anesthetized with ether, placed in the prone position on a flat table top, and nasoanal length measured in inches.

Sacrifice

Following at least 1 full day of being left completely undisturbed, all rats were sacrificed by decapitation within 30 sec of removal from the colony room between 9 and 11 AM. Blood from cervical wounds was collected into heparinized test tubes for radioimmunoassays[31] of resting levels of plasma growth hormone (GH) and prolactin. Brains were rapidly removed from the calvaria. The meninges and olfactory bulbs rostral to the most anterior coronal plane of the cortex were discarded; surface blood was rinsed away in cool (4° C) saline.

Pituitary and pineal glands were weighed, frozen over dry ice, and retained for radioimmunoassays of GH [31] and melatonin,[32] respectively. Brains were coronally sectioned midcollicularly, and hindbrains were discarded. A second coronal section was made through the hypothalamus at the anticipated center of MH injury, and tape-recorded comments were made about size and locus of lesions.[14] Both forebrain segments were then weighed and frozen in liquid nitrogen for fluorometric determinations [14] of endogenous 5-HT, 5-HIAA, NE, and dopamine (DA).

In addition to the above tissue collections, carcasses were subjected to autopsy for potential gastrointestinal pathology. Finally, the right tibia from all rats was removed and retained for linear measurement.

Results

5,7-Dihydroxytryptamine Effects

General Comments

Subcutaneous injection of Nembutal immediately after intracisternal 5,7-DHT administration did not completely prevent convulsions. Almost all rats so treated displayed at least mild tonic seizures within minutes of recovering from ether anesthesia. However, no gross tonic-clonic convulsions were observed. All forms of convulsive activity subsided within 24 hr of central 5,7-DHT injection.

Also noteworthy in drug-injected rats was the conspicuous lack of chronic and grossly exaggerated locomotory activity so characteristic of raphe-lesioned rats.[33-38] Indeed, to all outward appearances, central 5,7-DHT treatment seemed relatively innocuous, especially in comparison to the broad debilitory effects produced by similar doses of the catecholamine (CA) neurotoxin 6-hydroxydopamine (6-OHDA) by this same injection route.[14, 28]

Fatalities

Notwithstanding the seemingly benign nature of central 5,7-DHT treatment, eight of the 11 rats that died during the 51-day postinjection interval had received this drug. One such animal succumbed within 24 hr of injection, apparently because of respiratory complications caused by excessive anesthetization. In five additional rats, profound respiratory complications preceded and apparently caused death on Days 2–4 postinjection. Again, these complications might reflect adverse side effects of ether inhalation. The latter malady was not confined to 5,7-DHT-injected rats, because two controls presented with similar pathologic features prior to death. One drug-treated rat was hypophagic for several weeks and died emaciated on Day 42. Particularly curious was one death in each experimental group (Days 35 and 42) wherein rats presented grossly distended abdomens. It is unknown whether this swelling, which was observed a few weeks before, was contributory to death. However, as described below, this condition arose in other rats irrespective of groups at varying times throughout this study. It was this fact that led us to perform postmortem gastrointestinal examinations at the end of behavioral testing. Only data from rats that maintained good health throughout any particular phase of study were analyzed and are reported here.

Weight and Intake Changes

Two-way analyses of variance (two-way ANOVAs) revealed that BWs of both groups declined significantly ($p < 0.001$) during the first postinjection week (FIGURE 1). This effect was slightly but statistically more pronounced in rats that received 5,7-DHT ($p < 0.001$, day × drug interaction). After this first week, BWs of both groups rose ($p < 0.02$). However, the rise was less pronounced for 5,7-DHT-injected rats ($p < 0.001$, day × drug interaction).

Separate two-way ANOVAs on measures of food and water intake (FIGURE 1) revealed reliable depression with subsequent recovery across days for both groups ($p < 0.001$). However, no reliable drug effects were discernible by these analyses.

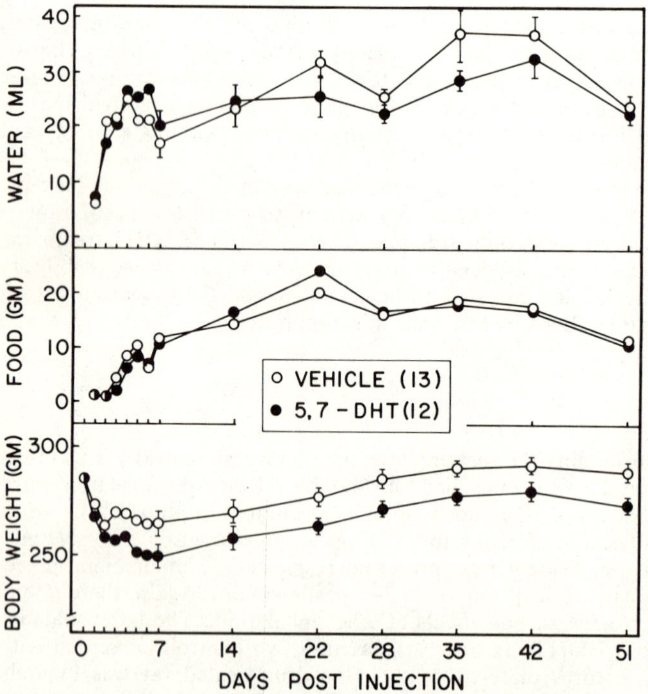

FIGURE 1. Measures of body weight (grams) and of 24-hr food (grams) and water (milliliters) intakes over 51 days after intracisternal injection of 200 μg of 5,7-DHT or its 1% ascorbic vehicle. Symbol definitions and numbers of rats per group are shown in the inset on the graph. Each symbol represents the group mean at the specified day postinjection. Vertical bars depict standard errors of the means at weekly intervals in cases where such variance exceeded that subtended by the symbol alone.

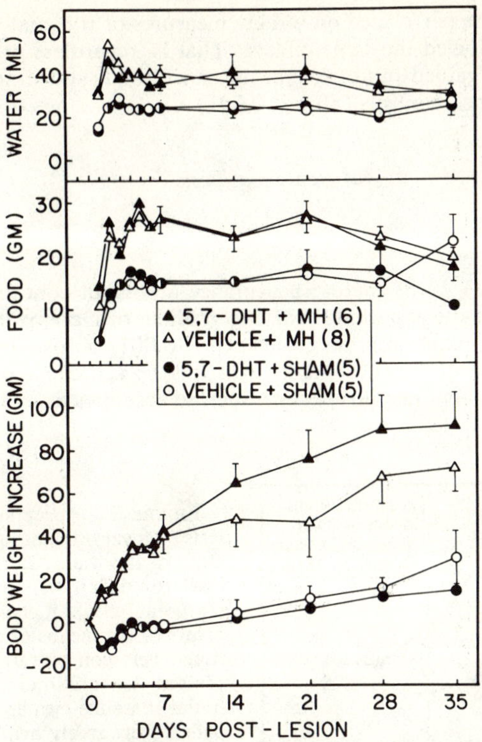

FIGURE 2. Measures of 24-hr food (grams) and water (milliliters) intakes and body weight (grams) increase above those registered at the time of MH lesions or sham surgery. Symbol definitions and numbers of rats per group are shown in the inset on the graph. Symbols and vertical bars represent group means and standard errors of the means as described in the legend to FIGURE 1.

Medial Hypothalamic Lesion Effects

At the time of MH surgery, 5,7-DHT-treated rats weighed reliably less ($p < 0.02$) than did vehicle-treated animals. Therefore, analyses of BW after MH versus sham lesions were performed on grams of BW change from operation weight.

Weight and Intake Changes

Three-way ANOVAs of BW change during the first week after stereotaxic surgery (FIGURE 2) revealed significant differences across days ($p < 0.001$). In particular, MH lesions produced highly reliable increases ($p < 0.001$) compared to sham surgery. This BW change was associated with elevated food ($p < 0.001$) and water ($p < 0.001$) intakes (FIGURE 2). However, no differences in BW or intakes could be ascribed to 5,7-DHT pretreatment.

Similar ANOVAs performed on weekly measures of BW and intake parameters (FIGURE 2) revealed the same effects. That is, regardless of drug pretreatment, lesioned rats gained more weight, ate more food, and drank more water than did sham-lesioned controls (all $p < 0.001$).

Acceptance Tests

Sucrose

Three-way ANOVAs of sucrose acceptance across all concentrations tested (FIGURE 3) revealed the expected [14] varying gradient of increments compared to baseline (water) drinking for all groups ($p < 0.001$, across concentrations). However, the degree of this acceptance was less ($p < 0.03$) for lesioned rats. Prior 5,7-DHT treatment had no effect on sucrose acceptance.

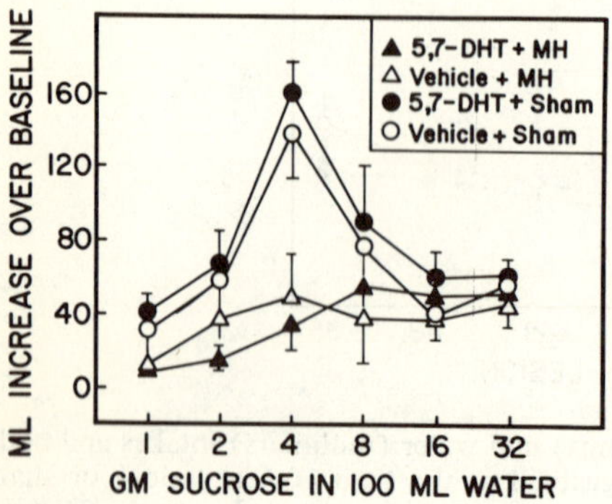

FIGURE 3. Acceptance (one-bottle tests) of varying sucrose solutions instead of tap water. Symbols (see graph inset for definition) and vertical bars represent group means and standard errors of the means of 24-hr consumptions per concentrations noted expressed as milliliter increases over initial baseline water intake. Food pellets were freely available during all tests; 48 hr of tap water replacement separated each test.

Quinine Sulfate

Similar three-way ANOVAs of quinine acceptance (FIGURE 4) revealed decreasing rejection across all groups ($p < 0.001$) as a function of concentration. However, lesioned rats were less willing ($p < 0.02$) than were sham-operated controls to accept quinine. Again, 5,7-DHT pretreatment had no reliable effect either alone or in interaction with lesions on this acceptance measure.

High-Fat Diet

All groups displayed reliable ($p < 0.001$) BW gains when given the high-fat diet (FIGURE 5) instead of food pellets. However, lesioned rats, which already weighed more than did sham-lesioned controls, gained considerably more ($p < 0.005$). This gain was associated with exaggerated ($p < 0.05$) food consumption (not shown graphically). On a kilocaloric basis, the magnitude of this increase was approximately twice that of nonlesioned groups. As in other

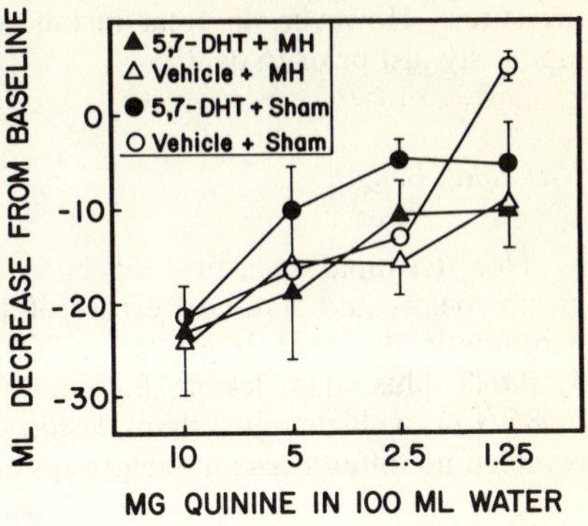

FIGURE 4. Acceptance of varying quinine sulfate solutions instead of tap water. Groups (see graph inset for definition) are the same as in FIGURE 3. Values plotted represent 24-hr decrements in milliliters of consumption compared to initial baseline water intake. Food availability and intertest intervals were as described in the legend to FIGURE 3.

acceptance tests, 5,7-DHT pretreatment did not modify the BW gain or intake levels of the high-fat diet compared to controls.

Somatic Variables

Fatalities After Medial Hypothalamic Surgery and Gastrointestinal Autopsies

Between the time of MH surgery and the termination of all behavioral testing, one rat from each of the four subgroups developed distended abdomens and died at varying times afterward. Gastrointestinal autopsies on one such rat (MH-lesioned) before impending death and on four others (two 5,7-DHT plus sham, two vehicle plus sham) that showed distension at the termination of the studies revealed grossly enlarged, atonic terminal ileums impacted with fecal material from 4 to 30 cm up to the ileocecal valve. In one instance, the cecum itself exhibited 4–5 cm long impaction.

It should be pointed out that other rats demonstrated earlier instances of such abdominal distension that remitted spontaneously. The presence of this pathologic feature in sublethal cases did not clearly influence feeding and BW

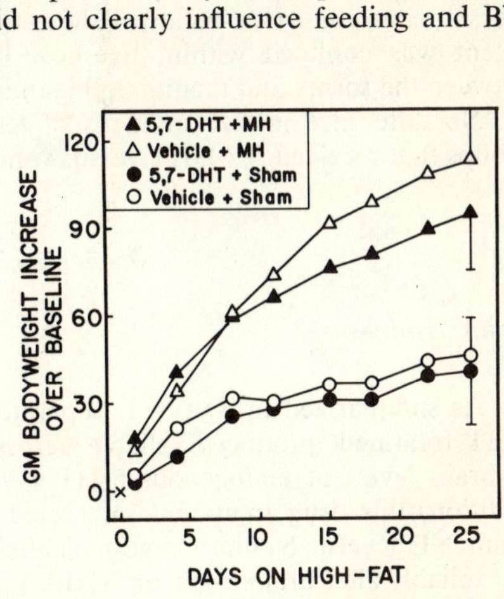

FIGURE 5. Mean increases in body weight (grams) for all subgroups (see graph inset for definition) after 25 days of exposure to a high-fat diet (33% Crisco plus 67% powdered chow) instead of pellet chow. Standard errors of the means are drawn for the last test day to permit visualization of maximal variance. Such increments in body weight reflect increased feeding (not shown).

measures. However, in fatal instances, intake parameters often dropped precipitously just prior to death.

Nasoanal Length

Near-terminal measures of body length (inches) revealed the following group means and standard errors of the means for the number of rats noted parenthetically in each group: 5,7-DHT plus MH lesion, 8.88 ± 0.085 (6); 5,7-DHT plus sham lesion, 8.89 ± 0.64 (4); vehicle plus MH lesion, 8.86 ± 0.085 (7); vehicle plus sham lesion, 8.75 ± 0.112 (4). Two-way ANOVAs revealed no differences among groups on these measures.

Tibial Length

By use of calipers to measure distance (millimeters) from the apex of the medial condyle to the juncture of the medial and lateral malleoli, the group means ± SEMs for the numbers of rats listed were: 5,7-DHT plus MH lesion, 38.8 ± 0.308 (6); 5,7-DHT plus sham lesion, 39.3 ± 0.629 (4); vehicle plus MH lesion, 39.6 ± 0.369 (7); vehicle plus sham lesion, 39.5 ± 0.289 (4). Again, no differences existed across these groups.

Medial Hypothalamic Lesions

The size and locus of MH lesion cavitation were indistinguishable from those reported previously from this laboratory[15] for the radiofrequency lesion parameters used. All lesions were large (2–3 mm in diameter), bilateral, and symmetrically centered near the posterior edge of the optic chiasma. Cavitation could be visualized from the caudal one half or one third of the chiasma to the medial or posterior planes of the ventromedial nuclei. Ventral intrusion never compromised the most basal 0.5 mm of MH tissue or the optic chiasma itself. Dorsal extension often included the reuniens nucleus of the thalamus. Lateral extent was confined within the most lateral edge of the dorsoventral plane between the fornix and mammalothalamic tracts.

No difference in size or locus of lesions could be discerned between subgroups that received 5,7-DHT versus vehicle pretreatment.

Biochemical Measures

Brain Amines

As summarized in TABLE 1, separate two-way ANOVAs revealed that 5,7-DHT treatment produced reliable across-group mean decrements ($p < 0.001$) in brain levels of endogenous 5-HT (−40.4%) and 5-HIAA (−39.6%). In addition, this drug treatment interacted with MH lesions ($p < 0.05$) to alter brain NE levels. Neither treatment affected brain DA. MH lesions alone did not reliably alter brain 5-HT or 5-HIAA.

TABLE 1

TERMINAL BRAIN MONOAMINE OR METABOLITE MEASURES

Compound *	5,7-DHT plus MH Lesion	5,7-DHT plus Sham Lesion	Vehicle plus MH Lesion	Vehicle plus Sham Lesion	2×2 ANOVA
5-HT	412±12.5 (6)†	373±10.1 (4)	654±18.5 (7)	685±41.4 (4)	drug ↓ $p < 0.001$
5-HIAA	270±17.8 (6)	236±11.0 (4)	421±19.6 (7)	411±26.9 (4)	drug ↓ $p < 0.001$
NE	402±15.5 (6)	389±16.3 (4)	368±11.5 (7)	438±44.2 (3)	drug×lesion $p < 0.05$
DA	1274±50.8 (6)	1170±34.4 (4)	1232±34.8 (7)	1270±32.0 (3)	ns

* All compounds were measured in nanograms per gram of tissue weight.
† All values represent the mean ± SEM, with the number of samples assayed shown in parentheses.

Neuroendocrine Indices

Separate two-way ANOVAs (Table 2) revealed reliable suppressant effects of MH lesions on resting levels of both plasma (−46.0%; $p < 0.01$) and pituitary (−38.1%; $p < 0.05$) GH. On the other hand, MH lesions elevated slightly the levels of pineal melatonin (+5.3%; $p < 0.01$). Central 5,7-DHT treatment had no effect on these hormones. No difference was detected across groups for resting levels of plasma prolactin.

Discussion

5,7-Dihydroxytryptamine, Eating, and Body Weight

Central injection of 5,7-DHT to DMI-pretreated adult female rats did not induce hyperphagia and/or enhanced BW gain throughout the 180+-day study. Indeed, the only clearly observable effect produced by such neurotoxin treatment was a small initial BW decline, which remained uncompensated for, and a persisting decrement in endogenous levels of brain 5-HT and 5-HIAA. Both of these findings are compatible with other recent observations [39–41] and are reminiscent of previous findings after midbrain raphe lesions.[21, 33–35] In these respects, the present data, along with those reported by Hoebel *et al.*,[23] fail to replicate Saller and Stricker's [27] observation of increased feeding to support enhanced linear growth (as in their juvenile male rats) or BW gain (as in their footnote regarding adult female rats).

It seemed conceivable that a dissociation could occur between variables that reflect somatic growth versus feeding and/or BW gain.[9] Therefore, we investigated the possibility that intracisternal 5,7-DHT treatment had altered the former but not the latter. However, indices of nasoanal length, tibial length, and resting levels of plasma and pituitary GH levels were all unchanged by the end of this study.

It might be argued [27] that the degree and/or specificity of brain 5-HT depletion induced by 5,7-DHT treatment was insufficient here to elicit overeating and BW increments. However, based on Saller and Stricker's description of their telencephalon and diencephalon dissections, we would estimate that 45–60% of forebrain 5-HT (the tissue index assessed here) had been depleted in their two hyperphagic groups. Quantitatively, this range encompasses the 46% mean depletion of our 5,7-DHT plus sham group relative to vehicle plus sham controls (see Table 1). On these grounds, the difference in magnitude of brain 5-HT depletion between these two studies seems more apparent than real. It also appears unlikely that our inability to replicate Saller and Stricker's findings [27] is due to nonspecific NE depletion. While our 5,7-DHT plus sham rats showed a mean 11% decrease in forebrain NE, inspection of individual data revealed considerable overlap with results obtained from vehicle plus sham controls.

It is also possible that the divergence in behavioral findings relates to the routes of 5,7-DHT injection used. Baumgarten and Björklund [42] have pointed out that the cisternal route (used here) seems less effective than the ventricular route (used by Saller and Stricker) in achieving maximal depletory effects. However, this criticism is more applicable to 5,6-dihydroxytryptamine, which can be oxidized easier than can 5,7-DHT.[42] Certainly, the biochemical considerations cited above argue against suggestions of substantial depletory differ-

TABLE 2

TERMINAL NEUROENDOCRINE MEASURES

Compound	5,7-DHT plus MH Lesion	5,7-DHT plus Sham Lesion	Vehicle plus MH Lesion	Vehicle plus Sham Lesion	2×2 ANOVA
Plasma GH *	35.3±3.45 (6)†	59.8±10.4 (4)	36.1±5.46 (6)	67.8±18.8 (4)	lesion ↓ $p < 0.01$
Pituitary GH ‡	33.3±3.58 (5)	44.2±3.02 (4)	38.6±2.59 (7)	45.5±5.84 (4)	lesion ↓ $p < 0.05$
Plasma prolactin *	55.2±13.3 (6)	25.7±12.4 (4)	28.1±9.54 (7)	9.60±2.26 (4)	ns
Pineal melatonin ‡	35.0±0.521 (6)	33.5±0.984 (4)	34.7±0.309 (7)	32.6±0.472 (4)	lesion ↑ $p < 0.01$

* Measured in nanograms per milliliter of plasma.
† All values represent the mean ± SEM, with the number of samples assayed shown in parentheses.
‡ Measured in nanograms per organ.

ences. Any possibly diminished potency of 5,7-DHT by our choice of the intracisternal route may have been compensated for by the use of 10 times more antioxidant (1% ascorbic acid vehicle) for the same [27] dose per volume of this neurotoxin. Finally, it should be pointed out that with the intraventricular route, Hoebel *et al.*[23] have also been unable to replicate Saller and Stricker's finding [27] of increased eating and BW gain in female rats. Of course, none of these studies to date, including the present report, can rule out nonspecific tissue damage, particularly to structures that surround the ventricular axis, as contributing to these behavioral discrepancies. There is clearly a need for additional work to better understand the effects of central 5,7-DHT administration as it may affect ingestive behaviors.

Before leaving this issue, it seems appropriate to discuss one other factor that might have contributed to Saller and Stricker's findings.[27] In both the present report and a footnote (number 13) in that paper, mention is made of apparent decreased gastrointestinal motility in some rats, which resulted in fecal retention and intestinal swelling. Saller and Stricker comment that such a pathologic feature may have reflected increased stress in their 5,7-DHT-treated rats, which subsequently prevented increased BW gain and/or growth. In that study, however, no control was made for the effects of intraperitoneal DMI pretreatment. Such control was employed in all rats studied here. As noted, the incidence of gastrointestinal pathology seemed equivocal across groups. Therefore, DMI pretreatment alone may be sufficient to evoke such intestinal atonia and/or decreased motility.§ Perhaps more germane to the issue of BW gain is our informal observation that such gastrointestinal distention is not always lethal and can be associated with occasionally elevated BW (though not increased food intake). This observation raises the possibility that some of the BW gain observed by Saller and Stricker [27] reflected fecal accumulation rather than exclusive increases in linear growth (in male rats) or fat deposition (in female rats). Notwithstanding this consideration, it seems clear that the magnitude of such an artifactual BW gain (e.g., 20–40 g) can in no way alone account for the 100–300 g range of the increments reported.[27]

Medial Hypothalamic Lesions After 5,7-Dihydroxytryptamine

The main purpose of the present study was to determine if chronic brain 5-HT depletion would selectively lower the level at which BW is regulated after MH lesions. This expectation was based on our recent finding [21] that 5-HT-depleting lesions of the midbrain raphe nuclei produced lower BW gain after MH injury. In this respect, while 5,7-DHT treatment seemed capable of producing chronic reductions in brain 5-HT, no parameter of intake or BW after MH lesions was affected differentially by such neurotoxin pretreatment.

The simplest conclusion to be drawn from this negative finding is that the capacity of 5-HT-depleting raphe lesions to reduce the BW set-point elevation

§ Aside from its ability to mimic sympathetic nerve activity, this possible action of DMI may be due to its well-known cholinergic blocking properties and/or its antagonism of ileum spasms induced by histamine or 5-HT. In addition, it seems possible that intraperitoneal deposition of DMI in its hydrochloride form could produce nonspecific peritoneal irritation; this effect, in turn, might cause reflex inhibition of intestinal motility.

that occurs after MH lesions [21] is unrelated to impaired 5-HT neurotransmission. Additional experiments are in progress to directly test the possibility that these earlier findings reflected damage to non-5-HT midbrain neurons.

Regardless of the outcome of these studies, there are several reasons to question the comparability of these experiments. These questions revolve around two major issues surrounding the main independent variables: the cross-validity in central modes of 5-HT depletion and the cross-reliability of MH lesioning.

Several findings imply that the functional consequences to brain 5-HT systems are different after raphe lesioning as compared to 5,7-DHT treatment.[39] First is the possibility that central 5,7-DHT induces denervation supersensitivity.[43] We are unaware of any comparable data that suggest that the same phenomenon might occur after midbrain raphe destruction. Second and perhaps related is the possible occurrence of functional regeneration in 5-HT systems originally impaired by central 5,7-DHT.[44] Perhaps relevant to either or both suggestions of recovered function are two findings here: (1) Pilot observations ‡ revealed that intracisternal 5,7-DHT after DMI pretreatment could deplete 60–70% of forebrain 5-HT 2 weeks after administration. However, with the same injection route, rat sex, and dose regimens, only 40–46% depletion was found 6 months later. This reduction in magnitude of depletion could reflect the effects of increased supersensitivity and/or regenerative growth, producing enhanced 5-HT synthesis or decreased 5-HT turnover with time. (2) Recent work [45] has shown that 45–50% brain 5-HT depletion up to 1 month after central 5,7-DHT injection is associated with subnormal plasma prolactin in rats. However, given the same magnitude of 5-HT depletion obtained here with the same dose of neurotoxin, no reliable difference in plasma prolactin levels could be detected. Aside from sex differences in rats studied as potentially accounting for this discrepancy, long-term changes in supersensitivity to 5-HT and/or regenerative growth could also account for these differences.

A final consideration regarding potential differences between raphe lesioning and 5,7-DHT modes of brain 5-HT depletion relates to the differential selectivity, uniformity, site, locus, and level of central nervous system affected. These issues have been discussed elsewhere.[17, 46]

It is also possible that the inability to replicate our earlier raphe plus MH lesion findings [21] relates to a lack of cross-reliability in MH lesion production. This possibility seems unlikely because vehicle plus MH-lesioned rats here displayed reliable hyperphagia, BW gain, and altered acceptance responding compatible with those seen before [21] in sham plus MH-lesioned rats. Also, there was no obvious difference between these studies in the size or locus of MH injury. In addition, unpublished observations of depressed GH levels in other MH-lesioned rats [15, 21] were confirmed here (see Table 2). On the other hand, endogenous levels of brain 5-HT were found to be normal in the present MH-lesioned groups. This finding is at variance with our consistent observations [14–17] of reliable 5-HT depletion after MH insult. Interestingly enough, the present vehicle plus MH-lesioned rats did not show the same magnitude of overeating and BW gain as that seen in these previous studies. This observation leaves open the fascinating possibility that the 5-HT depletory effects originally ascribed to MH lesion damage [14–17] might be a consequence rather than a cause of overeating. One mechanism by which such brain depletion could occur with overeating is by the enhanced dietary intake of proportionately more of the amino acids tyrosine, phenylalanine, leucine, isoleucine, and valine than of the natural 5-HT precursor tryptophan. Because all of these amino acids are believed to compete for the same transport carrier system into the brain,[47] this elevated amino acid intake, which would naturally follow hyperphagia on standard laboratory chow, could cause sufficient reductions in brain 5-HT precursor levels. These reductions, in turn, might diminish the quantity of 5-HT synthesized for putative neurotransmission. Studies are presently under way in conjunction with Dr. G. Harvey Anderson to assess this possibility. Should this hypothesis prove tenable, a major reevaluation of the potential role of brain 5-HT neurons in regulating feeding and BW gain[17, 18, 46] would be indicated.

Summary

The present study attempted to replicate and extend two recent studies that implicated aberrant brain 5-HT neurotransmission in the etiology of overeating and BW gain.[21, 27] Adult female rats received 25 mg/kg of desipramine hydrochloride 30–45 min prior to an intracisternal injection of 200 μg (free base) of 5,7-DHT creatinine sulfate or its 1% ascorbic acid aqueous vehicle. After 7 weeks of measuring food intake, water intake, and BW change, rats from both groups received radiofrequency lesions of the MH or sham surgery. After 5 additional weeks of intake and BW measurements, all rats were tested for 24-hr acceptance of varying sucrose and quinine solutions and for 25-day acceptance of a high-fat replacement diet. While 5,7-DHT depleted brain 5-HT by 45%, it did not induce overeating and BW gain alone[27] nor did it modify the overeating, obesity, or "finickiness" produced by hypothalamic injury.[21] Several factors that relate to specificity, sufficiency, and compatibility with other 5-HT depletory techniques were discussed, as were factors of similarity and dissimilarity between this and the previous experiments that we attempted to replicate.

Acknowledgments

The author thanks R. McArthur, P. Chan, C. Spegg, and E. Johansson for excellent technical assistance. Radioimmunoassays were performed by Dr. G. Brown's Section of Neuroendocrinology at the Clarke Institute. Gastrointestinal pathology was assessed by Dr. J. Kenyon, Department of Laboratory Animal Sciences, Faculty of Medicine, University of Toronto.

References

1. Brobeck, J. R. 1946. Mechanism of the development of obesity in animals with hypothalamic lesions. Physiol. Rev. **26:** 541–559.
2. Miller, N. E., C. J. Bailey & J. A. F. Stevenson. 1951. Decreased "hunger" but increased food intake resulting from hypothalamic lesions. Science **112:** 256–259.
3. Teitelbaum, P. 1955. Sensory control of hypothalamic hyperphagia. J. Comp. Physiol. Psychol. **48:** 156–163.
4. Graff, H. & E. Stellar. 1962. Hyperphagia, obesity, and finickiness. J. Comp. Physiol. Psychol. **55:** 418–424.

5. Corbit, J. D. & E. Stellar. 1964. Palatability, food intake, and obesity in normal and hyperphagic rats. J. Comp. Physiol. Psychol. **58:** 63–67.
6. Corbit, J. D. 1965. Hyperphagic hyperreactivity to adulteration of drinking water with quinine HCl. J. Comp. Physiol. Psychol. **60:** 123–124.
7. Bray, G. A. & T. F. Gallagher, Jr. 1975. Manifestations of hypothalamic obesity in man: a comprehensive investigation of eight patients and a review of the literature. Medicine **54:** 301–330.
8. Schacter, S. & J. Rodin. 1974. Obese Humans and Rats. John Wiley & Sons, Inc. New York, N.Y.
9. Bernardis, L. L. 1976. Origin of endocrine-metabolic changes in the weanling rat ventromedial syndrome. J. Neurosci. Res. **2:** 91–116.
10. Ungerstedt, U. 1971. Stereotaxic mapping of the monoamine pathways in the rat brain. Acta Physiol. Scand. Suppl. **367:** 1–48.
11. Ahlskog, J. E. & B. G. Hoebel. 1973. Overeating and obesity from damage to a noradrenergic system in the brain. Science **182:** 166–169.
12. Gold, R. M. 1973. Hypothalamic obesity: the myth of the ventromedial nucleus. Science **182:** 488–490.
13. Glick, S. D., S. Greenstein & D. H. Waters. 1973. Ventromedial hypothalamic lesions and brain catecholamines. Pharmacol. Biochem. Behav. **1:** 591–592.
14. Coscina, D. V., C. Rosenblum-Blinick, D. D. Godse & H. C. Stancer. 1973. Consummatory behaviors of hypothalamic hyperphagic rats after central injection of 6-hyroxydopamine. Pharmacol. Biochem. Behav. **1:** 629–642.
15. Coscina, D. V., D. D. Godse & H. C. Stancer. 1976. Neurochemical correlates of hypothalamic obesity in rats. Behav. Biol. **16:** 365–372.
16. Coscina, D. V. 1976. Relationship of goldthioglucose-induced obesity in mice to levels of brain biogenic amines. Paper presented at Eastern Psychological Association Meeting, New York.
17. Coscina, D. V. 1977. Brain amines in hypothalamic obesity. *In* Anorexia Nervosa. R. A. Vigersky, Ed. : 97–107. The Raven Press. New York, N.Y.
18. Hoebel, B. G. 1976. Satiety: hypothalamic stimulation, anorectic drugs, and neurochemical substrates. *In* Hunger, Basic Mechanisms and Clinical Implications. D. Novin, W. Wyrwicka & G. A. Bray, Eds. : 33–50. The Raven Press. New York, N.Y.
19. Ahlskog, J. E., P. K. Randall & B. G. Hoebel. 1975. Hypothalamic hyperphagia: dissociation from hyperphagia following destruction of noradrenergic neurons. Science **190:** 399–401.
20. Lorden, J., G. A. Oltmans & D. L. Margules. 1976. Central noradrenergic neurons: differential effects on body weight of electrolytic and 6-hydroxydopamine lesions in rats. J. Comp. Physiol. Psychol. **90:** 144–155.
21. Coscina, D. V. & H. C. Stancer. 1977. Selective blockade of hypothalamic hyperphagia and obesity in rats by serotonin-depleting midbrain lesions. Science **195:** 416–419.
22. Breisch, S. T., F. P. Zemlan & B. G. Hoebel. 1976. Hyperphagia and obesity following serotonin depletion by intraventricular p-chlorophenylalanine. Science **192:** 382–384.
23. Hoebel, B. G., F. P. Zemlan, M. E. Trulson, R. E. MacKenzie, R. P. DuCret & C. Norelli. This monograph.
24. Keesey, R. E. & T. L. Powley. 1975. Hypothalamic regulation of body weight. Amer. Sci. **63:** 558–565.
25. Björklund, A., H. G. Baumgarten & A. Rensch. 1975. 5,7-Dihydroxytryptamine: improvement of its selectivity for serotonin neurones in the CNS by pretreatment with desipramine. J. Neurochem. **24:** 833–835.
26. Gerson, S. & R. J. Baldessarini. 1975. Selective destruction of serotonin terminals in rat forebrain by high doses of 5,7-dihydroxytryptamine. Brain Res. **85:** 140–145.
27. Saller, C. F. & E. M. Stricker. 1976. Hyperphagia and increased growth in

rats after intraventricular injection of 5,7-dihydroxytryptamine. Science **192:** 385–387.
28. Coscina, D. V., J. Seggie, D. D. Godse & H. C. Stancer. 1973. Induction of rage in rats after central injection of 6-hydroxydopamine. Pharmacol. Biochem. Behav. **1:** 1–6.
29. Baumgarten, H. G. & L. Lachenmayer. 1972. 5,7-Dihydroxytryptamine: improvement in chemical lesioning of indoleamine neurons in the mammalian brain. Z. Zellforsch. **135:** 399–414.
30. Coscina, D. V. & R. A. McArthur. 1977. In preparation.
31. Brown, G. M. & J. B. Martin. 1974. Corticosterone, prolactin and growth hormone responses to handling and new environment in the rat. Psychosom. Med. **36:** 241–247.
32. Pang, S. F., G. M. Brown, L. J. Grota, J. W. Chambers & R. L. Rodman. 1977. Determination of n-acetylserotonin and melatonin activities in the pineal gland, retina, Harderian gland, brain and serum of rats and chickens. Neuroendocrinology **23:** 1–13.
33. Lorens, S. A., J. P. Sorensen & L. M. Yunger. 1971. Behavioral and neurochemical effects of lesions in the raphe system of the rat. J. Comp. Physiol. Psychol. **77:** 48–52.
34. Coscina, D. V., L. D. Grant, S. Balagura & S. P. Grossman. 1972. Hyperdipsia after serotonin-depleting midbrain lesions. Nature [New Biol.] **235:** 63–64.
35. Grant, L. D., D. V. Coscina, S. P. Grossman & D. X. Freedman. 1973. Muricide after serotonin depleting lesions of the midbrain raphe nuclei. Pharmacol. Biochem. Behav. **1:** 77–80.
36. Neill, D. B., L. D. Grant & S. P. Grossman. 1972. Selective potentiation of locomotor effects of amphetamine by midbrain raphe lesions. Physiol. Behav. **9:** 655–657.
37. Jacobs, B. L., W. D. Wise & K. M. Taylor. 1974. Differential behavioral and neurochemical effects following lesions of the dorsal or median raphe nuclei in rats. Brain Res. **79:** 353–361.
38. Lorens, S. A. & H. C. Guldberg. 1974. Regional 5-hydroxytryptamine following selective midbrain raphe lesions in the rat. Brain Res. **78:** 45–56.
39. Lorens, S. A. This monograph.
40. Breese, G. R. & B. S. Cooper. 1975. Behavioral and biochemical interactions of 5,7-dihydroxytryptamine with various drugs when administered intracisternally to adult and developing rats. Brain Res. **98:** 517–527.
41. Myers, R. D. This monograph.
42. Baumgarten, H. G. & A. Björklund. 1976. Neurotoxic indoleamines and monoamine neurons. Annu. Rev. Pharmacol. Toxicol. **16:** 101–111.
43. Trulson, M. E. & B. L. Jacobs. This monograph.
44. Nobin, A. & A. Björklund. This monograph.
45. Gil-Ad, I., F. Zambotti, M. O. Carruba, L. Vicentini & E E. Muller. 1976. Stimulatory role for brain serotonergic system on prolactin secretion in the male rat. Proc. Soc. Exp. Biol. Med. **151:** 512–518.
46. Blundell, J. 1977. Is there a role for serotonin (5-HT) in feeding? Int. J. Obesity **1:** 15–42.
47. Wurtman, R. J. & J. D. Fernstrom. 1974. Nutrition and the brain. *In* The Neurosciences: Third Study Program. F. O. Schmitt & F. G. Worden, Eds. Vol. **3:** 685–693. MIT Press. Cambridge, Mass.

LONG-TERM BIOCHEMICAL AND BEHAVIORAL EFFECTS OF *p*-CHLOROAMPHETAMINE IN THE RAT

Christer Köhler,* Svante B. Ross,†, Bolek Srebro,‡
and Sven-Ove Ögren †

*Institutes of *Psychology and ‡Physiology*
University of Bergen
5000 Bergen, Norway
and †Department of Pharmacology
Astra Läkemedel AB
S-151 85 Södertälje, Sweden

It is now well established that *p*-chloroamphetamine (*p*-CA) exerts long-term effects on central 5-hydroxytryptamine (5-HT, serotonin) neurons.[1, 2] *p*-Chloroamphetamine has been shown to reduce the cerebral levels of tryptophan hydroxylase, 5-HT, and 5-hydroxyindoleacetic acid (5-HIAA), synaptosomal high-affinity uptake of 5-HT, and 5-HT turnover for several months after its systemic administration in the rat.[2-7] These effects on 5-HT are similar to those observed after intraventricular administration of some of the hydroxylated tryptamines [5,6-dihydroxytryptamine (5,6-DHT), 5,7-dihydroxytryptamine (5,7-DHT)][8] and suggest a neurotoxic action of *p*-CA on brain indoleamine neurons. This hypothesis is supported indirectly by the findings of degenerative changes in perikarya located within midbrain regions rich in serotonergic cell bodies.[9]

The midbrain raphe nuclei give rise to separate 5-HT systems that innervate different regions of the central nervous system. It has been suggested that *p*-CA affects only some of these regions, preferentially those that innervate forebrain structures.[5] Several studies have demonstrated reduction in regional distribution[5] and synaptosomal uptake[6, 10] of 5-HT in different forebrain regions after *p*-CA treatment. In the present study, we have investigated the effect of *p*-CA on 5-HT synaptosomal uptake in regions where the serotonergic innervation is known to differ with regard to its midbrain origin.

It has been reported previously that blocking the 5-HT uptake mechanism may prevent the long-term biochemical effects of *p*-CA[11, 12] and *p*-CA-induced behavioral deficits.[13, 14] Prompted by indications that such protection may not be equally effective in all regions of the central nervous system,[13] we have investigated the effects of *p*-CA on regional synaptosomal uptake in animals pretreated with a new 5-HT selective uptake inhibitor, zimelidine.[12]

The fact that *p*-CA reduces central 5-HT but not catecholamines[2] for long periods after a single injection has made it an interesting pharmacologic agent in studies of the role of serotonin in behavior. Previous studies on the behavioral consequences of 5-HT depletion have suggested a role of serotonin in various forms of locomotor activity and aversive behaviors.[15] The present study reports some long-term effects on exploratory behavior and acquisition of an active avoidance response after *p*-CA treatment in the rat. To establish the serotonergic nature of these behavioral changes, some animals were pretreated with zimelidine before administration of *p*-CA.

0077-8923/78/0305-0645 $01.75/0

Biochemical Studies

Materials and Methods

Adult male albino rats (300–350 g) were used. They were housed individually in plastic cages [40 × 25 × 15 cm (height)] with food and water available ad lib.

(±)-*p*-Chloroamphetamine hydrochloride (*p*-CA) was obtained from Sigma Pharmaceuticals. Zimelidine [H102/09, (Z)-dimethylamino-1-(4-bromophenyl)-1-(3-pyridyl)propene] [12a] was obtained from Astra Pharmaceuticals. [3-^{14}C]-5-Hydroxytryptamine creatinine sulfate (sp act, 54 mCi/mmol) was purchased from the Radiochemical Centre, Amersham.

Measurement of 5-HT Levels in Some Brain Regions

Rats were treated either with *p*-CA (2 × 10 mg/kg, ip) or saline (1.0 ml/kg, ip) and sacrificed by decapitation 7 days after the second injection. 5-Hydroxytryptamine levels were measured in the cortex, including the hippocampus, and in the diencephalon, striatum, mesencephalon, and spinal cord. Samples from two animals were pooled. 5-Hydroxytryptamine content was measured fluorimetrically after cation exchange chromatography, according to Andén and Magnusson.[16]

Synaptosomal Accumulation of [^{14}C]-*5-HT*

Groups of four to eight animals each were treated with saline, *p*-CA (2 × 10 mg/kg, ip) or zimelidine (2 × 20 mg/kg, ip) 30 min before the *p*-CA injections. The animals were sacrificed by decapitation 7 days after the second injection, and the following brain areas were rapidly dissected: olfactory bulb, frontal cortex, entorhinal cortex, dorsal and ventral hippocampus, septum, striatum, hypothalamus, thalamus, superior and inferior colliculi, the remaining mesencephalon, and the spinal cord. Details of the dissection procedure will be reported elsewhere.[17] The brain samples were homogenized in 20 vol of 0.24 M sucrose. The accumulation of [^{14}C]-5-HT into synaptosomes was measured according to the method of Snyder and Coyle,[18] by use of a modified double-labeling technique.[19] The concentration of [^{14}C]-5-HT was 10^{-7} M. Active uptake was defined as the difference in accumulation at 37°C and at 0°C. An aliquot of the supernatant was taken for analysis of protein content of the synaptosomes, according to the procedure of Lowry *et al.*[20] Bovine serum albumin was used as the standard. [^{14}C]-5-Hydroxytryptamine uptake was expressed as picomoles per milligram of protein per minute and converted to the percentage of control values.

In parallel studies, regional catecholamine levels and synaptosomal uptake were studied after *p*-CA treatment. The results will be reported separately.

Results

Effects of p-*CA on 5-HT Levels in Some Brain Regions*

There was a clear difference in the effect of *p*-CA treatment on 5-HT levels in different brain regions (Figure 1). The strongest reduction in 5-HT levels was found in the cortex plus hippocampus (88%; $p < 0.001$, Student's *t* test) and in the striatum (92%; $p < 0.001$). 5-Hydroxytryptamine levels in the diencephalon were reduced by 57% ($p < 0.001$), while in the mesencephalon and in the striatum (92%; $p < 0.001$). 5-Hydroxytryptamine levels in the 23%; $p < 0.01$, respectively).

Brain catecholamines were not reduced 7 days after injection. In contrast to the effects of *p*-CA on the brain, the 5-HT content of the lung was not affected.

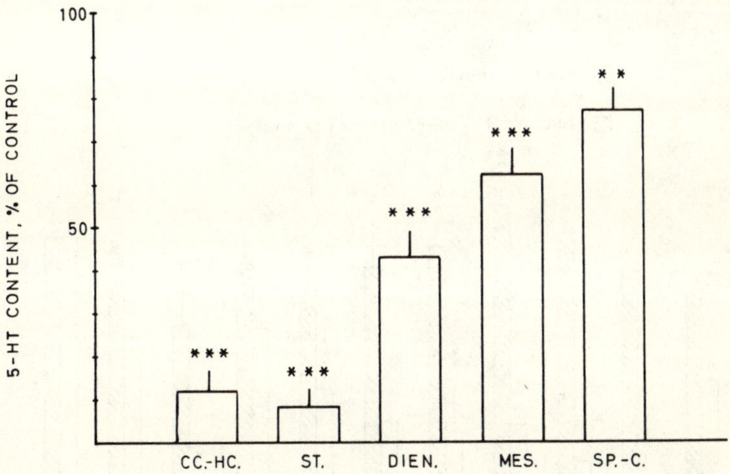

FIGURE 1. 5-Hydroxytryptamine (5-HT) content in various brain regions 7 days after *p*-CA treatment. Standard errors of the means are indicated. CC.-HC., Cortex plus hippocampus; ST., striatum; DIEN., diencephalon; MES., mesencephalon (including the colliculi), SP.-C., spinal cord. **, $p < 0.01$; ***, $p < 0.001$.

Synaptosomal Accumulation of [^{14}C]-5-HT in Selective Brain Regions

Large differences were found in the uptake of [^{14}C]-5-HT into synaptosomes prepared from several brain regions of control animals. Very high uptake of [^{14}C]-5-HT (expressed as picomoles of [^{14}C]-5-HT per milligram of protein per minute) was observed in the following regions: superior colliculi, 3.216; entorhinal cortex, 2.276; hypothalamus, 1.939; frontal cortex, 1.890; septum, 1.762. High uptake was found in the striatum (1.485), inferior colliculi (1.295), ventral hippocampus (1.281), olfactory bulb (1.132), and dorsal hippocampus (0.836). *p*-Chloroamphetamine treatment had different effects on [^{14}C]-5-HT uptake into homogenates of the selected brain regions (FIGURE 2). The largest reductions in uptake of [^{14}C]-5-HT were observed in the superior colliculi (76%; $p < 0.001$) and in the frontal (70%) and entorhinal (67%) cortices ($p < 0.001$ for both). Moderate reductions were found in the thalamus (65%; $p < 0.001$), the ventral (58%) and dorsal (56%) hippocampi ($p < 0.01$) for both, the septum (55%; $p < 0.05$), inferior colliculi

(48%; $p < 0.01$), striatum (45%; $p < 0.001$), and the olfactory bulb (42%; $p < 0.01$). However, in the hypothalamus (27%) and the mesencephalon (excluding the colliculi) (19%), *p*-CA treatment did not significantly reduce 5-HT uptake. The same was observed in the spinal cord. Pretreatment with zimelidine prevented the *p*-CA-induced reduction of [^{14}C]-5-HT uptake in most of the areas studied, particularly the colliculi, the frontal and entorhinal cortices, the dorsal and ventral hippocampi, the septum, and the thalamus. Zimelidine significantly blocked the effect of *p*-CA in the striatum and olfactory bulb, although less potently than in other regions. Lastly, synaptosomal uptake of [^{3}H]NA was not reduced in any of the regions studied after *p*-CA injections.

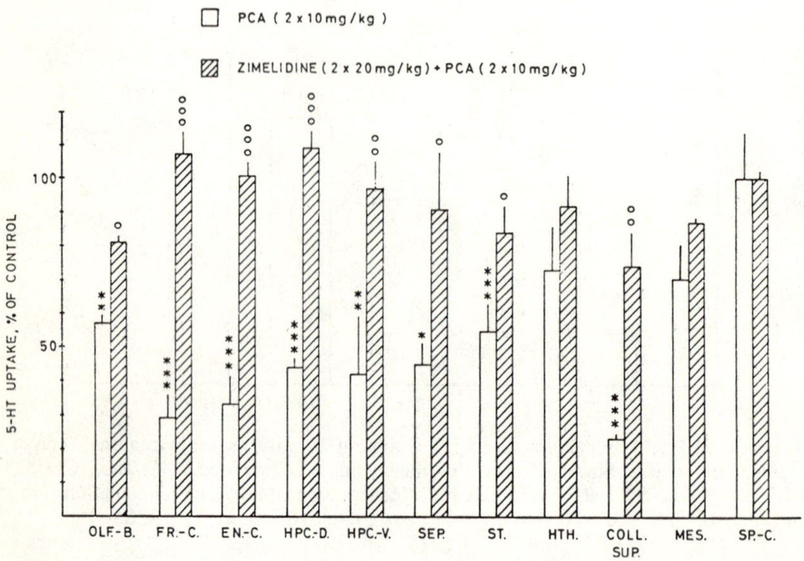

FIGURE 2. The effect of *p*-CA and zimelidine plus *p*-CA on synaptosomal uptake of 5-HT in different brain regions. Standard errors of the means are indicated. OLF.-B., Olfactory bulb; FR.-C., frontal cortex; EN.-C., entorhinal cortex; HPC.-D., hippocampus, dorsal part; HPC.-V., hippocampus, ventral part; SEP., septum; ST., striatum; HTH., hypothalamus; COLL. SUP., superior colliculus; MES., mesencephalon (excluding the colliculi); SP.-C., spinal cord. *p*-CA: ∗, $p < 0.05$; ∗∗, $p < 0.01$; ∗∗∗, $p < 0.001$. Zimelidine plus *p*-CA: ○, $p < 0.05$; ○○, $p < 0.01$; ○○○, $p < 0.001$.

Discussion

The present study confirms and extends previous findings [2, 6, 10] of long-term reductions in 5-HT levels and synaptosomal uptake in different brain regions after systemic *p*-CA treatment. The effect of *p*-CA on high-affinity uptake of [^{14}C]-5-HT could be blocked effectively by pretreatment with zimelidine in most areas studied. This observation supports previous findings that the long-term reduction in whole brain 5-HT and 5-HIAA could be prevented completely in rats pretreated with the 5-HT uptake inhibitors zimelidine [12] or Lilly 110,140.[21] These results support the suggestion that the neurotoxic effect of *p*-CA is dependent on an intact uptake mechanism of the 5-HT neurons.[12, 21]

Our findings indicate that at 7 days after injection, *p*-CA affects mainly the terminal regions of the ascending, but not the descending, 5-HT systems, without any detectable effect on mesencephalic samples. These observations are in agreement with the suggestion that *p*-CA affects forebrain terminal regions but do not confirm the suggestion that the degenerated perikarya observed in certain midbrain regions [9] represent serotonergic neurons.

The fact that *p*-CA decreases 5-HT uptake in regions innervated by either the B7 (striatum, thalamus), B8 (septum, hippocampus, entorhinal cortex), or the B9 (thalamus) cell groups indicates that *p*-CA affects all 5-HT systems that innervate the forebrain.

The weak effect of *p*-CA on 5-HT uptake in the hypothalamus is an interesting exception, because this region is known to be innervated by fibers from both the B7 and B8 and from the pontine B5 and B6 cell groups. Similarly, it has been reported that catecholaminergic terminals in the hypothalamus are affected little by the neurotoxic action of 6-hydroxydopamine.[22] The reason for this relative insensitivity of hypothalamic 5-HT to *p*-CA is still unknown.

Midbrain 5-HT uptake was found to be affected little by *p*-CA. This finding is in contrast to previous observations [6, 13] but could be explained by the fact that in the present study, the colliculi, which are rich in 5-HT terminals, had been removed from the mesencephalic sample. Similar to this finding, midbrain tryptophan hydroxylase has also been found [7] to be resistant to the long-term effects of *p*-CA. It seems, therefore, that *p*-CA causes degenerative effects restricted to terminals of the serotonergic projection that arise in the B7–B9 cell groups. This hypothesis is supported by recent fluorescence histochemical findings by Fuxe *et al.*,[23] who found that *p*-CA produced degeneration in forebrain 5-HT terminals and axons belonging to the mesencephalic raphe systems. However, in agreement with findings from our uptake studies, no evidence for degeneration in the hypothalamus was observed. Similarly, cell groups located in the mesencephalon or pons-medulla and in the descending bulbospinal projection remained intact after *p*-CA treatment. This differential and selective effect of *p*-CA is still unexplained.

Behavioral Studies

Materials and Methods

Five experiments are reported. Adult male albino rats were used. The animals were housed individually with food and water supplied ad lib. The testing was conducted in a sound- and light-controlled room, and the behavior of the animals was monitored by closed-circuit television (except for avoidance training).

The rats were assigned randomly to three equal groups ($n = 7$ or 8) that received on 2 consecutive days (at 3 PM) treatment with *p*-CA hydrochloride (Sigma, 10 mg/kg), zimelidine (20 mg/kg, 30 min before *p*-CA injections), or saline (1 ml/kg).

Behavioral testing started not earlier than 7 days after the second injection and was completed within 10–12 days. After the completion of testing, the animals were sacrificed by decapitation, and their brains were rapidly removed and dissected for estimation of forebrain 5-HT content.

Home Cage Activity and Exploration

In this experiment, the apparatus consisted of the home cage of each animal. The cage was placed in the testing room, its ceiling grid was removed, and the animal was left undisturbed. After three daily sessions (15 min), the number of rearing movements exhibited during each minute of the 15-min session was recorded. Twenty-four hours later, a novel object (a toy car, 12 cm long and 5 cm high) was placed in the home cage at the beginning of the session. In addition to the number of rearings, every contact with the novel object and its duration were recorded in 1-min intervals.

Habituation of the Orienting Response

The animals were tested in their home cages. A loudspeaker delivered 70 dB white noise pulses (5 sec; rise/fall time, 250 msec). The first stimulus was not delivered until the animal laid down and remained in this position for 5 min. The next stimuli were not delivered until the rat had again resumed a relaxed position and remained motionless for at least 30 sec. The rating scale described previously [24] was used to evaluate the rats' responses to each stimulus presentation. The response was divided into early and late components of the orienting response. The rating scale was used to evaluate the early component of the orienting response, while the length of the interstimulus intervals was used to measure the late component of the orienting response. The habituation criterion was three consecutive stimulus presentations without any visible response. Rehabituation was tested immediately and 24 hr later.

Open-Field Exploration

A wooden box elevated 15 cm above the floor was used as an open field [100 × 100 × 40 cm (height)]. The floor of the field was painted flat black, and the walls were covered with a transparent lacquer. White lines divided the field into 24 squares (20 × 20 cm). A rectangular hole (39 × 25 cm) was cut in the center of the field, so that the home cage could be inserted below the floor of the box. The floor grid of the home cage was supported by a cross that divided the floor area into four equal parts. Testing began after the cage had been placed in the apparatus and the ceiling grid had been removed to allow the rat to enter the field. Ambulation (crossing the floor line with half of the body), rearing (lifting both forepaws from the floor), and grooming (washing the head or body for more than 5 sec) were recorded continuously in 1-min intervals for a total of 15 min. All categories were recorded separately for the home cage and the field. The latencies of the first entry to the field and of the first grooming were noted.

Open-Field Activity

The apparatus had the same dimension and was painted identically as the one used in the previous experiment. The only difference was that its floor had no hole in the center of the field, and the corresponding area was divided into four parts by white lines. All other testing conditions were the same as in experiment 1. The categories recorded were the same as in the previous experiment. A novel object introduced on Day 4 was a toy car identical to the one described above. The number and duration of contacts with the object, in addition to the latency of the first approach, were recorded.

Acquisition of Two-Way Active Avoidance

The design, apparatus, and procedure have been described previously.[14] The experimental animals were preselected for acquisition training on the basis of escape performance to four unsignaled shocks. They were divided subsequently into three equal groups and received either *p*-CA, zimelidine plus *p*-CA, or saline (dose and sequence as described above).

The apparatus consisted of two compartments [25 × 25 × 25 cm (height)] separated by a partition with a 5 × 5 cm opening. The conditioned stimulus (CS) was a sine wave tone (60 dB) presented for 10 sec. A constant-current shock (1–2 mA for 5 sec) served as the unconditioned stimulus (UCS). Crossing to the opposite compartment during the CS was recorded as an avoidance and during the UCS as an escape response; latencies of both were recorded.

Eight daily sessions of 20 trials were given to all animals, starting 7 days after the second injection. Immediately after completion of the avoidance training (15 days after the injection), the animals were sacrificed, and their brains were removed and dissected for determination of 5-HT content.[16]

Results

Home Cage Activity and Exploration

Several forms of spontaneous activity have been shown to be affected by central 5-HT depletion in the rat.[25, 26] It was of interest, therefore, to study the long-term effect of *p*-CA on home cage activity. It is known that any novel stimulus of moderate intensity elicits orienting and exploration behaviors in the rat. An important determinant of this response is familiarity with the environment where the stimulus is presented. The present experiment investigated the effect of *p*-CA on reaction to a novel object introduced into the home cage.

There was no change in home cage activity 9 days after *p*-CA injections. The number of rearings that occurred in the home cage during the session that preceded the introduction of the novel object (baseline activity) was the same in all groups. When the object was presented, the *p*-CA group was found to rear significantly less than both the vehicle- ($t = 4.05$; 12 df; $p < 0.01$) and the zimelidine plus *p*-CA ($t = 2.39$, 12 df; $p < 0.05$) treated groups. Although the rats pretreated with zimelidine tended to rear less than did the vehicle

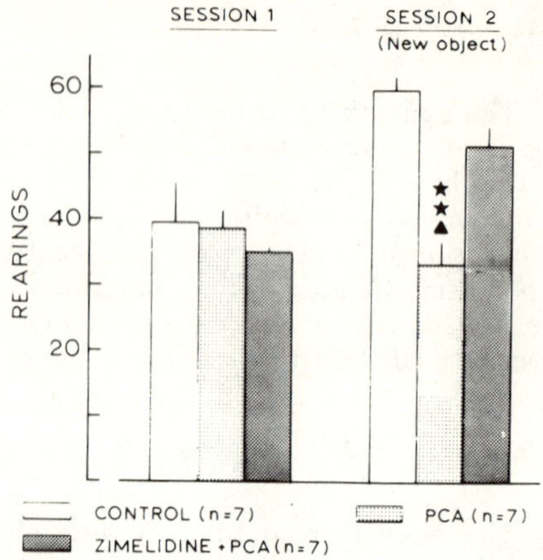

FIGURE 3. Mean rearing activity in the home cage. Standard errors of the means are indicated. ★★, Difference from control; ▲, difference from zimelidine plus *p*-CA.

controls, this difference did not reach statistical significance (FIGURE 3). Within-group comparisons of the two testing sessions revealed significant increases in rearing activity in both the control and zimelidine plus *p*-CA groups ($t = 3.79–4.97$, 6 df; $p < 0.02$). In contrast, the *p*-CA animals tended to decrease their rearing. There was no evidence of within-session habituation in any of the groups on Day 1, but on Day 2, this phenomenon was evident in the control and the zimelidine plus *p*-CA groups ($t = 3.09–4.39$, df; $p < 0.05$). The *p*-CA group made significantly less contacts with the novel object than did the control rats ($t = 2.25$, 12 df; $p < 0.05$). This was also evident for the zimelidine plus *p*-CA-treated group ($t = 2.23$, 12 df; $p < 0.05$). The *p*-CA and zimelidine plus *p*-CA-treated rats did not differ on this measure. No group difference was found with regard to the time spent in contact with the object (FIGURE 4). All groups decreased the number of contacts within the session ($p < 0.05$ for all groups).

In conclusion, there was no detectable difference in home cage activity between the control and the *p*-CA-treated animals before the novel object

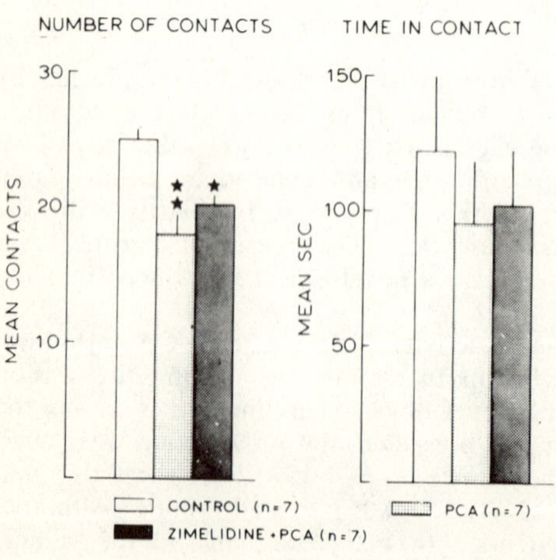

FIGURE 4. Mean numbers of contacts and of seconds in contact with the object introduced during session 2. ★★ and ★, $p < 0.05$.

was introduced into the cage. Alll animals clearly reacted when the object was placed in their cages, but only the control and zimelidine-pretreated rats significantly increased their rearing activity. Furthermore, only the control animals made more contacts with the novel object. It could be argued, however, that in a small cage, any increase in activity would enhance the probability of contact with the new object. All *p*-CA and zimelidine plus *p*-CA-treated rats made at least one contact with the novel object. When those animals approached the object, there was no difference in the duration of the contact. Thus, it cannot be concluded that *p*-CA injections affected exploration or investigatory behavior per se. These results seem to indicate that *p*-CA-treated rats are less aroused by a novel stimulus in the familiar environment. This effect was not prevented by pretreatment with zimelidine.

Habituation of the Orienting Response

The previous experiment showed that *p*-CA-treated animals reacted less to novel stimuli. In the present experiment, the orienting response to auditory stimuli and its habituation have been analyzed in the same rats.

The first stimulus was presented 5 min after the animals had resumed a relaxed position. On Day 1, there was no significant difference between the groups on this measure. On Day 2, the *p*-CA-injected animals laid down faster than did control and zimelidine plus *p*-CA-treated groups ($p < 0.05$, Student's t test). There was no difference in the rate of habituation as measured by the number of trials required to attain the criterion. This finding was true for both within- and between-session comparisons. The only clear-cut group difference was a rapid decline of response amplitudes observed in *p*-CA-injected animals during the first few trials ($T = 2$; $p < 0.05$, Wilcoxon's sign test). In contrast, the vehicle and the zimelidine plus *p*-CA-treated rats tended to show an initial increase of response amplitudes for both components of the orienting response.

It has been postulated [27] that responses to the repetitive presentation of a neutral stimulus are a net outcome of two independent processes: sensitization and habituation. Sensitization is manifested by an initial increase of response amplitude. The present observation of a rapid decrease of response amplitudes in the *p*-CA-injected rats suggests that *p*-CA may attenuate sensitization without detectable effects on habituation.

Open-Field Exploration

p-Chloroamphetamine-treated rats appeared to be less affected by a change in their home cage environment. However, the results were inconclusive as to the effect of *p*-CA on exploratory behavior. The present experiment investigated the long-term effects of *p*-CA on exploration of a novel environment.

Previous studies of exploration in rodents have employed tests in which the animal is removed from its home cage and placed in the testing apparatus. In the present experiment, we applied a test in which the rat had free access to the apparatus from its home cage.

The *p*-CA-injected group required significantly longer time to emerge from the home cage than did the control group on all 3 days of testing ($p < 12.05$

for all groups, Student's *t* test). The zimelidine-pretreated animals emerged later than did the control rats only on the first day ($p < 0.05$, Student's *t* test). When the first minute of the first session was analyzed, significantly more control than zimelidine plus *p*-CA-treated rats had entered the field ($p < 0.05$, Chi-square). The same was true for all days when the control and *p*-CA-injected rats were compared ($p < 0.05$ for all groups, Chi-square) (FIGURE 5).

Analysis of the field activity revealed that the *p*-CA-treated group made significantly less crossings than did both the vehicle ($t = 3.43–3.63$, 12 df; $p < 0.01$) and zimelidine plus *p*-CA- ($t = 2.33–4.46$, 12 df; $p < 0.05$) treated groups on all days of testing (FIGURE 6, *top*). This low level of locomotor activity was paralleled, on all days, by significantly reduced rearing in the *p*-CA-injected group as compared to both the control and the zimelidine plus *p*-CA-treated rats ($p < 0.01$ for all groups, Student's *t* test). The field activity

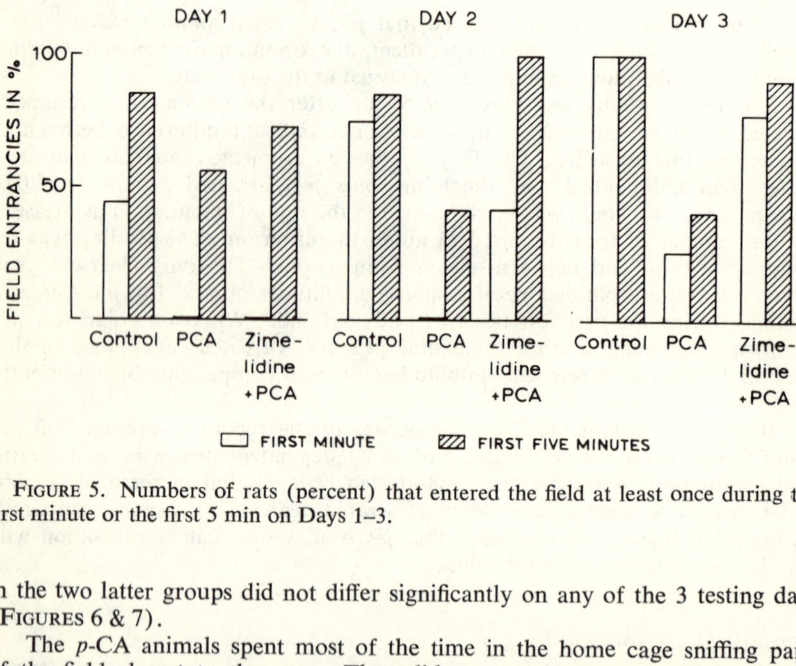

FIGURE 5. Numbers of rats (percent) that entered the field at least once during the first minute or the first 5 min on Days 1–3.

in the two latter groups did not differ significantly on any of the 3 testing days (FIGURES 6 & 7).

The *p*-CA animals spent most of the time in the home cage sniffing parts of the field closest to the cage. They did not manifest any gross locomotor disturbance that could have hindered climbing out of the cage. When the locomotor activity in the cage was analyzed, both the *p*-CA and the zimelidine plus *p*-CA-treated groups made more crossings than did the control group on Day 3 ($p < 0.05$, Student's *t* test; see FIGURE 6, *bottom*). Home cage rearing activity was significantly lower in the *p*-CA as compared to the control group on all days of testing ($p < 0.05$ for all groups, Student's *t* test). No other group difference was found with regard to this measure.

No significant within- or between-session decrease in locomotor or rearing activity was observed in any of the three groups. Lastly, on the first day of exposure to the field, the *p*-CA-treated group started to groom significantly

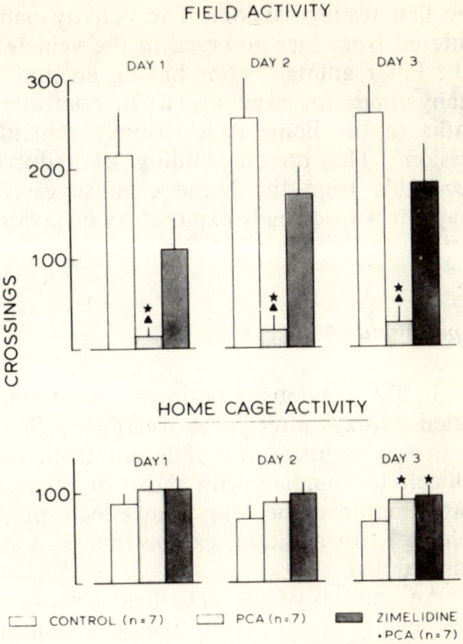

FIGURE 6. Mean numbers of squares crossed in the field (*top*) and the home cage (*bottom*) on all days of testing, ★, Difference from control; ▲, difference from zimelidine plus control.

faster than did the vehicle and the zimelidine plus *p*-CA-treated groups ($p < 0.05$ for all groups, Student's *t* test).

In summary, the *p*-CA-treated animals appeared to explore a novel environment accessible from the home cage less than did the control rats. It seems rather unlikely that this effect was due to an inability to enter the field, because all *p*-CA-injected rats did leave their home cage at least once during

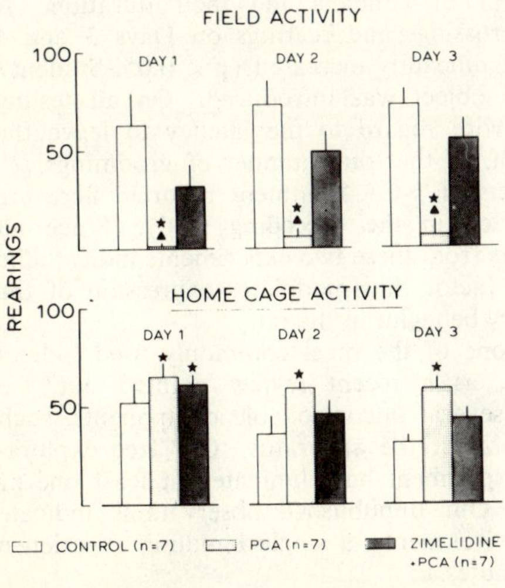

FIGURE 7. Mean numbers of rearings made in the field (*top*) and the home cage (*bottom*) on all days of testing. ★, Difference from control; ▲, difference from zimelidine plus control.

the first testing session. The activity pattern shown by the *p*-CA rats clearly differed from that observed in the vehicle or zimelidine plus *p*-CA-treated rats. The latter animals, after having entered the field, returned to the home cage many times for brief visits. In contrast, the *p*-CA rats, after retracing their paths to the home cage, usually remained there to the end of the testing session. The present finding of reduced activity in a novel environment accessible from the home cage suggests that the long-term effect of *p*-CA may affect specifically exploratory behavior in the rat.

Open-Field Activity

In the previous experiment, we found reduced open-field activity in rats tested 14 days after *p*-CA treatment. Since the open-field apparatus employed in that experiment was different from those commonly used, our results are difficult to compare with those of other reports. The last experiment, therefore, examined the long-term effects of *p*-CA treatment in a standard open-field test. In addition, exploration of a novel object introduced into the field was studied.

An analysis of the open-field activity showed that the *p*-CA group crossed significantly less squares than did both the control ($t = 2.46$–2.215, 12 df; $p < 0.05$) and the zimelidine plus *p*-CA treated ($t = 2.6$–2.26, 12 df; $p < 0.05$) groups on Days 3 and 4. The *p*-CA rats, furthermore, tended to be less active on the first 2 days of testing, but the group differences did not reach a statistically significant level (FIGURE 8).

Interestingly, only the *p*-CA group showed a significant within-session decrease in the number of crossings on Days 1 and 2 ($p < 0.05$, Student's t test). On Day 3, the *p*-CA-injected rats reared less than did the control ($t = 2.46$, 12 df; $p < 0.05$) but not the zimelidine plus *p*-CA group. No other group difference was found.

When the novel object was introduced on Day 4, there was no detectable difference with regard to the number of contacts and their duration. A comparison of the total number of crossings and rearings on Days 3 and 4 revealed that only the *p*-CA animals significantly increased ($p < 0.05$, Student's t test) their activity when the novel object was introduced. On all testing days, there was no group difference with regard to the latency to leave the central area, latency to the first groom, or the total number of groomings.

In conclusion, the behavioral effects of *p*-CA treatment reported here are strikingly different from those reported in the preceding study. Since all other conditions were similar, the results from these two experiments indicate the importance of testing conditions as a factor that modulates expression of the long-term effects of *p*-CA on exploratory behavior in the rat.

The open-field activity represents one of the most commonly used indices of exploration in rodents. However, as a recent review pointed out,[28] a standard open-field test has probably several uncontrollable components, such as handling, lighting conditions, and size of the apparatus. Our free exploration test employed in the preceding experiment has eliminated at least one of these components, namely, handling. Our unpublished observations indicate that, both in familiar and in novel environments, a single handling experience may affect rats' behavior for several minutes.

Two-Way Active Avoidance Acquisition

Electrolytic lesions in the midbrain raphe nuclei and systemic injections of *p*-chlorophenylalanine in the rat have been reported to alter responsiveness to electric shock and the acquisition of aversively motivated behaviors.[15] These effects were often related to a reduction in forebrain 5-HT levels. The present study examined the long-term effects of *p*-CA on the acquisition of a two-way (shuttle) avoidance response.

There was no apparent change in the initial response to electric shock or in the acquisition of escape response 7 days after the second *p*-CA injection. The same treatment, however, severely retarded the acquisition of two-way active avoidance. While the control and the zimelidine plus *p*-CA-injected groups

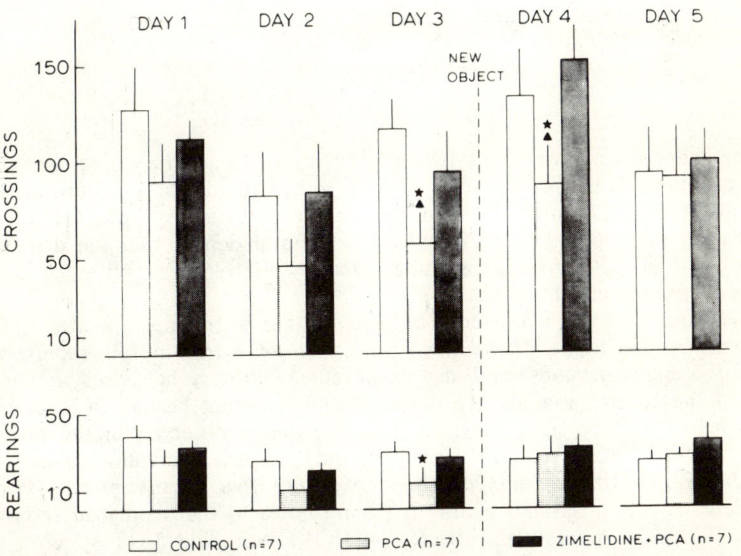

FIGURE 8. Mean numbers of crossings (*top*) and rearings (*bottom*) on Days 11–15. ★, Difference from control; ▲, difference from zimelidine plus control.

made 80–90% avoidance responses during the last two training sessions, the *p*-CA group showed an almost total absence of avoidance learning. The latter group made no more than 10% avoidances during eight testing sessions. The differences in avoidance performance of the *p*-CA and the two other groups was evident as early as during the third testing session ($p < 0.005$ for both comparisons, Fisher's test) (FIGURE 9).

There was no significant group difference on the first day of testing. The control and the zimelidine plus *p*-CA-treated groups significantly reduced their response latencies, mainly due to the improvement in avoidance performance. The *p*-CA-treated group, on the other hand, decreased their response latencies to the shock; at the end of testing, they exhibited short-latency escape responses. Lastly, the *p*-CA-treated animals did not display any evidence of increased freezing or intertrial activity.

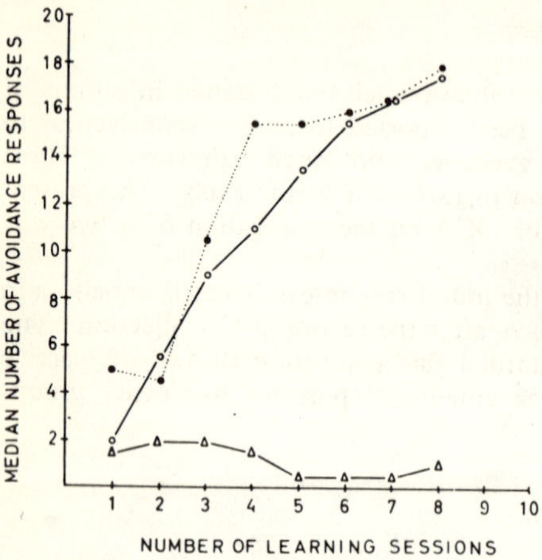

FIGURE 9. Number of avoidance responses of the *p*-CA (△), zimelidine plus *p*-CA (●) and vehicle (○) rats. (From Ögren *et al.*[14] By permission of *Neuroscience Letters.*)

The marked impairment of two-way avoidance acquisition observed after *p*-CA treatment in the present study appears to be caused by a long-term neurotoxic effect of *p*-CA on 5-HT neurons. This conclusion is supported by the observation that both the *p*-CA-induced behavioral deficits and the depletion of central 5-HT observed were antagonized by pretreatment with the 5-HT uptake inhibitor zimelidine.

The aspects of the avoidance contingency that were impaired after *p*-CA treatment remain to be determined. Since the *p*-CA-treated rats rapidly acquired the escape response and also displayed preparatory behaviors similar to those of the control animals, the observed deficit cannot be due to locomotor impairment. It is possible, however, that the deficit observed in the present study is specific to the paradigm used. In fact, recent observations[39] indicate that the escape training and/or shock exposure that occurs prior to drug treatment may be a critical factor that contributes to the avoidance response deficit. Since *p*-CA treatment affects mainly the ascending 5-HT systems, the present findings lend evidence for an important role of central 5-HT neurons in avoidance acquisition as measured in the present paradigm. It remains to be determined, however, whether the effect of *p*-CA on avoidance learning is due to its action on a particular 5-HT system.

Discussion

The major behavioral findings reported here have been summarized in TABLE 1. In brief, *p*-CA treatment reduced the response to a novel object introduced into the home cage but appeared to have no effect on general activity levels. Habituation of the orienting response was not affected by *p*-CA. The most striking effect of *p*-CA was an almost total elimination of open-field exploration. However, when open-field activity was tested, a decrease in activity after *p*-CA injection was observed only after 2 days of testing. Acquisition of a two-way active avoidance response was severely retarded in *p*-CA-treated animals; yet, there was no deficit in escape learning. In most of the tests, pretreatment with zimelidine prior to *p*-CA injection prevented these behavioral changes. Only in the exploration of a novel object and in the open field was there a transient effect in the zimelidine-pretreated animals.

In the rat, a commonly reported finding after interference with 5-HT neurons has been an alteration in locomotor activity. However, the effects reported appear to depend on the conditions and testing procedures employed and on the method of 5-HT depletion. For example, *p*-chlorophenylalanine (*p*-CPA), a potent 5-HT synthesis inhibitor, may increase [25, 29] locomotor activity of rats tested in familiar environments and decrease [30] or leave it unaffected [31] when rats are tested in novel environments. Furthermore, systemic administration of *p*-CA [26] or intracerebral application of the neurotoxin 5,7-DHT [32] produces reduced locomotor activity of rats exposed to novel environments. In contrast to reports of increased activity after *p*-CPA treatment, the present results indicate unchanged activity in a familiar environment after *p*-CA administration. This discrepancy may be due either to a difference in the action of the two drugs or, more likely, to the difference in the sampling methods employed.

The behavior of rats in novel environments (e.g., the open field) is determined by a complex interaction between curiosity and fear.[33] Thus, previously reported changes in locomotor activity after 5-HT depletion in rats may be secondary to alterations in sensory processes, reaction to novelty, habituation, or emotionality. Previous studies have focused mainly on the role of 5-HT

TABLE 1

SUMMARY OF THE EFFECTS OF *p*-CA ON EXPLORATION, HABITUATION, AND TWO-WAY ACTIVE AVOIDANCE ACQUISITION IN THE RAT

Test	*p*-CA	Zimelidine plus *p*-CA
Home Cage		
Baseline activity	0 *	0
Reaction to novel object		
Rearing	↓	0
Contact	↓	↓
Time in contact	0	0
Open-Field Activity		
Free exploration		
Rearing	↓	↓
Ambulation	↓	(↓)
Forced exploration		
Rearing	(↓)	0
Ambulation	(↓)	0
Orienting Response		
Amplitude	0	0
Habituation	0	0
Sensitization	↓	0
Two-Way Avoidance Acquisition		
Escape	0	0
Avoidance	↓	0

* 0, No effect; ↓, reduction; ( ), transient effect.

neurons in the control of locomotor activity without examining other aspects of open-field behavior, such as orienting, reactivity, and exploration. In the present study, we did not find any effect of *p*-CA on the first orienting response or on its habituation rate. The only detectable effect was a rapid decrease of the response amplitudes, suggesting an influence of *p*-CA on sensitization.

Little information exists regarding the effects of central 5-HT depletion on habituation of the orienting response in rats. Recently, File [34] has reported normal habituation and sensitization of the arrest reaction in rats treated with *p*-CPA. At variance with those results, and with the present findings, is Davis and Sheard's [35] report of enhanced sensitization of the startle reaction in *p*-CA-treated rats. An explanation for this discrepancy may reside in the different response systems under study, that is, orientation versus the startle response. In fact, it has been reported recently [36] that pharmacologic treatment may dissociate the rates of habituation of the startle response and of exploration, suggesting that separate neural systems may be responsible for different types of habituation.

It has been reported consistently that forebrain 5-HT-depleting lesions of the midbrain raphe nuclei produce enhanced locomotor activity in rats.[15, 37] This effect, furthermore, has been related specifically to a reduction in hippocampal 5-HT.[37] We have reported recently [38] that 5,7-DHT injections in the midbrain raphe nuclei did not enhance open-field locomotor activity despite 80% (forebrain) and 65% (hippocampal) reductions in 5-HT content. As we have already pointed out, the effect of systemic *p*-CPA administration on locomotor activity seems to depend on testing conditions. Several studies that employed this drug have demonstrated decreased locomotor activity. Recently, Vorhees *et al.*[26] reported reduced open-field activity in rats after *p*-CA treatment. The present results partly confirm this observation and indicate, furthermore, that attenuation of open-field activity may be due to the effects of *p*-CA on exploration.

The prominent reductions in septal, hippocampal, and entorhinal cortical 5-HT uptake after *p*-CA injections may implicate a role for 5-HT in these structures in exploratory behavior. We have observed recently that intraseptal injections of 5,7-DHT that deplete hippocampal 5-HT produce behavioral effects that resemble those observed here after *p*-CA treatment.

Concluding Remarks

p-Chloroamphetamine reduced 5-HT uptake in most forebrain regions studied but had little effect on the hypothalamus, mesencephalon, and spinal cord. These biochemical effects of *p*-CA could be prevented by pretreatment with the 5-HT uptake inhibitor zimelidine. Similarly, most of the long-term behavioral effects of *p*-CA could be antagonized by pretreatment with zimelidine. It can be concluded, therefore, that the behavioral changes observed were related to a neurotoxic effect of *p*-CA on 5-HT neurons in the forebrain.

Acknowledgments

It is a pleasure to thank Ms. Turid Mattsson for excellent technical assistance and Dr. O. J. Broch, Bergen for performing the biochemical analyses in some of the behavioral studies.

REFERENCES

1. FULLER, R. W. & B. B. MOLLOY. 1974. Recent studies with 4-chloroamphetamine and some analogues. *In* Advances in Biochemical Psychopharmacology. E. Costa, G. L. Gessa & M. Sandler, Eds. Vol. **10:** 195–205. The Raven Press. New York, N.Y.
2. SANDERS-BUSH, E., D. A. GALLAGER & F. SULSER. 1974. On the mechanism of brain 5-hydroxytryptamine depletion by p-chloroamphetamine and related drugs and the specificity of their action. *In* Advances in Biochemical Psychopharmacology. E. Costa, G. L. Gessa & M. Sandler, Eds. Vol. 10: 185–194. The Raven Press. New York, N.Y.
3. FULLER, R. W., H. D. SNODDY, B. W. ROUSH & B. B. MOLLOY 1973. Further structure activity studies on the lowering of brain 5-hydroxyindoles by 4-chloroamphetamine. Neuropharmacology **12:** 33–42.
4. GÁL, E. M., P. A. CHRISTIANSEN & L. M. YUNGER. 1975. Effect of p-chloroamphetamine on cerebral tryptophan-5-hydroxylase in vivo: a re-examination. Neuropharmacology **14:** 31–39.
5. NECKERS, L. M., L. BERTILSSON, S. H. KOSLOW & J. L. MEEK. 1976. Reduction of tryptophan hydroxylase activity and 5-hydroxytryptamine concentration in certain rat brain nuclei after p-chloroamphetamine. J. Pharmacol. Exp. Ther. **196:** 333–338.
6. SANDERS-BUSH, E., J. A. BUSHING & F. SULSER. 1975. Long-term effects of p-chloroamphetamine and related drugs on central serotonergic mechanisms. J. Pharmacol. Exp. Ther. **192:** 33–41.
7. SEKERKE, J. H., H. E. SMITH, J. A. BUSHING & E. SANDERS-BUSH. 1975. Correlation between brain levels and biochemical effects of the optical isomers of p-chloroamphetamine. J. Pharmacol. Exp. Ther. **193:** 835–844.
8. BJÖRKLUND, A., H. G. BAUMGARTEN & A. NOBIN. 1974. Chemical lesioning of central monoamine axons by means of 5,6-dihydroxytryptamine and 5,7-dihydroxytryptamine. *In* Advances in Biochemical Psychopharmacology. E. Costa, G. L. Gessa & M. Sandler, Eds. Vol. **10:** 13–33. The Raven Press. New York, N.Y.
9. HARVEY, J. A., S. E. MCMASTER & L. M. YUNGER. 1975. P-chloroamphetamine: selective neurotoxic action in brain. Science **187:** 841–843.
10. MASSARI, V. J. & E. SANDERS-BUSH. 1975. Synaptosomal uptake and levels of serotonin in rat brain areas after p-chloroamphetamine or B-9 lesions. Eur. J. Pharmacol. **33:** 419–422.
11. FULLER, R. W., K. W. PERRY & B. B. MOLLOY. 1975. Effect of 3-(p-trifluoromethylphenoxy)-N-methyl-3-phenylpropylamine on the depletion of brain serotonin by 4-chloroamphetamine. J. Pharmacol. Exp. Ther. **193:** 796–803.
12a. ROSS, S. B., S.-O. ÖGREN & A. L. RENYI. 1976. (z)-Dimethylamino-1-(4-bromophenyl)-1-(3-pyridyl)propene (HIO2/09), a new selective inhibitor of the neuronal 5-hydroxytryptamine uptake. Acta Pharmacol. Toxicol. **39:** 152–166.
12b. ROSS, S. B., S.-O. ÖGREN & A. L. RENYI. 1976. Antagonism of the acute and long-term biochemical effects of 4-chloroamphetamine on the 5-HT neurons in the rat brain by inhibition of the 5-hydroxytryptamine uptake. Acta Pharmacol. Toxicol. **39:** 456–476.
13. MESSING, R. B., L. PHEBUS, L. A. FISHER & L. D. LYTLE. 1976. Effects of p-chloroamphetamine on locomtor activity and brain 5-hydroxyindoles. Neuropharmacology **15:** 157–163.
14. ÖGREN, S.-O., CH. KÖHLER, S. B. ROSS & B. SREBRO. 1976. 5-Hydroxytryptamine depletion and avoidance acquisition in the rat. Antagonism of the long-term effects of p-chloroamphetamine with a selective inhibitor of 5-hydroxytryptamine uptake. Neurosci. Lett. **3:** 341–347.
15. SREBRO, B. & S. A. LORENS. 1975. Behavioral effects of selective midbrain raphe lesioins in the rat. Brain Res. **89:** 303–325.

16. Andén, N. E. & T. Magnusson. 1967. An improved method for fluorimetric determination of 5-hydroxytryptamine in tissue. Acta Physiol. Scand. **69:** 87–94.
17. Ögren, S.-O., Ch. Köhler, S. B. Ross & B. Srebro. 1977. The role of brain serotonin in one- and two-way active avoidance acquisition. Comparison of p-chloroamphetamine and p-chlorophenylalanine. In preparation.
18. Snyder, S. H. & J. T. Coyle. 1969. Regional differences in ^{3}H-norepinephrine and ^{3}H-dopamine uptake into brain homogenates. J. Pharmacol. Exp. Ther. **164:** 78–86.
19. Ross, S. B., A. L. Renyi & S.-O. Ögren. 1973. The inhibitory effect of various phenyl alkylamine derivatives on the uptake and of noradrenaline and 5-hydroxytryptamine in brain tissue. *In* Symposium on Pharmacological Agents on Biogenic Amines in the Central Nervous System. J. Knoll & K. Moggar, Eds. : 45–58. A Kodémiai Kiadó. Budapest, Hungary.
20. Lowry, O. H., N. J. Rosebrough, A. Farr & R. J. Randall. 1951. Protein measurement with the Folin phenol reagent. J. Biol. Chem. **193:** 265–275.
21. Wong, D. T., J. S. Horng, F. P. Bymaster, K. L. Hauser & B. B. Molloy. 1974. A selective inhibitor of serotonin uptake: Lilly 110140, 3-(p-trifluormethylyphenoxy)-N-methyl-3-phenylpropylamine. Life Sci. **15:** 471–479.
22. Fuxe, K., P. Eneroth, J. Å. Gustavsson, T. Hökfelt, G. Jonsson, A. Löfström & P. Skett. 1975. Effects of 6-OH-DA induced lesions of the ascending noradrenaline and adrenaline pathways to the tel and diencephalon on FSH, LH and prolactine secretion in the ovariectomized female rat. *In* Chemical Tools in Catecholamine Research. G. Jonsson, T. Malmors & Ch. Sachs, Eds. Vol. **1:** 273–283. North Holland Publishing Company. Amsterdam, The Netherlands.
23. Fuxe, K., S.-O. Ögren, L. F. Agnati & J. Å. Gustafsson. This monograph.
24. Köhler, Ch. 1976. Habituation of the orienting response after medial and lateral septal lesions in the albino rats. Behav. Biol. **16:** 63–72.
25. Fibiger, H. C. & B. A. Campbell. 1971. The effect of parachlorophenylalanine on spontaneous activity in the rat. Neuropharmacology **10:** 25–32.
26. Vorhees, C. V., G. J. Schaefer & R. J. Barrett. 1975. p-Chloroamphetamine: behavioral effects of reduced cerebral serotonin in rats. Pharmacol. Biochem. Behav. **3:** 279–284.
27. Groves, P. M. & R. F. Thompson. 1970. Habituation a dual process theory. Psychol. Rev. **77:** 419–450.
28. Walsh, R. N. & R. A. Cummins. 1976. The open-field test: a critical review. Psychol. Bull. **83:** 482–504.
29. Segal, D. S. 1976. Differential effects of para-chlorophenyl-alanine on amphetamine-induced locomotion and stereotypy. Brain Res. **116:** 267–276.
30. Marsden, C. A. & G. Curzon. 1976. Studies on the behavioral effects of tryptophan and p-chlorophenylalanine. Neuropharmacology **15:** 165–171.
31. Brody, J. F., Jr. 1970. Behavioral effects of serotonin depletion and of p-chlorophenylalanine (a serotonin depletor) in rats. Psychopharmacology **17:** 14–33.
32. Hole, K., K. Fuxe & G. Jonsson. 1976. Behavioral effects of 5,7-dihydroxytryptamine lesions of ascending 5-hydroxytryptamine pathways. Brain Res. **107:** 385–399.
33. Blanchard, R. J., M. J. Kelley & D. C. Blanchard. 1974. Defensive reactions and exploratory behavior in rats. J. Comp. Physiol. Psychol. **87:** 1129–1133.
34. File, S. E. 1975. Effects of parachlorophenylalanine and amphetamine on habituation of orienting. Pharmacol. Biochem. Behav. **3:** 979–983.
35. Davis, M. & M. H. Sheard. 1976. p-Chloroamphetamine (PCA): acute and chronic effects on habituation and sensitization of the acoustic startle response in rats. Eur. J. Pharmacol. **35:** 261–273.

36. WILLIAMS, J. M., L. M. HAMILTON & P. L. CARLTON. 1974. Pharmacological and anatomical dissociation of two types of habituation. J. Comp. Physiol. Psychol. **87:** 724–732.
37. JACOBS, B L. & A. COHEN. 1976. Differential behavioral effects of lesions of the median or dorsal raphe nuclei in rats open field and pain elicited aggression. J. Comp. Physiol. Psychol. **90:** 102–108.
38. LORENS, S. A., H. C. GULDBERG, K. HOLE, CH. KÖHLER & B. SREBRO. 1976. Activity, avoidance learning and regional 5-hydroxytryptamine following intra-brain stem 5,7-dihydroxytryptamine and electrolytic midbrain raphe lesions in the rat. Brain Res. **108:** 97–113.
39. ÖGREN, S.-O. 1977. Unpublished observations.

GENERAL DISCUSSION

G. R. Breese, *Moderator*

Department of Psychiatry
Biological Sciences Research Center
University of North Carolina School of Medicine
Chapel Hill, North Carolina 27514

Dr. J. A. Harvey: Dr. Jacobs, did you determine if responses to L-Dopa were altered?

Dr. B. L. Jacobs: We did not see any potentiation of L-Dopa responses.

Dr. Breese: I might add that our laboratory has found no evidence that L-Dopa-induced locomotor activity is increased in rats injected with 5,7-dihydroxytryptamine, as it is in 6-hydroxydopamine-treated rats. Furthermore, an increased sensitivity to 5-hydroxytryptamine is not apparent in 6-hydroxydopamine-treated rats.

Dr. Harvey: If you are pretreating rats with a monoamine oxidase inhibitor before administering tryptophan, how do you rule out the production of tryptamine?

Dr. Jacobs: I think in many cases you can't dismiss this possibility because tryptamine may, in fact, be formed and act on serotonin receptors. Therefore, I think the whole tryptamine question deserves more investigation.

Dr. J. F. Pujol: Dr. Jacobs, I'm surprised you discussed your functional results in terms of serotonergic mechanisms after such large doses of 5-hydroxytryptophan and LSD. It is well demonstrated that 5-HTP is decarboxylated in the brain by catecholaminergic serotonergic, and glial cells. Don't you agree that both neural and serotonergic mechanisms may be involved?

Dr. Jacobs: I feel that the syndrome under investigation is related to a specific serotonergic effect. While the precursor doses are high, such doses are necessary to produce an effect. The same is true for L-Dopa, and many of the effects observed after L-Dopa injection occur only at high doses.

Dr. M. E. Trulson: We have infused microgram quantities of serotonin intraventricularly into 5,7-dihydroxytryptamine-pretreated rats and reproduced the syndrome. I feel that this observation provides supportive evidence that the 5-HTP syndrome is induced by activating serotonergic receptors.

Dr. C. T. Giambalvo: Dr. Jacobs, could you clarify your statement that serotonin-containing fibers have no presynaptic receptor?

Dr. Jacobs: The point was that no one has presented evidence for a presynaptic terminal receptor on serotonin-containing neurons similar to the evidence accumulated for the presence of such a receptor on catecholamine-containing neurons.

Dr. J. H. Jacoby: Since chronic administration of *p*-chlorophenylalanine does not appear to induce a supersensitivity, is it possible that *p*-CPA somehow impairs the integrity of the development of supersensitivity?

Dr. Jacobs: If *p*-CPA does interfere with receptor integrity, it does so in such a way that all balancing factors permit responses exactly like those in control animals.

Dr. D. V. Coscina: Dr. Lorens, several years ago Dr. Grant and I showed that raphe lesions increased pain sensitivity using jump flinch indices. Dr. Harvey reported that, depending on the location of the raphe lesion, one may or may not have a hyperreactive preparation for the jump threshold. Could either you or Dr. Harvey comment on the location of this lesion?

DR. S. A. LORENS: With respect to pain sensitivity and electrolytic lesions, it has been reported that B3 lesions enhanced pain sensitivity; it is possible in those studies that reported an increased pain sensitivity after brain stem lesions that this nucleus was involved. I do not have an explanation for why electrolytic lesions of the raphe do not enhance pain sensitivity, as *p*-CPA does. It is possible that *p*-CPA influences the pain threshold by affecting peripheral stores of serotonin. Electrolytic lesions in the medial forebrain bundle or septum may alter the pain threshold by affecting the animal's unconditioned response to an uncomfortable stimulus.

DR. HARVEY: I would like to call everyone's attention to the fact that if changes in pain sensitivity occur with medial forebrain lesions or *p*-CPA, the effects are only seen during the day. The normal animal has a very low threshold at night; during the day, the threshold rises. We believe that this phenomenon is correlated with the circadian rhythm of serotonin content in the rat brain, which is low at night and high during the day. Thus, after the induction of a lesion that lowers serotonin, the pain threshold at night is very much like that of a normal animal. During the day, when the serotonin content increases and the pain threshold should increase, this effect is not present. In regard to raphe lesions, you will also find that this lesion produces changes that are different from the responses of controls at night but not during the day.

DR. G. JONSSON: Dr. Myers, is it possible that the hyperthermia produced by injection of 5,7-DHT is due to a release of prostaglandin?

DR. R. D. MYERS: We have tested this possibility and find that prostaglandin does not release 5-HT or noradrenalin and that neither 5-HT nor norepinephrine releases prostaglandin E_2.

DR. COSCINA: Do you have any biochemical evidence to support the possibility that 5,6-DHT injections are limited to just 5-HT systems or is it conceivable that some of the effects reflect alterations in other neural systems?

DR. MYERS: I believe that the initial incapacitation seen with 5,6-DHT treatment could be due to destruction of both catecholaminergic and serotonergic neurons. The time course of recovery is fascinating, because the animals recover their ability to regulate in the heat, which would be an expected result if the noradrenergic system were intact. However, they do not recover their ability to regulate in the cold, an expected response if the serotonergic system was incapacitated.

DR. HARVEY: Dr. Da Prada, did you say that you consider mounting behavior a form of sexual activity?

DR. M. DA PRADA: I feel that it is not a type of mating behavior.

DR. HARVEY: How, then, do you classify that behavior?

DR. DA PRADA: It is difficult to explain, but I think it is associated with limbic and hypothalamic structures. Furthermore, it can be observed in very young rats, suggesting that it is not a learned behavior.

DR. B. L. JACOBS: Two comments on Dr. Giambalvo's paper. First, if you depress raphe unit activity, you accelerate the activity of substantia nigra. Second, if you inject 5,7-DHT unilaterally into the medial forebrain bundle, you can get a serotonin turning model, that is, an animal that turns only in response to serotonergic manipulations and not in response to amphetamine, apomorphine, or L-Dopa.

Author Index

(Italicized page numbers refer to comments in Discussions)

Agnati, Luigi F., 346–369

Baldessarini, Ross J., 198–206
Baliga, B. Suren, 96–105
Barofsky, A.-L., *174*
Baumgarten, H. G., 3–24, 36–56, *56, 84, 94,* 107–117, *117–118, 133, 145, 171, 175–177, 189, 206, 258, 260–261,* 262–288, 423–436
Björklund, Anders, 3–24, 305–327, 370–384
Bobillier, P., 576–589
Bonetti, E. P., 595–620
Bosin, Talmage R., 134–145
Breese, G. R., 160–170, *171–174, 241, 260,* 664–665 (Moderator), *664*
Browning, Ronald A., 437–456, *456*
Bruckwick, Eleanor, 182–188

Campaigne, Ernest, 134–145
Campbell, Alexander, 198–206
Carruba, Michele O., *172,* 242–258, *258, 478*
Clemens, James A., 399–410, *410, 478*
Clineschmidt, Bradley V., 222–241, *241*
Cohen, Gerald, 74–84, *84*
Collu, Robert, 411–422, *422*
Coscina, Donald V., *172, 206, 241, 456, 479,* 627–644, *664–665*
Creveling, C. R., *56,* 57–73, *145*

Da Prada, M., *56, 118, 146, 173, 260, 385,* 595–620, *665*
Donelson, Alan C., 134–145
DuCret, R. P., 590–594

Fuller, Ray W., 147–157, *158–159, 172, 175–177,* 178–181, *241,* 259–261 (Moderator), *260*
Funderburk, W. H., *302–303*
Fuxe, Kjell, 346–369, *386*

Gál, E. M., *84, 105,* 119–127, *127, 158—159, 175–176, 241*
Gargiulo, Giuliana, 242–258
Gershon, M. D., *385*
Gerson, Sylvia C., 198–206
Giambalvo, C. T., *173–174, 189, 478,* 524–531, *589, 664*
Göthert, Manfred, 457–476, *477*
Growdon, John H., *159, 207,* 510–523
Gustafsson, Jan-Åke, 346–369

Hallman, H., 328–345
Hancke, J. L., 423–436
Harvey, John A., *259,* 289–302, *303–304,* 385–386 (Moderator), *385, 664–665*
Heikkila, Richard E., 74–84, *133*
Hoebel, B. G., *478–479,* 590–594
Hoffman, William E., 437–456
Höhn, K. G., 423–436
Horn, A. S., 128–133, *133, 158*

Jacobs, Barry L., *207, 479,* 497–509, *664*
Jacoby, Jacob H., 1–2, *146, 258, 261,* 387–398, 478–479 (Moderator), *478, 664*
Jonsson, G., 328–369, *665*
Jouvet, M., 576–589

Kaufman, Nathan, 480–496
Keane, P., 576–589
Keller, H. H., 595–620
Klemm, H. P., 3–24, 36–56, 107–117
Klupp, Niels, 457–476
Köhler, Christer, *385,* 645–663

Lachenmayer, L., 3–24, 107–117, 262–288, *385*
Lorens, Stanley A., 532–535, *665*
Lovenberg, W., 3–24, 182–188, *188–189, 221*
Lundberg, J. J., 262–288
Lytle, Loy D., 1–2, 175–177 (Moderator), 480–496

McGinty, J., *241, 385*
McGuffin, Jodie C., 222–241

MacKenzie, R. G., 590–594
Maickel, Roger P., 134–145, *145–146*
Massari, V. J., *259*
Meek, James L., 190–196, *259, 261*
Meshul, C. K., *84, 221*
Messing, Rita B., 480–496
Meyer, J., *410, 422*
Møllgård, K., 262–288
Moskowitz, Michael A., 96–105, *105–106, 259*
Mueller, R. A., 160–170
Munro, Hamish N., 96–105
Myers, R. D., *385, 478,* 556–575, *665*

Natelson, B., *479*
Nisticò, Giuseppe, 242–258
Nobin, Anders, 305–327, 370–384, *385*
Norelli, C., 590–594
Nowak, Thaddeus S., Jr., 96–105

Ögren, Sven-Ove, 346–369, 645–663

Pettibone, Douglas J., 480–496
Pflueger, A. Barbara, 222–241
Pieri, L., 595–620
Pieri, M., 595–620
Pollare, T., 328–345
Pujol, J. F., *478,* 576–589, *589, 664*

Radulovacki, M., *146, 196–197*
Rapport, Maurice M., 85–94
Reinhard, J. F., Jr., *476*
Renaud, B., 576–589
Rogers, Richard B., 134–145
Ross, Svante B., 645–663
Rotman, A., 57–73
Rubin, Deborah, 96–105

Sachs, Ch., 328–345
Sanders-Bush, Elaine, *127, 158, 175–176,* 208–221, *221,* 260
Sankar, Dr., *95*
Schlossberger, H. G., 3–35, 107–117
Sherman, A. D., 119–127
Silbergeld, E. K., *95*
Simonton, Ronald L., 437–456
Smith, G., *422*
Snodgrass, S. R., 524–531
Sperk, Günther, 198–206
Srebro, Bolek, 645–663
Stein, J., *260*
Steranka, Larry R., 208–221
Stewart, R. Malcolm, 198–206, *206–207, 260–261, 410*

Tamir, Hadassah, *56, 73,* 85–94, *94–95, 171, 188, 422*
Thomas, Ronald F., 387–398
Totaro, James A., 222–241
Trulson, Michael E., 497–509, 590–594, *664*

Uretsky, N. J., 524–531

Wiklund, Leif, 262–288, 370–384
Wishousky, Theodore I., 222–241
Wuttke, W., *410,* 423–436, *479*

Zacchei, Anthony G., 222–241
Zemlan, Frank P., 590–594, 621–626

Subject Index

Italicized page numbers refer to material in figures or tables.

The list of terms which precedes the index proper was adhered to strictly to bring about consistency in constructing the index. Under each chosen term are various synonyms, initialisms, etc. each preceded by "=". The reader is cautioned to check here for the various alternatives which may occur in the text itself, as different authors may use different terms. None of the alternatives appear in the index itself as references, because they may not have been needed alone as main headings and/or subheadings under other main headings.

N-Acetylnorfenfluramine
=FE_3
Adrenocorticotrophic hormone
=ACTH
Amine precursor uptake decarboxylase cells
=APUD cells
6-Aminodopamine
=6-ADA
3-(β-Aminoethyl)-5,6-dihydroxybenzo[*b*]-thiophene
=5,6-DHT-S
3-(β-Aminoethyl)-5,6-isopropylindendi-oxybenzo[*b*]thiophene
=IPST
Aurintricarboxylic acid
=ATA
Avoidance behavior
=CAR
=Conditioned avoidance response
Bacterial pyrogen
=Pyrifer® VII
p-Bromoamphetamine
=*p*-BA
=4-Bromoamphetamine
Catechol *O*-methyltransferase
=COMT
Catecholamines
=CA
p-Chloro-*N*-methylamphetamine
=*p*-CMA
p-Chloroamphetamine
=*p*-CA
=*p*-Chloro-α-methylphenylethylamine
=4-Chloroamphetamine
p-Chloronorephedrine
=*p*-Chloro-β-hydroxyamphetamine
=P_2
=PCNE
p-Chlorophenylalanine
=*p*-CPA
Chromatography, high-pressure liquid
=HPLC
Desmethylimipramine
=Desipramine hydrochloride
=DMI
=Norpramin®
Deuterium oxide
=D_2O
=Heavy water
5,6-Diacetoxytryptamine
=5,6-DAcOT
5,6-Dihydroxyindoleacetic acid
=DHIAA
5,6-Dihydroxytryptamine
=5,6-DHT
=5,6-HT
5,7-Dihydroxytryptamine
=5,7-DHT
=5,7-HT
3,4-Dimethoxyamphetamine
=3,4-DMA
=P_3
6,7-Dimethyl-5,6,7,8-tetrahydropterine
=$DMPH_4$
Dopa
=Dihydroxyphenylalanine
=DOPA
L-Dopa
=Levodopa
Dopa decarboxylase
=DDC
Dopamine
=DA
Dopamine β-hydroxylase
=DBH
Fenfluramine
=*N*-Ethyl-α-methyl-3-trifluoromethyl-β-phenylethylamine
=*N*-Ethyl-α-methyl-*m*-(trifluoro-methyl)phenylethylamine
=FE
=FEN
p-Fluoroamphetamine
=4-Fluoroamphetamine

Fluoxetine
=FLU
=Lilly 110,140
=3-(*p*-Trifluoromethylphenoxy)-*N*-methyl-3-phenylpropylamine
Follicle-stimulating hormone
=FSH
Glutathione, reduced
=GSH
Growth hormone
=GH
=Pituitary growth hormone
=Somatotropin
=STH
Harmaline
=HAR
Homovanillic acid
=HVA
5-Hydroxy-6-methoxytryptamine
=5-H-6-MT
6-Hydroxydopamine
=6-HD
=6-OHDA
=2,4,5-Trihydroxyphenylethylamine
5-Hydroxyindoleacetic acid
=5-HIAA
Hydroxyl radicals
=·OH
2-[*β*-(4-Hydroxyphenyl)ethylamino-methyl]tetralone
=HEAT
6-Hydroxytryptamine
=6-HT
5-Hydroxytryptophan
=5-HTP
Isoprenaline
=Isoproterenol
Lisuride maleate
=Lysenyl®
=Lysuride acid maleate
Lu 10-171
=1-(3-Dimethylamino)propyl-1-(*p*-fluorophenyl-5-phthalancarbonitrile hydrobromide
Luteinizing hormone
=LH
d-Lysergic acid diethylamide
=LSD
Maprotiline
=MAP
Maximal velocity
=V_{max}
Melanocyte-stimulating hormone
=MSH
5-Methoxy-*N*,*N*-dimethyltryptamine
=5-MDMT
3-Methoxy-4-hydroxyphenylethylene-glycol
=MOPEG
3-Methoxy-4-hydroxyphenylethylene-glycol sulfate
=3-Methoxy-4-hydroxyphenylglycol sulfate
=MOPEG-SO_4
5-Methoxy-6-hydroxytryptamine
=5-M-6-HT
α-Methyl-5-hydroxytryptamine
=α-Methyl-5-HT
=Serotonin, α-methyl
Michaelis constant
=$K_{m\ app}$
Monoamine oxidase
=MAO
Monoamine oxidase inhibitors
=MAOI
Norepinephrine
=NA
=NE
=Noradrenaline
Norfenfluramine
=NFE
=*m*-Trifluoromethylamphetamine
Pentylenetetrazol
=PTZ
Perchloric acid
=PCA
Peroxidase, horseradish
=HRP
1-Phenyl-3-(2-thiazolyl)-2-thiourea
=PTTU
Piperocaine
=Metycaine®
Prolactin
=LTH
=Luteotropin
=Pituitary lactogenic hormone
Reserpine
=RES
Ribonucleic acid
=Messenger RNA
=RNA
Serotonin
=Enteramine
=5-HT
=5-Hydroxytryptamine
Serotonin-binding protein
=SBP
Subcommissural organ
=SCO
Thyroid-stimulating hormone
=Thyrotrophin
=Thyrotropin
=TSH

m =Trifluoromethylbenzoylglycine
=FE_2
Tritium
=^3H
=T
Tryptophan
=TP
Tyrosine hydroxylase
=TH
Zimelidine
=(*z*)-3-Dimethylamino-1-(4-bromophenyl)-1-(3-pyridyl)propene
=H 102/09

Acetazolamide
effect of 5,7-dihydroxytryptamine or a raphe lesion on anticonvulsant action, 448—450, 452
effect on maximal electroshock-induced hindleg extension in rats, 449, *449*, 450
after a raphe nuclei lesion, 450, *450*
Acetylcholine, 519
effect of *p*-chloroamphetamine on, 260
raphe lesion lowering forebrain concentration, 453
Acetylcholinesterase, 538
N-Acetylnorfenfluramine, 120, 125—126
Activity level, 542—543
effects of
5,7-dihydroxytryptamine, 544
p-chloroamphetamine, 543—544
p-chlorophenylalanine, 543
selective raphe lesions, 543
novel vs. familiar environment, 543
Adenosine triphosphate
as inhibitor of serotonin-binding protein capacity, 91
Adrenal gland, 392
Adrenal-pituitary axis. *See* Pituitary-adrenal axis
Adrenal steroids
blocked by dexamethasone, 621, *622*
release stimulated by serotonin or 5-hydroxytryptophan, 392
Adrenalectomy, unilateral
and depletion of serotonin with *p*-chlorophenylalanine, 392
Adrenergic receptors, α- and β-, 391, 458
β-antagonists. *See* Pindolol
Adrenocorticotrophic hormone, 391—393
hypersecretion occurring after lesions of ascending serotonin pathways, as shown by increased basal corticosterone secretion, 362
Aggressive behavior
more prevalent in rats pretreated with *p*-chlorophenylalanine than 5-hydroxytryptophan, 486
Aldehyde dehydrogenase, 18
Amine precursor uptake decarboxylase cells, 284
Amines
influence of various antagonists on positive chronotropic effects in guinea pig right atria, 462, *464*
6-Aminodopamine, 76—77
mechanism of action, 80
3-(β-Aminoethyl)-5,6-dihydroxybenzo[*b*]-thiophene
ability to lower stored norepinephrine, 143
and behavioral changes in rats, 136, 139—140
stability, 145
3-(β-Aminoethyl)-5,6-isopropylindendi-oxybenzo[*b*]thiophene
biochemical and behavioral effects in rats, 140—141
d-Amphetamine, 479, 510, 527—531, 542
and brain serotonin concentration, 92
and protein synthesis, 105—106
effect on protein synthesis, 101—104
effects, 251—252, 254—256
hyperthermia induced by, 252, 254—256
increased responsiveness to, 385—386
Amphetamines, halogenated. *See* Halogenated amphetamines
Anorexia, 484—485. *See also* *p*-Chloroamphetamine; *p*-Chlorophenylalanine and serotonin; Fenfluramine; Fluoxetine
amphetamine-induced in rats, 593
fenfluramine-induced enhanced after 5,7-dihydroxytryptamine treatment in rats, 593—594, *593*
Antimycin A and oxygen consumption, 81
Apomorphine, 389, 391, 479, 510, 524, 527—531, 596, 605, 613, 615
inducing locomotor activity and stereotyped behavior in rats, 348
ineffective in producing myoclonic syndrome in rats, 511
stimulation of dopaminergic receptors to cause hypothermia or hyperthermia, 252
Aromatic amino acid decarboxylase inhibitors. *See* RO4-4602
Ascorbic acid depleted by serotonin or 5-hydroxytryptophan, 392
Atropine, 515, 518, 529
Aurintricarboxylic acid
effect on polysomal profiles, 102, *103*
effect on protein synthesis, 101—102, *101*
Avoidance behavior, 544—545
5,6-dihydroxytryptamine effects, 544
5,7-dihydroxytryptamine effects, 544—545
by rats, 657—658, *658*
discriminated, 545
one-way, signaled or unsignaled, 545
p-chloroamphetamine effects, 544—545
p-chlorophenylalanine effects, 544—545
selective raphe lesion effects, 544
two-way (shuttle), 544

Bacterial pyrogen, 254
Behavior. *See also* Aggressive behavior; Avoidance behavior; Behavior, rotational; *p*-Chloroamphetamine; Crossing behavior of rats; 5,6-Dihydroxytryptamine—and behavioral changes; Locomotor activity; *d*-Lysergic acid diethylamide-induced behavioral syndrome in rats; 5-Methoxy-6-hydroxytryptamine; Mounting behavior; Rearing activity; Seizure susceptibility changes; Sexual behavior
  as determined by injection of 5,7-dihydroxytryptamine in rats, 347—349, 366
  changes and percent reductions in regional serotonin concentrations after discrete electrolytic brain stem lesions, 533, *534*, 537, 542, 544
  contacts with objects by rats, 652, *652*
  effects of *p*-chloroamphetamine on rats, 658—660, *659*
  effects of serotonin depletion, *546*, 547
  exploration of field by rats, 654, *654*, *655*
  possibly related to route by which 5,7-dihydroxytryptamine was injected into rats, 638, 640
  scoring in rats, 514—515
  studies with rats, 497, 649—658
Behavior, rotational
  after nigrostriatal lesions in rats, 524
  after serotonin-pathway lesions in rats, 529
  after treatment with 5,6-dihydroxytryptamine and 5,7-dihydroxytryptamine in rats, 605, 608, 614—616
  correlation between size of lesion in the median raphe nucleus and vigor of the rotation, 526
  effects of serotonin neurotoxins in rats, 524—531
  in rats, 596
  injected rats tested with amphetamine, 524—526
  injected rats tested with apomorphine, 524—526
  time course, 526
Benserazide causing marked sexual stimulation in rats, 613
Benzodiazepine action over inhibitory γ-aminobutyric acidergic system and reduction of serotonin turnover, 363
Benztropine
  effect on retention of radioactivity in rat brain, 108, 110, 114—116
4-Benzyloxy-5-methoxyindole synthesis, 25—26, *26*
4-Benzyloxy-5-methoxytryptamine formiate ultraviolet spectra, 29, *31*
Blood platelets, 617
  after incubation in plasma at 37°C, 597
  from rabbits and guinea pigs, 615
  serotonin content after incubation in plasma with 5,6-dihydroxytryptamine, 5,7-dihydroxytryptamine, or metaraminol, from rats, 609, *609*
  storage pool deficiency, 616
  ultrastructural aspects after incubation in plasma with 5,6-dihydroxytryptamine, 5,7-dihydroxytryptamine, or metaraminol, from rats, 609, *610*, *611*, 616
Blood pressure, arterial
  hypertension in rats, 471—472
  pretreatment with neither reserpine nor cocaine decreased the pressor effect of 5,6-dihydroxytryptamine, 457, 459
  pressor effect of 5,7-dihydroxytryptamine considerably less than of 5,6-dihydroxytryptamine, 458
Body temperature. *See* Temperature, body
Body weight. *See* Weight, body
Brain. *See also* Hippocampus; Hypothalamus; Locus ceruleus; Serotonin—brain concentration; Serotonin—in brain
  5-hydroxyindoleacetic acid content in rats, 649
  areas B8 and B9, 623
  bulbospinal system of rats, 372
    regrowth, 377
  content of monoamine metabolites after 5,7-dihydroxytryptamine injected intraventricularly in rats pretreated with desmethylimipramine, 597, *600*, 613
  diencephalon functionally analyzed for serotonin system, 347—349
  dorsal accessory nucleus of the inferior olive of rats, 378, *380—381*
  enzymatic activity, 188—189
  fluorescence histochemical studies in rats, 376—379
  hypothalamus reactive to serotonin agonistic properties of quipazine, 390

Brain—(cont'd)
inferior olive of rats, 377—379, *378*
after administration of 5,6-dihydroxytryptamine, 379, *382*
medial hypothalamic lesions
after 5,7-dihydroxytryptamine in rats, 640—642
or sham surgery with measures of food and water intake and body weight in rats, 627, 633, *633*, 636, 638
midbrain raphe lesions in rats, 629, 641
midbrain stained with thionine showing lesion involving dorsal and median raphe nuclei in rats, 441, *441*
pathology in rats, 631
raphe-lesioned rats for monoamine analyses, 444, *446*, 447, 450, *451*
schematic representation in rats, 329, *329*
serotonin content, 665
decrease in rats, 503
in rats, 646—647, 649
in various regions after *p*-chloroamphetamine treatment in rats, 646—647, *647*
increase in rats, 665
of B9 region, 297
uptake in different regions affected by *p*-chloroamphetamine and zimelidine plus *p*-chloroamphetamine, in rats, 647, *648*
telencephalon functionally analyzed for serotonin system, 347—349
topography of area perfused with 5,6-dihydroxytryptamine in rats, 576, *577*
tryptophan concentrations and 3-(β-aminoethyl)-5,6-dihydroxybenzo[*b*]-thiophene, 146
Brain lesions. *See also* Behavior; Behavior, rotational; Dopamine; Norepinephrine
of medial forebrain bundle in rats, 529
of substantia nigra, 527
relation with dopamine neurons, 524
Brain metabolism of serotonin depletors, 119—126
Brain mitochondria
binding of radioactivity to after intraventricular injection of [^{14}C]-5,6-dihydroxytryptamine and [^{14}C]-5,7-dihydroxytryptamine, 11—12
interaction of 5,6-dihydroxytryptamine and 5,7-dihydroxytryptamine in rats, 74—75, 81—82, *81*
Brain stem serotonin and norepinephrine levels after 5,6-dihydroxytryptamine or 3-(β-aminoethyl)-5,6-dihydroxybenzo[*b*]thiophene in rats, 136, *138*, 139
p-Bromoamphetamine, 150, 158, 178, 289—291, 296—297
cf. *p*-chloroamphetamine in exerting neurotoxicity, 176
duration of reversibility of serotonin depletion, 148, *149*, 154
2-Bromolysergic acid diethylamide, 203, *204*
blocking myoclonic syndrome in rats, 511
p-Bromomethamphetamine, 150
Bufotenine as inhibitor of serotonin-binding protein capacity, 86, 88, 93

C

C cells of thyroid. *See* Thyroid C cells
Carbidopa to block 5-hydroxytryptophan effects, 410
Carbonic anhydrase inhibitors, 438, 452
Castration
did not affect mounting behavior in male rats, 603
of rats, 486
Catalase
inhibition of tryptamine conversion not antagonized, 41
Catechol *O*-methyltransferase, 17, 19, 36, 53
Catecholamine neurons, 600
effect of 5,6-dihydroxytryptamine on, 346
Catecholamines, 254—255, 540, 542, 586—587, 613, 616, 631
behavioral tolerance, 212
effect of *p*-chloroamphetamine on, 209
fluorometric estimation in brain, 244
histofluorescence, 244
metabolism, 212
uptake blockers. *See* Desmethylimipramine; Protriptyline
Centrally acting drugs. *See* Drugs, Centrally acting
Cerebrospinal fluid
circhoral luteinizing hormone pattern in ovariectomized injected rats, 427, *431*
serum luteinizing hormone, prolactin, and monoamine levels in ovariectomized rats treated in hypothalamus, 427, *428*
Chlorimipramine, 289, 372—373, *373*, *374—375*, 376

Chlorimipramine—(cont'd)
and serotonin-binding protein, 95
and serotonin uptake, 222
blocking of anorectic effects of fenfluramine, 241
blocking of uptake of 5,7-dihydroxytryptamine into adrenergic sites, 56
inhibiting serotonin transport, 20
inhibiting tyrosine hydroxylase activity increase in locus ceruleus by 5,6-dihydroxytryptamine treatment in cats, 584—585, *585*
pretreatment as a control for 5,7-dihydroxytryptamine in rats, 349
unavailable to protect serotonin fibers from 5,6-dihydroxytryptamine, 613
6-Chloro-2-aminotetralin, 153—154
p-Chloro-*N*-methylamphetamine, 147, 224, 546
m-Chloroamphetamine, 150, 154
o-Chloroamphetamine, 150
p-Chloroamphetamine, 147, 158—159, 178—179, 289—291, 296—297, 302—304, 385—386, 403, 481—486, 488—490, 511, 514—515, 529—530. *See also* in subheads under other main headings, e.g. Avoidance behavior—*p*-chloroamphetamine effects
abnormal behavior production, 212—213
administration, 123
and amino acid incorporation, 105—106
and concentration of serotonin and tryptophan hydroxylase in brain nuclei, 190
and glutathione levels, 215
and neurotoxicity, 153—154
and reduced glutathione in brain, 121, *123*
and sexual behavior in female rats, 621
and tryptophan 5-hydroxylase activity, 126
brain concentration after injection of *p*-chloroamphetamine or *p*-chloro-*N*-methylamphetamine in rats, 150—151, *151*
cf. fenfluramine, 239
correlation between turnover decrease in cortical serotonin and increase in striatal dopamine in rats injected unilaterally into median raphe nucleus or substantia nigra, 526—527, *527*
early changes in norepinephrine and dopamine and dynamics after injection, 209, *210*
effect in newborn animals, 172
effect of acute drug treatment on the postural syndrome induced in rats, 515, *517,* 519
effect of fluoxetine on serotonin decreases after injection, 214, *214*
effect of increasing doses on the postural syndrome induced in rats, 515, *518*
effect of pretreatment with cysteine or diethylmaleate on long-term effect, 215-216, *216*
effect on
mast cells and enterochromaffin cells, 385
mating behavior induced by estradiol benzoate in female rats, 621, *622*
proestrous surges of luteinizing hormone and prolactin in female rats, 402, *402*
serum levels of luteinizing hormone, prolactin, and growth hormone in male rats, 400, *401,* 409
effect on brain
B9 region in rats, 290—291, *292—293, 294*
B9 region in rats pretreated with fluoxetine, 292, *294*
serotonin, norepinephrine, and dopamine in rats, 291, *291*
effect when injected into median raphe nucleus in rats on serotonin and 5-hydroxyindoleacetic acid levels and serotonin and norepinephrine uptake, 525—526, *525*
fluorescence histochemistry of neurotoxic action, 357-358, *357*
hypothalamus and spinal cord resistant to depletion of serotonin after administration, 213
in discrete brain nuclei, 191—192
inducing anorexia, 478
influence of administration route on brain levels, 120, *120*
influence on brain concentration of serotonin and 5-hydroxyindoleacetic acid, 148, *148*
inhibition of protein synthesis, 103—104
injection effect on motor activity in rats, 210, *211*
injection increasing dopamine turnover, 589
injection of rats, 515, *516*

p-Chloroamphetamine—(cont'd)
interaction with fluoxetine in mice, 216, *217*
long-term biochemical and behavioral effects in rats, 645-660
long-term biochemical effects, 213-216
long-term neurotoxic effect on mating behavior and forebrain monamine levels in female rats, 621, *622*
mechanism of action, 177
metabolism, 126
metabolite accumulation after brain injection, 120, *121*
neurotoxic effects, 259, 480
neurotoxic metabolites and age of animals, 175—176
neurotoxicity, 125, 127, 175, 186—187, 261
producing decrease in brain serotonin and 5-hydroxyindoleacetic acid content, 289
reduction of brain serotonin and 5-hydroxyindoleacetic acid, 119
reduction of resting levels of thyroid-stimulating hormone in estrogen-primed ovariectomized rats, 394
selective neurotoxic damage to the brain, 351
Sidman avoidance, 211—212
species differences in mice and rats, 216
stereoisomers, 211
structure, 148, *149*
subcellular distribution of metabolites after brain injection, 121, *122*
synaptosomal uptake of serotonoin after single injection, 213, *213*
tolerance, 212
toxicity and metabolism, 215, 221
use to regulate body weight in human beings, 485
p-Chloroamphetamine analogs, 152, 154 *See also* 3-Phenylpiperidines, chlorinated
isomers, 158
structure, 148, *149*
p-Chloroamphetamine and serotonin actions on serotonin neurons, 141—142, 153—154
attempts to increase serotonin depletion by, 192—193
changes in central serotonin neurons, 284
depletion of serotonin, 259
duration of reversibility of serotonin depletion, 148, *149,* 154
effect on brain serotonin content when *p*-chloroamphetamine neonatally administered in rats, 163—164, *164*
effect on content of brain regions in rats, 292, *295*
effect on peripheral stores, 260
effect on serotonin neurons, 208
newborn rat brain serotonin, 241
reduction of brain serotonin in rats, 390—391
p-Chloroamphetamine homologs, 152
p-Chloromethamphetamine, 187
biochemical effects, 213, 217
tissue distributions of, 177
p-Chloronorephedrine, 126
metabolite of *p*-chloroamphetamine in brain, 120
none detectable in brain or liver after administration of iprindole, 124
p-Chlorophenylalanine, 145—146, 217, 388—389, 391—393, 405—406, 409, 444, 456, 487—488, 498, 515, 519, 532, 578, 590, 596, 664. *See also* in subheads under other main headings, e.g. Avoidance behavior —*p*-chlorophenylalanine effects
and supersensitivity, 258
chronic administration did not induce to various compounds, 503
lacking, 505
effect on brain monoamine concentrations in rabbits, 247, *249*
effect on growth hormone secretion in rats, 422
failure to block suckling-induced prolactin release, 433
in discrete brain nuclei, 193—194
induction of transient hyperphagia, 627
lowering of body temperature, 478
may increase locomotor activity in rats, 659
microinjection in rats, 571
mounting behavior in male, female, castrated, and ovariectomized rats alone or in combination with 5-hydroxytryptophan or haloperidol, 603, *603,* 614
pretreatment with neither enhances nor antagonizes 5,7-dihydroxytryptamine-induced hypothermia in rats, 489
preventing rather than enhancing audiogenic seizures in mice, 438

p-Chlorophenylalanine and serotonin
blockade of central nervous system serotonin synthesis, 242—243, 252, 254—256
blockade of central serotonin syndrome, 212
depleting brain serotonin content, 510
effect on serotonin in brain nuclei, 194, *195*
inducing reductions in brain serotonin associated with insomnia, 483
inhibiting serotonin synthesis, 423, 437
inhibiting tryptophan hydroxylase, depleting brain serotonin causing reduction in food intake and loss in body weight, 556
mechanism of action of serotonin neurons, 259
p-chlorophenylalanine decreasing brain serotonin biosynthesis and causing anorexia and hypodipsia, 484
reduction of brain serotonin in rats, 390
serotonin and tryptophan hydroxylase concentration in brain nuclei, 190
total insomnia following inhibition of serotonin biosynthesis, 576
p-Chlorophenylalanine methyl ester hydrochloride, 479
brains injected later with fluoxetine in rats, 229, *233*
Chlorpromazine and prevention of lethal hyperthermia, 261
Chlorpromazine effect on brain dopamine concentration, 173
Cholesterol, 623—624
Choline acetylase
effect of *p*-chloroamphetamine on, 260
Choline acetyltransferase, 538, 540
Chromatography, high-pressure liquid, 190—191
Cinanserin as putative serotonin receptor antagonist, 485
Cinanserin blockade of serotonin receptors, 623
Circadian rhythm, 587
Clonidine, 391, 393
Clorgyline, 48
Clozapine as inhibitor of serotonin-binding protein capacity, 92
Cocaine, 462, 468, 471
Cold environment, exposure to, 393—394
Corpus striatum synaptosomes from rats, 128
Corticosteroids
diurnal elevation in cats, rats, and man reduced by *p*-chloroamphetamine, 393
Corticosterone
elevation after lesions of the raphe nuclei induced with 5,6-dihydroxytryptamine, 392
suppressed by intraventricular or intracerebral injection of serotonin, 392
Covalent binding
of 5,6-dihydroxytryptamine to atrial protein, 65
of 5,7-dihydroxytryptamine to atrial protein, 65, *65*, 71
oxygen requirement for, 65, 67
p-chloroamphetamine or fenfluramine, 127
Covalent interaction of one or more oxidation products of an amine with intraneuronal proteins, 57
Cross-linking of protein
oxygen dependency on, 68
requirement for two active nucleophilic sites, 70
Crossing behavior of rats, 656, *657*
Cushing's disease therapy, 392—393
Cycloheximide, 203, 205, 207
N-Cyclopropyl-*p*-chloroamphetamine as inhibitor of monoamine oxidase, 150—152
Cyproheptadine, 284, 390, 462, 515, 518—519, 539
as blocker of serotonin receptors, 392
as inhibitor of serotonin-binding protein capacity, 91
as serotonin receptor antagonist, 484
pretreatment neither enhances nor antagonizes 5,7-dihydroxytryptamine-induced hypothermia in rats, 489
Cytochrome P-450, 221
Cytotoxicity
and mitochondrial damage, 71
and oxidative phosphorylation, 71
molecular mechanism involving autoxidation of quinone from 5,6-dihydroxytryptamine and superoxide and hydroxyl radicals, 85
selective target organelles for, 71
CZ-74 (drug) as inhibitor of serotonin-binding protein capacity, 86, 88, 93

D

Decarboxylase
inhibition by RO4-4602, 589
inhibitors, 437
Deoxycorticosterone acetate-saline hypertension in rats, 471

Desmethylimipramine, 118, 198, 324, 342—343, 388—390, 393, 400, 403, 418, 424, 426, 433, 515, 518, 595—596, 602—603, 605. *See also* in subheads under other main headings, e.g. 5,7-Dihydroxytryptamine—pretreatment of adult rats with desmethylimipramine
and serotonin-binding protein, 95
and uptake of various substances in rat brain tissue, 200, *201*
and/or 5,7-dihydroxytryptamine treatment of newborn and their effect on amine uptake *in vitro* in rat brain, 336—337, *336*
antagonizing accumulation of 5,7-dihydroxytryptamine into norepinephrine neurons, 16—17
as protective agent, 80
blocking brain norepinephrine transport sites, 107
blocking uptake of 5,7-dihydroxytryptamine into adrenergic sites, 56
effect of pretreatment on long-term depletion of brain and spinal cord serotonin and norepinephrine before injection of 5,7-dihydroxytryptamine, 314—315, *316—317*
effect on 5,7-dihydroxytryptamine-induced changes in serotonin and norepinephrine uptake *in vitro* in rat brain cortex, 334, *335*
effect on actions of 5,7-dihydroxytryptamine on newborn rat brain, 165, *165*
effect on retention of brain radioactivity in rats, 108, 110—111, 114—117
increasing the norepinephrine output evoked, 467
intraperitoneal injection in rats, 628—629
pretreatment, 541, 547
antagonizing brain norepinephrine, 4
before injection with 5,7-dihydroxytryptamine, 485
before intracerebroventricular injection of 5,7-dihydroxytryptamine, 472
in rats, 640—642
including amines in rat brain cortex, 200, *200*
with 5,7-dihydroxytryptamine intracisternally or intraventricularly, 13
used to protect norepinephrine neurons from neurotoxic effects of 5,7-dihydroxytryptamine, 440

Deuterium oxide, 29, 33
Development of serotonin neurons. *See* Serotonin neuron development
Developmental plasticity of serotonin neurons. *See* Serotonin neuron development
5,6-Diacetoxytryptamine as serotonin neurotoxin, 4, 306
Dibenamine pretreatment causing degenerative changes in the ultrastructure of cardiac nerves with 5,6-dihydroxytryptamine, 471
4,5-Dibenzyloxyindole synthesis, 25—26, *26*
Dibenzyloxyindoleacetonitrile synthesis, 29
4,5-Dibenzyloxytryptamine hydrogen oxalate ultraviolet spectra, *31*
5,6-Dibenzyloxytryptamine hydrogen oxalate ultraviolet spectra, *31*
5,7-Dibenzyloxytryptamine hydrogen oxalate ultraviolet spectra, *31*
β,β-Difluoro-4-chloroamphetamine, 289—291
β,β-Difluoro-*p*-chloroamphetamine, 152—154
5,7-Dihydroxyindoleacetaldehyde role in augmenting 5,7-dihydroxytryptamine toxicity, 54, 55
5,6-Dihydroxyindoleacetic acid, 17—18, 54
binding to rat hypothalamus homogenates, 38—39, 43—45, *52,* 53
identified by thin-layer chromatography, 46
5,7-Dihydroxyindoleacetic acid, 54, 118
binding to rat hypothalamus homogenates, 38—39, 43—45, *52,* 53
identified by thin-layer chromatography, 46
Dihydroxyindoleamines
effect on protein synthesis, 99, *100,* 103
Dihydroxymethoxyindoleacetaldehyde synthesis, 27
4,5-Dihydroxytryptamine
impairment of serotonin uptake, 306
rapid autoxidation preventing application as selective neurotoxic agent, 29
4,5-Dihydroxytryptamine hydrochloride synthesis, 28, *28*
5,6-Dihydroxytryptamine, 17, 107, 178—179, 198, 252, 254—256, 289, 297, 305, 385, 389, 393, 476, 485, 488, 566, 568, 570—573, 638, 665. *See* also in subheads under other main headings, e.g. Avoidance behavior —5,6-dihydroxytryptamine effects

5,6-Dihydroxytryptamine—(cont'd)
α methylation counteracting impairment of serotonin uptake, 307
and axonal degeneration after intraventricular injection, 370
and behavioral changes, 135—136, 145—146
in rats, 139—140
and cardiac lesions, 57
and cross-linking of protein, 60—61, *69*
and selective cytotoxicity, 57
as tool for morphologic studies of central indoleamine neurons, 322, *323,* 324
as tool for studying sleep mechanisms in cats, 576—587
binding to rat hypothalamus homogenates, 38—39, 43—45, *51*
biochemical effects when intracisternally injected, 199—200
brain lesions, 557—558
breakdown at biologic hydrogen ion concentration and 37°C, 53
chemical properties, 29, 33
chromatography, 60—61, *61,* 68—69
covalent binding to protein oxygen dependency, *60,* 68
effect alone and in combination with 5-hydroxytryptophan or 6-hydroxydopamine on brain amines and mounting behavior in male rats, 597, *598,* 614
effect of intraventricular injection on body weight in male rats, 414, *414*
brain levels of serotonin, norepinephrine, and dopamine in male rats, 412, *413,* 418
endocrine organ weights in male rats, 414, *415*
pituitary growth hormone levels in male rats, 417, *417*
plasma and pituitary follicle-stimulating hormone in male rats, 417, *418*
tail length in male rats, 414, *416*
testicular weight in rats, 414, *416,* 419
effect of methysergide on cumulative concentration response curve in isolated strips of rabbit aorta, 459, *460*
effect of *p*-chloroamphetamine on, 260
effect on brain biochemistry and behavior in rats, 141, *142*
effect on capacity of serotonin-binding protein from myenteric plexus, 89—90, *90*
effect on endogenous norepinephrine, dopamine, serotonin, and 5-hydroxyindoleacetic acid in different parts of brain in cats, 578, *580—581*
effect on monoamine brain concentrations in rabbits, 247, *249*
effects of intracerebral injections on monoamine uptake and content, 320, *321*
effects on the rat central nervous system after intraventricular or intracerebral application and on blood platelets *in vitro,* 595—617
enhancing release of melatonin and 5-methoxytryptophol, 388
indolic nitrogen of for serotonin neurotoxic effects, 143
inhibition of serotonin-binding protein capacity, 88—90, 93
injection can increase dopamine turnover, 589
intraventricular infusion in fowl, 248
irreversible binding to protein, 59, *59,* 60, 69
maximal velocity, 48-49
mechanism of action, 177
mechanism of neurotoxicity, 21
metabolized by monoamine oxidase in rat hypothalamus homogenates, 47
Michaelis constant, 47—48
microinjection in rats, 571
monoamine oxidase-catalyzed conversion of tryptamine, 41, *44*
inhibited, 41, *45*
monoamine oxidase inactivation
in brain homogenates, 42, *50*
in vitro, 42—43
monoamine oxidase inhibited *in vivo,* 43, *51*
not influencing pituitary stores or circulating growth hormone in prepubertal rats, 391
oxygen requirement for binding, 59
protein binding after intraventricular injection, 46, *52*
rats pretreated intraventricularly, 483
reaction with molecular oxygen, 19
reaction with sulfhydryl groups, 68
reactive nucleophiles, 68, *69*
retarding sexual maturation and massively increasing serum gonadotropins, 399
selectivity of action on central monoaminergic neurons, 17—19
synthesis, 1

5,7-Dihydroxytryptamine—(cont'd)
turning, contralateral, to lesioned side induced in rats by *d*-lysergic acid diethylamide bitartrate, apomorphine hydrochloride, and lisuride maleate, 605, *608*
uptake
inhibition by α-methyltyramine and serotonin, 62, *64,* 71
inhibition by cocaine and desmethylimipramine, 62, *64,* 71
temperature-dependent mechanism, 68
uptake by atria, 61, *62,* 68—69
temperature-dependent and irreversible binding, 65, *66,* 68—69, 71
with 5,7-dihydroxytryptamine and norepinephrine, 62, *64,* 68—69, 71
vasoconstrictor effect in isolated strips of rabbit aorta, 458—459
vasoconstrictor effect in rats and rabbits, 474
5,6-Dihydroxytryptamine analogs
effect on brain biochemistry and behavior in rats, 141, *142*
neurotoxicity, 160
5,6-Dihydroxytryptamine and serotonin
axonal swelling and accumulation of serotonin fluorescence, 213
cf. *p*-chloroamphetamine effects on brain, 645
degeneration of serotonergic neurons
fluorescent appearances in fowl, 248, 250—251, *252—253*
in rats and fowl, 242—243
reduction of brain levels in rabbits, 254
effect on development of neurons, 330
effects on synthesis and metabolism, 315
impairment of serotonin uptake, 306, *306,* 324
neuron systems after treatment, 315, 317, *317,* 319
neurotoxicity 4, 29, 36, 85, 175, 184—185
selective degeneration of neurons, 263, 285
selective neurotoxic action in brain serotonergic nerves, 480, 489
5,6-Dihydroxytryptamine creatinine sulfate·$2H_2O$ ultraviolet spectra, 29, *30,* 33
[1-^{14}C]-5,6-Dihydroxytryptamine
accumulation and retention in tissue slices, 308, 310—311, *310, 311*
brain regional patterns of radioactivity after intraventricular injection, 108, *109,* 114—117
in rats treated with benzotropine, 110, *110*
metabolism in rat brain *in vivo,* 17—18
radioactivity in hypothalamus, pons-medulla oblongata, and spinal cord after intraventricular injection, 5, *6*
radioactivity in rat perchloric acid-homogenized central nervous system, 7, 9, *10,* 11
radioactivity in rat urine after intraventricular injection, 5, *9,* 19
5,7-Dihydroxytryptamine, 107, 118, 178—179, 198, 254, 289, 297, 328, 343, 385, 388, 390, 400, 456, 476—477, 479, 485, 488, 530—531, 566, 568, 571, 573, 664. *See also* in subheads under other main headings, e.g. Brain mitochondria—interaction of 5,6-dihydroxytryptamine and 5,7-dihydroxytryptamine in rats
α-methylation counteracting impairment of serotonin uptake, 307
and forced extinction with intrabrain stem injections, 362—363, 365—366
and hyperactivity in rats, 478
and inhibition of monoamine oxidase, 58
and metabolites of neurotransmitter compounds, 173
and selective cytotoxicity, 57
and uptake of various substances by rat brain tissue, 200, *201*
appearance of rat brain dorsal raphe (B7) after intraventricular injection, 296, *298—299*
as tool for morphologic studies of central indoleamine neurons, 322, *323,* 324
better tool than 5,6-dihydroxytryptamine for mapping central serotonin neurons, 346
binding sites in brain, 118
binding sites in lumbar spinal cord, 118
binding to mitochondria, 67
binding to rat hypothalamus homogenates, 38—39, 43—45, *51*
biochemical effects when intracisternally injected, 199—200
blood-brain barrier for, 331

5,7-Dihydroxytryptamine—(cont'd)
brain content of 5-hydroxyindoleacetic acid, homovanillic acid, and 3-methoxyphenylglycol sulfate in male rats injected in the right medial forebrain bundle after pretreatment with desmethylimipramine, 605, *606,* 612
brain content of serotonin, dopamine, and norepinephrine after injection in the right medial forebrain bundle of rats pretreated intraperitoneally with desmethylimipramine or maprotiline, 603, *604,* 605, 612
brain lesions, 557—558
breakdown at biologic hydrogen ion concentration and 37°C, 53
cause of considerable degeneration of norepinephrine neurons and dopamine systems, 346
changes in brain monoamines and their metabolites induced in rats pretreated with harmaline or Lu 10-171, 597, 600, *601,* 613
changes in systolic blood pressure, pulse rate, and body weight after intracerebroventricular injection in spontaneously hypertensive rats, 472, *473*
chemical properties, 29, 33
chosen over 5,6-dihydroxytryptamine because of lower toxicity, 438
chromatography, 60, *61,* 68—69
continuous infusion into aortic cannula of an isolated rabbit heart, 465, *465*
correlation between decrease in cortical serotonin turnover and the increase in striatal dopamine turnover in rats injected unilaterally into median raphe nucleus or substantia nigra, 526—527, *527*
cysteinyl-tyrosyl bridge, 70
discrepancy between effects and those of a raphe lesion, 445
dopamine content after treatment of forebrain in male rats, 426, 427
duration of effects on brain monoamine oxidase content, 161—162, *162,* 164—165
effect of 1-phenyl-3-(2-thiazolyl)-2-thiourea, ethanol, and nialamide on the neurodegenerative action, 78, *79,* 81
effect of centrally acting drugs on brain monoamine oxidase content, 162—163, *163,* 165
effect of intracisternal injection on amines in rat cerebral cortex, 200, *200*
effect of *p*-chloroamphetamine on, 260
effect of various doses administered on brain monoamines in newborn rats, 163—164, *164*
effect on dopamine uptake, 528
effect on locomotor activity when neonatally injected into rats, 165—166, *166*
effect on mounting behavior of male and female rats pretreated with desmethylimipramine, harmaline, or Lu 10-171, 602, *602,* 614
effect on norepinephrine output, heart rate, and positive chronotropic effect of exogenous norepinephrine in isolated rabbit hearts, 465, *466,* 469
effect on peripheral norepinephrine neurons, 57
effect on proestrous surges of luteinizing hormone and prolactin in female rats, 403, *404*
effect on serum levels of luteinizing hormone, prolactin, and growth hormone in rats, 409
effect on tegmentum, 188
effect on tyrosine hydroxylase activity, 528
effect when injected
into median raphe nucleus in rats on serotonin and 5-hydroxyindoleacetic acid levels and serotonin and norepinephrine uptake, 525, *525*
into the medial forebrain bundle on the methiothepin-induced increase of monoamine metabolites in male rats, 605, *607,* 612
intraventricularly on increase of 5-hydroxyindoleacetic acid and homovanillic acid induced by methiopin in rats, 597, *600*
effects of intracerebral injections on monoamine uptake and content, 320, *321*
effects of intracerebroventricular injection on systolic blood pressure, pulse rate, and body weight of conscious spontaneously hypertensive rats, 472, *473*
effects of intraventricular injection on rat forebrain norepinephrine concentration, 439, *439*

5,7-Dihydroxytryptamine—(cont'd)
effects of intraventricular injection on rat forebrain norepinephrine, dopamine, and serotonin and brain stem serotonin, 439, *440*
effects of monoamine oxidase inhibitors on, 162, *162*
effects on serum values in rats on the afternoon of estrus, 402, *404*
effects on suckling-induced prolactin rise in female rats, 405, *405*
effects on the rat central nervous system after intraventricular or intracerebral application and on blood platelets *in vitro,* 595—617
efflux from atria, 65, *66,* 71
electroshock seizure threshold in rats, 444, *447*
formation of hydrogen peroxide from oxidative deamination, catalyzed by monoamine oxidase, 74, *75*
given to produce brain lesions in rats, 364—366
impairment of norepinephrine uptake, 307
in combination with a catecholamine uptake blocking agent, 497—498
influence on effect of electrical stimulation of postganglionic sympathetic nerves of an isolated rabbit heart, 465, *467,* 469—470
inhibition of serotonin-binding protein capacity, 88—89, 93
inhibition of uptake, 84
injection can increase dopamine turnover, 589
intracisternal injection in rats, 628
irreversible binding to protein, 59, *59, 60,* 69
keto-enol tautomeric forms, 29, *32*
maximal velocity, 48—49
measures of body weight and food and water intake after intracisternal injection in rats, 632, *632*
mechanism of action, 177
metabolized by monoamine oxidase in rat hypothalamus homogenates, 47
Michaelis constant, 47
microinjection in rats, 571
monoamine analyses
after anticonvulsant testing in treated rats, 450, *451*
in rats, 444, *445,* 447
monoamine oxidase and, 69
monoamine oxidase-catalyzed conversion of tryptamine, 41, *46*
inhibited, 41, *47*
monoamine oxidase inactivation
in brain homogenates, 43, *50*
in vitro, 42—43
in vivo, 43, 51
monoamines in rat central nervous system after given alone or to animals pretreated with uptake inhibitors, 597, *599,* 612
norepinephrine content of forebrain and brain stem in treated male rats, 426, *426*
o-quinone of. *See o*-Quinone of 5,7-dihydroxytryptamine
oscillatory luteinizing hormone pattern in ovariectomized rats after treatment, 427, *431*
p-quinoneimine of. *See p*-Quinoneimine of 5,7-dihydroxytryptamine
possible enhancement of toxicity of metabolism, 54—55
pretreatment of adult rats with desmethylimipramine, 524
pretreatment with, and locomotor activity, 173—174
protein binding after intraventricular injection, 46, *52*
rats pretreated intraventricularly with, 483
reaction product with cysteine and tyrosine residues, 70—71, *70*
reaction with molecular oxygen, 19
reactive aldehyde, 70, *70*
reducing luteinizing hormone levels in rats treated, 432
response of rats to maximal electroshock, 443, *444*
restored circhoral luteinizing hormone pattern in ovariectomized rats after treatment, 427, *431*
selectivity of section on central monoaminergic neurons, 17—18
serum luteinizing hormone and prolactin levels in treated proestrous rats, 430, *432*
serum luteinizing hormone, prolactin, and monoamine levels in hypothalamus of treated ovariectomized rats, 427, *428*
serum prolactin, luteinizing hormone, and follicle-stimulating hormone levels
in ether-stressed treated rats, 426, *430*
in mildly stressed male rats, 425, *425*
sex-related differences in response to, 174
specificity, 334, 336—337, 528
subcellular distribution, 67, *67,* 71

5,7-Dihydroxytryptamine—(cont'd)
supersensitivity produced by, 260
toxicity to norepinephrine neurons overcome by pretreatment of animals with protriptyline, 438
tritium overflow in brain slices of treated rats, 334, *335,* 336
tyrosyl derivative, 70, *70*
uptake
dose-dependent loss of with 6-hydroxydopamine or 5,7-dihydroxytryptamine, 71
inhibition by α-methyltyramine and serotonin, 62, *64*
inhibition by cocaine and desmethylimipramine, 62, *64,* 71
pretreatment with 6-hydroxydopamine, 5,7-dihydroxytryptamine, or 5,6-dihydroxytryptamine, 62, *64*
uptake and accumulation in norepinephrine neurons, 21
uptake by atria, 61, *63,* 71
plus 5,6-dihydroxytryptamine and norepinephrine, 62, *64,* 68—69, 71
uptake by brain, 176
vasconstrictor effect in
isolated strips of rabbit aorta, 458—459
rats and rabbits, 474
5,7-Dihydroxytryptamine and serotonin, 73
and brain concentration, 171—172
and regeneration within central pathways, 346
cf. *p*-chloroamphetamine effects on brain, 645
content in rat brain after treatment, 426, *429*
effect of neonatal administration to rats on *in vitro* serotonin
uptake, 331, *332*
with 6-hydroxytryptamine and desmethylimipramine on fluorescence histochemical demonstration of nerve terminals, 332, *333*
effect of pathway lesions in rat midbrain on
acquisition of one-way avoidance response, 352, *352*
forced extinction of one-way avoidance response, 352, *353*
retention of one-way avoidance response, 352, *353*
effect on central neurons, 3—4, 57
effect on development of neurons, 330
effect when administered to rats on *in vitro* serotonin uptake, 330—331, *331*
impairment of serotonin uptake, 306, 324
metabolism and neurotoxicity, 20—21
neurotoxicity, 4—5, 21, 29, 36, 85, 175, 186, 296, 480, 489
reducing brain serotonin and norepinephrine, 166
reducing brain serotonin content after pretreatment with desmethylimipramine, 434
reducing serotonin, 167—168
selective degeneration of serotonergic neurons, 263, 285
specificity in reducing serotonin content, 432—433
5,7-Dihydroxytryptamine creatinine sulfate
changes in ultraviolet absorption with hydrogen ion concentration, 29, *32*
ultraviolet spectra, 29, *30,* 33
after reaction with oxygen, 33, *33*
[^{14}C]-5,7-Dihydroxytryptamine
accumulation and retention in tissue slices, 308, 310—311, *310, 311*
affinity for brain serotonin and norepinephrine uptake sites, 16—17
brain regional patterns of radioactivity after intraventricular injection, 108, *109,* 114—117
in rats treated with desmethylimipramine, 111, *113*
in rats treated with pargyline, 110—111, *112*
in rats treated with 6-hydroxydopamine combined with nialamide, 111, *113*
comparison of intraventricular and intracisternal routes of administration, 12—16
effect of injection route on regional retention of radioactivity, 12, *13*
increase in fractions from caudal spinal cord, 117—118
radioactivity in hypothalamus, pons-medulla oblongata, and spinal cord after intraventricular injection, 5, *7, 8*
radioactivity in rat perchloric acid-homogenized central nervous system, 7, 9, *10,* 11
radioactivity in rat urine after intraventricular injection, 5, *9,* 19
6,7-Dihydroxytryptamine
biochemical effects when intracisternally injected, 199
impairment of norepinephrine uptake, 307

6,7-Dihydroxytryptamine—(cont'd)
impairment of serotonin uptake, 306
o-Dihydroxytryptamines and serotonin, 25, 33
3,4-Dimethoxyamphetamine, 126
as metabolite of *p*-chloroamphetamine, 176
in brain, 120, 124—125
as psychotoxic agent, 124
5,6-Dimethoxytryptamine
biochemical and behavioral effects in the rat, 140—141
6,7-Dimethyl-5,6,7,8-tetrahydropterine, 583
N,N-Dimethyltryptamine 487
o,ω-Dinitrostyrenes, 25—26
Dithiothreitol
antagonizing toxic action of 5,6-dihydroxytryptamine on monoamine oxidase, 53
preventing oxidation of 5,6-dihydroxytryptamine but not 5,7-dihydroxytryptamine, 54
Dopa, 525
accumulation, 221
determination in brain, 223—224
rat brain concentration after fenfluramine administration, 232—233, *237*
L-Dopa, 389—390, 479, 613, 664
ineffective in producing myoclonic syndrome in rats, 551
occasional hypersexuality seen in patients during therapy of parkinsonism, 613
Dopa decarboxylase, 583-585
Dopamine, 118, 252, 388—391, 412, 414, 498, 510, 519—520, 530—531, 576, 578, 602, 612—613, 616. *See also* in subheads under other main headings, e.g. 5,7-Dihydroxytryptamine—effect on dopamine uptake
analyzed in rat forebrain, 439—440
brain and spinal cord levels after treatment with 5,7-dihydroxytryptamine or *p*-chlorophenylalanine in rats, 498, *499,* 503
effect of 5,7-dihydroxytryptamine-induced lesions of serotonin pathways of midbrain on level
in front of the lesion in rats, *355,* 356
within cortex of rats, 354, *354,* 356
in brain, 423, 432, 540, 547
after intraventricular injection of 5,6-dihydroxytryptamine, 3
cortex of rats, 365
of female rats, 621
of rats, 631
telencephalon and diencephalon of rats, 349
in unilaterally lesioned rats, amphetamine thought to induce ipsilateral turning by releasing, 524
influence of 5,7-dihydroxytryptamine, α-methyl-5,7-dihydroxytryptamine, and desmethylimipramine on forebrain, 14—16, *15*
long-term depletion in brain, 20
mean concentrations after administration of 5,7-dihydroxytryptamine into rats, 590—591, *592*
measured by radioenzymatic catechol *O*-methyltransferase method, 424
role in luteinizing hormone release, 388
uptake by synaptosomes, 92
Dopamine β-hydroxylase, 182—184, 583—587
Dopamine denervation hypersensitivity, 406
Dopamine neurons, 107, 114—116, 342—343, 527, 530, 540
in rat brain, 336
regeneration after 5,7-dihydroxytryptamine treatment, 385
Dopamine receptor agonists, 595, 603, 605, 612—616. *See also* Apomorphine; *d*-Lysergic acid diethylamide; Ergocornine; Ergometrine; Piribedil
Dopamine releasers. *See d*-Methamphetamine; Methylphenidate
Dopaminergic-binding protein, 94
Drinking, 484—485
altered by 5-hydroxytryptophan administration in rats, 556
changes in monkeys and rats, 566, 568
deficits when 5,6-dihydroxytryptamine microinjected in monkeys and rats, 570—571
hypodipsia, 484—485. *See also p*-Chlorophenylalanine and serotonin
hypothalamic actions of serotonin neurotoxins, 573
in rats, 628—629
mean daily water intake for rats after serotonin injection and 5,6-dihydroxytryptamine, 566, *567*
regulated in rats, 628

Drugs, centrally acting
effect on actions of 5,7-dihydroxytryptamine in adult rats, 161—163
Drugs, intraventricular vs. intracisternal injections producing different effects, 171

Eating. *See* Feeding
β-Endorphin effect on growth hormone release in rats, 422
met-Enkephalin effect on growth hormone release in rats, 422
Enzymes. *See also* as main headings or in subheads under other main headings: Acetylcholinesterase; Adenyl cyclase; Aldehyde dehydrogenase; Brain—enzymatic activity; Catalase; Catechol *O*-methyltransferase; Choline acetylase; Choline acetyltransferase; Decarboxylase; Dopa decarboxylase; Dopamine β-hydroxylase; Hydroxyindole *O*-methyltransferase; Hydroxylases; Monoamine oxidase; Oxidase; Peroxidase, horseradish; Phenylalanine hydroxylase; Superoxide dismutase; Tryptophan hydroxylase; Tyrosine hydroxylase
activity in the locus ceruleus after intracisternal injection of 5,6-dihydroxytryptamine, 583, *584*
inhibitors. *See* Monoamine oxidase inhibitors; RO4-4602
Ergocornine, 605
Ergometrine, 605
Estradiol benzoate, 622—624
Estrous cycles in rats, 430, 433
Ethanol as hydroxyl radical scavenger, 82, 84
Ethylene-generating system for measuring flux of hydroxyl radicals, 75
Ethylene production by catalase, superoxide dismutase, and hydroxyl radical scavengers suppressed, 76—77, *76,* 80—81

Feeding, 484—485. *See also* Anorexia; Drinking; Hyperphagia
altered by 5-hydroxytryptophan administration in rats, 556
amino acids in rats, 641—642
changes in monkeys and rats, 566—568
deficits when 5,6-dihydroxytryptamine microinjected in monkeys and rats, 570—571
finickiness in rats, 627—628, 630, 634, 642
high-fat diet, 628,630
hypothalamic actions of serotonin neurotoxins, 573
intake after
intraventricular injection of *p*-chlorophenylalanine methyl ester in rats, 590, *591*
p-chlorophenylalanine- and sham-injected rats, 593, *593*
serotonin and 5,6-dihydroxytryptamine injection in rats, 566, *567*
intake in rats, 628—629, 638, 640
obesity, 484, 490
suppression by serotonin system, 594
Fenfluramine, 150, 178, 289—291, 296—297, 300, 302—304, 488—490, 498. *See also* in subheads under other main headings, e.g. 5-Hydroxyindoleacetic acid—brain concentration—after fenfluramine administration in newborn and adult rats
administration, 123
and newborn rats, 239
and tryptophan 5-hydroxylase activity, 126
brains in rats injected later with fluoxetine, 229, *233*
can induce anorexia, 478
determination in plasma and brain by gas chromatography, 224
mass spectrometry, 224—225
effect in newborn animals, 172
effect on brain B9 region in rats, 293, *295,* 296
hyperthermic effects, 255—256
influence of administration route on brain levels, 120, *120*
main site of action, 238
metabolism, 126
metabolite accumulation after brain injection, 120, *121*
method to differentiate from norfenfluramine, 241
neurotoxicity, 125, 127, 259, 480, 482, 484
plasma and brain concentration in rats, 225—226, *227, 228*
powerful anorectic agent, 556
subcellular distribution of metabolites after brain injection, 121, *122*
suppression of female mating, 621
time course for decrease in serotonin and 5-hydroxyindoleacetic acid in rat brain produced by, 225, *226*

Fenfluramine—(cont'd)
  used for regulation of body weight in human beings, 485
Fenfluramine and serotonin
  and reduction of brain serotonin and 5-hydroxyindoleacetic acid, 119
  and serotonin-binding protein capacity, 91
  effect of acute and chronic injection on brain serotonin content in rats, 296, *296*
  effect on brain serotonin content when fenfluramine neonatally administered in rats, 163—164, *164*
  effect on content in rat brain regions, 292, *295*
  effect on serotonergic terminal processes, 222
  possible serotonin deficiency, 1
  relation of concentration in brain to serotonin depletion, 222
  serotonin releasing, 621
p-Fluoroamphetamine, 150, 154, 289—291
Fluoxetine, 172, 289—290, 476, 515, 519, 649. *See also* in subheads under other main headings, e.g. *p*-Chloroamphetamine—interaction with fluoxetine in mice
  ability to reverse action of fenfluramine, 238
  and brain uptake and metabolism of amphetamines, 121
  and protein synthesis, 105—106
  and serotonin-binding protein, 95
  as inhibitor of serotonin-binding protein capacity, 92
  blocking of anorectic effects of fenfluramine, 241
  effects on *p*-chloroamphetamine, 158—159
  effects on serum prolactin and brain and hypothalamic serotonin and norepinephrine concentrations in female rats, 406, *407*
  high dosage with serotonin producing myoclonus, 207
  inducing analgesia, 487
  inhibition of mixed-function oxidase activity, 215
  metabolism in brain, 124
  potentiating action of 5-hydroxytryptophan 389, 392
  selectivity in rat brain, 229, *233*
  significantly blocking all aspects of the *p*-chloroamphetamine-induced myoclonic syndrome in rats, 518
  significantly decreasing serotonin reuptake, 503
  stimulant effects of *p*-chloroamphetamine blocked in pretreated animals, 482
  termination of norfenfluramine serotonin-reducing action, 236
  termination of serotonin-lowering action of fenfluramine, 226—227, *229*
  unavailable to protect serotonin fibers from 5,6-dihydroxytryptamine, 613
Follicle-stimulating hormone, 387—389, 412, 419—420, 434
  determined by radioimmunoassay, 424
  inhibitory effect of serotonin on release, 433
Food. *See* Feeding

Gastrointestinal system
  autopsies in rats, 635—636
  motility in rats, 640
  pathology in rats, 631, 640
Glutathione
  antagonized toxic action of 5,6-dihydroxytryptamine on monoamine oxidase, 53
  prevention of oxidation of 5,6-dihydroxytryptamine but not 5,7-dihydroxytryptamine, 54
Glutathione, reduced
  effect on brain serotonin, 121, *123*
Gonadotropin release elevated by progesterone, 624
Gramine synthesis of indole derivatives via, 27, *27*
Growth hormone, 390—391, 399—400, 412, 419—420
  in rats, 630, 638, 641
  plasma levels reduced by stress in rats, 422
  role of serotonin in regulating in primates, 391
Guanosine triphosphate as inhibitor of serotonin-binding protein capacity, 91

Habituation to novel stimuli by rats, 486—487
Halogenated amphetamines. *See also* *p*-Bromoamphetamine; *p*-Chloroamphetamine; β, β-Difluoro-4-chloroamphetamine; Fenfluramine; *p*-Fluoroamphetamine
  effect blocked by fluoxetine, 292

Halogenated amphetamines—(cont'd)
effect of chronic treatment, 292, 296
effect on brain serotonin, 291
neurotoxicity, 289—300
Haloperidol, 515, 518—519
blocking response to amphetamine and apomorphine re contralateral rotation in rats, 526
Harmaline, 596, 602, 613
Heart rate
increase induced by
bolus injections of 5,7-dihydroxytryptamine, 468, *468*
tryptamine derivatives and some other amines on right atrial strips of guinea pigs, 462, *463*
tryptamine derivatives on right atrial strips of guinea pigs, 461—462, *461*
isolated hearts of rabbits, 462, 465
positive chronotropic effect of 5,6-dihydroxytryptamine and 5,7-dihydroxytryptamine, 461—462
spontaneously beating right atria of guinea pigs. 461
Hippocampus
depletion of serotonin by *p*-chloroamphetamine, 196—197
loss of the serotonin innervating, 453
perfusion with serotonin increased seizure activity in rats, 437
Hippurates and enzymatic synthesis in rat brain homogenates, 122, *123*
Histamine, 462
Homovanillic acid, 118, 173, 596—597, 605, 612—613, 615—616
Hormones. *See also* as main headings or in subheads under other main headings: Adrenocorticotrophic hormone; Estrogen; Follicle-stimulating hormone; Gonadotropins; Growth hormone; Luteinizing hormone; Melanocyte-stimulating hormone; Progesterone; Prolactin; Thyroid-releasing hormone; Thyroid-stimulating hormone
used as terminal neuroendocrine measures, 638, *639*, 641
5-Hydroxy-7-chlorotryptamine, 175
5-Hydroxy-6-methoxyindoleacetic acid, 17
4-Hydroxy-5-methoxytryptamine
rapid autoxidation preventing application as selective neurotoxic agent, 29
4-Hydroxy-5-methoxytryptamine hydrochloride ultraviolet spectra, 29, *30*, 33
5-Hydroxy-6-methoxytryptamine, 18
and alteration of dopaminergic neurotransmission, 146
biochemical and behavioral effects in rats, 140
3-Hydroxybenzylhydrazine hydrochloride, 223, 225, 232—235, *237*
6-Hydroxydopamine, 53, 76—78, 107, 114—117, 254—255, 289, 297, 328, 339, 342, 362, 406, 436, 478, 530, 566, 568, 570, 594, 631, 649, 664. *See also* in subheads under other main headings, e.g. 5,6-Dihydroxytryptamine—effect alone and in combination with 5-hydroxytryptophan or 6-hydroxydopamine on brain amines and mounting behavior in male rats
1,4-*p*-quinone derivative of, 68
and neuroblastoma cells, 68
and regeneration in catecholamine-containing neurons, 370
autoxidation, 57
autoxidative and enzymatic formation from dopamine, 135
cytotoxicity, 57
degeneration of central catecholaminergic fibers by intraventricular and intracisternal injection, 19
effect of reducing agents on, 68
effect of polysomal profiles, 102, *103*
effect on protein synthesis, 98—99, *99*, *100*, 101—103, *101*
for lesioning central catecholamine neurons, 505
for lesioning rats, 605, 609, 614
formation of 2,5-dicysteinyl derivative, 68
formation of hydrogen peroxide from spontaneous reaction with oxygen, 74, *75*
inhibition of uptake, 84
introduction, 1
lesions, 526—527
long-lasting depletion of norepinephrine in peripheral organs, 36
mechanism of action, 80
neurotoxicity, 160, 184—185
paraquinone of, 57
paraquinone reaction with free sulfhydryl group, 68
quinoidal products, 80—82
shown after intracisternal injection, 342
uptake by atria
5,7-dihydroxytryptamine, 5,6-dihydroxytryptamine, and norepinephrine, 62, *64*, 68—69, 71

5-Hydroxyindole, 390
Hydroxyindole derivatives and serotonin-binding protein, 95
5,6-Hydroxyindoleacetaldehyde, 18
5-Hydroxyindoleacetic acid, 199, 208, 211, 213, 217, 239, 525, 530—531, 533, 576, 578—579, 596—597, 600, 605, 612—613, 616, 621. *See also* *p*-Chloroamphetamine—influence on brain concentration of serotonin and 5-hydroxyindolacetic acid
  brain concentration, 152
    after fenfluramine administration in newborn and adult rats, 231, *236*
    after norfenfluramine administration in rats, 229, *234*
    after *p*-chloroamphetamine intraperitoneally injected in rats, 515, *517*
    after treatment with fluoxetine in rats, 227, *228,* 229
  determination, 222—223
  effect of 5,6-dimethoxytryptamine on, 140—141
  formation from tryptophan in different central nervous system regions of rats, 315, *318*
  in brain of rats, 519—520, 631, 636, 638
  in central nervous system, 4
  increase caused by 5,6-dihydroxytryptamine and 3-(β-aminoethyl)-5,6-dihydroxybenzo[*b*]thiophene, 139
  increase or decrease caused by 5-hydroxy-6-methoxytryptamine, 140
  permanence of recovery in brain induced by fluoxetine in rats, 227, *231*
  probenecid-induced accumulation in cerebrospinal fluid and brain, 300, 304
  reduced in brain by *p*-chloroamphetamine in rats, 645
  spectrofluorometric assay from brain, *400*
  substantially blocking hyperphagia and obesity following medial hypothalamic lesion in rats, 628
[^{3}H]-5-Hydroxyindoleacetic acid
  metabolism from [^{3}H]-tryptophan, 373, *374—375,* 376
Hydroxyl radicals
  effect of scavenge on the neurodegenerative action of 6-hydroxydopamine and 6-aminodopamine, 77, *78,* 80—81
  from superoxide and hydrogen peroxide, *75,* 76—77, 80
  role of 6-hydroxydopamine in neurotoxicity, 74
  scavengers. *See* Ethanol; 1-Phenyl-3-(2-thiazolyl)-2-thiourea
Hydroxylases
  brain and spinal cord activities of tryptophan, tyrosine, and dopamine β-hydroxylases, 184, *184*
  effect of 5,6-dihydroxytryptamine on brain and spinal cord activities of tyrosine and dopamine β-hydroxylases, 185, *185*
Hydroxymethoxyindoleacetaldehyde synthesis, 27
2-[β-(4-Hydroxyphenyl)ethylamino-methyl]tetralone, 457—458
  antagonism of methysergide and toward the contractile response to vasoconstrictor amines in aorta of rabbits 360, *460*
  cf. methysergide and on cumulative concentration-response curve of 5,6-dihydroxytryptamine in aorta of rabbits, 460, *461*
6-Hydroxytryptamine
  as inhibitor of serotonin-binding protein capacity, 86, 88, 93
  neurotoxicity on catecholamine cell bodies, 300
5-Hydroxytryptophan, 254-256, 260, 284, 366, 389-390, 392-394, 437, 444, 481, 483, 485, 487, 525, 590, 602, 613, 615—616, 621, 627, 664. *See also* in subheads under other main headings, e.g. *p-Chlorophenylalanine*—mounting behavior in male, female, castrated, and ovariectomized rats alone or in combination with 5-hydroxytryptophan or haloperidol
  able to block preovulatory luteinizing hormone discharge in intact rats, 423
  administered intraperitoneally to rats pretreated with intracranial injections of 5,7-dihydroxytryptamine, 510, 519
  administration intraperitoneally elevating brain serotonin, 388
  and brain serotonin, 484
  and increase in circulating growth hormone level, 391
  and reduction of tryptophan 5-hydroxylase, 119
  blocking of ovulation, 399

5-Hydroxytryptophan—(cont'd)
brain concentration after fenfluramine administration in rats, 232—233, *237*
causing release of melanocyte-stimulating hormone after administration in frogs, 394
determination in brain, 223—224
effect on prolactin release in rats pretreated with
5,7-dihydroxytryptamine, 408, *408*
p-chloroamphetamine, 408—409, *408*
effects on serum prolactin, brain and hypothalamic serotonin, and brain and hypothalamic norepinephrine concentrations in female rats, 406, *407*
in brain serotonin biosynthesis, 483
increasing release of corticosteroids after total interruption of neural afferents to hypothalamus, 393
inducing prolactin rise in both *p*-chloroamphetamine- and 5,7-dihydroxytryptamine-treated animals, 410
injected rats, earlier injected with desmethylimipramine and intracranial 5,7-dihydroxytryptamine, 510—511, *512*
restoring cyclicity of luteinizing hormone, 389
supersensitivity to, 239
5-Hydroxytryptophan-induced neurologic syndrome
antiserotonin agents, 203
behavior after administration of desmethylimipramine plus 5,7-dihydroxytryptamine, 203, *205*
denervation supersensitivity, 205
myoclonus, 204—207
myoclonus after intracisternal injection of 5,7-dihydroxytryptamine, 201—203, *202*
myoclonus effect of drug treatment in 5,7-dihydroxytryptamine- plus desmethylimipramine-lesioned rats, 204, *205*
pharmacologic studies, 203
postsynaptic serotonin receptor mechanisms, 205
presynaptic mechanisms, 205
6-Hydroxytryptophan, 285
Hyperphagia. *See also* *p*-Chlorophenylalanine; 5-Hydroxyindoleacetic acid
5-hydroxytryptophan reversed *p*-chlorophenylalanine-induced in rats, 590, *591*
and obesity, 485
central injection of 5,7-dihydroxytryptamine to desmethylimipramine-pretreated adult female rats did not induce, 638
failure of 5,7-dihydroxytryptamine to produce, 590
in rats, 484, 642
induced by systemic injection of gold thioglucose in mice, 627
Hypertension. *See also* Deoxycorticosterone acetate-saline hypertension in rats
regulation by central serotonergic neurons in rats and rabbits, 471
regulation in rats and rabbits, 474
Hyperthermia, 251—252, 254—256, 558. *See also* *d*-Amphetamine; Apomorphine; Chlorpromazine and prevention of lethal hyperthermia; Fenfluramine; Pimozide; Sleeping and waking
and pretreatment effect of *p*-chlorophenylalanine or 5,6-dihydroxytryptamine in rabbits, 245—247
induced by
5-hydroxytryptophan plus phenylethylhydrazine in rabbits, 246, *248*
apomorphine in rabbits, 246, *246*, 255—256
bacterial pyrogen in rabbits, 246, *247*
d-amphetamine in rabbits, 245—246, *245*, 255—256
fenfluramine and serotonin in control and 5,6-dihydroxytryptamine-treated fowl, 251, *251*
fenfluramine in rabbits, 247, *250*
injection of 5,7-dihydroxytryptamine in rats, 665
injection of serotonin in cats, 242
Hypodipsia. *See* *p*-Chlorophenylalanine and serotonin; Drinking
Hypothalamohypophyseal portal capillaries, 387
Hypothalamus, 393—395
action of 5,6-dihydroxytryptamine on in monkeys and rats, 568, 570—572
dopamine levels, 429
fall in serotonin level, 392
lesion in rats, 572
mesencephalic norepinephrine fibers play key role in stimulating preovulatory luteinizing hormone surge, 423

Hypothalamus—(cont'd)
monoaminergic innervation, 423
norepinephrine levels, 429, 434
serotonin cell bodies, 436
serotonin content after 5,7-dihydroxytryptamine administration, 429
synaptosomes in rats, 128
Hypothermia. *See also* Apomorphine
fatal in monkeys, 563—564
induced by injection of norepinephrine in cats, 242

Indole derivatives
inhibition of serotonin-binding protein, 86
Indoleamines, 129—132, 470, 472
histofluorescence, 244
Indolectomy, chemical, 387—395
5,6-*o*-Indolquinone, 36
Iprindole administration, 124
Iprindole and brain uptake and metabolism of amphetamines, 121
Iproniazid, 470
uptake by atria of 5,7-dihydroxytryptamine, 5,6-dihydroxytryptamine, and norepinephrine, 62, *64*
Isoleucine, 642
Isoprenaline, 462

Kainic acid, 530

Leucine, 642
Lineweaver-Burk representations, 40—41, *41, 43, 44, 46,* 47, *48*
Lisuride maleate, 596, 605
given intraperitoneally and tested for inducing contralateral turning in rats, 605, *608,* 614—616
Locomotor activity, 480—482
effects of serotonin neurotoxic drugs on, 481—482
hyperactivity vs. hypoactivity, 481
hyperactivity, 482
hypoactivity, 482
in rats, 664
Locus ceruleus
and unilateral rotational behavior when lesioned, 589
in brain of cats, 582—587
Lordosis mediation by estrogen in female rats, 623—624
Lu 10-171, 596, 602, 613
preventing both serotonin depletion and the occurrence of mounting behavior in rats, 614
preventing destruction of serotonergic system, 56
Luteinizing hormone, 362—363, 387—389, 400, 423, 432—434
change of rhythm after ovariectomy plus 5,7-dihydroxytryptamine, 436
determined by radioimmunoassay, 424
inhibitory effects of serotonin infusion into cerebrospinal fluid, 399
surges induced by estrogen administration blocked by *p*-chlorophenylalanine injection in ovariectomized rats, 399
d-Lysergic acid diethylamide, 203, *204,* 207, 260, 349, 363, 487—488, 498, 596, 605, 614—616, 664
as inhibitor of serotonin-binding protein capacity, 91
binding *in vitro* in brain of rats, 337, *338,* 342—343
blocked myoclonic syndrome in rats, 511
d-Lysergic acid diethylamide-induced behavioral syndrome in rats, 406
[^{3}H]-*d*-Lysergic acid diethylamide
binding in rat cerebral cortex, 351, 358—360, *358, 359, 360, 361, 362*
after serotonin denervation of forebrain, 364—366

Maprotiline, 596, 603, 612—613
Mating behavior. *See* Sexual behavior
Melanocyte-stimulating hormone, 394
Melatonin
in rats, 631, 638
inhibitory influence on pituitary gonadotropin secretion, 388
Metaraminol, 597
d-Methamphetamine, 302, 605, 615
Methergoline, 515, 518—519
as blocker of serotonin receptors, 392
Methimazole structure, 77, *77,* 80—81
Methiopin, 389, 515, 518—519
as blocker of cerebral serotonin, dopamine, and norepinephrine receptors, 597, 605, 612
5-Methoxy-*N,N*-dimethyltryptamine, 254, 498
producing myoclonic syndrome in rats, 511
5-Methoxy-6-hydroxyindoleacetic acid, 17
3-Methoxy-4-hydroxyphenylethyleneglycol, 173
3-Methoxy-4-hydroxyphenylethyleneglycol sulfate, 582, 597, 605, 612—613, 616

5-Methoxy-6-hydroxytryptamine, 17
biochemical and behavioral effects in rats, 140
6-Methoxy-5-hydroxytryptamine, 17
5-Methoxytryptamine, 538
5-Methoxytryptophol inhibitory influence on pituitary gonadotropin secretion, 388
α-Methyl-5,6-dihydroxytryptamine, 69
α-Methyl-5,7-dihydroxytryptamine, 116
and cytotoxicity, 58
as serotonin neurotoxin, 4—5
cytotoxicity, lack of effect with, 69
intraventricular cf. intracisternal routes of administration, 12—16, *14, 15*
molar neurotoxicity, 20—21
toxic effects on norepinephrine- and serotonin-containing brain axons, 11
N-Methyl-5,6-dihydroxytryptamine impairment of norepinephrine uptake, 307
N-Methyl-5,7-dihydroxytryptamine impairment of norepinephrine uptake, 307
α-Methyl-5-hydroxytryptamine
as serotonergic false neurotransmitter, 194
measurement by chromatography, 191, *191*
use to deplete serotonin, 190
α-Methyl-5-hydroxytryptamine phosphate time course of changes in brain nuclei after injection, 194, *195*
α-Methyl-5-hydroxytryptophan, 194, 196
α-Methyl-*p*-tyrosine, 436, 515, 518—519
3-Methylcholanthrene induction of hepatic oxidase activity, 215
Methylphenidate, 605
Methylscopolamine, 515, 518
6-Methyltetrahydro-β-carboline and changes in central serotonergic neurons, 284
6-Methyltetrahydropterin as cofactor in assay of tryptophan and tyrosine hydroxylases, 183
Methysergide, 203, *204,* 207, 284, 390, 406, 409, 458—459, 462, 498, 504—505, 515, 518—519, 539
as inhibitor of serotonin-binding protein capacity, 91
blockade of serotonin receptors, 623
blocked myoclonic syndrome in rats, 511
effects on serum values on the afternoon of estrus in rats, 403, *405*
inhibition of estrous rise of serum prolactin, 403
pretreatment with neither enhances nor antagonizes 5,7-dihydroxytryptamine-induced hypothermia in rats, 489
putative serotonin receptor antagonist, 485
Monoamine oxidase, 11—12, 17—21, 470—471, 585. *See also* in subheads under other main headings, e.g. 5,7-Dihydroxytryptamine—metabolized by monoamine oxidase in rat hypothalamus homogenates
and clearance of 5,7-dihydroxytryptamine from the brain, 116
conversion of 5,6-dihydroxytryptamine, 40, *41,* 54—55
conversion of 5,7-dihydroxytryptamine, 40, *42,* 54—55
conversion of serotonin *in vitro,* 40, *43*
determination using [1-^{14}C]-5,6-dihydroxytryptamine and [1-^{14}C]-5,7-dihydroxytryptamine as substrates, 37
effect of intracisternal injection of 5,6-dihydroxytryptamine in locus ceruleus of cats, 583, *585*
effect of *p*-chloroamphetamine on, 260
inactivation by 5,6-dihydroxytryptamine or 5,7-dihydroxytryptamine, 53
inactivation *in vitro* and *in vivo,* 38
Michaelis constants and maximal velocities for conversion of 5,6-dihydroxytryptamine, 5,7-dihydroxytryptamine, and serotonin in rat hypothalamus, 40, *40*
time course conversion of 5,6-dihydroxytryptamine in rat hypothalamus, 39, *40*
Monoamine oxidase A, 48—49, 54
Monoamine oxidase B, 49
Monoamine oxidase inhibitors, 3—4, 19, 21, 74, 78, 81—82, 118, 254—256, 260, 307, 324, 347, 437—438, 481, 483, 486, 497, 520, 595, 600, 612—613, 664. *See also* Harmaline; 5-Hydroxytryptophan; Nialamide; Pargyline; Tranylcypromine
affecting cytotoxicity of 5,6-dihydroxytryptamine and 5,7-dihydroxytryptamine differently, 56
and neurotoxicity, 160

Monoamine oxidase inhibitors—(cont'd)
and uptake of 5,7-dihydroxytryptamine, 171-172
kinetics of 5,6-dihydroxytryptamine and 5,7-dihydroxytryptamine, 41—42
protection against action of 5,7-dihydroxytryptamine, 167
Monoamines, brain, 636, *637*. *See also* in subheads under other main headings, e.g. 5,7-Dihydroxytryptamine —changes in brain monoamines and their metabolites induced in rats pretreated with harmaline or Lu 10-171
and their main metabolites, 616
catecholamines. *See* Dopamine; Norepinephrine
indoleamines. *See* 5-Hydroxyindoleacetic acid; Serotonin
Morphine, 462, 488
effect on growth hormone release in rats, 422
Morphine analgesia, effects of
5,6-dihydroxytryptamine, 541
5,7-dihydroxytryptamine, 541—542
6-hydroxydopamine, 541
fluoxetine, 541
in rats, 348
p-chlorophenylalanine, 541
selective electrolytic raphe lesions, 540—541
zimelidine, 541
Mounting behavior. *See also* *p*-Chlorophenylalanine; Lu 10-171
after 5,6-dihydroxytryptamine or 5,7-dihydroxytryptamine in rats, 602, 616
after *p*-chlorophenylalanine in rats, 603, 613—614
classified as sexual behavior or not, 665
in rats, 596
serotonin neurons shown to originate in, 394
Myoclonus
and cerebellar function, 260—261
drugs interfering with dopaminergic, norepinephrinergic, and cholinergic neurotransmission did not block syndrome in rats, 511
effect of serotonin antagonists produced by 5-hydroxytryptophan after 5,7-dihydroxytryptamine administration in rats, 511, *514*
induced by *p*-chloroamphetamine, 518—520
record induced by 5-hydroxytryptophan in rats, 510, *513*
syndrome related specifically to central indoleamine neurons, 511

Nasoanal length in rats, 630, 636, 638
Neuroblastoma cells. *See* 6-Hydroxydopamine—and neuroblastoma cells
Neuroendocrine regulation and role of serotonin, 387, 394—395
Neurohumoral amine metabolic pathways, 182, *183*
Neurons. *See* Catecholamine neurons; Dopamine neurons; Norepinephrine neurons; Serotonin neurons
Neurotoxic indoleamines. *See also* 5,6-Diacetoxytryptamine
cardiovascular effects, 457—474
degenerative effects on central monoamine neurons, 305—324
effects of various doses on brain and spinal cord serotonin and brain and spinal cord catecholamine content, 312—313, *312, 314*
effects on monoamine levels, 312—313
effects on serotonin and norepinephrine uptake, 313—314
in vitro, 307, *307*
fluorescence morphology, 315—317, 319
local intracerebral injections, 320—322
morphologic studies, 322—324
uptake site affinity and inhibition *in vitro,* 307—308, *309,* 313
Neurotoxins. *See* 5,6-Diacetoxytryptamine; 5,6-Dihydroxytryptamine; 5,7-Dihydroxytryptamine; *o*-Dihydroxytryptamine; 6-Hydroxydopamine; α-Methyl-5,7-dihydroxytryptamine; Neurotoxic indoleamines; Serotonin-like neurotoxins
Nialamide, 114, 117
as potentiating agent, 80
decreasing lordosis response frequency in female rats, 485
inhibiting conversion of 5,6-dihydroxytryptamine and 5,7-dihydroxytryptamine by monoamine oxidase, 47, 54
Nicotinic drugs and nicotinic receptors, 468
Nigrostriatal pathway, 524
model for interaction with other tracts, 530

Nigrostriatal pathway—(cont'd)
modulation by striatal interneurons, 530
motor efferents, 530
o-Nitrotoluene, 25
Nociception
analgesia, 488
hyperalgesia, 487—488
Norepinephrine, 390, 393, 412, 418, 467, 477, 481, 490, 519—520, 576, 578, 582, 595, 602, 612—613, 624, 627, 665. *See also* in subheads under other main headings, e.g. Desmethylimipramine — blocking brain norepinephrine transport sites
5,7-dihydroxytryptamine effect on brain and spinal cord concentration in rats, 200
analyzed in forebrain of rats, 439—440
and specificity of 5,7-dihydroxytryptamine, 334, 336
appearing to be caused by an activation of presynaptic serotonin receptors on sympathetic nerve terminals, 462, 465
as neurotransmitter, 134
brain and spinal cord levels after treatment with 5,7-dihydroxytryptamine or *p*-chlorophenylalanine in rats, 498, *499*, 503
catabolism, 21
concentration after administration of 5,7-dihydroxytryptamine in rats, 590—591, *592*
concentration-response curve, 460
depletion by reserpine, 3
depletion in the heart, 135
effect of 5,6-dimethoxytryptamine on, 140
effect of 5,7-dihydroxytryptamine-induced lesions of serotonin pathways of midbrain on level
in front of the lesion in rats, *355*, 356
within cortex in rats, 354, *354*, 356
effects of various drugs
and of calcium ion omission on output from isolated hearts and on the positive chronotropic effect evoked by 5,7-dihydroxytryptamine bolus injection in rabbits, 468, *469*
on output from isolated hearts of rabbits, 468, *468*
in brain, 402, 432—433, 540—541, 547
after intraventricular injection of 5,6-dihydroxytryptamine, 3—4, 146
cortex of rats, 365
hypothalamus, 406
long-term depletion, 20
of female rats, 621
of rats, 631, 638
telencephalon and diencephalon of rats, 349
turnover in rats, 252
influence of 5,7-dihydroxytryptamine
on output and on increase in heart rate evoked by electrical stimulation, 469—470, *470*
on removal of exogenous, by isolated hearts of rabbits, 469, *470*
plus α-methyl-5,7-dihydroxytryptamine and desmethylimipramine on central nervous system, 14—16, 15
measured by radioenzymatic catechol *O*-methyltransferase method, 424
role in luteinizing hormone release, 388
substrate for monoamine oxidase, 48
uptake by atria in mice, 74—75
Norepinephrine neurons, 107, 114—116, 337, 339—340, 342—343
lesions by 6-hydroxydopamine and regeneration of, 347
pathways in rats, 605, 612—613
pathways unaffected in rats, 603
regeneration after 5,7-dihydroxytryptamine or 5,6-dihydroxytryptamine treatment, 379, 383
regeneration after 5,7-dihydroxytryptamine treatment, 385
Norfenfluramine, 120, 125, 150. *See also* in subheads under other main headings, e.g. 5-Hydroxyindoleacetic acid—brain concentration—after norfenfluramine administration in rats
as major metabolite of fenfluramine, 236
depletion of brain serotonin, 5-hydroxyindoleacetic acid, and tryptophan hydroxylase in rats, 150, 151
determination in plasma and brain by gas chromatography and gas chromatography-mass spectrometry, 224—225
mediator of long-term effect of fenfluramine, 236
method to differentiate from fenfluramine, 241

Norfenfluramine—(cont'd)
plasma and brain concentration in rats, 225—226, *227, 228*
relation of brain concentration to depletion of serotonin, 222

Ontogenesis of serotonin neurons. *See* Serotonin neuron development
Osmiophilia of serotonin organelles, 609, *609, 610*
Ovariectomy not affecting mounting behavior in female rats, 603
Ovulation blocking by 5-hydroxytryptophan. *See* 5-Hydroxytryptophan—blocking of ovulation
Oxygen consumption. *See also* Antimycin A and oxygen consumption
cyanide effect in brain mitochondria of rats, 79, *79,* 81
dopamine, serotonin, and boiled mitochondrial preparation, 79, *79,* 81
in brain mitochondria of rats, 75
in brain mitochondria stimulated by 5,6-dihydroxytryptamine and 5,7-dihydroxytryptamine in rats, 79, *79,* 81
Oxygen reduction to form superoxide radical and ionized peroxide, 75, *75,* 80

Pain sensitivity, effects of
5-hydroxytryptophan, 539—540
5,6-dihydroxytryptamine, 539
5,7-dihydroxytryptamine, 539—540
and raphe lesions, 664—665
electrolytic lesions in medial forebrain bundle, 539—540
fluoxetine, 540
p-chloroamphetamine, 540
p-chlorophenylalanine, 539—541
quipazine, 540
tryptophan, 540
Pancreas protection against alloxan by thiourea, 80
Paraquinone of 6-hydroxydopamine. *See* 6-Hydroxydopamine
Pargyline, 4, 17, 20, 203, *204,* 502
decreasing lordosis response frequency in female rats, 485
effect on actions of 5,7-dihydroxytryptamine on newborn brain in rats, 165, *165*
effect on retention of rat brain radioactivity, 108, 110—111, 114, 116
Parkinsonism, therapy for. *See* L-Dopa
Pentobarbital effect on growth hormone release in rats, 422
Pentylenetetrazol inducing convulsions in mice, 437
Pentylenetetrazol response after treatment with 5,7-dihydroxytryptamine or a raphe nuclei lesions in rats, 442, *443*
Perchloric acid, 5, 7, 9, 11—12, 20, 29, 38—39, *52,* 53—54, 108, 110—111, 115—116
Peroxidase, horseradish, 537
Phenethylamine derivatives, autoxidizable. *See* 6-Hydroxydopamine
Phenobarbital, 438
and β,β-difluoro-*p*-chloroamphetamine, 153
and half-life of *p*-chloroamphetamine, 221
effect of 5,7-dihydroxytryptamine or a raphe lesion on anticonvulsant action, 448—450, 452
effect on maximal electroshock-induced hindleg extension in rats, 449, *449,* 450
induction of hepatic oxidase activity, 215
Phenothiazines reducing receptor availability of neurotransmitters, 389
Phenoxybenzamine, 515, 518
Phentolamine, 458
1-Phenyl-3-(2-thiazolyl)-2-thiourea
as copper chelator, 84
as hydroxyl radical scavenger, 82
as protective agent, 80
structure, 77, *77,* 80—81
Phenylalanine, 546, 642
Phenylalanine hydroxylase, 540, 546
Phenylethylamine compound effect on protein synthesis, 96, 99, *100,* 101
3-Phenylpiperidines, chlorinated, 153
Phenytoin, 438
effect of 5,7-dihydroxytryptamine or a raphe lesion on anticonvulsant action, 448—450, 452
effect on maximal electroshock-induced hindleg extension
after a raphe nuclei lesion in rats, 450, *450*
in rats, 449, *449,* 450
Pimozide, 515, 518
counteraction of hyperthermic effect of *d*-amphetamine, 252
Pindolol, 467
Piperocaine, 623—624

Piperonyl butoxide
effect on reduction of serotonin and tryptophan hydroxylase by *p*-chloroamphetamine, 208, *209*
inhibition of hepatic oxidase activity, 215
Piribedil, 605
Pituitary-adrenal axis suppression by intraventricular injection of indoles, 392
Pituitary, anterior, regulation of secretion, 387
Platelets, blood. *See* Blood platelets
Postural syndrome and *p*-chloroamphetamine-induced abnormalities, tremor, and myoclonus in rats, 511, 514
Probenecid, 17
Progesterone location of implants in female rats, 624, *624*
Progesterone uptake by all brain regions in rats and guinea pigs, 623
Prolactin, 362—363, 389—390, 394, 400, 434
in rat plasma, 436
in rats, 630, 638, 641
serum estrogen-induced surges blocked by *p*-chlorophenylalanine or methysergide, 399
serum values on the day of vaginal estrus in rats, 402, *403*
stimulation of release after intracerebroventricular serotonin injection, after administration of 5-hydroxytryptophan and treatment with serotonin agonists, 399
stress-induced release governed primarily by the norepinephrine system, 433
Propranolol, 462, 515, 518
Prostaglandins, 665
Protein synthesis inhibition by neurotoxins, 96—104
Protriptyline, 438, 595
as norepinephrine uptake blocker, 346
plus 5,7-dihydroxytryptamine as a means of selective lesioning serotonergic neurons, 439, 452
pretreatment with, 543
preventing the effects of 5,7-dihydroxytryptamine on norepinephrine neurons, 439

Quinine sulfate
acceptance of varying solutions instead of tap water by rats, 634, *635*
in drinking water for rats, 630
Quinoid structures, 18, 29, 54
of 6-hydroxydopamine, 36
1,4-*p*-Quinone derivative of 6-hydroxydopamine. *See* 6-Hydroxydopamine—1,4-*p*-quinone derivative of
o-Quinone
of 5,7-dihydroxytryptamine, 21
sulfhydryl group reduction, 103
p-Quinone
sulfhydryl group reduction, 103
p-Quinoneimine of 5,7-dihydroxytryptamine, 19—21
neurotoxicity, 55
Quinoneimines from 5,6-dihydroxytryptamine or 5,7-dihydroxytryptamine, 53
Quinones
from 5,6-dihydroxytryptamine or 5,7-dihydroxytryptamine, 53—54
neurotoxic effects, 175
Quipazine, 207, 487
producing myoclonic syndrome in rats, 511
stimulation of serotonin receptors stimulating prolactin release, 389—390

Raphe nuclei
dorsal raphe nucleus, 482
in midbrain, 484—485
in rats, 488—490
lesions. *See also* Acetazolamide; Acetylcholine; Activity level; Avoidance behavior; Behavior, rotational; Brain; 5,7-Dihydroxytryptamine; Pain sensitivity, effects of; Pentylenetetrazol; Phenytoin; Receptor supersensitivity; Spatial reversal learning
producing behavioral and electrophysiologic signs of insomnia in cats, 483
locomotor activity, 481
median or dorsal, 489—490
nucleus raphe dorsalis or nucleus raphe magnus, 487
Rearing activity
behavior of rats, 652, *652,* 654, *655,* 656, *657*
Receptor supersensitivity
absence after median raphe nucleus lesions, 527
due to serotonin or dopamine neurons in rat brain, 526
Ungerstedt's model, 526—527

Rectal temperature
effect of 5-hydroxytryptophan in rats pretreated with desmethylimipramine and 5,7-dihydroxytryptamine, 511, *513*
Reserpine, 302, 437, 462, 467, 578. *See also* in subheads under other main headings, e.g. Norepinephrine—depletion by reserpine
amine depletion with, 392
and serotonin-binding protein, 95
as inhibitor of serotonin-binding protein capacity, 86
brain injected later with fluoxetine in rats, 229, *233*
reduction of thyroid-stimulating hormone with, 394
uptake by atria
5,7-dihydroxytryptamine, 5,6-dihydroxytryptamine, and norepinephrine, 62, *64,* 71
Reticulocyte ribosomes, 97
Ribonucleic acid, globin messenger, 97—98
RO4–4602, 203, *204*
correlation between the decrease in cortical 5-hydroxytryptophan and the decrease in striatal dopamine after injection of rats, 528, *528*
preventing conversion of 5-hydroxytryptophan to serotonin, 392
to measure the turnover of dopamine and serotonin in rats, 525
Rotational behavior. *See* Behavior, rotational
Rotenone and oxygen consumption, 81

Salicylates and brain serotonin concentration, 92
Schizophreniform symptoms, fenfluramine-induced, 490
Seizure susceptibility changes, 456
after intracerebral treatment with 5,7-dihydroxytryptamine, 437—453
effect of 5,7-dihydroxytryptamine, 438—445, 447—448
effect of a raphe lesion, 438—445, *446,* 447—448
electroshock seizures by corneal stimulation in rats, 442
forebrain norepinephrine and dopamine concentrations unaltered in 5,7-dihydroxytryptamine-treated rats, 452
induced by pentylenetetrazol by intraperitoneal injection in rats, 441—442, 445, 452
role of serotonin, 453
Serotonergic elements, 242
receptor supersensitivity, 243
selective destruction, 263
Serotonergic motor syndromes, 510—520
Serotonergic nerve terminals, supersensitivity after destruction of. *See* Supersensitivity after destruction of serotonergic nerve terminals
Serotonergic raphe inputs to the nigrostriatal pathway diagrammatic model, 529, *529*
Serotonin, 254—256, 422, 437—440, 442, 444—445, 447—448, 450, 452—453, 472, 497, 504—505, 530—531, 556, 568, 613, 616—617, 627. *See also* other main headings beginning "Seroton-" or with the word "serotonin" (e.g. 5,7-Dihydroxytryptamine and serotonin); *also* subheads throughout the subject index may refer to serotonin and should be sought by referring to related concept, as they are too numerous to list here (e.g. Brain—serotonin content; Hydroxyindole derivatives—and serotonin-binding protein)
abnormal motor response complex induced by *p*-chloroamphetamine, 209, 211
alterations in postnatal development of serotonin neurons after administration at birth, 339, *341*
analyzed in forebrain, brain stem, and spinal cord of rats, 439
and mediation of hyperthermic response to *d*-amphetamine and apomorphine in rabbits, 252
as neurotransmitter, 134, 483
binding sites in lumbar spinal cord, 118
brain and spinal cord levels after *p*-chlorophenylalanine and 5,7-dihydroxytryptamine administration, 511, 514, *514*
brain and spinal cord levels after treatment with 5,7-dihydroxytryptamine or *p*-chlorophenylalanine in rats, 498, *499,* 503
brain concentration, 152, 171
after fenfluramine administration in rats, 231, *235*
after intraperitoneal injection of *p*-chloroamphetamine in rats, 515, 517

Serotonin—(cont'd)
brain concentration—(cont'd)
after norfenfluramine administration in rats, 229, *234*
after treatment with fluoxetine in rats, 227, *228*, 229
in forebrain and midbrain after fenfluramine administration in rats, 231, *237*
in rats, 519
measured, 87—88
permanence of recovery induced by fluoxetine in rats, 227, *230*
spectrofluorometric assay, 400
brain concentration depletion
after injection of fenfluramine, 222
after injection of *p*-chloroamphetamine or *p*-chloro-*N*-methylamphetamine in rats, 150—151, *151*
after intracisternal injection of 5,7-dihydroxytryptamine, 4
and recovery in rat brain regions after single intraventricular injection of 5,6-dihydroxytryptamine, 370, *371*, 372
by *p*-chloroamphetamine in rats, 645
in whole brain by *p*-chlorophenylalanine and reversed by 5-hydroxytryptophan in rats, 590, *592*
long-term, 20
related to route of administration used, 171
causing release of melanocyte-stimulating hormone after administration in frogs, 394
chronic receptor blockade, 498
competition of unlabeled vs. tritium-labeled *in vitro* in brain of rats, 337, *339*
competition vs. *d*-lysergic acid diethylamide for [^{3}H]-*d*-lysergic acid diethylamide *in vitro* in brain of rats, 337, *340*
depletion, 119
effect of 5,7-dihydroxytryptamine-induced lesions of serotonin pathways of midbrain on level
in front of the lesion in rats, *355*, 356
within cortex in rats, 364, *354*, 356
endogenous in bulbospinal system given injection of 5,6-dihydroxytryptamine intraventricularly in rats, 372, *372*
failure of depletion to prevent induced prolactin rise in females treated with 5,7-dihydroxytryptamine, 478
formation from tryptophan in different central nervous system regions in rats, 315, *318*
history of isolation and characterization, 1
importance of interval between injection of fenfluramine and fluoxetine on recovery in brain of rats, 227, *232*
in brain, 402, 411, 418—420, 432—434
activating the secretion of adrenocorticotrophic hormone, prolactin, and growth hormone, 411
after injection of 5,6-dihydroxytryptamine, 3—5
binding *in vitro* in rats, 337, *338*
cortex in rats, 365
cortical uptake after fenfluramine administration in rats, 231, *235*
depletion having anorexigenic effect, 418—419
hypothalamus, 406
in rats, 631, 636, 638, 640—642
inhibits secretion of gonadotropins, 411
recovery produced by fluoxetine in rats treated with norfenfluramine, 229, *230*
reduced with *p*-chlorophenylalanine plus 5,6-dihydroxytryptamine, 436
telencephalon and diencephalon in rats, 349
time course of changes in nuclei after injection of α-methyl-5-hydroxytryptamine, 194, *195*
time course of changes in nuclei after injection of *p*-chloroamphetamine, 192, *193*
tryptophan-induced elevation, 392
uptake in rats, 334, *335*
whole brain, 406
influence of 5,7-dihydroxytryptamine, α-methyl-5,7-dihydroxytryptamine, and desmethylimipramine on central nervous system, 13—14, *14*
inhibition of surgery-induced adrenal steroid secretion with intraventricular doses in dogs, 392
intraventricular injection shown to inhibit gonadotropin release and to stimulate prolactin release in male rats, 423
long-lasting effect of fenfluramine, 234—235
maximal velocity, 48—49

Serotonin—(cont'd)
mean concentrations after administration of 5,7-dihydroxytryptamine into rats, 590—591, *592*
measured by radioenzymatic hydroxyindole *O*-methyltransferase method, 424
measurement by chromatography, 191, *191*
metabolic effects, 394
Michaelis constant, 48
microinjection into hypothalamus at room temperature raises body temperature in monkeys, 488—489
monoamine oxidase-catalyzed conversion of tryptamine, 41, *48*
inhibited, 41, *49*
nervous system regional differences, 213—214
reduction of cerebrospinal serotonin accompanied by enhancement of sexual activity in male rats, 595
release, 561
role in mechanism of sleep in cats, 576
storage and/or transport, 86—87
synthesis rate, 234
toxic inhibitory tone on gonadotropin release, 388
uptake, 133
by synaptosomes, 92
Serotonin analogs, 129—132
Serotonin-binding protein
5,6-dihydroxytryptamine and 5,7-dihydroxytryptamine as inhibitors of, 85
binding of newly synthesized serotonin, 92, *93*
characteristics of capacity, 87
ferrous ion-enhancing capacity, 86
inhibition of serotonin binding by neurotoxins, 88—89, *88*
location in synaptosomal cytoplasm and cytosol of brain and peripheral nervous tissue, 86
role of sulfhydryl group, 86
Serotonin cell group (B1–B9) anatomy, 532—533, *533, 535, 536—537,* 537—538, 544
afferent connections, 538—539
efferent projections from
dorsal raphe to amygdala, 537
dorsal raphe to caudate-putamen, 537
median raphe to individual hypothalamic nuclei, 537
median raphe to septum-hippocampus, 537
Serotonin denervation hypersensitivity, 406
Serotonin depletion methods
differences in neurochemical effects, 532
importance in determining behavioral effects, 532
Serotonin inhibitors. *See* Lisuride maleate; Lu 10-171; Methysergide
Serotonin-like neurotoxins, 129—132
Serotonin neuron development, 329, 338—343, 346—350, 366
axonal sprouting in rats, 376
effects of neonatal 5,7-dihydroxytryptamine treatment in rats, 330—332, 334
growth regulation, 342
intracisternal 5,7-dihydroxytryptamine injection, 341—342
not associated with collateral sprouting of nerves not damaged by 5,7-dihydroxytryptamine, 342
route of neurotoxin administration, 328—329
serotonin receptors, 337, 342
Serotonin neuron ontogenesis. *See* Serotonin neuron development
Serotonin neurons, 107, 114—116, 573, 576, 582, 659—660
affecting behavior including sleep, feeding, temperature control, and pain sensitivity, 510
effect of 5,6-dihydroxytryptamine on, 346
effect of *p*-chloroamphetamine on in rats, 645, 649
evaluation of degeneration, 350—351
false neurotransmitter as tool for study of, 190
hyperresponsivity in rats after lesions, 363—364
in hypothalamus, 410
in rat brain, 642
in rats, 602, 605, 609
negative reenforcement, 363
regeneration after 5,7-dihydroxytryptamine or 5,6-dihydroxytryptamine treatment, 379, 383
regeneration in brain after
5,6-dihydroxytryptamine-induced axotomy in rats, 370—383
5,7-dihydroxytryptamine-induced axotomy in rats, 370
supersensitivity disappearing, as shown by reduced responsiveness to 5-methoxydimethyltryptamine and to tryptophan, 347

Serotonin neurons—(cont'd)
terminals as primary site of action of fenfluramine, 238
Serotonin neurotoxins, 394—395, 409. *See also* 5,6-Diacetoxytryptamine; 5,6-Dihydroxytryptamine; 5,7-Dihydroxytryptamine; *o*-Dihydroxytryptamine; 6-Hydroxydopamine; α-Methyl-5,7-dihydroxytryptamine; Neurotoxic indoleamines; Serotonin-like neurotoxins
as tools to study thyroid function, 394
Serotonin receptors, 457, 664
newly synthesized serotonin, 86
supersensitivity, 593
Serotonin-secreting neurons, central, 457
Serotonin system, ascending and progesterone, 623—624
[^{3}H]Serotonin
accumulation and retention in tissue slices, 308, 310—311, *310, 311*
binding in brain cortex after serotonin denervation of forebrain in rats, 364—366
binding in brain cortex of rats, 351, 358—360, *358, 359, 360, 361,* 362
metabolism from [^{3}H]tryptophan, 373, *374—375,* 376
uptake in cortex and hypothalamus after 5,7-dihydroxytryptamine-induced lesions in rats, 356
uptake in medulla oblongata and spinal cord after 5,6-dihydroxytryptamine treatment in rats, 372, *373*
uptake of total tritium and unchanged [^{3}H]serotonin by brain slices and spinal cord in rats, 370, *371,* 372
Sexual behavior. *See also* *p*-Chloroamphetamine; L-Dopa; Fenfluramine; Mounting behavior
dependent on ovarian hormones estrogen and progesterone appearing mediated by a serotonergic system, 624
effect of intracranial hormone administration in female rats, 623, *623*
facilitated by norepinephrine in female rats, 622
increased after injection of *p*-chlorophenylalanine in rats, 485—486
suppressed by serotonin in female rats, 622
Sleeping and waking, 481—484. *See also* *p*-Chlorophenylalanine; 5,6-Dihydroxytryptamine—as tool for studying sleep mechanisms in cats; Serotonin—role in mechanism of sleep in cats
amount of paradoxical sleep after injection of 5,6-dihydroxytryptamine in cats, 578, *579*
amount of stage-2 sleep after injection of 5,6-dihydroxytryptamine in cats, 578, *579*
and hyperthermia in serotonin-infused fowl, 247—248
and lesions of serotonin pathways with 5,7-dihydroxytryptamine in rats, 349, 365
arousal, 480, 487
brain serotonin neurons may play important role in mediating, 480
rapid eye movement, 483—484
sedation and depression, 481—482
slow-wave sleep, 484
Spatial reversal learning effects of midbrain raphe lesions, 545
Spinal cord
decrease in serotonin level in rats, 503
in rats, 377
Spleen
norepinephrine level after single dose of 5,6-dihydroxytryptamine or 3-(β-aminoethyl)-5,6-dihydroxybenzo[*b*]thiophene in rats, 136, *137*
serotonin level after single dose of 5,6-dihydroxytryptamine or 3-(β-aminoethyl)-5,6-dihydroxybenzo[*b*]thiophene in rats, 136, *137*
Subcommissural organ
after injection of 5,6-dihydroxytryptamine in rats, 267, *268,* 279, *280—281,* 283
after intraventricular injection of 5,7-dihydroxytryptamine in rats, *272—273,* 275, *276, 277*
and serotonergic synapses, 262—264
apical process in contact with cerebrospinal fluid after intraventricular injection of 5,7-dihydroxytryptamine in rats, 273, *275*
autoradiography, 265
definition and location, 262
detection of somatostatin, luteinizing hormone-releasing factor, thyrotropin-releasing factor, and arginine vasotocin in, 283
electron microscopic observations in rats, 265, 269, *270—271,* 273
fluorescence histochemistry, 264
fluorescence micrograph of saggital section from rats, 266, *266*

Subcommissural organ—(cont'd)
histochemical demonstration of nucleoproteins, glycoproteins, and neurosecretory proteins, 265
increased extracellular space in hypendyma after injection of 5,6-dihydroxytryptamine in rats, 275, *278*
indoleaminergic innervation confirmed, 284—285
inhibitory effects of serotonin on protein synthesis, 284
innervation, 282—283
of rats as model for neuron-target cell interaction, 282—284
perinuclear region of cells after intraventricular injection of 5,6-dihydroxytryptamine in rats, 273, *274*
protein and glycoprotein synthesis inhibited by serotonergic input, 264
secretory function, 279—280, 282—283
serotonergic terminals in the ependymal cells, 385
serotonin terminals, 282—284
varicose fluorescent fibers after intraventricular injection of 6-hydroxytryptamine in rats, 267, *268*
Subsensitivity after destruction of serotonergic nerve terminals affected by various doses of fenfluramine in injected rats, 501—502, *502*
Sucrose
acceptance of varying solutions instead of tap water by rats, 634, *634*
in drinking water for rats, 630
Sulfhydryl groups, 54, 56. *See also* 5,6-Dihydroxytryptamine—reaction with sulfhydryl groups; Serotonin-binding protein
in peptides and proteins, 18
Superoxide dismutase
inhibition of tryptamine conversion not antagonized, 41
Superoxides
sites of action from neurotoxins, 84
Supersensitivity after destruction of serotonergic nerve terminals
5-methoxy-*N,N*-dimethyltryptamine, 497—498
5,7-dihydroxytryptamine, 498, 500—502, 504—506
adenyl cyclase activity, 506
administration of L-tryptophan (after monoamine oxidase inhibition) and L-hydroxytryptophan both producing the syndrome, 497, 504—505
effect of various doses of 5-methoxy-*N,N*-dimethyltryptamine in injected rats, 500—501, *501,* 504—505
effect of various doses of *d*-lysergic acid diethylamide in injected rats, 500—501, *501,* 504—505
effect of various doses of L-5-hydroxytryptophan in injected rats, 500, *500,* 503, *503*
effect of various doses of L-tryptophan in rats, 501, *502*
fenfluramnie, 497, 504—505
p-chloroamphetamine, 497
presynaptic, 504
serotonin, 497
Sympathetic nerves of the heart, 457
decrease in norepinephrine content, 471
degeneration of the terminals, 471
in rabbits, 462
norepinephrine uptake, 469
serotonin causing norepinephrine release from, 468
Synaptosomal accumulation of serotonin in various brain regions in rats, 647—648

Temperature, body. *See also p*-Chlorophenylalanine; Hyperthermia; Hypothermia; Rectal temperature
average deviation in response to exposure to cold or warm air in rats, 559, *561*
behavioral thermoregulation in monkeys, 564—566
deviation from a baseline after
injection of 5,6-dihydroxytryptamine in monkeys, 562, *563,* 565, *565*
injection of saline and 5,6-dihydroxytryptamine in rats, 558, *560*
intrahypothalamic injection of serotonin in monkeys, 558, *559*
risk of exposure to cold and warm air after serotonin and saline injection and after 5,6-dihydroxytryptamine injection into anterior hypothalamus in rats, 561, *562*
hypothalamic actions of serotonin neurotoxins, 573
maximum rise induced by serotonin injected into hypothalamus before and after 5,6-dihydroxytryptamine injections at the same site in rats, 558, *560*
of monkeys after injection of 5,6-dihydroxytryptamine, 564, *564*
regulation, 488—489

Temperature, body—(cont'd)
thermoregulatory mechanism
after hypothalamic injection with 5,6-dihydroxytryptamine, 559, 561
deficits in monkeys and rats, 571—572
in hypothalamus, 556—557
Testosterone, 486
Thiourea
as protective agent, 80—81
structure, 77, *77*
Thyroid C cells
uptake of 6-hydroxydopamine, 56
Thyroid-releasing hormone, 389
Thyroid-stimulating hormone, 393—394
Tibial length, right, in rats, 631, 636, 638
Transmitter turnover
correlations with rotational behavior, 526
dopamine turnover, 526
serotonin turnover, 526
Tranylcypromine, 510, 519
Trazodone as inhibitor of serotonin-binding protein capacity, 91
m-Trifluoromethylbenzoylglycine
recovery from rat brain as fenfluramine metabolite, 120, 122, 125—126
Triton® X-100
and oxygen consumption, 81
Tryptamine, 387, 538, 540, 546, 664
and excess concentration of 5,6-dihydroxytryptamine, 5,7-dihydroxytryptamine, or serotonin to inhibit conversion, 49, 54
in brain, 481
noncompetitive inhibition by 5,6-dihydroxytryptamine or 5,7-dihydroxytryptamine, 53
Tryptamine derivatives
acute toxicity, 458
in isolated strips of rabbit aorta, 459, *459*, 461
numbering system, 128, *129*
Tryptamines
synthesis, 25, 27
Tryptamines, hydroxylated
effect on serotonin, norepinephrine, and dopamine uptake into hypothalamic homogenates, 128—129, *130*, 131
effect on serotonin uptake into hypothalamic homogenates, 128—129, *129*
Tryptophan, 199, 203—205, 254, 389, 391, 483, 519, 532, 576, 578, 642, 664
as precursor for serotonin, 510
essential amino acid for serotonin biosynthesis, 481
in central nervous system, 4
in diet, 487
loading to increase neuronal serotonin levels, 394
pretreatment with neither enhances nor antagonizes 5,7-dihydroxytryptamine-induced hypothermia in rats, 489
Tryptophan hydroxylase, 153—154, 158—159, 182—184, 204, 259, 297, 437, 532—533, 537—538, 540, 546, 586, 621
and *p*-chloroamphetamine, 208, 211, 213—215, 217
in rat brain, 649
inactivated by *p*-chlorophenylalanine, 519
measurement by high-pressure liquid chromatography, 190
reduced in brain by *p*-chloroamphetamine in rats, 645
time course of changes in brain nuclei after injection of *p*-chloroamphetamine, 192, *193*
Tryptophan-induced neurologic syndrome myoclonus, 206—207
Tryptophan-poor diets, 539—540
Tubulin relation to serotonin-binding protein, 90—91
Turning behavior. *See* Behavior, rotational
Tyrosine, 583, 642
in brain of cats, 578
Tyrosine hydroxylase, 182—184, 221, 255, 478, 540, 546, 589, 621
effect of intracisternally injected 5,6-dihydroxytryptamine
in locus ceruleus of cats, 583—584, *585*
on activity in the locus and frontal cortex in cats, 582—583, *583*, *584*
increase in activity in the locus ceruleus after bilateral injection of 5,6-dihydroxytryptamine in cats, 585—586, *586*

Urethane anesthetizing of rats, 458

Valine, 642
injections, 487
Vinblastine
as inhibitor of serotonin-binding protein capacity, 86
effect on capacity of serotonin-binding protein in brain, 90, *91*

Water intake. *See* Drinking
Weight, body. *See also* *p*-Chloroamphetamine; 5,6-Dihydroxytryptamine; Fenfluramine
after exposure to a high-fat diet instead of pellet chow in rats, 634, *635*
average loss or gain in rats injected intraventricularly with 5,6-dihydroxytryptamine, 6-hydroxydopamine, or artificial cerebrospinal fluid and some given 5,7-dihydroxytryptamine similarly, 568, *570*
changes in monkeys and rats, 566, 568
ethanol preference test obtained for 5,6-dihydroxytryptamine- and 6-hydroxydopamine-injected rats, 571, *572*
gain accompanying overeating in rats, 571
gain in rats, 642
in rats, 627, 638, 640—642
loss or gain in rats after single injection of 5,6-dihydroxytryptamine, 5,7-dihydroxytryptamine, 6-hydroxydopamine, or cerebrospinal fluid in brain ventricle, 568, *569*
obesity in rats, 627
regulation in rats, 628, 640, 642

Zimelidine, 352, *352*, *353*, 354—356, *355*, 362, 645—646, 648—649, 651—660. *See* also in subheads under other main headings, e.g. Morphine analgesia, effects of—zimelidine
as potent serotonin uptake-blocking agent